Vibration of Continuous Systems

Vibration of Continuous Systems

Singiresu S. Rao

Professor and Chairman
Department of Mechanical and Aerospace Engineering
University of Miami
Coral Gables, Florida

JOHN WILEY & SONS, INC.

Library of Congress Cataloging-in-Publication Data:

Rao, S. S.
    Vibration of Continuous Systems / Singiresu S. Rao.
        p. cm.
    Includes index.
    ISBN-13 978-0-471-77171-5 (cloth)
    ISBN-10 0-471-77171-6 (cloth)
    1. Vibration–Textbooks. 2. Structural dynamics–Textbooks.
    I. Title.
    TA355.R378 2007
    624.1'71–dc22

2006008775

To Lord Sri Venkateswara

Contents

Preface

This book covers analytical methods of vibration analysis of continuous structural systems, including strings, bars, shafts, beams, circular rings and curved beams, membranes, plates, and shells. The propagation of elastic waves in structures and solid bodies is also introduced. The objectives of the book are (1) to make a methodical and comprehensive presentation of the vibration of various types of structural elements, (2) to present the exact analytical and approximate analytical methods of analysis, and (3) to present the basic concepts in a simple manner with illustrative examples.

Continuous structural elements and systems are encountered in many branches of engineering, such as aerospace, architectural, chemical, civil, ocean, and mechanical engineering. The design of many structural and mechanical devices and systems requires an accurate prediction of their vibration and dynamic performance characteristics. The methods presented in the book can be used in these applications. The book is intended to serve as a textbook for a dual-level or first graduate-level course on vibrations or structural dynamics. More than enough material is included for a one-semester course. The chapters are made as independent and self-contained as possible so that a course can be taught by selecting appropriate chapters or through equivalent self-study. A successful vibration analysis of continuous structural elements and systems requires a knowledge of mechanics of materials, structural mechanics, ordinary and partial differential equations, matrix methods, variational calculus, and integral equations. Applications of these techniques are presented throughout. The selection, arrangement, and presentation of the material has been made based on the lecture notes for a course taught by the author. The contents of the book permit instructors to emphasize a variety of topics, such as basic mathematical approaches with simple applications, bars and beams, beams and plates, or plates and shells. The book will also be useful as a reference book for practicing engineers, designers, and vibration analysts involved in the dynamic analysis and design of continuous systems.

Organization of the Book

The book is organized into 17 chapters and two appendixes. The basic concepts and terminology used in vibration analysis are introduced in Chapter 1. The importance, origin, and a brief history of vibration of continuous systems are presented. The difference between discrete and continuous systems, types of excitations, description of harmonic functions, and basic definitions used in the theory of vibrations and representation of periodic functions in terms of Fourier series and the Fourier integral are discussed. Chapter 2 provides a brief review of the theory and techniques used in the vibration analysis of discrete systems. Free and forced vibration of single- and multidegree-of-freedom systems are outlined. The eigenvalue problem and its role in the modal analysis used in the free and forced vibration analysis of discrete systems are discussed.

Various methods of formulating vibration problems associated with continuous systems are presented in Chapters 3, 4, and 5. The equilibrium approach is presented in Chapter 3. Use of Newton's second law of motion and D'Alembert's principle is outlined, with application to different types of continuous elements. Use of the variational approach in deriving equations of motion and associated boundary conditions is described in Chapter 4. The basic concepts of calculus of variations and their application to extreme value problems are outlined. The variational methods of solid mechanics, including the principles of minimum potential energy, minimum complementary energy, stationary Reissner energy, and Hamilton's principle, are presented. The use of Hamilton's principle in the formulation of continuous systems is illustrated with torsional vibration of a shaft and transverse vibration of a thin beam. The integral equation approach for the formulation of vibration problems is presented in Chapter 5. A brief outline of integral equations and their classification, and the derivation of integral equations, are given together with examples. The solution of integral equations using iterative, Rayleigh–Ritz, Galerkin, collocation, and numerical integration methods is also discussed in this chapter.

The common solution procedure based on eigenvalue and modal analyses for the vibration analysis of continuous systems is outlined in Chapter 6. The orthogonality of eigenfunctions and the role of the expansion theorem in modal analysis are discussed. The forced vibration response of viscously damped systems are also considered in this chapter. Chapter 7 covers the solution of problems of vibration of continuous systems using integral transform methods. Both Laplace and Fourier transform techniques are outlined together with illustrative applications.

The transverse vibration of strings is presented in Chapter 8. This problem finds application in guy wires, electric transmission lines, ropes and belts used in machinery, and the manufacture of thread. The governing equation is derived using equilibrium and variational approaches. The traveling-wave solution and separation of variables solution are outlined. The free and forced vibration of strings are considered in this chapter. The longitudinal vibration of bars is the topic of Chapter 9. Equations of motion based on simple theory are derived using the equilibrium approach as well as Hamilton's principle. The natural frequencies of vibration are determined for bars with different end conditions. Free vibration response due to initial excitation and forced vibration of bars are both presented, as is response using modal analysis. Free and forced vibration of bars using Rayleigh and Bishop theories are also outlined in Chapter 9. The torsional vibration of shafts plays an important role in mechanical transmission of power in prime movers and other high-speed machinery. The torsional vibration of uniform and nonuniform rods with both circular and noncircular cross sections is described in Chapter 10. The equations of motion and free and forced vibration of shafts with circular cross section are discussed using the elementary theory. The Saint-Venant and Timoshenko–Gere theories are considered in deriving the equations of motion of shafts with noncircular cross sections. Methods of determining the torsional rigidity of noncircular shafts are presented using the Prandtl stress function and Prandtl membrane analogy.

Chapter 11 deals with the transverse vibration of beams. Starting with the equation of motion based on Euler–Bernoulli or thin beam theory, natural frequencies and mode shapes of beams with different boundary conditions are determined. The free vibration response due to initial conditions, forced vibration under fixed and moving loads, response under axial loading, rotating beams, continuous beams, and beams on an elastic foundation are presented using the Euler–Bernoulli theory. The effects of rotary inertia (Rayleigh theory) and rotary inertia and shear deformation (Timoshenko theory) on the transverse vibration of beams are also considered. Finally, coupled bending–torsional vibration of beams is discussed toward the end of Chapter 11. In-plane flexural and coupled twist-bending vibration of circular rings and curved beams is considered in Chapter 12. The equations of motion and free vibration solutions are presented first using a simple theory. Then the effects of rotary inertia and shear deformation are considered. The vibration of rings is important in a study of the vibration of ring-stiffened shells used in aerospace applications, gears, and stators of electrical machines.

The transverse vibration of membranes is the topic of Chapter 13. Membranes find application in drums and microphone condensers. The equation of motion of membranes is derived using both the equilibrium and variational approaches. The free and forced vibration of rectangular and circular membranes are both discussed in this chapter. Chapter 14 covers the transverse vibration of plates. The equation of motion and the free and forced vibration of both rectangular and circular plates are presented. The vibration of plates subjected to in-plane forces, plates on elastic foundation, and plates with variable thickness is also discussed. Finally, the effect of rotary inertia and shear deformation on the vibration of plates is outlined according to Mindlin's theory. The vibration of shells is the topic of Chapter 15. First the theory of surfaces is presented using shell coordinates. Then the strain–displacement relations according to Love's approximations, stress–strain, and force and moment resultants are given. Then the equations of motion are derived from Hamilton's principle. The equations of motion of circular cylindrical shells and their natural frequencies are considered using Donnel–Mushtari–Vlasov and Love's theories. Finally, the effect of rotary inertia and shear deformation on the vibration of shells is considered.

Wave propagation in elastic solids is considered in Chapter 16. The one-dimensional wave equation and the traveling-wave solution are presented. The wave motion in strings and wave propagation in a semi-infinite medium, along with reflection and transmission of waves at fixed and free boundaries, are discussed. The differences between compressional or P waves and shear or S waves are discussed. The flexural waves in beams and the propagation of dilatational and distortional waves is considered in an infinite elastic medium. Rayleigh or surface waves are also discussed. Finally, Chapter 17 is devoted to the approximate analytical methods useful for vibration analysis. The computational details of the Rayleigh, Rayleigh–Ritz, assumed modes, weighted residual, Galerkin, collocation, subdomain collocation, and least squares methods are presented along with numerical examples. Appendix A presents the basic equations of elasticity. Laplace and Fourier transform pairs associated with some simple and commonly used functions are summarized in Appendix B.

Acknowledgments

I would like to thank the many graduate students who offered constructive suggestions when drafts of this book were used as class notes. I would also like to express my gratitude to my wife, Kamala, for her infinite patience, encouragement, and moral support in completing this book. She shares with me the fun and pain associated with the writing of the book. Finally, and most importantly, I would like to acknowledge the many intangible contributions made by my granddaughters, Siriveena and Samanthaka, that helped the completion of this work in a timely fashion.

S. S. Rao

srao@miami.edu
Miami, November 2005

Symbols

a	radius of a circular membrane or plate
a, b	dimensions of a membrane or plate along the x and y directions
A	cross-sectional area; area of a plate; amplitude
A, B, C, D	constants
c	velocity of wave propagation; damping constant
c_1, c_2	damping constants of dampers
$[c], [c_{ij}]$	damping matrix
C	torsional rigidity
C_1, C_2, C_3, C_4	constants
$[d], [d_{ij}]$	damping matrix
D	flexural rigidity of a plate or shell; domain
E	Young's modulus
EA	axial stiffness of a bar
EI	bending stiffness of a beam
f	linear frequency (Hz)
$\vec{f}$	vector of forces
f_0	uniform load; amplitude of force
$f(t), F(t)$	force
$f(x, t)$	external force per unit length
$f(x, y, t)$	external transverse force per unit area on a membrane or plate
F_0	concentrated force
$F(m, n, t)$	Fourier transform of $f(x, y, t)$
$F(s)$	Laplace transform of $f(t)$
$F(\omega)$	Fourier transform of $f(t)$
$F_j(t)$	concentrated force at point $\vec{X}_j$
G	shear modulus
GJ	torsional rigidity
h	thickness of a plate or shell
$H(t)$	Heaviside function
i	$\sqrt{-1}$
I	area moment of inertia of cross section of a beam; functional
I_0	mass polar moment of inertia per unit length
I_m, K_m	modified Bessel functions of order m of the first and second kind
I_p, J	polar moment of inertia of cross section

J_m, Y_m	Bessel functions of order m of the first and second kind
k	shear correction factor; spring stiffness
k_1, k_2	stiffnesses of springs
$[k], [k_{ij}]$	stiffness matrix
l	length of a string, bar, shaft or beam
L	Laplace transform operator; operator for stiffness distribution; length of a beam; Lagrangian
L^{-1}	inverse Laplace transform operator
m	mass
m_1, m_2	masses
$[m], [m_{ij}]$	mass matrix
M	mass; operator for mass distribution
M_x, M_y, M_{xy}	moment resultants in a plate or shell
n	number of degrees of freedom
$N_i(t)$	generalized force corresponding to $\eta_i(t)$
N_x, N_y, N_{xy}	in-plane force resultants in a plate or shell
P	tension in a string; tension per unit length in a membrane; axial force
$Q_n(t)$	generalized force corresponding to $\eta_n(t)$
Q_x, Q_y	shear force resultants in a plate or shell
r, θ	polar coordinates
R	radius; Rayleigh's quotient
s	number of concentrated forces
S	boundary
t	time
T	kinetic energy; function of time t
$T^*_{\max}$	reference kinetic energy
u, v, w	displacement components along the x, y, and z directions
$u(x, t)$	axial displacement
$u_0(x), \dot{u}_0(x)$	initial values of $u(x, t)$ and $\dot{u}(x, t)$
U	potential energy
$U_n(x)$	nth mode of vibration or eigenfunction
v	velocity
V	domain; volume; shear force
$w(x, t)$	transverse deflection of a string or beam
$w(x, y, t)$	transverse deflection of a membrane or plate
$w_0(x), \dot{w}_0(x)$	initial values of $w(x, t)$ and $\dot{w}(x, t)$
$w_0(x, y), \dot{w}_0(x, y)$	initial values of $w(x, y, t)$ and $\dot{w}(x, y, t)$
W	work done by external forces
$W(m, n, t)$	Fourier transform of $w(x, y, t)$
$W_i(x)$	ith normal mode shape of a string or beam
$W_{mn}(x, y)$	mode shape of a membrane or plate
$W_0(a), \dot{W}_0(a)$	Fourier transforms of $w_0(x)$ and $\dot{w}_0(x)$

$W_0(m, n)$, $\dot{W}_0(m, n)$	Fourier transforms of $w_0(x, y)$ and $\dot{w}_0(x, y)$
$\overline{W}(p, s)$	Laplace transform of $W(x, s)$
$\dot{x}$	time derivative of $x(t)$
$x_i(t)$	displacement of ith mass
$\vec{x}(t)$	vector of displacements
$\vec{X}$	vector of amplitudes in $\vec{x}(t)$
$\vec{X}^{(i)}$	ith eigenvector
$[X]$	modal matrix
α, β	curvilinear coordinates
δ	variation operator
$\delta(x - x_0)$	Dirac delta function
δ_{ij}	Kronecker delta
$\varepsilon, \varepsilon_{xx}$	axial strain in a bar or beam
$\varepsilon_{xx}, \varepsilon_{yy}, \varepsilon_{zz}, \varepsilon_{xy}, \varepsilon_{yz}, \varepsilon_{zx}$	components of strain
$\vec{\varepsilon}$	vector of strains
$\eta_i(t)$	ith modal or generalized coordinate
$\eta_i(0) = \eta_{i0}$	initial value of $\eta_i(t)$
$\vec{\eta}(t)$	vector of modal or generalized coordinates
θ	angular coordinate
λ, λ_i	eigenvalue; ith eigenvalue
ν	Poisson's ratio
π	strain energy
π_0	strain energy density
π_p	potential energy
ρ	mass density; mass per unit length of a string; mass per unit area of a membrane
$\rho, \rho(\theta), \rho(x)$	radius of curvature of a curved beam
σ, σ_{xx}	axial stress in a bar or beam
$\sigma_{xx}, \sigma_{yy}, \sigma_{zz}, \sigma_{xy}, \sigma_{yz}, \sigma_{zx}$	components of stress
$\vec{\sigma}$	vector of stresses
τ	time period
ϕ, ϕ_0	phase angle
ω	frequency of vibration (rad/s); forcing frequency
ω_n	natural frequency; nth natural frequency
Ω	forcing frequency
$\nabla^2 = \frac{\partial^2}{\partial x^2} + \frac{\partial^2}{\partial y^2}$	harmonic or Laplace operator

1

Introduction: Basic Concepts and Terminology

1.1 CONCEPT OF VIBRATION

Any repetitive motion is called *vibration* or *oscillation*. The motion of a guitar string, motion felt by passengers in an automobile traveling over a bumpy road, swaying of tall buildings due to wind or earthquake, and motion of an airplane in turbulence are typical examples of vibration. The theory of vibration deals with the study of oscillatory motion of bodies and the associated forces. The oscillatory motion shown in Fig. 1.1(a) is called *harmonic motion* and is denoted as

$$x(t) = X \cos \omega t \tag{1.1}$$

where X is called the *amplitude of motion*, ω is the *frequency of motion*, and t is the time. The motion shown in Fig. 1.1(b) is called *periodic motion*, and that shown in Fig. 1.1(c) is called *nonperiodic* or *transient motion*. The motion indicated in Fig. 1.1(d) is *random* or *long-duration nonperiodic vibration*.

The phenomenon of vibration involves an alternating interchange of potential energy to kinetic energy and kinetic energy to potential energy. Hence, any vibrating system must have a component that stores potential energy and a component that stores kinetic energy. The components storing potential and kinetic energies are called a *spring* or *elastic element* and a *mass* or *inertia element*, respectively. The elastic element stores potential energy and gives it up to the inertia element as kinetic energy, and vice versa, in each cycle of motion. The repetitive motion associated with vibration can be explained through the motion of a mass on a smooth surface, as shown in Fig. 1.2. The mass is connected to a linear spring and is assumed to be in equilibrium or rest at position 1. Let the mass m be given an initial displacement to position 2 and released with zero velocity. At position 2, the spring is in a maximum elongated condition, and hence the potential or strain energy of the spring is a maximum and the kinetic energy of the mass will be zero since the initial velocity is assumed to be zero. Because of the tendency of the spring to return to its unstretched condition, there will be a force that causes the mass m to move to the left. The velocity of the mass will gradually increase as it moves from position 2 to position 1. At position 1, the potential energy of the spring is zero because the deformation of the spring is zero. However, the kinetic energy and hence the velocity of the mass will be maximum at position 1 because of conservation of energy (assuming no dissipation of energy due to damping or friction). Since the velocity is maximum at position 1, the mass will

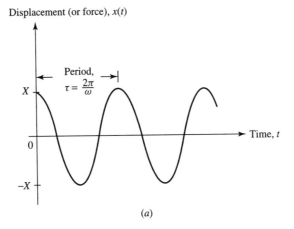

(a)

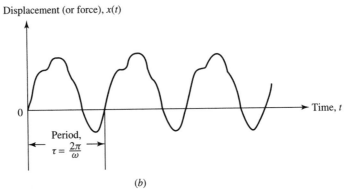

(b)

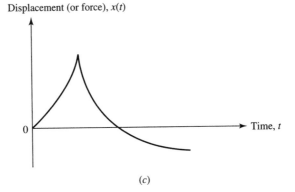

(c)

Figure 1.1   Types of displacements (or forces): (*a*) periodic simple harmonic; (*b*) periodic, nonharmonic; (*c*) nonperiodic, transient; (*d*) nonperiodic, random.

Displacement (or force), $x(t)$

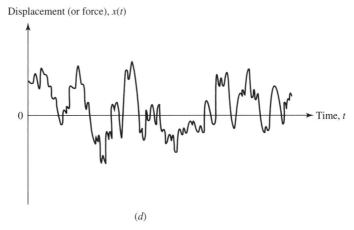

(d)

Figure 1.1 (*continued*)

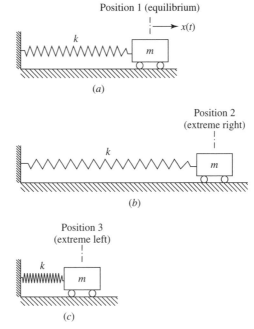

Figure 1.2 Vibratory motion of a spring–mass system: (*a*) system in equilibrium (spring undeformed); (*b*) system in extreme right position (spring stretched); (*c*) system in extreme left position (spring compressed).

continue to move to the left, but against the resisting force due to compression of the spring. As the mass moves from position 1 to the left, its velocity will gradually decrease until it reaches a value of zero at position 3. At position 3 the velocity and hence the kinetic energy of the mass will be zero and the deflection (compression) and hence the potential energy of the spring will be maximum. Again, because of the tendency of the spring to return to its uncompressed condition, there will be a force that causes the mass *m* to move to the right from position 3. The velocity of the mass will increase gradually as it moves from position 3 to position 1. At position 1, all of the potential energy of the spring has been converted to the kinetic energy of the mass, and hence the velocity of the mass will be maximum. Thus, the mass continues to move to the right against increasing spring resistance until it reaches position 2 with zero velocity. This completes one cycle of motion of the mass, and the process repeats; thus, the mass will have oscillatory motion.

The initial excitation to a vibrating system can be in the form of initial displacement and/or initial velocity of the mass element(s). This amounts to imparting potential and/or kinetic energy to the system. The initial excitation sets the system into oscillatory motion, which can be called *free vibration*. During free vibration, there will be exchange between potential and kinetic energies. If the system is conservative, the sum of potential energy and kinetic energy will be a constant at any instant. Thus, the system continues to vibrate forever, at least in theory. In practice, there will be some damping or friction due to the surrounding medium (e.g., air), which will cause loss of some energy during motion. This causes the total energy of the system to diminish continuously until it reaches a value of zero, at which point the motion stops. If the system is given only an initial excitation, the resulting oscillatory motion eventually will come to rest for all practical systems, and hence the initial excitation is called *transient excitation* and the resulting motion is called *transient motion*. If the vibration of the system is to be maintained in a steady state, an external source must replace continuously the energy dissipated due to damping.

1.2   IMPORTANCE OF VIBRATION

Any body having mass and elasticity is capable of oscillatory motion. In fact, most human activities, including hearing, seeing, talking, walking, and breathing, also involve oscillatory motion. Hearing involves vibration of the eardrum, seeing is associated with the vibratory motion of light waves, talking requires oscillations of the laryng (tongue), walking involves oscillatory motion of legs and hands, and breathing is based on the periodic motion of lungs. In engineering, an understanding of the vibratory behavior of mechanical and structural systems is important for the safe design, construction, and operation of a variety of machines and structures.

The failure of most mechanical and structural elements and systems can be associated with vibration. For example, the blade and disk failures in steam and gas turbines and structural failures in aircraft are usually associated with vibration and the resulting fatigue. Vibration in machines leads to rapid wear of parts such as gears and bearings, loosening of fasteners such as nuts and bolts, poor surface finish during metal cutting, and excessive noise. Excessive vibration in machines causes not only the failure of components and systems but also annoyance to humans. For example, imbalance in diesel engines can cause ground waves powerful enough to create a nuisance in urban areas. Supersonic aircraft create sonic booms that shatter doors and windows. Several spectacular failures of bridges, buildings, and dams are associated with wind-induced vibration, as well as oscillatory ground motion during earthquakes.

In some engineering applications, vibrations serve a useful purpose. For example, in vibratory conveyors, sieves, hoppers, compactors, dentist drills, electric toothbrushes, washing machines, clocks, electric massaging units, pile drivers, vibratory testing of materials, vibratory finishing processes, and materials processing operations such as casting and forging, vibration is used to improve the efficiency and quality of the process.

1.3 ORIGINS AND DEVELOPMENTS IN MECHANICS AND VIBRATION

The earliest human interest in the study of vibration can be traced to the time when the first musical instruments, probably whistles or drums, were discovered. Since that time, people have applied ingenuity and critical investigation to study the phenomenon of vibration and its relation to sound. Although certain very definite rules were observed in the art of music, even in ancient times, they can hardly be called science. The ancient Egyptians used advanced engineering concepts such as the use of dovetailed cramps and dowels in the stone joints of major structures such as the pyramids during the third and second millennia B.C.

As far back as 4000 B.C., music was highly developed and well appreciated in China, India, Japan, and perhaps Egypt [1, 6]. Drawings of stringed instruments such as harps appeared on the walls of Egyptian tombs as early as 3000 B.C. The British Museum also has a nanga, a primitive stringed instrument from 155 B.C. The present system of music is considered to have arisen in ancient Greece.

The scientific method of dealing with nature and the use of logical proofs for abstract propositions began in the time of Thales of Miletos (640–546 B.C.), who introduced the term *electricity* after discovering the electrical properties of yellow amber. The first person to investigate the scientific basis of musical sounds is considered to be the Greek mathematician and philosopher Pythagoras (582–507 B.C.). Pythagoras established the Pythagorean school, the first institute of higher education and scientific research. Pythagoras conducted experiments on vibrating strings using an apparatus called the monochord. Pythagoras found that if two strings of identical properties but different lengths are subject to the same tension, the shorter string produces a higher note, and in particular, if the length of the shorter string is one-half that of the longer string, the shorter string produces a note an octave above the other. The concept of pitch was known by the time of Pythagoras; however, the relation between the pitch and the frequency of a sounding string was not known at that time. Only in the sixteenth century, around the time of Galileo, did the relation between pitch and frequency become understood [2].

Daedalus is considered to have invented the pendulum in the middle of the second millennium B.C. One initial application of the pendulum as a timing device was made by Aristophanes (450–388 B.C.). Aristotle wrote a book on sound and music around 350 B.C. and documents his observations in statements such as "the voice is sweeter than the sound of instruments" and "the sound of the flute is sweeter than that of the lyre." Aristotle recognized the vectorial character of forces and introduced the concept of vectorial addition of forces. In addition, he studied the laws of motion, similar to those of Newton. Aristoxenus, who was a musician and a student of Aristotle, wrote a three-volume book called *Elements of Harmony*. These books are considered the oldest books available on the subject of music. Alexander of Afrodisias introduced the ideas of potential and kinetic energies and the concept of conservation of energy. In about 300 B.C., in addition to his contributions to geometry, Euclid gave a brief description of music in a treatise called *Introduction to Harmonics*. However, he did not discuss the physical nature of sound in the book. Euclid was distinguished for his teaching ability, and his greatest work, the *Elements*, has seen numerous editions and remains one of the most influential books of mathematics of all time. Archimedes (287–212 B.C.) is called by some scholars the father of mathematical physics. He developed the rules of statics. In his *On Floating Bodies*, Archimedes developed major rules of fluid pressure on a variety of shapes and on buoyancy.

China experienced many deadly earthquakes in ancient times. Zhang Heng, a historian and astronomer of the second century A.D., invented the world's first seismograph to measure earthquakes in A.D. 132 [3]. This seismograph was a bronze vessel in the form of a wine jar, with an arrangement consisting of pendulums surrounded by a group of eight lever mechanisms pointing in eight directions. Eight dragon figures, with a bronze ball in the mouth of each, were arranged outside the jar. An earthquake in any direction would tilt the pendulum in that direction, which would cause the release of the bronze ball in that direction. This instrument enabled monitoring personnel to know the direction, time of occurrence, and perhaps, the magnitude of the earthquake.

The foundations of modern philosophy and science were laid during the sixteenth century; in fact, the seventeenth century is called the *century of genius* by many. Galileo (1564–1642) laid the foundations for modern experimental science through his measurements on a simple pendulum and vibrating strings. During one of his trips to the church in Pisa, the swinging movements of a lamp caught Galileo's attention. He measured the period of the pendulum movements of the lamp with his pulse and was amazed to find that the time period was not influenced by the amplitude of swings. Subsequently, Galileo conducted more experiments on the simple pendulum and published his findings in *Discourses Concerning Two New Sciences* in 1638. In this work, he discussed the relationship between the length and the frequency of vibration of a simple pendulum, as well as the idea of sympathetic vibrations or resonance [4].

Although the writings of Galileo indicate that he understood the interdependence of the parameters—length, tension, density and frequency of transverse vibration—of a string, they did not offer an analytical treatment of the problem. Marinus Mersenne (1588–1648), a mathematician and theologian from France, described the correct behavior of the vibration of strings in 1636 in his book *Harmonicorum Liber*. For the first time, by knowing (measuring) the frequency of vibration of a long string, Mersenne was able to predict the frequency of vibration of a shorter string having the same density and tension. He is considered to be the first person to discover the laws of vibrating strings. The truth was that Galileo was the first person to conduct experimental studies on vibrating strings; however, publication of his work was prohibited until 1638, by order of the Inquisitor of Rome. Although Galileo studied the pendulum extensively and discussed the isochronism of the pendulum, Christian Huygens (1629–1695) was the person who developed the pendulum clock, the first accurate device developed for measuring time. He observed deviation from isochronism due to the nonlinearity of the pendulum, and investigated various designs to improve the accuracy of the pendulum clock.

The works of Galileo contributed to a substantially increased level of experimental work among many scientists and paved the way to the establishment of several professional organizations, such as the Academia Naturae in Naples in 1560, Academia dei Lincei in Rome in 1606, Royal Society in London in 1662, the French Academy of Sciences in 1766, and the Berlin Academy of Science in 1770.

The relation between the pitch and frequency of vibration of a taut string was investigated further by Robert Hooke (1635–1703) and Joseph Sauveur (1653–1716). The phenomenon of mode shapes during the vibration of stretched strings, involving no motion at certain points and violent motion at intermediate points, was observed independently by Sauveur in France (1653–1716) and John Wallis in England (1616–1703). Sauveur called points with no motion *nodes* and points with violent motion, *loops*. Also, he observed that vibrations involving nodes and loops had higher frequencies than those involving no nodes. After observing that the values of the higher frequencies were integral multiples of the frequency of simple vibration with no nodes, Sauveur termed the frequency of simple vibration the *fundamental frequency* and the higher frequencies, the *harmonics*. In addition, he found that the vibration of a stretched string can contain several harmonics simultaneously. The phenomenon of beats was also observed by Sauveur when two organ pipes, having slightly different pitches, were sounded together. He also tried to compute the frequency of vibration of a taut string from the measured sag of its middle point. Sauveur introduced the word *acoustics* for the first time for the science of sound [7].

Isaac Newton (1642–1727) studied at Trinity College, Cambridge and later became professor of mathematics at Cambridge and president of the Royal Society of London. In 1687 he published the most admired scientific treatise of all time, *Philosophia Naturalis Principia Mathematica.* Although the laws of motion were already known in one form or other, the development of differential calculus by Newton and Leibnitz made the laws applicable to a variety of problems in mechanics and physics. Leonhard Euler (1707–1783) laid the groundwork for the calculus of variations. He popularized the use of free-body diagrams in mechanics and introduced several notations, including $e = 2.71828\ldots$, $f(x)$, $\sum$, and $i = \sqrt{-1}$. In fact, many people believe that the current techniques of formulating and solving mechanics problems are due more to Euler than to any other person in the history of mechanics. Using the concept of inertia force, Jean D'Alembert (1717–1783) reduced the problem of dynamics to a problem in statics. Joseph Lagrange (1736–1813) developed the variational principles for deriving the equations of motion and introduced the concept of generalized coordinates. He introduced *Lagrange equations* as a powerful tool for formulating the equations of motion for lumped-parameter systems. Charles Coulomb (1736–1806) studied the torsional oscillations both theoretically and experimentally. In addition, he derived the relation between electric force and charge.

Claude Louis Marie Henri Navier (1785–1836) presented a rigorous theory for the bending of plates. In addition, he considered the vibration of solids and presented the continuum theory of elasticity. In 1882, Augustin Louis Cauchy (1789–1857) presented a formulation for the mathematical theory of continuum mechanics. William Hamilton (1805–1865) extended the formulation of Lagrange for dynamics problems and presented a powerful method (Hamilton's principle) for the derivation of equations of motion of continuous systems. Heinrich Hertz (1857–1894) introduced the terms *holonomic* and *nonholonomic* into dynamics around 1894. Jules Henri Poincaré

(1854–1912) made many contributions to pure and applied mathematics, particularly to celestial mechanics and electrodynamics. His work on nonlinear vibrations in terms of the classification of singular points of nonlinear autonomous systems is notable.

1.4 HISTORY OF VIBRATION OF CONTINUOUS SYSTEMS

The precise treatment of the vibration of continuous systems can be associated with the discovery of the basic law of elasticity by Hooke, the second law of motion by Newton, and the principles of differential calculus by Leibnitz. Newton's second law of motion is used routinely in modern books on vibrations to derive the equations of motion of a vibrating body.

Strings   A theoretical (dynamical) solution of the problem of the vibrating string was found in 1713 by the English mathematician Brook Taylor (1685–1731), who also presented the famous Taylor theorem on infinite series. He applied the fluxion approach, similar to the differential calculus approach developed by Newton and Newton's second law of motion, to an element of a continuous string and found the true value of the first natural frequency of the string. This value was found to agree with the experimental values observed by Galileo and Mersenne. The procedure adopted by Taylor was perfected through the introduction of partial derivatives in the equations of motion by Daniel Bernoulli, Jean D'Alembert, and Leonhard Euler. The fluxion method proved too clumsy for use with more complex vibration analysis problems. With the controversy between Newton and Leibnitz as to the origin of differential calculus, patriotic Englishmen stuck to the cumbersome fluxions while other investigators in Europe followed the simpler notation afforded by the approach of Leibnitz.

In 1747, D'Alembert derived the partial differential equation, later referred to as the *wave equation*, and found the wave travel solution. Although D'Alembert was assisted by Daniel Bernoulli and Leonhard Euler in this work, he did not give them credit. With all three claiming credit for the work, the specific contribution of each has remained controversial.

The possibility of a string vibrating with several of its harmonics present at the same time (with displacement of any point at any instant being equal to the algebraic sum of displacements for each harmonic) was observed by Bernoulli in 1747 and proved by Euler in 1753. This was established through the dynamic equations of Daniel Bernoulli in his memoir, published by the Berlin Academy in 1755. This characteristic was referred to as the *principle of the coexistence of small oscillations*, which is the same as the *principle of superposition* in today's terminology. This principle proved to be very valuable in the development of the theory of vibrations and led to the possibility of expressing any arbitrary function (i.e., any initial shape of the string) using an infinite series of sine and cosine terms. Because of this implication, D'Alembert and Euler doubted the validity of this principle. However, the validity of this type of expansion was proved by Fourier (1768–1830) in his *Analytical Theory of Heat* in 1822.

It is clear that Bernoulli and Euler are to be credited as the originators of the modal analysis procedure. They should also be considered the originators of the Fourier expansion method. However, as with many discoveries in the history of science, the persons credited with the achievement may not deserve it completely. It is often the person who publishes at the right time who gets the credit.

The analytical solution of the vibrating string was presented by Joseph Lagrange in his memoir published by the Turin Academy in 1759. In his study, Lagrange assumed that the string was made up of a finite number of equally spaced identical mass particles, and he established the existence of a number of independent frequencies equal to the number of mass particles. When the number of particles was allowed to be infinite, the resulting frequencies were found to be the same as the harmonic frequencies of the stretched string. The method of setting up the differential equation of motion of a string (called the *wave equation*), presented in most modern books on vibration theory, was developed by D'Alembert and described in his memoir published by the Berlin Academy in 1750.

Bars   Chladni in 1787, and Biot in 1816, conducted experiments on the longitudinal vibration of rods. In 1824, Navier, presented an analytical equation and its solution for the longitudinal vibration of rods.

Shafts   Charles Coulomb did both theoretical and experimental studies in 1784 on the torsional oscillations of a metal cylinder suspended by a wire [5]. By assuming that the resulting torque of the twisted wire is proportional to the angle of twist, he derived an equation of motion for the torsional vibration of a suspended cylinder. By integrating the equation of motion, he found that the period of oscillation is independent of the angle of twist. The derivation of the equation of motion for the torsional vibration of a continuous shaft was attempted by Caughy in an approximate manner in 1827 and given correctly by Poisson in 1829. In fact, Saint-Venant deserves the credit for deriving the torsional wave equation and finding its solution in 1849.

Beams   The equation of motion for the transverse vibration of thin beams was derived by Daniel Bernoulli in 1735, and the first solutions of the equation for various support conditions were given by Euler in 1744. Their approach has become known as the *Euler–Bernoulli* or *thin beam theory*. Rayleigh presented a beam theory by including the effect of rotary inertia. In 1921, Stephen Timoshenko presented an improved theory of beam vibration, which has become known as the *Timoshenko* or *thick beam theory*, by considering the effects of rotary inertia and shear deformation.

Membranes   In 1766, Euler, derived equations for the vibration of rectangular membranes which were correct only for the uniform tension case. He considered the rectangular membrane instead of the more obvious circular membrane in a drumhead, because he pictured a rectangular membrane as a superposition of two sets of strings laid in perpendicular directions. The correct equations for the vibration of rectangular and circular membranes were derived by Poisson in 1828. Although a solution corresponding to axisymmetric vibration of a circular membrane was given by Poisson, a nonaxisymmetric solution was presented by Pagani in 1829.

Plates   The vibration of plates was also being studied by several investigators at this time. Based on the success achieved by Euler in studying the vibration of a rectangular membrane as a superposition of strings, Euler's student James Bernoulli, the grand-nephew of the famous mathematician Daniel Bernoulli, attempted in 1788 to derive an equation for the vibration of a rectangular plate as a gridwork of beams. However, the resulting equation was not correct. As the torsional resistance of the plate was not considered in his equation of motion, only a resemblance, not the real agreement, was noted between the theoretical and experimental results.

The method of placing sand on a vibrating plate to find its mode shapes and to observe the various intricate modal patterns was developed by the German scientist Chladni in 1802. In his experiments, Chladni distributed sand evenly on horizontal plates. During vibration, he observed regular patterns of modes because of the accumulation of sand along the nodal lines that had no vertical displacement. Napoléon Bonaparte, who was a trained military engineer, was present when Chladni gave a demonstration of his experiments on plates at the French Academy in 1809. Napoléon was so impressed by Chladni's demonstration that he gave a sum of 3000 francs to the French Academy to be presented to the first person to give a satisfactory mathematical theory of the vibration of plates. When the competition was announced, only one person, Sophie Germain, entered the contest by the closing date of October 1811 [8]. However, an error in the derivation of Germain's differential equation was noted by one of the judges, Lagrange. In fact, Lagrange derived the correct form of the differential equation of plates in 1811. When the academy opened the competition again, with a new closing date of October 1813, Germain entered the competition again with a correct form of the differential equation of plates. Since the judges were not satisfied, due to the lack of physical justification of the assumptions she made in deriving the equation, she was not awarded the prize. The academy opened the competition again with a new closing date of October 1815. Again, Germain entered the contest. This time she was awarded the prize, although the judges were not completely satisfied with her theory. It was found later that her differential equation for the vibration of plates was correct but the boundary conditions she presented were wrong. In fact, Kirchhoff, in 1850, presented the correct boundary conditions for the vibration of plates as well as the correct solution for a vibrating circular plate.

The great engineer and bridge designer Navier (1785–1836) can be considered the originator of the modern theory of elasticity. He derived the correct differential equation for rectangular plates with flexural resistance. He presented an exact method that transforms the differential equation into an algebraic equation for the solution of plate and other boundary value problems using trigonometric series. In 1829, Poisson extended Navier's method for the lateral vibration of circular plates.

Kirchhoff (1824–1887) who included the effects of both bending and stretching in his theory of plates published in his book *Lectures on Mathematical Physics*, is considered the founder of the extended plate theory. Kirchhoff's book was translated into French by Clebsch with numerous valuable comments by Saint-Venant. Love extended Kirchhoff's approach to thick plates. In 1915, Timoshenko presented a solution for circular plates with large deflections. Foppl considered the nonlinear theory of plates in 1907; however, the final form of the differential equation for the large deflection of plates was developed by von Kármán in 1910. A more rigorous plate theory that considers the effects of transverse shear forces was presented by Reissner. A plate theory that includes the effects of both rotatory inertia and transverse shear deformation, similar to the Timoshenko beam theory, was presented by Mindlin in 1951.

Shells   The derivation of an equation for the vibration of shells was attempted by Sophie Germain, who in 1821 published a simplified equation, with errors, for the vibration of a cylindrical shell. She assumed that the in-plane displacement of the neutral surface of a cylindrical shell was negligible. Her equation can be reduced to the correct form for a rectangular plate but not for a ring. The correct equation for the vibration of a ring had been given by Euler in 1766.

Aron, in 1874, derived the general shell equations in curvilinear coordinates, which were shown to reduce to the plate equation when curvatures were set to zero. The equations were complicated because no simplifying assumptions were made. Lord Rayleigh proposed different simplifications for the vibration of shells in 1882 and considered the neutral surface of the shell either extensional or inextensional. Love, in 1888, derived the equations for the vibration of shells by using simplifying assumptions similar to those of beams and plates for both in-plane and transverse motions. Love's equations can be considered to be most general in unifying the theory of vibration of continuous structures whose thickness is small compared to other dimensions. The vibration of shells, with a consideration of rotatory inertia and shear deformation, was presented by Soedel in 1982.

Approximate Methods   Lord Rayleigh published his book on the theory of sound in 1877; it is still considered a classic on the subject of sound and vibration. Notable among the many contributions of Rayleigh is the method of finding the fundamental frequency of vibration of a conservative system by making use of the principle of conservation of energy—now known as *Rayleigh's method*. Ritz (1878–1909) extended Rayleigh's method for finding approximate solutions of boundary value problems. The method, which became known as the *Rayleigh–Ritz method*, can be considered to be a variational approach. Galerkin (1871–1945) developed a procedure that can be considered a weighted residual method for the approximate solution of boundary value problems.

Until about 40 years ago, vibration analyses of even the most complex engineering systems were conducted using simple approximate analytical methods. Continuous systems were modeled using only a few degrees of freedom. The advent of high-speed digital computers in the 1950s permitted the use of more degrees of freedom in modeling engineering systems for the purpose of vibration analysis. Simultaneous development of the finite element method in the 1960s made it possible to consider thousands of degrees of freedom to approximate practical problems in a wide spectrum of areas, including machine design, structural design, vehicle dynamics, and engineering mechanics. Notable contributions to the theory of the vibration of continuous systems are summarized in Table 1.1.

1.5   DISCRETE AND CONTINUOUS SYSTEMS

The *degrees of freedom* of a system are defined by the minimum number of independent coordinates necessary to describe the positions of all parts of the system at any instant of time. For example, the spring–mass system shown in Fig. 1.2 is a single-degree-of-freedom system since a single coordinate, $x(t)$, is sufficient to describe the position of the mass from its equilibrium position at any instant of time. Similarly, the simple pendulum shown in Fig. 1.3 also denotes a single-degree-of-freedom system. The reason is that the position of a simple pendulum during motion can be described by using a single angular coordinate, θ. Although the position of a simple pendulum can be stated in terms of the Cartesian coordinates x and y, the two coordinates x and y are not independent; they are related to one another by the constraint $x^2 + y^2 = l^2$, where l is the

Table 1.1   Notable Contributions to the Theory of Vibration of Continuous Systems

Period	Scientist	Contribution
582–507 B.C.	Pythagoras	Established the first school of higher education and scientific research. Conducted experiments on vibrating strings. Invented the monochord.
384–322 B.C.	Aristotle	Wrote a book on acoustics. Studied laws of motion (similar to those of Newton). Introduced vectorial addition of forces.
Third century B.C.	Alexander of Afrodisias	Kinetic and potential energies. Idea of conservation of energy.
325–265 B.C.	Euclid	Prominent mathematician. Published a treatise called *Introduction to Harmonics*.
A.D.		
1564–1642	Galileo Galilei	Experiments on pendulum and vibration of strings. Wrote the first treatise on modern dynamics.
1642–1727	Isaac Newton	Laws of motion. Differential calculus. Published the famous *Principia Mathematica*.
1653–1716	Joseph Sauveur	Introduced the term *acoustics*. Investigated harmonics in vibration.
1685–1731	Brook Taylor	Theoretical solution of vibrating strings. Taylor's theorem.
1700–1782	Daniel Bernoulli	Principle of angular momentum. Principle of superposition.
1707–1783	Leonhard Euler	Principle of superposition. Beam theory. Vibration of membranes. Introduced several mathematical symbols.
1717–1783	Jean D'Alembert	Dynamic equilibrium of bodies in motion. Inertia force. Wave equation.
1736–1813	Joseph Louis Lagrange	Analytical solution of vibrating strings. Lagrange's equations. Variational calculus. Introduced the term *generalized coordinates*.
1736–1806	Charles Coulomb	Torsional vibration studies.
1756–1827	E. F. F. Chladni	Experimental observation of mode shapes of plates.
1776–1831	Sophie Germain	Vibration of plates.
1785–1836	Claude Louis Marie Henri Navier	Bending vibration of plates. Vibration of solids. Originator of modern theory of elasticity.
1797–1872	Jean Marie Duhamel	Studied partial differential equations applied to vibrating strings and vibration of air in pipes. Duhamel's integral.
1805–1865	William Hamilton	Principle of least action. Hamilton's principle.

Table 1.1   (*continued*)

Period	Scientist	Contribution
1824–1887	Gustav Robert Kirchhoff	Presented extended theory of plates. Kirchhoff's laws of electrical circuits.
1842–1919	John William Strutt (Lord Rayleigh)	Energy method. Effect of rotatory inertia. Shell equations.
1874	H. Aron	Shell equations in curvilinear coordinates.
1888	A. E. H. Love	Classical theory of thin shells.
1871–1945	Boris Grigorevich Galerkin	Approximate solution of boundary value problems with application to elasticity and vibration.
1878–1909	Walter Ritz	Extended Rayleigh's energy method for approximate solution of boundary value problems.
1956	Turner, Clough, Martin, and Topp	Finite element method.

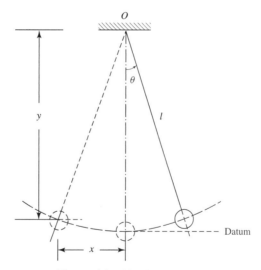

Figure 1.3   Simple pendulum.

constant length of the pendulum. Thus, the pendulum is a single-degree-of-freedom system. The mass–spring–damper systems shown in Fig. 1.4(*a*) and (*b*) denote two- and three-degree-of-freedom systems, respectively, since they have, two and three masses that change their positions with time during vibration. Thus, a multidegree-of-freedom system can be considered to be a system consisting of point masses separated by springs and dampers. The parameters of the system are discrete sets of finite numbers. These systems are also called *lumped-parameter*, *discrete*, or *finite-dimensional systems*.

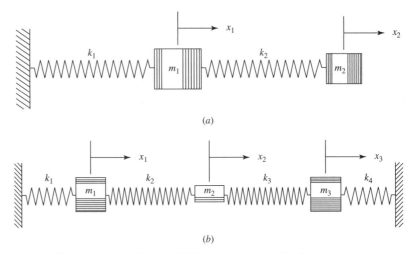

Figure 1.4   (*a*) Two- and (b) three-degree-of-freedom systems.

On the other hand, in a continuous system, the mass, elasticity (or flexibility), and damping are distributed throughout the system. During vibration, each of the infinite number of point masses moves relative to each other point mass in a continuous fashion. These systems are also known as *distributed, continuous*, or *infinite-dimensional systems*. A simple example of a continuous system is the cantilever beam shown in Fig. 1.5. The beam has an infinite number of mass points, and hence an infinite number of coordinates are required to specify its deflected shape. The infinite number of coordinates, in fact, define the elastic deflection curve of the beam. Thus, the cantilever beam is considered to be a system with an infinite number of degrees of freedom. Most mechanical and structural systems have members with continuous elasticity and mass distribution and hence have infinite degrees of freedom.

The choice of modeling a given system as discrete or continuous depends on the purpose of the analysis and the expected accuracy of the results. The motion of an *n*-degree-of-freedom system is governed by a system of *n* coupled second-order ordinary differential equations. For a continuous system, the governing equation of motion is in the form of a partial differential equation. Since the solution of a set of ordinary differential equations is simple, it is relatively easy to find the response of a discrete system that is experiencing a specified excitation. On the other hand, solution of a partial differential equation is more involved, and closed-form solutions are available for only a few continuous systems that have a simple geometry and simple, boundary conditions and excitations. However, the closed-form solutions that are available will often provide insight into the behavior of more complex systems for which closed-form solutions cannot be found.

For an *n*-degree-of-freedom system, there will be, at most, *n* distinct natural frequencies of vibration with a mode shape corresponding to each natural frequency. A continuous system, on the other hand, will have an infinite number of natural frequencies, with one mode shape corresponding to each natural frequency. A continuous system can be approximated as a discrete system, and its solution can be obtained in a simpler manner. For example, the cantilever beam shown in Fig. 1.5(*a*) can be

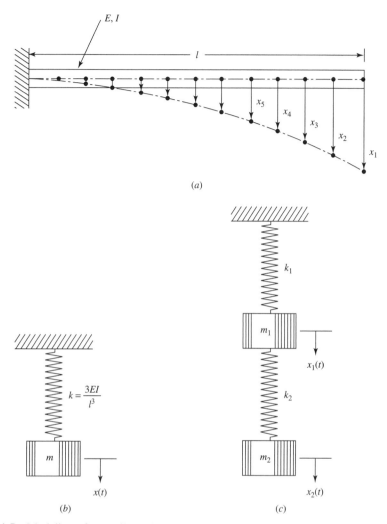

Figure 1.5   Modeling of a cantilever beam as (*a*) a continuous system, (*b*) a single-degree-of-freedom system, and (*c*) a two-degree-of-freedom system.

approximated as a single degree of freedom by assuming the mass of the beam to be a concentrated point mass located at the free end of the beam and the continuous flexibility to be approximated as a simple linear spring as shown in Fig. 1.5(*b*). The accuracy of approximation can be improved by using a two-degree-of-freedom model as shown in Fig. 1.5(*c*), where the mass and flexibility of the beam are approximated by two point masses and two linear springs.

1.6   VIBRATION PROBLEMS

Vibration problems may be classified into the following types [9]:

   1. *Undamped and damped vibration.*   If there is no loss or dissipation of energy due to friction or other resistance during vibration of a system, the system is said to be *undamped*. If there is energy loss due to the presence of damping, the system is called *damped*. Although system analysis is simpler when neglecting damping, a consideration of damping becomes extremely important if the system operates near resonance.

2. *Free and forced vibration.* If a system vibrates due to an initial disturbance (with no external force applied after time zero), the system is said to undergo *free vibration*. On the other hand, if the system vibrates due to the application of an external force, the system is said to be under *forced vibration*.

3. *Linear and nonlinear vibration.* If all the basic components of a vibrating system (i.e., the mass, the spring, and the damper) behave linearly, the resulting vibration is called *linear vibration*. However, if any of the basic components of a vibrating system behave nonlinearly, the resulting vibration is called *nonlinear vibration*. The equation of motion governing linear vibration will be a linear differential equation, whereas the equation governing nonlinear vibration will be a nonlinear differential equation. Most vibratory systems behave nonlinearly as the amplitudes of vibration increase to large values.

1.7 VIBRATION ANALYSIS

A vibratory system is a dynamic system for which the response (output) depends on the excitations (inputs) and the characteristics of the system (e.g., mass, stiffness, and damping) as indicated in Fig. 1.6. The excitation and response of the system are both time dependent. Vibration analysis of a given system involves determination of the response for the excitation specified. The analysis usually involves mathematical modeling, derivation of the governing equations of motion, solution of the equations of motion, and interpretation of the response results.

The purpose of mathematical modeling is to represent all the important characteristics of a system for the purpose of deriving mathematical equations that govern the behavior of the system. The mathematical model is usually selected to include enough details to describe the system in terms of equations that are not too complex. The mathematical model may be linear or nonlinear, depending on the nature of the system characteristics. Although linear models permit quick solutions and are simple to deal with, nonlinear models sometimes reveal certain important behavior of the system which cannot be predicted using linear models. Thus, a great deal of engineering judgment is required to develop a suitable mathematical model of a vibrating system. If the mathematical model of the system is linear, the principle of superposition can be used. This means that if the responses of the system under individual excitations $f_1(t)$ and $f_2(t)$ are denoted as $x_1(t)$ and $x_2(t)$, respectively, the response of the system would be

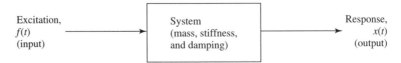

Figure 1.6 Input–output relationship of a vibratory system.

$x(t) = c_1 x_1(t) + c_2 x_2(t)$ when subjected to the excitation $f(t) = c_1 f_1(t) + c_2 f_2(t)$, where c_1 and c_2 are constants.

Once the mathematical model is selected, the principles of dynamics are used to derive the equations of motion of the vibrating system. For this, the free-body diagrams of the masses, indicating all externally applied forces (excitations), reaction forces, and inertia forces, can be used. Several approaches, such as D'Alembert's principle, Newton's second law of motion, and Hamilton's principle, can be used to derive the equations of motion of the system. The equations of motion can be solved using a variety of techniques to obtain analytical (closed-form) or numerical solutions, depending on the complexity of the equations involved. The solution of the equations of motion provides the displacement, velocity, and acceleration responses of the system. The responses and the results of analysis need to be interpreted with a clear view of the purpose of the analysis and the possible design implications.

1.8   EXCITATIONS

Several types of excitations or loads can act on a vibrating system. As stated earlier, the excitation may be in the form of initial displacements and initial velocities that are produced by imparting potential energy and kinetic energy to the system, respectively. The response of the system due to initial excitations is called *free vibration*. For real-life systems, the vibration caused by initial excitations diminishes to zero eventually and the initial excitations are known as *transient excitations*.

In addition to the initial excitations, a vibrating system may be subjected to a large variety of external forces. The origin of these forces may be environmental, machine induced, vehicle induced, or blast induced. Typical examples of environmentally induced dynamic forces include wind loads, wave loads, and earthquake loads. *Machine-induced loads* are due primarily to imbalance in reciprocating and rotating machines, engines, and turbines, and are usually periodic in nature. *Vehicle-induced loads* are those induced on highway and railway bridges from speeding trucks and trains crossing them. In some cases, dynamic forces are induced on bodies and equipment located inside vehicles due to the motion of the vehicles. For example, sensitive navigational equipment mounted inside the cockpit of an aircraft may be subjected to dynamic loads induced by takeoff, landing, or in-flight turbulence. *Blast-induced loads* include those generated by explosive devices during blast operations, accidental chemical explosions, or terrorist bombings.

The nature of some of the dynamic loads originating from different sources is shown in Fig. 1.1. In the case of rotating machines with imbalance, the induced loads will be harmonic, as shown in Fig. 1.1(*a*). In other types of machines, the loads induced due to the unbalance will be *periodic*, as shown in Fig. 1.1(*b*). A blast load acting on a vibrating structure is usually in the form of an overpressure, as shown in Fig. 1.1(*c*). The blast overpressure will cause severe damage to structures located close to the explosion. On the other hand, a large explosion due to underground detonation may even affect structures located far away from the explosion. Earthquake-, wave-, and wind-, gust-, or turbulence-, induced loads will be random in nature, as indicated in Fig. 1.1(*d*).

It can be seen that harmonic force is the simplest type of force to which a vibrating system can be subjected. The harmonic force also plays a very important role in the study of vibrations. For example, any periodic force can be represented as an infinite sum of harmonic forces using Fourier series. In addition, any nonperiodic force can be represented (by considering its period to be approaching infinity) in terms of harmonic forces using the Fourier integral. Because of their importance in vibration analysis, a detailed discussion of harmonic functions is given in the following section.

1.9 HARMONIC FUNCTIONS

In most practical applications, harmonic time dependence is considered to be same as sinusoidal vibration. For example, the harmonic variations of alternating current and electromagnetic waves are represented by sinusoidal functions. As an application in the area of mechanical systems, the motion of point S in the action of the Scotch yoke mechanism shown in Fig. 1.7 is simple harmonic. In this system, a crank of radius A rotates about point O. It can be seen that the amplitude is the maximum value of $x(t)$ from the zero value, either positively or negatively, so that $A = \max |x(t)|$. The frequency is related to the period τ, which is the time interval over which $x(t)$ repeats such that $x(t + \tau) = x(t)$.

The other end of the crank (P) slides in the slot of the rod that reciprocates in the guide G. When the crank rotates at the angular velocity ω, endpoint S of the slotted link is displaced from its original position. The displacement of endpoint S in time t is given by

$$x = A \sin \theta = A \sin \omega t \tag{1.2}$$

and is shown graphically in Fig. 1.7. The velocity and acceleration of point S at time t are given by

$$\frac{dx}{dt} = \omega A \cos \omega t \tag{1.3}$$

$$\frac{d^2 x}{dt^2} = -\omega^2 A \sin \omega t = -\omega^2 x \tag{1.4}$$

Equation (1.4) indicates that the acceleration of point S is directly proportional to the displacement. Such motion, in which the acceleration is proportional to the displacement and is directed toward the mean position, is called *simple harmonic motion*. According to this definition, motion given by $x = A \cos \omega t$ will also be simple harmonic.

1.9.1 Representation of Harmonic Motion

Harmonic motion can be represented by means of a vector $\vec{OP}$ of magnitude A rotating at a constant angular velocity ω, as shown in Fig. 1.8. It can be observed that the projection of the tip of the vector $\vec{X} = \vec{OP}$ on the vertical axis is given by

$$y = A \sin \omega t \tag{1.5}$$

and its projection on the horizontal axis by

$$x = A \cos \omega t \tag{1.6}$$

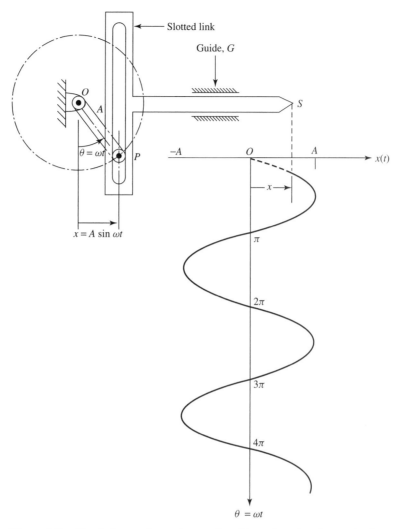

Figure 1.7   Simple harmonic motion produced by a Scotch yoke mechanism.

Equations (1.5) and (1.6) both represent simple harmonic motion. In the vectorial method of representing harmonic motion, two equations, Eqs. (1.5) and (1.6), are required to describe the vertical and horizontal components. Harmonic motion can be represented more conveniently using complex numbers. Any vector $\vec{X}$ can be represented as a complex number in the xy plane as

$$\vec{X} = a + ib \tag{1.7}$$

where $i = \sqrt{-1}$ and a and b denote the x and y components of $\vec{X}$, respectively, and can be considered as the *real* and *imaginary parts* of the vector $\vec{X}$. The vector $\vec{X}$ can also be expressed as

$$\vec{X} = A\,(\cos\theta + i\,\sin\theta) \tag{1.8}$$

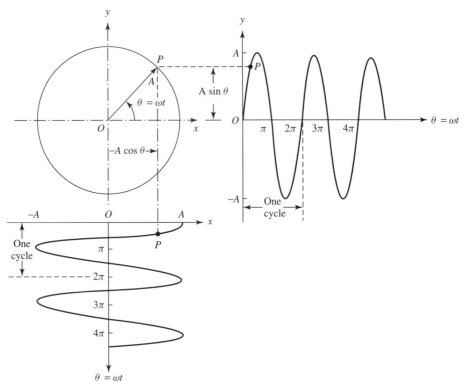

Figure 1.8   Harmonic motion: projection of a rotating vector.

where

$$A = (a^2 + b^2)^{1/2} \tag{1.9}$$

denotes the modulus or magnitude of the vector $\vec{X}$ and

$$\theta = \tan^{-1} \frac{b}{a} \tag{1.10}$$

indicates the argument or the angle between the vector and the x axis. Noting that

$$\cos\theta + i\sin\theta = e^{i\theta} \tag{1.11}$$

Eq. (1.8) can be expressed as

$$\vec{X} = A(\cos\theta + i\sin\theta) = Ae^{i\theta} \tag{1.12}$$

Thus, the rotating vector $\vec{X}$ of Fig. 1.8 can be written, using complex number representation, as

$$\vec{X} = Ae^{i\omega t} \tag{1.13}$$

where ω denotes the circular frequency (rad/sec) of rotation of the vector $\vec{X}$ in the counterclockwise direction. The harmonic motion given by Eq. (1.13) can be differentiated with respect to time as

$$\frac{d\vec{X}}{dt} = \frac{d}{dt}(Ae^{i\omega t}) = i\omega Ae^{i\omega t} = i\omega\vec{X} \tag{1.14}$$

$$\frac{d^2\vec{X}}{dt^2} = \frac{d}{dt}(i\omega Ae^{i\omega t}) = -\omega^2 Ae^{i\omega t} = -\omega^2\vec{X} \tag{1.15}$$

Thus, if $\vec{X}$ denotes harmonic motion, the displacement, velocity, and acceleration can be expressed as

$$x(t) = \text{displacement} = \text{Re}[Ae^{i\omega t}] = A\cos\omega t \tag{1.16}$$

$$\dot{x}(t) = \text{velocity} = \text{Re}[i\omega Ae^{i\omega t}] = -\omega A\sin\omega t = \omega A\cos(\omega t + 90°) \tag{1.17}$$

$$\ddot{x}(t) = \text{acceleration} = \text{Re}[-\omega^2 Ae^{i\omega t}] = -\omega^2 A\cos\omega t = \omega^2 A\cos(\omega t + 180°) \tag{1.18}$$

where Re denotes the real part, or alternatively as

$$x(t) = \text{displacement} = \text{Im}[Ae^{i\omega t}] = A\sin\omega t \tag{1.19}$$

$$\dot{x}(t) = \text{velocity} = \text{Im}[i\omega Ae^{i\omega t}] = \omega A\cos\omega t = \omega A\sin(\omega t + 90°) \tag{1.20}$$

$$\ddot{x}(t) = \text{acceleration} = \text{Im}[-\omega^2 Ae^{i\omega t}] = -\omega^2 A\sin\omega t = \omega^2 A\sin(\omega t + 180°) \tag{1.21}$$

where Im denotes the imaginary part. Eqs. (1.16)–(1.21) are shown as rotating vectors in Fig. 1.9. It can be seen that the acceleration vector leads the velocity vector by 90°, and the velocity vector leads the displacement vector by 90°.

1.9.2 Definitions and Terminology

Several definitions and terminology are used to describe harmonic motion and other periodic functions. The motion of a vibrating body from its undisturbed or equilibrium position to its extreme position in one direction, then to the equilibrium position, then

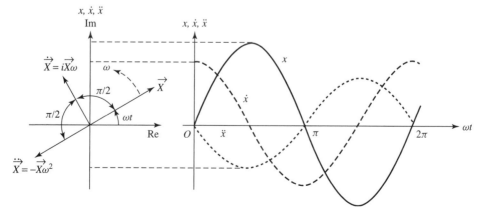

Figure 1.9 Displacement (x), velocity ($\dot{x}$), and acceleration ($\ddot{x}$) as rotating vectors.

to its extreme position in the other direction, and then back to the equilibrium position is called a *cycle* of vibration. One rotation or an angular displacement of 2π radians of pin P in the Scotch yoke mechanism of Fig. 1.7 or the vector $\vec{OP}$ in Fig. 1.8 represents a cycle.

The *amplitude* of vibration denotes the maximum displacement of a vibrating body from its equilibrium position. The amplitude of vibration is shown as A in Figs. 1.7 and 1.8. The *period* of oscillation represents the time taken by the vibrating body to complete one cycle of motion. The period of oscillation is also known as the *time period* and is denoted by τ. In Fig. 1.8, the time period is equal to the time taken by the vector $\vec{OP}$ to rotate through an angle of 2π. This yields

$$\tau = \frac{2\pi}{\omega} \tag{1.22}$$

where ω is called the *circular frequency*. The frequency of oscillation or *linear frequency* (or simply the *frequency*) indicates the number of cycles per unit time. The frequency can be represented as

$$f = \frac{1}{\tau} = \frac{\omega}{2\pi} \tag{1.23}$$

Note that ω is called the *circular frequency* and is measured in radians per second, whereas f is called the *linear frequency* and is measured in cycles per second (hertz). If the sine wave is not zero at time zero (i.e., at the instant we start measuring time), as shown in Fig. 1.10, it can be denoted as

$$y = A\sin(\omega t + \phi) \tag{1.24}$$

where $\omega t + \phi$ is called the *phase* of the motion and ϕ the *phase angle* or initial phase. Next, consider two harmonic motions denoted by

$$y_1 = A_1\sin\omega t \tag{1.25}$$

$$y_2 = A_2\sin(\omega t + \phi) \tag{1.26}$$

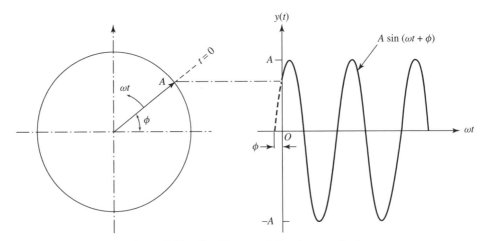

Figure 1.10   Significance of the phase angle ϕ.

Since the two vibratory motions given by Eqs. (1.25) and (1.26) have the same frequency ω, they are said to be *synchronous motions*. Two synchronous oscillations can have different amplitudes, and they can attain their maximum values at different times, separated by the time $t = \phi/\omega$, where ϕ is called the phase angle or *phase difference*. If a system (a single-degree-of-freedom system), after an initial disturbance, is left to vibrate on its own, the frequency with which it oscillates without external forces is known as its *natural frequency* of vibration. A discrete system having n degrees of freedom will have, in general, n distinct natural frequencies of vibration. A continuous system will have an infinite number of natural frequencies of vibration.

As indicated earlier, several harmonic motions can be combined to find the resulting motion. When two harmonic motions with frequencies close to one another are added or subtracted, the resulting motion exhibits a phenomenon known as *beats*. To see the phenomenon of beats, consider the difference of the motions given by

$$x_1(t) = X \sin \omega_1 t \equiv X \sin \omega t \tag{1.27}$$

$$x_2(t) = X \sin \omega_2 t \equiv X \sin(\omega - \delta)t \tag{1.28}$$

where δ is a small quantity. The difference of the two motions can be denoted as

$$x(t) = x_1(t) - x_2(t) = X[\sin \omega t - \sin(\omega - \delta)t] \tag{1.29}$$

Noting the relationship

$$\sin A - \sin B = 2 \sin \frac{A - B}{2} \cos \frac{A + B}{2} \tag{1.30}$$

the resulting motion $x(t)$ can be represented as

$$x(t) = 2X \sin \frac{\delta t}{2} \cos \left(\omega - \frac{\delta}{2} \right) t \tag{1.31}$$

The graph of $x(t)$ given by Eq. (1.31) is shown in Fig. 1.11. It can be observed that the motion, $x(t)$, denotes a cosine wave with frequency $(\omega_1 + \omega_2)/2 = \omega - \delta/2$, which is approximately equal to ω, and with a slowly varying amplitude of

$$2X \sin \frac{\omega_1 - \omega_2}{2} t = 2X \sin \frac{\delta t}{2}$$

Whenever the amplitude reaches a maximum, it is called a *beat*. The frequency δ at which the amplitude builds up and dies down between 0 and $2X$ is known as the *beat frequency*. The phenomenon of beats is often observed in machines, structures, and electric power houses. For example, in machines and structures, the beating phenomenon occurs when the forcing frequency is close to one of the natural frequencies of the system.

Example 1.1   Find the difference of the following harmonic functions and plot the resulting function for $A = 3$ and $\omega = 40$ rad/s: $x_1(t) = A \sin \omega t$, $x_2(t) = A \sin 0.95\omega t$.

SOLUTION   The resulting function can be expressed as

$$x(t) = x_1(t) - x_2(t) = A \sin \omega t - A \sin 0.95\omega t$$

$$= 2A \sin 0.025\omega t \cos 0.975\omega t \tag{E1.1.1}$$

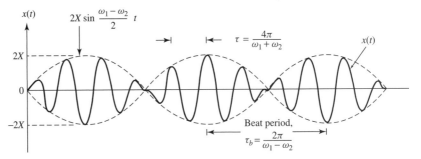

Figure 1.11   Beating phenomenon.

The plot of the function $x(t)$ is shown in Fig. 1.11. It can be seen that the function exhibits the phenomenon of beats with a beat frequency of $\omega_b = 1.00\omega - 0.95\omega = 0.05\omega = 2$ rad/s.

1.10   PERIODIC FUNCTIONS AND FOURIER SERIES

Although harmonic motion is the simplest to handle, the motion of many vibratory systems is not harmonic. However, in many cases the vibrations are periodic, as indicated, for example, in Fig. 1.1(b). Any periodic function of time can be represented as an infinite sum of sine and cosine terms using Fourier series. The process of representing a periodic function as a sum of harmonic functions (i.e., sine and cosine functions) is called *harmonic analysis*. The use of Fourier series as a means of describing periodic motion and/or periodic excitation is important in the study of vibration. Also, a familiarity with Fourier series helps in understanding the significance of experimentally determined frequency spectrums. If $x(t)$ is a periodic function with period τ, its Fourier series representation is given by

$$x(t) = \frac{a_0}{2} + a_1 \cos \omega t + a_2 \cos 2\omega t + \cdots + b_1 \sin \omega t + b_2 \sin 2\omega t + \cdots$$

$$= \frac{a_0}{2} + \sum_{n=1}^{\infty}(a_n \cos n\omega t + b_n \sin n\omega t) \tag{1.32}$$

where $\omega = 2\pi/\tau$ is called the *fundamental frequency* and $a_0, a_1, a_2, \ldots, b_1, b_2, \ldots$ are constant coefficients. To determine the coefficients a_n and b_n, we multiply Eq. (1.32) by $\cos n\omega t$ and $\sin n\omega t$, respectively, and integrate over one period $\tau = 2\pi/\omega$: for example, from 0 to $2\pi/\omega$. This leads to

$$a_0 = \frac{\omega}{\pi} \int_0^{2\pi/\omega} x(t)\, dt = \frac{2}{\tau} \int_0^{\tau} x(t)\, dt \tag{1.33}$$

$$a_n = \frac{\omega}{\pi} \int_0^{2\pi/\omega} x(t) \cos n\omega t\, dt = \frac{2}{\tau} \int_0^{\tau} x(t) \cos n\omega t\, dt \tag{1.34}$$

$$b_n = \frac{\omega}{\pi} \int_0^{2\pi/\omega} x(t) \sin n\omega t\, dt = \frac{2}{\tau} \int_0^{\tau} x(t) \sin n\omega t\, dt \tag{1.35}$$

Equation (1.32) shows that any periodic function can be represented as a sum of harmonic functions. Although the series in Eq. (1.32) is an infinite sum, we can approximate most periodic functions with the help of only a first few harmonic functions.

Fourier series can also be represented by the sum of sine terms only or cosine terms only. For example, any periodic function $x(t)$ can be expressed using cosine terms only as

$$x(t) = d_0 + d_1 \cos(\omega t - \phi_1) + d_2 \cos(2\omega t - \phi_2) + \cdots \tag{1.36}$$

where

$$d_0 = \frac{a_0}{2} \tag{1.37}$$

$$d_n = (a_n^2 + b_n^2)^{1/2} \tag{1.38}$$

$$\phi_n = \tan^{-1} \frac{b_n}{a_n} \tag{1.39}$$

The Fourier series, Eq. (1.32), can also be represented in terms of complex numbers as

$$x(t) = e^{i(0)\omega t} \left(\frac{a_0}{2} - \frac{ib_0}{2} \right)$$

$$+ \sum_{n=1}^{\infty} \left[e^{in\omega t} \left(\frac{a_n}{2} - \frac{ib_n}{2} \right) + e^{-in\omega t} \left(\frac{a_n}{2} + \frac{ib_n}{2} \right) \right] \tag{1.40}$$

where $b_0 = 0$. By defining the complex Fourier coefficients c_n and c_{-n} as

$$c_n = \frac{a_n - ib_n}{2} \tag{1.41}$$

$$c_{-n} = \frac{a_n + ib_n}{2} \tag{1.42}$$

Eq. (1.40) can be expressed as

$$x(t) = \sum_{n=-\infty}^{\infty} c_n e^{in\omega t} \tag{1.43}$$

The Fourier coefficients c_n can be determined, using Eqs. (1.33)–(1.35), as

$$c_n = \frac{a_n - ib_n}{2} = \frac{1}{\tau} \int_0^{\tau} x(t)(\cos n\omega t - i \sin n\omega t)\, dt$$

$$= \frac{1}{\tau} \int_0^{\tau} x(t) e^{-in\omega t}\, dt \tag{1.44}$$

The harmonic functions $a_n \cos n\omega t$ or $b_n \sin n\omega t$ in Eq. (1.32) are called the *harmonics of order n* of the periodic function $x(t)$. A harmonic of order n has a period τ/n. These harmonics can be plotted as vertical lines on a diagram of amplitude (a_n and b_n or d_n and ϕ_n) versus frequency ($n\omega$), called the *frequency spectrum* or *spectral diagram*.

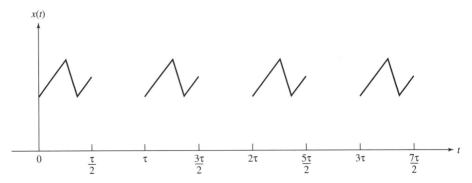

Figure 1.12   Typical periodic function.

1.11   NONPERIODIC FUNCTIONS AND FOURIER INTEGRALS

As shown in Eqs. (1.32), (1.36), and (1.43), any periodic function can be represented by a Fourier series. If the period τ of a periodic function increases indefinitely, the function $x(t)$ becomes nonperiodic. In such a case, the Fourier integral representation can be used as indicated below.

Let the typical periodic function shown in Fig. 1.12 be represented by a complex Fourier series as

$$x(t) = \sum_{n=-\infty}^{\infty} c_n e^{in\omega t}, \qquad \omega = \frac{2\pi}{\tau} \tag{1.45}$$

where

$$c_n = \frac{1}{\tau} \int_{-\tau/2}^{\tau/2} x(t) e^{-in\omega t}\, dt \tag{1.46}$$

Introducing the relations

$$n\,\omega = \omega_n \tag{1.47}$$

$$(n+1)\omega - n\,\omega = \omega = \frac{2\pi}{\tau} = \Delta\omega_n \tag{1.48}$$

Eqs. (1.45) and (1.46) can be expressed as

$$x(t) = \sum_{n=-\infty}^{\infty} \frac{1}{\tau}(\tau c_n)e^{i\omega_n t} = \frac{1}{2\pi}\sum_{n=-\infty}^{\infty} (\tau c_n)e^{i\omega_n t}\,\Delta\omega_n \tag{1.49}$$

$$\tau c_n = \int_{-\tau/2}^{\tau/2} x(t) e^{-i\omega_n t}\, dt \tag{1.50}$$

As $\tau \to \infty$, we drop the subscript n on ω, replace the summation by integration, and write Eqs. (1.49) and (1.50) as

$$x(t) = \lim_{\substack{\tau \to \infty \\ \Delta\omega_n \to 0}} \frac{1}{2\pi} \sum_{n=-\infty}^{\infty} (\tau c_n) e^{i\omega_n t} \Delta\omega_n = \frac{1}{2\pi} \int_{-\infty}^{\infty} X(\omega) e^{i\omega t} \, d\omega \qquad (1.51)$$

$$X(\omega) = \lim_{\substack{\tau \to \infty \\ \Delta\omega_n \to 0}} (\tau c_n) = \int_{-\infty}^{\infty} x(t) e^{-i\omega t} \, dt \qquad (1.52)$$

Equation (1.51) denotes the Fourier integral representation of $x(t)$ and Eq. (1.52) is called the *Fourier transform* of $x(t)$. Together, Eqs. (1.51) and (1.52) denote a *Fourier transform pair*. If $x(t)$ denotes excitation, the function $X(\omega)$ can be considered as the spectral density of excitation with $X(\omega)\,d\omega$ denoting the contribution of the harmonics in the frequency range ω to $\omega + d\omega$ to the excitation $x(t)$.

Example 1.2   Consider the nonperiodic rectangular pulse load $f(t)$, with magnitude f_0 and duration s, shown in Fig. 1.13(a). Determine its Fourier transform and plot the amplitude spectrum for $f_0 = 200$ lb, $s = 1$ sec, and $t_0 = 4$ sec.

SOLUTION   The load can be represented in the time domain as

$$f(t) = \begin{cases} f_0, & t_0 < t < t_0 + s \\ 0, & t_0 > t > t_0 + s \end{cases} \qquad (E1.2.1)$$

The Fourier transform of $f(t)$ is given by, using Eq. (1.52),

$$F(\omega) = \int_{-\infty}^{\infty} f(t) e^{-i\omega t} \, dt = \int_{t_0}^{t_0+s} f_0 e^{-i\omega t} \, dt$$

$$= f_0 \frac{i}{\omega} (e^{-i\omega(t_0+s)} - e^{-i\omega t_0})$$

$$= \frac{f_0}{\omega} \{[\sin \omega(t_0 + s) - \sin \omega t_0] + i[\cos \omega(t_0 + s) - \cos \omega t_0]\} \qquad (E1.2.2)$$

The amplitude spectrum is the modulus of $F(\omega)$:

$$|F(\omega)| = |F(\omega)F^*(\omega)|^{1/2} \qquad (E1.2.3)$$

where $F^*(\omega)$ is the complex conjugate of $F(\omega)$:

$$F^*(\omega) = \frac{f_0}{\omega} \{[\sin \omega(t_0 + s) - \sin \omega t_0] - i[\cos(\omega t_0 + s) - \cos \omega t_0]\} \qquad (E1.2.4)$$

By substituting Eqs. (E1.2.2) and (E1.2.4) into Eq. (E1.2.3), we can obtain the amplitude spectrum as

$$|F(\omega)| = \frac{f_0}{|\omega|} (2 - 2\cos \omega s)^{1/2} \qquad (E1.2.5)$$

or

$$\frac{|F(\omega)|}{f_0} = \frac{1}{|\omega|} (2 - 2\cos \omega)^{1/2} \qquad (E1.2.6)$$

The plot of Eq. (E1.2.6) is shown in Fig. 1.13(b).

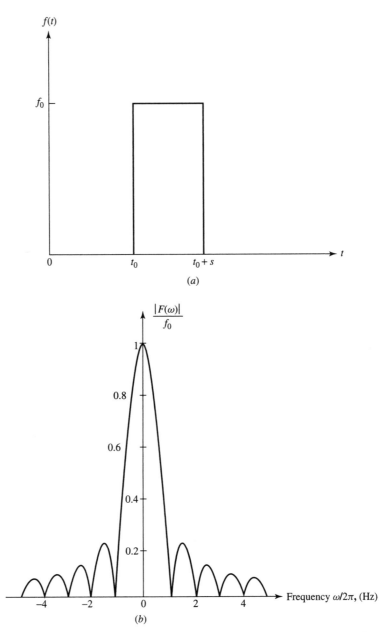

Figure 1.13 Fourier transform of a nonperiodic function: (*a*) rectangular pulse; (*b*) amplitude spectrum.

1.12 LITERATURE ON VIBRATION OF CONTINUOUS SYSTEMS

Several textbooks, monographs, handbooks, encyclopedia, vibration standards, books dealing with computer programs for vibration analysis, vibration formulas, and specialized topics as well as journals and periodicals are available in the general area of vibration of continuous systems. Among the large number of textbooks written on the subject of vibrations, the books by Magrab [10], Fryba [11], Nowacki [12], Meirovitch [13], and Clark [14] are devoted specifically to the vibration of continuous systems. Monographs by Leissa on the vibration of plates and shells [15, 16] summarize the results available in the literature on these topics. A handbook edited by Harris and Piersol [17] gives a comprehensive survey of all aspects of vibration and shock. A handbook on viscoelastic damping [18] describes the damping characteristics of polymeric materials, including rubber, adhesives, and plastics, in the context of design of machines and structures. An encyclopedia edited by Braun et al. [19] presents the current state of knowledge in areas covering all aspects of vibration along with references for further reading.

Pretlove [20], gives some computer programs in BASIC for simple analyses, and Rao [9] gives computer programs in Matlab, C++, and Fortran for the vibration analysis of a variety of systems and problems. Reference [21] gives international standards for acoustics, mechanical vibration, and shock. References [22–24] basically provide all the known formulas and solutions for a large variety of vibration problems, including those related to beams, frames, and arches. Several books have been written on the vibration of specific systems, such as spacecraft [25], flow-induced vibration [26], dynamics and control [27], foundations [28], and gears [29]. The practical aspects of vibration testing, measurement, and diagnostics of instruments, machinery, and structures are discussed in Refs. [30–32].

The most widely circulated journals that publish papers relating to vibrations are the *Journal of Sound and Vibration, ASME Journal of Vibration and Acoustics, ASME Journal of Applied Mechanics, AIAA Journal, ASCE Journal of Engineering Mechanics, Earthquake Engineering and Structural Dynamics, Computers and Structures, International Journal for Numerical Methods in Engineering, Journal of the Acoustical Society of America, Bulletin of the Japan Society of Mechanical Engineers, Mechanical Systems and Signal Processing, International Journal of Analytical and Experimental Modal Analysis, JSME International Journal Series III, Vibration Control Engineering, Vehicle System Dynamics,* and *Sound and Vibration.* In addition, *the Shock and Vibration Digest, Noise and Vibration Worldwide,* and *Applied Mechanics Reviews* are abstract journals that publish brief discussions of recently published vibration papers.

REFERENCES

1. D. C. Miller, *Anecdotal History of the Science of Sound*, Macmillan, New York, 1935.
2. N. F. Rieger, The quest for $\sqrt{k/m}$: notes on the development of vibration analysis, Part I, Genius awakening, *Vibrations*, Vol. 3, No. 3–4, pp. 3–10, 1987.
3. Chinese Academy of Sciences, *Ancient China's Technology and Science*, Foreign Languages Press, Beijing, 1983.
4. R. Taton, Ed., *Ancient and Medieval Science: From the Beginnings to 1450*, translated by A. J. Pomerans, Basic Books, New York, 1957.
5. S. P. Timoshenko, *History of Strength of Materials*, McGraw-Hill, New York, 1953.

6. R. B. Lindsay, The story of acoustics, *Journal of the Acoustical Society of America,* Vol. 39, No. 4, pp. 629–644, 1966.

7. J. T. Cannon and S. Dostrovsky, *The Evolution of Dynamics: Vibration Theory from 1687 to 1742*, Springer-Verlag, New York, 1981.

8. L. L. Bucciarelli and N. Dworsky, *Sophie Germain: An Essay in the History of the Theory of Elasticity*, D. Reidel, Dordrecht, The Netherlands, 1980.

9. S. S. Rao, *Mechanical Vibrations*, 4th ed., Prentice Hall, Upper Saddle River, NJ, 2004.

10. E. B. Magrab, *Vibrations of Elastic Structural Members*, Sijthoff & Noordhoff, Alphen aan den Rijn, The Netherlands, 1979.

11. L. Fryba, *Vibration of Solids and Structures Under Moving Loads*, Noordhoff International Publishing, Groningen, The Netherlands, 1972.

12. W. Nowacki, *Dynamics of Elastic Systems*, translated by H. Zorski, Wiley, New York, 1963.

13. L. Meirovitch, *Analytical Methods in Vibrations*, Macmillan, New York, 1967.

14. S. K. Clark, *Dynamics of Continuous Elements*, Prentice-Hall, Englewood Cliffs, NJ, 1972.

15. A. W. Leissa, *Vibration of Plates*, NASA SP-160, National Aeronautics and Space Administration, Washington, DC, 1969.

16. A. W. Leissa, *Vibration of Shells*, NASA SP-288, National Aeronautics and Space Administration, Washington, DC, 1973.

17. C. M. Harris and A. G. Piersol, Eds., *Harris' Shock and Vibration Handbook*, 5th ed., McGraw-Hill, New York, 2002.

18. D. I. G. Jones, *Handbook of Viscoelastic Vibration Damping*, Wiley, Chichester, West Sussex, England, 2001.

19. S. G. Braun, D. J. Ewans, and S. S. Rao, Eds., *Encyclopedia of Vibration*, 3 vol., Academic Press, San Diego, CA, 2002.

20. A. J. Pretlove, *BASIC Mechanical Vibrations*, Butterworths, London, 1985.

21. International Organization for Standardization, *Acoustics, Vibration and Shock: Handbook of International Standards for Acoustics, Mechanical Vibration and Shock*, Standards Handbook 4, ISO, Geneva, Switzerland, 1980.

22. R. D. Blevins, *Formulas for Natural Frequency and Mode Shape*, Van Nostrand Reinhold, New York, 1979.

23. I. A. Karnovsky and O. I. Lebed, *Free Vibrations of Beams and Frames: Eigenvalues and Eigenfunctions*, McGraw-Hill, New York, 2004.

24. I. A. Karnovsky and O. I. Lebed, *Non-classical Vibrations of Arches and Beams: Eigenvalues and Eigenfunctions*, McGraw-Hill, New York, 2004.

25. J. Wijker, *Mechanical Vibrations in Spacecraft Design*, Springer-Verlag, Berlin, 2004.

26. R. D. Blevins, *Flow-Induced Vibration*, 2nd ed., Krieger, Melbourne, FL, 2001.

27. H. S. Tzou and L. A. Bergman, Eds., *Dynamics and Control of Distributed Systems*, Cambridge University Press, Cambridge, 1998.

28. J. P. Wolf and A. J. Deaks, *Foundation Vibration Analysis: A Strength of Materials Approach*, Elsevier, Amsterdam, 2004.

29. J. D. Smith, *Gears and Their Vibration: A Basic Approach to Understanding Gear Noise*, Marcel Dekker, New York, 1983.

30. J. D. Smith, *Vibration Measurement Analysis*, Butterworths, London, 1989.

31. G. Lipovszky, K. Solyomvari, and G. Varga, *Vibration Testing of Machines and Their Maintenance*, Elsevier, Amsterdam, 1990.

32. S. Korablev, V. Shapin, and Y. Filatov, in *Vibration Diagnostics in Precision Instruments*, Engl. ed., E. Rivin, Ed., Hemisphere Publishing, New York, 1989.

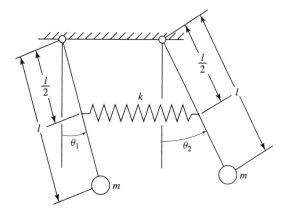

Figure 1.14   Two simple pendulums connected by a spring.

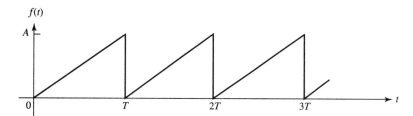

Figure 1.15   Sawtooth function.

PROBLEMS

1.1   Express the following function as a sum of sine and cosine functions:

$$f(t) = 5\sin(10t - 2.5)$$

1.2   Consider the following harmonic functions:

$$x_1(t) = 5\sin 20t \quad \text{and} \quad x_2(t) = 8\cos\left(20t + \frac{\pi}{3}\right)$$

Express the function $x(t) = x_1(t) + x_2(t)$ as (**a**) a cosine function with a phase angle, and (**b**) a sine function with a phase angle.

1.3   Find the difference of the harmonic functions $x_1(t) = 6\sin 30t$ and $x_2(t) = 4\cos(30t + \pi/4)$ (**a**) as a sine function with a phase angle, and (**b**) as a cosine function with a phase angle.

1.4   Find the sum of the harmonic functions $x_1(t) = 5\cos\omega t$ and $x_2(t) = 10\cos(\omega t + 1)$ using (**a**) trigonometric relations, (**b**) vectors, and (**c**) complex numbers.

1.5   The angular motions of two simple pendulums connected by a soft spring of stiffness k are described by (Fig. 1.14)

$$\theta_1(t) = A\cos\omega_1 t\cos\omega_2 t, \qquad \theta_2(t) = A\sin\omega_1 t\sin\omega_2 t$$

where A is the amplitude of angular motion and ω_1 and ω_2 are given by

$$\omega_1 = \frac{k}{8m}\sqrt{\frac{l}{g}}, \qquad \omega_2 = \sqrt{\frac{g}{l}} + \omega_1$$

Plot the functions $\theta_1(t)$ and $\theta_2(t)$ for $0 \le t \le 13.12$ s and discuss the resulting motions for the following data: $k = 1$ N/m, $m = 0.1$ kg, $l = 1$ m, and $g = 9.81$ m/s^2.

1.6   Find the Fourier cosine and sine series expansion of the function shown in Fig. 1.15 for $A = 2$ and $T = 1$.

1.7   Find the Fourier cosine and sine series representation of a series of half-wave rectified sine pulses shown in Fig. 1.16 for $A = \pi$ and $T = 2$.

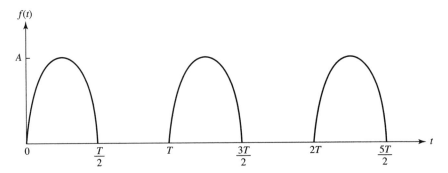

Figure 1.16   Half sine pulses.

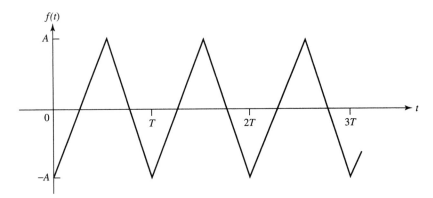

Figure 1.17   Triangular wave.

1.8   Find the complex Fourier series expansion of the sawtooth function shown in Fig. 1.15.

1.9   Find the Fourier series expansion of the triangular wave shown in Fig. 1.17.

1.10 Find the complex Fourier series representation of the function $f(t) = e^{-2t}$, $-\pi < t < \pi$.

1.11 Consider a transient load, $f(t)$, given by

$$f(t) = \begin{cases} 0, & t < 0 \\ e^{-t}, & t \geq 0 \end{cases}$$

Find the Fourier transform of $f(t)$.

1.12 The Fourier sine transform of a function $f(t)$, denoted by $F_s(\omega)$, is defined as

$$F_s(\omega) = \int_0^\infty f(t) \sin \omega t \, dt, \qquad \omega > 0$$

and the inverse of the transform $F_s(\omega)$ is defined by

$$f(t) = \frac{2}{\pi} \int_0^\infty F_s(\omega) \sin \omega t \, d\omega, \qquad t > 0$$

Using these definitions, find the Fourier sine transform of the function $f(t) = e^{-at}$, $a > 0$.

1.13 Find the Fourier sine transform of the function $f(t) = te^{-t}$, $t \geq 0$.

1.14 Find the Fourier transform of the function

$$f(t) = \begin{cases} e^{-at}, & t \geq 0 \\ 0, & t < 0 \end{cases}$$

2

Vibration of Discrete Systems: Brief Review

2.1 VIBRATION OF A SINGLE-DEGREE-OF-FREEDOM SYSTEM

The number of degrees of freedom of a vibrating system is defined by the minimum number of displacement components required to describe the configuration of the system during vibration. Each system shown in Fig. 2.1 denotes a single-degree-of-freedom system. The essential features of a vibrating system include (1) a mass m, producing an inertia force: $m\ddot{x}$; (2) a spring of stiffness k, producing a resisting force: kx; and (3) a damping mechanism that dissipates the energy. If the equivalent viscous damping coefficient is denoted as c, the damping force produced is $c\dot{x}$.

2.1.1 Free Vibration

In the absence of damping, the equation of motion of a single-degree-of-freedom system is given by

$$m\ddot{x} + kx = f(t) \tag{2.1}$$

where $f(t)$ is the force acting on the mass and $x(t)$ is the displacement of the mass m. The free vibration of the system, in the absence of the forcing function $f(t)$, is governed by the equation

$$m\ddot{x} + kx = 0 \tag{2.2}$$

The solution of Eq. (2.2) can be expressed as

$$x(t) = x_0 \cos \omega_n t + \frac{\dot{x}_0}{\omega_n} \sin \omega_n t \tag{2.3}$$

where ω_n is the natural frequency of the system, given by

$$\omega_n = \sqrt{\frac{k}{m}} \tag{2.4}$$

$x_0 = x(t = 0)$ is the initial displacement and $\dot{x}_0 = dx(t = 0)/dt$ is the initial velocity of the system. Equation (2.3) can also be expressed as

$$x(t) = A \, \cos(\omega_n t - \phi) \tag{2.5}$$

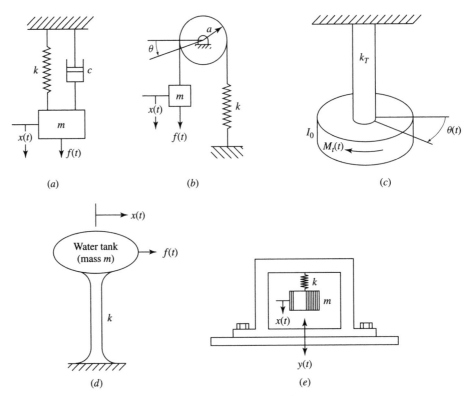

Figure 2.1 Single-degree-of-freedom systems.

or

$$x(t) = A \; \sin(\omega_n t + \phi_0) \qquad (2.6)$$

where

$$A = \left[x_0^2 + \left(\frac{\dot{x}_0}{\omega_n} \right)^2 \right]^{1/2} \qquad (2.7)$$

$$\phi = \tan^{-1} \frac{\dot{x}_0}{x_0 \omega_n} \qquad (2.8)$$

$$\phi_0 = \tan^{-1} \frac{x_0 \omega_n}{\dot{x}_0} \qquad (2.9)$$

The free vibration response of the system indicated by Eq. (2.5) is shown graphically in Fig. 2.2.

The equation of motion for the vibration of a viscously damped system is given by

$$m\ddot{x} + c\dot{x} + kx = f(t) \qquad (2.10)$$

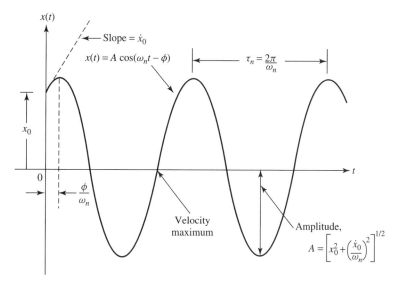

Figure 2.2 Free vibration response.

By dividing throughout by m, Eq. (2.10) can be rewritten as

$$\ddot{x} + 2\zeta\omega_n\dot{x} + \omega_n^2 x = F(t) \tag{2.11}$$

where ζ is the damping ratio, given by

$$\zeta = \frac{c}{2m\omega_n} = \frac{c}{c_c} \tag{2.12}$$

where c_c is known as the critical damping constant:

$$c_c = 2m\omega_n = 2\sqrt{km} \tag{2.13}$$

and

$$F(t) = \frac{f(t)}{m} \tag{2.14}$$

The system is considered to be underdamped, critically damped, and overdamped if the value of the damping ratio is less than 1, equal to 1, and greater than 1, respectively. The free vibration of a damped system is governed by the equation

$$\ddot{x} + 2\zeta\omega_n\dot{x} + \omega_n^2 x = 0 \tag{2.15}$$

The free vibration response of the system [i.e., the solution of Eq. (2.15)], with different levels of damping can be expressed as follows:

1. Underdamped system ($\zeta < 1$):

$$x(t) = e^{-\zeta\omega_n t}\left(x_0 \cos \omega_d t + \frac{\dot{x}_0 + \zeta\omega_n x_0}{\omega_d} \sin \omega_d t\right) \tag{2.16}$$

where $x_0 = x(t = 0)$ is the initial displacement, $\dot{x}_0 = dx(t = 0)/dt$ is the initial velocity, and ω_d is the frequency of the damped vibration given by

$$\omega_d = \sqrt{1 - \zeta^2}\,\omega_n \tag{2.17}$$

2. Critically damped system ($\zeta = 1$):

$$x(t) = [x_0 + (\dot{x}_0 + \omega_n x_0)\,t]\,e^{-\omega_n t} \tag{2.18}$$

3. Overdamped system ($\zeta > 1$):

$$x(t) = C_1 e^{(-\zeta + \sqrt{\zeta^2 - 1})\omega_n t} + C_2 e^{(-\zeta - \sqrt{\zeta^2 - 1})\omega_n t} \tag{2.19}$$

where

$$C_1 = \frac{x_0 \omega_n (\zeta + \sqrt{\zeta^2 - 1}) + \dot{x}_0}{2\omega_n \sqrt{\zeta^2 - 1}} \tag{2.20}$$

$$C_2 = \frac{-x_0 \omega_n (\zeta - \sqrt{\zeta^2 - 1}) - \dot{x}_0}{2\omega_n \sqrt{\zeta^2 - 1}} \tag{2.21}$$

The motions indicated by Eqs. (2.16), (2.18), and (2.19) are shown graphically in Fig. 2.3.

2.1.2 Forced Vibration under Harmonic Force

For an undamped system subjected to the harmonic force $f(t) = f_0 \cos \omega t$, the equation of motion is

$$m\ddot{x} + kx = f_0 \cos \omega t \tag{2.22}$$

where f_0 is the magnitude and ω is the frequency of the applied force. The steady-state solution or the particular integral of Eq. (2.22) is given by

$$x_p(t) = X \cos \omega t \tag{2.23}$$

where

$$X = \frac{f_0}{k - m\omega^2} = \frac{\delta_{st}}{1 - (\omega/\omega_n)^2} \tag{2.24}$$

denotes the maximum amplitude of the steady-state response and

$$\delta_{st} = \frac{f_0}{k} \tag{2.25}$$

indicates the static deflection of the mass under the force f_0. The ratio

$$\frac{X}{\delta_{st}} = \frac{1}{1 - (\omega/\omega_n)^2} \tag{2.26}$$

represents the ratio of the dynamic to static amplitude of motion and is called the *amplification factor*, *magnification factor*, or *amplitude ratio*. The variation of the amplitude

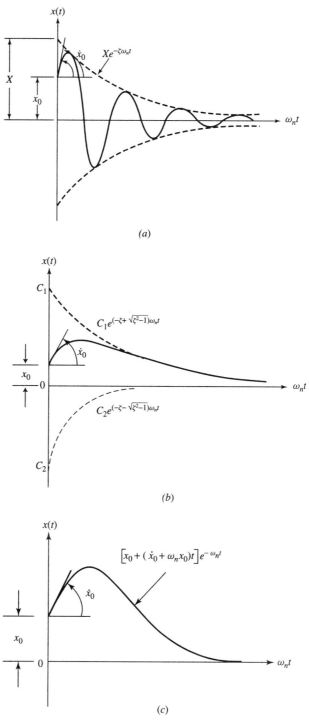

Figure 2.3   Damped free vibration response: (*a*) underdamped vibration ($\zeta < 1$); (*b*) over-damped vibration ($\zeta > 1$); (*c*) critically damped vibration ($\zeta = 1$).

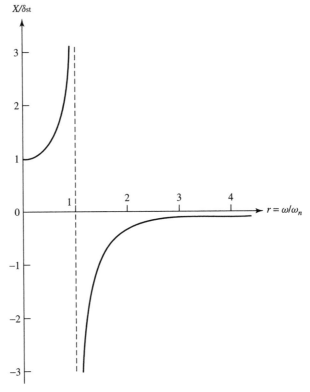

Figure 2.4   Magnification factor of an undamped system.

ratio with frequency ratio is shown in Fig. 2.4. The total solution of Eq.(2.22), including the homogeneous solution and the particular integral, is given by

$$x(t) = \left(x_0 - \frac{f_0}{k - m\omega^2}\right) \cos \omega_n t + \frac{\dot{x}_0}{\omega_n} \sin \omega_n t + \frac{f_0}{k - m\omega^2} \cos \omega t \qquad (2.27)$$

At resonance, $\omega/\omega_n = 1$, and the solution given by Eq. (2.27) can be expressed as

$$x(t) = x_0 \cos \omega_n t + \frac{\dot{x}_0}{\omega_n} \sin \omega_n t + \frac{\delta_{st} \omega_n t}{2} \sin \omega_n t \qquad (2.28)$$

This solution can be seen to increase indefinitely, with time as shown in Fig. 2.5. When a viscously damped system is subjected to the harmonic force $f(t) = f_0 \cos \omega t$, the equation of motion becomes

$$m\ddot{x} + c\dot{x} + kx = f_0 \cos \omega t \qquad (2.29)$$

The particular solution of Eq. (2.29) can be expressed as

$$x_p(t) = X \cos(\omega t - \phi) \qquad (2.30)$$

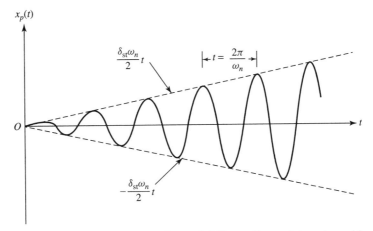

Figure 2.5   Response when $r = \omega/\omega_n = 1$ (effects of x_0 and $\dot{x}_0$ not considered).

where X is the amplitude and ϕ is the phase angle, given by

$$X = \frac{f_0}{[(k - m\omega^2)^2 + c^2\omega^2]^{1/2}} = \frac{\delta_{st}}{[(1 - r^2)^2 + (2\zeta r)^2]^{1/2}} \tag{2.31}$$

$$\phi = \tan^{-1}\frac{c\omega}{k - m\omega^2} = \tan^{-1}\frac{2\zeta r}{1 - r^2} \tag{2.32}$$

where

$$\delta_{st} = \frac{f_0}{k} \tag{2.33}$$

denotes the static deflection under the force f_0,

$$r = \frac{\omega}{\omega_n} \tag{2.34}$$

indicates the frequency ratio, and

$$\zeta = \frac{c}{c_c} = \frac{c}{2\sqrt{mk}} = \frac{c}{2m\omega_n} \tag{2.35}$$

represents the damping ratio. The variations of the amplitude ratio or magnification factor

$$\frac{X}{\delta_{st}} = \frac{1}{\sqrt{(1 - r^2)^2 + (2\zeta r)^2}} \tag{2.36}$$

and the phase angle, ϕ, given by Eq. (2.32), with the frequency ratio, r, are as shown in Fig. 2.6.

The total solution of Eq. (2.29), including the homogeneous solution and the particular integral, in the case of an underdamped system can be expressed as

$$x(t) = X_0 e^{-\zeta\omega_n t}\cos(\omega_d t - \phi_0) + X\cos(\omega t - \phi) \tag{2.37}$$

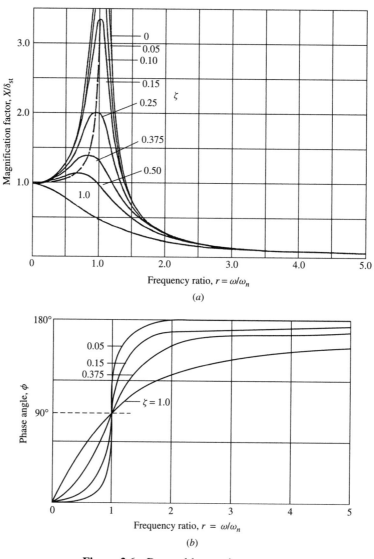

Figure 2.6 Damped harmonic response.

where ω_d is the frequency of damped vibration given by Eq. (2.17), X and ϕ are given by Eqs. (2.31) and (2.32), respectively, and X_0 and ϕ_0 can be determined from the initial conditions.

For example, if the initial conditions are given by $x(t = 0) = x_0$ and $dx(t = 0)/dt = \dot{x}_0$, Eq. (2.37) yields

$$x_0 = X_0 \cos \phi_0 + X \cos \phi \tag{2.38}$$

$$\dot{x}_0 = -\zeta \omega_n X_0 \cos \phi_0 + \omega_d X_0 \sin \phi_0 + \omega X \sin \phi \tag{2.39}$$

The solution of Eqs. (2.38) and (2.39) gives X_0 and ϕ_0.

If the harmonic force acting on the system is denoted in complex form, the equation of motion of the system becomes

$$m\ddot{x} + c\dot{x} + kx = f_0 e^{i\omega t} \tag{2.40}$$

where $i = \sqrt{-1}$. In this case, the particular solution of Eq. (2.40) can be expressed as

$$x_p(t) = X e^{i\omega t} \tag{2.41}$$

where X is a complex constant given by

$$X = \frac{f_0}{k - m\omega^2 + ic\omega} \tag{2.42}$$

which can be rewritten in the form

$$\frac{kX}{f_0} = H(i\omega) \equiv \frac{1}{1 - r^2 + i2\zeta r} \tag{2.43}$$

where $H(i\omega)$ is called the *complex frequency response* of the system. Equation (2.41) can be rewritten as

$$x_p(t) = \frac{f_0}{k} |H(i\omega)| e^{i(\omega t - \phi)} \tag{2.44}$$

where $|H(i\omega)|$ denotes the absolute value of $H(i\omega)$:

$$|H(i\omega)| = \left| \frac{kX}{f_0} \right| = \frac{1}{[(1 - r^2)^2 + (2\zeta r)^2]^{1/2}} \tag{2.45}$$

and ϕ indicates the phase angle:

$$\phi = \tan^{-1} \frac{c\omega}{k - m\omega^2} = \tan^{-1} \frac{2\zeta r}{1 - r^2} \tag{2.46}$$

2.1.3   Forced Vibration under General Force

For a general forcing function, $f(t)$, the solution of Eq. (2.11) can be found by taking Laplace transforms of the various terms using the relations

$$L[x(t)] = \overline{x}(s) \tag{2.47}$$

$$L[\dot{x}(t)] = s\overline{x}(s) - x(0) \tag{2.48}$$

$$L[\ddot{x}(t)] = s^2\overline{x}(s) - sx(0) - \dot{x}(0) \tag{2.49}$$

$$L[F(t)] = \overline{F}(s) \tag{2.50}$$

where $\overline{X}(s)$ and $F(s)$ are the Laplace transforms of $x(t)$ and $F(t)$, respectively. Thus, Eq. (2.11) becomes

$$[s^2\overline{x}(s) - sx(0) - \dot{x}(0)] + 2\zeta\omega_n[s\overline{x}(s) - x(0)] + \omega_n^2\overline{x}(s) = \overline{F}(s) \tag{2.51}$$

or

$$\overline{x}(s) = \frac{1}{\Delta}[\overline{F}(s) + (s + 2\zeta\omega_n)x_0 + \dot{x}_0] \tag{2.52}$$

where $x_0 = x(0)$ and $\dot{x}_0 = \dot{x}(0)$, and

$$\Delta = s^2 + 2\zeta\omega_n s + \omega_n^2 = (s + \zeta\omega_n)^2 + \omega_d^2 \tag{2.53}$$

By virtue of the inverse transforms

$$L^{-1}\left[\frac{1}{\Delta}\right] = \frac{e^{-\zeta\omega_n t}}{\omega_d}\sin\omega_d t \tag{2.54}$$

$$L^{-1}\left[\frac{s + \zeta\omega_n}{\Delta}\right] = e^{-\zeta\omega_n t}\cos\omega_d t \tag{2.55}$$

$$L^{-1}\left[\frac{\overline{F}(s)}{\Delta}\right] = \frac{1}{\omega_d}\int_0^t F(\tau)e^{-\zeta\omega_n(t-\tau)}\sin\omega_d(t-\tau)\, d\tau \tag{2.56}$$

The solution can be expressed as

$$x(t) = \int_0^t F(\tau)h(t-\tau)\, d\tau + g(t)x_0 + h(t)\dot{x}_0 \tag{2.57}$$

where

$$h(t) = \frac{1}{\omega_d}e^{-\zeta\omega_n t}\sin\omega_d t \tag{2.58}$$

$$g(t) = e^{-\zeta\omega_n t}\left(\cos\omega_d t + \frac{\zeta\omega_n}{\omega_d}\sin\omega_d t\right) \tag{2.59}$$

The first term in Eq. (2.57) is called the *convolution integral* or *Duhamel's integral*, and the second and third terms are called *transients* because of the presence of $e^{-\zeta\omega_n t}$, which is a decaying function of time. Note that in Eq. (2.57), the condition for an oscillatory solution is that $\zeta < 1$.

Example 2.1   Find the response of an underdamped spring–mass–damper system to a unit impulse by assuming zero initial conditions.

SOLUTION   The equation of motion can be expressed as

$$m\ddot{x} + c\dot{x} + kx = \delta(t) \tag{E2.1.1}$$

where $\delta(t)$ denotes the unit impulse. By taking the Laplace transform of both sides of Eq. (E2.1.1) and using the initial conditions $x_0 = \dot{x}_0 = 0$, we obtain

$$(ms^2 + cs + k)\,\overline{x}(s) = 1 \tag{E2.1.2}$$

or

$$\overline{x}(s) = \frac{1}{ms^2 + cs + k} = \frac{1/m}{s^2 + 2\zeta\omega_n s + \omega_n^2} \tag{E2.1.3}$$

Since $\zeta < 1$, the inverse transform yields

$$x(t) = \frac{1/m}{\omega_n\sqrt{1-\zeta^2}}e^{-\zeta\omega_n t}\sin\left(\omega_n\sqrt{1-\zeta^2}t\right) = \frac{1}{m\omega_d}e^{-\zeta\omega_n t}\sin\omega_d t \qquad \text{(E2.1.4)}$$

2.2 VIBRATION OF MULTIDEGREE-OF-FREEDOM SYSTEMS

A typical n-degree-of-freedom system is shown in Fig. 2.7(a). For a multidegree-of-freedom system, it is more convenient to use matrix notation to express the equations of motion and describe the vibrational response. Let x_i denote the displacement of mass m_i measured from its static equilibrium position; $i = 1, 2, \ldots, n$. The equations of motion of the n-degree-of-freedom system shown in Fig. 2.7(a) can be derived from the free-body diagrams of the masses shown in Fig. 2.7(b) and can be expressed in matrix form as

$$[m]\ddot{\vec{x}} + [c]\dot{\vec{x}} + [k]\vec{x} = \vec{f} \qquad (2.60)$$

where $[m]$, $[c]$, and $[k]$ denote the mass, damping, and stiffness matrices, respectively:

$$[m] = \begin{bmatrix} m_1 & 0 & 0 & \cdots & 0 \\ 0 & m_2 & 0 & \cdots & 0 \\ 0 & 0 & m_3 & \cdots & 0 \\ \vdots & & & \ddots & \\ 0 & 0 & 0 & \cdots & m_n \end{bmatrix} \qquad (2.61)$$

$$[c] = \begin{bmatrix} c_1+c_2 & -c_2 & 0 & \cdots & 0 & 0 \\ -c_2 & c_2+c_3 & -c_3 & \cdots & 0 & 0 \\ 0 & -c_3 & c_3+c_4 & \cdots & 0 & 0 \\ \vdots & \vdots & \vdots & \vdots & \vdots & \vdots \\ 0 & 0 & 0 & \cdots & -c_{n-1} & c_n \end{bmatrix} \qquad (2.62)$$

$$[k] = \begin{bmatrix} k_1+k_2 & -k_2 & 0 & \cdots & 0 & 0 \\ -k_2 & k_2+k_3 & -k_3 & \cdots & 0 & 0 \\ 0 & -k_3 & k_3+k_4 & \cdots & 0 & 0 \\ \vdots & \vdots & \vdots & & \vdots & \vdots \\ 0 & 0 & 0 & \cdots & -k_{n-1} & k_n \end{bmatrix} \qquad (2.63)$$

The vectors $\vec{x}$, $\dot{\vec{x}}$, and $\ddot{\vec{x}}$ indicate, respectively, the vectors of displacements, velocities, and accelerations of the various masses, and $\vec{f}$ represents the vector of forces acting on the masses:

$$\vec{x} = \begin{Bmatrix} x_1 \\ x_2 \\ x_3 \\ \vdots \\ x_n \end{Bmatrix}, \quad \dot{\vec{x}} = \begin{Bmatrix} \dot{x}_1 \\ \dot{x}_2 \\ \dot{x}_3 \\ \vdots \\ \dot{x}_n \end{Bmatrix}, \quad \ddot{\vec{x}} = \begin{Bmatrix} \ddot{x}_1 \\ \ddot{x}_2 \\ \ddot{x}_3 \\ \vdots \\ \ddot{x}_n \end{Bmatrix}, \quad \vec{f} = \begin{Bmatrix} f_1 \\ f_2 \\ f_3 \\ \vdots \\ f_n \end{Bmatrix} \qquad (2.64)$$

where a dot over x_i represents a time derivative of x_i.

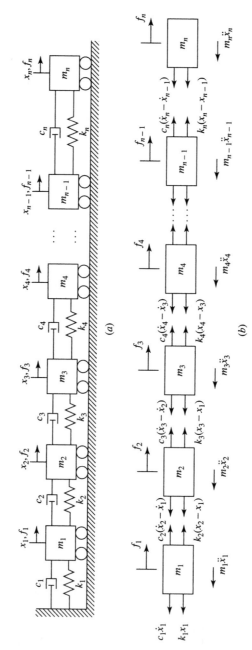

Figure 2.7 (a) An n-degree-of-freedom system; (b) free-body diagrams of the masses.

Note that the spring–mass–damper system shown in Fig. 2.7 is a particular case of a general n-degree-of-freedom system. In their most general form, the mass, damping, and stiffness matrices in Eq. (2.60) are fully populated and can be expressed as

$$[m] = \begin{bmatrix} m_{11} & m_{12} & m_{13} & \cdots & m_{1n} \\ m_{12} & m_{22} & m_{23} & \cdots & m_{2n} \\ \cdot & & & & \\ \cdot & & & & \\ m_{1n} & m_{2n} & m_{3n} & \cdots & m_{nn} \end{bmatrix} \tag{2.65}$$

$$[c] = \begin{bmatrix} c_{11} & c_{12} & c_{13} & \cdots & c_{1n} \\ c_{12} & c_{22} & c_{23} & \cdots & c_{2n} \\ \cdot & & & \cdots & \cdot \\ \cdot & & & \cdots & \cdot \\ \cdot & & & \cdots & \cdot \\ c_{1n} & c_{2n} & c_{3n} & \cdots & c_{nn} \end{bmatrix} \tag{2.66}$$

$$[k] = \begin{bmatrix} k_{11} & k_{12} & k_{13} & \cdots & k_{1n} \\ k_{12} & k_{22} & k_{23} & \cdots & k_{2n} \\ \cdot & & & \cdots & \cdot \\ \cdot & & & \cdots & \cdot \\ \cdot & & & \cdots & \cdot \\ k_{1n} & k_{2n} & k_{3n} & \cdots & k_{nn} \end{bmatrix} \tag{2.67}$$

Equation (2.60) denotes a system of n coupled second-order ordinary differential equations. These equations can be decoupled using a procedure called *modal analysis*, which requires the natural frequencies and normal modes or natural modes of the system. To determine the natural frequencies and normal modes, the eigenvalue problem corresponding to the vibration of the undamped system is to be solved.

2.2.1 Eigenvalue Problem

The free vibration of the undamped system is governed by the equation

$$[m]\ddot{\vec{x}} + [k]\vec{x} = \vec{0} \tag{2.68}$$

The solution of Eq. (2.68) is assumed to be harmonic as

$$\vec{x} = \vec{X} \sin(\omega t + \phi) \tag{2.69}$$

so that

$$\ddot{\vec{x}} = -\omega^2 \vec{X} \sin(\omega t + \phi) \tag{2.70}$$

where $\vec{X}$ is the vector of amplitudes of $\vec{x}(t)$, ϕ is the phase angle, and ω is the frequency of vibration. Substituting Eqs. (2.69) and (2.70) into Eq. (2.68), we obtain

$$[[k] - \omega^2[m]]\vec{X} = \vec{0} \tag{2.71}$$

Equation (2.71) represents a system of n algebraic homogeneous equations in unknown coefficients $X_1, X_2, \ldots, X_n$ (amplitudes of $x_1, x_2, \ldots, x_n$) with ω^2 playing the role of a parameter. For a nontrivial solution of the vector of coefficients $\vec{X}$, the determinant of the coefficient matrix must be equal to zero:

$$||[k] - \omega^2[m]|| = 0 \tag{2.72}$$

Equation (2.72) is a polynomial equation of nth degree in ω^2 (ω^2 is called the *eigenvalue*) and is called the *characteristic equation* or *frequency equation*.

The roots of the polynomial give the n eigenvalues, $\omega_1^2, \omega_2^2, \ldots, \omega_n^2$. The positive square roots of the eigenvalues yield the natural frequencies of the system, $\omega_1, \omega_2, \ldots, \omega_n$. The natural frequencies are usually arranged in increasing order of magnitude, so that $\omega_1 \leq \omega_2 \leq \ldots \leq \omega_n$. The lowest frequency ω_1 is referred to as the *fundamental frequency*. For each natural frequency ω_i, a corresponding nontrivial vector $\vec{X}^{(i)}$ can be obtained from Eq. (2.71):

$$[[k] - \omega_i^2[m]]\vec{X}^{(i)} = \vec{0} \tag{2.73}$$

The vector $\vec{X}^{(i)}$ is called the *eigenvector, characteristic vector, modal vector*, or *normal mode* corresponding to the natural frequency ω_i.

Of the n homogeneous equations represented by Eq. (2.73), any set of $n - 1$ equations can be solved to express any $n - 1$ quantities out of $X_1^{(i)}, X_2^{(i)}, \ldots, X_n^{(i)}$ in terms of the remaining $X^{(i)}$. Since Eq. (2.73) denotes a system of homogeneous equations, if $\vec{X}^{(i)}$ is a solution of Eq. (2.73), then $c_i \vec{X}^{(i)}$ is also a solution, where c_i is an arbitrary constant. This indicates that the shape of a natural mode is unique, but not its amplitude. Usually, a magnitude is assigned to the eigenvector $\vec{X}^{(i)}$ to make it unique using a process called *normalization*. A common normalization procedure, called *normalization with respect to the mass matrix*, consists of setting

$$\vec{X}^{(i)^\mathrm{T}}[m]\vec{X}^{(i)} = 1, \qquad i = 1, 2, \ldots, n \tag{2.74}$$

where the superscript T denotes the transpose.

2.2.2 Orthogonality of Modal Vectors

The modal vectors possess an important property known as *orthogonality* with respect to the mass matrix $[m]$ as well as the stiffness matrix $[k]$ of the system. To see this property, consider two distinct eigenvalues ω_i^2 and ω_j^2 and the corresponding eigenvectors $\vec{X}^{(i)}$ and $\vec{X}^{(j)}$. These solutions satisfy Eq. (2.71), so that

$$[k]\vec{X}^{(i)} = \omega_i^2[m]\vec{X}^{(i)} \tag{2.75}$$

$$[k]\vec{X}^{(j)} = \omega_j^2[m]\vec{X}^{(j)} \tag{2.76}$$

Premultiplication of both sides of Eq. (2.75) by $\vec{X}^{(j)^\mathrm{T}}$ and Eq. (2.76) by $\vec{X}^{(i)^\mathrm{T}}$ leads to

$$\vec{X}^{(j)^\mathrm{T}}[k]\vec{X}^{(i)} = \omega_i^2 \vec{X}^{(j)^\mathrm{T}}[m]\vec{X}^{(i)} \tag{2.77}$$

$$\vec{X}^{(i)^\mathrm{T}}[k]\vec{X}^{(j)} = \omega_j^2 \vec{X}^{(i)^\mathrm{T}}[m]\vec{X}^{(j)} \tag{2.78}$$

Noting that the matrices $[k]$ and $[m]$ are symmetric, we transpose Eq. (2.78) and subtract the result from Eq. (2.77), to obtain

$$(\omega_i^2 - \omega_j^2)\vec{X}^{(j)^\mathrm{T}}[m]\vec{X}^{(i)} = 0 \tag{2.79}$$

Since the eigenvalues are distinct, $\omega_i^2 \neq \omega_j^2$ and Eq. (2.79) leads to

$$\vec{X}^{(j)^{\mathrm{T}}}[m]\vec{X}^{(i)} = 0, \qquad i \neq j \tag{2.80}$$

Substitution of Eq. (2.80) in Eq. (2.77) results in

$$\vec{X}^{(j)^{\mathrm{T}}}[k]\vec{X}^{(i)} = 0, \qquad i \neq j \tag{2.81}$$

Equations (2.80) and (2.81) denote the orthogonality property of the eigenvectors with respect to the mass and stiffness matrices, respectively. When $j = i$, Eqs. (2.77) and (2.78) become

$$\vec{X}^{(i)^{\mathrm{T}}}[k]\vec{X}^{(i)} = \omega_i^2 \vec{X}^{(i)^{\mathrm{T}}}[m]\vec{X}^{(i)} \tag{2.82}$$

If the eigenvectors are normalized according to Eq. (2.74), Eq. (2.82) gives

$$\vec{X}^{(i)^{\mathrm{T}}}[k]\vec{X}^{(i)} = \omega_i^2 \tag{2.83}$$

By considering all the eigenvectors, Eqs. (2.74) and (2.83) can be written in matrix form as

$$[X]^{\mathrm{T}}[m][X] = [I] = \begin{bmatrix} 1 & & & 0 \\ & 1 & & \\ & & \ddots & \\ 0 & & & 1 \end{bmatrix} \tag{2.84}$$

$$[X]^{\mathrm{T}}[k][X] = [\omega_i^2] = \begin{bmatrix} \omega_1^2 & & & 0 \\ & \omega_2^2 & & \\ & & \ddots & \\ 0 & & & \omega_n^2 \end{bmatrix} \tag{2.85}$$

where the $n \times n$ matrix $[X]$, called the *modal matrix*, contains the eigenvectors $\vec{X}^{(1)}$, $\vec{X}^{(2)}, \ldots, \vec{X}^{(n)}$ as columns:

$$[X] = \begin{bmatrix} \vec{X}^{(1)} & \vec{X}^{(2)} & \cdots & \vec{X}^{(n)} \end{bmatrix} \tag{2.86}$$

2.2.3   Free Vibration Analysis of an Undamped System Using Modal Analysis

The free vibration of an undamped n-degree-of-freedom system is governed by the equations

$$[m]\ddot{\vec{x}} + [k]\vec{x} = \vec{0} \tag{2.87}$$

The n coupled second-order homogeneous differential equations represented by Eq. (2.87) can be uncoupled using modal analysis. In the analysis the solution, $\vec{x}(t)$, is expressed as a superposition of the normal modes $\vec{X}^{(i)}, i = 1, 2, \ldots, n$:

$$\vec{x}(t) = \sum_{i=1}^{n} \eta_i(t)\vec{X}^{(i)} = [X]\vec{\eta}(t) \tag{2.88}$$

where $[X]$ is the modal matrix, $\eta_i(t)$ are unknown functions of time, known as *modal coordinates* (or *generalized coordinates*), and $\vec{\eta}(t)$ is the vector of modal coordinates:

$$\vec{\eta}(t) = \begin{Bmatrix} \eta_1(t) \\ \eta_2(t) \\ \vdots \\ \eta_n(t) \end{Bmatrix} \tag{2.89}$$

Equation (2.88) represents the *expansion theorem* and is based on the fact that eigenvectors are orthogonal and form a basis in n-dimensional space. This implies that any vector, such as $\vec{x}(t)$, in n-dimensional space can be generated by a linear combination of a set of linearly independent vectors, such as the eigenvectors $\vec{X}^{(i)}, i = 1, 2, \ldots, n$. Substitution of Eq. (2.88) into Eq. (2.87) gives

$$[m][X]\ddot{\vec{\eta}} + [k][X]\vec{\eta} = \vec{0} \tag{2.90}$$

Premultiplication of Eq. (2.90) by $[X]^T$ leads to

$$[X]^T[m][X]\ddot{\vec{\eta}} + [X]^T[k][X]\vec{\eta} = \vec{0} \tag{2.91}$$

In view of Eqs. (2.84) and (2.85), Eq. (2.91) reduces to

$$\ddot{\vec{\eta}} + [\omega_i^2]\vec{\eta} = \vec{0} \tag{2.92}$$

which denotes a set of n uncoupled second-order differential equations:

$$\frac{d^2\eta_i(t)}{dt^2} + \omega_i^2\eta_i(t) = 0, \quad i = 1, 2, \ldots, n \tag{2.93}$$

If the initial conditions of the system are given by

$$\vec{x}(t = 0) = \vec{x}_0 = \begin{Bmatrix} x_{1,0} \\ x_{2,0} \\ \vdots \\ x_{n,0} \end{Bmatrix} \tag{2.94}$$

$$\dot{\vec{x}}(t = 0) = \dot{\vec{x}}_0 = \begin{Bmatrix} \dot{x}_{1,0} \\ \dot{x}_{2,0} \\ \vdots \\ \dot{x}_{n,0} \end{Bmatrix} \tag{2.95}$$

the corresponding initial conditions on $\vec{\eta}(t)$ can be determined as follows. Premultiply Eq. (2.88) by $[X]^T[m]$ and use Eq. (2.84) to obtain

$$\vec{\eta}(t) = [X]^T[m]\vec{x}(t) \tag{2.96}$$

Thus,

$$\left\{ \begin{array}{c} \eta_1(0) \\ \eta_2(0) \\ \vdots \\ \eta_n(0) \end{array} \right\} = \vec{\eta}(0) = [X]^{\mathrm{T}}[m]\vec{x}_0 \tag{2.97}$$

$$\left\{ \begin{array}{c} \dot{\eta}_1(0) \\ \dot{\eta}_2(0) \\ \vdots \\ \dot{\eta}_n(0) \end{array} \right\} = \dot{\vec{\eta}}(0) = [X]^{\mathrm{T}}[m]\dot{\vec{x}}_0 \tag{2.98}$$

The solution of Eq. (2.93) can be expressed as [see Eq. (2.3)]

$$\eta_i(t) = \eta_i(0)\cos\omega_i t + \frac{\dot{\eta}_i(0)}{\omega_i}\sin\omega_i t, \qquad i = 1, 2, \ldots, n \tag{2.99}$$

where $\eta_i(0)$ and $\dot{\eta}_i(0)$ are given by Eqs. (2.97) and (2.98) as

$$\eta_i(0) = \vec{X}^{(i)^{\mathrm{T}}}[m]\vec{x}_0 \tag{2.100}$$

$$\dot{\eta}_i(0) = \vec{X}^{(i)^{\mathrm{T}}}[m]\dot{\vec{x}}_0 \tag{2.101}$$

Once $\eta_i(t)$ are determined, the free vibration solution, $\vec{x}(t)$, can be found using Eq. (2.88).

Example 2.2   Find the free vibration response of the two-degree-of-freedom system shown in Fig. 2.8 using modal analysis for the following data: $m_1 = 2$ kg, $m_2 = 5$ kg, $k_1 = 10$N/m, $k_2 = 20$N/m, $k_3 = 5$N/m, $x_1(0) = 0.1$ m, $x_2(0) = 0$, $\dot{x}_1(0) = 0$, and $\dot{x}_2(0) = 5$ m/s.

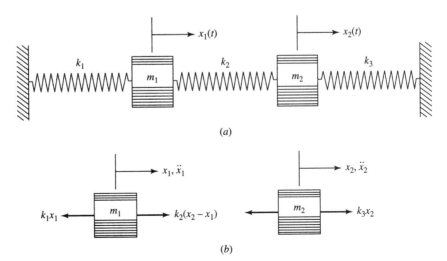

Figure 2.8   Two-degree-of-freedom system: (*a*) system in equilibrium; (*b*) free-body diagrams.

SOLUTION   The equations of motion can be expressed as

$$\begin{bmatrix} m_1 & 0 \\ 0 & m_2 \end{bmatrix} \begin{Bmatrix} \ddot{x}_1 \\ \ddot{x}_2 \end{Bmatrix} + \begin{bmatrix} k_1 + k_2 & -k_2 \\ -k_2 & k_2 + k_3 \end{bmatrix} \begin{Bmatrix} x_1 \\ x_2 \end{Bmatrix} = \begin{Bmatrix} 0 \\ 0 \end{Bmatrix} \tag{E2.2.1}$$

For free vibration, we assume harmonic motion as

$$x_i(t) = X_i \cos(\omega t + \phi), \qquad i = 1, 2 \tag{E2.2.2}$$

where X_i is the amplitude of $x_i(t)$, ω is the frequency, and ϕ is the phase angle. Substitution of Eq. (E2.2.2) into Eq. (E2.2.1) leads to the eigenvalue problem

$$\begin{bmatrix} -\omega^2 m_1 + k_1 + k_2 & -k_2 \\ -k_2 & -\omega^2 m_2 + k_2 + k_3 \end{bmatrix} \begin{Bmatrix} X_1 \\ X_2 \end{Bmatrix} = \begin{Bmatrix} 0 \\ 0 \end{Bmatrix} \tag{E2.2.3}$$

Using the known data, Eq. (E2.2.3) can be written as

$$\begin{bmatrix} -2\omega^2 + 30 & -20 \\ -20 & -5\omega^2 + 25 \end{bmatrix} \begin{Bmatrix} X_1 \\ X_2 \end{Bmatrix} = \begin{Bmatrix} 0 \\ 0 \end{Bmatrix} \tag{E2.2.4}$$

For a nontrivial solution of X_1 and X_2, the determinant of the coefficient matrix in Eq. (E2.2.4) is set equal to zero to obtain the frequency equation:

$$\begin{vmatrix} -2\omega^2 + 30 & -20 \\ -20 & -5\omega^2 + 25 \end{vmatrix} = 0$$

or

$$\omega^4 - 20\omega^2 + 35 = 0 \tag{E2.2.5}$$

The roots of Eq. (E2.2.5) give the natural frequencies of vibration of the system as

$$\omega_1 = 1.392028 \text{ rad/s}, \qquad \omega_2 = 4.249971 \text{ rad/s} \tag{E2.2.6}$$

Substitution of $\omega = \omega_1 = 1.392028$ in Eq. (E2.2.4) leads to $X_2^{(1)} = 1.306226 X_1^{(1)}$, while $\omega = \omega_2 = 4.249971$ in Eq. (E2.2.4) yields $X_2^{(2)} = -0.306226 X_1^{(2)}$. Thus, the mode shapes or eigenvectors of the system are given by

$$\vec{X}^{(1)} = \begin{Bmatrix} X_1^{(1)} \\ X_2^{(1)} \end{Bmatrix} = \begin{Bmatrix} 1 \\ 1.306226 \end{Bmatrix} X_1^{(1)} \tag{E2.2.7}$$

$$\vec{X}^{(2)} = \begin{Bmatrix} X_1^{(2)} \\ X_2^{(2)} \end{Bmatrix} = \begin{Bmatrix} 1 \\ -0.306226 \end{Bmatrix} X_1^{(2)} \tag{E2.2.8}$$

where $X_1^{(1)}$ and $X_1^{(2)}$ are arbitrary constants. By normalizing the mode shapes with respect to the mass matrix, we can find the values of $X_1^{(1)}$ and $X_1^{(2)}$ as

$$\vec{X}^{(1)^T} [m] \vec{X}^{(1)} = (X_1^{(1)})^2 \{ 1 \quad 1.306226 \} \begin{bmatrix} 2 & 0 \\ 0 & 5 \end{bmatrix} \begin{Bmatrix} 1 \\ 1.306226 \end{Bmatrix} = 1$$

or $X_1^{(1)} = 0.30815$, and

$$\vec{X}^{(2)^{\mathrm{T}}}[m]\vec{X}^{(2)} = (X_1^{(2)})^2\{\, 1 \quad -0.306226 \,\} \begin{bmatrix} 2 & 0 \\ 0 & 5 \end{bmatrix} \begin{Bmatrix} 1 \\ -0.306226 \end{Bmatrix} = 1$$

or $X_1^{(2)} = 0.63643$. Thus, the modal matrix becomes

$$[X] = [\, \vec{X}^{(1)} \quad \vec{X}^{(2)} \,] = \begin{bmatrix} 0.30815 & 0.63643 \\ 0.402513 & -0.19489 \end{bmatrix} \qquad (E2.2.9)$$

Using

$$\vec{x}(t) = [X]\vec{\eta}(t) \qquad (E2.2.10)$$

Eq. (E2.2.1) can be expressed in scalar form as

$$\frac{d^2\eta_i(t)}{dt^2} + \omega_i^2\eta_i(t) = 0, \qquad i = 1, 2 \qquad (E2.2.11)$$

The initial conditions of $\eta_i(t)$ can be determined using Eqs. (2.100) and (2.101) as

$$\eta_i(0) = \vec{X}^{(i)^{\mathrm{T}}}[m]\vec{x}(0) \quad \text{or} \quad \vec{\eta}(0) = [X]^{\mathrm{T}}[m]\vec{x}(0) \qquad (E2.2.12)$$

$$\dot{\eta}_i(0) = \vec{X}^{(i)^{\mathrm{T}}}[m]\dot{\vec{x}}(0) \quad \text{or} \quad \dot{\vec{\eta}}(0) = [X]^{\mathrm{T}}[m]\dot{\vec{x}}(0) \qquad (E2.2.13)$$

$$\vec{\eta}(0) = \begin{bmatrix} 0.30815 & 0.63643 \\ 0.402513 & -0.19489 \end{bmatrix}^{\mathrm{T}} \begin{bmatrix} 2 & 0 \\ 0 & 5 \end{bmatrix} \begin{Bmatrix} 0.1 \\ 0 \end{Bmatrix} = \begin{Bmatrix} 0.61630 \\ 1.27286 \end{Bmatrix} \qquad (E2.2.14)$$

$$\dot{\vec{\eta}}(0) = \begin{bmatrix} 0.30815 & 0.63643 \\ 0.402513 & -0.19489 \end{bmatrix}^{\mathrm{T}} \begin{bmatrix} 2 & 0 \\ 0 & 5 \end{bmatrix} \begin{Bmatrix} 0 \\ 5 \end{Bmatrix} = \begin{Bmatrix} 10.06282 \\ -4.87225 \end{Bmatrix} \qquad (E2.2.15)$$

The solution of Eq. (E2.2.11) is given by Eq. (2.99):

$$\eta_i(t) = \eta_i(0)\cos\omega_i t + \frac{\dot{\eta}_i(0)}{\omega_i}\sin\omega_i t, \qquad i = 1, 2 \qquad (E2.2.16)$$

Using the initial conditions of Eqs. (E2.2.14) and (E2.2.15), we find that

$$\eta_1(t) = 0.061630\cos 1.392028t + 7.22889\sin 1.392028t \qquad (E2.2.17)$$

$$\eta_2(t) = 0.127286\cos 4.249971t - 1.14642\sin 4.24997t \qquad (E2.2.18)$$

The displacements of the masses m_1 and m_2, in meters, can be determined from Eq. (E2.2.10) as

$$\vec{x}(t) = \begin{bmatrix} 0.30815 & 0.63643 \\ 0.402513 & -0.19489 \end{bmatrix} \begin{Bmatrix} 0.061630\cos 1.392028t + 7.22889\sin 1.392028t \\ 0.127286\cos 4.249971t - 1.14642\sin 4.24997t \end{Bmatrix}$$

$$= \begin{Bmatrix} 0.018991\cos 1.392028t + 2.22758\sin 1.392028t + 0.081009\cos 4.24997t \\ \quad - 0.72962\sin 4.24997t \\ 0.024807\cos 1.392028t + 2.909722\sin 1.392028t - 0.024807\cos 4.24997t \\ \quad + 0.223426\sin 4.24997t \end{Bmatrix}$$

$$(E2.2.19)$$

2.2.4   Forced Vibration Analysis of an Undamped System Using Modal Analysis

The equations of motion can be expressed as

$$[m]\ddot{\vec{x}} + [k]\vec{x} = \vec{f}(t) \tag{2.102}$$

The eigenvalues ω_i^2 and the corresponding eigenvectors $\vec{X}^{(i)}$, $i = 1, 2, \ldots, n$, of the system are assumed to be known. The solution of Eq. (2.102) is assumed to be given by a linear combination of the eigenvectors as

$$\vec{x}(t) = \sum_{i=1}^{n} \eta_i(t) \vec{X}^{(i)} = [X]\vec{\eta}(t) \tag{2.103}$$

where $\eta_i(t)$ denote modal coordinates and $[X]$ represents the modal matrix. Substituting Eq. (2.103) into Eq. (2.102) and premultiplying the result by $[X]^\mathrm{T}$ results in

$$[X]^\mathrm{T}[m][X]\ddot{\vec{\eta}} + [X]^\mathrm{T}[k][X]\vec{\eta} = [X]^\mathrm{T}\vec{f} \tag{2.104}$$

Using Eqs. (2.84) and (2.85), Eq. (2.104) can be written as

$$\ddot{\vec{\eta}} + [\omega_i^2]\vec{\eta} = \vec{Q} \tag{2.105}$$

where $\vec{Q}$ is called the *vector of modal forces* (or *generalized forces*) given by

$$\vec{Q}(t) = [X]^\mathrm{T}\vec{f}(t) \tag{2.106}$$

The n uncoupled differential equations indicated by Eq. (2.105) can be expressed in scalar form as

$$\frac{d^2\eta_i(t)}{dt^2} + \omega_i^2\eta_i(t) = Q_i(t), \qquad i = 1, 2, \ldots, n \tag{2.107}$$

where

$$Q_i(t) = \vec{X}^{(i)^\mathrm{T}}\vec{f}(t), \qquad i = 1, 2, \ldots, n \tag{2.108}$$

Each of the equations in (2.107) can be considered as the equation of motion of an undamped single-degree-of-freedom system subjected to a forcing function. Hence, the solution of Eq. (2.107) can be expressed, using $\eta_i(t)$, $Q_i(t)$, $\eta_{i,0}$, and $\dot{\eta}_{i,0}$ in place of $x(t)$, $F(t)$, x_0, and $\dot{x}_0$, respectively, and setting $\omega_d = \omega_i$ and $\zeta = 0$ in Eqs. (2.57)–(2.59), as

$$\eta_i(t) = \int_0^t Q_i(\tau)h(t - \tau)\, d\tau + g(t)\eta_{i,0} + h(t)\dot{\eta}_{i,0} \tag{2.109}$$

with

$$h(t) = \frac{1}{\omega_i}\sin\omega_i t \tag{2.110}$$

$$g(t) = \cos\omega_i t \tag{2.111}$$

The initial values $\eta_{i,0}$ and $\dot{\eta}_{i,0}$ can be determined from the known initial conditions $\vec{x}_0$ and $\dot{\vec{x}}_0$, using Eqs. (2.97) and (2.98).

2.2.5   Forced Vibration Analysis of a System with Proportional Damping

In proportional damping, the damping matrix $[c]$ in Eq. (2.60) can be expressed as a linear combination of the mass and stiffness matrices as

$$[c] = \alpha[m] + \beta[k] \tag{2.112}$$

where α and β are known constants. Substitution of Eq. (2.112) into Eq. (2.60) yields

$$[m]\ddot{\vec{x}} + (\alpha[m] + \beta[k])\dot{\vec{x}} + [k]\vec{x} = \vec{f} \tag{2.113}$$

As indicated earlier, in modal analysis, the solution of Eq. (2.113) is assumed to be of the form

$$\vec{x}(t) = [X]\vec{\eta}(t) \tag{2.114}$$

Substituting Eq. (2.114) into Eq. (2.113) and premultiplying the result by $[X]^{\mathrm{T}}$ leads to

$$[X]^{\mathrm{T}}[m][X]\ddot{\vec{\eta}} + (\alpha[X]^{\mathrm{T}}[m][X]\dot{\vec{\eta}} + \beta[X]^{\mathrm{T}}[k][X]\dot{\vec{\eta}}) + [X]^{\mathrm{T}}[k][X]\vec{\eta} = [X]^{\mathrm{T}}\vec{f} \tag{2.115}$$

When Eqs. (2.84) and (2.85) are used, Eq. (2.115) reduces to

$$\ddot{\vec{\eta}} + (\alpha[I] + \beta[\omega_i^2])\dot{\vec{\eta}} + [\omega_i^2]\vec{\eta} = \vec{Q} \tag{2.116}$$

where

$$\vec{Q} = [X]^{\mathrm{T}}\vec{f} \tag{2.117}$$

By defining

$$\alpha + \beta\omega_i^2 = 2\zeta_i\omega_i, \qquad i = 1, 2, \ldots, n \tag{2.118}$$

where ζ_i is called the *modal viscous damping factor* in the ith mode, Eq. (2.116) can be rewritten in scalar form as

$$\frac{d^2\eta_i(t)}{dt^2} + 2\zeta_i\omega_i\frac{d\eta_i(t)}{dt} + \omega_i^2\eta_i(t) = Q_i(t), \qquad i = 1, 2, \ldots, n \tag{2.119}$$

Each of the equations in (2.119) can be considered as the equation of motion of a viscously damped single-degree-of-freedom system whose solution is given by Eqs. (2.57)–(2.59). Thus, the solution of Eq. (2.119) is given by

$$\eta_i(t) = \int_0^t Q_i(\tau)h(t - \tau)\,d\tau + g(t)\eta_{i,0} + h(t)\dot{\eta}_{i,0} \tag{2.120}$$

where

$$h(t) = \frac{1}{\omega_{di}}e^{-\zeta_i\omega_i t}\sin\omega_{di}t \tag{2.121}$$

$$g(t) = e^{-\zeta_i\omega_i t}\left(\cos\omega_{di}t + \frac{\zeta_i\omega_i}{\omega_{di}}\sin\omega_{di}t\right) \tag{2.122}$$

and ω_{di} is the ith frequency of damped vibration:

$$\omega_{di} = \sqrt{1 - \zeta_i^2}\,\omega_i \tag{2.123}$$

2.2.6   Forced Vibration Analysis of a System with General Viscous Damping

The equations of motion of an n-degree-of-freedom system with arbitrary viscous damping can be expressed in the form of Eq. (2.60):

$$[m]\ddot{\vec{x}} + [c]\dot{\vec{x}} + [k]\vec{x} = \vec{f} \tag{2.124}$$

In this case, the modal matrix will not diagonalize the damping matrix, and an analytical solution is not possible in the configuration space. However, it is possible to find an analytical solution in the state space if Eq. (2.124) is expressed in state-space form. For this, we add the identity $\dot{\vec{x}}(t) = \dot{\vec{x}}(t)$ to an equivalent form of Eq. (2.124) as

$$\dot{\vec{x}}(t) = \dot{\vec{x}}(t) \tag{2.125}$$

$$\ddot{\vec{x}}(t) = -[m]^{-1}[c]\dot{\vec{x}}(t) - [m]^{-1}[k]\vec{x}(t) + [m]^{-1}\vec{f} \tag{2.126}$$

By defining a $2n$-dimensional state vector $\vec{y}(t)$ as

$$\vec{y}(t) = \begin{Bmatrix} \vec{x}(t) \\ \dot{\vec{x}}(t) \end{Bmatrix} \tag{2.127}$$

Eqs. (2.125) and (2.126) can be expressed in state form as

$$\dot{\vec{y}}(t) = [A]\vec{y}(t) + [B]\vec{f}(t) \tag{2.128}$$

where the coefficient matrices $[A]$ and $[B]$, of order $2n \times 2n$ and $2n \times n$, respectively, are given by

$$[A] = \begin{bmatrix} [0] & [I] \\ -[m]^{-1}[k] & -[m]^{-1}[c] \end{bmatrix} \tag{2.129}$$

$$[B] = \begin{bmatrix} [0] \\ [m]^{-1} \end{bmatrix} \tag{2.130}$$

Modal Analysis in State Space   For the modal analysis, first we consider the free vibration problem with $\vec{f} = \vec{0}$ so that Eq. (2.128) reduces to

$$\dot{\vec{y}}(t) = [A]\vec{y}(t) \tag{2.131}$$

This equation denotes a set of $2n$ first-order ordinary differential equations with constant coefficients. The solution of Eq. (2.131) is assumed to be of the form

$$\vec{y}(t) = \vec{Y}e^{\lambda t} \tag{2.132}$$

where $\vec{Y}$ is a constant vector and λ is a constant scalar. By substituting Eq. (2.132) into Eq. (2.131), we obtain, by canceling the term $e^{\lambda t}$ on both sides,

$$[A]\vec{Y} = \lambda \vec{Y} \tag{2.133}$$

Equation (2.133) can be seen to be a standard algebraic eigenvalue problem with a nonsymmetric real matrix, $[A]$. The solution of Eq. (2.133) gives the eigenvalues λ_i and the corresponding eigenvectors $\vec{Y}^{(i)}$, $i = 1, 2, \ldots, 2n$. These eigenvalues and eigenvectors can be real or complex. If λ_i is a complex eigenvalue, it can be shown that its complex conjugate $(\bar{\lambda}_i)$ will also be an eigenvalue. Also, the eigenvectors $\vec{Y}^{(i)}$ and $\bar{\vec{Y}}^{(i)}$, corresponding to λ_i and $\bar{\lambda}_i$, will also be complex conjugates to one another. The eigenvectors $\vec{Y}^{(i)}$ corresponding to the eigenvalue problem, Eq. (2.133), are called the *right eigenvectors* of the matrix $[A]$. The eigenvectors corresponding to the transpose of the matrix are called the *left eigenvectors* of $[A]$. Thus, the left eigenvectors, corresponding to the eigenvalues λ_i, are obtained by solving the eigenvalue problem

$$[A]^{\mathrm{T}}\vec{Z} = \lambda\vec{Z} \tag{2.134}$$

Since the determinants of the matrices $[A]$ and $[A]^{\mathrm{T}}$ are equal, the characteristic equations corresponding to Eqs. (2.133) and (2.134) will be identical:

$$\|[A] - \lambda[I]\| \equiv \|[A]^{\mathrm{T}} - \lambda[I]\| = 0 \tag{2.135}$$

Thus, the eigenvalues of Eqs. (2.133) and (2.134) will be identical. However, the eigenvectors of $[A]$ and $[A]^{\mathrm{T}}$ will be different. To find the relationship between $\vec{Y}^{(i)}$, $i = 1, 2, \ldots, 2n$ and $\vec{Z}^{(j)}$, $j = 1, 2, \ldots, 2n$, the eigenvalue problems corresponding to $\vec{Y}^{(i)}$ and $\vec{Z}^{(j)}$ are written as

$$[A]\vec{Y}^{(i)} = \lambda_i\vec{Y}^{(i)} \quad \text{and} \quad [A]^{\mathrm{T}}\vec{Z}^{(j)} = \lambda_j\vec{Z}^{(j)} \tag{2.136}$$

or

$$\vec{Z}^{(j)^{\mathrm{T}}}[A] = \lambda_j\vec{Z}^{(j)^{\mathrm{T}}} \tag{2.137}$$

Premultiplying the first of Eq. (2.136) by $\vec{Z}^{(j)^{\mathrm{T}}}$ and postmultiplying Eq. (2.137) by $\vec{Y}^{(i)}$, we obtain

$$\vec{Z}^{(j)^{\mathrm{T}}}[A]\vec{Y}^{(i)} = \lambda_i\vec{Z}^{(j)^{\mathrm{T}}}\vec{Y}^{(i)} \tag{2.138}$$

$$\vec{Z}^{(j)^{\mathrm{T}}}[A]\vec{Y}^{(i)} = \lambda_j\vec{Z}^{(j)^{\mathrm{T}}}\vec{Y}^{(i)} \tag{2.139}$$

Subtracting Eq. (2.139) from Eq. (2.138) gives

$$(\lambda_i - \lambda_j)\vec{Z}^{(j)^{\mathrm{T}}}\vec{Y}^{(i)} = 0 \tag{2.140}$$

Assuming that $\lambda_i \neq \lambda_j$, Eq. (2.140) yields

$$\vec{Z}^{(j)^{\mathrm{T}}}\vec{Y}^{(i)} = 0, \qquad i, \ j = 1, 2, \ldots, 2n \tag{2.141}$$

which show that the ith right eigenvector of $[A]$ is orthogonal to the jth left eigenvector of $[A]$, provided that the corresponding eigenvalues λ_i and λ_j are distinct. By substituting Eq. (2.141) into Eq. (2.138) or Eq. (2.139), we find that

$$\vec{Z}^{(j)^{\mathrm{T}}}[A]\,\vec{Y}^{(i)} = 0, \qquad i, \ j = 1, 2, \ldots, 2n \tag{2.142}$$

By setting $i = j$ in Eq. (2.138) or Eq. (2.139), we obtain

$$\vec{Z}^{(i)^{\mathrm{T}}}[A]\vec{Y}^{(i)} = \lambda_i\vec{Z}^{(i)^{\mathrm{T}}}\vec{Y}^{(i)}, \qquad i = 1, 2, \ldots, 2n \tag{2.143}$$

When the right and left eigenvectors of $[A]$ are normalized as

$$\vec{Z}^{(i)^{\mathrm{T}}}\vec{Y}^{(i)} = 1, \qquad i = 1, 2, \ldots, 2n \tag{2.144}$$

Eq. (2.143) gives

$$\vec{Z}^{(i)^{\mathrm{T}}}[A]\vec{Y}^{(i)} = \lambda_i, \qquad i = 1, 2, \ldots, 2n \tag{2.145}$$

Equations (2.144) and (2.145) can be expressed in matrix form as

$$[Z]^{\mathrm{T}}[Y] = [I] \tag{2.146}$$

$$[Z]^{\mathrm{T}}[A][Y] = [\lambda_i] \tag{2.147}$$

where the matrices of right and left eigenvectors are defined as

$$[Y] \equiv \begin{bmatrix} \vec{Y}^{(1)} & \vec{Y}^{(2)} & \cdots & \vec{Y}^{(2n)} \end{bmatrix} \tag{2.148}$$

$$[Z] \equiv \begin{bmatrix} \vec{Z}^{(1)} & \vec{Z}^{(2)} & \cdots & \vec{Z}^{(2n)} \end{bmatrix} \tag{2.149}$$

and the diagonal matrix of eigenvalues is given by

$$[\lambda_i] = \begin{bmatrix} \lambda_1 & & & 0 \\ & \lambda_2 & & \\ & & \ddots & \\ 0 & & & \lambda_{2n} \end{bmatrix} \tag{2.150}$$

In the modal analysis, the solution of the state equation, Eq. (2.128), is assumed to be a linear combination of the right eigenvectors as

$$\vec{y}(t) = \sum_{i=1}^{2n} \eta_i(t)\vec{Y}^{(i)} = [Y]\,\vec{\eta}(t) \tag{2.151}$$

where $\eta_i(t)$, $i = 1, 2, \ldots, 2n$, are modal coordinates and $\vec{\eta}(t)$ is the vector of modal coordinates:

$$\vec{\eta}(t) = \begin{Bmatrix} \eta_1(t) \\ \eta_2(t) \\ \vdots \\ \eta_{2n}(t) \end{Bmatrix} \tag{2.152}$$

Substituting Eq. (2.151) into Eq. (2.128) and premultiplying the result by $[Z]^{\mathrm{T}}$, we obtain

$$[Z]^{\mathrm{T}}[Y]\dot{\vec{\eta}}(t) = [Z]^{\mathrm{T}}[A]\,[Y]\,\vec{\eta}(t) + [Z]^{\mathrm{T}}[B]\,\vec{f}(t) \tag{2.153}$$

In view of Eqs. (2.146) and (2.147), Eq. (2.153) reduces to

$$\dot{\vec{\eta}}(t) = [\lambda_i]\,\vec{\eta}(t) + \vec{Q}(t) \tag{2.154}$$

which can be written in scalar form as

$$\frac{d\eta_i(t)}{dt} = \lambda_i \eta_i(t) + Q_i(t), \qquad i = 1, 2, \ldots, 2n \tag{2.155}$$

where the vector of modal forces is given by

$$\vec{Q}(t) = [Z]^{\mathrm{T}}[B] \, \vec{f}(t) \tag{2.156}$$

and the ith modal force by

$$Q_i(t) = \vec{Z}^{(i)^{\mathrm{T}}}[B]\vec{f}(t), \qquad i = 1, \, 2, \, \ldots, \, 2n \tag{2.157}$$

The solutions of the first-order ordinary differential equations, Eq. (2.155), can be expressed as

$$\eta_i(t) = \int_0^t e^{\lambda_i(t-\tau)} Q_i(\tau) \, d\tau + e^{\lambda_i t} \eta_i(0), \qquad i = 1, \, 2, \, \ldots, \, 2n \tag{2.158}$$

which can be written in matrix form as

$$\vec{\eta}(t) = \int_0^t e^{[\lambda_i](t-\tau)} \vec{Q}(\tau) \, d\tau + e^{[\lambda_i]t} \vec{\eta}(0) \tag{2.159}$$

where $\vec{\eta}(0)$ denotes the initial value of $\vec{\eta}(t)$. To determine $\vec{\eta}(0)$, we premultiply Eq. (2.151) by $\vec{Z}^{(i)^{\mathrm{T}}}$ to obtain

$$\vec{Z}^{(i)^{\mathrm{T}}} \vec{y}(t) = \vec{Z}^{(i)^{\mathrm{T}}}[Y] \, \vec{\eta}(t) \tag{2.160}$$

In view of the orthogonality relations, Eq. (2.141), Eq. (2.160) gives

$$\eta_i(t) = \vec{Z}^{(i)^{\mathrm{T}}} \vec{y}(t), \qquad i = 1, \, 2, \, \ldots, \, 2n \tag{2.161}$$

By setting $t = 0$ in Eq. (2.161), the initial value of $\eta_i(t)$ can be found as

$$\eta_i(0) = \vec{Z}^{(i)^{\mathrm{T}}} \vec{y}(0), \qquad i = 1, \, 2, \, \ldots, \, 2n \tag{2.162}$$

Finally, the solution of Eq. (2.128) can be expressed, using Eqs. (2.151) and (2.159), as

$$\vec{y}(t) = \int_0^t [Y]e^{[\lambda_i](t-\tau)} \vec{Q}(\tau) \, d\tau + [Y]e^{[\lambda_i]t}\vec{\eta}(0) \tag{2.163}$$

Example 2.3   Find the forced response of the viscously damped two-degree-of-freedom system shown in Fig. 2.9 using modal analysis for the following data: $m_1 = 2$ kg, $m_2 = 5$ kg, $k_1 = 10$ N/m, $k_2 = 20$ N/m, $k_3 = 5$ N/m, $c_1 = 2$ N $\cdot$ s/m, $c_2 = 3$ N $\cdot$ s/m, $c_3 = 1.0$ N $\cdot$ s/m, $f_1(t) = 0$, $f_2(t) = 5$ N, and $t \geq 0$. Assume the initial conditions to be zero.

SOLUTION   The equations of motion of the system are given by

$$[m]\ddot{\vec{x}} + [c]\dot{\vec{x}} + [k]\vec{x} = \vec{f} \tag{E2.3.1}$$

where

$$[m] = \begin{bmatrix} m_1 & 0 \\ 0 & m_2 \end{bmatrix} = \begin{bmatrix} 2 & 0 \\ 0 & 5 \end{bmatrix} \tag{E2.3.2}$$

$$[c] = \begin{bmatrix} c_1 + c_2 & -c_2 \\ -c_2 & c_2 + c_3 \end{bmatrix} = \begin{bmatrix} 5 & -3 \\ -3 & 4 \end{bmatrix} \tag{E2.3.3}$$

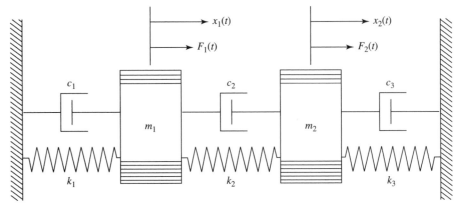

Figure 2.9   Viscously damped two-degree-of-freedom system.

$$[k] = \begin{bmatrix} k_1 + k_2 & -k_2 \\ -k_2 & k_2 + k_3 \end{bmatrix} = \begin{bmatrix} 30 & -20 \\ -20 & 25 \end{bmatrix} \qquad (E2.3.4)$$

$$\vec{x} = \begin{Bmatrix} x_1 \\ x_2 \end{Bmatrix}, \qquad \dot{\vec{x}} = \begin{Bmatrix} \dot{x}_1 \\ \dot{x}_2 \end{Bmatrix}, \qquad \ddot{\vec{x}} = \begin{Bmatrix} \ddot{x}_1 \\ \ddot{x}_2 \end{Bmatrix}, \qquad \vec{f} = \begin{Bmatrix} f_1 \\ f_2 \end{Bmatrix} \qquad (E2.3.5)$$

The equations of motion can be stated in state form as

$$\dot{\vec{y}} = [A]\vec{y} + [B]\vec{f} \qquad (E2.3.6)$$

where

$$[A] = \begin{bmatrix} [0] & [I] \\ -[m]^{-1}[k] & -[m]^{-1}[c] \end{bmatrix}$$

$$= \begin{bmatrix} 0 & 0 & 1 & 0 \\ 0 & 0 & 0 & 1 \\ -15 & 10 & -\frac{5}{2} & \frac{3}{2} \\ 4 & -5 & \frac{3}{5} & -\frac{4}{5} \end{bmatrix} \qquad (E2.3.7)$$

$$[B] = \begin{bmatrix} [0] \\ [m]^{-1} \end{bmatrix} = \begin{bmatrix} 0 & 0 \\ 0 & 0 \\ \frac{1}{2} & 0 \\ 0 & \frac{1}{5} \end{bmatrix} \qquad (E2.3.8)$$

$$\vec{y} = \begin{Bmatrix} x_1 \\ x_2 \\ \dot{x}_1 \\ \dot{x}_2 \end{Bmatrix} \qquad (E2.3.9)$$

The solution of the eigenvalue problem

$$[A]\vec{Y} = \lambda \vec{Y}$$

or

$$\begin{bmatrix} 0 & 0 & 1 & 0 \\ 0 & 0 & 0 & 1 \\ -15 & 10 & -\frac{5}{2} & \frac{3}{2} \\ 4 & -5 & \frac{3}{5} & -\frac{4}{5} \end{bmatrix} \begin{Bmatrix} Y_1 \\ Y_2 \\ Y_3 \\ Y_4 \end{Bmatrix} = \lambda \begin{Bmatrix} Y_1 \\ Y_2 \\ Y_3 \\ Y_4 \end{Bmatrix} \tag{E2.3.10}$$

is given by

$$\begin{aligned} \lambda_1 &= -1.4607 + 3.9902i \\ \lambda_2 &= -1.4607 - 3.9902i \\ \lambda_3 &= -0.1893 + 1.3794i \\ \lambda_4 &= -0.1893 - 1.3794i \end{aligned} \tag{E2.3.11}$$

$$[Y] \equiv \begin{bmatrix} \vec{Y}^{(1)} & \vec{Y}^{(2)} & \vec{Y}^{(3)} & \vec{Y}^{(4)} \end{bmatrix}$$

$$= \begin{bmatrix} -0.0754 - 0.2060i & -0.0754 + 0.2060i & -0.0543 - 0.3501i & -0.0543 + 0.3501i \\ 0.0258 + 0.0608i & 0.0258 - 0.0608i & -0.0630 - 0.4591i & -0.0630 + 0.4591i \\ 0.9321 & 0.9321 & 0.4932 - 0.0085i & 0.4932 + 0.0085i \\ -0.2803 + 0.0142i & -0.2803 - 0.0142i & 0.6452 & 0.6452 \end{bmatrix} \tag{E2.3.12}$$

The solution of the eigenvalue problem

$$[A]^{\mathrm{T}}\vec{Z} = \lambda \vec{Z}$$

or

$$\begin{bmatrix} 0 & 0 & -15 & 4 \\ 0 & 0 & 10 & -5 \\ 1 & 0 & -\frac{5}{2} & \frac{3}{5} \\ 0 & 1 & \frac{3}{2} & -\frac{4}{5} \end{bmatrix} \begin{Bmatrix} Z_1 \\ Z_2 \\ Z_3 \\ Z_4 \end{Bmatrix} = \lambda \begin{Bmatrix} Z_1 \\ Z_2 \\ Z_3 \\ Z_4 \end{Bmatrix} \tag{E2.3.13}$$

gives λ_i as indicated in Eq. (E2.3.11) and $\vec{Z}^{(i)}$ as

$$[Z] \equiv \begin{bmatrix} \vec{Z}^{(1)} & \vec{Z}^{(2)} & \vec{Z}^{(3)} & \vec{Z}^{(4)} \end{bmatrix}$$

$$= \begin{bmatrix} 0.7736 & 0.7736 & 0.2337 - 0.0382i & 0.2337 + 0.0382i \\ -0.5911 + 0.0032i & -0.5911 - 0.0032i & 0.7775 & 0.7775 \\ 0.0642 - 0.1709i & 0.0642 + 0.1709i & 0.0156 - 0.1697i & 0.0156 + 0.1697i \\ -0.0418 + 0.1309i & -0.0418 - 0.1309i & 0.0607 - 0.5538i & 0.0607 + 0.5538i \end{bmatrix} \tag{E2.3.14}$$

The vector of modal forces is given by

$$\vec{Q}(t) = [Z]^{T}[B]\vec{f}(t)$$

$$= \begin{bmatrix} 0.7736 & -0.5911+0.0032i & 0.0642-0.1709i & -0.0418+0.1309i \\ 0.7736 & -0.5911-0.0032i & 0.0642+0.1709i & -0.0418-0.1309i \\ 0.2337-0.0382i & 0.7775 & 0.0156-0.1697i & 0.0607-0.5538i \\ 0.2337+0.0382i & 0.7775 & 0.0156+0.1697i & 0.0607+0.5538i \end{bmatrix}$$

$$\cdot \begin{bmatrix} 0 & 0 \\ 0 & 0 \\ 0.5 & 0 \\ 0 & 0.2 \end{bmatrix} \begin{Bmatrix} 0 \\ 5 \end{Bmatrix} = \begin{Bmatrix} -0.0418+0.1309i \\ -0.0418-0.1309i \\ 0.0607-0.5538i \\ 0.0607+0.5538i \end{Bmatrix} \qquad \text{(E2.3.15)}$$

Since the initial values, $x_1(0)$, $x_2(0)$, $\dot{x}_1(0)$, and $\dot{x}_2(0)$, are zero, all $\eta_i(0) = 0, i = 1, 2, 3, 4$, from Eq. (2.162). Thus, the values of $\eta_i(t)$ are given by

$$\eta_i(t) = \int_0^t e^{\lambda_i(t-\tau)} Q_i(\tau)\,d\tau, \qquad i = 1, 2, 3, 4 \qquad \text{(E2.3.16)}$$

since $Q_i(\tau)$ is a constant (complex quantity), Eq. (E2.3.16) gives

$$\eta_i(t) = \frac{Q_i}{\lambda_i}(e^{\lambda_i t} - 1), \qquad i = 1, 2, 3, 4 \qquad \text{(E2.3.17)}$$

Using the values of Q_i and λ_i from Eqs. (E2.3.15) and (E2.3.11), $\eta_i(t)$ can be expressed as

$$\begin{aligned} \eta_1(t) &= (0.0323 - 0.0014i)\,[e^{(-1.4607+3.9902i)t} - 1] \\ \eta_2(t) &= (0.0323 + 0.0014i)\,[e^{(-1.4607-3.9902i)t} - 1] \\ \eta_3(t) &= (-0.4 + 0.0109i)\,[e^{(-0.1893+1.3794i)t} - 1] \\ \eta_4(t) &= (-0.4 - 0.0109i)\,[e^{(-0.1893-1.3794i)t} - 1] \end{aligned} \qquad \text{(E2.3.18)}$$

Finally, the state variables can be found from Eq. (2.151) as

$$\vec{y}(t) = [Y]\vec{\eta}(t) \qquad \text{(E2.3.19)}$$

In view of Eqs. (E2.3.12) and (E2.3.18), Eq. (E2.3.19) gives

$$\begin{aligned} y_1(t) &= 0.0456\,[e^{(-1.4607+3.9902i)t} - 1] & \text{m} \\ y_2(t) &= 0.0623\,[e^{(-1.4607-3.9902i)t} - 1] & \text{m} \\ y_3(t) &= -0.3342\,[e^{(-0.1893+1.3794i)t} - 1] & \text{m/s} \\ y_4(t) &= -0.5343\,[e^{(-0.1893-1.3794i)t} - 1] & \text{m/s} \end{aligned} \qquad \text{(E2.3.20)}$$

2.3 RECENT CONTRIBUTIONS

Single-Degree-of-Freedom Systems   Anderson and Ferri [5] investigated the properties of a single-degree-of-freedom system damped with generalized friction laws. The system was studied first by using an exact time-domain method and then by using first-order harmonic balance. It was observed that the response amplitude can be increased or decreased by the addition of amplitude-dependent friction. These results suggest that in situations where viscous damping augmentation is difficult or impractical, as in the case of space structures and turbomachinery bladed disks, beneficial damping properties can be achieved through the redesign of frictional interfaces. Bishop et al. [6] gave an elementary explanation of the Duhamel integral as well as Fourier and Laplace transform techniques in linear vibration analysis. The authors described three types of receptances and explained the relationships between them.

Multidegree-of-Freedom Systems   The dynamic absorbers play a major role in reducing vibrations of machinery. Soom and Lee [7] studied the optimal parameter design of linear and nonlinear dynamic vibration absorbers for damped primary systems. Shaw et al. [8] showed that the presence of nonlinearities can introduce dangerous instabilities, which in some cases may result in multiplication rather than reduction of the vibration amplitudes. For systems involving a large number of degrees of freedom, the size of the eigenvalue problem is often reduced using a model reduction or dynamic condensation process to find an approximate solution rapidly. Guyan reduction is a popular technique used for model reduction [9]. Lim and Xia [10] presented a technique for dynamic condensation based on iterated condensation. The quantification of the extent of nonproportional viscous damping in discrete vibratory systems was investigated by Prater and Singh [11]. Lauden and Akesson derived an exact complex dynamic member stiffness matrix for a damped second-order Rayleigh–Timoshenko beam vibrating in space [12].

The existence of classical real normal modes in damped linear vibrating systems was investigated by Caughey and O'Kelly [13]. They showed that the necessary and sufficient condition for a damped system governed by the equation of motion

$$[I]\ddot{\vec{x}}(t) + [A]\dot{\vec{x}}(t) + [B]\vec{x}(t) = \vec{f}(t) \tag{2.164}$$

to possess classical normal modes is that matrices $[A]$ and $[B]$ be commutative; that is, $[A][B] = [B][A]$. The scope of this criterion was reexamined and an alternative form of the condition was investigated by other researchers [14]. The settling time of a system can be defined as the time for the envelope of the transient part of the system response to move from its initial value to some fraction of the initial value. An expression for the settling time of an underdamped linear multidegree-of-freedom system was derived by Ross and Inman [15].

REFERENCES

1. S. S. Rao, *Mechanical Vibrations*, 4th ed., Prentice Hall, Upper Saddle River, NJ, 2004.
2. L. Meirovitch, *Fundamentals of Vibrations*, McGraw-Hill, New York, 2001.
3. D. J. Inman, *Engineering Vibration*, 2nd ed., Prentice Hall, Upper Saddle River, NJ, 2001.
4. A. K. Chopra, *Dynamics of Structures: Theory and Applications to Earthquake Engineering*, Prentice Hall, Englewood Cliffs, NJ, 1995.
5. J. R. Anderson and A. A. Ferri, Behavior of a single-degree-of-freedom system with a generalized friction law, *Journal of Sound and Vibration*, Vol. 140, No. 2, pp. 287–304, 1990.

6. R. E. D. Bishop, A. G. Parkinson, and J. W. Pendered, Linear analysis of transient vibration, *Journal of Sound and Vibration*, Vol. 9, No. 2, pp. 313–337, 1969.

7. A. Soom and M. Lee, Optimal design of linear and nonlinear vibration absorbers for damped system, *Journal of Vibrations, Acoustics, Stress and Reliability in Design*, Vol. 105, No. 1, pp. 112–119, 1983.

8. J. Shah, S. W. Shah, and A. G. Haddow, On the response of the nonlinear vibration absorber, *Journal of Non-linear Mechanics*, Vol. 24, pp. 281–293, 1989.

9. R. J. Guyan, Reduction of stiffness and mass matrices, *AIAA Journal*, Vol. 3, No. 2, p. 380, 1965.

10. R. Lim and Y. Xia, A new eigensolution of structures via dynamic condensation, *Journal of Sound and Vibration*, Vol. 266, No. 1, pp. 93–106, 2003.

11. G. Prater and R. Singh, Quantification of the extent of non-proportional viscous damping in discrete vibratory systems, *Journal of Sound and Vibration*, Vol. 104, No. 1, pp. 109–125, 1986.

12. R. Lauden and B. Akesson, Damped second-order Rayleigh–Timoshenko beam vibration in space: an exact complex dynamic member stiffness matrix, *International Journal for Numerical Methods in Engineering*, Vol. 19, No. 3, pp. 431–449, 1983.

13. T. K. Caughey and M. E. J. O'Kelly, Classical normal modes in damped systems, *Journal of Applied Mechanics*, Vol. 27, pp. 269–271, 1960.

14. A. S. Phani, On the necessary and sufficient conditions for the existence of classical normal modes in damped linear dynamic systems, *Journal of Sound and Vibration*, Vol. 264, No. 3, pp. 741–745, 2002.

15. A. D. S. Ross and D. J. Inman, Settling time of underdamped linear lumped parameter systems, *Journal of Sound and Vibration*, Vol. 140, No. 1, pp. 117–127, 1990.

PROBLEMS

2.1 A building frame with four identical columns that have an effective stiffness of k and a rigid floor of mass m is shown in Fig. 2.10. The natural period of vibration of the frame in the horizontal direction is found to be 0.45 s. When a heavy machine of mass 500 kg is mounted (clamped) on the floor, its natural period of vibration in the horizontal direction is found to be 0.55 s. Determine the effective stiffness k and mass m of the building frame.

2.2 The propeller of a wind turbine with four blades is shown in Fig. 2.11. The aluminum shaft AB on which the blades are mounted is a uniform hollow shaft of outer diameter 2 in., inner diameter 1 in., and length 10 in. If

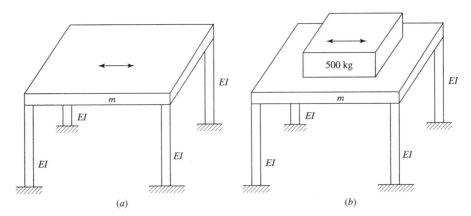

(a)   (b)

Figure 2.10

each blade has a mass moment of inertia of 0.5 lb-in.-sec^2, determine the natural frequency of vibration of the blades about the y-axis. [*Hint*: The torsional stiffness k_t of a shaft of length l is given by $k_t = GI_0/l$, where G is the shear modulus ($G = 3.8 \times 10^6$ psi for aluminum) and I_0 is the polar moment of inertia of the cross section of the shaft.]

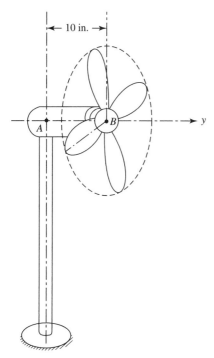

Figure 2.11

2.3 What is the difference between the damped and undamped natural frequencies and natural time periods for a damping ratio of 0.5?

2.4 A spring–mass system with mass 1 kg is found to vibrate with a natural frequency of 10 Hz. The same system when immersed in an oil is observed to vibrate with a natural frequency of 9 Hz. Find the damping constant of the oil.

2.5 Find the response of an undamped spring–mass system subjected to a constant force F_0 applied during $0 \le t \le \tau$ using a Laplace transform approach. Assume zero initial conditions.

2.6 A spring–mass system with mass 10 kg and stiffness 20,000 N/m is subjected to the force shown in

Fig. 2.12. Determine the response of the mass using the convolution integral.

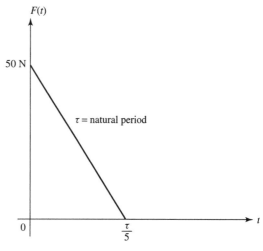

Figure 2.12

2.7 Find the response of a spring–mass system subjected to the force $F(t) = F_0 e^{i\omega t}$ using the method of Laplace transforms. Assume the initial conditions to be zero.

2.8 Consider a spring–mass system with $m = 10$ kg and $k = 5000$ N/m subjected to a harmonic force $F(t) = 400 \cos 10t$ N. Find the total system response with the initial conditions $x_0 = 0.1$ m and $\dot{x}_0 = 5$ m/s.

2.9 Consider a spring–mass–damper system with $m = 10$ kg, $k = 5000$ N/m, and $c = 200$ N·s/m subjected to a harmonic force $F(t) = 400 \cos 10t$ N. Find the steady-state and total system response with the initial conditions $x_0 = 0.1$ m and $\dot{x}_0 = 5$ m/s.

2.10 A simplified model of an automobile and its suspension system is shown in Fig. 2.13 with the following data: mass $m = 1000$ kg, radius r of gyration about the center of mass $G = 1.0$ m, spring constant of front suspension $k_f = 20$ kN/m, and spring constant of rear suspension $k_r = 15$ kN/m.

(a) Derive the equations of motion of an automobile by considering the vertical displacement of the center of mass y and rotation of the body about the center of mass θ as the generalized coordinates.

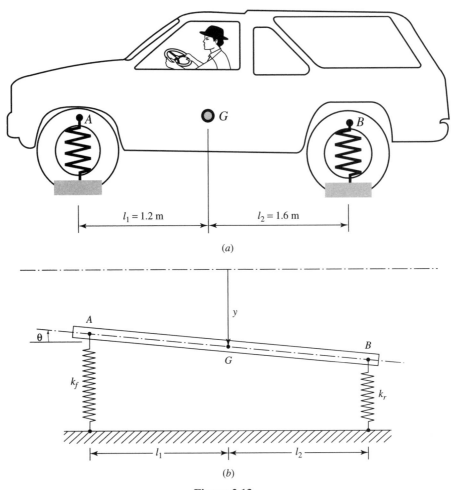

(a)

(b)

Figure 2.13

(b) Determine the natural frequencies and mode shapes of the automobile in bounce (up-and-down motion) and pitch (angular motion) modes.

2.11 Find the natural frequencies and the m-orthogonal mode shapes of the system shown in Fig. 2.9(a) for the following data: $k_1 = k_2 = k_3 = k$ and $m_1 = m_2 = m$.

2.12 Determine the natural frequencies and the m-orthogonal mode shapes of the system shown in Fig. 2.14.

2.13 Find the free vibration response of the system shown in Fig. 2.8(a) using modal analysis. The data are as follows: $m_1 = m_2 = 10$ kg, $k_1 = k_2 =$

$k_3 = 500$ N/m, $x_1(0) = 0.05$ m, $x_2(0) = 0.10$ m, and $\dot{x}_1(0) = \dot{x}_2(0) = 0$.

2.14 Consider the following data for the two-degree-of-freedom system shown in Fig. 2.9: $m_1 = 1$ kg, $m_2 = 2$ kg, $k_1 = 500$ N/m, $k_2 = 100$ N/m, $k_3 = 300$ N/m, $c_1 = 3$ N·s/m, $c_2 = 1$ N·s/m, and $c_3 = 2$ N·s/m.

(a) Derive the equations of motion.

(b) Discuss the nature of error involved if the off-diagonal terms of the damping matrix are neglected in the equations derived in part (a).

(c) Find the responses of the masses resulting from the initial conditions $x_1(0) = 5$ mm, $x_2(0) = 0$, $\dot{x}_1(0) = 1$ m/s, and $\dot{x}_2(0) = 0$.

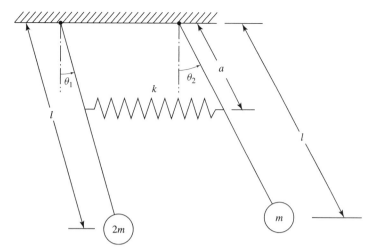

Figure 2.14

2.15 Determine the natural frequencies and m-orthogonal mode shapes of the three-degree-of-freedom system shown in Fig. 2.15 for the following data: $m_1 = m_3 = m$, $m_2 = 2\,m$, $k_1 = k_4 = k$, and $k_2 = k_3 = 2\,k$.

2.16 Find the free vibration response of the system described in Problem 2.14 using modal analysis

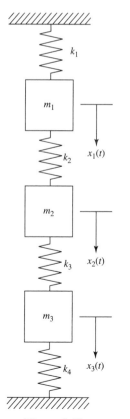

Figure 2.15

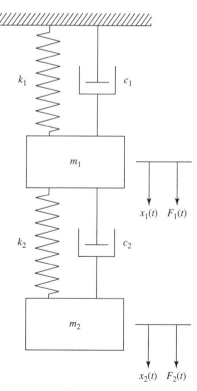

Figure 2.16

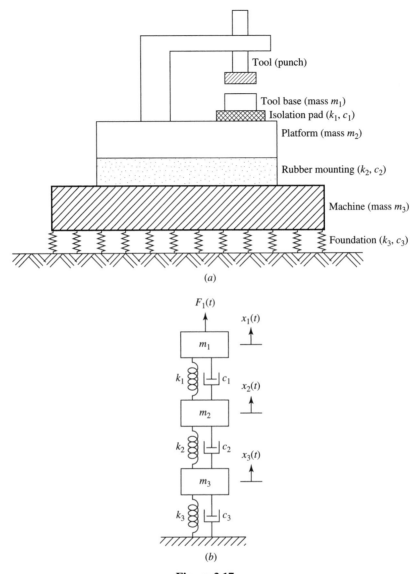

Figure 2.17

with the following data: $m = 2$ kg, $k = 100$ N/m, $x_1(0) = 0.1$ m, and $x_2(0) = x_3(0) = \dot{x}_1(0) = \dot{x}_2(0) = \dot{x}_3(0) = 0$.

2.17 Consider the two-degree-of-freedom system shown in Fig. 2.16 with the following data: $m_1 = 10$ kg, $m_2 = 1$ kg, $k_1 = 100$ N/m, $k_2 = 10$ N/m, and dampers c_1 and c_2 corresponding to proportional damping with $\alpha = 0.1$ and $\beta = 0.2$. Find the steady-state response of the system.

2.18 A punch press mounted on a foundation as shown in Fig. 2.17(a) has been modeled as a three-degree-of-freedom system as indicated in Fig. 2.17(b). The data are as follows: $m_1 = 200$ kg, $m_2 = 2000$ kg, $m_3 = 5000$ kg, $k_1 = 2 \times 10^5$ N/m, $k_2 = 1 \times 10^5$ N/m, and $k_3 = 5 \times 10^5$ N/m. The damping constants c_1, c_2, and c_3 correspond to modal damping ratios of $\zeta_1 = 0.02$, $\zeta_2 = 0.04$, and $\zeta_3 = 0.06$ in the first, second, and third modes of the system, respectively.

Find the response of the system using modal analysis when the tool base m_1 is subjected to an impact force $F_1(t) = 500\delta(t)$ N.

2.19 A spring–mass–damper system with $m = 0.05$ lb-sec^2/in., k $= 50$ lb/in., and $c = 1$ lb-sec/in., is subjected to a harmonic force of magnitude 20 lb. Find the resonant amplitude and the maximum amplitude of the steady-state motion.

2.20 A machine weighing 25 lb is subjected to a harmonic force of amplitude 10 lb and frequency 10 Hz. If the maximum displacement of the machine is to be restricted to 1 in., determine the necessary spring constant of the foundation for the machine. Assume the damping constant of the foundation to be 0.5 lb-sec/in.

3

Derivation of Equations: Equilibrium Approach

3.1 INTRODUCTION

The equations of motion of a vibrating system can be derived by using the dynamic equilibrium approach, variational method, or integral equation formulation. The dynamic equilibrium approach is considered in this chapter. The variational and integral equation approaches are presented in Chapters 4 and 5, respectively. The dynamic equilibrium approach can be implemented by using either Newton's second law of motion or D'Alembert's principle.

3.2 NEWTON'S SECOND LAW OF MOTION

Newton's second law of motion can be used conveniently to derive the equations of motion of a system under the following conditions:

1. The system undergoes either pure translation or pure rotation.
2. The motion takes place in a single plane.
3. The forces acting on the system either have a constant orientation or are oriented parallel to the direction along which the point of application moves.

If these conditions are not satisfied, application of Newton's second law of motion becomes complex, and other methods, such as the variational and integral equation approaches, can be used more conveniently. *Newton's second law of motion* can be stated as follows: The rate of change of the linear momentum of a system is equal to the net force acting on the system. Thus, if several forces $\vec{F}_1, \vec{F}_2, \ldots$ act on the system, the resulting force acting on the system is given by $\sum_i \vec{F}_i$ and Newton's second law of motion can be expressed as

$$\sum_i \vec{F}_i = \frac{d}{dt}(m\vec{v}) = m\vec{a} \tag{3.1}$$

where m is the constant mass, $\vec{v}$ is the linear velocity, $\vec{a}$ is the linear acceleration, and $m\vec{v}$ is the linear momentum. Equation (3.1) can be extended to angular motion. For the planar motion of a body, the angular momentum about the center of mass can be expressed as $I\omega$, where I is the constant mass moment of inertia of the body about an axis perpendicular to the plane of motion and passing through the centroid (centroidal axis) and ω is the angular velocity of the body. Then Newton's second law of motion states that the rate of change of angular momentum is equal to the net moment acting about the centroidal axis of the body:

$$\sum_i M_i = \frac{d}{dt}(I\omega) = I\dot{\omega} = I\alpha \tag{3.2}$$

where $M_1, M_2, \ldots$ denote the moments acting about the centroidal axis of the body and $\dot{\omega} = d\omega/dt = \alpha$, the angular acceleration of the body.

3.3 D'ALEMBERT'S PRINCIPLE

D'Alembert's principle is just a restatement of Newton's second law of motion. For the linear motion of a mass, Newton's second law of motion, Eq. (3.1), can be rewritten as

$$\sum_i \vec{F}_i - m\vec{a} = \vec{0} \tag{3.3}$$

Equation (3.3) can be considered as an equilibrium equation in which the sum of all forces, including the force $-m\vec{a}$ is in equilibrium. The term $-m\vec{a}$ represents a fictitious force called the *inertia force* or *D'Alembert force*. Equation (3.3) denotes D'Alembert's principle, which can be stated in words as follows: The sum of all external forces, including the inertia force, keeps the body in a state of dynamic equilibrium. Note that the minus sign associated with the inertia force in Eq. (3.3) denotes that when $\vec{a} = d\vec{v}/dt > 0$, the force acts in the negative direction. As can be seen from Eqs. (3.1) and (3.3), Newton's second law of motion and D'Alembert's principle are equivalent. However, Newton's second law of motion is more commonly used in deriving the equations of motion of vibrating bodies and systems. The equations of motion of the axial vibration of a bar, transverse vibration of a thin beam, and the transverse vibration of a thin plate are derived using the equilibrium approach in the following sections.

3.4 EQUATION OF MOTION OF A BAR IN AXIAL VIBRATION

Consider an elastic bar of length l with varying cross-sectional area $A(x)$, as shown in Fig. 3.1. The axial forces acting on the cross sections of a small element of the bar of length dx are given by P and $P + dP$ with

$$P = \sigma A = EA\frac{\partial u}{\partial x} \tag{3.4}$$

where σ is the axial stress, E is Young's modulus, u is the axial displacement, and $\partial u/\partial x$ is the axial strain. If $f(x, t)$ denotes the external force per unit length, the resulting force acting on the bar element in the x direction is

$$(P + dP) - P + f\,dx = dP + f\,dx$$

The application of Newton's second law of motion gives

$$\text{mass} \times \text{acceleration} = \text{resultant force}$$

or

$$\rho A\,dx\,\frac{\partial^2 u}{\partial t^2} = dP + f\,dx \tag{3.5}$$

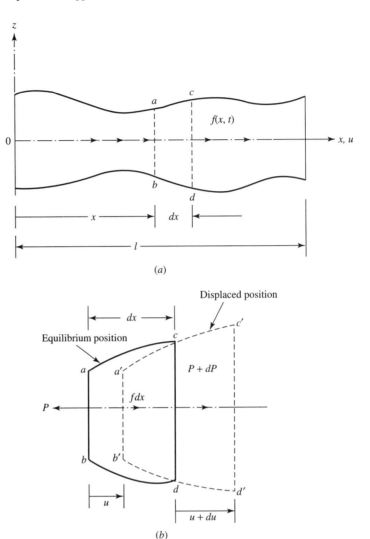

Figure 3.1   Longitudinal vibration of a bar.

where ρ is the mass density of the bar. By using the relation $dP = (\partial P / \partial x)\, dx$ and Eq. (3.4), the equation of motion for the forced longitudinal vibration of a nonuniform bar, Eq. (3.5), can be expressed as

$$\frac{\partial}{\partial x}\left[EA(x)\frac{\partial u(x,t)}{\partial x}\right] + f(x,t) = \rho(x)A(x)\frac{\partial^2 u}{\partial t^2}(x,t) \tag{3.6}$$

For a uniform bar, Eq. (3.6) reduces to

$$EA\frac{\partial^2 u}{\partial x^2}(x,t) + f(x,t) = \rho A\frac{\partial^2 u}{\partial t^2}(x,t) \tag{3.7}$$

Equation (3.6) or (3.7) can be solved using the appropriate initial and boundary conditions of the bar. For example, if the bar is subjected to a known initial displacement $u_0(x)$ and initial velocity $\dot{u}_0(x)$ the initial conditions can be stated as

$$u(x, t = 0) = u_0(x), \qquad 0 \le x \le l \tag{3.8}$$

$$\frac{\partial u}{\partial t}(x, t = 0) = \dot{u}_0(x), \qquad 0 \le x \le l \tag{3.9}$$

If the bar is fixed at $x = 0$ and free at $x = l$, the boundary conditions can be stated as follows. At the fixed end:

$$u(0, t) = 0, \qquad t > 0 \tag{3.10}$$

At the free end:

$$\text{axial force} = AE \frac{\partial u}{\partial x}(l, t) = 0$$

or

$$\frac{\partial u}{\partial x}(l, t) = 0, \qquad t > 0 \tag{3.11}$$

Other possible boundary conditions of the bar are discussed in Chapter 9.

3.5   EQUATION OF MOTION OF A BEAM IN TRANSVERSE VIBRATION

A thin beam subjected to a transverse force is shown in Fig. 3.2(a). Consider the free-body diagram of an element of a beam of length dx shown in Fig. 3.2(b), where $M(x, t)$ is the bending moment, $V(x, t)$ is the shear force, and $f(x, t)$ is the external transverse force per unit length of the beam. Since the inertia force (mass of the element times the acceleration) acting on the element of the beam is

$$\rho A(x) \, dx \frac{\partial^2 w}{\partial t^2}(x, t)$$

the force equation of motion in the z direction gives

$$-(V + dV) + f(x, t) \, dx + V = \rho A(x) \, dx \frac{\partial^2 w}{\partial t^2}(x, t) \tag{3.12}$$

where ρ is the mass density and $A(x)$ is the cross-sectional area of the beam. The moment equilibrium equation about the y axis passing through point P in Fig. 3.2 leads to

$$(M + dM) - (V + dV) \, dx + f(x, t) \, dx \frac{dx}{2} - M = 0 \tag{3.13}$$

Writing

$$dV = \frac{\partial V}{\partial x} \, dx \quad \text{and} \quad dM = \frac{\partial M}{\partial x} \, dx$$

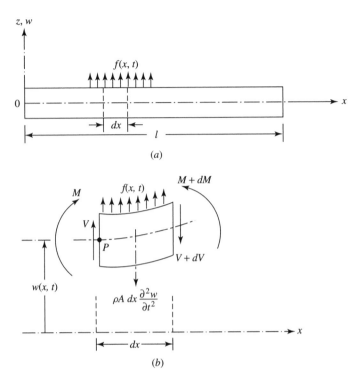

Figure 3.2 Transverse vibration of a thin beam.

and disregarding terms involving second powers in dx, Eqs. (3.12) and (3.13) can be written as

$$-\frac{\partial V}{\partial x}(x, t) + f(x, t) = \rho A(x)\frac{\partial^2 w}{\partial t^2}(x, t) \tag{3.14}$$

$$\frac{\partial M}{\partial x}(x, t) - V(x, t) = 0 \tag{3.15}$$

By using the relation $V = \partial M/\partial x$ from Eqs. (3.15), (3.14) becomes

$$-\frac{\partial^2 M}{\partial x^2}(x, t) + f(x, t) = \rho A(x)\frac{\partial^2 w}{\partial t^2}(x, t) \tag{3.16}$$

From the elementary theory of bending of beams (also known as the *Euler–Bernoulli* or *thin beam theory*), the relationship between bending moment and deflection can be expressed as [1, 2]

$$M(x, t) = EI(x)\frac{\partial^2 w}{\partial x^2}(x, t) \tag{3.17}$$

where E is Young's modulus and $I(x)$ is the moment of inertia of the beam cross section about the y axis. Inserting Eq. (3.17) into Eq. (3.16), we obtain the equation of motion for the forced lateral vibration of a nonuniform beam:

$$\frac{\partial^2}{\partial x^2}\left[EI(x)\frac{\partial^2 w}{\partial x^2}(x,t)\right] + \rho A(x)\frac{\partial^2 w}{\partial t^2}(x,t) = f(x,t) \tag{3.18}$$

For a uniform beam, Eq. (3.18) reduces to

$$EI\frac{\partial^4 w}{\partial x^4}(x,t) + \rho A\frac{\partial^2 w}{\partial t^2}(x,t) = f(x,t) \tag{3.19}$$

Equation (3.19) can be solved using the proper initial and boundary conditions. For example, if the beam is given an initial displacement $w_0(x)$ and an initial velocity $\dot{w}_0(x)$, the initial conditions can be expressed as

$$w(x, t=0) = w_0(x), \qquad 0 \le x \le l \tag{3.20}$$

$$\frac{\partial w}{\partial t}(x, t=0) = \dot{w}_0(x), \qquad 0 \le x \le l \tag{3.21}$$

If the beam is fixed at $x=0$ and pinned at $x=l$, the deflection and slope will be zero at $x=0$ and the deflection and the bending moment will be zero at $x=l$. Hence, the boundary conditions are given by

$$w(x=0, t) = 0, \qquad t > 0 \tag{3.22}$$

$$\frac{\partial w}{\partial x}(x=0, t) = 0, \qquad t > 0 \tag{3.23}$$

$$w(x=l, t) = 0, \qquad t > 0 \tag{3.24}$$

$$\frac{\partial^2 w}{\partial x^2}(x=l, t) = 0, \qquad t > 0 \tag{3.25}$$

Other possible boundary conditions of the beam are given in Chapter 11.

3.6   EQUATION OF MOTION OF A PLATE IN TRANSVERSE VIBRATION

The following assumptions are made in deriving the differential equation of motion of a transversely vibrating plate:

1. The thickness h of the plate is small compared to its other dimensions.
2. The middle plane of the plate does not undergo in-plane deformation (i.e., the middle plane is a neutral surface).
3. The transverse deflection w is small compared to the thickness of the plate.
4. The influence of transverse shear deformation is neglected (i.e., straight lines normal to the middle surface before deformation remain straight and normal after deformation).
5. The effect of rotary inertia is neglected.

The plate is referred to a system of orthogonal coordinates xyz. The middle plane of the plate is assumed to coincide with the xy plane before deformation, and the deflection of the middle surface is defined by $w(x, y, t)$, as shown in Fig. 3.3(a).

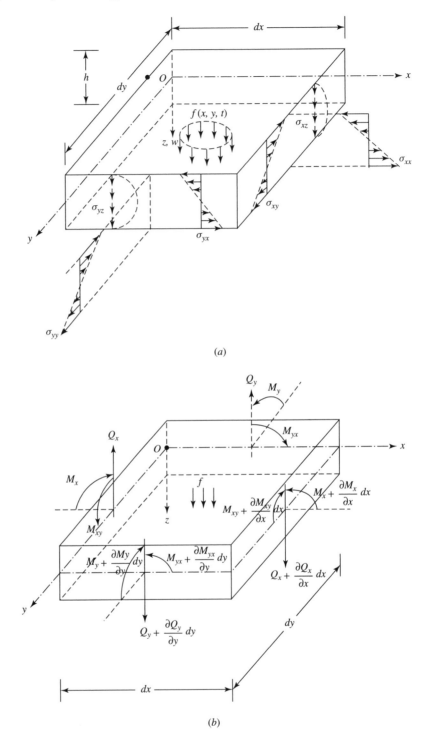

(a)

(b)

Figure 3.3   (a) Stresses in a plate; (b) forces and induced moment resultants in an element of a plate.

3.6.1 State of Stress

For thin plates subjected to bending forces (i.e., transverse loads and bending moments), the direct stress in the z direction (σ_{zz}) is usually neglected. Thus, the nonzero stress components are σ_{xx}, σ_{yy}, σ_{xy}, σ_{yz}, and σ_{xz}. As we are considering flexural (bending) deformations only, there will be no resulting force in the x and y directions; that is,

$$\int_{-h/2}^{h/2} \sigma_{xx}\, dz = 0, \qquad \int_{-h/2}^{h/2} \sigma_{yy}\, dz = 0 \tag{3.26}$$

It can be noted that in beams, which can be considered as one-dimensional analogs of plates, the shear stress σ_{xy} will not be present. As in beam theory, the stresses σ_{xx} (and σ_{yy}) and σ_{xz} (and σ_{yz}) are assumed to vary linearly and parabolically, respectively, over the thickness of the plate, as indicated in Fig. 3.3(a). The shear stress σ_{xy} is assumed to vary linearly over the thickness of the plate, as shown in Fig. 3.3(a). The stresses σ_{xx}, σ_{yy}, σ_{xy}, σ_{yz}, and σ_{xz} are used in defining the following force and moment resultants per unit length:

$$M_x = \int_{-h/2}^{h/2} \sigma_{xx} z\, dz$$

$$M_y = \int_{-h/2}^{h/2} \sigma_{yy} z\, dz$$

$$M_{xy} = \int_{-h/2}^{h/2} \sigma_{xy} z\, dz = M_{yx} \qquad \text{since} \quad \sigma_{yx} = \sigma_{xy} \tag{3.27}$$

$$Q_x = \int_{-h/2}^{h/2} \sigma_{xz}\, dz$$

$$Q_y = \int_{-h/2}^{h/2} \sigma_{yz}\, dz$$

These force and moment resultants are shown in Fig. 3.3(b).

3.6.2 Dynamic Equilibrium Equations

By considering an element of the plate, the differential equation of motion in terms of force and moment resultants can be derived. For this we consider the bending moments and shear forces to be functions of x, y, and t, so that if M_x acts on one side of the element, $M_x + dM_x = M_x + (\partial M_x/\partial x)dx$ acts on the opposite side. The resulting equations of motion can be written as follows.

Dynamic equilibrium of forces in the z direction:

$$\left(Q_x + \frac{\partial Q_x}{\partial x}\, dx \right) dy + \left(Q_y + \frac{\partial Q_y}{\partial y}\, dy \right) dx + f\, dx\, dy - Q_x\, dy - Q_y\, dx$$

$$= \text{mass of element} \times \text{acceleration in the } z \text{ direction}$$

$$= \rho h\, dx\, dy \frac{\partial^2 w}{\partial t^2}$$

or

$$\frac{\partial Q_x}{\partial x} + \frac{\partial Q_y}{\partial y} + f(x, y, t) = \rho h \frac{\partial^2 w}{\partial t^2} \tag{3.28}$$

where $f(x,y,t)$ is the intensity of the external distributed load and ρ is the density of the material of the plate.

Equilibrium of moments about the x axis:

$$\left(Q_y + \frac{\partial Q_y}{\partial y} dy\right) dx\, dy = \left(M_y + \frac{\partial M_y}{\partial y} dy\right) dx + \left(M_{xy} + \frac{\partial M_{xy}}{\partial x} dx\right) dy$$
$$- M_y\, dx - M_{xy}\, dy - f\, dx\, dy \frac{dy}{2}$$

By neglecting terms involving products of small quantities, this equation can be written as

$$Q_y = \frac{\partial M_y}{\partial y} + \frac{\partial M_{xy}}{\partial x} \tag{3.29}$$

Equilibrium of moments about the y axis:

$$\left(Q_x + \frac{\partial Q_x}{\partial x} dx\right) dy\, dx = \left(M_x + \frac{\partial M_x}{\partial x} dx\right) dy + \left(M_{yx} + \frac{\partial M_{yx}}{\partial y} dy\right) dx$$
$$- M_x\, dy - M_{yx}\, dx - f\, dx\, dy \frac{dx}{2}$$

or

$$Q_x = \frac{\partial M_x}{\partial x} + \frac{\partial M_{xy}}{\partial y} \tag{3.30}$$

3.6.3   Strain–Displacement Relations

To derive the strain–displacement relations, consider the bending deformation of a small element (by neglecting shear deformation), as shown in Fig. 3.4. In the edge view of the element (in the xz plane), $PQRS$ is the undeformed position and $P'Q'R'S'$ is the deformed position of the element. Due to the assumption that "normals to the middle plane of the undeformed plate remain straight and normal to the middle plane after deformation," line AB will become $A'B'$ after deformation. Thus, points such as K will have in-plane displacements u and v (parallel to the x and y axes), due to rotation of the normal AB about the y and x axes, respectively. The in-plane displacements of K can be expressed as (Fig. 3.4b and c)

$$u = -z \frac{\partial w}{\partial x}$$
$$v = -z \frac{\partial w}{\partial y} \tag{3.31}$$

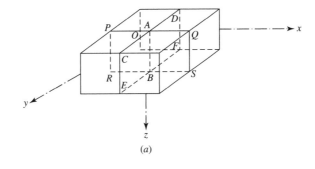

(a)

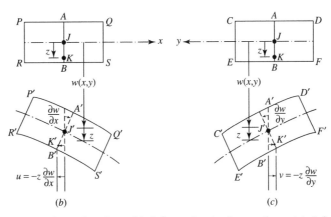

(b)   (c)

Figure 3.4 (a) Edge view of a plate; (b) deformation in the xz plane; (c) deformation in the yz plane.

The linear strain–displacement relations are given by

$$\varepsilon_{xx} = \frac{\partial u}{\partial x}, \qquad \varepsilon_{yy} = \frac{\partial v}{\partial y}, \qquad \varepsilon_{xy} = \frac{\partial u}{\partial y} + \frac{\partial v}{\partial x} \tag{3.32}$$

where ε_{xx} and ε_{yy} are normal strains parallel to the x and y axes, respectively, and ε_{xy} is the shear strain in the xy plane. Equations (3.31) and (3.32) yield

$$\varepsilon_{xx} = \frac{\partial u}{\partial x} = \frac{\partial}{\partial x}\left(-z\frac{\partial w}{\partial x}\right) = -z\frac{\partial^2 w}{\partial x^2}$$

$$\varepsilon_{yy} = \frac{\partial v}{\partial y} = \frac{\partial}{\partial y}\left(-z\frac{\partial w}{\partial y}\right) = -z\frac{\partial^2 w}{\partial y^2} \tag{3.33}$$

$$\varepsilon_{xy} = \frac{\partial u}{\partial y} + \frac{\partial v}{\partial x} = \frac{\partial}{\partial y}\left(-z\frac{\partial w}{\partial x}\right) + \frac{\partial}{\partial x}\left(-z\frac{\partial w}{\partial y}\right) = -2z\frac{\partial^2 w}{\partial x\partial y}$$

Equations (3.31) show that the transverse displacement $w(x, y, t)$ completely describes the deformation state of the plate.

3.6.4   Moment–Displacement Relations

We assume the plate to be in a state of plane stress. Thus, the stress–strain relations can be expressed as

$$\sigma_{xx} = \frac{E}{1-v^2}\varepsilon_{xx} + \frac{vE}{1-v^2}\varepsilon_{yy}$$

$$\sigma_{yy} = \frac{E}{1-v^2}\varepsilon_{yy} + \frac{vE}{1-v^2}\varepsilon_{xx} \tag{3.34}$$

$$\sigma_{xy} = G\varepsilon_{xy}$$

where E is Young's modulus, G is the shear modulus, and v is Poisson's ratio. By substituting Eq. (3.33) into Eq. (3.34) and the resulting stress into the first three equations of (3.27), we obtain, after integration,

$$M_x = -D\left(\frac{\partial^2 w}{\partial x^2} + v\frac{\partial^2 w}{\partial y^2}\right)$$

$$M_y = -D\left(\frac{\partial^2 w}{\partial y^2} + v\frac{\partial^2 w}{\partial x^2}\right) \tag{3.35}$$

$$M_{xy} = M_{yx} = -(1-v)D\frac{\partial^2 w}{\partial x \partial y}$$

where D, the *flexural rigidity of the plate*, is given by

$$D = \frac{Eh^3}{12(1-v^2)} \tag{3.36}$$

The flexural rigidity D is analogous to the flexural stiffness of a beam (EI). In fact, $D = EI$ for a plate of unit width when v is taken as zero. The use of Eqs. (3.35) in Eqs. (3.29) and (3.30) lead to the relations

$$Q_x = -D\frac{\partial}{\partial x}\left(\frac{\partial^2 w}{\partial x^2} + \frac{\partial^2 w}{\partial y^2}\right)$$

$$Q_y = -D\frac{\partial}{\partial y}\left(\frac{\partial^2 w}{\partial x^2} + \frac{\partial^2 w}{\partial y^2}\right) \tag{3.37}$$

3.6.5   Equation of Motion in Terms of Displacement

By substituting Eqs. (3.35) and (3.37) into Eqs. (3.28)–(3.30), we notice that moment equilibrium equations (3.29) and (3.30) are satisfied automatically, and Eq. (3.28) gives the desired equation of motion as

$$D\left(\frac{\partial^4 w}{\partial x^4} + 2\frac{\partial^4 w}{\partial x^2 \partial y^2} + \frac{\partial^4 w}{\partial y^4}\right) + \rho h\frac{\partial^2 w}{\partial t^2} = f(x, y, t) \tag{3.38}$$

If $f(x, y, t) = 0$, we obtain the free vibration equation as

$$D\left(\frac{\partial^4 w}{\partial x^4} + 2\frac{\partial^4 w}{\partial x^2 \partial y^2} + \frac{\partial^4 w}{\partial y^4}\right) + \rho h\frac{\partial^2 w}{\partial t^2} = 0 \tag{3.39}$$

Equations (3.38) and (3.39) can be written in a more general form:

$$D\nabla^4 w + \rho h \frac{\partial^2 w}{\partial t^2} = f \tag{3.40}$$

$$D\nabla^4 w + \rho h \frac{\partial^2 w}{\partial t^2} = 0 \tag{3.41}$$

where $\nabla^4 = \nabla^2 \nabla^2$, the *biharmonic operator*, is given by

$$\nabla^4 = \frac{\partial^4}{\partial x^4} + 2\frac{\partial^4}{\partial x^2 \partial y^2} + \frac{\partial^4}{\partial y^4} \tag{3.42}$$

in Cartesian coordinates.

3.6.6   Initial and Boundary Conditions

As the equation of motion, Eq. (3.38) or (3.39), involves fourth-order partial derivatives with respect to x and y, and second-order partial derivatives with respect to t, we need to specify four conditions in terms of each of x and y (i.e., two conditions for any edge) and two conditions in terms of t (usually, in the form of initial conditions) to find a unique solution of the problem. If the displacement and velocity of the plate at $t = 0$ are specified as $w_0(x, y)$ and $\dot{w}_0(x, y)$, the initial conditions can be expressed as

$$w(x, y, 0) = w_0(x, y) \tag{3.43}$$

$$\frac{\partial w}{\partial t}(x, y, 0) = \dot{w}_0(x, y) \tag{3.44}$$

The general boundary conditions that are applicable for any type of geometry of the plate can be stated as follows. Let n and s denote the coordinates in the directions normal and tangential to the boundary. At a fixed edge, the deflection and the slope along the normal direction must be zero:

$$w = 0 \tag{3.45}$$

$$\frac{\partial w}{\partial n} = 0 \tag{3.46}$$

For a simply supported edge, the deflection and the bending moment acting on the edge about the s direction must be zero; that is,

$$w = 0 \tag{3.47}$$

$$M_n = 0 \tag{3.48}$$

where the expression for M_n in terms of normal and tangential coordinates is given by [5]

$$M_n = -D\left[\nabla^2 w - (1 - \nu)\left(\frac{1}{R}\frac{\partial w}{\partial n} + \frac{\partial^2 w}{\partial s^2}\right)\right] \tag{3.49}$$

where R denotes the radius of curvature of the edge. For example, if the edge with $y = b = $ constant of a rectangular plate is simply supported, Eqs. (3.47) and (3.48) become

$$w(x, b) = 0, \qquad 0 \le x \le a \tag{3.50}$$

$$M_y = -D \left(\frac{\partial^2 w}{\partial y^2} + v \frac{\partial^2 w}{\partial x^2} \right)(x, b) = 0, \quad 0 \le x \le a \tag{3.51}$$

where the dimensions of the plate are assumed to be a and b parallel to the x and y axes, respectively. The other possible boundary conditions of the plate are discussed in Chapter 14.

3.7   ADDITIONAL CONTRIBUTIONS

In the equilibrium approach, the principles of equilibrium of forces and moments are used by considering an element of the physical system. This gives the analyst a physical feel of the problem. Hence, the approach has been used historically by many authors to derive equations of motion. For example, Love [6] considered the free-body diagram of a curved rod to derive coupled equations of motion for the vibration of a curved rod or beam. Timoshenko and Woinowsky-Krieger derived equations of motion for the vibration of plates and cylindrical shells [7]. Static equilibrium equations of symmetrically loaded shells of revolution have been derived using the equilibrium approach, and the resulting equations have subsequently been specialized for spherical, conical, circular cylindrical, toroidal, and ellipsoidal shells by Ugural [3] for determining the membrane stresses. The approach was also used to derive equilibrium equations of axisymmetrically loaded circular cylindrical and general shells of revolution by including the bending behavior.

In the equilibrium approach, the boundary conditions are developed by considering the physics of the problem. Although the equilibrium and variational approaches can give the same equations of motion, the variational methods have the advantage of yielding the exact form of the boundary conditions automatically. Historically, the development of plate theory, in terms of the correct forms of the governing equation and the boundary conditions, has been associated with the energy (or variational) approach. Several investigators, including Bernoulli, Germain, Lagrange, Poisson, and Navier, have attempted to present a satisfactory theory of plates but did not succeed completely. Later, Kirchhoff [8] derived the correct governing equations for plates using minimization of the (potential) energy and pointed out that there exist only two boundary conditions on a plate edge. Subsequently, Lord Kelvin and Tait [9] gave physical insight to the boundary conditions given by Kirchhoff by converting twisting moments along the edge of the plate into shearing forces. Thus, the edges are subject to only two forces: shear and moment.

REFERENCES

1. F. P. Beer, E. R. Johnston, Jr., and J. T. DeWolf, *Mechanics of Materials*, 3rd ed., McGraw-Hill, New York, 2002.

2. S. S. Rao, *Mechanical Vibrations*, 4th ed., Prentice Hall, Upper Saddle River, NJ, 2004.

3. A. C. Ugural, *Stresses in Plates and Shells*, McGraw-Hill, New York, 1981.

4. S. S. Rao, *The Finite Element Method in Engineering*, 3rd ed., Butterworth-Heinemann, Boston, 1999.

5. E. Ventsel and T. Krauthammer, *Thin Plates and Shells: Theory, Analysis, and Applications*, Marcel Dekker, New York, 2001.

6. A. E. H. Love, *A Treatise on the Mathematical Theory of Elasticity*, 4th ed, Dover, New York, 1944.

7. S. P. Timoshenko and S. Woinowsky-Krieger, *Theory of Plates and Shells*, McGraw-Hill, New York, 1959.

8. G. R. Kirchhoff, Über das Gleichgewicht und die Bewegung einer elastishen Scheibe, *Journal fuer die Reine und Angewandte Mathematik*, Vol. 40, pp. 51–88, 1850.

9. Lord Kelvin and P. G. Tait, *Treatise on Natural Philosophy*, Vol. 1, Clarendon Press, Oxford, 1883.

10. W. Flügge, *Stresses in Shells,* 2nd ed., Springer-Verlag, New York, 1973.

PROBLEMS

3.1 The system shown in Fig. 3.5 consists of a cylinder of mass M_0 and radius R that rolls without slipping on a horizontal surface. The cylinder is connected to a viscous damper of damping constant c and a spring of stiffness k. A uniform bar of length l and mass M is pin-connected to the center of the cylinder and is subjected to a force F at the other end. Derive the equations of motion of the two-degree-of-freedom system using the equilibrium approach.

3.2 Consider a prismatic bar with one end (at $x = 0$) connected to a spring of stiffness K_0 and the other end (at $x = l$) attached to a mass M_0 as shown in Fig. 3.6. The bar has a length of l, cross-sectional area A, mass density ρ, and modulus of elasticity E. Derive the equation of motion for the axial vibration of the bar and the boundary conditions using the equilibrium approach.

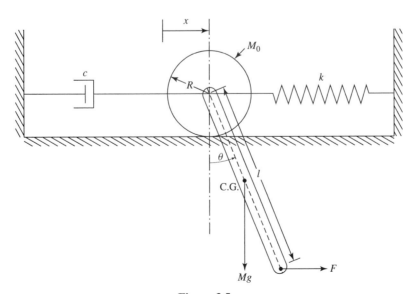

Figure 3.5

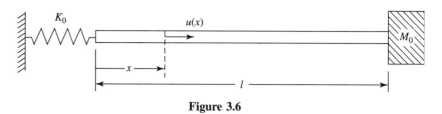

Figure 3.6

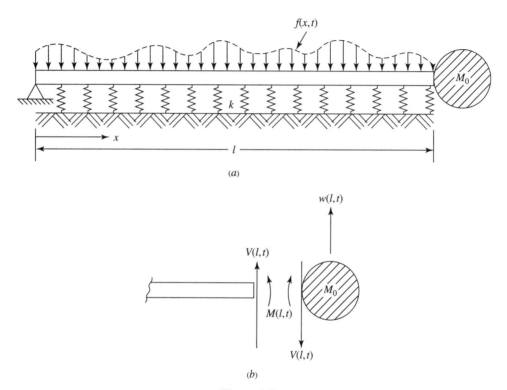

Figure 3.7

3.3   A beam resting on an elastic foundation and subjected to a distributed transverse force $f(x, t)$ is shown in Fig. 3.7(a). One end of the beam (at $x = 0$) is simply supported and the other end (at $x = l$) carries a mass M_0. The free-body diagram of the end mass M_0 is shown in Fig. 3.7(b).

(a) Derive the equation of motion of the beam using the equilibrium approach.

(b) Find the boundary conditions of the beam.

3.4   Consider a differential element of a membrane under uniform tension T in a polar coordinate system as shown in Fig. 3.8. Derive the equation of motion for the transverse vibration of a circular membrane of radius R using the equilibrium approach. Assume that the membrane has a mass of m per unit area.

3.5   Consider a differential element of a circular plate subjected to the transverse distributed force $f(r, \theta, t)$ as shown in Fig. 3.9. Noting that Q_t and M_{rt} vanish

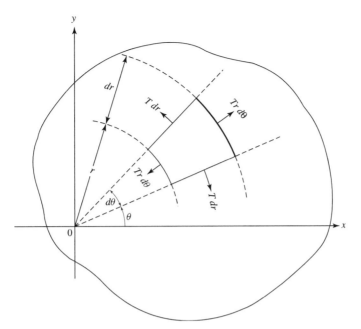

Figure 3.8

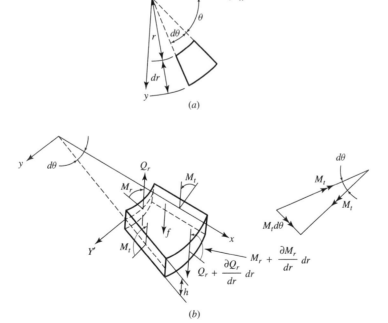

Figure 3.9

due to the symmetry, derive an equation of motion for the transverse vibration of a circular plate using the equilibrium approach.

3.6   Consider a rectangular plate resting on an elastic foundation with a foundation modulus k so that the resisting force offered by the foundation to a transverse deflection of the plate w is given by kw per unit area. The plate is subjected to a transverse force $f(x, y, t)$ per unit area. Derive a differential equation of motion governing the transverse vibration of the plate using the equilibrium approach.

4

Derivation of Equations: Variational Approach

4.1 INTRODUCTION

As stated earlier, vibration problems can be formulated using an equilibrium, a variational, or an integral equation approach. The variational approach is considered in this chapter. In the variational approach, the conditions of extremization of a functional are used to derive the equations of motion. The variational methods offer the following advantages:

1. Forces that do no work, such as forces of constraint on masses, need not be considered.
2. Accelerations of masses need not be considered; only velocities are needed.
3. Mathematical operations are to be performed on scalars, not on vectors, in deriving the equations of motion.

Since the variational methods make use of the principles of calculus of variations, the basic concepts of calculus of variations are presented. However, a brief review of the calculus of a single variable is given first to indicate the similarity of the concepts.

4.2 CALCULUS OF A SINGLE VARIABLE

To understand the principles of calculus of variations, we start with the extremization of a function of a single variable from elementary calculus [2]. For this, consider a continuous and differentiable function of one variable, defined in the interval (x_1, x_2), with extreme points at a, b, and c as shown in Fig. 4.1. In this figure the point $x = a$ denotes a local minimum with $f(a) \leq f(x)$ for all x in the neighborhood of a. Similarly, the point $x = b$ represents a local maximum with $f(b) \geq f(x)$ for all x in the neighborhood of b. The point $x = c$ indicates a stationary or inflection point with $f(c) \leq f(x)$ on one side and $f(c) \geq f(x)$ on the other side of the neighborhood of c. To establish the conditions of extreme values of the function $f(x)$, consider a Taylor series expansion of the function about an extreme point such as $x = a$:

$$f(x) = f(a) + \frac{df}{dx}\bigg|_a (x-a) + \frac{1}{2!}\frac{d^2f}{dx^2}\bigg|_a (x-a)^2 + \frac{1}{3!}\frac{d^3f}{dx^3}\bigg|_a (x-a)^3 + \cdots$$

$$(4.1)$$

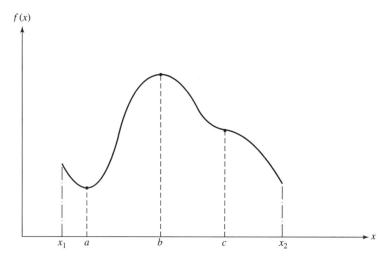

$f(x)$

x_1 a b c x_2 x

Figure 4.1   Extreme values of a function of one variable.

which can be rewritten as

$$f(x) - f(a) = \frac{df}{dx}\Big|_a (x - a) + \frac{1}{2!}\frac{d^2f}{dx^2}\Big|_a (x - a)^2 + \cdots \tag{4.2}$$

If $x = a$ is a local minimum, the quantity on the right-hand side of Eq. (4.2) must be positive for all values of x in the neighborhood of a. Since the value of $x - a$ can be positive, zero, or negative in the neighborhood of a, the necessary condition is that $df/dx|_a = 0$ and a sufficient condition is that $d^2f/dx^2|_a > 0$ for $f(a)$ to be a local minimum. A similar procedure can be used to establish the conditions of local maximum at $x = b$ and stationary point at $x = c$. Conditions for the extreme values of $f(x)$ can be summarized as follows. Local minimum at $x = a$:

$$\frac{df}{dx}\Big|_a = 0, \qquad \frac{d^2f}{dx^2}\Big|_a > 0 \tag{4.3}$$

Local maximum at $x = b$:

$$\frac{df}{dx}\Big|_b = 0, \qquad \frac{d^2f}{dx^2}\Big|_b < 0 \tag{4.4}$$

Stationary point at $x = c$:

$$\frac{df}{dx}\Big|_c = 0, \qquad \frac{d^2f}{dx^2}\Big|_c = 0 \tag{4.5}$$

4.3   CALCULUS OF VARIATIONS

The calculus of variations deals with the determination of extreme (minima, maxima, or stationary) values of functionals. A *functional* is defined as a function of one or more other functions. A simple problem in calculus of variations can be stated as follows. Find the function $\phi(x)$ that satisfies the conditions

$$\phi(x_1) = \phi_1, \qquad \phi(x_2) = \phi_2 \qquad (4.6)$$

and makes the integral functional

$$I = \int_{x_1}^{x_2} f(x, \phi, \phi_x) \, dx \qquad (4.7)$$

stationary. Here x_1, x_2, ϕ_1, and ϕ_2 are given, x is the independent variable, ϕ is the unknown function of x, $\phi_x = d\phi(x)/dx$, and $f(x, \phi, \phi_x)$ is a known function of x, ϕ, and ϕ_x. To find the true solution $\phi(x)$ that extremizes the functional I, we consider a family of trial functions $\overline{\phi}(x)$ defined by

$$\overline{\phi}(x) = \phi(x) + \varepsilon\eta(x) \qquad (4.8)$$

where ε is a parameter and $\eta(x)$ is an arbitrary differentiable function with

$$\eta(x_1) = \eta(x_2) = 0 \qquad (4.9)$$

Thus, for any specified function $\eta(x)$, there is a family of functions given by Eq. (4.8) with each value of ε designating a member of that family. Equation (4.9) ensures that the trial functions satisfy the end conditions specified:

$$\overline{\phi}(x_1) = \phi(x_1) = \phi_1$$
$$\overline{\phi}(x_2) = \phi(x_2) = \phi_2 \qquad (4.10)$$

Geometrically, the family of curves $\phi(x) = \overline{\phi}(x)$ connect the points (x_1,ϕ_1) and (x_2,ϕ_2) as shown in Fig. 4.2. The minimizing curve $\phi(x)$ is a member of the family for $\varepsilon = 0$. The difference between the curves $\phi(x)$ and $\overline{\phi}(x)$ is given by $\varepsilon\eta(x)$. Using $\overline{\phi}$ and $\overline{\phi}_x = d\overline{\phi}/dx$ for ϕ and $\phi_x = d\phi/dx$, respectively, in $f(x, \phi, \phi_x)$, the integral over the trial curve can be expressed as

$$\overline{I}(\varepsilon) = \int_{x_1}^{x_2} f(x, \overline{\phi}, \overline{\phi}_x) \, dx = \int_{x_1}^{x_2} f(x, \phi + \varepsilon\eta, \phi_x + \varepsilon\eta_x) \, dx \qquad (4.11)$$

where $\eta_x = d\eta/dx$. As in the case of the calculus of one variable, we expand the functional $\overline{I}(\varepsilon)$ about $\varepsilon = 0$:

$$\overline{I}(\varepsilon) = \overline{I}\big|_{\varepsilon=0} + \frac{d\overline{I}}{d\varepsilon}\bigg|_{\varepsilon=0} \varepsilon + \frac{1}{2!}\frac{d^2\overline{I}}{d\varepsilon^2}\bigg|_{\varepsilon=0} \varepsilon^2 + \cdots \qquad (4.12)$$

which can be rewritten as

$$\overline{I}(\varepsilon) - \overline{I}\big|_{\varepsilon=0} = \frac{d\overline{I}}{d\varepsilon}\bigg|_{\varepsilon=0} \varepsilon + \frac{1}{2!}\frac{d^2\overline{I}}{d\varepsilon^2}\bigg|_{\varepsilon=0} \varepsilon^2 + \cdots \qquad (4.13)$$

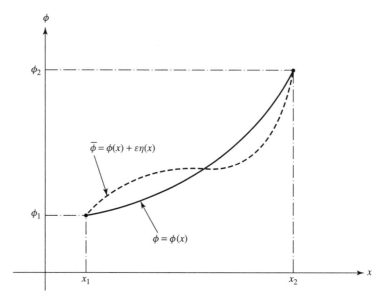

Figure 4.2 Exact and trial solutions.

It can be observed that the necessary condition for the extremum of $\overline{I}$ is that

$$\frac{d\overline{I}}{d\varepsilon}\bigg|_{\varepsilon=0} = \overline{I}_\varepsilon(0) = 0 \tag{4.14}$$

Using differentiation of an integral[1] and noting that both $\overline{\phi}$ and $\overline{\phi}_x$ are functions of ε, we obtain

$$\overline{I}_\varepsilon = \frac{d\overline{I}}{d\varepsilon} = \int_{x_1}^{x_2} \left(\frac{\partial f}{\partial \overline{\phi}}\frac{\partial \overline{\phi}}{\partial \varepsilon} + \frac{\partial f}{\partial \overline{\phi}_x}\frac{\partial \overline{\phi}_x}{\partial \varepsilon} \right) dx = \int_{x_1}^{x_2} \left(\frac{\partial f}{\partial \overline{\phi}}\eta + \frac{\partial f}{\partial \overline{\phi}_x}\eta_x \right) dx \tag{4.15}$$

When ε is set equal to zero, $(\overline{\phi}, \overline{\phi}_x)$ are replaced by (ϕ, ϕ_x) and Eq. (4.15) reduces to

$$\overline{I}_\varepsilon(0) = \frac{d\overline{I}}{d\varepsilon}(0) = \int_{x_1}^{x_2} \left(\frac{\partial f}{\partial \phi}\eta + \frac{\partial f}{\partial \phi_x}\eta_x \right) dx = 0 \tag{4.16}$$

---

[1]If

$$I = I(\varepsilon) = \int_{x_1(\varepsilon)}^{x_2(\varepsilon)} f(x, \varepsilon)\, d\varepsilon \tag{a}$$

then

$$I_\varepsilon = \frac{dI}{d\varepsilon} = f(x_2, \varepsilon)\frac{dx_2}{d\varepsilon} - f(x_1, \varepsilon)\frac{dx_1}{d\varepsilon} + \int_{x_1(\varepsilon)}^{x_2(\varepsilon)} \frac{\partial f}{\partial \varepsilon}\, dx \tag{b}$$

If x_1 and x_2 are constants, Eq. (b) reduces to

$$I_\varepsilon = \frac{dI}{d\varepsilon} = \int_{x_1}^{x_2} \frac{\partial f}{\partial \varepsilon}\, dx \tag{c}$$

Integrating the second term of the integral in Eq. (4.16) by parts, we obtain

$$\overline{I}_\varepsilon(0) = \frac{\partial f}{\partial \phi_x}\eta\Big|_{x_1}^{x_2} + \int_{x_1}^{x_2}\left[\frac{\partial f}{\partial \phi} - \frac{d}{dx}\left(\frac{\partial f}{\partial \phi_x}\right)\right]\eta\,dx = 0 \tag{4.17}$$

In view of Eq. (4.9), Eq. (4.17) gives

$$\overline{I}_\varepsilon(0) = \int_{x_1}^{x_2}\left[\frac{\partial f}{\partial \phi} - \frac{d}{dx}\left(\frac{\partial f}{\partial \phi_x}\right)\right]\eta\,dx = 0 \tag{4.18}$$

Since Eq. (4.18) must hold for all η, we have

$$\frac{\partial f}{\partial \phi} - \frac{d}{dx}\left(\frac{\partial f}{\partial \phi_x}\right) = 0 \tag{4.19}$$

This equation, known as the *Euler–Lagrange equation*, is, in general, a second-order differential equation. The solution of Eq. (4.19) gives the function $\phi(x)$ that makes the integral I stationary.

4.4 VARIATION OPERATOR

Equation (4.17) can also be derived using a variation operator δ, defined as

$$\delta\phi = \overline{\phi}(x) - \phi(x) \tag{4.20}$$

where $\phi(x)$ is the true function of x that extremizes I, and $\overline{\phi}(x)$ is another function of x which is infinitesimally different from $\phi(x)$ at every point x in the interval $x_1 < x < x_2$. The variation of a function $\phi(x)$ denotes an infinitesimal change in the function at a given value of x. The change is virtual and arbitrary. The variation differs from the usual differentiation, which denotes a measure of the change in a function (such as ϕ) resulting from a specified change in an independent variable (such as x). In view of Eq. (4.8), Eq. (4.20) can be represented as

$$\delta\phi(x) = \overline{\phi}(x) - \phi(x) = \varepsilon\eta(x) \tag{4.21}$$

where the parameter ε tends to zero. The variation operator has the following important properties, which are useful in the extremization of the functional I.

1. Since the variation operator is defined to cause an infinitesimal change in the function ϕ for a fixed value of x, we have

$$\delta x = 0 \tag{4.22}$$

   and hence the independent variable x will not participate in the variation process.

2. The variation operator is commutative with respect to the operation of differentiation. For this, consider the derivative of a variation:

$$\frac{d}{dx}\delta\phi = \frac{d}{dx}\varepsilon\eta(x) = \varepsilon\frac{d\eta(x)}{dx} \tag{4.23}$$

Next, consider the operation of the variation of a derivation, $\delta(d\phi/dx)$. Using the definition of Eq. (4.21),

$$\delta\frac{d\phi}{dx} = \frac{d\overline{\phi}}{dx} - \frac{d\phi}{dx} = \frac{d}{dx}(\overline{\phi} - \phi) = \frac{d}{dx}\varepsilon\eta(x) = \varepsilon\frac{d\eta(x)}{dx} \tag{4.24}$$

Thus, Eqs. (4.23) and (4.24) indicate that the operations of differentiation and variation are commutative:

$$\frac{d}{dx}\delta\phi = \delta\frac{d\phi}{dx} \tag{4.25}$$

3. The variation operator is commutative with respect to the operation of integration. For this, consider the variation of an integral:

$$\delta\int_{x_1}^{x_2}\phi(x)\,dx = \int_{x_1}^{x_2}\overline{\phi}(x)\,dx - \int_{x_1}^{x_2}\phi(x)\,dx$$

$$= \int_{x_1}^{x_2}[\overline{\phi}(x) - \phi(x)]\,dx = \int_{x_1}^{x_2}\delta\phi(x)\,dx \tag{4.26}$$

Equation (4.26) establishes that the operations of integration and variation are commutative:

$$\delta\int_{x_1}^{x_2}\phi(x)\,dx = \int_{x_1}^{x_2}\delta\phi(x)\,dx \tag{4.27}$$

For the extremization of the functional I of Eq. (4.7), we follow the procedure used for the extremization of a function of a single variable and define the functional I to be stationary if the first variation is zero:

$$\delta I = 0 \tag{4.28}$$

Using Eq. (4.7) and the commutative property of Eq. (4.27), Eq. (4.28) can be written as

$$\delta I = \int_{x_1}^{x_2}\delta f\,dx = 0 \tag{4.29}$$

where the variation of f is caused by the varying function $\phi(x)$:

$$\delta f = f(x,\overline{\phi},\overline{\phi}_x) - f(x,\phi,\phi_x) = f(x,\phi + \varepsilon\eta, \phi_x + \varepsilon\eta_x) - f(x,\phi,\phi_x) \tag{4.30}$$

The expansion of the function $f(x,\phi + \varepsilon\eta, \phi_x + \varepsilon\eta_x)$ about (x,ϕ,ϕ_x) gives

$$f(x,\phi + \varepsilon\eta, \phi_x + \varepsilon\eta_x) = f(x,\phi,\phi_x) + \frac{\partial f}{\partial\phi}\varepsilon\eta + \frac{\partial f}{\partial\phi_x}\varepsilon\eta_x + \cdots \tag{4.31}$$

Since ε is assumed to be small, we neglect terms of higher order in ε in Eq. (4.31), so that

$$\delta f = \varepsilon\left(\frac{\partial f}{\partial\phi}\eta + \frac{\partial f}{\partial\phi_x}\eta_x\right)$$

Thus,

$$\delta I = \int_{x_1}^{x_2} \delta f \, dx = \varepsilon \int_{x_1}^{x_2} \left(\frac{\partial f}{\partial \phi} \eta + \frac{\partial f}{\partial \phi_x} \eta_x \right) dx = 0 \tag{4.32}$$

for all functions $\eta(x)$. The second term in the integral, with ε in Eq. (4.32), can be integrated by parts as

$$\int_{x_1}^{x_2} \frac{\partial f}{\partial \phi_x} \eta_x \, dx = - \int_{x_1}^{x_2} \eta \frac{d}{dx} \left(\frac{\partial f}{\partial \phi_x} \right) dx + \eta \frac{\partial f}{\partial \phi_x} \bigg|_{x_1}^{x_2} \tag{4.33}$$

Using Eq. (4.33) in (4.32), we obtain

$$\frac{\delta I}{\varepsilon} = \int_{x_1}^{x_2} \left[\frac{\partial f}{\partial \phi} - \frac{d}{dx} \left(\frac{\partial f}{\partial \phi_x} \right) \right] \eta(x) \, dx + \eta \frac{\partial f}{\partial \phi_x} \bigg|_{x_1}^{x_2} = 0 \tag{4.34}$$

Since the function $\eta(x)$ is arbitrary, Eq. (4.34) will be satisfied for all possible values of $\eta(x)$ only if

$$\frac{\partial f}{\partial \phi} - \frac{d}{dx} \left(\frac{\partial f}{\partial \phi_x} \right) = 0 \tag{4.35}$$

$$\eta \frac{\partial f}{\partial \phi_x} \bigg|_{x_1}^{x_2} = 0 \tag{4.36}$$

Equation (4.35) can be seen to be the Euler–Lagrange equation, and Eq. (4.36) denotes the boundary conditions. Since the function $\phi(x)$ is specified or fixed at the endpoints, as $\phi(x_1) = \phi_1$ and $\phi(x_2) = \phi_2$ [see Eq. (4.6)], $\eta(x_1) = \eta(x_2) = 0$ [see Eq. (4.9)] and hence no variation is permitted in $\phi(x)$ at the endpoints. Thus, Eq. (4.36) will be satisfied automatically.

4.5   FUNCTIONAL WITH HIGHER-ORDER DERIVATIVES

The extremization of functionals involving higher-order derivatives will be useful in deriving the equations of motion of several continuous systems. To illustrate the methodology, we consider the extremization of a functional (I) involving second derivatives:

$$I = \int_{x_1}^{x_2} f(x, \phi, \phi_x, \phi_{xx}) \, dx \tag{4.37}$$

where $\phi = \phi(x)$, $\phi_x = d\phi/dx$, and $\phi_{xx} = d^2\phi/dx^2$. Let $\phi(x)$ denote the true function that extremizes the functional I and $\overline{\phi}(x)$ a tentative solution:

$$\overline{\phi}(x) = \phi(x) + \varepsilon \eta(x) \tag{4.38}$$

When Eq. (4.38) is used for ϕ in Eq. (4.37), we obtain

$$\overline{I} = \int_{x_1}^{x_2} f(x, \overline{\phi}, \overline{\phi}_x, \overline{\phi}_{xx}) \, dx \tag{4.39}$$

By proceeding as in Section 4.3, the necessary condition for the extremum of $\overline{I}$ can be expressed as

$$\left. \frac{d\overline{I}}{d\varepsilon} \right|_{\varepsilon=0} = \overline{I}_{\varepsilon}(0) = 0 \tag{4.40}$$

where

$$\frac{d\overline{I}}{d\varepsilon} = \frac{d}{d\varepsilon} \int_{x_1}^{x_2} f(x, \overline{\phi}, \overline{\phi}_x, \overline{\phi}_{xx}) \, dx = \int_{x_1}^{x_2} \left(\frac{\partial f}{\partial \overline{\phi}} \frac{\partial \overline{\phi}}{\partial \varepsilon} + \frac{\partial f}{\partial \overline{\phi}_x} \frac{\partial \overline{\phi}_x}{\partial \varepsilon} + \frac{\partial f}{\partial \overline{\phi}_{xx}} \frac{\partial \overline{\phi}_{xx}}{\partial \varepsilon} \right) dx$$

$$= \int_{x_1}^{x_2} \left(\frac{\partial f}{\partial \overline{\phi}} \eta + \frac{\partial f}{\partial \overline{\phi}_x} \eta_x + \frac{\partial f}{\partial \overline{\phi}_{xx}} \eta_{xx} \right) dx \tag{4.41}$$

By setting $\varepsilon = 0$, $\overline{\phi}$ becomes ϕ and the condition of Eq. (4.40) becomes

$$\int_{x_1}^{x_2} \left(\frac{\partial f}{\partial \phi} \eta + \frac{\partial f}{\partial \phi_x} \eta_x + \frac{\partial f}{\partial \phi_{xx}} \eta_{xx} \right) dx = 0 \tag{4.42}$$

The second and third terms of the integral in Eq. (4.42) can be integrated by parts as

$$\int_{x_1}^{x_2} \frac{\partial f}{\partial \phi_x} \eta_x \, dx = - \int_{x_1}^{x_2} \frac{d}{dx} \left(\frac{\partial f}{\partial \phi_x} \right) \eta \, dx + \left. \frac{\partial f}{\partial \phi_x} \eta \right|_{x_1}^{x_2} \tag{4.43}$$

$$\int_{x_1}^{x_2} \frac{\partial f}{\partial \phi_{xx}} \eta_{xx} \, dx = - \int_{x_1}^{x_2} \frac{d}{dx} \left(\frac{\partial f}{\partial \phi_{xx}} \right) \eta_x \, dx + \left. \frac{\partial f}{\partial \phi_{xx}} \eta_x \right|_{x_1}^{x_2}$$

$$= + \int_{x_1}^{x_2} \frac{d^2}{dx^2} \left(\frac{\partial f}{\partial \phi_{xx}} \right) \eta \, dx - \left. \frac{d}{dx} \left(\frac{\partial f}{\partial \phi_{xx}} \right) \eta \right|_{x_1}^{x_2} + \left. \frac{\partial f}{\partial \phi_{xx}} \eta_x \right|_{x_1}^{x_2} \tag{4.44}$$

Using Eqs. (4.43) and (4.44), Eq. (4.42) can be written as

$$\int_{x_1}^{x_2} \left[\frac{\partial f}{\partial \phi} - \frac{d}{dx} \left(\frac{\partial f}{\partial \phi_x} \right) + \frac{d^2}{dx^2} \left(\frac{\partial f}{\partial \phi_{xx}} \right) \right] \eta \, dx$$

$$+ \left. \frac{\partial f}{\partial \phi_x} \eta \right|_{x_1}^{x_2} - \left. \frac{d}{dx} \left(\frac{\partial f}{\partial \phi_{xx}} \right) \eta \right|_{x_1}^{x_2} + \left. \frac{\partial f}{\partial \phi_{xx}} \eta_x \right|_{x_1}^{x_2} = 0 \tag{4.45}$$

If the function $\phi(x)$ and its first derivative $\phi_x(x) = d\phi(x)/dx$ are specified or fixed at the endpoints x_1 and x_2, both $\eta(x)$ and $\eta_x(x) = d\eta(x)/dx$ will be zero at x_1 and x_2, and hence each of the terms

$$\left. \left[\frac{\partial f}{\partial \phi_x} - \frac{d}{dx} \left(\frac{\partial f}{\partial \phi_{xx}} \right) \right] \eta \right|_{x_1}^{x_2}, \qquad \left. \frac{\partial f}{\partial \phi_{xx}} \eta_x \right|_{x_1}^{x_2}$$

will be zero. Hence, the necessary condition for the extremization of the functional I, also known as the *Euler–Lagrange equation*, can be obtained from Eq. (4.45) as

$$\frac{\partial f}{\partial \phi} - \frac{d}{dx} \left(\frac{\partial f}{\partial \phi_x} \right) + \frac{d^2}{dx^2} \left(\frac{\partial f}{\partial \phi_{xx}} \right) = 0 \tag{4.46}$$

Note that if the functional I involves derivatives of higher than second order, so that

$$I = \int_{x_1}^{x_2} f(x, \phi^{(0)}, \phi^{(1)}, \phi^{(2)}, \ldots, \phi^{(j)}) \, dx \tag{4.47}$$

where $\phi^{(j)}$ denotes the jth-order derivative of ϕ,

$$\phi^{(j)} = \frac{d^j \phi}{dx^j}, \qquad j = 1, 2, \ldots \tag{4.48}$$

the corresponding Euler–Lagrange equation can be derived as

$$\sum_{j=0}^{n} (-1)^{n-j} \frac{d^{n-j}}{dx^{n-j}} \left(\frac{\partial f}{\partial \phi^{(n-j)}} \right) = 0 \tag{4.49}$$

4.6   FUNCTIONAL WITH SEVERAL DEPENDENT VARIABLES

In some applications, such as the vibration of a multidegree-of-freedom system, the functional will contain a single independent variable (such as time) but several dependent variables (such as the displacements of individual masses). To consider the extremization of such functionals, let

$$I = \int_{x_1}^{x_2} f(x, \phi_1, \phi_2, \ldots, \phi_n, (\phi_1)_x, (\phi_2)_x, \ldots, (\phi_n)_x) \, dx \tag{4.50}$$

where $(\phi_i)_x = d\phi_i / dx$, $i = 1, 2, \ldots, n$. To find the functions $\phi_1(x), \phi_2(x), \ldots, \phi_n(x)$ with specified end conditions, $\phi_i(x_1) = \phi_{i1}$ and $\phi_i(x_2) = \phi_{i2}$ that extremize the functional I of Eq. (4.50), we assume a set of tentative differentiable functions $\overline{\phi}_i(x)$ as

$$\overline{\phi}_i(x) = \phi_i(x) + \varepsilon \eta_i(x), \qquad i = 1, 2, \ldots, n \tag{4.51}$$

where ε is a parameter and $\eta_i(x)$ are arbitrary differentiable functions with

$$\eta_i(x_1) = \eta_i(x_2) = 0 \tag{4.52}$$

Using $\overline{\phi}_i$ and $(\overline{\phi}_i)_x = d\overline{\phi}_i / dx$ for ϕ_i and $(\phi_i)_x = d\phi_i / dx$ in Eq. (4.50), we obtain

$$\overline{I}(\varepsilon) = \int_{x_1}^{x_2} f(x, \overline{\phi}_1, \overline{\phi}_2, \ldots, \overline{\phi}_n, (\overline{\phi}_1)_x, (\overline{\phi}_2)_x, \ldots, (\overline{\phi}_n)_x) \, dx \tag{4.53}$$

By proceeding as in Section 4.3, the necessary condition for the extremum of $\overline{I}$ can be expressed as

$$\left. \frac{d\overline{I}}{d\varepsilon} \right|_{\varepsilon=0} = \overline{I}_\varepsilon(0) = 0 \tag{4.54}$$

where

$$\frac{d\bar{I}}{d\varepsilon} = \int_{x_1}^{x_2} \left[\frac{\partial f}{\partial \bar{\phi}_1} \frac{\partial \bar{\phi}_1}{\partial \varepsilon} + \cdots + \frac{\partial f}{\partial \bar{\phi}_n} \frac{\partial \bar{\phi}_n}{\partial \varepsilon} + \frac{\partial f}{\partial (\bar{\phi}_1)_x} \frac{\partial (\bar{\phi}_1)_x}{\partial \varepsilon} + \cdots + \frac{\partial f}{\partial (\bar{\phi}_n)_x} \frac{\partial (\bar{\phi}_n)_x}{\partial \varepsilon} \right] dx$$

$$= \int_{x_1}^{x_2} \left[\frac{\partial f}{\partial \bar{\phi}_1} \eta_1 + \cdots + \frac{\partial f}{\partial \bar{\phi}_n} \eta_n + \frac{\partial f}{\partial (\bar{\phi}_1)_x} \eta_{1x} + \cdots + \frac{\partial f}{\partial (\bar{\phi}_n)_x} \eta_{nx} \right] dx \qquad (4.55)$$

For $\varepsilon = 0$, $\bar{\phi}_i = \phi_i$ and $(\bar{\phi}_i)_x = (\phi_i)_x$, $i = 1, 2, \ldots, n$, and the necessary condition of Eq. (4.54) becomes

$$\int_{x_1}^{x_2} \left[\frac{\partial f}{\partial \phi_1} \eta_1 + \cdots + \frac{\partial f}{\partial \phi_n} \eta_n + \frac{\partial f}{\partial (\phi_1)_x} \eta_{1x} + \cdots + \frac{\partial f}{\partial (\phi_n)_x} \eta_{nx} \right] dx = 0 \qquad (4.56)$$

By using the relation

$$\int_{x_1}^{x_2} \frac{\partial f}{\partial (\phi_i)_x} \eta_{ix} \, dx = - \int_{x_1}^{x_2} \frac{d}{dx} \left(\frac{\partial f}{\partial (\phi_i)_x} \right) \eta_i \, dx + \frac{\partial f}{\partial (\phi_i)_x} \eta_i \bigg|_{x_1}^{x_2}, \qquad i = 1, 2, \ldots, n$$

$$(4.57)$$

and noting that $\eta_i = 0$ at x_1 and x_2 from Eq. (4.52), Eq. (4.56) can be expressed as

$$\int_{x_1}^{x_2} \left\{ \left[\frac{\partial f}{\partial \phi_1} - \frac{d}{dx} \left(\frac{\partial f}{\partial (\phi_1)_x} \right) \right] \eta_1 + \cdots + \left[\frac{\partial f}{\partial \phi_n} - \frac{d}{dx} \left(\frac{\partial f}{\partial (\phi_n)_x} \right) \right] \eta_n \right\} dx = 0$$

$$(4.58)$$

Since $\eta_1(x), \ldots, \eta_n(x)$ are arbitrary functions of x, we assume a particular $\eta_i(x)$ to be arbitrary and all the remaining $\eta_j(x) = 0$ $(j = 1, 2, \ldots, i-1, i+1, \ldots, n)$ so that Eq. (4.58) leads to the necessary conditions, also known as the Euler–Lagrange equations:

$$\frac{\partial f}{\partial \phi_i} - \frac{d}{dx} \left(\frac{\partial f}{\partial (\phi_i)_x} \right) = 0, \qquad i = 1, 2, \ldots, n \qquad (4.59)$$

Note that if the functional involves the second derivatives of the functions $\phi_i(x)$ as

$$I = \int_{x_1}^{x_2} f(x, \phi_1, \phi_2, \ldots, \phi_n, (\phi_1)_x, (\phi_2)_x, \ldots, (\phi_n)_x, (\phi_1)_{xx}, (\phi_2)_{xx}, \ldots, (\phi_n)_{xx}) \, dx$$

$$(4.60)$$

the Euler–Lagrange equations can be derived as

$$\frac{d^2}{dx^2} \left(\frac{\partial f}{\partial (\phi_i)_{xx}} \right) - \frac{d}{dx} \left(\frac{\partial f}{\partial (\phi_i)_x} \right) + \frac{\partial f}{\partial \phi_i} = 0, \qquad i = 1, 2, \ldots, n \qquad (4.61)$$

In general, if the functional I involves derivatives of higher than the second order, so that

$$I = \int_{x_1}^{x_2} f(x, \phi_1^{(0)}, \ldots, \phi_n^{(0)}, \phi_1^{(1)}, \ldots, \phi_n^{(1)}, \phi_1^{(2)}, \ldots, \phi_n^{(2)}, \ldots, \phi_1^{(j)}, \ldots, \phi_n^{(j)}) \, dx$$

$$(4.62)$$

where $\phi_i^{(j)}$ denotes the jth-order derivative of ϕ_i,

$$\phi_i^{(j)} = \frac{d^j \phi_i(x)}{dx^j} \tag{4.63}$$

the corresponding Euler–Lagrange equations can be derived as

$$\sum_{j=0}^{n} (-1)^{n-j} \frac{d^{n-j}}{dx^{n-j}} \left(\frac{\partial f}{\partial \phi_i^{(n-j)}} \right) = 0, \qquad i = 1, 2, \ldots, n \tag{4.64}$$

4.7   FUNCTIONAL WITH SEVERAL INDEPENDENT VARIABLES

Many problems involve extremization of a functional involving more than one independent variable. Hence, we consider the extremization of a functional in the form of a multiple integral:

$$I = \iiint_V f(x, y, z, \phi, \phi_x, \phi_y, \phi_z)\, dV \tag{4.65}$$

where x, y, and z are the independent variables and ϕ is the dependent variable with $\phi = \phi(x, y, z)$, $\phi_i = \partial\phi(i = x, y, z)/\partial i$, and V is the volume or domain of integration bounded by a surface S. We assume that the function $\phi(x, y, z)$ is specified on the surface S. If $\phi(x, y, z)$ is the true function that extremizes the functional I, we consider a trial function $\overline{\phi}(x, y, z)$ that differs infinitesimally from ϕ in volume V as

$$\overline{\phi}(x, y, z) = \phi(x, y, z) + \varepsilon \eta(x, y, z) \tag{4.66}$$

All the trial functions are assumed to attain the same value on the boundary S, so that

$$\eta(x, y, z) = 0 \qquad \text{on } S \tag{4.67}$$

When $\overline{\phi}$ is used for ϕ in Eq. (4.65), we obtain[2]

$$\overline{I} = \int_V f(x, y, z, \phi + \varepsilon\eta, \phi_x + \varepsilon\eta_x, \phi_y + \varepsilon\eta_y, \phi_z + \varepsilon\eta_z)\, dV \tag{4.68}$$

The necessary condition for the extremum of $\overline{I}$ can be expressed as

$$\frac{d\overline{I}}{d\varepsilon}\bigg|_{\varepsilon=0} = \overline{I}_\varepsilon(0) = 0 \tag{4.69}$$

where

$$\frac{d\overline{I}}{d\varepsilon} = \int_V \left(\frac{\partial f}{\partial \overline{\phi}} \frac{\partial \overline{\phi}}{\partial \varepsilon} + \frac{\partial f}{\partial \overline{\phi}_x} \frac{\partial \overline{\phi}_x}{\partial \varepsilon} + \frac{\partial f}{\partial \overline{\phi}_y} \frac{\partial \overline{\phi}_y}{\partial \varepsilon} + \frac{\partial f}{\partial \overline{\phi}_z} \frac{\partial \overline{\phi}_z}{\partial \varepsilon} \right) dV$$

$$= \int_V \left(\frac{\partial f}{\partial \overline{\phi}}\eta + \frac{\partial f}{\partial \overline{\phi}_x}\eta_x + \frac{\partial f}{\partial \overline{\phi}_y}\eta_y + \frac{\partial f}{\partial \overline{\phi}_z}\eta_z \right) dV \tag{4.70}$$

---

[2]For simplicity, the multiple integral is written as $\int_V$.

By setting $\varepsilon = 0$, $\overline{\phi}$ becomes ϕ and the condition of Eq. (4.69) becomes

$$\int_V \left(\frac{\partial f}{\partial \phi} \eta + \frac{\partial f}{\partial \phi_x} \eta_x + \frac{\partial f}{\partial \phi_y} \eta_y + \frac{\partial f}{\partial \phi_z} \eta_z \right) dV = 0 \qquad (4.71)$$

Applying Green's theorem, Eq. (4.71) can be expressed as

$$-\int_V \left[\frac{d}{dx} \left(\frac{\partial f}{\partial \phi_x} \right) + \frac{d}{dy} \left(\frac{\partial f}{\partial \phi_y} \right) + \frac{d}{dz} \left(\frac{\partial f}{\partial \phi_z} \right) \right] \eta \, dV$$

$$+ \int_S \left(\frac{\partial f}{\partial \phi_x} l_x + \frac{\partial f}{\partial \phi_y} l_y + \frac{\partial f}{\partial \phi_z} l_z \right) \eta \, dS = 0 \qquad (4.72)$$

where l_x, l_y, and l_z denote the cosines of the angle between the normal to the surface S and the x, y, and z axes, respectively. Since $\eta = 0$ on S according to Eq. (4.67), and $\eta(x, y, z)$ is arbitrary in V, the necessary condition for extremization or the Euler–Lagrange equation becomes

$$\frac{\partial f}{\partial \phi} - \frac{d}{dx} \left(\frac{\partial f}{\partial \phi_x} \right) - \frac{d}{dy} \left(\frac{\partial f}{\partial \phi_y} \right) - \frac{d}{dz} \left(\frac{\partial f}{\partial \phi_z} \right) = 0 \qquad (4.73)$$

4.8   EXTREMIZATION OF A FUNCTIONAL WITH CONSTRAINTS

In some cases the extremization of a functional subject to a condition is desired. The best known case, called the *isoperimetric problem*, involves finding the closed curve of a given perimeter for which the enclosed area is a maximum. To demonstrate the procedure involved, consider the problem of finding a continuously differentiable function $\phi(x)$ that extremizes the functional

$$I = \int_{x_1}^{x_2} f(x, \phi, \phi_x) \, dx \qquad (4.74)$$

while making the functional

$$J = \int_{x_1}^{x_2} g(x, \phi, \phi_x) \, dx \qquad (4.75)$$

assume a prescribed value and with both $\phi(x_1) = \phi_1$ and $\phi(x_2) = \phi_2$ prescribed. If $\phi(x)$ denotes the true solution of the problem, we consider a two-parameter family of trial solutions $\overline{\phi}(x)$ as

$$\overline{\phi}(x) = \phi(x) + \varepsilon_1 \eta_1(x) + \varepsilon_2 \eta_2(x) \qquad (4.76)$$

where ε_1 and ε_2 are parameters and $\eta_1(x)$ and $\eta_2(x)$ are arbitrary differentiable functions with

$$\eta_1(x_1) = \eta_1(x_2) = \eta_2(x_1) = \eta_2(x_2) = 0 \qquad (4.77)$$

Equation (4.77) ensures that $\overline{\phi}(x_1) = \phi(x_1) = \phi_1$ and $\overline{\phi}(x_2) = \phi(x_2) = \phi_2$. Note that $\overline{\phi}(x)$ cannot be expressed as merely a one-parameter family of functions because any change in the value of the single parameter, in general, will alter the value of J, whose value must be maintained as prescribed.

We can use the method of Lagrange multipliers to solve the problem. When $\bar{\phi}$ is substituted for ϕ in Eqs. (4.74) and (4.75), we obtain $\bar{I}(\varepsilon_1, \varepsilon_2)$ and $\bar{J}(\varepsilon_1, \varepsilon_2)$. Hence, we define a new function, L, as

$$L = \bar{I}(\varepsilon_1, \varepsilon_2) + \lambda \bar{J}(\varepsilon_1, \varepsilon_2) = \int_{x_1}^{x_2} F(x, \bar{\phi}, \bar{\phi}_x)\, dx \tag{4.78}$$

where λ is an undetermined constant, called a *Lagrange multiplier*, and

$$F(x, \bar{\phi}, \bar{\phi}_x) = f(x, \bar{\phi}, \bar{\phi}_x) + \lambda g(x, \bar{\phi}, \bar{\phi}_x) \tag{4.79}$$

The necessary conditions for the extremum of L, which also correspond to the solution of the original constrained problem, can be expressed as

$$\left. \frac{\partial L}{\partial \varepsilon_1} \right|_{\varepsilon_1 = \varepsilon_2 = 0} = \left. \frac{\partial L}{\partial \varepsilon_2} \right|_{\varepsilon_1 = \varepsilon_2 = 0} = 0 \tag{4.80}$$

From Eqs. (4.78) and (4.79), we obtain

$$\frac{\partial L}{\partial \varepsilon_j} = \int_{x_1}^{x_2} \left(\frac{\partial F}{\partial \bar{\phi}} \frac{\partial \bar{\phi}}{\partial \varepsilon_j} + \frac{\partial F}{\partial \bar{\phi}_x} \frac{\partial \bar{\phi}_x}{\partial \varepsilon_j} \right) dx = \int_{x_1}^{x_2} \left[\frac{\partial F}{\partial \bar{\phi}} \eta_j + \frac{\partial F}{\partial \bar{\phi}_x} (\eta_j)_x \right] dx, \qquad j = 1, 2 \tag{4.81}$$

where

$$(\eta_j)_x = \frac{d\eta_j(x)}{dx} \tag{4.82}$$

Setting $\varepsilon_1 = \varepsilon_2 = 0$, $(\bar{\phi}, \bar{\phi}_x)$ will be replaced by (ϕ, ϕ_x), so that the conditions of Eq. (4.80) become

$$\left. \frac{\partial F}{\partial \varepsilon_j} \right|_{\varepsilon_1 = \varepsilon_2 = 0} = \int_{x_1}^{x_2} \left[\frac{\partial F}{\partial \phi} \eta_j + \frac{\partial F}{\partial \phi_x} (\eta_j)_x \right] dx = 0, \qquad j = 1, 2 \tag{4.83}$$

Integrating the second term of the integral in Eq. (4.83) by parts leads to

$$\int_{x_1}^{x_2} \left[\frac{\partial F}{\partial \phi} - \frac{d}{dx} \left(\frac{\partial F}{\partial \phi_x} \right) \right] \eta_j\, dx = 0, \qquad j = 1, 2 \tag{4.84}$$

Since the functions $\eta_1(x)$ and $\eta_2(x)$ are arbitrary, the necessary condition or Euler–Lagrange equation can be expressed as

$$\frac{\partial F}{\partial \phi} - \frac{d}{dx} \left(\frac{\partial F}{\partial \phi_x} \right) = 0 \tag{4.85}$$

Note: Solution of the second-order Euler–Lagrange equation (4.85) yields a function $\phi(x)$ with three unknown quantities: two constants of integration and one Lagrange multiplier [see Eq. (4.79)]. For a given isoperimetric problem, the two end conditions specified, $\phi(x_1) = \phi_1$ and $\phi(x_2) = \phi_2$, and the prescribed value of J can be used to find the three unknown quantities. The following example illustrates the procedure.

Example 4.1 Determine the shape of a perfectly flexible rope of uniform cross section that hangs at rest in a vertical plane with its endpoints fixed. The length of the rope is specified as l.

SOLUTION Since the rope is to be in a static equilibrium position, the potential energy of the system is to be minimized subject to the constraint stated on the length of the rope. Let (x_1, y_1) and (x_2, y_2) denote the fixed endpoints of the rope in the xy (vertical) plane with $x_1 < x_2$ (see Fig. 4.3). If the mass of the rope per unit length is denoted by ρ, the potential energy of an elemental length of the rope (ds) at (x, y) is given by $\rho\, ds\, g_0 y$, where g_0 denotes the acceleration due to gravity. Thus, the total potential energy to be minimized is given by

$$I = \int_{s=0}^{l} \rho\, ds\, g_0 y \tag{E4.1.1}$$

Using the relation $(dx)^2 + (dy)^2 = (ds)^2$, we obtain

$$ds = \sqrt{1 + \left(\frac{dy}{dx}\right)^2}\, dx \tag{E4.1.2}$$

Thus, the variational problem can be stated as follows: Determine the curve (function) $y(x)$ that passes through points (x_1, y_1) and (x_2, y_2), which minimizes

$$I = \rho g_0 \int_{x_1}^{x_2} y \sqrt{1 + \left(\frac{dy}{dx}\right)^2}\, dx \tag{E4.1.3}$$

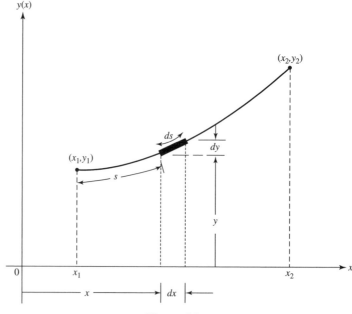

Figure 4.3

with the constraint

$$J = \int_{s=0}^{l} ds = \int_{x_1}^{x_2} g\, dx \equiv \int_{x_1}^{x_2} \sqrt{1 + \left(\frac{dy}{dx}\right)^2}\, dx = l \tag{E4.1.4}$$

where

$$g = \sqrt{1 + \left(\frac{dy}{dx}\right)^2}\, dx \tag{E4.1.5}$$

The function, F, can be expressed as

$$F = f + \lambda g = \rho g_0 y \sqrt{1 + y_x^2} + \lambda \sqrt{1 + y_x^2} \tag{E4.1.6}$$

Noting that F is independent of x, the Euler–Lagrange equation can be expressed as [Eq. (4.85) with y in place of ϕ]

$$\frac{d}{dx}\left[(\rho g_0 y + \lambda)\left(\frac{y_x^2}{\sqrt{1 + y_x^2}} - \sqrt{1 + y_x^2}\right)\right] = 0 \tag{E4.1.7}$$

The integration of Eq. (E4.1.7) with respect to x yields

$$(\rho g_0 y + \lambda)\left(\frac{y_x^2}{\sqrt{1 + y_x^2}} - \sqrt{1 + y_x^2}\right) = c_1 \tag{E4.1.8}$$

where c_1 is a constant. Equation (E4.1.8) can be rearranged to obtain

$$y_x^2 - (1 + y_x^2) = \frac{c_1 \sqrt{1 + y_x^2}}{\rho g_0 y + \lambda} \tag{E4.1.9}$$

Squaring both sides of Eq. (E4.1.9) and rearranging yields

$$dx = c_1 \frac{dy}{\sqrt{(\rho g_0 y + \lambda)^2 - c_1^2}} \tag{E4.1.10}$$

or

$$x = c_1 \int \frac{dy}{\sqrt{(\rho g_0 y + \lambda)^2 - c_1^2}} + c_2 \tag{E4.1.11}$$

By using the transformation

$$z = \rho g_0 y + \lambda \tag{E4.1.12}$$

Eq. (E4.1.11) can be written as

$$x = \frac{c_1}{\rho g_0} \int \frac{dz}{\sqrt{z^2 - c_1^2}} + c_2 = -\frac{c_1}{\rho g_0} \cosh^{-1}\frac{z}{c_1} + c_2 \tag{E4.1.13}$$

Thus, the solution of the isoperimetric problem is given by

$$y(x) = -\frac{\lambda}{\rho g_0} - \frac{c_1}{\rho g_0} \cosh \frac{\rho g_0 (x - c_2)}{c_1} \tag{E4.1.14}$$

This solution indicates that the shape of a hanging rope is a catenary with vertical axis. The constants c_1, c_2, and λ can be determined by making the catenary pass through the specified points (x_1, y_1) and (x_2, y_2) and using the constraint equation (E4.1.4).

4.9 BOUNDARY CONDITIONS

In all previous sections the necessary condition for the extremum of a given functional was derived by assuming that the variation is zero on the boundary (at the endpoints x_1 and x_2 in the case of a single independent variable). This assumption is equivalent to asserting a specific value of the function on the boundary and is not subject to variation. However, there are many applications where the function to be varied is not specified on the boundary, but other equally valid boundary conditions are imposed. It is to be noted that investigation of the boundary conditions is an integral part of the variational approach, and any alteration of the boundary conditions causes a corresponding change in the extremum value of the functional. If the function is not specified on the boundary, the proper type of boundary conditions that can be imposed will be supplied by the variational method. In fact, one of the attractive features of the variational approach for complex problems is that it gives not only the governing differential equation(s) of motion, in the form of Euler–Lagrange equation(s), but also the correct boundary conditions of the problem. Consider the extremization of the functional

$$I = \int_{x_1}^{x_2} f(x, \phi, \phi_x, \phi_{xx}) \, dx \tag{4.86}$$

The necessary condition for the extremum of I can be expressed as [see Eq. (4.45)].

$$\int_{x_1}^{x_2} \left[\frac{\partial f}{\partial \phi} - \frac{d}{dx}\left(\frac{\partial f}{\partial \phi_x}\right) + \frac{d^2}{dx^2}\left(\frac{\partial f}{\partial \phi_{xx}}\right) \right] \eta \, dx + \frac{\partial f}{\partial \phi_{xx}} \eta_x \Big|_{x_1}^{x_2}$$

$$+ \left[\frac{\partial f}{\partial \phi_x} - \frac{d}{dx}\left(\frac{\partial f}{\partial \phi_{xx}}\right) \right] \eta \Big|_{x_1}^{x_2} = 0 \tag{4.87}$$

To satisfy Eq. (4.87) for any arbitrary function $\eta(x)$ in $x_1 < x < x_2$, we need to satisfy all the following equations:

$$\frac{\partial f}{\partial \phi} - \frac{d}{dx}\left(\frac{\partial f}{\partial \phi_x}\right) + \frac{d^2}{dx^2}\left(\frac{\partial f}{\partial \phi_{xx}}\right) = 0 \tag{4.88}$$

$$\frac{\partial f}{\partial \phi_{xx}} \eta_x \Big|_{x_1}^{x_2} = 0 \tag{4.89}$$

$$\left[\frac{\partial f}{\partial \phi_x} - \frac{d}{dx}\left(\frac{\partial f}{\partial \phi_{xx}}\right) \right] \eta \Big|_{x_1}^{x_2} = 0 \tag{4.90}$$

As seen earlier, Eq. (4.88) denotes the governing differential equation (Euler–Lagrange equation), while Eqs. (4.89) and (4.90) indicate the boundary conditions to be satisfied. It is not necessary to specify the values of η and $\eta_x = d\eta/dx$ at x_1 and x_2 in order to satisfy Eqs. (4.89) and (4.90). We can satisfy these equations by specifying, alternately, the following:

$$\left. \frac{\partial f}{\partial \phi_{xx}} \right|_{x_1}^{x_2} = 0 \tag{4.91}$$

$$\left. \frac{\partial f}{\partial \phi_x} - \frac{d}{dx}\left(\frac{\partial f}{\partial \phi_{xx}} \right) \right|_{x_1}^{x_2} = 0 \tag{4.92}$$

The conditions specified by Eqs. (4.91) and (4.92) are called *natural boundary conditions* because they come out naturally from the extremization process (if they are satisfied, they are called *free boundary conditions*). The conditions

$$\delta\phi|_{x_1}^{x_2} = 0 \quad \text{or} \quad \eta|_{x_1}^{x_2} = 0 \tag{4.93}$$

$$\delta\phi_x|_{x_1}^{x_2} = 0 \quad \text{or} \quad \eta_x|_{x_1}^{x_2} = 0 \tag{4.94}$$

are called *geometric* or *kinematic* or *forced boundary conditions*. It can be seen that Eqs. (4.89) and (4.90) can be satisfied by any combination of natural and geometric boundary conditions at each of the endpoints x_1 and x_2:

$$\text{specify value of } \phi_x \text{ (so that } \eta_x = 0\text{) or specify } \frac{\partial f}{\partial \phi_{xx}} = 0 \tag{4.95}$$

or

$$\text{specify value of } \phi \text{ (so that } \eta = 0\text{) or specify } \frac{\partial f}{\partial \phi_x} - \frac{d}{dx}\left(\frac{\partial f}{\partial \phi_{xx}} \right) = 0 \tag{4.96}$$

The physical significance of the natural and geometric boundary conditions is discussed for a beam deflection problem in the following example.

Example 4.2   A uniform elastic cantilever beam of length l is loaded uniformly as shown in Fig. 4.4. Derive the governing differential equation and the proper boundary conditions of the beam. Also find the deflection of the beam.

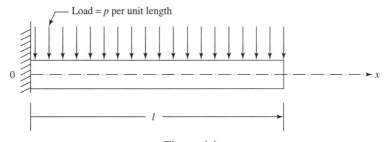

Figure 4.4

SOLUTION   The principle of minimum potential energy is applicable for elastic bodies subject to static loads. This principle states that of all possible displacement configurations a body can assume that satisfy compatibility and given displacement boundary conditions, the configuration that satisfies the equilibrium conditions makes the potential energy assume a minimum value. Hence, if U denotes the potential energy of a body, U must be minimum for the true equilibrium state. Thus, δU must be zero.

The potential energy of a body is given by the strain energy minus the work done by the external loads. In the present case, the strain energy due to the bending of the beam (π) is given by

$$\pi = \frac{EI}{2} \int_0^l \left(\frac{d^2u}{dx^2}\right)^2 dx$$

where E is Young's modulus, I is the moment of inertia of the cross section of the beam about the neutral axis, and $u(x)$ is the transverse displacement of the beam. The work done by the external loads (W) is given by

$$W = \int_0^l pu\, dx$$

where p denotes distributed load (load per unit length) on the beam. Thus, the potential energy functional becomes

$$U = \int_0^l \left[\frac{EI}{2}(u'')^2 - pu\right] dx \tag{E4.2.1}$$

where a prime indicates differentiation with respect to x once. By comparing Eq. (E4.2.1) with the general form in Eq. (4.86), we obtain

$$f(x, u, u', u'') = \frac{EI}{2}(u'')^2 - pu \tag{E4.2.2}$$

Noting that

$$\frac{\partial f}{\partial u} = -p$$

$$\frac{\partial f}{\partial u'} = 0 \tag{E4.2.3}$$

$$\frac{\partial f}{\partial u''} = EIu''$$

the Euler–Lagrange equation or the differential equation of equilibrium of the beam can be obtained from Eq. (4.46) as

$$EIu'''' - p = 0 \tag{E4.2.4}$$

The boundary conditions indicated in Eq. (4.45) take the form

$$EIu'''\delta u|_{x_1}^{x_2} = 0 \tag{E4.2.5}$$

$$EIu''\delta u'|_{x_1}^{x_2} = 0 \tag{E4.2.6}$$

From elementary strength of materials, we notice that EIu''' is the shear force, EIu'' is the bending moment, u' is the rotation or slope, and u is the transverse displacement. Equation (E4.2.5) states that either the shear force or the variation of displacement must be zero at each end. Similarly, Eq. (E4.2.6) states that either the bending moment or the variation of slope must be zero at each end of the beam. In the present case, at $x = x_1 = 0$, the beam is fixed and hence $u = u' = 0$, and at $x = x_2 = l$, the beam is free and hence $EIu'' = EIu''' = 0$. Thus, we have the geometric (displacement) boundary conditions at the fixed end and free or natural boundary conditions at the free end of the beam.

The deflection of the beam, $u(x)$, can be found by solving Eq. (E4.2.4):

$$EIu'''' = p = \text{constant} \tag{E4.2.7}$$

Integrating this equation, we obtain

$$EIu''' = px + c_1 \tag{E4.2.8}$$

where c_1 is a constant. Since $EIu''' = 0$ at $x = l$, we can find that

$$c_1 = -pl \tag{E4.2.9}$$

$$EIu''' = px - pl \tag{E4.2.10}$$

Integration of Eq. (E4.2.10) gives

$$EIu'' = \frac{px^2}{2} - pxl + c_2 \tag{E4.2.11}$$

where c_2 is a constant. As $EIu'' = 0$ at $x = l$,

$$c_2 = \frac{pl^2}{2} \tag{E4.2.12}$$

This gives

$$EIu'' = \frac{p}{2}(x^2 - 2xl + l^2) \tag{E4.2.13}$$

Integrating this again, we have

$$EIu' = \frac{p}{2}\left(\frac{x^3}{3} - x^2l + xl^2\right) + c_3 \tag{E4.2.14}$$

where the constant c_3 can be found by using the condition that $u' = 0$ at $x = 0$. This leads to $c_3 = 0$ and

$$EIu' = \frac{p}{2}\left(\frac{x^3}{3} - x^2l + xl^2\right) \tag{E4.2.15}$$

Integration of this equation gives

$$EIu = \frac{p}{2}\left(\frac{x^4}{12} - \frac{x^3l}{3} + \frac{x^2l^2}{2}\right) + c_4 \tag{E4.2.16}$$

Since $u = 0$ at $x = 0$, we have $c_4 = 0$, and the deflection of the beam is given by

$$u(x) = \frac{p}{2EI} \left(\frac{x^4}{12} - \frac{x^3 l}{3} + \frac{x^2 l^2}{2} \right)$$ (E4.2.17)

4.10 VARIATIONAL METHODS IN SOLID MECHANICS

Several variational methods can be used to derive the governing differential equations of an elastic body. The principles of minimum potential energy, minimum complementary energy, and stationary Reissner energy can be used to formulate static problems. The variational principle valid for dynamics of systems of particles, rigid bodies, or deformable solids is called *Hamilton's principle*. All these variational principles are discussed in this section, with more emphasis placed on Hamilton's principle.

4.10.1 Principle of Minimum Potential Energy

The potential energy of an elastic body (U) is defined as

$$U = \pi - W_P$$ (4.97)

where π is the strain energy and W_P is the work done on the body by the external forces ($-W_P$ is also called the *potential energy* of the applied loads). The principle of minimum potential energy can be stated as follows: Of all possible displacement states a body can assume (u, v, and w for a three-dimensional body) that satisfy compatibility and specified kinematic or displacement boundary conditions, the state that satisfies the equilibrium equations makes the potential energy assume a minimum value. If the potential energy is expressed in terms of the displacement components u, v, and w, the principle of minimum potential energy gives, at the equilibrium state,

$$\delta U(u, v, w) = \delta \pi(u, v, w) - \delta W_P(u, v, w) = 0$$ (4.98)

where the variation is to be taken with respect to the displacement in Eq. (4.98), while the forces and stresses are assumed constant. The strain energy of a linear elastic body is given by

$$\pi = \frac{1}{2} \iiint_V \vec{\varepsilon}^{\mathrm{T}} \vec{\sigma} \, dV$$ (4.99)

where $\vec{\varepsilon}$ is the strain vector, $\vec{\sigma}$ is the stress vector, V is the volume of the body, and the superscript T denotes the transpose. By using the stress–strain relations

$$\vec{\sigma} = [D]\vec{\varepsilon}$$ (4.100)

where $[D]$ is the elasticity matrix, the strain energy can be expressed as

$$\pi = \frac{1}{2} \iiint_V \vec{\varepsilon}^{\mathrm{T}} [D] \vec{\varepsilon} \, dV$$ (4.101)

Note that the initial strains are assumed to be absent in Eq. (4.101). If initial strains are present, with the initial strain vector given by $\vec{\varepsilon}_0$, the strain energy of the body becomes

$$\pi = \tfrac{1}{2} \iiint_V \vec{\varepsilon}^{\mathrm{T}}[D]\vec{\varepsilon}\, dV - \iiint_V \vec{\varepsilon}^{\mathrm{T}}[D]\vec{\varepsilon}_0\, dV \qquad (4.102)$$

The work done by the external forces can be expressed as

$$W_P = \iiint_V (\overline{\phi}_x u + \overline{\phi}_y v + \overline{\phi}_z w)\, dV + \iint_{S_2} (\overline{\Phi}_x u + \overline{\Phi}_y v + \overline{\Phi}_z w)\, dS_2 \qquad (4.103)$$

where S_2 is the surface of the body on which surface forces (tractions) are prescribed. Denoting the known body force vector $\vec{\phi}$, the prescribed surface force (traction) vector $\vec{\Phi}$, and the displacement vector $\vec{u}$ as

$$\vec{\phi} = \begin{Bmatrix} \overline{\phi}_x \\ \overline{\phi}_y \\ \overline{\phi}_z \end{Bmatrix}, \quad \vec{\Phi} = \begin{Bmatrix} \overline{\Phi}_x \\ \overline{\Phi}_y \\ \overline{\Phi}_z \end{Bmatrix}, \quad \vec{u} = \begin{Bmatrix} u \\ v \\ w \end{Bmatrix}$$

Eq. (4.103) can be written equivalently as

$$W_P = \iiint_V \vec{\phi}^{\mathrm{T}} \vec{u}\, dV + \iint_{S_2} \vec{\Phi}^{\mathrm{T}} \vec{u}\, dS_2 \qquad (4.104)$$

Using Eqs. (4.102) and (4.104), the potential energy of the body can be expressed as

$$U(u, v, w) = \tfrac{1}{2} \iiint_V \vec{\varepsilon}^{\mathrm{T}}[D](\vec{\varepsilon} - 2\vec{\varepsilon}_0)\, dV - \iiint_V \vec{\phi}^{\mathrm{T}} \vec{u}\, dV - \iint_{S_2} \vec{\Phi}^{\mathrm{T}} \vec{u}\, dS_2 \qquad (4.105)$$

Thus, according to the principle of minimum potential energy, the displacement field $\vec{u}(x, y, z)$ that minimizes U and satisfies all the boundary conditions is the one that satisfies the equilibrium equations. In the principle of minimum potential energy, we minimize the functional U, and the resulting equations denote the equilibrium equations while the compatibility conditions are satisfied identically.

4.10.2   Principle of Minimum Complementary Energy

The complementary energy of an elastic body (U_c) is defined as

$$U_c = \text{complementary strain energy in terms of stresses } (\overline{\pi})$$

$$- \text{ work done by the applied loads during stress changes } (\overline{W}_P)$$

The principle of the minimum complementary energy can be stated as follows: Of all possible stress states that satisfy the equilibrium equations and the stress boundary conditions, the state that satisfies the compatibility conditions will make the complementary energy assume a minimum value. By expressing the complementary energy

U_c in terms of the stresses σ_{ij}, the principle of minimum complementary energy gives, for compatibility,

$$\delta U_c(\sigma_{xx}, \sigma_{yy}, \ldots, \sigma_{zx}) = \delta \overline{\pi}((\sigma_{xx}, \sigma_{yy}, \ldots, \sigma_{zx}) - \delta \overline{W}_P(\sigma_{xx}, \sigma_{yy}, \ldots, \sigma_{zx}) = 0 \tag{4.106}$$

where the variation is taken with respect to the stress components in Eq. (4.106) while the displacements are assumed constant. The complementary strain energy of a linear elastic body can be expressed as

$$\overline{\pi} = \frac{1}{2} \iiint_V \vec{\sigma}^T \vec{\varepsilon} \, dV \tag{4.107}$$

In the presence of known initial strains $\vec{\varepsilon}_0$, the complementary strain energy becomes

$$\overline{\pi} = \frac{1}{2} \iiint_V \vec{\sigma}^T([C]\vec{\sigma} + 2\vec{\varepsilon}_0) \, dV \tag{4.108}$$

where the strain–stress relations are assumed to be of the form

$$\vec{\varepsilon} = [C]\vec{\sigma} \tag{4.109}$$

The work done by the applied loads during stress change, also known as *complementary work*, is given by

$$\overline{W}_P = \iint_{S_1} (\phi_x \overline{u} + \phi_y \overline{v} + \phi_z \overline{w}) \, dS_1 = \iint_{S_1} \vec{\phi}^T \vec{u} \, dS_1 \tag{4.110}$$

where S_1 is the part of the surface of the body on which the values of displacements are prescribed as

$$\vec{\overline{u}} = \begin{Bmatrix} \overline{u} \\ \overline{v} \\ \overline{w} \end{Bmatrix}$$

Thus, the complementary energy of the body can be expressed, using Eqs. (4.108) and (4.110), as

$$U_c(\sigma_{xx}, \sigma_{yy}, \ldots, \sigma_{zx}) = \frac{1}{2} \iiint_V \vec{\sigma}([C]\vec{\sigma} + 2\vec{\varepsilon}_0) \, dV - \iint_{S_1} \vec{\phi}^T \vec{u} \, dS_1 \tag{4.111}$$

In the principle of minimum complementary energy, the functional U_c is minimized and the resulting equations denote the compatibility equations, while the equilibrium equations are satisfied identically.

4.10.3 Principle of Stationary Reissner Energy

In the principle of minimum potential energy, the potential energy U is expressed in terms of displacements, and variations of u, v, and w are permitted. Similarly, in the case of the principle of minimum complementary energy, the complementary energy U_c is expressed in terms of stresses, and variations of $\sigma_{xx}, \sigma_{yy}, \ldots, \sigma_{zx}$ are permitted. In the present case, the Reissner energy U_r is expressed in terms of both displacements and stresses, and variations with respect to both displacements and stresses are permitted. The Reissner energy for a linearly elastic body is defined as

$$U_r = \iiint\limits_V (\text{internal stresses} \times \text{strains expressed in terms of displacements}$$

$$- \text{complementary strain energy in terms of stresses})\, dV$$

$$- \text{work done by applied forces}$$

$$= \iiint\limits_V \left\{ \left[\sigma_{xx} \frac{\partial u}{\partial x} + \sigma_{yy} \frac{\partial v}{\partial y} + \cdots + \sigma_{zx} \left(\frac{\partial w}{\partial x} + \frac{\partial u}{\partial z} \right) \right] - \pi \right\} dV$$

$$- \iiint\limits_V (\bar{\phi}_x u + \bar{\phi}_y v + \bar{\phi}_z w)\, dV - \iint\limits_{S_2} (\bar{\Phi}_x u + \bar{\Phi}_y v + \bar{\Phi}_z w)\, dS_2$$

$$- \iint\limits_{S_1} \left[(u - \bar{u})\Phi_x + (v - \bar{v})\Phi_y + (w - \bar{w})\Phi_z \right] dS_1$$

$$= \iiint\limits_V (\vec{\sigma}^{\mathrm{T}} \vec{\varepsilon} - \tfrac{1}{2}\vec{\sigma}^{\mathrm{T}}[C]\vec{\sigma} - \vec{\bar{\Phi}}^{\mathrm{T}} \vec{u})\, dV - \iint\limits_{S_2} \vec{u}^{\mathrm{T}} \vec{\bar{\Phi}}\, dS_2 - \iint\limits_{S_1} (\vec{u} - \vec{\bar{u}})^{\mathrm{T}} \vec{\Phi}\, dS_1$$

$$(4.112)$$

When the variation of U_r is set equal to zero by considering variations in both displacements and stresses, we obtain

$$\delta U_r = \sum_{ij} \frac{\partial U_r}{\partial \sigma_{ij}} \delta \sigma_{ij} + \left(\frac{\partial U_r}{\partial u} \delta u + \frac{\partial U_r}{\partial v} \delta v + \frac{\partial U_r}{\partial w} \delta w \right) = 0 \qquad (4.113)$$

where the subscripts i and j are used to include all the components of stress $\sigma_{xx}, \sigma_{yy}, \ldots, \sigma_{zx}$. The first term on the right-hand side of Eq. (4.113) gives the stress–displacement relations, and the second term gives the equilibrium equations and boundary conditions. The principle of stationary Reissner energy can be stated in words as follows: Of all possible stress and displacement states a body can have, the particular set that makes the Reissner energy stationary gives the correct stress–displacement and equilibrium equations, along with the boundary conditions.

4.10.4   Hamilton's Principle

The variational principle that can be used for dynamic problems is called Hamilton's principle. According to this principle, variation of the functional is taken with respect to time. The functional used in Hamilton's principle, similar to U, U_c, and U_r, is called the *Lagrangian* (L) and is defined as

$$L = T - U = \text{kinetic energy} - \text{potential energy} \qquad (4.114)$$

Development of Hamilton's principle for discrete as well as continuous systems is presented in the following sections.

Hamilton's Principle for Discrete Systems   Let a discrete system (system with a finite number of degrees of freedom) be composed of n masses or particles. First, we consider a single particle of mass m, at the position vector $\vec{r}$, subjected to a force $\vec{f}(\vec{r})$. The position of the particle, $\vec{r}$, at any time t is given by Newton's second law of motion:

$$m\frac{d^2\vec{r}}{dt^2} - \vec{f}(\vec{r}) = \vec{0} \tag{4.115}$$

If the true path of the particle is $\vec{r}(t)$, we define a varied path as $\vec{r} + \delta\vec{r}$, where $\delta\vec{r}$ denotes the variation of the path at any fixed time t. We assume that the true path and the varied path are same at two distinct times t_1 and t_2, so that

$$\delta\vec{r}(t_1) = \delta\vec{r}(t_2) = 0 \tag{4.116}$$

By taking the dot product of Eq. (4.115) with $\delta\vec{r}$ and integrating with respect to time from t_1 to t_2 yields

$$\int_{t_1}^{t_2} \left[m\frac{d^2\vec{r}}{dt^2} - \vec{f}(\vec{r}) \right] \cdot \delta\vec{r}\, dt = 0 \tag{4.117}$$

The first term of the integral in Eq. (4.117) can be integrated by parts as

$$\int_{t_1}^{t_2} m\frac{d^2\vec{r}}{dt^2} \cdot \delta\vec{r}\, dt = - \int_{t_1}^{t_2} m\frac{d\vec{r}}{dt} \cdot \frac{d(\delta\vec{r})}{dt}\, dt + m\frac{d\vec{r}}{dt} \cdot \delta\vec{r}\Big|_{t_1}^{t_2} \tag{4.118}$$

In view of Eq. (4.116), the second term on the right side of Eq. (4.118) will be zero and Eq. (4.117) becomes

$$\int_{t_1}^{t_2} \left[m\frac{d\vec{r}}{dt} \cdot \frac{d}{dt}\delta\vec{r} + \vec{f}(r) \cdot \delta\vec{r} \right] dt = 0 \tag{4.119}$$

The kinetic energy of the particle (T) is given by

$$T = \frac{1}{2}m\frac{d\vec{r}}{dt} \cdot \frac{d\vec{r}}{dt} \tag{4.120}$$

and hence

$$\delta T = m\frac{d\vec{r}}{dt} \cdot \delta\frac{d\vec{r}}{dt} = m\frac{d\vec{r}}{dt} \cdot \frac{d\delta\vec{r}}{dt} \tag{4.121}$$

Using Eq. (4.121) in Eq. (4.119), the general form of Hamilton's principle for a single mass (particle) can be expressed as

$$\int_{t_1}^{t_2} (\delta T + \vec{f} \cdot \delta\vec{r})\, dt = 0 \tag{4.122}$$

Conservative Systems   For a conservative system, the sum of the potential and kinetic energies is a constant, and the force $\vec{f}$ can be derived from the potential energy U as

$$\vec{f} = -\nabla U \tag{4.123}$$

where ∇ denotes the gradient operator. Noting that

$$\nabla U \cdot \delta \vec{r} = \frac{\partial U}{\partial x} \delta x + \frac{\partial U}{\partial y} \delta y + \frac{\partial U}{\partial z} \delta z = \delta U(x, y, z) \tag{4.124}$$

Hamilton's principle for a single mass in a conservative system is given by

$$\delta I = \delta \int_{t_1}^{t_2} L \, dt = 0 \tag{4.125}$$

where

$$L = T - U \tag{4.126}$$

is called the *Lagrangian function*. Thus, Hamilton's principle for a particle acted by a conservative force can be stated as follows: Of all possible paths that the particle could take from its position at time t_1 to its position at time t_2, the true path will be the one that extremizes the integral

$$I = \int_{t_1}^{t_2} L \, dt \tag{4.127}$$

Use of Generalized Coordinates   If the position of the path at any time t, $\vec{r}$, is expressed in terms of the generalized coordinates q_1, q_2, and q_3 (instead of x, y, and z), the Lagrangian L can be expressed as

$$L = L(q_1, q_2, q_3, \dot{q}_1, \dot{q}_2, \dot{q}_3) \tag{4.128}$$

where $\dot{q}_i = dq_i/dt$ ($i = 1, 2, 3$) denotes the ith generalized velocity. Then the necessary condition for the extremization of I can be written as

$$\delta I = \delta \int_{t_1}^{t_2} L(q_1, q_2, q_3, \dot{q}_1, \dot{q}_2, \dot{q}_3) \, dt = \int_{t_1}^{t_2} \sum_{i=1}^{3} \left[\frac{\partial L}{\partial q_i} - \frac{d}{dt} \left(\frac{\partial L}{\partial \dot{q}_i} \right) \right] \delta q_i \, dt = 0 \tag{4.129}$$

If q_i are linearly independent, with no constraints among q_i, all δq_i are independent, and hence Eq. (4.129) leads to

$$\frac{\partial L}{\partial q_i} - \frac{d}{dt} \left(\frac{\partial L}{\partial \dot{q}_i} \right) = 0, \qquad i = 1, 2, 3 \tag{4.130}$$

Equations (4.130) denote the Euler–Lagrange equations that correspond to the extremization of I and are often called the *Lagrange equations of motion*.

Nonconservative Systems   If the forces are not conservative, the general form of Hamilton's principle, given by Eq. (4.122), can be rewritten as

$$\delta \int_{t_1}^{t_2} T \, dt + \int_{t_1}^{t_2} \delta W_{\text{nc}} \, dt = 0 \tag{4.131}$$

where

$$\delta W_{\text{nc}} = \vec{f} \cdot \delta \vec{r}$$

denotes the virtual work done by the nonconservative force $\vec{f}$. In this case, a functional I does not exist for extremization. If the virtual work δW_{nc} is expressed in terms of generalized coordinates (q_1, q_2, q_3) and generalized forces (Q_1, Q_2, Q_3) as

$$\delta W_{\text{nc}} = \sum_{i=1}^{3} Q_i \delta q_i \tag{4.132}$$

where δq_i is the virtual generalized displacement, Eq. (4.131) can be expressed as

$$\int_{t_1}^{t_2} \sum_{i=1}^{3} \left[\frac{\partial T}{\partial q_i} - \frac{d}{dt} \left(\frac{\partial T}{\partial \dot{q}_i} \right) + Q_i \right] \delta q_i \, dt = 0 \tag{4.133}$$

Thus, the Euler–Lagrange equations corresponding to Eq. (4.133) are given by

$$\frac{\partial T}{\partial q_i} - \frac{d}{dt} \left(\frac{\partial T}{\partial \dot{q}_i} \right) + Q_i = 0, \qquad i = 1, 2, 3 \tag{4.134}$$

System of Masses   If a system of n mass particles or rigid bodies with masses m_i and position vectors $\vec{r}_i$ are considered, Hamilton's principle can be expressed as follows. For conservative forces,

$$\delta \int_{t_1}^{t_2} L(q_1, q_2, \ldots, \dot{q}_1, \dot{q}_2, \ldots) \, dt = 0 \tag{4.135}$$

which is a generalization of Eq. (4.125). For nonconservative forces,

$$\delta \int_{t_1}^{t_2} T \, dt + \int_{t}^{t_2} \delta W_{\text{nc}} \, dt = 0 \tag{4.136}$$

which is a generalization of Eq. (4.131). The kinetic energy and the virtual work in Eqs. (4.135) and (4.136) are given by

$$T = \frac{1}{2} \sum_{i=1}^{n} m_i \frac{d\vec{r}_i}{dt} \cdot \frac{d\vec{r}_i}{dt} \tag{4.137}$$

$$\delta W (\delta W_{\text{nc}}) = \sum_{i=1}^{n} \vec{f}_i \cdot \delta \vec{r}_i \tag{4.138}$$

As can be seen from Eqs. (4.125) and (4.131), Hamilton's principle reduces the problems of dynamics to the study of a scalar integral that does not depend on the coordinates used. Note that Hamilton's principle yields merely the equations of motion of the system but not the solution of the dynamics problem.

Hamilton's Principle for Continuous Systems   For a continuous system, the kinetic energy of the body, T, can be expressed as

$$T = \tfrac{1}{2} \iiint\limits_{V} \rho \dot{\vec{u}}^{\mathrm{T}} \dot{\vec{u}}\, dV \tag{4.139}$$

where ρ is the density of the material and $\dot{\vec{u}}$ is the vector of velocity components at any point in the body:

$$\dot{\vec{u}} = \left\{ \begin{array}{c} \dot{u} \\ \dot{v} \\ \dot{w} \end{array} \right\}$$

Thus, the Lagrangian can be written as

$$L = \frac{1}{2} \iiint\limits_{V} (\rho \dot{\vec{u}}^{\mathrm{T}} \dot{\vec{u}} - \vec{\varepsilon}^{\mathrm{T}}[D]\vec{\varepsilon} + 2\vec{u}^{\mathrm{T}}\vec{\phi})\, dV + \iint\limits_{S_2} \vec{u}^{\mathrm{T}}\vec{\Phi}\, dS_2 \tag{4.140}$$

Hamilton's principle can be stated in words as follows: Of all possible time histories of displacement states that satisfy the compatibility equations and the constraints or the kinematic boundary conditions and that also satisfy the conditions at initial and final times t_1 and t_2, the history corresponding to the actual solution makes the Lagrangian functional a minimum. Hamilton's principle can thus be expressed as

$$\delta \int_{t_1}^{t_2} L\, dt = 0 \tag{4.141}$$

Generalized Hamilton's Principle   For an elastic body in motion, the equations of dynamic equilibrium for an element of the body can be written, using Cartesian tensor notation, as

$$\sigma_{ij,j} + \phi_i = \rho \frac{\partial^2 u_i}{\partial t^2}, \qquad i = 1, 2, 3 \tag{4.142}$$

where ρ is the density of the material, ϕ_i is the body force per unit volume acting along the x_i direction, u_i is the component of displacement along the x_i direction, the σ_{ij} denotes the stress tensor

$$\sigma_{ij} = \begin{bmatrix} \sigma_{11} & \sigma_{12} & \sigma_{13} \\ \sigma_{21} & \sigma_{22} & \sigma_{23} \\ \sigma_{31} & \sigma_{32} & \sigma_{33} \end{bmatrix} \equiv \begin{bmatrix} \sigma_{xx} & \sigma_{xy} & \sigma_{xz} \\ \sigma_{xy} & \sigma_{yy} & \sigma_{yz} \\ \sigma_{xz} & \sigma_{yz} & \sigma_{zz} \end{bmatrix} \tag{4.143}$$

and

$$\sigma_{ij,j} = \frac{\partial \sigma_{i1}}{\partial x_1} + \frac{\partial \sigma_{i2}}{\partial x_2} + \frac{\partial \sigma_{i3}}{\partial x_3} \tag{4.144}$$

with $x_1 = x$, $x_2 = y$, $x_3 = z$ and $u_1 = u$, $u_2 = v$, $u_3 = w$.

The solid body is assumed to have a volume V with a bounding surface S. The bounding surface S is assumed to be composed of two parts, S_1 and S_2, where the displacements u_i are prescribed on S_1 and surface forces (tractions) are prescribed on S_2. Consider a set of virtual displacements δu_i of the vibrating body which vanishes over the boundary surface S_1, where values of displacements are prescribed, but are arbitrary over the rest of the boundary surface S_2, where surface tractions are prescribed. The virtual work done by the body and surface forces is given by

$$\iiint_V \phi_i \delta u_i \, dV + \iint_S \Phi_i \delta u_i \, dS \tag{4.145}$$

where Φ_i indicates the prescribed surface force along the direction u_i. Although the surface integral is expressed over S in Eq. (4.145), it needs to be integrated only over S_2, since δu_i vanishes over the surface S_1, where the boundary displacements are prescribed. The surface forces Φ_i can be represented as

$$\Phi_i = \sigma_{ij} \nu_j \equiv \sum_{j=1}^{3} \sigma_{ij} \nu_j, \qquad i = 1, 2, 3 \tag{4.146}$$

where $\vec{\nu} = \{\nu_1 \quad \nu_2 \quad \nu_3\}^T$ is the unit vector along the outward normal of the surface S with ν_1, ν_2, and ν_3 as its components along the x_1, x_2, and x_3 directions, respectively. By substituting Eq. (4.146), the second term on the right-hand side of Eq. (4.145) can be written as

$$\iint_S \sigma_{ij} \delta u_i \nu_j \, dS \tag{4.147}$$

Using Gauss's theorem [7], expression (4.147) can be rewritten in terms of the volume integral as

$$\iint_S \Phi_i \delta u_i \, dS = \iint_S \sigma_{ij} \delta u_i \nu_j \, dS = \iiint_V (\sigma_{ij} \delta u_i)_{,j} \, dV$$

$$= \iiint_V \sigma_{ij,j} \delta u_i \, dV + \iiint_V \sigma_{ij} \delta u_{i,j} \, dV \tag{4.148}$$

Because of the symmetry of the stress tensor, the last term in Eq. (4.148) can be written as

$$\iiint_V \sigma_{ij} \delta u_{i,j} \, dV = \iiint_V \sigma_{ij} \left[\tfrac{1}{2} (\delta u_{i,j} + \delta u_{j,i}) \right] dV = \iiint_V \sigma_{ij} \delta \varepsilon_{ij} \, dV \tag{4.149}$$

where ε_{ij} denotes the strain tensor:

$$\varepsilon_{ij} = \begin{bmatrix} \varepsilon_{11} & \varepsilon_{12} & \varepsilon_{13} \\ \varepsilon_{21} & \varepsilon_{22} & \varepsilon_{23} \\ \varepsilon_{31} & \varepsilon_{32} & \varepsilon_{33} \end{bmatrix} \equiv \begin{bmatrix} \varepsilon_{xx} & \varepsilon_{xy} & \varepsilon_{xz} \\ \varepsilon_{xy} & \varepsilon_{yy} & \varepsilon_{yz} \\ \varepsilon_{xz} & \varepsilon_{yz} & \varepsilon_{zz} \end{bmatrix} \tag{4.150}$$

In view of the equations of dynamic equilibrium, Eq. (4.142), the first integral on the right hand side of Eq. (4.148), can be expressed as

$$\iiint_V \sigma_{ij,j} \delta u_i \, dV = \iiint_V \left(\rho \frac{\partial^2 u_i}{\partial t^2} - \phi_i \right) \delta u_i \, dV \tag{4.151}$$

Thus, the second term of expression (4.145) can be written as

$$\iint_S \Phi_i \delta u_i \, dS = \iiint_V \sigma_{ij} \delta \varepsilon_{ij} \, dV + \iiint_V \left(\rho \frac{\partial^2 u_i}{\partial t^2} - \phi_i \right) \delta u_i \, dV \qquad (4.152)$$

This gives the variational equation of motion

$$\iiint_V \sigma_{ij} \delta \varepsilon_{ij} \, dV = \iiint_V \left(\phi_i - \rho \frac{\partial^2 u_i}{\partial t^2} \right) \delta u_i \, dV + \iint_S \Phi_i \delta u_i \, dS \qquad (4.153)$$

This equation can be stated more concisely by introducing different levels of restrictions. If the body is perfectly elastic, Eq. (4.153) can be stated in terms of the strain energy density π_0 as

$$\delta \iiint_V \pi_0 \, dV = \iiint_V \left(\phi_i - \rho \frac{\partial^2 u_i}{\partial t^2} \right) \delta u_i \, dV + \iint_S \Phi_i \delta u_i \, dS \qquad (4.154)$$

or

$$\delta \iiint_V \left(\pi_0 + \rho \frac{\partial^2 u_i}{\partial t^2} \delta u_i \right) dV = \iiint_V \phi_i \delta u_i \, dV + \iint_S \Phi_i \delta u_i \, dS \qquad (4.155)$$

If the variations δu_i are identified with the actual displacements $(\partial u_i / \partial t) dt$ during a small time interval dt, Eq. (4.155) states that in an arbitrary time interval, the sum of the energy of deformation and the kinetic energy increases by an amount that is equal to the work done by the external forces during the same time interval.

Treating the virtual displacements δu_i as functions of time and space not identified with the actual displacements, the variational equation of motion, Eq. (4.154), can be integrated between two arbitrary instants of time t_1 and t_2 and we obtain

$$\int_{t_1}^{t_2} \iiint_V \delta \pi_0 \, dV \, dt = \int_{t_1}^{t_2} dt \iiint_V \phi_i \delta u_i \, dV + \int_{t_1}^{t_2} dt \iint_S \Phi_i \delta u_i \, dS$$

$$- \int_{t_1}^{t_2} dt \iiint_V \rho \frac{\partial^2 u_i}{\partial t^2} \delta u_i \, dV \qquad (4.156)$$

Denoting the last term in Eq. (4.156) as A, inverting the order of integration, and integrating by parts leads to

$$A = \iiint_V \rho \frac{\partial u_i}{\partial t} \delta u_i \, dV \Big|_{t_1}^{t_2} - \iiint_V dV \int_{t_1}^{t_2} \frac{\partial u_i}{\partial t} \left(\rho \frac{\partial \delta u_i}{\partial t} + \frac{\partial \rho}{\partial t} \delta u_i \right) dt \qquad (4.157)$$

In most problems, the time rate of change of the density of the material, $\partial \rho / \partial t$, can be neglected. Also, we consider δu_i to be zero at all points of the body at initial and final times t_1 and t_2, so that

$$\delta u_i(t_1) = \delta u_i(t_2) = 0 \qquad (4.158)$$

In view of Eq. (4.158), Eq. (4.157) can be rewritten as

$$A = -\int_{t_1}^{t_2} \iiint_V \rho \frac{\partial u_i}{\partial t} \frac{\partial \delta u_i}{\partial t} \, dV \, dt = -\int_{t_1}^{t_2} \iiint_V \rho \frac{\partial u_i}{\partial t} \delta \frac{\partial u_i}{\partial t} \, dV \, dt$$

$$= -\int_{t_1}^{t_2} \delta \iiint_V \frac{1}{2} \rho \frac{\partial u_i}{\partial t} \frac{\partial u_i}{\partial t} \, dV \, dt = -\int_{t_1}^{t_2} \delta T \, dt \qquad (4.159)$$

where

$$T = \frac{1}{2} \iiint_V \rho \frac{\partial u_i}{\partial t} \frac{\partial u_i}{\partial t} \, dV \qquad (4.160)$$

is the kinetic energy of the vibrating body. Thus, Eq. (4.156) can be expressed as

$$\int_{t_1}^{t_2} \delta(\pi - T) \, dt = \int_{t_1}^{t_2} \iiint_V \phi_i \delta u_i \, dV \, dt + \int_{t_1}^{t_2} \iint_{S_2} \Phi_i \delta u_i \, dS \, dt \qquad (4.161)$$

where π denotes the total strain energy of the solid body:

$$\pi = \iiint_V \pi_0 \, dV \qquad (4.162)$$

If the external forces acting on the body are such that the sum of the integrals on the right-hand side of Eq. (4.161) denotes the variation of a single function W (known as the potential energy of loading), we have

$$\iiint_V \phi_i \delta u_i \, dV + \iint_{S_2} \Phi_i \delta u_i \, dS = -\delta W \qquad (4.163)$$

Then Eq. (4.161) can be expressed as

$$\delta \int_{t_1}^{t_2} L \, dt \equiv \int_{t_1}^{t_2} (\pi - T + W) \, dt = 0 \qquad (4.164)$$

where

$$L = \pi - T + W \qquad (4.165)$$

is called the Lagrangian function and Eq. (4.164) is known as Hamilton's principle. Note that a negative sign is included, as indicated in Eq. (4.163), for the potential energy of loading (W). *Hamilton's principle* can be stated in words as follows: The time integral of the Lagrangian function between the initial time t_1 and the final time t_2 is an extremum for the actual displacements (motion) with respect to all admissible virtual displacements that vanish throughout the entire time interval: first, at all points of the body at the instants t_1 and t_2, and second, over the surface S_1, where the displacements are prescribed.

Hamilton's principle can be interpreted in another way by considering the displacements $u_i(x_1, x_2, x_3, t)$, $i = 1, 2, 3$, to constitute a dynamic path in space. Then *Hamilton's principle* states: Among all admissible dynamic paths that satisfy the prescribed geometric boundary conditions on S_1 at all times and the prescribed conditions at two arbitrary instants of time t_1 and t_2 at every point of the body, the actual dynamic path (solution) makes the Lagrangian function an extremum.

4.11 APPLICATIONS OF HAMILTON'S PRINCIPLE

4.11.1 Equation of Motion for Torsional Vibration of a Shaft (Free Vibration)

Strain Energy   To derive a general expression for the strain energy of a shaft, consider the shaft to be of variable cross section under a torsional load as shown in Fig. 4.5. If $\theta(x, t)$ denotes the angular displacement of the cross section at x, the angular displacement of the cross section at $x + dx$ can be denoted as $\theta(x, t) + [\partial\theta(x, t)/\partial x]\, dx$, due to the distributed torsional load $m_t(x, t)$. The shear strain at a radial distance r is given by $\gamma = r(\partial\theta/\partial x)$. The corresponding shear stress can be represented as $\tau = G\gamma = Gr(\partial\theta/\partial x)$, where G is the shear modulus. The strain energy density π_0 can be represented as $\pi_0 = \frac{1}{2}\tau\gamma = \frac{1}{2}Gr^2\,(\partial\theta/\partial x)^2$. The total strain energy of the shaft can be determined as

$$\pi = \iiint_V \pi_0\, dV = \int_0^L \iint_A \frac{1}{2}Gr^2 \left(\frac{\partial\theta}{\partial x}\right)^2 dA\, dx = \frac{1}{2}\int_0^L GJ \left(\frac{\partial\theta}{\partial x}\right)^2 dx \quad (4.166)$$

where V is the volume, L is the length, A is the cross-sectional area, and $J = I_p$ polar moment of inertia (for a uniform circular shaft) of the shaft.

Kinetic Energy   The kinetic energy of a shaft with variable cross section can be expressed as

$$T = \frac{1}{2}\int_0^L I_0(x) \left(\frac{\partial\theta(x, t)}{\partial t}\right)^2 dx \quad (4.167)$$

where $I_0(x) = \rho I_p(x)$ is the mass moment of inertia per unit length of the shaft and ρ is the density. By using Eqs. (4.166) and (4.167), Hamilton's principle can be used to obtain

$$\delta \int_{t_1}^{t_2} (T - \pi)\, dt = \delta \int_0^L \left[\frac{1}{2}\int_0^L I_0 \left(\frac{\partial\theta}{\partial t}\right)^2 dx - \frac{1}{2}\int_0^L GJ \left(\frac{\partial\theta}{\partial x}\right)^2 dx\right] dt = 0 \quad (4.168)$$

By carrying out the variation operation, the various terms in Eq. (4.168) can be rewritten, noting that δ and $\partial/\partial t$ as well as δ and $\partial/\partial x$ are commutative, as

$$\delta \int_{t_1}^{t_2} \left[\frac{1}{2}\int_0^L I_0 \left(\frac{\partial\theta}{\partial t}\right)^2 dx\right] dt = -\int_{t_1}^{t_2} \int_0^L I_0 \frac{\partial^2\theta}{\partial t^2}\delta\theta\, dx\, dt \quad (4.169)$$

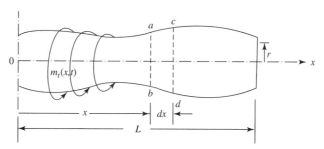

Figure 4.5   Torsional vibration of a shaft.

assuming that θ is prescribed at t_1 and t_2 so that $\delta\theta = 0$ at t_1 and t_2. Similarly,

$$\delta \int_{t_1}^{t_2} \left[\frac{1}{2} \int_0^L GJ \left(\frac{\partial\theta}{\partial x} \right)^2 dx \right] dt = \int_{t_1}^{t_2} \left[GJ \frac{\partial\theta}{\partial x} \delta\theta \bigg|_0^L - \int_0^L \frac{\partial}{\partial x} \left(GJ \frac{\partial\theta}{\partial x} \right) \delta\theta \, dx \right] dt \tag{4.170}$$

Thus, Eq. (4.168) becomes

$$\int_{t_1}^{t_2} \left\{ \int_0^L \left[\frac{\partial}{\partial x} \left(GJ \frac{\partial\theta}{\partial x} \right) - I_0 \frac{\partial^2\theta}{\partial t^2} \right] \delta\theta \, dx - GJ \frac{\partial\theta}{\partial x} \delta\theta \bigg|_{t_1}^{t_2} \right\} dt = 0 \tag{4.171}$$

Assuming that $\delta\theta = 0$ at $x = 0$ and $x = L$, and $\delta\theta$ is arbitrary in $0 < x < L$, Eq. (4.171) requires that

$$\frac{\partial}{\partial x} \left(GJ \frac{\partial\theta}{\partial x} \right) - I_0 \frac{\partial^2\theta}{\partial t^2} = 0, \qquad 0 < x < L \tag{4.172}$$

$$\left(GJ \frac{\partial\theta}{\partial x} \right) \delta\theta = 0 \qquad \text{at} \quad x = 0 \text{ and } x = L \tag{4.173}$$

Equation (4.172) denotes the equation of motion of the shaft, and Eq. (4.173) indicates the boundary conditions. The boundary conditions require that either $GJ(\partial\theta/\partial x) = 0$ (stress is zero) or $\delta\theta = 0$ (θ is specified) at $x = 0$ and $x = L$.

4.11.2   Transverse Vibration of a Thin Beam

Consider an element of a thin beam in bending as shown in Fig. 4.6. If w denotes the deflection of the beam at any point x along the length of the beam, the slope of the deflected centerline is given by $\partial w/\partial x$. Since a plane section of the beam remains plane after deformation according to simple (thin) beam theory, the axial displacement of a fiber located at a distance z from the neutral axis u due to the transverse displacement w can be expressed as (point A moves to A')

$$u = -z \frac{\partial w}{\partial x} \tag{4.174}$$

Thus, the axial strain can be expressed as

$$\varepsilon_x = \frac{\partial u}{\partial x} = -z \frac{\partial^2 w}{\partial x^2} \tag{4.175}$$

and the axial stress as

$$\sigma_x = E\varepsilon_x = -Ez \frac{\partial^2 w}{\partial x^2} \tag{4.176}$$

The strain energy density of the beam element (π_0) is given by

$$\pi_0 = \frac{1}{2} \sigma_x \varepsilon_x = \frac{1}{2} E z^2 \left(\frac{\partial^2 w}{\partial x^2} \right)^2 \tag{4.177}$$

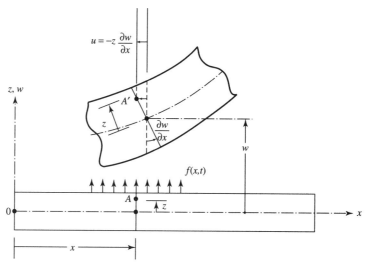

Figure 4.6   Beam in bending.

and hence the strain energy of the beam (π) can be expressed as

$$\pi = \iiint_V \pi_0 \, dV = \int_0^L \frac{1}{2} E \iint_A [dA(x)z^2] \left(\frac{\partial^2 w}{\partial x^2}\right)^2 dx = \frac{1}{2} \int_0^L EI(x) \left(\frac{\partial^2 w}{\partial x^2}\right)^2 dx$$

(4.178)

where $I(x)$ denotes the area moment of inertia of the cross section of the beam at x:

$$I(x) = \iint_A dA(x)z^2$$

(4.179)

The kinetic energy of the beam can be expressed as

$$T = \frac{1}{2} \int_0^L m(x) \left[\frac{\partial w(x,t)}{\partial t}\right]^2 dx$$

(4.180)

where $m(x) = \rho A(x)$ is the mass per unit length of the beam and ρ is the density of the beam. The virtual work of the applied distributed force, $f(x,t)$, is given by

$$\delta W(t) = \int_0^L f(x,t)\delta w(x,t) \, dx$$

(4.181)

Noting that the order of integrations with respect to t and x can be interchanged and the operators δ and d/dx or δ and d/dt are commutative, the variations of π and T can be written as

$$\delta\pi = \delta \int_{t_1}^{t_2} \int_0^L \frac{1}{2} EI \left(\frac{\partial^2 w}{\partial x^2}\right)^2 dx \, dt = \int_{t_1}^{t_2} \int_0^L EI \frac{\partial^2 w}{\partial x^2} \delta\left(\frac{\partial^2 w}{\partial x^2}\right) dx \, dt$$

$$= \int_{t_1}^{t_2} \left[EI\frac{\partial^2 w}{\partial x^2} \delta\frac{\partial w}{\partial x}\Big|_0^L - \int_0^L \frac{\partial}{\partial x}\left(EI\frac{\partial^2 w}{\partial x^2}\right)\delta\frac{\partial w}{\partial x} \, dx \right] dt$$

(4.182)

Since

$$
\int_{t_1}^{t_2} \int_0^L \frac{\partial}{\partial x} \left(EI \frac{\partial^2 w}{\partial x^2} \right) \delta \left(\frac{\partial w}{\partial x} \right) dx \, dt
$$

$$
= \int_{t_1}^{t_2} \left[\frac{\partial}{\partial x} \left(EI \frac{\partial^2 w}{\partial x^2} \right) \delta w \Big|_0^L - \int_0^L \frac{\partial^2}{\partial x^2} \left(EI \frac{\partial^2 w}{\partial x^2} \right) \delta w \, dx \right] dt \qquad (4.183)
$$

Eq. (4.182) becomes

$$
\delta \pi = \int_{t_1}^{t_2} \left[EI \frac{\partial^2 w}{\partial x^2} \delta \left(\frac{\partial w}{\partial x} \right) \Big|_0^L - \frac{\partial}{\partial x} \left(EI \frac{\partial^2 w}{\partial x^2} \right) \delta w \Big|_0^L \right.
$$

$$
\left. + \int_0^L \frac{\partial^2}{\partial x^2} \left(EI \frac{\partial^2 w}{\partial x^2} \right) \delta w \, dx \right] dt \qquad (4.184)
$$

$$
\delta T = \delta \int_{t_1}^{t_2} \int_0^L \frac{1}{2} m(x) \left(\frac{\partial w}{\partial t} \right)^2 dx \, dt = \int_{t_1}^{t_2} \int_0^L m(x) \frac{\partial w}{\partial t} \delta \left(\frac{\partial w}{\partial t} \right) dx \, dt
$$

$$
= \int_0^L \left[\left(m \frac{\partial w}{\partial t} \right) \delta w \Big|_{t_1}^{t_2} - \int_{t_1}^{t_2} \frac{\partial}{\partial t} \left(m \frac{\partial w}{\partial t} \right) \delta w \, dt \right] dx
$$

$$
- \int_0^L \left[\int_{t_1}^{t_2} \frac{\partial}{\partial t} \left(m \frac{\partial w}{\partial t} \right) \delta w \, dt \right] dx \qquad (4.185)
$$

because δw is zero at $t = t_1$ and $t = t_2$. Thus, Hamilton's principle can be stated as

$$
\delta \int_{t_1}^{t_2} (T - \pi + W) \, dt = 0
$$

or

$$
\int_{t_1}^{t_2} \left[- \int_0^L m \frac{\partial^2 w}{\partial t^2} \delta w \, dx - EI \frac{\partial^2 w}{\partial x^2} \delta \left(\frac{\partial w}{\partial x} \right) \Big|_0^L + \frac{\partial}{\partial x} \left(EI \frac{\partial^2 w}{\partial x^2} \right) \delta w \Big|_0^L \right.
$$

$$
\left. - \int_0^L \frac{\partial^2}{\partial x^2} \left(EI \frac{\partial^2 w}{\partial x^2} \right) \delta w \, dx + \int_0^L f \delta w \, dt \right] dt = 0 \qquad (4.186)
$$

Equation (4.186) leads to the following equations:

$$
\frac{\partial^2}{\partial x^2} \left(EI \frac{\partial^2 w}{\partial x^2} \right) + m \frac{\partial^2 w}{\partial x^2} - f = 0, \qquad 0 < x < L \qquad (4.187)
$$

$$
EI \frac{\partial^2 w}{\partial x^2} \delta \left(\frac{\partial w}{\partial x} \right) \Big|_o^L = 0 \qquad (4.188)
$$

$$
\frac{\partial}{\partial x} \left(EI \frac{\partial^2 w}{\partial x^2} \right) \delta w \Big|_0^L = 0 \qquad (4.189)
$$

Equation (4.187) denotes the equation of motion for the transverse vibration of the beam, and Eqs. (4.188) and (4.189) represent the boundary conditions. It can be seen that Eq. (4.188) requires that either

$$EI\frac{\partial^2 w}{\partial x^2} = 0 \quad \text{or} \quad \delta\left(\frac{\partial w}{\partial x}\right) = 0 \qquad \text{at } x = 0 \text{ and } x = L \tag{4.190}$$

while Eq. (4.189) requires that either

$$\frac{\partial}{\partial x}\left(EI\frac{\partial^2 w}{\partial x^2}\right) = 0 \quad \text{or} \quad \delta w = 0 \qquad \text{at } x = 0 \text{ and } x = L \tag{4.191}$$

Thus, Eqs. (4.188) and (4.189) can be satisfied by the following common boundary conditions:

1. Fixed or clamped end:

$$w = \text{transverse deflection} = 0, \qquad \frac{\partial w}{\partial x} = \text{bending slope} = 0 \tag{4.192}$$

2. Pinned or hinged end:

$$w = \text{transverse deflection} = 0, \qquad EI\frac{\partial^2 w}{\partial x^2} = \text{bending moment} = 0 \tag{4.193}$$

3. Free end:

$$EI\frac{\partial^2 w}{\partial x^2} = \text{bending moment} = 0, \qquad \frac{\partial}{\partial x}\left(EI\frac{\partial^2 w}{\partial x^2}\right) = \text{shear force} = 0 \tag{4.194}$$

4.12   RECENT CONTRIBUTIONS

Nagem et al. [11] observed that the Hamiltonian formulation of the damped oscillator can be used to model dissipation in quantum mechanics, to analyze low-temperature thermal fluctuations in *RLC* circuits, and to establish Pontryagin control theory for damped systems. Sato examined the governing equations used for the vibration and stability of a Timoshenko beam from the point of view of Hamilton's principle [12]. He derived the governing equations using an extended Hamilton's principle by considering the deviation of the external force following the deflection of the beam at its tip in terms of the angle ε measured from the x axis (which is taken to be along the length of the beam).

The variational finite difference method was presented for the vibration of sector plates by Singh and Dev [13]. Conventional finite difference techniques are normally applied to discretize the differential formulation either by approximating the field variable directly or by replacing the differentials by appropriate difference quotients. In general, the boundary conditions pose difficulties, particularly in problems with complex geometric configurations. The difficulties of conventional finite difference analysis can be overcome by using an integral-based finite difference approach in which the principle of virtual work or minimum potential energy is used. Reference [13] demonstrates the application of the variational finite difference method to vibration problems.

Gladwell and Zimmermann [14] presented the variational formulations of the equations governing the harmonic vibration of structural and acoustic systems. Two formulations, one involving displacements only and the other involving forcelike quantities only, were presented along with a discussion of the dual relationship. The principles were applied to the vibration of membranes and plates, to coupled air-membrane and air-plate vibrations, and to the vibration of isotropic elastic solid.

REFERENCES

1. S. S. Rao, Theory of vibration: variational methods, in *Encyclopedia of Vibration*, S. Braun, D. Ewins, and S. S. Rao, Eds., Academic Press, London, 2002, Vol. 2, pp. 1344–1360.

2. J. S. Rao, *Advanced Theory of Vibration*, Wiley Eastern, New Delhi, India, 1992.

3. K. M. Liew, C. M. Wang, Y. Xiang, and S. Kitipornchai, *Vibration of Mindlin Plates: Programming the p-Version Ritz Method*, Elsevier Science, Oxford, 1998.

4. R. D. Mindlin, Influence of rotatory inertia and shear on flexural motions of isotropic, elastic plates, *Journal of Applied Mechanics*, Vol. 18, No. 1, pp. 31–38, 1951.

5. R. Weinstock, *Calculus of Variations: With Applications to Physics and Engineering*, McGraw-Hill, New York, 1952.

6. R. R. Schechter, *The Variational Method in Engineering*, McGraw-Hill, New York, 1967.

7. Y. C. Fung, *Foundations of Solid Mechanics*, Prentice-Hall, Englewood Cliffs, NJ, 1965.

8. J. N. Reddy, *Energy and Variational Methods in Applied Mechanics*, Wiley, New York, 1984.

9. S. S. Rao, *Engineering Optimization: Theory and Practice*, 3rd ed., Wiley, New York, 1996.

10. K. Washizu, *Variational Methods in Elasticity and Plasticity*, Pergamon Press, Oxford, 1968.

11. R. Nagem, B. A. Rhodes, and G. V. H. Sandri, Hamiltonian mechanics of the damped harmonic oscillator, *Journal of Sound and Vibration*, Vol. 144, No. 3, pp. 536–538, 1991.

12. K. Sato, On the governing equations for vibration and stability of a Timoshenko beam: Hamilton's principle," *Journal of Sound and Vibration*, Vol. 145, No. 2, pp. 338–340, 1991.

13. J. P. Singh and S. S. Dev, Variational finite difference method for the vibration of sector plates, *Journal of Sound and Vibration*, Vol. 136, No. 1, pp. 91–104, 1990.

14. G. M. L. Gladwell and G. Zimmermann, On energy and complementary energy formulations of acoustic and structural vibration problems, *Journal of Sound and Vibration*, Vol. 3, No. 3, pp. 233–241, 1966.

PROBLEMS

4.1 Formulate the problem of finding a plane curve of smallest arc length $y(x)$ that connects points (x_1, y_1) and (x_2, y_2).

4.2 Solve the problem formulated in Problem 4.1 and show that the shortest distance between points (x_1, y_1) and (x_2, y_2) is a straight line.

4.3 A plane curve $y(x)$ is used to connect points (x_1, y_1) and (x_2, y_2) with $x_1 < x_2$. The curve $y(x)$ is rotated about the x axis to generate a surface of revolution in the range $x_1 \leq x \leq x_2$ (Fig. 4.7). Formulate the problem of finding the curve $y(x)$ that corresponds to minimum area of the surface of revolution in the xy plane.

Figure 4.7

4.4 Solve the problem formulated in Problem 4.3.

4.5 Given two points $A = (x_1, y_1)$ and $B = (x_2, y_2)$ in the xy plane, consider an arc defined by $y = y(x) > 0$, $x_1 \leq x \leq x_2$, that passes through A and B whose rotation about the x axis generates a surface of revolution. Find the arc $y = y(x)$ such that the area included in $x_1 \leq x \leq x_2$ is a minimum.

4.6 Consider the Lagrangian functional L, given by

$$L = \int_0^l \frac{\rho A}{2} \left(\frac{\partial u}{\partial t} \right)^2 dx - \int_0^l \frac{AE}{2} \left(\frac{\partial u}{\partial x} \right)^2 dx$$
$$+ \int_0^l fu\, dx + Fu(l)$$

This functional corresponds to the axial vibration of a bar where $u(x, t)$ denotes the axial displacement.

(a) Find the first variation of the functional L with $\delta u(0, t) = \delta u(x, t_1) = \delta u(x, t_2) = 0$.

(b) Derive the Euler–Lagrange equations by setting the coefficients of δu in $(0, l)$ and at $x = l$ in the result of part (a) to zero separately.

4.7 Consider a solid body of revolution obtained by rotating a curve $y = y(x)$ in the xy plane passing through the origin $(0,0)$, about the x axis as shown in Fig. 4.8. When this body of revolution moves in the $-x$ direction at a velocity v in a fluid of density ρ, the normal pressure acting on an element of the surface of the body is given by

$$p = 2\rho v^2 \sin^2 \theta \qquad (4.1)$$

where θ is the angle between the direction of the velocity of the fluid and the tangent to the surface. The drag force on the body, P, can be found by integrating the x component of the force acting on the surface of a slice of the body shown in Fig. 4.8(b)[9]:

$$P = 4\pi\rho v^2 \int_0^L \left(\frac{dy}{dx} \right)^3 y\, dx \qquad (4.2)$$

Find the curve $y = y(x)$ that minimizes the drag on the body of revolution, given by Eq. (4.2), subject to the condition that $y(x)$ satisfies the end conditions $y(x = 0) = 0$ and $y(x = L) = R$.

4.8 Consider the functional $I(w)$ that arises in the transverse bending of a thin rectangular plate resting on an elastic foundation:

$$I(w) = \frac{D}{2} \int_{x=0}^a \int_{y=0}^b \left[\left(\frac{\partial^2 w}{\partial x^2} \right)^2 + \left(\frac{\partial^2 w}{\partial y^2} \right)^2 \right.$$
$$+ 2\nu \frac{\partial^2 w}{\partial x^2} \frac{\partial^2 w}{\partial y^2} + 2(1 - \nu) \left(\frac{\partial^2 w}{\partial x \partial y} \right)^2 + kw^2 \right]$$
$$\times dx\, dy - \int_{x=0}^a \int_{y=0}^b f_0 w\, dx\, dy$$

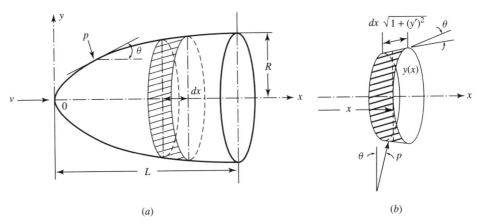

Figure 4.8   Solid body of revolution moving in a fluid.

where $w(x, y)$ denotes the transverse deflection, D the bending rigidity, ν the Poisson ratio, k the foundation modulus, and f_0 the transverse distributed load acting on the plate with $w = 0$ on the boundary of the plate.

(a) Find the first variation of the functional $I(w)$ with respect to w.

(b) Derive the Euler–Lagrange equation corresponding to the functional $I(w)$.

(c) Identify the natural and forced boundary conditions of the problem.

4.9 Consider the problem of minimizing the functional $I(y)$ given by

$$I(y) = \int_{x=a}^{b} y(x)\, dx, \qquad y(a) = A, \quad y(b) = B$$

subject to the constraint

$$\int_{x=a}^{b} \sqrt{1 + \left(\frac{dy}{dx}\right)^2}\, = l$$

Derive the Euler–Lagrange equations of the problem using Lagrange multipliers.

4.10 The transverse deflection of a membrane of area A (in the xy plane), subjected to a distributed transverse load $f(x, y)$, gives rise to the functional

$$I(w) = \frac{1}{2} \iint_A \left[\left(\frac{\partial w}{\partial x}\right)^2 + \left(\frac{\partial w}{\partial y}\right)^2 - 2fw \right] dx\, dy$$

Derive the governing differential equation and the boundary conditions by minimizing the functional $I(w)$.

4.11 The potential energy of a thin beam, $I(w)$, lying along the x axis subjected to a distributed transverse load $f(x)$ per unit length, a bending moment M_1 and a shear force V_1 at the end $x = 0$, and a bending moment M_2 and a shear force V_2 at the end $x = l$ is given by

$$I(w) = \frac{1}{2} \int_{x=0}^{l} EI \left(\frac{d^2 w}{dx^2}\right)^2 dx - \int_{x=0}^{l} fw\, dx$$
$$+ M_1 \left(\frac{dw}{dx}\right)_{x=0} - M_2 \left(\frac{dw}{dx}\right)_{x=l}$$
$$- V_1 w|_{x=0} + V_2 w|_{x=l}$$

where $w(x)$ denotes the transverse deflection and EI the bending stiffness of the beam. Derive the governing differential equation and the boundary conditions of the beam by minimizing the potential energy.

5

Derivation of Equations: Integral Equation Approach

5.1 INTRODUCTION

In this chapter we describe the integral formulation of the equations of motion governing the vibration of continuous systems. An *integral equation* is an equation in which the unknown function appears under one or more signs of integration. The general form of an integral equation is given by

$$\int_a^b K(t,\xi)\phi(\xi)\,d\xi + a_0(t)\phi(t) = f(t) \tag{5.1}$$

where $K(t,\xi)$ is a known function of the variables t and ξ and is called the *kernal* or *nucleus*, $\phi(\xi)$ is an unknown function, $a_0(t)$ and $f(t)$ are known functions, and a and b are known limits of integration. The function $\phi(t)$ which satisfies Eq. (5.1) is called the *solution* of the integral Eq. (5.1). Physically, Eq. (5.1) relates the present value of the function $\phi(t)$ to the sum or integral of what had happened to all its previous values, $\phi(\xi)$, from the previous state, a, to the present state, b. The first and second terms on the left-hand side of Eq. (5.1) are called the *regular* and *exceptional parts* of the equation, respectively, while the term on the right-hand side, $f(t)$, is called the *disturbance function*. In some cases, the integral equation may contain the derivatives of the unknown function $\phi(\xi)$ as

$$\int_a^b K(t,\xi)\phi(\xi)\,d\xi + a_0(t)\phi(t) + a_1(t)\phi^{(1)}(t) + \cdots + a_n(t)\phi^{(n)}(t) = f(t) \tag{5.2}$$

where $a_1(t), \ldots, a_n(t)$ are known functions of t and $\phi^{(i)}(t) = d^i\phi/dt^i$, $i = 1, 2, \ldots, n$. Equation (5.2) is called an *integrodifferential equation*.

5.2 CLASSIFICATION OF INTEGRAL EQUATIONS

Integral equations can be classified in a variety of ways, as indicated below.

5.2.1 Classification Based on the Nonlinear Appearance of $\phi(t)$

If the unknown function $\phi(t)$ appears nonlinearly in the regular and/or exceptional parts, the equation is said to be a *nonlinear integral equation*. For example, the equation

$$G[\phi(t)] - \int_a^b H[t, \xi, \phi(\xi)] \, d\xi = f(t) \tag{5.3}$$

where G and/or H are nonlinear functions of $\phi(t)$, is called a *nonlinear integral equation*. On the other hand, if both G and H in Eq. (5.3) are linear in terms of $\phi(t)$, the equation is said to be a *linear integral equation*. Thus, Eq. (5.1) is called a linear integral equation, while Eq. (5.2) is said to be a linear integrodifferential equation.

5.2.2   Classification Based on the Location of Unknown Function $\phi(t)$

Based on the location of the unknown function, the integral equations are said to be of the first, second, or third kind. For example, if the unknown function appears under the integral sign only, the equation is said to be of the *first kind*. If the unknown function appears both under the integral sign and outside the integral, the equation is considered to be of the second or third kind. In the *second kind* of integral equation, the unknown function, appearing outside the integral sign, appears alone, whereas in the *third kind*, it appears in the form of a product $a_0(t)\phi(t)$, where $a_0(t)$ is a known function of t. According to this classification, Eq. (5.1) is an integral equation of the third kind. The corresponding equations of the second and first kinds can be expressed, respectively, as

$$\phi(t) - \lambda \int_a^b K(t, \xi)\phi(\xi) \, d\xi = f(t) \tag{5.4}$$

and

$$\int_a^b K(t, \xi)\phi(\xi) \, d\xi = f(t) \tag{5.5}$$

If $f(t) = 0$ in Eq. (5.4), we obtain

$$\phi(t) = \lambda \int_a^b K(t, \xi)\phi(\xi) \, d\xi \tag{5.6}$$

which is called a *homogeneous integral equation*. Note that the λ in Eqs. (5.4) and (5.6) denotes a constant and can be incorporated into the kernel $K(t, \xi)$. However, in many applications, this constant represents a significant parameter that may assume several values. Hence, it is included as a separate parameter in these equations.

5.2.3   Classification Based on the Limits of Integration

Based on the type of integral in the regular part, the integral equations are classified as Fredholm- or Volterra-type equations. If the integral is over finite limits with fixed endpoints (definite integral), the equation is said to be of *Fredholm type*. On the other hand, if the integration limits are variable (indefinite integral), the integral equation is said to be of *Volterra type*. It can be seen that in Eqs. (5.1) to (5.6), the regular parts involve definite integrals and hence they are considered to be of Fredholm type.

If $K(t, \xi) = 0$ for $\xi > t$, the regular parts of Eqs. (5.1) to (5.6) can be expressed as indefinite integrals as

$$\int_a^t K(t, \xi)\phi(\xi) \, d\xi$$

and hence the resulting equations will be of Volterra type Thus, Volterra-type integral equations of the third, second, and first kind can be expressed, in sequence, as

$$\int_a^t K(t, \xi)\phi(\xi)\, d\xi + a_0(t)\phi(t) = f(t) \tag{5.7}$$

$$\phi(t) - \int_a^t K(t, \xi)\phi(\xi)\, d\xi = f(t) \tag{5.8}$$

$$\int_a^t K(t, \xi)\phi(\xi)\, d\xi = f(t) \tag{5.9}$$

Similar to Eq. (5.6), the Volterra-type homogeneous integral equation can be written as

$$\phi(t) = \int_a^t K(t, \xi)\phi(\xi)\, d\xi \tag{5.10}$$

5.2.4 Classification Based on the Proper Nature of an Integral

If the regular part of the integral equation contains a singular integral, the equation is called a *singular integral equation*. Otherwise, the equation is called a *normal integral equation*. The singularity in the integral may be due to either an infinite range of integration or a nonintegrable or unbounded kernel which causes the integrand to become infinite at some point in the range of integration. Thus, the following equations are examples of singular integral equations:

$$\int_0^\infty K(t, \xi)\phi(\xi)\, d\xi = f(t) \tag{5.11}$$

$$\phi(t) - \lambda \int_{-\infty}^\infty K(t, \xi)\phi(\xi)\, d\xi = f(t) \tag{5.12}$$

5.3 DERIVATION OF INTEGRAL EQUATIONS

5.3.1 Direct Method

The direct method of deriving integral equations is illustrated through the following example.

Example 5.1 Load Distribution on a String Consider the problem of finding the load distribution on a tightly stretched string, which results in a specified deflection shape of the string. Let a string of length L be under tension P. When a concentrated load F is applied to the string at point ξ, the string will deflect as shown in Fig. 5.1. Let the transverse displacement of the string at ξ due to F be δ. Then the displacement $w(x)$ at any other point x can be expressed as

$$w(x) = \begin{cases} \dfrac{\delta x}{\xi}, & 0 \leq x \leq \xi \\[2mm] \delta\dfrac{L - x}{L - \xi}, & \xi \leq x \leq L \end{cases} \tag{E5.1.1}$$

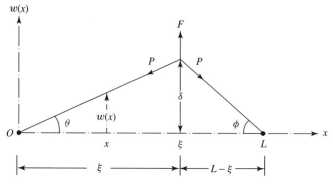

Figure 5.1   Tightly stretched string subject to a force F.

For small displacements δ, the conditions for the equilibrium of horizontal and vertical forces can be written as

$$P \cos \theta = P \cos \phi \tag{E5.1.2}$$

$$F = P \sin \theta + P \sin \phi$$

$$\approx P(\tan \theta + \tan \phi)$$

$$= P \frac{L\delta}{\xi(L - \xi)} \tag{E5.1.3}$$

Equation (E5.1.3) can be solved for δ, which upon substitution in Eq. (E5.1.3) results in

$$w(x) = \frac{F}{P} g(x, \xi) \tag{E5.1.4}$$

where $g(x, \xi)$ is the *impulse response function*, also known as *Green's function*, given by

$$g(x, \xi) = \begin{cases} \dfrac{x(L - \xi)}{L}, & x \le \xi \\[2ex] \dfrac{\xi(L - x)}{L}, & \xi \le x \end{cases} \tag{E5.1.5}$$

If the external load applied to the string is distributed with a magnitude of $f(\xi)$ per unit length, the transverse displacement of the string can be expressed as

$$w(x) = \frac{1}{P} \int_0^L g(x, \xi) f(\xi) \, d\xi \tag{E5.1.6}$$

If the displacement variation $w(x)$ is specified, Eq. (E5.1.6) becomes an integral equation of the first kind for the unknown force distribution $f(x)$. For free vibration, the force per unit length, due to inertia, is given by

$$f(x, t) = -\rho(x) \frac{\partial^2 w(x, t)}{\partial t^2} \tag{E5.1.7}$$

where $\rho(x)$ is the mass density (mass per unit length) of the string at x. Using Eq. (E5.1.7), Eq. (E5.1.6) can be written as

$$w(x, t) = -\frac{1}{P} \int_0^L g(x, \xi)\rho(\xi)\frac{\partial^2 w(\xi, t)}{\partial t^2} \, d\xi \qquad (E5.1.8)$$

When $\rho(x)$ is known, Eq. (E5.1.8) denotes the governing integrodifferential equation for the displacement $w(x, t)$. Assuming a simple harmonic solution with frequency ω,

$$w(x, t) = W(x) \sin \omega t \qquad (E5.1.9)$$

where $W(x)$ denotes the amplitude of displacement of the string at x, Eq. (E5.1.8) becomes

$$W(x) = \frac{\omega^2}{P} \int_0^L g(x, \xi)\rho(\xi)W(\xi) \, d\xi \qquad (E5.1.10)$$

which can be seen to be an integral equation of the second kind for $W(x)$.

5.3.2   Derivation from the Differential Equation of Motion

The equation of motion for the free vibration of a string can be expressed as (see Eq. (8.9))

$$c^2\frac{\partial^2 w}{\partial x^2} = \frac{\partial^2 w}{\partial t^2} \qquad (5.13)$$

where

$$c^2 = \frac{P}{\rho} \qquad (5.14)$$

ρ is the mass density of the string per unit length and P is the tension. If the string is fixed at both ends, the boundary conditions are given by

$$w(0, t) = 0$$
$$w(L, t) = 0 \qquad (5.15)$$

If the string is given an initial displacement $f(x)$ and initial velocity $g(x)$, we have

$$w(x, 0) = f(x)$$
$$\frac{\partial w}{\partial t}(x, 0) = g(x) \qquad (5.16)$$

Using the separation of variables technique, $w(x, t)$ can be expressed as

$$w(x, t) = X(x)T(t) \qquad (5.17)$$

where X is a function of x and T is a function of t. Using Eq. (5.17), Eq. (5.13) can be rewritten as

$$\frac{d^2T/dt^2}{T} = c^2\frac{d^2X/dx^2}{X} = -\lambda c^2 \qquad (5.18)$$

where $-\lambda c^2$ is a constant. Equation (5.18) yields two ordinary differential equations:

$$\frac{d^2 T}{dt^2} + \lambda c^2 T = 0 \tag{5.19}$$

$$\frac{d^2 X}{dx^2} + \lambda X = 0 \tag{5.20}$$

Since $T(t) \neq 0$, the boundary conditions can be expressed as

$$X(0) = 0$$

$$X(L) = 0 \tag{5.21}$$

Integration of Eq. (5.20) gives

$$\frac{dX}{dx} = -\lambda \int_0^x X \, d\xi + c_1 \tag{5.22}$$

where c_1 is a constant. Integration of Eq. (5.22) leads to

$$X = -\lambda \int_0^x d\eta \int_0^x X \, d\xi + c_1 x + c_2 \tag{5.23}$$

where c_2 is a constant. Changing the order of integration, Eq. (5.23) can be rewritten as

$$X(x) = -\lambda \int_0^x X(\xi) \, d\xi \int_\xi^x d\eta + c_1 x + c_2$$

$$= -\lambda \int_0^x (x - \xi) X(\xi) \, d\xi + c_1 x + c_2 \tag{5.24}$$

The use of the boundary conditions, Eq. (5.21), results in

$$c_2 = 0$$

$$c_1 = \frac{1}{L} \int_0^L (L - \xi) X(\xi) \, d\xi \tag{5.25}$$

The differential Eq. (5.13) can thus be expressed as an equivalent integral equation in $X(x)$ as

$$X(x) + \lambda \int_0^x (x - \xi) X(\xi) \, d\xi - \lambda \frac{x}{L} \int_0^L (L - \xi) X(\xi) \, d\xi = 0 \tag{5.26}$$

Introducing

$$K(x, \xi) = \begin{cases} \xi(L - x), & 0 \leq \xi < x \\ x(L - \xi), & x \leq \xi \leq L \end{cases} \tag{5.27}$$

Eq. (5.26) can be rewritten as

$$X(x) = \lambda \int_0^L K(x, \xi) X(\xi) \, d\xi \tag{5.28}$$

This equation can be seen to be of the same form as Eq. (E5.1.10) (Volterra-type homogeneous integral equation). Equation (5.28) can be solved using the procedure outlined in the following example.

Example 5.2 Find the solution of Eq. (5.28).

SOLUTION We rewrite Eq. (5.28) as

$$X(x) + \lambda \int_0^x (x - \xi) X(\xi) \, d\xi - \lambda x \, d = 0 \qquad \text{(E5.2.1)}$$

where

$$d = \frac{1}{L} \int_0^L (L - \xi) X(\xi) \, d\xi \qquad \text{(E5.2.2)}$$

Taking the Laplace transform of Eq. (E5.2.1), we obtain

$$\overline{X}(s) + \lambda \frac{1}{s^2} \overline{X}(s) - \lambda \frac{1}{s^2} d = 0 \qquad \text{(E5.2.3)}$$

where $\overline{X}(s)$ is the Laplace transform of $X(x)$. Equation (E5.2.3) yields

$$\overline{X}(s) = \frac{\lambda d}{s^2 + \lambda} \qquad \text{(E5.2.4)}$$

and the inverse transform of Eq. (E5.2.4) gives

$$X(x) = \sqrt{\lambda} \, d \, \sin \sqrt{\lambda} x \qquad \text{(E5.2.5)}$$

Substitution of Eq. (E5.2.5) into Eq. (E5.2.2) gives

$$d \int_0^L (L - \xi) \sqrt{\lambda} \sin \sqrt{\lambda} \xi \, d\xi = dL \qquad \text{(E5.2.6)}$$

which can be satisfied when $d = 0$ or

$$\int_0^L (L - \xi) \sqrt{\lambda} \sin \sqrt{\lambda} \xi \, d\xi = L \qquad \text{(E5.2.7)}$$

Since $d = 0$ leads to the trivial solution $X(x) = 0$ and $w(x, t) = 0$, Eq. (E5.2.7) must be satisfied. Equation (E5.2.7) yields

$$L - \frac{\sin \sqrt{\lambda} L}{\sqrt{\lambda}} = L \qquad \text{(E5.2.8)}$$

or

$$\sin \sqrt{\lambda} L = 0 \qquad \text{(E5.2.9)}$$

or

$$\lambda = \frac{n^2 \pi^2}{L^2}, \qquad n = 1, 2, \ldots \qquad \text{(E5.2.10)}$$

Equations (E5.2.5) and (E5.2.10) lead to

$$X(x) = a \sin \frac{n \pi x}{L} \tag{E5.2.11}$$

where a is a constant.

5.4 GENERAL FORMULATION OF THE EIGENVALUE PROBLEM

5.4.1 One-Dimensional Systems

For a one-dimensional continuous system, the displacement $w(x, t)$ can be expressed as

$$w(x, t) = \int_0^L a(x, \xi) f(\xi, t) \, d\xi \tag{5.29}$$

where $a(x, \xi)$ is the flexibility influence function that satisfies the boundary conditions of the system and $f(\xi, t)$ is the distributed load at point ξ at time t. For a system undergoing free vibration, the load represents the inertia force, so that

$$f(x, t) = -m(x) \frac{\partial^2 w(x, t)}{\partial t^2} \tag{5.30}$$

where $m(x)$ is the mass per unit length. Assuming a harmonic motion of frequency ω during free vibration,

$$w(x, t) = W(x) \cos \omega t \tag{5.31}$$

Eq. (5.30) can be expressed as

$$f(x, t) = \omega^2 m(x) W(x) \cos \omega t \tag{5.32}$$

Substituting Eqs. (5.31) and (5.32) into Eq. (5.29) results in

$$W(x) = \omega^2 \int_0^L a(x, \xi) m(\xi) W(\xi) \, d\xi \tag{5.33}$$

It can be seen that Eq. (5.33) is a homogeneous integral equation of the second kind and represents the eigenvalue problem of the system in integral form.

Example 5.3 Free Transverse Vibration of a Membrane   Consider a membrane of area A whose equilibrium shape lies in the xy plane. Let the membrane be fixed at its boundary, S, and subjected to a uniform tension P (force per unit length). Let the transverse displacement of point $Q(x, y)$ due to the transverse load $f(\xi, \eta) \, d\xi \, d\eta$ applied at the point $R(\xi, \eta)$ be $w(Q)$. By considering the equilibrium of a small element of area $dx \, dy$ of the membrane, the differential equation can be derived as

$$\frac{\partial^2 w}{\partial x^2} + \frac{\partial^2 w}{\partial y^2} = -\frac{f(x, y)}{P} \tag{E5.3.1}$$

The Green's function of the membrane, $K(x, y; \xi, \eta)$, is given by [4]

$$K(Q, R) \equiv K(x, y; \xi, \eta) = \log \frac{1}{r} - h(Q, R) \tag{E5.3.2}$$

where r denotes the distance between two points Q and R in the domain of the membrane:

$$r = \sqrt{(x - \xi)^2 + (y - \eta)^2} \tag{E5.3.3}$$

and $h(Q, R)$ is a harmonic function whose values on the boundary of the membrane, S, are the same as those of $\log(1/r)$ so that $K(Q, R)$ will be zero on S. For example, if the membrane is circular with center at $(0,0)$ and radius a, the variation of the function $K(Q, R)$ will be as shown in Fig. 5.2. Since the membrane is fixed along its boundary S, the transverse displacement of point Q can be expressed as

$$w(Q) = \frac{1}{2\pi P} \iint\limits_{A} K(Q, R) f(R) \, dA \tag{E5.3.4}$$

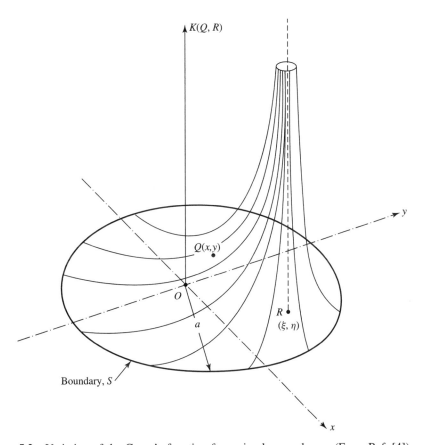

Figure 5.2 Variation of the Green's function for a circular membrane. (From Ref. [4]).

From this static relation, the free vibration relation can be obtained by substituting $-\rho(R)[\partial^2 w(R)/\partial t^2]$ for $f(R)$ in Eq. (E5.3.4) so that

$$w(Q) = -\frac{1}{2\pi P} \iint_A K(Q, R)\rho(R)\frac{\partial^2 w}{\partial t^2}(R)\,dA \qquad \text{(E5.3.5)}$$

Assuming harmonic motion with frequency ω, we have

$$w(Q) = W(Q)e^{i\omega t} \qquad \text{(E5.3.6)}$$

where $W(Q)$ denotes the amplitude of vibration at point Q. Substitution of Eq. (E5.3.6) in Eq. (E5.3.5) yields the relation

$$W(Q) = \frac{\omega^2}{2\pi P} \iint_A K(Q, R)\rho(R)W(R)\,dA \qquad \text{(E5.3.7)}$$

5.4.2   General Continuous Systems

The general form of Eq. (5.33), valid for any continuous system, can be expressed as

$$W(x) = \lambda \int_V g(x, \xi)m(\xi)W(\xi)\,dV(\xi) \qquad \text{(5.34)}$$

where $W(x)$ and $W(\xi)$ denote the displacements at points x and ξ, respectively. Depending on the dimensionality of the problem, points x and ξ may be defined by one, two, or three spatial coordinates. The general flexibility influence function $g(x, \xi)$, also known as the *Green's function*, is symmetric in x and ξ, [i.e., $g(x, \xi) = g(\xi, x)$] for a self-adjoint problem.

Note that the kernel, $g(x, \xi)m(\xi)$, in Eq. (5.34) is not symmetric unless $m(\xi)$ is a constant. However, the kernel can be made symmetric by noting the fact that $m(\xi) > 0$ and introducing the function $\phi(x)$:

$$\phi(x) = \sqrt{m(x)}W(x) \qquad \text{(5.35)}$$

By multiplying both sides of Eq. (5.34) by $\sqrt{m(x)}$ and using Eq. (5.35), we obtain

$$\phi(x) = \lambda \int_V K(x, \xi)\phi(\xi)\,dV(\xi) \qquad \text{(5.36)}$$

where the kernel

$$K(x, \xi) = \sqrt{m(x)m(\xi)}g(x, \xi) \equiv K(\xi, x) \qquad \text{(5.37)}$$

can be seen to be symmetric. An advantage of the transformation above is that a symmetric kernel usually possesses an infinite number of eigenvalues, λ, for which Eq. (5.36) will have nonzero solutions. On the other hand, a nonsymmetric kernel may or may not have eigenvalues [1]. For any specific eigenvalue λ_i, Eq. (5.36) has a nontrivial solution $\phi_i(x)$, which is related to $W_i(x)$ by Eq. (5.35). The function $W_i(x)$ represents the eigenfunction corresponding to the eigenvalue λ_i of the system.

5.4.3   Orthogonality of Eigenfunctions

It can be shown that the eigenfunctions $\phi_i(x)$ are orthogonal in the usual sense, while the functions $W_i(x)$ are orthogonal with respect to the functions $m(x)$. For this, consider Eq. (5.36), corresponding to two distinct eigenvalues λ_i and λ_j:

$$\phi_i(x) = \lambda_i \int_V K(x, \xi)\phi_i(\xi)\, dV(\xi) \tag{5.38}$$

$$\phi_j(x) = \lambda_j \int_V K(x, \xi)\phi_j(\xi)\, dV(\xi) \tag{5.39}$$

Multiply Eq. (5.38) by $\phi_j(x)$, integrate over the domain V, and use Eq. (5.39) to obtain

$$
\begin{aligned}
\int_V \phi_i(x)\phi_j(x)\, dV(x) &= \lambda_i \int_V \phi_j(x) \left[\int_V K(x, \xi)\phi_i(\xi)\, dV(\xi) \right] dV(x) \\
&= \lambda_i \int_V \phi_i(\xi) \left[\int_V K(\xi, x)\phi_j(\xi)\, dV(x) \right] dV(\xi) \\
&= \frac{\lambda_i}{\lambda_j} \int_V \phi_i(\xi)\phi_j(\xi)\, dV(\xi)
\end{aligned}
\tag{5.40}
$$

which yields

$$(\lambda_i - \lambda_j) \int_V \phi_i(x)\phi_j(x)\, dV(x) = 0 \tag{5.41}$$

Since λ_i and λ_j are distinct, $\lambda_i \neq \lambda_j$, Eq. (5.41) leads to the orthogonality relation

$$\int_V \phi_i(x)\phi_j(x)\, dV(x) = 0, \qquad \lambda_i \neq \lambda_j \tag{5.42}$$

When Eq. (5.35) is used in Eq. (5.42), we obtain the orthogonality relation for the eigenfunctions $W_i(x)$ as

$$\int_V m(x)W_i(x)W_j(x)\, dV(x) = \begin{cases} 0 & \text{for } \lambda_i \neq \lambda_j \\ 1 & \text{for } \lambda_i = \lambda_j \end{cases} \tag{5.43}$$

5.5   SOLUTION OF INTEGRAL EQUATIONS

Several methods, both exact and approximate methods, can be used to find the solutions of integral equations [1, 4–6]. The method of undetermined coefficients and the Rayleigh–Ritz, Galerkin, collocation, and numerical integration methods are considered in this section.

5.5.1 Method of Undetermined Coefficients

In this method the unknown function is assumed to be in the form of a power series of a finite number of terms. The assumed function is then substituted into the integral equation and the regular part is integrated. This results in a set of simultaneous equations in terms of the unknown coefficients. Solution of these simultaneous equations yields the solution of the integral equation.

Example 5.4 Find the solution of the integral equation

$$2 \int_0^1 (1 - \xi + x\xi)\phi(\xi)\,d\xi = -x + 1 \tag{E5.4.1}$$

SOLUTION   Assume the solution of $\phi(x)$ in a power series of two terms as

$$\phi(x) = c_1 + c_2 x \tag{E5.4.2}$$

where c_1 and c_2 are constants to be determined. Substitute Eq. (E5.4.2) into Eq. (E5.4.1) and carry out the integration to obtain

$$2 \int_0^1 (1 - \xi + x\xi)(c_1 + c_2\xi)\,d\xi = -x + 1 \tag{E5.4.3}$$

Upon integration, Eq. (E5.4.3) becomes

$$\left(c_1 + \tfrac{4}{3}c_2\right) + x \left(c_1 + \tfrac{2}{3}c_2\right) = -x + 1 \tag{E5.4.4}$$

Equating similar terms on both sides of Eq. (E5.4.4), we obtain

$$c_1 + \tfrac{4}{3}c_2 = 1$$
$$c_1 + \tfrac{2}{3}c_2 = -1 \tag{E5.4.5}$$

Equations (E5.4.5) yield $c_1 = -3$ and $c_2 = 3$. Thus, the solution of the integral Eq. (E5.4.1) is given by

$$\phi(x) = -3 + 3x \tag{E5.4.6}$$

5.5.2 Iterative Method

An iterative method similar to the matrix iteration method for the solution of a matrix eigenvalue problem can be used for the solution of the integral Eq. (5.34). The iteration method assumes that the eigenvalues are distinct and well separated such that $\lambda_1 < \lambda_2 < \lambda_3 \cdots$. In addition, the iteration method is based on the expansion theorem related to the eigenfunctions $W_i(x)$. Similar to the expansion theorem of the matrix eigenvalue problem, the expansion theorem related to the integral formulation of the eigenvalue problem can be stated as

$$W(x) = \sum_{i=1}^{\infty} c_i W_i(x) \tag{5.44}$$

where the coefficients c_i are determined as

$$c_i = \int_V m(x) W(x) W_i(x) \, dV(x) \tag{5.45}$$

Equation (5.44) indicates that any function $W(x)$ that satisfies the boundary conditions of the system can be represented as a linear combination of the eigenfunctions $W_i(x)$ of the system.

First Eigenfunction   The iteration method starts with the selection of a trial function $W_1^{(1)}(x)$ as an approximation to the first eigenfunction or mode shape, $W_1(x)$. Substituting $W_1^{(1)}(x)$ for $W(x)$ on the right-hand side of Eq. (5.34) and evaluating the integral, the next (improved) approximation to the eigenfunction $W_1(x)$ can be obtained:

$$W_1^{(2)}(x) = \int_V g(x, \xi) m(\xi) W_1^{(1)}(\xi) \, dV(\xi) \tag{5.46}$$

Using Eq. (5.44), Eq. (5.46) can be expressed as

$$
\begin{aligned}
W_1^{(2)}(x) &= \sum_{i=1}^{\infty} c_i \int_V g(x, \xi) m(\xi) W_i(\xi) \, dV(\xi) \\
&= \sum_{i=1}^{\infty} \frac{c_i W_i(x)}{\lambda_i}
\end{aligned}
\tag{5.47}
$$

The definition of the eigenvalue problem, Eq. (5.34), yields

$$W_i(x) = \lambda_i \int_V g(x, \xi) m(\xi) W_i(\xi) \, dV(\xi) \tag{5.48}$$

Using $W_1^{(2)}(x)$ as the trial function on the right-hand side of Eq. (5.48), we obtain the new approximation, $W_1^{(3)}(x)$, as

$$
\begin{aligned}
W_1^{(3)}(x) &= \int_V g(x, \xi) m(\xi) W_1^{(2)}(\xi) \, dV(\xi) \\
&= \sum_{i=1}^{\infty} \frac{c_i W_i(x)}{\lambda_i^2}
\end{aligned}
\tag{5.49}
$$

The continuation of the process leads to

$$W_1^{(n)}(x) = \sum_{i=1}^{\infty} \frac{c_i W_i(x)}{\lambda_i^{n-1}}, \qquad n = 2, 3, \ldots \tag{5.50}$$

Since the eigenvalues are assumed to satisfy the relation $\lambda_1 < \lambda_2 \cdots$, the first term on the right-hand side of Eq. (5.50) becomes large compared to the other terms and as

$n \to \infty$, Eq. (5.50) yields

$$\lim_{n\to\infty} W_1^{(n-1)}(x) = \frac{c_1 W_1(x)}{\lambda_1^{n-2}} \tag{5.51}$$

$$\lim_{n\to\infty} W_1^{(n)}(x) = \frac{c_1 W_1(x)}{\lambda_1^{n-1}} \tag{5.52}$$

Equations (5.51) and (5.52) yield the converged eigenvalue λ_1 as

$$\lambda_1 = \lim_{n\to\infty} \frac{W_1^{(n-1)}(x)}{W_1^{(n)}(x)} \tag{5.53}$$

and the converged eigenvector can be taken as

$$W_1(x) = \lim_{n\to\infty} W_1^{(n)}(x) \tag{5.54}$$

Higher Eigenfunctions To determine the second eigenfunction, the trial function $W_2^{(1)}(x)$ used must be made completely free of the first eigenfunction, $W_1(x)$. For this we use any arbitrary trial function $\tilde{W}_2^{(1)}(x)$ to generate $W_2^{(1)}(x)$ as

$$W_2^{(1)}(x) = \tilde{W}_2^{(1)}(x) - a_1 W_1(x) \tag{5.55}$$

where a_1 is a constant that can be determined from the orthogonality condition of the eigenfunctions:

$$\int_V m(x) W_2^{(1)}(x) W_1(x) \, dV(x)$$

$$= \int_V m(x) \tilde{W}_2^{(1)}(x) W_1(x) \, dV(x) - a_1 \int_V m(x)[W_1(x)]^2 \, dV(x) = 0 \tag{5.56}$$

or

$$a_1 = \frac{\int_V m(x) \tilde{W}_2^{(1)}(x) W_1(x) \, dV(x)}{\int_V m(x)[W_1(x)]^2 \, dV(x)} \tag{5.57}$$

When $W_1(x)$ is normalized according to Eq. (5.43),

$$\int_V m(x)[W_1(x)]^2 \, dV(x) = 1 \tag{5.58}$$

Eq. (5.57) becomes

$$a_1 = \int_V m(x) \tilde{W}_2^{(1)}(x) W_1(x) \, dV(x) \tag{5.59}$$

Once a_1 is determined, we substitute Eq. (5.55) for $W(x)$ on the right-hand side of Eq. (5.34), evaluate the integral, and denote the result as $\tilde{W}_2^{(2)}(x)$, the next (improved) approximation to the true eigenfunction $W_2(x)$:

$$\tilde{W}_2^{(2)}(x) = \int_V g(x, \xi) m(\xi) W_2^{(1)}(\xi) \, dV(\xi) \tag{5.60}$$

For the next iteration, we generate $W_2^{(2)}(x)$ that is free of $W_1(x)$ as

$$W_2^{(2)}(x) = \tilde{W}_2^{(2)}(x) - a_2 W_1(x) \tag{5.61}$$

where a_2 can be found using an equation similar to Eq.(5.59) as

$$a_2 = \int_V m(x) \tilde{W}_2^{(2)}(x) W_1(x) \, dV(x) \tag{5.62}$$

when $W_1(x)$ is normalized according to Eq. (5.43).

When the iterative process is continued, we obtain, as $n \to \infty$, the converged result as

$$\lambda_2 = \lim_{n \to \infty} \frac{W_2^{(n-1)}(x)}{W_2^{(n)}(x)} \tag{5.63}$$

$$W_2(x) = \lim_{n \to \infty} W_2^{(n)}(x) \tag{5.64}$$

To find the third eigenfunction of the system, we start with any arbitrary trial function $\tilde{W}_3^{(1)}(x)$ and generate the function $W_3^{(1)}(x)$ that is completely free of the first and second eigenfunctions $W_1(x)$ and $W_2(x)$ as

$$W_3^{(1)}(x) = \tilde{W}_3^{(1)}(x) - a_1 W_1(x) - a_2 W_2(x) \tag{5.65}$$

where the constants a_1 and a_2 can be found by making $W_3^{(1)}(x)$ orthogonal to both $W_1(x)$ and $W_2(x)$. The procedure used in finding the second eigenfunction can be used to find the converged solution for λ_3 and $W_3(x)$. In fact, a similar process can be used to find all other higher eigenvalues and eigenfunctions.

Example 5.5 Find the first eigenvalue and the corresponding eigenfunction of a tightly stretched string under tension using the iterative method with the trial function

$$W_1^{(1)}(x) = \frac{x(L - x)}{L^2}$$

SOLUTION Let the mass of the string be m per unit length and the tension in the string be P. The Green's function or the flexibility influence function, $g(x, \xi)$, can be derived by applying a unit load at point ξ and finding the resulting deflection at point x as shown in Fig. 5.3. For vertical force equilibrium, we have

$$P \frac{a}{\xi} + P \frac{a}{L - \xi} = 1 \tag{E5.5.1}$$

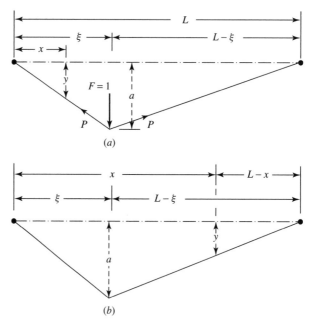

Figure 5.3

which yields

$$a = \frac{\xi(L - \xi)}{PL} \tag{E5.5.2}$$

Thus, the Green's function is given by

$$g(x, \xi) = \begin{cases} \dfrac{ax}{\xi}, & \xi > x \\ \dfrac{a(L - x)}{L - \xi}, & \xi < x \end{cases} \tag{E5.5.3}$$

which can be expressed as

$$g(x, \xi) = \begin{cases} \dfrac{x(L - \xi)}{PL}, & \xi > x \\ \dfrac{\xi(L - x)}{PL}, & \xi < x \end{cases} \tag{E5.5.4}$$

Using the trial function

$$W_1^{(1)}(x) = \frac{x(L - x)}{L^2} \tag{E5.5.5}$$

for $W_1(x)$ on the right-hand side of Eq. (5.34), we obtain the new trial function $W_1^{(2)}(x)$ as

$$
\begin{aligned}
W_1^{(2)}(x) &= \int_{\xi=0}^{L} g(x,\xi)m(\xi)W_1^{(1)}(\xi)\,d\xi \\
&= \int_{\xi=0}^{x} g(x,\xi)m(\xi)W_1^{(1)}(\xi)\,d\xi + \int_{\xi=x}^{L} g(x,\xi)m(\xi)W_1^{(1)}(\xi)\,d\xi \\
&= \int_{\xi=0}^{x} \frac{\xi(L-x)}{PL}(m)\frac{\xi(L-\xi)}{L^2}\,d\xi \\
&\quad + \int_{\xi=x}^{L} \frac{x(L-\xi)}{PL}(m)\frac{\xi(L-\xi)}{L^2}\,d\xi
\end{aligned}
\tag{E5.5.6}
$$

Equation (E5.5.6) can be simplified as

$$
W_1^{(2)}(x) = \frac{m}{12PL^2}(L^3x - 2Lx^3 + x^4)
\tag{E5.5.7}
$$

Using Eqs. (E5.5.5) and (E5.5.7), we obtain

$$
\omega_1^2 \approx \frac{W_1^{(1)}(x)}{W_1^{(2)}(x)} = \frac{12P}{mL^2}\frac{x/L - x^2/L^2}{x/L - 2(x^3/L^3) + x^4/L^4}
\tag{E5.5.8}
$$

or

$$
\omega_1^2 \approx \frac{12P}{mL^2}
\tag{E5.5.9}
$$

or

$$
\omega_1 \approx 3.4641\sqrt{\frac{P}{mL^2}}
\tag{E5.5.10}
$$

This approximate solution can be seen to be quite good compared to the exact value of the first natural frequency, $\omega_1 = \pi\sqrt{P/mL^2}$.

5.5.3   Rayleigh–Ritz Method

In the *Rayleigh–Ritz method*, also known as the *assumed modes method*, the solution of the free vibration problem is approximated by a linear combination of n admissible functions, $u_i(x)$, as

$$
w(x,t) = \sum_{i=1}^{n} u_i(x)\eta_i(t)
\tag{5.66}
$$

where $\eta_i(t)$ are time-dependent generalized coordinates. The kinetic energy of the system, $T(t)$, can be expressed as

$$
\begin{aligned}
T(t) &= \tfrac{1}{2}\int_{x=0}^{L} m(x)[\dot{w}(x,t)]^2\,dx \\
&= \tfrac{1}{2}\sum_{i=1}^{n}\sum_{j=1}^{n} m_{ij}\dot{\eta}_i(t)\dot{\eta}_j(t)
\end{aligned}
\tag{5.67}
$$

where $m(x)$ denotes the mass per unit length, a dot above a symbol represents a time derivative, and

$$m_{ij} = m_{ji} = \int_{x=0}^{L} m(x)u_i(x)u_j(x)\,dx \tag{5.68}$$

The potential energy of the system, $U(t)$, can be expressed in terms of the flexibility influence function $g(x, \xi)$ and the distributed load $f(x, t)$ as

$$U(t) = \tfrac{1}{2} \int_{x=0}^{L} f(x, t) \left[\int_0^L g(x, \xi) f(\xi, t)\,d\xi \right] dx \tag{5.69}$$

Assuming that

$$f(x, t) = \omega^2 m(x) W(x, t) \tag{5.70}$$

for free vibration, Eq. (5.69) can be expressed as

$$U(t) = \frac{\omega^4}{2} \int_0^L m(x) W(x, t) \left[\int_0^L g(x, \xi) m(\xi) W(\xi, t)\,d\xi \right] dx \tag{5.71}$$

Substitution of Eq. (5.66) into (5.71) leads to

$$U(t) = \tfrac{1}{2}\tilde{\lambda}^2 \sum_{i=1}^{n} \sum_{j=1}^{n} k_{ij}\eta_i(t)\eta_j(t) \tag{5.72}$$

where

$$k_{ij} = k_{ji} = \int_0^L m(x)u_i(x) \left[\int_0^L g(x, \xi) m(\xi) u_j(\xi)\,d\xi \right] dx \tag{5.73}$$

and $\tilde{\lambda}$ denotes an approximation of ω^2. Equations (5.66), (5.67), and (5.72) essentially approximate the continuous system by an n-degree-of-freedom system.

Lagrange's equations for an n-degree-of-freedom conservative system are given by

$$\frac{d}{dt}\left(\frac{\partial T}{\partial \dot{\eta}_k}\right) - \frac{\partial T}{\partial \eta_k} + \frac{\partial U}{\partial \eta_k} = 0, \qquad k = 1, 2, \dots, n \tag{5.74}$$

where η_k is the generalized displacement and $\dot{\eta}_k$ is the generalized velocity. Substitution of Eqs. (5.67) and (5.72) into (5.74) yields the following equations of motion:

$$\sum_{i=1}^{n} m_{ki}\ddot{\eta}_i + \tilde{\lambda}^2 \sum_{i=1}^{n} k_{ki}\eta_i = 0, \qquad k = 1, 2, \dots, n \tag{5.75}$$

For harmonic variation of $\eta_i(t)$,

$$\ddot{\eta}_i = -\tilde{\lambda}\eta_i \tag{5.76}$$

and Eqs. (5.75) lead to the matrix eigenvalue problem

$$\tilde{\lambda}[k]\vec{\eta} = [m]\vec{\eta} \tag{5.77}$$

where $[k] = [k_{ij}]$ and $[m] = [m_{ij}]$ are symmetric matrices and $\tilde{\lambda} = \omega^2$. The problem of Eq. (5.77) can be solved readily to find the eigenvalues $\tilde{\lambda}$ and the corresponding eigenvectors $\vec{\eta}$. The eigenfunctions of the continuous system can then be determined using Eq. (5.66).

Example 5.6 Find the natural frequencies of a tightly stretched string under tension using the Rayleigh–Ritz method.

SOLUTION Let the mass of the string be m per unit length and the tension in the string be P. The Green's function or flexibility influence function of the string can be expressed as (see Example 5.1)

$$g(x, \xi) = g(\xi, x) = \begin{cases} \dfrac{\xi(L - x)}{LP}, & \xi < x \\ \dfrac{x(L - \xi)}{LP}, & \xi > x \end{cases} \tag{E5.6.1}$$

We assume a two-term solution for the deflection of the string as

$$w(x, t) = u_1(x)\eta_1(t) + u_2(x)\eta_2(t) \tag{E5.6.2}$$

where the admissible functions $u_1(x)$ and $u_2(x)$ are chosen as

$$u_1(x) = \frac{x}{L}\left(1 - \frac{x}{L}\right) \tag{E5.6.3}$$

$$u_2(x) = \frac{x^2}{L^2}\left(1 - \frac{x}{L}\right) \tag{E5.6.4}$$

and $\eta_1(t)$ and $\eta_2(t)$ are the time-dependent generalized coordinates to be determined. The elements of the matrix $[m]$ can be determined as

$$m_{ij} = \int_{x=0}^{L} m(x)u_i(x)u_j(x)\,dx \tag{E5.6.5}$$

Equation (E5.6.5) gives

$$m_{11} = m\int_{x=0}^{L} u_1^2(x)\,dx = \frac{m}{L^2}\int_{x=0}^{L} x^2\left(1 - \frac{x}{L}\right)^2 dx = \frac{mL^3}{30L^2} = \frac{mL}{30}$$

$$m_{12} = m_{21} = m\int_{x=0}^{L} u_1(x)u_2(x)\,dx = \frac{m}{L^3}\int_{x=0}^{L} x^3\left(1 - \frac{x}{L}\right)^2 dx = \frac{mL^4}{60L^3} = \frac{mL}{60}$$

$$m_{22} = m\int_{x=0}^{L} u_2^2(x)\,dx = \frac{m}{L^4}\int_{x=0}^{L} x^4\left(1 - \frac{x}{L}\right)^2 dx = \frac{mL^5}{105L^4} = \frac{mL}{105}$$

The elements of the matrix $[k]$ can be found from

$$k_{ij} = k_{ji} = \int_0^{L} m(x)u_i(x)\left[\int_0^{L} g(x, \xi)m(\xi)u_j(\xi)\,d\xi\right] dx \tag{E5.6.6}$$

In Eq. (E5.6.6) the inside integral for k_{11} can be evaluated as follows:

$$\int_0^L g(x,\xi)m(\xi)u_1(\xi)\,d\xi$$

$$= m\int_0^x \frac{\xi(L-x)}{LP}\frac{\xi}{L}\left(1-\frac{\xi}{L}\right)d\xi + m\int_x^L \frac{x(L-\xi)}{LP}\frac{\xi}{L}\left(1-\frac{\xi}{L}\right)d\xi$$

$$= \frac{m}{L^2P}\left(-\frac{L}{6}x^3 + \frac{1}{12}x^4 + \frac{L^3}{12}x\right) \tag{E5.6.7}$$

Thus, Eq. (E5.6.6) yields

$$k_{11} = \int_0^L m\left(\frac{x}{L}-\frac{x^2}{L^2}\right)\frac{m}{L^2P}\left(-\frac{L}{6}x^3 + \frac{1}{12}x^4 + \frac{L^3}{12}x\right)dx$$

$$= \frac{m^2}{L^4P}\int_0^L \frac{Lx-x^2}{12}(-2Lx^3 + x^4 + L^3x)\,dx$$

$$= \frac{17m^2L^3}{5040P} \tag{E5.6.8}$$

Similarly, we can obtain from Eq.(E5.6.6)

$$k_{12} = k_{21} = \int_0^L m\left(\frac{x^2}{L^2}-\frac{x^3}{L^3}\right)\frac{m}{L^2P}\left(-\frac{L}{6}x^3 + \frac{x^4}{12} + \frac{L^3}{12}x\right)dx$$

$$= \frac{m^2L^3}{P}\left(\frac{17}{10,080}\right) \tag{E5.6.9}$$

The inside integral in Eq. (E5.6.6) for k_{22} can be evaluated as

$$\int_0^L g(x,\xi)m(\xi)u_2(\xi)\,d\xi$$

$$= m\int_0^L \frac{\xi(L-x)}{LP}\left(\frac{\xi^2}{L^2}-\frac{\xi^3}{L^3}\right)d\xi + m\int_x^L \frac{x(L-\xi)}{LP}\left(\frac{\xi^2}{L^2}-\frac{\xi^3}{L^3}\right)d\xi$$

$$= \frac{m}{L^4P}\frac{-5L^2x^4 + 3Lx^5 + 2L^5x}{60} \tag{E5.6.10}$$

Thus, Eq. (E5.6.6) gives k_{22} as

$$k_{22} = \int_0^L m(x)u_2(x)\frac{m}{60L^4P}(-5L^2x^4 + 3Lx^5 + 2L^5x)\,dx$$

$$= \frac{m^2}{60L^4P}\int_0^L \left(\frac{x^2}{L^2}-\frac{x^3}{L^3}\right)(-5L^2x^4 + 3Lx^5 + 2L^5x)\,dx$$

$$= \frac{m^2L^3}{P}\left(\frac{11}{12,600}\right) \tag{E5.6.11}$$

Thus, the eigenvalue problem can be expressed as

$$\frac{mL}{420}\begin{bmatrix} 14 & 7 \\ 7 & 4 \end{bmatrix}\vec{X} = \frac{m^2L^3\lambda}{50,400}\begin{bmatrix} 170 & 85 \\ 85 & 44 \end{bmatrix}\vec{X} \tag{E5.6.12}$$

or

$$\begin{bmatrix} 14 & 7 \\ 7 & 4 \end{bmatrix} \vec{X} = \tilde{\lambda} \begin{bmatrix} 170 & 85 \\ 85 & 44 \end{bmatrix} \vec{X} \qquad \text{(E5.6.13)}$$

where

$$\tilde{\lambda} = \frac{mL^2\lambda}{120P} \qquad \text{(E5.6.14)}$$

$$\lambda = \omega^2 \qquad \text{(E5.6.15)}$$

is the eigenvalue and $\vec{X}$ is the eigenvector (mode shape). The solution of Eq. (E5.6.13) is given by

$$\tilde{\lambda}_1 = 0.0824, \qquad \tilde{\lambda}_2 = 0.3333 \qquad \text{(E5.6.16)}$$

with

$$\vec{X}^{(1)} = \begin{Bmatrix} 1.0000 \\ 0.0 \end{Bmatrix}, \qquad \vec{X}^{(2)} = \begin{Bmatrix} -0.4472 \\ 0.8944 \end{Bmatrix} \qquad \text{(E5.6.17)}$$

5.5.4 Galerkin's Method

In the Galerkin method, the function ϕ is approximated by a linear combination of n comparison functions, $u_i(x)$, as

$$\phi(x, t) = \sum_{i=1}^{n} u_i(x)\eta_i \qquad \text{(5.78)}$$

where η_i are coefficients (or generalized coordinates) to be determined. Consider the eigenvalue problem of the continuous system in the integral form

$$w(x) = \lambda \int_V g(x, \xi) m(\xi) w(\xi) \, dV(\xi) \qquad \text{(5.79)}$$

where $w(x)$ is the displacement at point x and $g(x, \xi)$ is the symmetric Green's function or flexibility influence function. By introducing

$$\phi(x) = \sqrt{m(x)} w(x) \qquad \text{(5.80)}$$

and multiplying both sides of Eq. (5.79) by $\sqrt{m(x)}$, we obtain

$$\phi(x) = \lambda \int_V K(x, \xi) \phi(\xi) \, dV(\xi) \qquad \text{(5.81)}$$

where $K(x, \xi)$ denotes the symmetric kernel:

$$K(x, \xi) = g(x, \xi)\sqrt{m(x)}\sqrt{m(\xi)} \qquad \text{(5.82)}$$

When the approximate solution, Eq. (5.78), is substituted into Eq. (5.81), the equality will not hold; hence an error function, $\varepsilon(x)$, also known as the *residual*, can be defined as

$$\varepsilon(x) = \phi(x) - \tilde{\lambda} \int_V K(x, \xi)\phi(\xi)\, dV(\xi) \tag{5.83}$$

where $\tilde{\lambda}$ denotes an approximate value of λ and V indicates the domain of the system. To determine the coefficients η_k, the weighted integral of the error function over the domain of the system is set equal to zero. Using the functions $u_1(x), u_2(x), \ldots, u_n(x)$ as the weighting functions, n equations can be derived:

$$\int_V \varepsilon(x) u_k(x)\, dV(x) = 0, \qquad k = 1, 2, \ldots, n \tag{5.84}$$

Substituting Eq. (5.83) into (5.84), we obtain

$$\sum_{i=1}^n \eta_i \int_V u_k(x) u_j(x)\, dV(x)$$

$$- \tilde{\lambda} \sum_{i=1}^n \eta_i \int_V \eta_k(x) \left[\int_V K(x, \xi) u_i(\xi)\, dV(\xi) \right] dV(x) = 0 \tag{5.85}$$

Defining

$$u_i(x) = \sqrt{m(x)}\tilde{u}_i(x), \qquad i = 1, 2, \ldots, n \tag{5.86}$$

$$k_{ik} = k_{ki} = \int_V u_k(x) \left[\int_V K(x, \xi) u_i(\xi)\, dV(\xi) \right] dV(x)$$

$$= \int_V m(x)\tilde{u}_k(x) \left[\int_V g(x, \xi) m(\xi)\tilde{u}_i(\xi)\, dV(\xi) \right] dV(x) \tag{5.87}$$

$$m_{ik} = m_{ki} = \int_V u_k(x) u_i(x)\, dV(x)$$

$$= \int_V m(x)\tilde{u}_k(x)\tilde{u}_i(x)\, dV(x) \tag{5.88}$$

Eq. (5.85) can be expressed as

$$\tilde{\lambda}[k]\vec{\eta} = [m]\vec{\eta} \tag{5.89}$$

which can be seen to be similar to Eq. (5.77).

5.5.5 Collocation Method

Consider the eigenvalue problem of a continuous system in integral form:

$$w(x) = \int_V g(x, \xi) m(\xi) w(\xi)\, dV(\xi) \tag{5.90}$$

The solution of the free vibration problem is approximated by a linear combination of n comparison functions, $u_i(x)$, as

$$w(x) = \sum_{i=1}^{n} u_i(x)\eta_i \tag{5.91}$$

where η_i are coefficients or generalized coordinates to be determined. When Eq. (5.91) is substituted into Eq. (5.90), the equality will not hold; hence an error function or residual $\varepsilon(x)$ can be defined as

$$\varepsilon(x) = w(x) - \tilde{\lambda} \int_V g(x, \xi) m(\xi) \, dV(\xi) \tag{5.92}$$

By substituting Eq. (5.91) into (5.90), the error function can be expressed as

$$\varepsilon(x) = w(x) - \tilde{\lambda} \int_V g(x, \xi) m(\xi) w(\xi) \, dV(\xi)$$
$$= \sum_{i=1}^{n} \eta_i u_i(x) - \tilde{\lambda} \sum_{i=1}^{n} \eta_i \int_V g(x, \xi) m(\xi) u_i(\xi) \, dV(\xi) \tag{5.93}$$

To determine the coefficients η_k, the error function is set equal to zero at n distinct points. By setting the error, Eq. (5.93), equal to zero at the points $x_k (k = 1, 2, \ldots, n)$, we obtain

$$\varepsilon(x_k) = 0, \qquad k = 1, 2, \ldots, n \tag{5.94}$$

Equations (5.93) and (5.94) lead to the eigenvalue problem

$$\sum_{i=1}^{n} (m_{ki} - \tilde{\lambda} k_{ki})\eta_i = 0, \qquad k = 1, 2, \ldots, n \tag{5.95}$$

which can be expressed in matrix form as

$$[m]\vec{\eta} = \tilde{\lambda}[k]\vec{\eta} \tag{5.96}$$

where the elements of the matrices $[m]$ and $[k]$ are given by

$$m_{ki} = u_i(x_k) \tag{5.97}$$

$$k_{ki} = \int_V g(x_k, \xi) m(\xi) u_i(\xi) \, dV(\xi) \tag{5.98}$$

It is to be noted that the matrices $[m]$ and $[k]$ are, in general, not symmetric. The solution of the eigenvalue problem with nonsymmetric matrices $[m]$ and $[k]$ is more complex than the one with symmetric matrices [3].

5.5.6   Numerical Integration Method

In the numerical integration method, the regular part of the integral equation is decomposed into the form of a sum, and the equation is then reduced to a set of simultaneous linear equations with the values of the unknown function at some points in the domain of integration treated as the unknown quantities. The procedure is illustrated through the following example.

Example 5.7   Find the solution of the integral equation

$$\phi(x) + \int_0^1 (1 + x\xi)\phi(\xi)\, d\xi = f(x) \equiv x^2 - \frac{23}{24}x + \frac{4}{3} \tag{E5.7.1}$$

numerically and compare the result with the exact solution

$$\phi(x) = x^2 - 2x + 1 \tag{E5.7.2}$$

SOLUTION   We use the Gauss integration method for the numerical solution of Eq. (E5.7.1).

In Gauss integration, the integral is evaluated by using the formula

$$\int_{-1}^1 g(t)\, dt = \sum_{i=1}^n w_i g(t_i) \tag{E5.7.3}$$

where n is called the *number of Gauss points*, w_i are called *weights*, and t_i are the specified values of t in the range of integration. For any specified n, the values of w_i and t_i are chosen so that the formula will be exact for polynomials up to and including degree $2n - 1$. Since the range of integration in Eq. (E5.7.3) for x is -1 to $+1$, the formula can be made applicable to a general range of integration using a transformation of the variable. Thus, an integral of the form $\int_a^b f(x)\, dx$ can be evaluated, using the Gauss integration method, as

$$\int_a^b f(x)\, dx = \frac{b-a}{2} \sum_{i=1}^n w_i f(x_i) \tag{E5.7.4}$$

where the coordinate transformation

$$x = \frac{(b-a)t + a + b}{2} \tag{E5.7.5}$$

is used so that

$$x_i = \frac{(b-a)t_i + a + b}{2} \tag{E5.7.6}$$

Using $n = 4$, the corresponding values of w_i and t_i are given by [2]

$$w_1 = w_4 = 0.34785\,48451\,47454$$
$$w_2 = w_3 = 0.65214\,51548\,62546 \qquad \text{(E5.7.7)}$$

$$t_1 = -0.861136311594053$$
$$t_2 = -0.339981043584856$$
$$t_3 = -t_2 \qquad \text{(E5.7.8)}$$
$$t_4 = -t_1$$

The values of the variable x_i given by Eq. (E5.7.6) for $a = 0$ and $b = 1$ are

$$
\begin{aligned}
x_1 &= 0.06943184 \\
x_2 &= 0.33000946 \\
x_3 &= 0.66999054 \\
x_4 &= 0.93056816
\end{aligned}
\qquad \text{(E5.7.9)}
$$

Treating the values of $\phi(x)$ at the Gaussian points x_i as unknowns, Eq. (E5.7.1) can be expressed as

$$\phi_i + \tfrac{1}{2}[w_1(1 + x_i\xi_1)\phi_1 + w_2(1 + x_i\xi_2)\phi_2 + w_3(1 + x_i\xi_3)\phi_3 + w_4(1 + x_i\xi_4)\phi_4]$$
$$= x_i^2 - \tfrac{23}{24}x_i + \tfrac{4}{3}, \qquad i = 1, 2, 3, 4 \qquad \text{(E5.7.10)}$$

where $\phi_i = \phi(x_i)$, $x_i \equiv \xi_i$ are given by Eq. (E5.7.9), and w_i are given by Eq.(E5.7.7). The four linear equations indicated by Eq. (E5.7.10) are given by

$$1.17476588\phi_1 + 0.33354393\phi_2 + 0.34124103\phi_3$$
$$+ 0.18516506\phi_4 = 1.27161527$$

$$0.17791265\phi_1 + 1.36158392\phi_2 + 0.39816827\phi_3$$
$$+ 0.22733989\phi_4 = 1.12598062$$
$$\qquad\qquad\qquad\qquad\qquad\qquad\qquad\qquad\qquad \text{(E5.7.11)}$$
$$0.18201829\phi_1 + 0.39816827\phi_2 + 1.47244242\phi_3$$
$$+ 0.28236628\phi_4 = 1.14014649$$

$$0.18516506\phi_1 + 0.42620826\phi_2 + 0.52936965\phi_3$$
$$+ 1.32454112\phi_4 = 1.30749607$$

The solution of Eq. (E5.7.11) is given by $\phi_1 = 0.8660$, $\phi_2 = 0.4489$, $\phi_3 = 0.1089$, and $\phi_4 = 0.0048$, which can be seen to be same as the exact solution given by Eq. (E5.7.2) with four-decimal-place accuracy.

5.6   RECENT CONTRIBUTIONS

Strings and Bars   Laura and Gutierrez determined the fundamental frequency coefficient of vibrating systems using Rayleigh's optimization concept when solving integral equations by means of the Ritz method [13]. The authors considered the transverse vibration of a string of variable density, the longitudinal vibration of a rod with a nonuniform cross section, and the transverse vibration of a beam with ends elastically restrained against rotation as illustrative examples.

Beams An integral equation approach was used by many investigators for the solution of a vibrating beam [13]. In Ref. [9], Bergman and McFarland used Green's functions to study the free vibrations of an Euler–Bernoulli beam with homogeneous boundary conditions, supported in its interior by arbitrarily located pin supports and translational and rotational springs. A method of determining the dynamic response of prismatic damped Euler–Bernoulli beams subjected to distributed and concentrated loads using dynamic Green's functions was presented by Abu-Hilal [8]. The method gives exact solutions in closed form and can be used for single- and multispan beams, single- and multiloaded beams, and statically determinate or indeterminate beams. The responses of a statically indeterminate cantilevered beam and a cantilevered beam with elastic support are considered as example problems. The use of Green's functions in the frequency analysis of Timoshenko beams with oscillators was considered by Kukla [7].

Membranes Spence and Horgan [10] derived bounds on the natural frequencies of composite circular membranes using an integral equation method. The membrane was assumed to have a stepped radial density. Although such problems, involving discontinuous coefficients in the differential equation, can be treated using classical variational methods, it was shown that an eigenvalue estimation technique based on an integral formulation is more efficient. Gutierrez and Laura [11] analyzed the transverse vibrations of composite membranes using the integral equation method and Rayleigh's optimization suggestion. Specifically, the fundamental frequency of vibration of membranes of nonuniform density was determined.

Plates Bickford and Wu [12] considered the problem of finding upper and lower bounds on the natural frequencies of free vibration of a circular plate with stepped radial density. The problem, which involves discontinuous coefficients in the governing differential equation, has been formatted with an integral equation by using a Green's function and the basic theory of linear integral equations.

REFERENCES

1. J. Kondo, *Integral Equations*, Kodansha, Tokyo, and Clarendon Press, Oxford, 1991.

2. S. S. Rao, *Applied Numerical Methods for Engineers and Scientists*, Prentice Hall, Upper Saddle River, NJ, 2002.

3. S. S. Rao, *Mechanical Vibrations*, 4th ed., Prentice Hall, Upper Saddle River, NJ, 2004.

4. F. G. Tricomi, *Integral Equations*, Interscience, New York, 1957.

5. J. A. Cochran, *The Analysis of Linear Integral Equations*, McGraw-Hill, New York, 1972.

6. L. Meirovitch, *Analytical Methods in Vibrations*, Macmillan, New York, 1967.

7. S. Kukla, Application of Green functions in frequency analysis of Timoshenko beams with oscillators, *Journal of Sound and Vibration*, Vol. 205, No. 3, pp. 355–363, 1997.

8. M. Abu-Hilal, Forced vibration of Euler–Bernoulli beams by means of dynamic Green functions, *Journal of Sound and Vibration*, Vol. 267, No. 2, pp. 191–207, 2003.

9. L. A. Bergman and D. M. McFarland, On the vibration of a point- supported linear distributed system, *Journal of Vibration, Acoustics, Stress, and Reliability in Design*, Vol. 110, No. 4, pp. 485–492, 1988.

10. J. P. Spence and C. O. Horgan, Bounds on natural frequencies of composite circular membranes: integral equation methods, *Journal of Sound and Vibration*, Vol. 87, pp. 71–81, 1983.

11. R. H. Gutierrez and P. A. A. Laura, Analysis of transverse vibrations of composite membranes using the integral equation method and Rayleigh's optimization suggestion, *Journal of Sound and Vibration*, Vol. 147, No. 3, pp. 515–518, 1991.

12. W. B. Bickford and S. Y. Wu, Bounds on natural frequencies of composite circular plates: integral equation methods, *Journal of Sound and Vibration*, Vol. 126, No. 1, pp. 19–36, 1988.

13. P. A. A. Laura and R. H. Gutierrez, Rayleigh's optimization approach for the approximate solution of integral equations for vibration problems, *Journal of Sound and Vibration*, Vol. 139, No. 1, pp. 63–70, 1990.

PROBLEMS

5.1 Classify the following integral equations:

(a) $\phi(x) + \lambda \int_0^1 (x + y)\phi^2(y)\,dy = 0$

(b) $\phi(x) - \lambda \int_0^1 e^{x-y}\phi(y)\,dy = f(x)$

(c) $z(x, t) = \int_0^t G(x, y)\left[p(y) - \mu(y)\frac{\partial^2 z}{\partial t^2}\right]dy,$
$0 \leq x \leq l$

5.2 Classify the equation

$$\phi(x) = x + \int_0^x (\xi - x)\phi(\xi)\,d\xi \qquad (5.1)$$

Show that the function $\phi(x) = \sin x$ is a solution of Eq. (5.1).

5.3 Consider the integral equation

$$\phi(x) = \omega^2 \int_0^1 K(x, \xi)\phi(\xi)\,d\xi, \qquad 0 \leq x \leq 1 \quad (5.1)$$

where

$$K(x, \xi) = \begin{cases} x(1 - \xi), & x \leq \xi \\ \xi(1 - x), & \xi \leq x \end{cases} \qquad (5.2)$$

Determine the condition(s) under which the function $\phi(x) = \sin \omega x$ satisfies Eq. (5.1).

5.4 Consider the integral equation [1]

$$\phi(t) - \int_{-\infty}^t e^{-(t-\xi)}(t - \xi)\phi(\xi) = te^t \qquad (5.1)$$

and the function

$$\phi(t) = c_1 te^t + c_2 e^t \qquad (5.2)$$

Show that the function $\phi(t)$ given by Eq. (5.2) is a solution of Eq. (5.1).

5.5 Solve the equation

$$\phi(x) - 6 \int_0^1 x\xi^2\phi(\xi)\,d\xi = 2e^x - x + 1$$

using the method of undetermined coefficients by assuming the solution $\phi(x)$ to be of the form

$$\phi(x) = 2e^x + c_1 x + 1$$

5.6 Consider the integral equation

$$\phi(x) = \lambda \int_0^1 (1 + x\xi)\phi(\xi)\,d\xi$$

Solve this equation using the method of undetermined coefficients by assuming the solution as

$$\phi(x) = c_1 + c_2 x$$

5.7   Find the solution of the equation

$$\phi(x) - \lambda \int_0^1 (x - \xi)^2 \phi(\xi) \, d\xi = x$$

by assuming a solution of the form

$$\phi(x) = c_1 + c_2 x + c_3 x^2$$

Use the method of undetermined coefficients.

5.8   Solve the integral equation

$$\phi(x) - \int_0^1 (x - \xi) \phi(\xi) \, d\xi = 4x + x^2$$

using the method of undetermined coefficients. Assume the solution $\phi(x)$ as

$$\phi(x) = c_1 + c_2 x + c_3 x^2$$

5.9   Find the solution of the following equation using a numerical method [1]:

$$\phi(t) - \frac{1}{4} \int_0^{\pi/2} t \xi \phi(\xi) \, d\xi = \sin t - \frac{t}{4}$$

Compare your solution with the exact solution, $\phi(t) = \sin t$.

5.10 Find the solution of the following integral equation (eigenvalue problem) using the Gaussian integration method:

$$\phi(x) = \lambda \int_0^1 (x + \xi) \phi(\xi) \, d\xi$$

(a) with two Gaussian points; **(b)** with four Gaussian points; and **(c)** with six Gaussian points.
[*Hint*: The locations and weights corresponding to different number of Gauss points are given in Table 5.1.]:

Table 5.1   Data for Problem 5.10

Number of points, n	Locations, x_i	Weights, w_i
1	0.00000 00000 00000	2. 00000 00000 00000
2	±0.57735 02691 89626	1. 00000 00000 00000
3	±0.77459 66692 41483	0.55555 55555 55555
	0.00000 00000 00000	0.88888 88888 88889
4	±0.86113 63115 94053	0.34785 48451 47454
	±0.33998 10435 84856	0.65214 51548 62546
5	±0.90617 98459 38664	0.23692 68850 56189
	±0.53846 93101 05683	0.47862 86704 99366
	0.00000 00000 00000	0.56888 88888 88889
6	±0.93246 95142 03152	0.17132 44923 79170
	±0.66120 93864 66265	0.36076 15730 48139
	±0.23861 91860 83197	0.46791 39345 72691

Source: Ref. [2].

6

Solution Procedure: Eigenvalue and Modal Analysis Approach

6.1 INTRODUCTION

The equations of motion of many continuous systems are in the form of nonhomogeneous linear partial differential equations of order 2 or higher subject to boundary and initial conditions. The boundary conditions may be homogeneous or nonhomogeneous. The initial conditions are usually stated in terms of the values of the field variable and its time derivative at time zero. The solution procedure basically involves two steps. In the first step, the nonhomogeneous part of the equation of motion is neglected and the homogeneous equation is solved using the separation-of-variables technique. This leads to an eigenvalue problem whose solution yields an infinite set of eigenvalues and the corresponding eigenfunctions. The eigenfunctions are orthogonal and form a complete set in the sense that any function $\tilde{f}(\vec{X})$ that satisfies the boundary conditions of the problem can be represented by a linear combination of the eigenfunctions. This property constitutes what is known as the *expansion theorem*. In the second step, the solution of the nonhomogeneous equation is assumed to be a sum of the products of the eigenfunctions and time-dependent generalized coordinates using the expansion theorem. This process leads to a set of second-order ordinary differential equations in terms of the generalized coordinates. These equations are solved using the initial conditions of the problem. Once the generalized coordinates are known, complete solution of the problem can be determined from the expansion theorem.

6.2 GENERAL PROBLEM

The equation of motion of an undamped continuous system is in the form of a partial differential equation which can be expressed as

$$M(\vec{X})\frac{\partial^2 w(\vec{X}, t)}{\partial t^2} + L[w(\vec{X}, t)] = f(\vec{X}, t) + \sum_{j=1}^{s} F_j(t)\delta(\vec{X} - \vec{X}_j), \qquad \vec{X} \in V \quad (6.1)$$

where $\vec{X}$ is a typical point in the domain of the system (V), $M(\vec{X})$ is the mass distribution, $w(\vec{X}, t)$ is the field variable or displacement of the system that depends on the spatial variables $(\vec{X})$ and time (t), $L[w(\vec{X}, t)]$ is the stiffness distribution of the system, $f(\vec{X}, t)$ is the distributed force acting on the system, $F_j(t)$ is the concentrated force acting at the point $\vec{X} = \vec{X}_j$ of the system, s is the number of concentrated forces acting on the system, and $\delta(\vec{X} - \vec{X}_j)$ is the Dirac delta function, defined as

$$\delta(\vec{X} - \vec{X}_j) = 0, \qquad \vec{X} \neq \vec{X}_j$$

$$\int_V \delta(\vec{X} - \vec{X}_j) \, dV = 1$$

(6.2)

Note that the vector $\vec{X}$ will be identical to x for one-dimensional systems, includes x and y for two-dimensional systems, and consists of x, y, and z for three-dimensional systems. In Eq. (6.1), L and M are linear homogeneous differential operators involving derivatives with respect to the spatial variables $\vec{X}$ (but not with respect to time, t) up to the orders $2p$ and $2q$, respectively, where p and q are integers with $p > q$. For example, for a two-dimensional problem in a Cartesian coordinate system, the operator L can be expressed for $p = 1$ as

$$L[w] = c_1 w + c_2 \frac{\partial w}{\partial x} + c_3 \frac{\partial w}{\partial y} + c_4 \frac{\partial^2 w}{\partial x^2} + c_5 \frac{\partial^2 w}{\partial y^2} + c_6 \frac{\partial^2 w}{\partial x \partial y}$$

(6.3)

and the linearity of L implies that

$$L[c_1 w_1 + c_2 w_2] = c_1 L[w_1] + c_2 L[w_2]$$

(6.4)

where $c_1, c_2, \ldots, c_6$ are constants. In the case of the transverse vibration of a string having a mass distribution of $\rho(x)$ per unit length and subjected to a constant tension P, the operators M and L are given by [see Eq. (8.8)]

$$M = \rho(x)$$

(6.5)

$$L = P \frac{\partial^2}{\partial x^2}$$

(6.6)

In the case of the torsional vibration of a shaft, the operators M and L are given by [see Eq. (10.19)]

$$M = I_0(x)$$

(6.7)

$$L = \frac{\partial}{\partial x} \left(G I_P \frac{\partial}{\partial x} \right)$$

(6.8)

where $I_0(x)$ is the mass polar moment of inertia of the shaft per unit length, G is the shear modulus, and $I_P(x)$ is the polar moment of inertia of the cross section of the shaft. Similarly, in the case of the transverse vibration of a uniform plate in bending, the operators M and L are given by [see Eq. (14.8)]

$$M = \rho h$$

(6.9)

$$L = D \left(\frac{\partial^4}{\partial x^4} + 2 \frac{\partial^4}{\partial x^2 \partial y^2} + \frac{\partial^4}{\partial y^4} \right)$$

(6.10)

where ρ is the density, h is the thickness, and D is the flexural rigidity of the plate.

The governing differential equation (6.1) is subject to p boundary conditions at every point of boundary S of domain V of the system. The boundary conditions can be expressed as

$$A_i[w] = \lambda B_i[w], \qquad i = 1, 2, \ldots, p \tag{6.11}$$

where A_i and B_i are linear homogeneous differential operators involving derivatives of w, with respect to the normal and tangential directions of the boundary, up to the order $2p - 1$, and λ is a parameter known as the eigenvalue of the system. In some problems, the boundary conditions do not involve the eigenvalue λ, in which case Eq. (6.11) reduces to

$$A_i[w] = 0, \qquad i = 1, 2, \ldots, p \tag{6.12}$$

We shall consider mostly boundary conditions of the type given by Eq. (6.12) in further discussions. In the case of free vibration, f and all F_j will be zero, and Eq. (6.1) reduces to the homogeneous form

$$M(\vec{X})\frac{\partial^2 w(\vec{X}, t)}{\partial t^2} + L[w(\vec{X}, t)] = 0, \qquad \vec{X} \in V \tag{6.13}$$

6.3   SOLUTION OF HOMOGENEOUS EQUATIONS: SEPARATION-OF-VARIABLES TECHNIQUE

The separation-of-variables technique is applicable to the solution of homogeneous second- and higher-order linear partial differential equations with constant coefficients subject to homogeneous boundary conditions. The partial differential equations may represent initial or boundary value problems. To illustrate the method of separation of variables, we consider a homogeneous hyperbolic equation of the form

$$\rho(x)\frac{\partial^2 w(x, t)}{\partial t^2} + L[w(x, t)] = 0, \qquad x \in G, \quad t > 0 \tag{6.14}$$

where G denotes a bounded region such as $[0, l]$, x is the spatial variable, t is time, $\rho(x)$ is positive and independent of t, L is a linear differential operator, and $w(x, t)$ is an unknown function to be determined. The homogeneous boundary conditions can be stated as

$$A_1 w(0, t) + B_1 \frac{\partial w}{\partial x}(0, t) = 0 \tag{6.15}$$

$$A_2 w(l, t) + B_2 \frac{\partial w}{\partial x}(l, t) = 0 \tag{6.16}$$

The initial conditions for Eq. (6.14) can be expressed as

$$w(x, 0) = f(x), \qquad x \in G \tag{6.17}$$

$$\frac{\partial w}{\partial t}(x, 0) = g(x), \qquad x \in G \tag{6.18}$$

The separation-of-variables technique replaces the partial differential equation in two parameters, x and t, Eq. (6.14), by two ordinary differential equations. The solution of Eq. (6.14) is assumed to be a product of two functions, each depending on only one of parameters x and t as

$$w(x, t) = W(x)T(t) \tag{6.19}$$

where $W(x)$ is required to satisfy the boundary conditions, Eqs. (6.15) and (6.16), and $T(t)$ is required to satisfy the initial conditions, Eqs. (6.17) and (6.18). Substituting Eq. (6.19) into Eq. (6.14) and dividing throughout by $\rho(x)W(x)T(t)$ yields

$$\frac{T''(t)}{T(t)} = -\frac{L[W(x)]}{\rho(x)W(x)} \tag{6.20}$$

where $T''(t) = d^2T(t)/dt^2$. Since the left and right sides of Eq. (6.20) depend on different variables, they cannot be functions of their respective variables. Thus, each side of Eq. (6.20) must equal a constant. By denoting this constant by $-\lambda$, we obtain the following equations for $W(x)$ and $T(t)$ (see Problem 6.1):

$$L[W(x)] = \lambda\rho(x)W(x) \tag{6.21}$$

$$T''(t) + \lambda T(t) = 0 \tag{6.22}$$

The ordinary differential equations (6.21) and (6.22) can be solved by satisfying the boundary and initial conditions to find $W(x)$ and $T(t)$, respectively. Consequently, $w(x, t) = W(x)T(t)$ will satisfy both the boundary and initial conditions. Since both the equation and the boundary conditions for $W(x)$ are homogeneous, $W(x) = 0$ will be a solution of the problem (called the *trivial solution*). However, we require nonzero solutions for $W(x)$ in the boundary value problem, and such solutions exist only for certain values of the parameter λ in Eq. (6.21). The problem of determining the nonzero $W(x)$ and the corresponding value of the parameter λ is known as an *eigenvalue problem*. Here λ is called an *eigenvalue* and $W(x)$ is called an *eigenfunction*. The eigenvalue problem is known as the *Sturm–Liouville problem* in the mathematical literature and is discussed in the following section.

6.4   STURM–LIOUVILLE PROBLEM

The mathematical models for the vibration of some continuous systems are in the form of a certain type of two-point boundary value problem known as the Sturm–Liouville problem. The Sturm–Liouville problem is a one-dimensional eigenvalue problem whose governing equation is of the general form

$$\frac{d}{dx}\left[p(x)\frac{dw}{dx}\right] + [q(x) + \lambda r(x)]w(x) = 0, \qquad a < x < b$$

or

$$(pw')' + (q + \lambda r)w = 0, \qquad a < x < b \tag{6.23}$$

with boundary conditions in the form

$$A_1 w(a) + B_1 w'(a) = 0 \tag{6.24}$$

$$A_2 w(b) + B_2 w'(b) = 0 \tag{6.25}$$

where a prime denotes a derivative with respect to x, and $p(x)$, $q(x)$, and $r(x)$ are continuous functions defined in the closed interval $a \le x \le b$, with $p(x) > 0$, $r(x) > 0$ and $A_i \ge 0$, $B_i \ge 0$ with $A_i + B_i > 0$ for $i = 1, 2$. The boundary conditions of Eqs. (6.24) and (6.25) are said to be homogeneous because a linear combination of $w(x)$ and $w'(x)$ at $x = a$ and $x = b$ are both equal to zero.

6.4.1 Classification of Sturm–Liouville Problems

Based on the nature of the boundary conditions and the behavior of $p(x)$ at the boundaries, Sturm–Liouville problems can be classified as regular, periodic, or singular. The problem defined by Eqs. (6.23)–(6.25) is called a *regular Sturm–Liouville problem*. Note that $p(x) > 0$ and is continuous in the interval $a \le x \le b$ with constants A_i and B_i not equal to zero simultaneously in the ith boundary condition, $i = 1, 2$ [Eqs. (6.24) and (6.25)]. In this case the problem involves finding constant values of λ corresponding to each of which a nontrivial solution $w(x)$ can be found for Eq. (6.23) while satisfying the boundary conditions of Eqs. (6.24) and (6.25).

If the function $p(x)$ and the boundary conditions involving $w(x)$ and $w'(x)$ are periodic over the interval $a \le x \le b$, the problem is called a *periodic Sturm–Liouville problem*. In this case, the problem involves finding constant values of λ corresponding to each of which a nontrivial solution can be found for Eq. (6.23) while satisfying the periodic boundary conditions given by

$$p(a) = p(b), \qquad w(a) = w(b), \qquad w'(a) = w'(b) \tag{6.26}$$

If the functions $p(x)$ or $r(x)$ or both are zero at any one or both of the boundary points a and b, the problem is called a *singular Sturm–Liouville problem*. In this case the problem involves finding constant values of λ corresponding to each of which a nontrivial solution $w(x)$ can be found to satisfy Eq. (6.23) and the boundary conditions of Eqs. (6.24) and (6.25). For example, if the singular point is located at either $x = a$ or $x = b$ so that either $p(a) = 0$ or $p(b) = 0$, the boundary condition that is often imposed at the singular point basically requires the solution $w(x)$ to be bounded at that point.

Note that Eq. (6.23) always has the solution $w(x) = 0$, called the *trivial solution*. For nontrivial solutions (solutions that are not identically zero) that satisfy the specified boundary conditions at $x = a$ and $x = b$, the parameter λ cannot be arbitrary. Thus, the problem involves finding constant values of λ for which nontrivial solutions exist that satisfy the specified boundary conditions.

Each value of λ for which a nontrivial solution can be found is called an *eigenvalue* of the problem and the corresponding solution $w(x)$ is called an *eigenfunction* of the problem. Because the Sturm–Liouville problem is homogeneous, it follows that the eigenfunctions are not unique. The eigenfunction corresponding to any eigenvalue can be multiplied by any constant factor, and the resulting function remains an eigenfunction to the same eigenvalue λ.

Example 6.1 Regular Sturm–Liouville Problem   Find the solution to the eigenvalue problem

$$\frac{d^2 W(x)}{dx^2} + \lambda W(x) = 0 \tag{E6.1.1}$$

with the boundary conditions

$$W(0) - \frac{dW}{dx}(0) = 0 \tag{E6.1.2}$$

$$W(1) + \frac{dW}{dx}(1) = 0 \tag{E6.1.3}$$

SOLUTION   Equation (E6.1.1) can be identified as a Sturm–Liouville problem, Eq. (6.23), with $p(x) = 1, q(x) = 0$, and $r(x) = 1$. Theoretically, we can consider three cases: $\lambda < 0$, $\lambda = 0$, and $\lambda > 0$.

When $\lambda < 0$, we set $\lambda = -\alpha^2$ and write Eq. (E6.1.1) as

$$\frac{d^2W(x)}{dx^2} - \alpha^2 W(x) = 0 \tag{E6.1.4}$$

which has the general solution

$$W(x) = c_1 e^{\alpha x} + c_2 e^{-\alpha x} \tag{E6.1.5}$$

The boundary conditions, Eqs. (E6.1.2) and (E6.1.3), become

$$c_1(1 - \alpha) + c_2(1 + \alpha) = 0 \tag{E6.1.6}$$

$$c_1(1 + \alpha)e^{\alpha} + c_2(1 - \alpha)e^{-\alpha} = 0 \tag{E6.1.7}$$

It can be shown (see Problem 6.2) that for $\alpha > 0$, c_1 and c_2 do not have nonzero solution; the only solution is the trivial solution, $c_1 = 0$ and $c_2 = 0$. Hence, the problem has no negative eigenvalue.

When $\lambda = 0$, Eq. (E6.1.1) reduces to

$$\frac{d^2W(x)}{dx^2} = 0 \tag{E6.1.8}$$

which has the general solution

$$W(x) = c_1 + c_2 x \tag{E6.1.9}$$

The boundary conditions, Eqs. (E6.1.2) and (E6.1.3), become

$$c_1 - c_2 = 0 \tag{E6.1.10}$$

$$c_1 + 2c_2 = 0 \tag{E6.1.11}$$

The only solution to Eqs. (E6.1.10) and (E6.1.11) is the trivial solution $c_1 = 0$ and $c_2 = 0$. Hence, $\lambda = 0$ is not an eigenvalue of the problem

When $\lambda > 0$, we set $\lambda = \alpha^2$ and write Eq. (E6.1.1) as

$$\frac{d^2W(x)}{dx^2} + \alpha^2 W(x) = 0 \tag{E6.1.12}$$

which has the general solution

$$W(x) = c_1 \cos \alpha x + c_2 \sin \alpha x \tag{E6.1.13}$$

The boundary conditions, Eqs. (E6.1.2) and (E6.1.3), become

$$c_1 - \alpha c_2 = 0 \tag{E6.1.14}$$

$$c_1 \cos \alpha + c_2 \sin \alpha - c_1 \alpha \sin \alpha + c_2 \alpha \cos \alpha = 0 \tag{E6.1.15}$$

The solution of Eqs. (E6.1.14) and (E6.1.15) is given by

$$c_1 = \alpha c_2 \tag{E6.1.16}$$

$$c_2[(1 - \alpha^2) \sin \alpha + 2\alpha \cos \alpha] = 0 \tag{E6.1.17}$$

If c_2 is zero in Eq. (E6.1.17), c_1 will also be zero from Eq. (E6.1.16). This will be a trivial solution. Hence, for a nontrivial solution, we should have

$$(1 - \alpha^2) \sin \alpha + 2\alpha \cos \alpha = 0 \tag{E6.1.18}$$

Equation (E6.1.18) is a transcendental equation whose roots are given by

$$\tan \alpha_i = \frac{2\alpha_i}{\alpha_i^2 - 1}, \qquad i = 1, 2, \ldots \tag{E6.1.19}$$

Equation (E6.1.19) can be solved numerically to find α_i, hence the eigenvalues are given by $\lambda_i = \alpha_i^2$ and the corresponding eigenfunctions are given by

$$W_i(x) = c_2(\alpha_i \cos \alpha_i x + \sin \alpha_i x) \tag{E6.1.20}$$

Example 6.2 Periodic Sturm–Liouville Problem   Find the solution of the boundary value problem

$$\frac{d^2 W(x)}{dx^2} + \lambda W(x) = 0, \qquad 0 < x < 1 \tag{E6.2.1}$$

subject to the boundary conditions

$$W(0) = W(1) \tag{E6.2.2}$$

$$\frac{dW}{dx}(0) = \frac{dW}{dx}(1) \tag{E6.2.3}$$

SOLUTION   Equation (E6.2.1) can be identified as the Sturm–Liouville problem of Eq. (6.23) with $p(x) = 1$, $q(x) = 0$, and $r(x) = 1$. We can consider three cases: $\lambda < 0$, $\lambda = 0$, and $\lambda > 0$.

When $\lambda < 0$, we set $\lambda = -\alpha^2$ and write Eq. (E6.2.1) as

$$\frac{d^2 W(x)}{dx^2} - \alpha^2 W(x) = 0 \tag{E6.2.4}$$

which has the general solution

$$W(x) = c_1 e^{\alpha x} + c_2 e^{-\alpha x} \tag{E6.2.5}$$

The boundary conditions, Eqs. (E6.2.2) and (E6.2.3), become

$$c_1(1 - e^\alpha) + c_2(1 - e^{-\alpha}) = 0 \tag{E6.2.6}$$

$$c_1\alpha(1 - e^\alpha) - c_2\alpha(1 - e^{-\alpha}) = 0 \tag{E6.2.7}$$

Equation (E6.2.6) gives

$$c_2 = -c_1\frac{1 - e^\alpha}{1 - e^{-\alpha}} \tag{E6.2.8}$$

Substitution of Eq.(E6.2.8) in Eq. (E6.2.7) yields

$$2c_1\alpha(1 - e^\alpha) = 0 \tag{E6.2.9}$$

Since $\alpha > 0$, Eq. (E6.2.9) gives $c_1 = 0$ and hence $c_2 = 0$ [from Eq. (E6.2.8)]. This is a trivial solution and hence the problem has no negative eigenvalue.

When $\lambda = 0$, Eq. (E6.2.1) reduces to

$$\frac{d^2W(x)}{dx^2} = 0 \tag{E6.2.10}$$

which has the general solution

$$W(x) = c_1 + c_2x \tag{E6.2.11}$$

The boundary conditions, Eqs. (E6.2.2) and (E6.2.3), become

$$c_1 = c_1 + c_2 \tag{E6.2.12}$$

$$c_2 = c_2 \tag{E6.2.13}$$

Equations (E6.2.12) and (E6.2.13) imply that $c_2 = 0$ and hence the solution, Eq. (E6.2.11), reduces to

$$W(x) = c_1 \tag{E6.2.14}$$

where c_1 is any nonzero constant. This shows that $\lambda = 0$ is an eigenvalue of the problem with the corresponding eigenfunction given by Eq. (E6.2.14).

When $\lambda > 0$, we set $\lambda = \alpha^2$ and write Eq. (E6.2.1) as

$$\frac{d^2W(x)}{dx^2} + \alpha^2W(x) = 0 \tag{E6.2.15}$$

which has the general solution

$$W(x) = c_1\cos\alpha x + c_2\sin\alpha x \tag{E6.2.16}$$

The boundary conditions, Eqs. (E6.2.2) and (E6.2.3), become

$$c_1 = c_1\cos\alpha + c_2\sin\alpha \tag{E6.2.17}$$

$$c_2\alpha = -c_1\alpha\sin\alpha + c_2\alpha\cos\alpha \tag{E6.2.18}$$

The solution of Eqs. (E6.2.17) and (E6.2.18) yields

$$c_2 = c_1 \frac{1 - \cos \alpha}{\sin \alpha} \tag{E6.2.19}$$

and

$$c_2 = -c_1 \frac{\sin \alpha}{1 - \cos \alpha} \tag{E6.2.20}$$

which imply that either $c_1 = 0$ or

$$\cos \alpha = 1 \tag{E6.2.21}$$

Equation (E6.2.21) implies that α is zero or an integer multiple of 2π, so that

$$\alpha_m = \pm m \cdot 2\pi, \qquad m = 0, 1, 2, \dots \tag{E6.2.22}$$

Thus, the eigenvalues of the problem are given by

$$\lambda_m = \alpha_m^2 = 4m^2\pi^2, \qquad m = 0, 1, 2, \dots \tag{E6.2.23}$$

with the corresponding eigenfunctions given by Eq. (E6.2.16):

$$W_m(x) = c_1 \cos 2m\pi x + c_2 \sin 2m\pi x, \qquad m = 0, 1, 2, \dots \tag{E6.2.24}$$

Example 6.3 Singular Sturm–Liouville Problem The free transverse vibration of a circular membrane of radius a, clamped around the edge, is governed by the equation

$$r^2 \frac{d^2 W(r)}{dr^2} + r \frac{dW(r)}{dr} + (\omega^2 - m^2) W(r) = 0, \qquad 0 \le r \le a \tag{E6.3.1}$$

subject to the requirement of a bounded solution with the condition

$$W(a) = 0 \tag{E6.3.2}$$

where r denotes the radial direction, $W(r)$ is the transverse displacement, ω is the natural frequency (ω^2 is called the eigenvalue), and m is an integer [see Eq. (13.126)]. Find the solution of the problem.

SOLUTION Equation (E6.3.1) can be identified as Bessel's differential equation of order m with the parameter ω. The equation can be rewritten in the form of a Sturm–Liouville equation:

$$\frac{d}{dr}\left(r \frac{dW}{dr}\right) + \left(\omega^2 r^2 - \frac{m^2}{r}\right) W = 0, \qquad 0 \le r \le a \tag{E6.3.3}$$

which can be compared to Eq. (6.23) with the notations $p(x) = r$, $q(x) = -m^2/r$, $r(x) = r^2$, and $\lambda = \omega^2$. It can be seen that Eq. (E6.3.3) denotes a singular Sturm–Liouville problem because $p(0) = 0$. The solution of Eq. (E6.3.3) is given by

$$W(r) = B_1 J_m(\omega r) + B_2 Y_m(\omega r) \tag{E6.3.4}$$

where B_1 and B_2 are constants and J_m and Y_m are Bessel functions of the first and second kind, respectively. The solution is required to be bounded, but $W(r)$ approaches infinity at $r = 0$. Thus, B_2 must be set equal to zero to ensure a bounded solution. Thus, Eq. (E6.3.4) reduces to

$$W(r) = B_1 J_m(\omega r) \qquad \text{(E6.3.5)}$$

The use of the boundary condition of Eq. (E6.32) in Eq. (E6.3.5) yields

$$J_m(\omega a) = 0 \qquad \text{(E6.3.6)}$$

Equation (E6.3.6) has infinite roots $\omega_i a$, $i = 1, 2, \ldots$. Thus, the ith eigenvalue of the membrane is given by $\lambda_i = \omega_i^2$, and the corresponding eigenfunction by

$$W_i(r) = B_1 J_m(\omega_i r) \qquad \text{(E6.3.7)}$$

where the constant B_1 can be selected arbitrarily.

6.4.2   Properties of Eigenvalues and Eigenfunctions

The fundamental properties of eigenvalues and eigenfunctions of Sturm–Liouville problems are given below.

1. Regular and periodic Sturm–Liouville problems have an infinite number of distinct real eigenvalues $\lambda_1, \lambda_2, \ldots$ which can be arranged as

$$\lambda_1 < \lambda_2 < \cdots$$

The smallest eigenvalue λ_1 is finite and the largest one is infinity:

$$\lim_{n \to \infty} \lambda_n = \infty$$

2. A unique eigenfunction exists, except for an arbitrary multiplicative constant, for each eigenvalue of a regular Sturm–Liouville problem.
3. The infinite sequence of eigenfunctions $w_1(x), w_2(x), \ldots$ defined over the interval $a \leq x \leq b$ are said to be orthogonal with respect to a weighting function $r(x) \geq 0$ if

$$\int_a^b r(x) w_m(x) w_n(x) \, dx = 0, \qquad m \neq n \qquad (6.27)$$

When $m = n$, Eq. (6.27) defines the *norm* of $w_n(x)$, denoted $||w_n(x)||$, as

$$||w_n(x)||^2 = \int_a^b r(x) w_m^2(x) \, dx > 0 \qquad (6.28)$$

By normalizing the function $w_m(x)$ as

$$\overline{w}_m(x) = \frac{w_m(x)}{||w_m(x)||}, \qquad m = 1, 2, \ldots \qquad (6.29)$$

we obtain the *orthonormal* functions $\overline{w}_m(x)$ with the properties

$$\int_a^b r(x)\overline{w}_m(x)\overline{w}_n(x)\,dx = 0, \qquad m \neq n \tag{6.30}$$

$$\int_a^b r(x)\overline{w}_m(x)\overline{w}_n(x)\,dx = 1, \qquad m = n \tag{6.31}$$

4. *Expansion theorem.* The orthogonality of the eigenfunctions $w_1(x), w_2(x), \ldots$ over the interval $a \leq x \leq b$ with respect to a weighting function $r(x)$ permits them to be used to represent any function $\tilde{f}(x)$ over the same interval as a linear combination of $w_m(x)$ as

$$\tilde{f}(x) = \sum_{m=1}^{\infty} c_m w_m(x) = c_1 w_1(x) + c_2 w_2(x) + \cdots \tag{6.32}$$

where $c_1, c_2, \ldots$ are constants known as *coefficients of the expansion.* Equation (6.32) denotes the eigenfunction expansion of $\tilde{f}(x)$ and is known as the *expansion theorem.* To determine the coefficients c_m, we multiply Eq. (6.32) by $r(x)w_n(x)$ and integrate the result with respect to x from a to b:

$$\int_a^b r(x)\tilde{f}(x)w_n(x)\,dx = \sum_{m=1}^{\infty}\left[\int_a^b c_m r(x)w_m(x)w_n(x)\,dx\right] \tag{6.33}$$

When Eqs. (6.27) and (6.28) are used, Eq. (6.33) reduces to

$$\int_a^b r(x)\tilde{f}(x)w_n(x)\,dx = c_n \int_a^b r(x)[w_n(x)]^2\,dx = c_n\|w_n(x)\|^2 \tag{6.34}$$

Equation (6.34) gives the coefficients c_n as

$$c_n = \frac{\int_a^b r(x)\tilde{f}(x)w_n(x)\,dx}{\|w_n(x)\|^2}, \qquad n = 1, 2, \ldots \tag{6.35}$$

5. *Orthogonality of eigenfunctions.* If the functions $p(x), q(x), r(x)$, and $r(x)$ are real valued and continuous with $r(x) > 0$ on the interval $a \leq x \leq b$, the eigenfunctions $w_m(x)$ and $w_n(x)$ corresponding to different eigenvalues λ_m and λ_n, respectively, are orthogonal with respect to the weighting function $r(x)$. This property can be proved as follows.

Since the eigenfunctions satisfy the Sturm–Liouville equation (6.23), we have

$$(pw_m')' + (q + \lambda_m r)w_m = 0, \qquad a < x < b \tag{6.36}$$

$$(pw_n')' + (q + \lambda_n r)w_n = 0, \qquad a < x < b \tag{6.37}$$

Multiply Eq. (6.36) by $w_n(x)$ and Eq. (6.37) by $w_m(x)$ and subtract the resulting equations one from the other to obtain

$$(\lambda_m - \lambda_n)rw_m w_n = w_m(pw_n')' - w_n(pw_m')' = (pw_n'w_m - pw_m'w_n)' \tag{6.38}$$

Integration of Eq. (6.38) with respect to x from a to b results in

$$(\lambda_m - \lambda_n) \int_a^b r w_m(x) w_n(x)\, dx = [p(w_n'(x) w_m(x) - w_m'(x) w_n(x)]_a^b$$

$$= p(b)[w_n'(b) w_m(b) - w_m'(b) w_n(b)]$$

$$- p(a)[w_n'(a) w_m(a) - w_m'(a) w_n(a)] \qquad (6.39)$$

Based on whether $p(x)$ is zero or not at $x = a$ or $x = b$, we need to consider the following cases:

(a) $p(a) = 0$ and $p(b) = 0$: In this case, the expression on the right-hand side of Eq. (6.39) is zero. Since λ_m and λ_n are distinct, we have

$$\int_a^b r(x) w_m(x) w_n(x)\, dx = 0, \qquad m \neq n \qquad (6.40)$$

Note that Eq. (6.40) is valid irrespective of the boundary conditions of Eqs. (6.24) and (6.25).

(b) $p(b) = 0$ and $p(a) \neq 0$: In this case, the expression on the right-hand side of Eq. (6.39) reduces to

$$- p(a)[w_n'(a) w_m(a) - w_m'(a) w_n(a)] \qquad (6.41)$$

The boundary condition at $x = a$ can be written as

$$A_1 w_m(a) + B_1 w_m'(a) = 0 \qquad (6.42)$$

$$A_1 w_n(a) + B_1 w_n'(a) = 0 \qquad (6.43)$$

Multiply Eq. (6.42) by $w_n(a)$ and Eq. (6.43) by $w_m(a)$ and subtract the resulting equations one from the other to obtain

$$B_1[w_n'(a) w_m(a) - w_m'(a) w_n(a)] = 0 \qquad (6.44)$$

Assuming that $B_1 \neq 0$, the expression in brackets in Eq. (6.44) must be zero. This means that the expression in (6.41) is zero. Hence, the orthogonality condition given in Eq. (6.40) is valid. Note that if $B_1 = 0$, then $A_1 \neq 0$ by assumption, and a similar argument proves the orthogonality condition in Eq. (6.40).

(c) $p(a) = 0$ and $p(b) \neq 0$: By using a procedure similar to that of case (b) with the boundary condition of Eq. (6.25), the orthogonality condition in Eq. (6.40) can be proved.

(d) $p(a) \neq 0$ and $p(b) \neq 0$: In this case we need to use both the procedures of cases (b) and (c) to establish the validity of Eq. (6.40).

(e) $p(a) = p(b)$: In this case the right-hand side of Eq. (6.39) can be written as

$$p(b)[w_n'(b) w_m(b) - w_m'(b) w_n(b) - w_n'(a) w_m(a) + w_m'(a) w_n(a)] \qquad (6.45)$$

By using the boundary condition of Eq. (6.24) as before, we can prove that the expression in brackets in Eq. (6.45) is zero. This proves the orthogonality condition given in Eq. (6.40).

6.5   GENERAL EIGENVALUE PROBLEM

The eigenvalue problem considered in Section 6.4, also known as the *Sturm–Liouville problem*, is valid only for one-dimensional systems. A general eigenvalue problem applicable to one-, two-, and three-dimensional systems is discussed in this section.

In the case of free vibration, f and all F_j will be zero and Eq. (6.1) reduces to the homogeneous form

$$M(\vec{X})\frac{\partial^2 w(\vec{X}, t)}{\partial t^2} + L[w(\vec{X}, t)] = 0, \qquad \vec{X} \in V \tag{6.46}$$

For the natural frequencies of vibration, we assume the displacement $w(\vec{X}, t)$ to be a harmonic function as

$$w(\vec{X}, t) = W(\vec{X})e^{i\omega t} \tag{6.47}$$

where $W(\vec{X})$ denotes the mode shape (also called the *eigenfunction* or *normal mode*) and ω indicates the natural frequency of vibration. Using Eq. (6.47), Eq. (6.46) can be represented as

$$L[W] = \lambda M[W] \tag{6.48}$$

where $\lambda = \omega^2$ is also called the *eigenvalue* of the system. Equation (6.48), along with the boundary conditions of Eq. (6.11) or (6.12), defines the eigenvalue problem of the system. The solution of the eigenvalue problem yields an infinite number of eigenvalues $\lambda_1, \lambda_2, \ldots$ and the corresponding eigenfunctions $W_1(\vec{X}), W_2(\vec{X}), \ldots$. The eigenvalue problem is said to be homogeneous, and the amplitudes of the eigenfunctions $W_i(\vec{X}), i = 1, 2, \ldots$, are arbitrary. Thus, only the shapes of the eigenfunctions can be determined uniquely.

6.5.1   Self-Adjoint Eigenvalue Problem

Before defining a problem known as the *self-adjoint eigenvalue problem*, two types of functions, called *admissible* and *comparison functions*, are introduced. These functions are used in certain approximate methods of solving the eigenvalue problem. As seen in Chapter 4, the boundary conditions of Eq. (6.12) are composed of geometric (or forced) and natural (or free) boundary conditions.

A function $u(\vec{X})$ is said to be an *admissible function* if it is p times differentiable over the domain V and satisfies only the geometric boundary conditions of the eigenvalue problem. Note that an admissible function does not satisfy the natural boundary conditions as well as the governing differential equation of the eigenvalue problem. A function $u(\vec{X})$ is said to be a *comparison function* if it is $2p$ times differentiable over the domain V and satisfies all the boundary conditions (both geometric and natural) of the eigenvalue problem. Note that a comparison function does not satisfy the governing differential equation of the eigenvalue problem. On the other hand, the *eigenfunctions* $W_i(\vec{X}), i = 1, 2, \ldots$ satisfy the governing differential equation as well as all the boundary conditions of the eigenvalue problem.

Definition   The eigenvalue problem defined by Eqs. (6.48) and (6.12) is said to be *self-adjoint* if for any two arbitrary comparison functions $u_1(\vec{X})$ and $u_2(\vec{X})$, the following relations are valid:

$$\int_V u_1(\vec{X})L[u_2(\vec{X})]\,dV = \int_V u_2(\vec{X})L[u_1(\vec{X})]\,dV \tag{6.49}$$

$$\int_V u_1(\vec{X})M[u_2(\vec{X})]\,dV = \int_V u_2(\vec{X})M[u_1(\vec{X})]\,dV \tag{6.50}$$

Positive Definite Problem   An eigenvalue problem, defined by Eqs. (6.48) and (6.12), is said to be *positive definite* if the operators L and M are both positive definite. The operator L is considered *positive* if for any comparison function $u(\vec{X})$, the following relation is valid:

$$\int_V u(\vec{X})L[u(\vec{X})]\,dV \geq 0 \tag{6.51}$$

The operator L is considered *positive definite* if the integral in Eq. (6.51) is zero only when $u(\vec{X})$ is identically equal to zero. Similar definitions are valid for the operator M. The eigenvalue problem is said to be *semidefinite* if the operator L is only positive and the operator M is positive definite. It is to be noted that the eigenvalue problems corresponding to most continuous systems considered in subsequent discussions are self-adjoint, as implied by Eqs. (6.49) and (6.50). In most cases, the operator $M(\vec{X})$ denotes the distributed mass of the system, and hence the positive definiteness of $M(\vec{X})$ is ensured.

If the system or the eigenvalue problem is positive definite, all the eigenvalues λ_i will be positive. If the system is semidefinite, some λ_i will be zero. It can be seen that these properties are similar to that of a discrete system.

Example 6.4   The free axial vibration of a uniform bar fixed at both the ends $x = 0$ and $x = L$ is governed by the equation

$$EA\frac{\partial^2 u(x,t)}{\partial x^2} + m\frac{\partial^2 u(x,t)}{\partial t^2} = 0 \tag{E6.4.1}$$

where E is Young's modulus, A is the cross sectional, area, m is the mass per unit length, and $u(x,t)$ is the axial displacement of the bar. Show that the eigenvalue problem, obtained with

$$u(x,t) = U(x)\cos\omega t \tag{E6.4.2}$$

in Eq. (E6.4.1), is self-adjoint. Consider the following comparison functions:

$$U_1(x) = C_1 x(L - x), \qquad U_2(x) = C_2\sin\frac{\pi x}{L} \tag{E6.4.3}$$

SOLUTION   The eigenvalue problem corresponding to Eq. (E6.4.1) is given by

$$EA\frac{d^2 U(x)}{dx^2} = \lambda m U(x) \tag{E6.4.4}$$

where $\lambda = \omega^2$ is the eigenvalue. Comparing Eq. (E6.4.4) with Eq. (6.48), we identify the operators L and M as

$$L = EA\frac{\partial^2}{\partial x^2}, \qquad M = m \tag{E6.4.5}$$

The eigenvalue problem will be self-adjoint if the following conditions hold true:

$$\int_0^L U_1(x)L[U_2(x)]\,dx = \int_0^L U_2(x)L[U_1(x)]\,dx \tag{E6.4.6}$$

$$\int_0^L U_1(x)M[U_2(x)]\,dx = \int_0^L U_2(x)M[U_1(x)]\,dx \tag{E6.4.7}$$

The boundary conditions of the bar can be expressed as

$$U(0) = 0, \qquad U(L) = 0 \tag{E6.4.8}$$

The comparison functions given by Eq. (E6.4.3) can be seen to satisfy the boundary conditions, Eq. (E6.4.8). Using $U_1(x)$ and $U_2(x)$, we find that

$$\int_0^L U_1(x)L[U_2(x)]\,dx = \int_0^L C_1 x(L-x)EA\frac{d^2}{dx^2}\left(C_2 \sin\frac{\pi x}{L}\right)dx$$
$$= -\frac{4C_1 C_2 EAL}{\pi} \tag{E6.4.9}$$

$$\int_0^L U_2(x)L[U_1(x)]\,dx = \int_0^L C_2 \sin\frac{\pi x}{L}EA\frac{d^2}{dx^2}(C_1 xL - C_1 x^2)\,dx$$
$$= -\frac{4C_1 C_2 EAL}{\pi} \tag{E6.4.10}$$

It can be seen that Eq. (E6.4.6) is satisfied. Similarly, Eq. (E6.4.7) can also be shown to be satisfied. Thus, the eigenvalue problem is self-adjoint.

6.5.2 Orthogonality of Eigenfunctions

The orthogonality property, proved for the Sturm–Liouville problem in Section 6.4.2, can also be established for a general eigenvalue problem. For this, let λ_i and λ_j denote two distinct eigenvalues, with $W_i = W_i(\vec{X})$ and $W_j = W_j(\vec{X})$ indicating the corresponding eigenfunctions. Then

$$L[W_i] = \lambda_i M[W_i] \tag{6.52}$$

$$L[W_j] = \lambda_j M[W_j] \tag{6.53}$$

Multiply Eq. (6.52) by W_j and Eq. (6.53) by W_i and subtract the resulting equations from each other:

$$W_j L[W_i] - W_i L[W_j] = \lambda_i W_j M[W_i] - \lambda_j W_i M[W_j] \tag{6.54}$$

Integrate both sides of Eq. (6.54) over the domain V of the system to obtain

$$\int_V (W_j L[W_i] - W_i L[W_j]) \, dV = \int_V (\lambda_i W_j M[W_i]) - \lambda_j W_i M[W_j] \, dV \tag{6.55}$$

If the eigenvalue problem is assumed to be self-adjoint, then

$$\int_V W_j L[W_i] \, dV = \int_V W_i L[W_j] \, dV \tag{6.56}$$

and

$$\int_V W_j M[W_i] \, dV = \int_V W_i M[W_j] \, dV \tag{6.57}$$

In view of Eqs. (6.56) and (6.57), Eq. (6.55) reduces to

$$(\lambda_i - \lambda_j) \int_V W_i M[W_j] \, dV = 0 \tag{6.58}$$

Since λ_i and λ_j are distinct, Eq. (6.58) yields

$$\int_V W_i M[W_j] \, dV = 0 \qquad \text{for } \lambda_i \neq \lambda_j \tag{6.59}$$

Equations (6.59) and (6.53) can be used to obtain

$$\int_V W_i L[W_j] \, dV = 0 \qquad \text{for } \lambda_i \neq \lambda_j \tag{6.60}$$

Equations (6.59) and (6.60) are known as the *generalized orthogonality conditions* and the eigenfunctions $W_i(\vec{X})$ and $W_j(\vec{X})$ are considered to be orthogonal in a generalized sense. The eigenfunctions $W_i(\vec{X})$ can be normalized with respect to $M[W_i]$ by setting

$$\int_V W_i(\vec{X}) M[W_i(\vec{X})] \, dV = 1, \qquad i = 1, 2, \ldots \tag{6.61}$$

Equation (6.61) basically specifies the amplitude of the eigenfunction $W_i(\vec{X})$; without this normalization, the amplitude of the function $W_i(\vec{X})$ remains arbitrary. If $M[W_i(\vec{X})]$ denotes the mass distribution $M(\vec{X})$, the orthogonality condition can be written as

$$\int_V M(\vec{X}) W_i(\vec{X}) W_j(\vec{X}) \, dV = 0 \qquad \text{for } \lambda_i \neq \lambda_j \tag{6.62}$$

In this case, the functions $\sqrt{M(\vec{X})} W_i(\vec{X})$ and $\sqrt{M(\vec{X})} W_j(\vec{X})$ are considered to be orthogonal in the usual sense.

6.5.3 Expansion Theorem

As in the case of the Sturm–Liouville problem, the eigenfunctions constitute a complete set in the sense that any function $\tilde{f}(\vec{X})$ that satisfies the homogeneous boundary conditions of the problem can be represented by a linear combination of the eigenfunctions $W_m(\vec{X})$ of the problem as

$$\tilde{f}(\vec{X}) = \sum_{m=1}^{\infty} c_m W_m(\vec{X}) \tag{6.63}$$

where the coefficients c_m can be determined as in the case of Eq. (6.35) as

$$c_m = \frac{\int_V \tilde{f}(\vec{X}) M[W_m(\vec{X})]\, dV}{\|W_m(\vec{X})\|^2}, \qquad m = 1, 2, \ldots \tag{6.64}$$

Equation (6.63), also known as the *expansion theorem*, plays an important role in vibration analysis and is commonly used to find the forced vibration response of a system by modal analysis.

6.6   SOLUTION OF NONHOMOGENEOUS EQUATIONS

The equation of motion of a continuous system subjected to external forces leads to a nonhomogeneous partial differential equation given by Eq. (6.1):

$$M(\vec{X})\frac{\partial^2 w(\vec{X}, t)}{\partial t^2} + L[w(\vec{X}, t)] = f(\vec{X}, t) + \sum_{j=1}^{s} F_j(t)\delta(\vec{X} - \vec{X}_j), \qquad \vec{X} \in V \tag{6.65}$$

subject to the boundary conditions indicated in Eq. (6.12):

$$A_i[w(\vec{X}, t)] = 0, \qquad i = 1, 2, \ldots, p \tag{6.66}$$

and initial conditions similar to those given by Eqs. (6.17) and (6.18):

$$w(\vec{X}, 0) = f(\vec{X}) \tag{6.67}$$

$$\frac{\partial w}{\partial t}(\vec{X}, 0) = g(\vec{X}) \tag{6.68}$$

To find the solution or response of the system, $w(\vec{X}, t)$, we use a procedure known as *modal analysis*. This procedure involves the following steps:

1. Solve the eigenvalue problem associated with Eqs. (6.65) and (6.66). The eigenvalue problem consists of the differential equation

$$L[W(\vec{X})] = \lambda M[W(\vec{X})], \qquad \vec{X} \in V \tag{6.69}$$

with the boundary conditions

$$A_i[W(\vec{X})] = 0, \qquad i = 1, 2, \ldots, p \tag{6.70}$$

where $\lambda = \omega^2$ is the eigenvalue, $W(\vec{X})$ is the eigenfunction, and ω is the natural frequency of the system. The solution of Eqs. (6.69) and (6.70) yields an infinite set of eigenvalues $\lambda_1, \lambda_2, \ldots$ and the corresponding eigenfunctions, also known as *mode shapes*, $W_1(\vec{X}), W_2(\vec{X}), \ldots$. The eigenfunctions are orthogonal, so that

$$\int_V M(\vec{X}) W_m(\vec{X}) W_n(\vec{X})\, dV = \delta_{mn} \tag{6.71}$$

2. Normalize the eigenfunctions so that

$$\int_V W_m(\vec{X}) L[W_n(\vec{X})]\, dV = \omega_m^2 \delta_{mn} \tag{6.72}$$

3. Express the forced response of the system [i.e., the solution of the problem in Eqs. (6.65)–(6.68)] using the expansion theorem as

$$w(\vec{X}, t) = \sum_{m=1}^{\infty} W_m(\vec{X}) \eta_m(t) \tag{6.73}$$

where the $\eta_m(t)$ are known as the time-dependent generalized coordinates. In Eq. (6.73), the eigenfunctions $W_m(\vec{X})$ are known from steps 1 and 2, while the generalized coordinates $\eta_m(t)$ are unknown and to be determined by satisfying the equation of motion and the initial conditions of Eqs. (6.67) and (6.68). To determine $\eta_m(t)$, we substitute Eq. (6.73) in Eq. (6.65) to obtain

$$M(\vec{X}) \frac{\partial^2}{\partial t^2} \left[\sum_{m=1}^{\infty} W_m(\vec{X}) \eta_m(t) \right] + L \left[\sum_{m=1}^{\infty} W_m(\vec{X}) \eta_m(t) \right]$$

$$= f(\vec{X}, t) + \sum_{j=1}^{s} F_j(t) \delta(\vec{X} - \vec{X}_j) \tag{6.74}$$

which can be rewritten as

$$\sum_{m=1}^{\infty} \ddot{\eta}_m(t) M(\vec{X}) W_m(\vec{X}) + \sum_{m=1}^{\infty} \eta_m(t) L[W_m(\vec{X})]$$

$$= f(\vec{X}, t) + \sum_{j=1}^{s} F_j(t) \delta(\vec{X} - \vec{X}_j) \tag{6.75}$$

where $\ddot{\eta}_m(t) = d^2 \eta_m(t)/dt^2$. By multiplying Eq. (6.75) by $W_n(\vec{X})$ and integrating the result over the domain V, we obtain

$$\sum_{m=1}^{\infty} \ddot{\eta}_m(t) \int_V W_n(\vec{X}) M(\vec{X}) W_m(\vec{X}) \, dV$$

$$+ \sum_{m=1}^{\infty} \eta_m(t) \int_V W_n(\vec{X}) L[W_m(\vec{X})] \, dV$$

$$= \int_V W_n(\vec{X}) f(\vec{X}, t) \, dV + \sum_{j=1}^{s} \int_V W_n(\vec{X}) F_j(t) \delta(\vec{X} - \vec{X}_j) \, dV \tag{6.76}$$

Using the property of Dirac delta function given by Eq. (6.2), the last term of Eq. (6.76) can be simplified as

$$\sum_{j=1}^{s} W_n(\vec{X}_j) F_j(t) \tag{6.77}$$

In view of Eqs. (6.71), (6.72), and (6.77), Eq. (6.76) can be rewritten as

$$\ddot{\eta}_m(t) + \omega_m^2 \eta_m(t) = Q_m(t), \qquad m = 1, 2, \ldots \tag{6.78}$$

where $Q_m(t)$ is called the mth *generalized force*, given by

$$Q_m(t) = \int_V W_m(\vec{X}) f(\vec{X}, t) \, dV + \sum_{j=1}^{s} W_m(\vec{X}_j) F_j(t) \qquad (6.79)$$

Equation (6.78) denotes an infinite set of uncoupled second-order ordinary differential equations. A typical equation in (6.78) can be seen to be similar to the equation of a single-degree-of-freedom system [see Eq. (2.107)]. The solution of Eq. (6.78) can be expressed as [see Eq. (2.109)]

$$\eta_m(t) = \frac{1}{\omega_m} \int_0^t Q_m(\tau) \sin \omega_m(t - \tau) \, d\tau$$

$$+ \eta_m(0) \cos \omega_m t + \dot{\eta}_m(0) \frac{\sin \omega_m t}{\omega_m}, \qquad m = 1, 2, \ldots \qquad (6.80)$$

where $\eta_m(0)$ and $\dot{\eta}_m(0)$ are the initial values of the generalized coordinate (generalized displacement) $\eta_m(t)$ and the time derivative of the generalized coordinate (generalized velocity) $\dot{\eta}_m(t) = d\eta_m(t)/dt$. Using the initial conditions of Eqs. (6.67) and (6.68), the values of $\eta_m(0)$ and $\dot{\eta}_m(0)$ can be determined as

$$\eta_m(0) = \int_V M(\vec{X}) W_m(\vec{X}) w(\vec{X}, 0) \, dV$$

$$= \int_V M(\vec{X}) W_m(\vec{X}) f(\vec{X}) \, dV \qquad (6.81)$$

$$\dot{\eta}_m(0) = \int_V M(\vec{X}) W_m(\vec{X}) \dot{w}(\vec{X}, 0) \, dV$$

$$= \int_V M(\vec{X}) W_m(\vec{X}) g(\vec{X}) \, dV \qquad (6.82)$$

Finally, the solution of the problem (i.e., the forced response of the system) can be found using Eqs. (6.80) and (6.73).

6.7 FORCED RESPONSE OF VISCOUSLY DAMPED SYSTEMS

Consider the vibration of a viscously damped continuous system. We assume the damping force, F_d, resisting the motion of the system to be proportional to the velocity and opposite to the direction of the velocity, similar to the case of a discrete system:

$$F_d(\vec{X}, t) = -C \frac{\partial w(\vec{X}, t)}{\partial t} = -\frac{\partial}{\partial t} C[w(\vec{X}, t)] \qquad (6.83)$$

where C is a linear homogeneous differential operator, similar to the operator L, composed of derivatives with respect to the spatial coordinates $\vec{X}$ (but not with respect to time t) of order up to $2p$. Thus, the equation of motion of the viscously damped system can be expressed, similar to Eq. (6.65), as [3, 4]

$$M(\vec{X})\frac{\partial^2 w(\vec{X}, t)}{\partial t^2} + \frac{\partial}{\partial t}C[w(\vec{X}, t)] + L[w(\vec{X}, t)]$$

$$= f(\vec{X}, t) + \sum_{j=1}^{s} F_j(t)\delta(\vec{X} - \vec{X}_j), \qquad \vec{X} \in V \qquad (6.84)$$

subject to the homogeneous boundary conditions

$$A_i[w(\vec{X}, t)] = 0, \qquad i = 1, 2, \ldots, p \qquad (6.85)$$

and the initial conditions

$$w(\vec{X}, 0) = f(\vec{X}) \qquad (6.86)$$

$$\frac{\partial w}{\partial t}(\vec{X}, 0) = g(\vec{X}) \qquad (6.87)$$

For the undamped system, we find the eigenvalues λ_m and the corresponding eigenfunctions $W_m(\vec{X})$ by solving the eigenvalue problem

$$L[W(\vec{X})] = \lambda M[W(\vec{X})], \qquad \vec{X} \in V \qquad (6.88)$$

subject to the boundary conditions

$$A_i[W(\vec{X})] = 0, \qquad i = 1, 2, \ldots, p \qquad (6.89)$$

The orthogonal eigenfunctions are assumed to be normalized according to Eqs. (6.71) and (6.72). As in the case of an undamped system, the damped response of the system is assumed to be a sum of the products of eigenfunctions and time-dependent generalized coordinates $\eta_m(t)$, using the expansion theorem, as

$$w(\vec{X}, t) = \sum_{m=1}^{\infty} W_m(\vec{X})\eta_m(t) \qquad (6.90)$$

Substitution of Eq. (6.90) into Eq. (6.84) leads to

$$\sum_{m=1}^{\infty} \ddot{\eta}_m(t)M(\vec{X})W_m(\vec{X}) + \sum_{m=1}^{\infty} \dot{\eta}_m(t)C[W_m(\vec{X})] + \sum_{m=1}^{\infty} \eta_m(t)L[W_m(\vec{X})]$$

$$= f(\vec{X}, t) + \sum_{j=1}^{s} F_j(t)\delta(\vec{X} - \vec{X}_j), \qquad \vec{X} \in V \qquad (6.91)$$

By multiplying Eq. (6.91) by $W_n(\vec{X})$ and integrating the result over the domain of the system V, and using Eqs. (6.71) and (6.72), we obtain

$$\ddot{\eta}_m(t) + \sum_{n=1}^{\infty} c_{mn}\dot{\eta}_m(t) + \omega_m^2\eta_m(t) = Q_m(t), \qquad m = 1, 2, \ldots \qquad (6.92)$$

where c_{mn}, known as the *viscous damping coefficients*, are given by

$$c_{mn} = \int_V W_m(\vec{X})C[W_n(\vec{X})]\,dV \qquad (6.93)$$

and $Q_m(t)$, called the *generalized forces*, are given by

$$Q_m(t) = \int_V W_m(\vec{X}) f(\vec{X}, t) \, dV + \sum_{j=1}^{s} W_m(\vec{X}_j) F_j(t) \tag{6.94}$$

In many practical situations, the viscous damping operator C is not known. To simplify the analysis, the operator C is assumed to be a linear combination of the operator L and the mass function M:

$$C = \alpha_1 L + \alpha_2 M \tag{6.95}$$

where α_1 and α_2 are constants. With this assumption, the viscous damping coefficients can be expressed as

$$c_{mn} = c_{mn}\delta_{mn} = 2\zeta_m \omega_m \delta_{mn} \tag{6.96}$$

where ζ_m is called the *damping ratio*. Introducing Eq. (6.96) into Eq. (6.92), we obtain a set of uncoupled second-order ordinary differential equations:

$$\ddot{\eta}_m(t) + 2\zeta_m \omega_m \dot{\eta}_m(t) + \omega_m^2 \eta_m(t) = Q_m(t), \qquad m = 1, 2, \ldots \tag{6.97}$$

Equations (6.97) are similar to those of a viscously damped single-degree-of-freedom system [see Eq. (2.119)]. The solution of Eq. (6.97) is given by [see Eq. (2.120)]

$$\eta_m(t) = \int_0^t Q_m(\tau) h(t - \tau) \, d\tau + g(t)\eta_m(0) + h(t)\dot{\eta}_m(0) \tag{6.98}$$

where

$$h(t) = \frac{1}{\omega_{dm}} e^{-\zeta_m \omega_m t} \sin \omega_{dm} t \tag{6.99}$$

$$g(t) = e^{-\zeta_m \omega_m t} \left(\cos \omega_{dm} t + \frac{\zeta_m \omega_m}{\omega_{dm}} \sin \omega_{dm} t \right) \tag{6.100}$$

and ω_{dm} is the mth frequency of damped vibration given by

$$\omega_{dm} = \sqrt{1 - \zeta_m^2} \, \omega_m \tag{6.101}$$

where the system is assumed to be underdamped. Once the $\eta_m(t)$ are known, the solution of the original equation (6.84) can be found from Eq. (6.90).

6.8   RECENT CONTRIBUTIONS

A discussion of the various methods of physical modeling and a brief survey of the direct solution techniques for a class of linear vibration systems, including discrete as well as distributed parameter systems, have been presented by Chen [5]. The discussion of discrete systems includes close-coupled, far-coupled, and branched systems. The discussion of continuous systems includes one-dimensional problems of vibrating strings and beams and two-dimensional problems of vibrating membranes and plates. Anderson and Thomas discussed three methods for solving boundary value problems that have both time derivatives of the dependent variable and known time-dependent functions in the boundary conditions [6]. In these methods, the time dependence is eliminated from the boundary conditions by decomposing the solution into a quasistatic part and a dynamic part. The boundary conditions containing the time derivatives of the dependent variable are satisfied identically by imposing special requirements on the quasistatic portion of the complete solution. The method is illustrated with a problem that deals with the forced thickness–stretch vibrations of an elastic plate.

Anderson [7] investigated the forced vibrations of two elastic bodies having a surface contact within the framework of the classical linear theory of elasticity. The generalized orthogonality condition and a simple form of the generalized forces are derived. The procedure is illustrated by considering the example of the forced thickness–stretch vibration of a two-layer plate system.

As indicated earlier, the modal analysis, based on eigenfunction expansion, is a commonly used technique for the transient analysis of continuous systems. However, the conventional modal expansion is not directly applicable to non-self-adjoint systems whose eigenfunctions are nonorthogonal. In Ref. [8], an exact closed-form solution method was presented for transient analysis of general one-dimensional distributed systems that have non-self-adjoint operators, and eigenvalue-dependent boundary conditions are subject to arbitrary external, initial, and boundary disturbances. In this reference, an eigenfunction series solution is derived through introduction of augmented spatial operators and through application of the modal expansion theorem given in Ref. [9]. The method is demonstrated by considering a cantilever beam with end mass, viscous damper, and spring.

Structural intensity can be used to describe the transfer of vibration energy. The spatial distribution of structural intensity within a structure offers information on energy transmission paths and positions of sources and sinks of mechanical energy. Gavric et al. [10] presented a method for the measurement of structural intensity using a normal mode approach. The method is tested on an assembly of two plates.

REFERENCES

1. E. Kreyszig, *Advanced Engineering Mathematics*, 8th ed., Wiley, New York, 1999.

2. E. Zauderer, *Partial Differential Equations of Applied Mathematics*, 2nd ed., Wiley, New York, 1989.

3. L. Meirovitch, *Analytical Methods in Vibrations*, Macmillan, New York, 1967.

4. R. Courant and D. Hilbert, *Methods of Mathematical Physics*, Interscience, New York, Vol. I, 1937, Vol. II, 1962.

5. F. Y. Chen, On modeling and direct solution of certain free vibration systems, *Journal of Sound and Vibration*, Vol. 14, No. 1, pp. 57–79, 1971.

6. G. L. Anderson and C. R. Thomas, A forced vibration problem involving time derivatives in the boundary conditions, *Journal of Sound and Vibration*, Vol. 14, No. 2, pp. 193–214, 1971.

7. G. L. Anderson, On the forced vibrations of elastic bodies in contact, *Journal of Sound and Vibration*, Vol. 16, No. 4, pp. 533–549, 1971.

8. B. Yang and X. Wu, Transient response of one-dimensional distributed systems: a closed form eigenfunction expansion realization, *Journal of Sound and Vibration*, Vol. 208, No. 5, pp. 763–776, 1997.

9. B. Yang, Integral formulas for non-self-adjoint distributed dynamic systems, *AIAA Journal*, Vol. 34, No. 10, pp. 2132–2139, 1996.

10. L. Gavric, U. Carlsson, and L. Feng, Measurement of structural intensity using a normal mode approach, *Journal of Sound and Vibration*, Vol. 206, No. 1, pp. 87–101, 1997.

PROBLEMS

6.1 Prove that the left and right sides of Eq. (6.20) must be equal to a negative constant.

6.2 Show that for $\alpha > 0$, Eqs. (E6.1.6) and (E6.1.7) do not have a nonzero solution for c_1 and c_2.

6.3 Convert the following differential equation to Sturm–Liouville form, Eq. (6.23):

$$(1 - x^2)^2 \frac{d^2 w}{dx^2} - 2x(1 - x^2)\frac{dw}{dx}$$
$$+ [\lambda(1 - x^2) - n^2]w = 0$$

6.4 Find the eigenvalues and eigenfunctions of the equation

$$\frac{d^2 w}{dx^2} + \lambda w = 0, \qquad w(0) = 0, \quad w(2\pi) = 0$$

6.5 Determine whether the following functions are orthogonal in the interval $0 \le x \le l$:

$$W_i(x) = \sin\frac{i\pi x}{l}, \qquad i = 1, 2, \ldots$$

6.6 Determine whether the following differential equation is self-adjoint:

$$x^2 \frac{d^2 w}{dx^2} + x\frac{dw}{dx} + (x^2 - m^2)w = 0$$

6.7 Determine whether the following differential equation is self-adjoint:

$$(1 - x^2)\frac{d^2 w}{dx^2} - 2x\frac{dw}{dx} - \lambda w = 0$$

6.8 Consider the Sturm–Liouville equation

$$(xw')' + \left(\omega^2 x - \frac{1}{x}\right)w = 0, \qquad 0 \le x \le 1$$

Determine the bounded solution of the equation subject to the condition $w(1) = 0$.

6.9 Consider the differential equation corresponding to the transverse vibration of a string fixed at $x = 0$ and $x = l$:

$$\frac{d^2 W(x)}{dx^2} + \alpha^2 W(x) = 0, \qquad 0 \le x \le l$$

where $\alpha^2 = \omega^2 \rho/P$, ρ is the mass per unit length and P is the tension in the string. Determine whether each of the following functions is an admissible, comparison, or eigenfunction:

(a) $W(x) = c\sin(\pi x/l)$

(b) $W(x) = cx(x - l)$

(c) $W(x) = cx(2x - l)$

6.10 The eigenvalue problem corresponding to the transverse vibration of a uniform beam is given by

$$EI\frac{d^4 W(x)}{dx^4} = \lambda m W(x)$$

where EI is the bending stiffness, m is the mass per unit length, $W(x)$ is the transverse displacement (eigenfunction or mode shape), and $\lambda = \omega^2$ is the eigenvalue. Assuming the beam to be simply supported at both ends $x = 0$ and $x = L$, show that the problem is self-adjoint by considering the following comparison functions:

$$W_1(x) = C_1\sin\frac{\pi x}{L}, \qquad W_2(x) = C_2 x(2Lx^2 - x^3 - L^3)$$

6.11 A uniform shaft with torsional rigidity GJ is fixed at $x = 0$ and carries a rigid disk of mass polar moment of inertia I_0 at $x = L$. State the boundary conditions of the shaft in torsional vibration at $x = 0$ and $x = L$ and establish that one of the boundary conditions depends on the natural frequency of vibration of the shaft.

7

Solution Procedure: Integral Transform Methods

7.1 INTRODUCTION

Integral transforms are considered to be operational methods or operational calculus methods that are developed for the efficient solution of differential and integral equations. In these methods, the operations of differentiation and integration are symbolized by algebraic operators. Oliver Heaviside (1850–1925) was the first person to develop and use the operational methods for solution of the telegraph equation and the second-order hyperbolic partial differential equations with constant coefficients in 1892 [1]. However, his operational methods were based mostly on intuition and lacked mathematical rigor. Although subsequently, the operational methods have developed into one of most useful mathematical methods, contemporary mathematicians hardly recognized Heaviside's work on operational methods, due to its lack of mathematical rigor.

Subsequently, many mathematicians tried to interpret and justify Heaviside's work. For example, Bromwich and Wagner tried to justify Heaviside's work on the basis of contour integration [2, 3]. Carson attempted to derive the operational method using an infinite integral of the Laplace type [4]. Van der Pol and other mathematicians tried to derive the operational method by employing complex variable theory [5]. All these attempts proved successful in establishing the mathematical validity of the operational method in the early part of the twentieth century. As such, the modern concept of the operational method has a rigorous mathematical foundation and is based on the functional transformation provided by Laplace and Fourier integrals.

In general, if a function $f(t)$, defined in terms of the independent variable t, is governed by a differential equation with certain initial or boundary conditions, the integral transforms convert $f(t)$ into $F(s)$ defined by

$$F(s) = \int_{t_1}^{t_2} f(t) K(s, t) \, dt \tag{7.1}$$

where s is a parameter, $K(s, t)$ is called the *kernal* of the transformation, and t_1 and t_2 are the limits of integration. The transform is said to be finite if t_1 and t_2 are finite. Equation (7.1) is called the *integral transformation* of $f(t)$. It converts a differential equation into an algebraic equation in terms of the new, transformed function $F(s)$. The initial or boundary conditions will be accounted for automatically in the process of conversion to an algebraic equation. The resulting algebraic equation can be solved for $F(s)$ without much difficulty. Once $F(s)$ is known, the original function $f(t)$ can be found by using the inverse integral transformation.

If a function f, defined in terms of two independent variables, is governed by a partial differential equation, the integral transformation reduces the number of independent variables by one. Thus, instead of a partial differential equation, we need to solve only an ordinary differential equation, which is much simpler. A major task in using the integral transform method involves carrying out the inverse transformation. The transform and its inverse are called the *transform pair*. The most commonly used integral transforms are the Fourier and Laplace transforms. The application of both these transforms for the solution of vibration problems is considered in this chapter.

7.2 FOURIER TRANSFORMS

7.2.1 Fourier Series

In Section 1.10 we saw that the Fourier series expansion of a function $f(t)$ that is periodic with period τ and contains only a finite number of discontinuities is given by

$$f(t) = \frac{a_0}{2} + \sum_{n=1}^{\infty} \left(a_n \cos \frac{2n\pi, t}{\tau} + b_n \sin \frac{2n\pi t}{\tau} \right) \tag{7.2}$$

where the coefficients a_n and b_n are given by

$$a_0 = \frac{2}{\tau} \int_{-\tau/2}^{\tau/2} f(t) \, dt$$

$$a_n = \frac{2}{\tau} \int_{-\tau/2}^{\tau/2} f(t) \cos \frac{2n\pi t}{\tau} \, dt, \qquad n = 1, 2, \ldots \tag{7.3}$$

$$b_n = \frac{2}{\tau} \int_{-\tau/2}^{\tau/2} f(t) \sin \frac{2n\pi t}{\tau} \, dt, \qquad n = 1, 2, \ldots.$$

Using the identities

$$\cos \frac{2\pi t}{\tau} = \frac{e^{i(2\pi t/\tau)} + e^{-i(2\pi t/\tau)}}{2}, \qquad \sin \frac{2\pi t}{\tau} = \frac{e^{i(2\pi t/\tau)} - e^{-i(2\pi t/\tau)}}{2i} \tag{7.4}$$

Eq. (7.2) can be expressed as

$$f(t) = \frac{a_0}{2} + \sum_{n=1}^{\infty} \left[a_n \frac{e^{i(n \cdot 2\pi t/\tau)} + e^{-i(n \cdot 2\pi t/\tau)}}{2} + b_n \frac{e^{i(n \cdot 2\pi t/\tau)} - e^{-i(n \cdot 2\pi t/\tau)}}{2i} \right]$$

$$= e^{i(0)(2\pi t/\tau)} \left(\frac{a_0}{2} - \frac{i b_0}{2} \right) + \sum_{n=1}^{\infty} \left[e^{in \cdot 2\pi t/\tau} \left(\frac{a_n}{2} - \frac{i b_n}{2} \right) + e^{-in \cdot 2\pi t/\tau} \left(\frac{a_n}{2} + \frac{i b_n}{2} \right) \right] \tag{7.5}$$

where $b_0 = 0$. By defining the complex Fourier coefficients c_n and c_{-n} as

$$c_n = \frac{a_n - ib_n}{2}, \qquad c_{-n} = \frac{a_n + ib_n}{2} \tag{7.6}$$

Eq. (7.5) can be expressed as

$$f(t) = \sum_{n=-\infty}^{\infty} c_n e^{in \cdot 2\pi t/\tau} \tag{7.7}$$

where the Fourier coefficients c_n can be determined using Eqs. (7.3) as

$$
\begin{aligned}
c_n &= \frac{a_n - ib_n}{2} \\
&= \frac{1}{\tau} \int_{-\tau/2}^{\tau/2} f(t) \left(\cos \frac{n \cdot 2\pi t}{\tau} - i \sin \frac{n \cdot 2\pi t}{\tau} \right) dt = \frac{1}{\tau} \int_{-\tau/2}^{\tau/2} f(t) e^{-in(2\pi t/\tau)} \, dt
\end{aligned}
\tag{7.8}
$$

Using Eq. (7.8), Eq. (7.7) can be written as

$$f(t) = \sum_{n=-\infty}^{\infty} \frac{e^{in \cdot 2\pi t/\tau}}{\tau} \int_{-\tau/2}^{\tau/2} f(t) e^{-in(2\pi t/\tau)} \, dt \tag{7.9}$$

7.2.2 Fourier Transforms

When the period of the periodic function $f(t)$ in Eq. (7.9) is extended to infinity, the expansion will be applicable to nonperiodic functions as well. For this, let $\omega_n = n \cdot 2\pi/\tau$ and $\Delta\omega_n = n\omega_0 - (n-1)\omega_0 = 2\pi/\tau$. As $\tau \to \infty$, $\Delta\omega_n \to d\omega \to 0$ and the subscript n need not be used since the discrete value of ω_n becomes continuous. By using the relations $n\omega_0 = \omega$ and $d\omega = 2\pi/\tau$ as $\tau \to \infty$, Eq. (7.9) becomes

$$
\begin{aligned}
f(t) &= \lim_{\tau \to \infty} \sum_{n=-\infty}^{\infty} \frac{1}{\tau} e^{in(2\pi t/\tau)} \int_{-\tau/2}^{\tau/2} f(t) e^{-in(2\pi t/\tau)} \, dt \\
&= \frac{1}{2\pi} \int_{-\infty}^{\infty} e^{-i\omega t} \int_{-\infty}^{\infty} f(t) e^{i\omega t} \, dt \, d\omega
\end{aligned}
\tag{7.10}
$$

Equation (7.10), called the *Fourier integral*, is often expressed in the form of the following Fourier transform pair:

$$F(\omega) = \int_{-\infty}^{\infty} f(t) e^{i\omega t} \, dt \tag{7.11}$$

$$f(t) = \frac{1}{2\pi} \int_{-\infty}^{\infty} F(\omega) e^{-i\omega t} \, d\omega \tag{7.12}$$

where $F(\omega)$ is called the *Fourier transform* of $f(t)$ and $f(t)$ is called the *inverse Fourier transform* of $F(\omega)$. In Eq. (7.12), $F(\omega) \, d\omega$ can be considered as the harmonic contribution of the function $f(t)$ in the frequency range ω to $\omega + d\omega$. This also denotes the limiting value of c_n as $\tau \to \infty$, as indicated by Eq. (7.8). Thus, Eq. (7.12) denotes an infinite sum of harmonic oscillations in which all frequencies from $-\infty$ to ∞ are represented.

Notes

1. By rewriting Eq. (7.10) as

$$f(t) = \frac{1}{\sqrt{2\pi}} \int_{-\infty}^{\infty} \left[\frac{1}{\sqrt{2\pi}} \int_{-\infty}^{\infty} f(t)e^{-i\omega t} \, dt \right] e^{i\omega t} \, d\omega \tag{7.13}$$

the Fourier transform pair can be defined in a symmetric form as

$$F(\omega) = \frac{1}{\sqrt{2\pi}} \int_{-\infty}^{\infty} f(t)e^{-i\omega t} \, dt \tag{7.14}$$

$$f(t) = \frac{1}{\sqrt{2\pi}} \int_{-\infty}^{\infty} F(\omega)e^{i\omega t} \, d\omega \tag{7.15}$$

It is also possible to define the Fourier transform pair as

$$F(\omega) = \frac{1}{\sqrt{2\pi}} \int_{-\infty}^{\infty} f(t)e^{i\omega t} \, dt \tag{7.16}$$

$$f(t) = \frac{1}{\sqrt{2\pi}} \int_{-\infty}^{\infty} F(\omega)e^{-i\omega t} \, d\omega \tag{7.17}$$

2. The Fourier transform pair corresponding to an even function $f(t)$ can be defined as follows:

$$F(\omega) = \int_{0}^{\infty} f(t) \cos \omega t \, dt \tag{7.18}$$

$$f(t) = \frac{2}{\pi} \int_{0}^{\infty} F(\omega) \cos \omega t \, d\omega \tag{7.19}$$

The Fourier sine transform pair corresponding to an odd function $f(t)$ can be defined as

$$F(\omega) = \int_{0}^{\infty} f(t) \sin \omega t \, dt \tag{7.20}$$

$$f(t) = \frac{2}{\pi} \int_{0}^{\infty} F(\omega) \sin \omega t \, d\omega \tag{7.21}$$

3. The Fourier transform pair is applicable only to functions $f(t)$ that satisfy Dirichlet's conditions in the range $(-\infty, \infty)$. A function $f(t)$ is said to satisfy Dirichlet's conditions in the interval (a, b) if (a) $f(t)$ has only a finite number of maxima and minima in (a, b) and (b) $f(t)$ has only a finite number of finite discontinuities with no infinite discontinuity in (a, b). As an example, the function $f(t) = t/(1 + t^2)$ satisfies Dirichlet's conditions in the interval $(-\infty, \infty)$, whereas the function $f(t) = 1/(1 - t)$ does not satisfy Dirichlet's conditions in any interval containing the point $t = 1$ because $f(t)$ has an infinite discontinuity at $t = 1$.

7.2.3   Fourier Transform of Derivatives of Functions

Let the Fourier transform of the jth derivative of the function $f(t)$ be denoted as $F^{(j)}(\omega)$. Then, by using the definition of Eq. (7.11),

$$F^{(j)}(\omega) = \int_{-\infty}^{\infty} \frac{d^j f(t)}{dt^j} e^{i\omega t}\, dt = e^{i\omega t} \frac{d^{j-1} f(t)}{dt^{j-1}} \bigg|_{-\infty}^{\infty} - i\,\omega\, F^{(j-1)}(\omega) \tag{7.22}$$

Assuming that the $(j-1)$st derivative of $f(t)$ is zero as $t \to \pm\infty$, Eq. (7.22) reduces to

$$F^{(j)}(\omega) = -i\,\omega\, F^{(j-1)}(\omega) \tag{7.23}$$

Again assuming that all derivatives of order $1, 2, \ldots, j-1$ are zero as $t \to \pm\infty$, Eq. (7.23) yields

$$F^{(j)}(\omega) = (-i\,\omega)^j\, F(\omega) \tag{7.24}$$

where $F(\omega)$ is the complex Fourier transform of $f(t)$ given by Eq. (7.11).

7.2.4   Finite Sine and Cosine Fourier Transforms

The Fourier series expansion of a function $f(t)$ in the interval $0 \le t \le \pi$ is given by [using Eq. (1.32)]

$$f(t) = \frac{a_0}{\pi} + \frac{2}{\pi} \sum_{n=1}^{\infty} a_n \cos nt \tag{7.25}$$

where

$$a_n = \int_0^{\pi} f(t) \cos nt\, dt \tag{7.26}$$

Using Eqs. (7.25) and (7.26), the finite cosine Fourier transform pair is defined as

$$F(n) = \int_0^{\pi} f(t) \cos nt\, dt \tag{7.27}$$

$$f(t) = \frac{F(0)}{\pi} + \frac{2}{\pi} \sum_{n=1}^{\infty} F(n) \cos nt \tag{7.28}$$

A similar procedure can be used to define the finite sine Fourier transforms. Starting with the Fourier sine series expansion of a function $f(t)$ defined in the interval $0 \le t \le \pi$ [using Eq. (7.32)], we obtain

$$f(t) = \frac{2}{\pi} \sum_{n=1}^{\infty} b_n \sin nt \tag{7.29}$$

where

$$b_n = \int_0^{\pi} f(t) \sin nt\, dt \tag{7.30}$$

the finite sine Fourier transform pair is defined as

$$F(n) = \int_0^\pi f(t) \sin nt \, dt \tag{7.31}$$

$$f(t) = \frac{2}{\pi} \sum_{n=1}^\infty F(n) \sin nt \tag{7.32}$$

When the independent variable t is defined in the range $(0,a)$ instead of $(0,\pi)$, the finite cosine transform is defined as

$$F(n) = \int_0^a f(t) \cos \xi t \, dt \tag{7.33}$$

where ξ is yet unspecified. Defining a new variable y as $y = \pi t/a$ so that $dy = (\pi/a)dt$, Eq. (7.33) can be rewritten as

$$F(n) = \frac{a}{\pi} \int_0^\pi \overline{f}(y) \cos \left(\xi \frac{ya}{\pi} \right) dy \tag{7.34}$$

where

$$\overline{f}(y) = f \left(\frac{ya}{\pi} \right) \tag{7.35}$$

If $\xi a/\pi = n$ or $\xi = n\pi/a$, then

$$\frac{a}{\pi} \overline{f}(y) = \frac{1}{\pi} F(0) + \frac{2}{\pi} \sum_{n=1}^\infty F(n) \cos ny \tag{7.36}$$

Returning to the original variable t, we define the finite cosine transform pair as

$$F(n) = \int_0^a f(t) \cos \frac{n\pi t}{a} \, dt \tag{7.37}$$

$$f(t) = \frac{F(0)}{a} + \frac{2}{a} \sum_{n=1}^\infty F(n) \cos \frac{n\pi t}{a} \tag{7.38}$$

Similarly, the finite sine Fourier transform pair is defined as

$$F(n) = \int_0^a f(t) \sin \frac{n\pi t}{a} \, dt \tag{7.39}$$

$$f(t) = \frac{2}{a} \sum_{n=1}^\infty F(n) \sin \frac{n\pi t}{a} \tag{7.40}$$

Example 7.1 Find the Fourier transform of the function

$$f(x) = \begin{cases} a, & 0 < x < a \\ 0, & x > a \end{cases} \tag{E7.1.1}$$

SOLUTION   The Fourier transform of $f(x)$, as defined by Eq. (7.14), is given by

$$F(\omega) = \frac{1}{\sqrt{2\pi}} \int_{-\infty}^{\infty} f(x) e^{-i\omega x} \, dx$$

$$= \frac{1}{\sqrt{2\pi}} \int_{0}^{a} a e^{-i\omega x} \, dx = \frac{a}{\sqrt{2\pi}} \left(\frac{e^{-i\omega x}}{-i\omega} \right)_{0}^{a}$$

$$= \frac{a(1 - e^{-i\omega a})}{i\omega\sqrt{2\pi}} \tag{E7.1.2}$$

Example 7.2   Find the Fourier transform of the function

$$f(x) = c_1 f_1(x) + c_2 f_2(x)$$

where c_1 and c_2 are constants.

SOLUTION   The Fourier transform of $f(x)$ can be found using Eq. (7.14) as

$$F(\omega) = \frac{1}{\sqrt{2\pi}} \int_{-\infty}^{\infty} f(x) e^{-i\omega x} \, dx$$

$$= \frac{c_1}{\sqrt{2\pi}} \int_{-\infty}^{\infty} f_1(x) e^{-i\omega x} \, dx + \frac{c_2}{\sqrt{2\pi}} \int_{-\infty}^{\infty} f_2(x) e^{-i\omega x} \, dx$$

$$= c_1 F_1(\omega) + c_2 F_2(\omega) \tag{E7.2.1}$$

This shows that Fourier transform is a linear operation; that is, the Fourier transform of a linear sum of a set of functions is equal to the linear sum of the Fourier transforms of the individual functions.

Example 7.3   Find the Fourier transform of the function $f(ax)$, where a is a positive constant.

SOLUTION   The Fourier transform of $f(ax)$ is given by [Eq. (7.14)]

$$\frac{1}{\sqrt{2\pi}} \int_{-\infty}^{\infty} f(ax) e^{-i\omega x} \, dx \tag{E7.3.1}$$

By introducing a new variable t as $t = ax$ so that $dt = a\,dx$, the expression (E7.3.1) can be rewritten as

$$\frac{1}{a\sqrt{2\pi}} \int_{-\infty}^{\infty} f(t) e^{-i\omega t/a} \, dt \tag{E7.3.2}$$

Thus, the Fourier transform of $f(ax)$ is given by

$$\frac{1}{a} F\left(\frac{\omega}{a}\right), \qquad a > 0 \tag{E7.3.3}$$

7.3 FREE VIBRATION OF A FINITE STRING

Consider a string of length l under tension P and fixed at the two endpoints $x = 0$ and $x = l$. The equation of motion governing the transverse vibration of the string is given by

$$c^2 \frac{\partial^2 w(x, t)}{\partial x^2} = \frac{\partial^2 w(x, t)}{\partial t^2}; 0 \le x \le l \tag{7.41}$$

By redefining the spatial coordinate x in terms of p as

$$p = \frac{x\pi}{l} \tag{7.42}$$

Eq. (7.41) can be rewritten as

$$\frac{\pi^2 c^2}{l^2} \frac{\partial^2 w(p, t)}{\partial x^2} = \frac{\partial^2 w(p, t)}{\partial t^2}, \qquad 0 \le p \le \pi \tag{7.43}$$

We now take finite sine transform of Eq. (7.43). According to Eq. (7.31), we multiply Eq. (7.43) by $\sin np$ and integrate with respect to p from 0 to π:

$$\frac{\pi^2 c^2}{l^2} \int_0^\pi \frac{\partial^2 w}{\partial p^2} \sin np \, dp = \int_0^\pi \frac{\partial^2 w}{\partial t^2} \sin np \, dp \tag{7.44}$$

where

$$\int_0^\pi \frac{\partial^2 w(p, t)}{\partial p^2} \sin np \, dp = \left(\frac{\partial w}{\partial p} \sin np - nw \cos np \right) \Big|_0^\pi - n^2 \int_0^\pi w \sin np \, dp \tag{7.45}$$

Since the string is fixed at $p = 0$ and $p = \pi$, the first term on the right-hand side of Eq. (7.45) vanishes, so that

$$\int_0^\pi \frac{\partial^2 w(p, t)}{\partial p^2} \sin np \, dp = -n^2 \int_0^\pi w \sin np \, dp \tag{7.46}$$

Thus, Eq. (7.44) becomes

$$-\frac{n^2 \pi^2 c^2}{l^2} \int_0^\pi w \sin np \, dp = \frac{\partial^2}{\partial t^2} \int_0^\pi w \sin np \, dp \tag{7.47}$$

Defining the finite Fourier sine transform of $w(p, t)$ as [see Eq. (7.31)]

$$W(n, t) = \int_0^\pi w(p, t) \sin np \, dp \tag{7.48}$$

Eq. (7.47) can be expressed as an ordinary differential equation as

$$\frac{d^2 W(n, t)}{dt^2} + \frac{\pi^2 c^2 n^2}{l^2} W(n, t) = 0 \tag{7.49}$$

The solution of Eq. (7.49) is given by

$$W(n, t) = \tilde{C}_1 e^{i(\pi cn/l)t} + \tilde{C}_2 e^{-i(\pi cn/l)t}$$

or

$$W(n, t) = C_1 \cos \frac{\pi c n t}{l} + C_2 \sin \frac{\pi c n t}{l} \qquad (7.50)$$

where the constants $\tilde{C}_1$ and $\tilde{C}_2$ or C_1 and C_2 can be determined from the known initial conditions of the string.

Let the initial conditions of the string be given by

$$w(x, t = 0) = w_0(x) \qquad (7.51)$$

$$\frac{\partial w}{\partial t}(x, t = 0) = \dot{w}_0(x) \qquad (7.52)$$

In terms of the finite Fourier sine transform $W(n, t)$ defined by Eq. (7.48), Eqs. (7.51) and (7.52) can be expressed as

$$W(n, t = 0) = W_0(n) \qquad (7.53)$$

$$\frac{dW}{dt}(n, t = 0) = \dot{W}_0(n) \qquad (7.54)$$

where

$$W_0(n) = \int_0^\pi w_0(p) \sin np \, dp \qquad (7.55)$$

or

$$W_0(n) = \frac{\pi}{l} \int_0^l w_0(\xi) \sin \frac{n\pi\xi}{l} \, d\xi \qquad (7.56)$$

$$\dot{W}_0(n) = \int_0^\pi \dot{w}_0(p) \sin np \, dp \qquad (7.57)$$

or

$$\dot{W}_0(n) = \frac{\pi}{l} \int_0^l \dot{w}_0(\xi) \sin \frac{n\pi\xi}{l} \, d\xi \qquad (7.58)$$

Equations (7.53), (7.54), and (7.50) lead to

$$C_1 = W_0(n) \qquad (7.59)$$

$$C_2 = \frac{l}{nc\pi} \dot{W}_0(n) \qquad (7.60)$$

Thus, the solution, Eq. (7.50), becomes

$$W(n, t) = W_0(n) \cos \frac{n\pi ct}{l} + \frac{l}{n\pi c} \dot{W}_0(n) \sin \frac{n\pi ct}{l} \qquad (7.61)$$

The inverse finite Fourier sine transform of $W(n, t)$ is given by [see Eq. (7.32)]

$$w(p, t) = \frac{2}{\pi} \sum_{n=1}^{\infty} W(n, t) \sin np \qquad (7.62)$$

Substituting Eq. (7.61) into (7.62), we obtain

$$w(p, t) = \frac{2}{\pi} \sum_{n=1}^{\infty} W_0(n) \cos \frac{n\pi ct}{l} \sin np + \frac{2l}{\pi^2 c} \sum_{n=1}^{\infty} \frac{1}{n} \dot{W}_0(n) \sin \frac{n\pi ct}{l} \sin np \quad (7.63)$$

Using Eqs. (7.56) and (7.58), Eq. (7.63) can be expressed in terms of x and t as

$$w(x, t) = \frac{2}{l} \sum_{n=1}^{\infty} \sin \frac{n\pi x}{l} \cos \frac{n\pi ct}{l} \int_0^l w_0(\xi) \sin \frac{n\pi \xi}{l} \, d\xi$$

$$+ \frac{2}{\pi c} \sum_{n=1}^{\infty} \frac{1}{n} \sin \frac{n\pi x}{l} \sin \frac{n\pi ct}{l} \int_0^l \dot{w}_0(\xi) \sin \frac{n\pi \xi}{l} \, d\xi \quad (7.64)$$

7.4   FORCED VIBRATION OF A FINITE STRING

Consider a string of length l under tension P, fixed at the two endpoints $x = 0$ and $x = l$, and subjected to a distributed transverse force $\tilde{f}(x, t)$. The equation of motion of the string is given by [see Eq. (8.7)]

$$P \frac{\partial^2 w(x, t)}{\partial x^2} + \tilde{f}(x, t) = \rho \frac{\partial^2 w(x, t)}{\partial t^2} \quad (7.65)$$

or

$$c^2 \frac{\partial^2 w}{\partial x^2} + f(x, t) = \frac{\partial^2 w}{\partial t^2} \quad (7.66)$$

where

$$f(x, t) = \frac{\tilde{f}(x, t)}{\rho} \quad (7.67)$$

As in Section 7.3 we change the spatial variable x to p as

$$p = \frac{x\pi}{l} \quad (7.68)$$

so that Eq. (7.66) can be written as

$$\frac{\pi^2}{l^2} \frac{\partial^2 w(p, t)}{\partial p^2} + f\left(\frac{lp}{\pi}, t\right) = \frac{1}{c^2} \frac{\partial^2 w}{\partial t^2} \quad (7.69)$$

By proceeding as in the case of free vibration (Section 7.3), Eq. (7.69) can be expressed as an ordinary differential equation:

$$\frac{d^2 W(n, t)}{dt^2} + \frac{\pi^2 c^2 n^2}{l^2} W(n, t) = c^2 F(n, t) \quad (7.70)$$

where

$$W(n, t) = \int_0^{\pi} w(p, t) \sin np \, dp \quad (7.71)$$

$$F(n, t) = \int_0^{\pi} f\left(\frac{lp}{n}, t\right) \sin np \, dp \quad (7.72)$$

Assuming the initial conditions of the string to be zero, the steady-state solution of Eq. (7.70) can be expressed as

$$W(n, t) = \frac{cl}{\pi n} \int_0^t F(n, \tau) \sin \frac{n\pi c}{l}(t - \tau)\, d\tau \tag{7.73}$$

The inverse finite Fourier sine transform of $W(n, t)$ is given by [see Eq. (7.32)]

$$w(p, t) = \frac{2}{\pi} \sum_{n=1}^{\infty} W_n(n, t) \sin np$$

$$= \frac{2cl}{\pi^2} \sum_{n=1}^{\infty} \frac{1}{n} \sin np \int_0^t F(n, \tau) \sin \frac{n\pi c}{l}(t - \tau)\, d\tau \tag{7.74}$$

or

$$w(x, t) = \frac{2cl}{\pi^2} \sum_{n=1}^{\infty} \frac{1}{n} \sin \frac{n\pi x}{l} \int_0^t F(n, \tau) \sin \frac{n\pi c}{l}(t - \tau)\, d\tau \tag{7.75}$$

Example 7.4 Find the response of a string of length l, fixed at $x = 0$ and $x = l$, under the action of the harmonic force $f(x, t) = \tilde{f}_0(x)e^{i\omega t}$, where ω is the forcing frequency. Assume the initial displacement and velocity of the string to be zero.

SOLUTION Since the force is harmonic, the response of the string is assumed to be harmonic as

$$w(x, t) = u(x)e^{i\omega t} \tag{E7.4.1}$$

and the equation of motion, Eq. (7.66), becomes

$$c^2 \frac{d^2u(x)}{dx^2} + \omega^2 u(x) + f_0(x) = 0 \tag{E7.4.2}$$

where

$$f_0(x) = \frac{\tilde{f}_0(x)}{\rho} \tag{E7.4.3}$$

By introducing the new spatial variable $p = \pi x/l$ [as defined in Eq. (7.42)], Eq. (E7.4.2) can be written as

$$\frac{\pi^2 c^2}{l^2} \frac{d^2u(p)}{dp^2} + \omega^2 u(p) + f_0\left(\frac{lp}{\pi}\right) = 0 \tag{E7.4.4}$$

We now take the finite Fourier sine transform of Eq. (E7.4.4). According to Eq. (7.31), we multiply Eq. (E7.4.4) by $\sin np$ and integrate with respect to p from 0 to π:

$$\int_0^\pi \left[\frac{\pi^2 c^2}{l^2} \frac{d^2u}{dp^2} + \omega^2 u + f_0\left(\frac{lp}{\pi}\right) \right] \sin np\, dp = 0 \tag{E7.4.5}$$

where

$$\int_0^\pi \frac{d^2u}{dp^2} \sin np \, dp = \left(\frac{du}{dp} \sin np - nu \cos np \right) \Bigg|_0^\pi - n^2 \int_0^\pi u \sin np \, dp \quad \text{(E7.4.6)}$$

The first term on the right-hand side of Eq. (E7.4.6) will be zero because $u(0) = u(\pi) = 0$ (since the string is fixed at $p = 0$ and $p = \pi$) and $\sin 0 = \sin n\pi = 0$. We define the finite Fourier sine transforms of $u(p)$ and $f_0(lp/\pi)$ as $U(n)$ and $F_0(n)$:

$$U(n) = \int_0^\pi u(p) \sin np \, dp \quad \text{(E7.4.7)}$$

$$F_0(n) = \int_0^\pi f_0\left(\frac{lp}{\pi}\right) \sin np \, dp \quad \text{(E7.4.8)}$$

The inverse finite Fourier sine transforms of Eqs. (E7.4.7) and (E7.4.8) yield

$$u(p) = \frac{2}{\pi} \sum_{n=1}^\infty U(n) \sin np \quad \text{(E7.4.9)}$$

$$f_0\left(\frac{lp}{\pi}\right) = \frac{2}{\pi} \sum_{n=1}^\infty F_0(n) \sin np \quad \text{(E7.4.10)}$$

Thus, Eq. (E7.4.5) can be rewritten as

$$-\frac{\pi^2 c^2 n^2}{l^2} U(n) + \omega^2 U(n) + F_0(n) = 0$$

or

$$U(n) = \frac{F_0(n)}{(\pi^2 c^2 n^2 / l^2)(1 - \omega^2 / \omega_n^2)} \quad \text{(E7.4.11)}$$

where

$$\omega_n^2 = \frac{\pi^2 c^2 n^2}{l^2} \quad \text{(E7.4.12)}$$

denotes the natural frequency of the string. Finally, by taking the inverse finite Fourier sine transform of Eq. (E7.4.9) using Eqs. (E7.4.9) and (E7.4.10), we obtain the steady-state forced response of the string as

$$w(x, t) = \frac{2e^{i\omega t}}{Pl} \sum_{n=1}^\infty \frac{\sin(n\pi x / l) \int_0^l \tilde{f}_0(y) \sin(n\pi y / l) y \, dy}{(n^2 \pi^2 / l^2)(1 - \omega^2 / \omega_n^2)} \quad \text{(E7.4.13)}$$

7.5 FREE VIBRATION OF A BEAM

Consider a uniform beam of length l simply supported at $x = 0$ and $x = l$. The equation of motion governing the transverse vibration of the beam is given by [see Eq. (3.19)]

$$\frac{\partial^4 w}{\partial x^4} + \frac{1}{c^2} \frac{\partial^2 w}{\partial t^2} = 0 \quad \text{(7.76)}$$

where

$$c^2 = \frac{EI}{\rho A} \tag{7.77}$$

The boundary conditions can be expressed as

$$w(x, t) = 0 \qquad \text{at } x = 0, x = l \tag{7.78}$$

$$\frac{\partial^2 w}{\partial x^2}(x, t) = 0 \qquad \text{at } x = 0, x = l \tag{7.79}$$

We take finite Fourier sine transform of Eq. (7.76). For this, we multiply Eq. (7.76) by $\sin(n\pi x/l)$ and integrate with respect to x from 0 to l:

$$\int_0^l \frac{\partial^4 w}{\partial x^4} \sin \frac{n\pi x}{l}\, dx + \frac{1}{c^2} \int_0^l \frac{\partial^2 w}{\partial t^2} \sin \frac{n\pi x}{l}\, dx = 0 \tag{7.80}$$

Here

$$\int_0^l \frac{\partial^4 w}{\partial x^4} \sin \frac{n\pi x}{l}\, dx = \frac{\partial^3 w}{\partial x^3} \sin \frac{n\pi x}{l}\Big|_0^l - \int_0^l \frac{n\pi}{l} \frac{\partial^3 w}{\partial x^3} \cos \frac{n\pi x}{l}\, dx$$

$$= \frac{\partial^3 w}{\partial x^3} \sin \frac{n\pi x}{l}\Big|_0^l - \frac{n\pi}{l} \frac{\partial^2 w}{\partial x^2} \cos \frac{n\pi x}{l}\Big|_0^l$$

$$- \left(\frac{n\pi}{l}\right)^2 \int_0^l \frac{\partial^2 w}{\partial x^2} \sin \frac{n\pi x}{l}\, dx \tag{7.81}$$

In view of the boundary conditions of Eq. (7.79), Eq. (7.81) reduces, to

$$\int_0^l \frac{\partial^4 w}{\partial x^4} \sin \frac{n\pi x}{l}\, dx = -\left(\frac{n\pi}{l}\right)^2 \int_0^l \frac{\partial^2 w}{\partial x^2} \sin \frac{n\pi x}{l}\, dx \tag{7.82}$$

Again using integration by parts, the integral on the right-hand side of Eq. (7.82) can be expressed as

$$\int_0^l \frac{\partial^2 w}{\partial x^2} \sin \frac{n\pi x}{l}\, dx = \frac{\partial w}{\partial x} \sin \frac{n\pi x}{l}\Big|_0^l - \int_0^l \frac{n\pi}{l} \frac{\partial w}{\partial x} \cos \frac{n\pi x}{l}\, dx$$

$$= \frac{\partial w}{\partial x} \sin \frac{n\pi x}{l}\Big|_0^l - \frac{n\pi}{l} w \cos \frac{n\pi x}{l}\Big|_0^l - \left(\frac{n\pi}{l}\right)^2 \int_0^l w \sin \frac{n\pi x}{l}\, dx$$

$$= -\left(\frac{n\pi}{l}\right)^2 \int_0^l w \sin \frac{n\pi x}{l}\, dx \tag{7.83}$$

in view of the boundary conditions of Eq. (7.78). Thus, Eq. (7.80) can be expressed as

$$\left(\frac{n\pi}{l}\right)^4 \int_0^l w \sin \frac{n\pi x}{l}\, dx + \frac{1}{c^2} \frac{\partial^2}{\partial t^2} \int_0^l w \sin \frac{n\pi x}{l}\, dx = 0 \tag{7.84}$$

Defining the finite Fourier sine transform of $w(x, t)$ as [see Eq. (7.39)]

$$W(n, t) = \int_0^l w(x, t) \sin \frac{n\pi x}{l} \, dx \qquad (7.85)$$

Eq (7.84) reduces to the ordinary differential equation

$$\frac{d^2 W(n, t)}{dt^2} + \frac{c^2 n^4 \pi^4}{l^4} W(n, t) = 0 \qquad (7.86)$$

The solution of Eq. (7.86) can be expressed as

$$W(n, t) = \overline{C}_1 e^{i(cn^2\pi^2/l^2)t} + \overline{C}_2 e^{-i(cn^2\pi^2/l^2)t} \qquad (7.87)$$

or

$$W(n, t) = C_1 \cos \frac{cn^2\pi^2}{l^2} t + C_2 \sin \frac{cn^2\pi^2}{l^2} t \qquad (7.88)$$

Assuming the initial conditions of the beam as

$$w(x, t = 0) = w_0(x) \qquad (7.89)$$

$$\frac{dw}{dt}(x, t = 0) = \dot{w}_0(x) \qquad (7.90)$$

the finite Fourier sine transforms of Eqs. (7.89) and (7.90) yield

$$W(n, t = 0) = W_0(n) \qquad (7.91)$$

$$\frac{dW}{dt}(n, t = 0) = \dot{W}_0(n) \qquad (7.92)$$

where

$$W_0(n) = \int_0^l w_0(x) \sin \frac{n\pi x}{l} \, dx \qquad (7.93)$$

$$\dot{W}_0(n) = \int_0^l \dot{w}_0(x) \sin \frac{n\pi x}{l} \, dx \qquad (7.94)$$

Using initial conditions of Eqs. (7.93) and (7.94), Eq. (7.88) can be expressed as

$$W(n, t) = W_0(n) \cos \frac{cn^2\pi^2}{l^2} t + \frac{l^2}{cn^2\pi^2} \dot{W}_0(n) \sin \frac{cn^2\pi^2}{l^2} t \qquad (7.95)$$

Finally, the transverse displacement of the beam, $w(x, t)$, can be determined by using the finite inverse Fourier sine transform of Eq. (7.95) as

$$w(n, t) = \frac{2}{l} \sum_{n=1}^{\infty} W(n, t) \sin \frac{n\pi x}{l} \qquad (7.96)$$

which can be rewritten, using Eqs. (7.93) and (7.94), as

$$w(x, t) = \frac{2}{l} \sum_{n=1}^{\infty} \sin \frac{n\pi x}{l} \cos \frac{cn^2\pi^2}{l^2} t \int_{\xi=0}^{l} w_0(\xi) \sin \frac{n\pi\xi}{l} d\xi$$

$$+ \frac{2l}{c\pi^2} \sum_{n=1}^{\infty} \frac{1}{n^2} \sin \frac{n\pi x}{l} \sin \frac{cn^2\pi^2}{l^2} t \int_{\xi=0}^{l} \dot{w}_0(\xi) \sin \frac{n\pi\xi}{l} d\xi \qquad (7.97)$$

7.6   LAPLACE TRANSFORMS

The Laplace transform technique is an operational method that can be used conveniently for solving linear ordinary differential equations with constant coefficients. The method can also be used for the solution of linear partial differential equations that govern the response of continuous systems. Its advantage lies in the fact that differentiation of the time function corresponds to multiplication of the transform by a complex variable s. This reduces a differential equation in time t to an algebraic equation in s. Thus, the solution of the differential equation can be obtained by using either a Laplace transform table or the partial fraction expansion method. An added advantage of the Laplace transform method is that during the solution process, the initial conditions of the differential equation are taken care of automatically, so that both the homogeneous (complementary) solution and the particular solution can be obtained simultaneously.

The Laplace transformation of a time-dependent function, $f(t)$, denoted as $F(s)$, is defined as

$$L[f(t)] = F(s) = \int_0^{\infty} f(t)e^{-st} dt \qquad (7.98)$$

where L is an operational symbol denoting that the quantity upon which it operates is to be transformed by the Laplace integral

$$\int_0^{\infty} e^{-st} dt \qquad (7.99)$$

The inverse or reverse process of finding the function $f(t)$ from the Laplace transform $F(s)$, known as the *inverse Laplace transform*, is donated as

$$L^{-1}[F(s)] = f(t) = \frac{1}{2\pi i} \int_{c-i\infty}^{c+i\infty} F(s)e^{st} ds, \qquad t > 0 \qquad (7.100)$$

Certain conditions are to be satisfied for the existence of the Laplace transform of the function $f(t)$. One condition is that the absolute value of $f(t)$ must be bounded as

$$|f(t)| \leq Ce^{\alpha t} \qquad (7.101)$$

for some constants C and α. This means that if the values of the constants C and α can be found such that

$$|e^{-st} f(t)| \leq Ce^{(\alpha - s)t} \qquad (7.102)$$

then

$$L[f(t)] = \int_0^\infty e^{-st} f(t)\, dt \le C \int_0^\infty e^{(\alpha-s)t}\, dt = \frac{C}{s-\alpha} \qquad (7.103)$$

Another condition is that the function $f(t)$ must be piecewise continuous. This means that in a given interval, the function $f(t)$ has a finite number of finite discontinuities and no infinite discontinuity.

7.6.1   Properties of Laplace Transforms

Some of the important properties of Laplace transforms are indicated below.

1. *Linearity property*.   If c_1 and c_2 are any constant and $f_1(t)$ and $f_2(t)$ are functions of t with Laplace transforms $F_1(s)$ and $F_2(s)$, respectively, then

$$L[c_1 f_1(t) + c_2 f_2(t)] = c_1 L[f_1(t)] + c_2 L[f_2(t)]$$
$$= c_1 F_1(s) + c_2 F_2(s) \qquad (7.104)$$

The validity of Eq. (7.104) can be seen from the definition of the Laplace transform. Because of this property, the operator L can be seen to be a linear operator.

2. *First translation or shifting property*.   If $L[f(t)] = F(s)$ for $s > \beta$, then

$$L[e^{at} f(t)] = F(s-a) \qquad (7.105)$$

where $s - a > \beta$ and a may be a real or complex number. To see the validity of Eq. (7.105), we use the definition of the Laplace transform

$$L[e^{at} f(t)] = \int_0^\infty e^{at} e^{-st} f(t)\, dt = \int_0^\infty e^{-(s-a)t} f(t)\, dt = F(s-a) \qquad (7.106)$$

Equation (7.105) shows that the effect of multiplying $f(t)$ by e^{at} in the real domain is to shift the transform of $f(t)$ by an amount a in the s-domain.

3. *Second translation or shifting property*.   If

$$L[f(t)] = F(s) \quad \text{and} \quad g(t) = \begin{cases} f(t-a), & t > a \\ 0, & t < a \end{cases}$$

then

$$L[g(t)] = e^{-as} F(s) \qquad (7.107)$$

4. *Laplace transformation of derivatives*.   If $L[f(t)] = F(s)$, then

$$L[f'(t)] = L\left[\frac{df(t)}{dt}\right] = sF(s) - f(0) \qquad (7.108)$$

To see the validity of Eq. (7.108), we use the definition of Laplace transform as

$$L\left[\frac{df(t)}{dt}\right] = \int_0^\infty e^{-st} \frac{df(t)}{dt}\, dt \qquad (7.109)$$

Integrating the right-hand side of Eq. (7.109) by parts, we obtain

$$e^{-st} f(t)|_0^\infty - \int_0^\infty (-se^{-st}) f(t) \, dt = -f(0) + sF(s) \tag{7.110}$$

The property of Eq. (7.108) can be extended to the nth derivative of $f(t)$ to obtain

$$L\left[\frac{d^{(n)} f(t)}{dt^n}\right] = L[f^{(n)(t)}]$$

$$= -f^{(n-1)}(0) - sf^{(n-2)}(0) - s^2 f^{(n-3)}(0) - \cdots - s^{(n-1)} f(0) - s^{(n)} F(s) \tag{7.111}$$

where

$$f^{(n-i)}(0) = \frac{d^{n-i} f(t)}{dt^{n-i}}\bigg|_{t=0} \tag{7.112}$$

5. *Convolution theorem.* Let the Laplace transforms of the functions $f(t)$ and $g(t)$ be given by $F(s)$ and $G(s)$, respectively. Then

$$L[(f * g)(t)] = F(s) * G(s) \tag{7.113}$$

where $F * G$ is called the *convolution* or the *faltung* of F and G. Equation (7.113) can be expressed equivalently as

$$L\left[\int_0^t f(\tau)g(t-\tau) \, d\tau\right] = F(s)G(s) \tag{7.114}$$

or conversely,

$$L^{-1}[F(s)G(s)] = \int_0^t f(\tau)g(t-\tau) \, d\tau \tag{7.115}$$

To prove the validity of Eqs. (7.113) to (7.115), consider the definition of the Laplace transform and the convolution operation as

$$L[(f * g)(t)] = \int_0^\infty e^{-st} \left[\int_0^t f(\tau)g(t-\tau) \, d\tau\right] dt \tag{7.116}$$

From the region of integration shown in Fig. 7.1, the integral in Eq. (7.116) can be rewritten, by interchanging the order of integration, as

$$L[(f * g)(t)] = \int_0^\infty f(\tau) \left[\int_\tau^\infty e^{-st} g(t-\tau) \, dt\right] d\tau \tag{7.117}$$

By using the second property, the inner integral can be written as $e^{-st}G(s)$, so that Eq. (7.117) can be expressed as

$$L[(f * g)(t)] = \int_0^\infty G(s)e^{-s\tau} f(\tau) \, d\tau = G(s) \int_0^\infty e^{-s\tau} f(\tau) \, d\tau$$

$$= G(s)F(s) \tag{7.118}$$

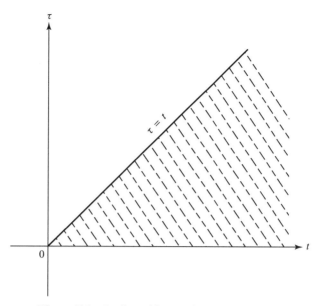

Figure 7.1 Region of integration in Eq. (7.116).

The converse result can be stated as

$$L^{-1}[F(s)G(s)] = \int_0^t f(\tau)g(t - \tau)\,d\tau$$
$$= f(t) * g(t) \tag{7.119}$$

7.6.2 Partial Fraction Method

In the Laplace transform method, sometimes we need to find the inverse transformation of the function

$$F(s) = \frac{P(s)}{Q(s)} \tag{7.120}$$

where $P(s)$ and $Q(s)$ are polynomials in s with the degree of $P(s)$ less than that of $Q(s)$. Let the polynomial $Q(s)$ be of order n with roots $a_1, a_2, a_3, \ldots, a_n$, so that

$$Q(s) = (s - a_1)(s - a_2)(s - a_3) \cdots (s - a_n) \tag{7.121}$$

First, let us consider the case in which all the n roots $a_1, a_2, a_3, \ldots, a_n$ are distinct, so that Eq. (7.120) can be expressed as

$$F(s) = \frac{P(s)}{Q(s)} = \frac{c_1}{s - a_1} + \frac{c_2}{s - a_2} + \frac{c_3}{s - a_3} + \cdots + \frac{c_n}{s - a_n} \tag{7.122}$$

where c_i are coefficients. The points $a_1, a_2, a_3, \ldots, a_n$ are called *simple poles* of $F(s)$.

The poles denote points at which the function $F(s)$ becomes infinite. The coefficients c_i in Eq. (7.122) can be found as

$$c_i = \lim_{s \to a_i} [(s - a_i)F(s)] = \left. \frac{P(s)}{Q'(s)} \right|_{s=a_i} \tag{7.123}$$

where $Q'(s)$ is the derivative of $Q(s)$ with respect to s. Using the result

$$L^{-1}\left[\frac{1}{s - a_i} \right] = e^{a_i t} \tag{7.124}$$

the inverse transform of Eq. (7.122) can be found as

$$f(t) = L^{-1}[F(s)] = c_1 e^{a_1 t} + c_2 e^{a_2 t} + \cdots + c_n e^{a_n t} = \sum_{i=1}^{n} c_i e^{a_i t}$$

$$= \sum_{i=1}^{n} \lim_{s \to a_i} [(s - a_i)F(s)e^{st}]$$

$$= \sum_{i=1}^{n} \left. \frac{P(s)}{Q'(s)} e^{st} \right|_{s=a_i} \tag{7.125}$$

Next, let us consider the case in which $Q(s)$ has a multiple root of order k, so that

$$Q(s) = (s - a_1)^k (s - a_2)(s - a_3) \cdots (s - a_{n-k}) \tag{7.126}$$

In this case, Eq. (7.120) can be expressed as

$$F(s) = \frac{P(s)}{Q(s)} = \frac{c_{11}}{s - a_1} + \frac{c_{12}}{(s - a_2)^2} + \cdots + \frac{c_{1k}}{(s - a_1)^k}$$

$$+ \frac{c_2}{s - a_2} + \frac{c_3}{s - a_3} + \cdots + \frac{c_{n-k}}{s - a_{n-k}} \tag{7.127}$$

Note that the coefficients c_{1j} can be determined as

$$c_{1j} = \frac{1}{(k-j)!} \frac{d^{k-j}}{ds^{k-j}} [(s - a_1)^k F(s)]|_{s=a_1}, \qquad j = 1, 2, 3, \ldots, k \tag{7.128}$$

while the coefficients c_i, $i = 2, 3, \ldots, n - k$, can be found as in Eq. (7.125). Since

$$L^{-1}\left[\frac{1}{(s - a_1)^j} \right] = \frac{t^{j-1}}{(j-1)!} e^{a_1 t} \tag{7.129}$$

the inverse of Eq. (7.127) can be expressed as

$$f(t) = \left[c_{11} + c_{12}t + c_{13}\frac{t^2}{2!} + \cdots + c_{1k}\frac{t^{k-1}}{(k-1)!} \right] e^{a_1 t}$$

$$+ c_2 e^{a_2 t} + c_3 e^{a_3 t} + c_4 e^{a_4 t} + \cdots + c_{n-k} e^{a_{n-k} t} \tag{7.130}$$

7.6.3 Inverse Transformation

The inverse Laplace transformation, denoted as $L^{-1}[F(s)]$, is also defined by the complex integration formula

$$L^{-1}[F(s)] = f(t) = \frac{1}{2\pi i} \int_{\alpha-i\infty}^{\alpha+i\infty} e^{st} F(s) \, ds \qquad (7.131)$$

where α is a suitable real constant, in Eq. (7.131), the path of the integration is a line parallel to the imaginary axis that crosses the real axis at Re $s = \alpha$ and extends from $-\infty$ to $+\infty$. We assume that $F(s)$ is an analytic function of the complex variable s in the right half-plane Re $s > \alpha$ and all the poles lie to the left of the line $x = \alpha$. This condition is usually satisfied for all physical problems possessing stability since the poles to the right of the imaginary axis denote instability. The details of evaluation of Eq. (7.131) depend on the nature of the singularities of $F(s)$.

The path of the integration is the straight line $\overline{L}$ as shown in Fig. 7.2 in the complex s plane, with equation $s = \alpha + iR$, $-\infty < R < +\infty$ and Re $s = \alpha$ is chosen so that all the singularities of the integrand of Eq. (7.131) lie to the left of the line $\overline{L}$. The Cauchy-residue theorem is used to evaluate the contour integral as

$$\int_C e^{st} F(s) \, ds = \int_{\overline{L}} e^{st} F(s) \, ds + \int_\Gamma e^{st} F(s) \, ds$$

$$= 2\pi i [\text{sum of the residues of } e^{st} F(s) \text{ at the poles inside } C] \qquad (7.132)$$

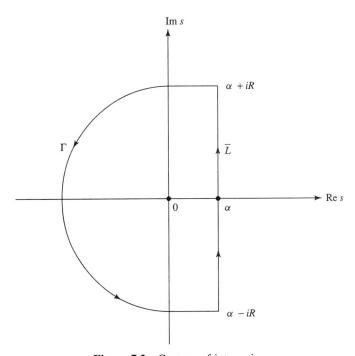

Figure 7.2 Contour of integration.

where $R \to \infty$ and the integral over Γ tends to zero in most cases. Thus, Eq. (7.131) reduces to the form

$$\lim_{R \to \infty} \frac{1}{2\pi i} \int_{\alpha-iR}^{\alpha+iR} e^{st} F(s) \, ds$$

$$= \text{sum of the residues of } e^{st} F(s) \text{ at the poles of } F(s) \qquad (7.133)$$

The following example illustrates the procedure of contour integration.

Example 7.5   Find the inverse Laplace transform of the function

$$F(s) = \frac{s}{s^2 + c^2} \qquad (E7.5.1)$$

SOLUTION   The inverse transform is given by

$$f(t) = \frac{1}{2\pi i} \int_{\alpha-i\infty}^{\alpha+i\infty} e^{st} F(s) \, ds$$

$$= \frac{1}{2\pi i} \int_{\alpha-i\infty}^{\alpha+i\infty} e^{st} \frac{s}{s^2 + c^2} \, ds \qquad (E7.5.2)$$

The integrand in Eq. (E7.5.2) has two simple poles at $s = \pm ic$, and residues at these poles are given by

$$R_1 = \text{residue of } e^{st} F(s) \text{ at } s = -ic$$

$$= \lim_{s \to -ic} (s + ic) \frac{s e^{st}}{s^2 + c^2} = \frac{1}{2} e^{-ict} \qquad (E7.5.3)$$

$$R_2 = \text{residue of } e^{st} F(s) \text{ at } s = ic$$

$$= \lim_{s \to +ic} (s - ic) \frac{s e^{st}}{s^2 + c^2} = \frac{1}{2} e^{ict} \qquad (E7.5.4)$$

Hence,

$$f(t) = \frac{1}{2\pi i} \int_{\alpha-i\infty}^{\alpha+i\infty} e^{st} F(s) \, ds = R_1 + R_2 = \frac{1}{2}(e^{ict} + e^{-ict}) = \cos ct \qquad (E7.5.5)$$

7.7   FREE VIBRATION OF A STRING OF FINITE LENGTH

In this case the equation of motion is

$$c^2 \frac{\partial^2 w}{\partial x^2} - \frac{\partial^2 w}{\partial t^2} = 0 \qquad (7.134)$$

If the string is fixed at $x = 0$ and $x = l$, the boundary conditions are

$$w(0, t) = 0 \qquad (7.135)$$

$$w(l, t) = 0 \qquad (7.136)$$

Let the initial conditions of the string be given by

$$w(x, t = 0) = w_0(x) \tag{7.137}$$

$$\frac{\partial w}{\partial t}(x, t = 0) = \dot{w}_0(x) \tag{7.138}$$

Applying Laplace transforms to Eq. (7.134), we obtain

$$c^2 \frac{d^2 W(x, s)}{dx^2} - s^2 W(x, s) + s w_0(x) + \dot{w}_0(x) = 0 \tag{7.139}$$

where

$$W(x, s) = \int_0^\infty w(x, t) e^{-st} \, dt \tag{7.140}$$

Taking finite Fourier sine transform of Eq. (7.139), we obtain

$$(s^2 + c^2 p_n^2) \overline{W}(p_n, s) = s \overline{W}_0(p_n) + \dot{\overline{W}}_0(p_n) \tag{7.141}$$

where

$$\overline{W}(p_n, s) = \int_0^l W(x, s) \sin p_n x \, dx \tag{7.142}$$

$$\overline{W}_0(p_n) = \int_0^l w_0(x) \sin p_n x \, dx \tag{7.143}$$

$$\dot{\overline{W}}_0(p_n) = \int_0^l \dot{w}_0(x) \sin p_n x \, dx \tag{7.144}$$

with

$$p_n = \frac{n\pi}{l} \tag{7.145}$$

Equation (7.141) gives

$$\overline{W}(p_n, s) = \frac{s \overline{W}_o(p_n) + \dot{\overline{W}}_0(p_n)}{s^2 + c^2 p_n^2} \tag{7.146}$$

Performing the inverse finite Fourier sine transform of Eq. (7.146) yields

$$W(x, s) = \frac{2}{l} \sum_{n=1}^\infty \frac{\sin p_n x}{s^2 + c^2 p_n^2} \int_0^l [s w_0(\xi) + \dot{w}_0(\xi)] \sin p_n \xi \, d\xi \tag{7.147}$$

Finally, by taking the inverse Laplace transform of $W(x, s)$ in Eq. (7.147), we obtain

$$w(x, t) = \frac{2}{l} \sum_{n=1}^\infty \sin p_n x \left[\cos \frac{cn\pi t}{l} \right.$$

$$\left. \cdot \int_0^l w_0(\xi) \sin p_n \xi \, d\xi + \frac{\sin(cn\pi t/l)}{cn\pi/l} \int_0^l \dot{w}_0(\xi) \sin p_n \xi \, d\xi \right] \tag{7.148}$$

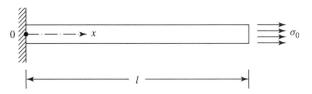

Figure 7.3   Axial stress at the end of a bar.

Example 7.6   A uniform bar is fixed at $x = 0$ and subjected to an axial stress σ_0 at $x = l$ as shown in Fig. 7.3. Assuming the bar to be at rest initially, determine the axial vibration response of the bar.

SOLUTION   The equation governing the longitudinal vibration of a bar is giving by [see Eq. (9.15)]:

$$c^2 \frac{\partial^2 u(x, t)}{\partial x^2} = \frac{\partial^2 u(x, t)}{\partial t^2} \tag{E7.6.1}$$

The boundary conditions are given by

$$u(0, t) = 0 \tag{E7.6.2}$$

$$\frac{\partial u}{\partial x}(l, x) = \varepsilon_x(l, t) = \frac{\sigma_0}{E} \tag{E7.6.3}$$

and the initial conditions by

$$u(x, 0) = 0 \tag{E7.6.4}$$

$$\frac{\partial u}{\partial t}(x, 0) = 0 \tag{E7.6.5}$$

By taking Laplace transform of Eq. (E7.6.1) with respect to t, we obtain

$$s^2 U(x, s) - su(x, 0) - \frac{\partial u}{\partial t}(x, 0) = c^2 \frac{d^2 U(x, s)}{dx^2} \tag{E7.6.6}$$

which in view of the initial conditions of Eqs. (E7.6.4) and (E7.6.5), reduces to

$$\frac{d^2 U}{dx^2} - \frac{s^2}{c^2} U = 0 \tag{E7.6.7}$$

where

$$U(x, s) = L[u(x, t)] \tag{E7.6.8}$$

Noting that $L(1) = 1/s$, the Laplace transforms of Eqs. (E7.6.2) and (E7.6.3) can be written as

$$U(0, s) = 0 \tag{E7.6.9}$$

$$\frac{dU}{dx}(l, s) = \frac{\sigma_0}{Es} \tag{E7.6.10}$$

The solution of Eq. (E7.6.7) is given by

$$U(x, s) = c_1 \cosh \frac{sx}{c} + c_2 \sinh \frac{sx}{c} \tag{E7.6.11}$$

where the constants c_1 and c_2 can be found using Eqs. (E7.6.9) and (E7.6.10) as

$$c_1 = 0 \tag{E7.6.12}$$

$$c_2 \frac{s}{c} \cosh \frac{sl}{c} = \frac{\sigma_0}{Es} \quad \text{or} \quad c_2 = \frac{c\sigma_0}{Es^2 \cosh(sl/c)} \tag{E7.6.13}$$

Thus, the solution, $U(x, s)$, becomes

$$U(x, s) = \frac{c\sigma_0}{E} \frac{\sinh(sx/c)}{s^2 \cosh(sl/c)} \tag{E7.6.14}$$

By taking the inverse Laplace transform of Eq. (E7.6.14), we obtain the axial displacement of the bar as [8]

$$u(x, t) = \frac{\sigma_0}{E} \left[x + \frac{8l}{\pi^2} \sum_{n=1}^{\infty} \frac{(-1)^n}{(2n-1)^2} \sin \frac{(2n-1)\pi x}{2l} \cos \frac{(2n-1)\pi ct}{2l} \right] \tag{E7.6.15}$$

7.8 FREE VIBRATION OF A BEAM OF FINITE LENGTH

The equation of motion for the transverse vibration of a beam is given by

$$c^2 \frac{\partial^4 \overline{w}(x, t)}{\partial x^4} + \frac{\partial^2 \overline{w}(x, t)}{\partial t^2} = 0 \tag{7.149}$$

where

$$c^2 = \frac{EI}{\rho A} \tag{7.150}$$

For free vibration, $\overline{w}(x, t)$ is assumed to be harmonic with frequency ω:

$$\overline{w}(x, t) = w(x)e^{i\omega t} \tag{7.151}$$

so that Eq. (7.149) reduces to an ordinary differential equation:

$$\frac{d^4 w(x)}{dx^4} - \beta^4 w(x) = 0 \tag{7.152}$$

where

$$\beta^4 = \frac{\omega^2}{c^2} = \frac{\omega^2 \rho A}{EI} \tag{7.153}$$

By taking Laplace transforms of Eq. (7.152), we obtain

$$s^4 W(s) - s^3 w(0) - s^2 w'(0) - sw''(0) - w'''(0) - \beta^4 W(s) = 0 \tag{7.154}$$

or

$$W(s) = \frac{1}{s^4 - \beta^4} \left[s^3 w(0) + s^2 w'(0) + sw''(0) + w'''(0) \right] \tag{7.155}$$

where $w(0)$, $w'(0)$, $w''(0)$, and $w'''(0)$ denote the deflection and its first, second, and third derivative, respectively, at $x = 0$. By noting that

$$L^{-1}\left[\frac{s^3}{s^4 - \beta^4}\right] = \frac{1}{2}(\cosh \beta x + \cos \beta x) \tag{7.156}$$

$$L^{-1}\left[\frac{s^2}{s^4 - \beta^4}\right] = \frac{1}{2\beta}(\sinh \beta x + \sin \beta x) \tag{7.157}$$

$$L^{-1}\left[\frac{s}{s^4 - \beta^4}\right] = \frac{1}{2\beta^2}(\cosh \beta x - \cos \beta x) \tag{7.158}$$

$$L^{-1}\left[\frac{1}{s^4 - \beta^4}\right] = \frac{1}{2\beta^3}(\sinh \beta x - \sin \beta x) \tag{7.159}$$

the inverse Laplace transform of Eq.(7.155) gives

$$w(x) = \frac{1}{2}(\cosh \beta x + \cos \beta x)w(0) + \frac{1}{2\beta}(\sinh \beta x + \sin \beta x)w'(0)$$

$$+ \frac{1}{2\beta^2}(\cosh \beta x - \cos \beta x)w''(0) + \frac{1}{2\beta^3}(\sinh \beta x - \sin \beta x)w'''(0) \tag{7.160}$$

7.9   FORCED VIBRATION OF A BEAM OF FINITE LENGTH

The governing equation is given by

$$EI\frac{\partial^4 w(x, t)}{\partial x^4} + \rho A\frac{\partial^2 w}{\partial t^2} = f(x, t) \tag{7.161}$$

where $f(x, t)$ denotes the time-varying distributed force. Let the initial deflection and velocity be given by $w_0(x)$ and $\dot{w}_0(x)$, respectively. The Laplace transform of Eq. (7.161), with respect to t with s as the subsidiary variable, yields

$$\frac{d^4 W(x, s)}{dx^4} + \frac{\rho A}{EI}s^2 W(x, s) = \frac{\rho A}{EI}[sw_0(x) + \dot{w}_0(x)] + \frac{1}{EI}F(x, s) \tag{7.162}$$

or

$$\frac{d^4 W(x, s)}{dx^4} - \beta^4 W(x, s) = G(x, s) \tag{7.163}$$

where

$$\beta^4 = -\frac{\rho A s^2}{EI} \tag{7.164}$$

$$G(x, s) = \frac{\rho A}{EI}[sw_0(x) + \dot{w}_0(x)] + \frac{1}{EI}F(x, s) \tag{7.165}$$

Again, by taking Laplace transform of Eq. (7.163) with respect to x with p as the subsidiary variable, we obtain

$$(p^4 - \beta^4)\overline{W}(p, s) = \overline{G}(p, s) + p^3 \overline{W}(0, s)$$

$$+ p^2 \overline{W}_x'(0, s) + p \overline{W}_x''(0, s) + \overline{W}_x'''(0, s)$$

or

$$\overline{W}(p, s) = \frac{\overline{G}(p, s)}{p^4 - \beta^4} + \frac{p^3 \overline{W}(0, s) + p^2 \overline{W}'_x(0, s) + p \overline{W}''_x(0, s) + \overline{W}'''_x(0, s)}{p^4 - \beta^4}$$

(7.166)

where $\overline{W}(0, s)$, $\overline{W}'_x(0, s)$, $\overline{W}''_x(0, s)$, and $\overline{W}'''_x(0, s)$ denote the Laplace transforms with respect to t of $w(x, t)$, $(\partial w/\partial x)(x, t)$, $(\partial^2 w/\partial x^2)(x, t)$ and $(\partial^3 w/\partial x^3)(x, t)$ respectively, at $x = 0$. Next, we perform the inverse Laplace transform of Eq. (7.166) with respect to x. For this, we use Eqs. (7.156)–(7.159) and express the inverse transform of Eq. (7.166) as

$$W(x, s) = \frac{1}{2\beta^3} \int_0^x G(\eta, s)[\sinh \beta(x - \eta) - \sin \beta(x - \eta)] \, d\eta$$

$$+ \frac{1}{2} \overline{W}(0, s)(\cosh \beta x + \cos \beta x) + \frac{1}{2\beta} \overline{W}'_x(0, s)(\sinh \beta x + \sin \beta x)$$

$$+ \frac{1}{2\beta^2} \overline{W}''_x(0, s)(\cosh \beta x - \cos \beta x) + \frac{1}{2\beta^3} \overline{W}'''_x(0, s)(\sinh \beta x - \sin \beta x)$$

(7.167)

Finally, we perform the inverse Laplace transform of Eq. (7.167) with respect to t to find the desired solution, $w(x, t)$. The procedure is illustrated in the following example.

Example 7.7   A uniform beam of length l is subjected to a concentrated harmonic force $f_0 \sin \omega t$ at $x = \xi, 0 < \xi < l$. Assuming the end conditions of the beam to be simple supports and the initial conditions to be zero, determine the response of the beam.

SOLUTION   The boundary conditions of the beam can be expressed as

$$w(0, t) = 0 \tag{E7.7.1}$$

$$\frac{\partial^2 w}{\partial x^2}(0, t) = 0 \tag{E7.7.2}$$

$$w(l, t) = 0 \tag{E7.7.3}$$

$$\frac{\partial^2 w}{\partial x^2}(l, t) = 0 \tag{E7.7.4}$$

Taking Laplace transforms, Eqs. (E7.7.1)–(E7.7.4) can be written as

$$W(0, s) = 0 \tag{E7.7.5}$$

$$W''_x(0, s) = 0 \tag{E7.7.6}$$

$$W(l, s) = 0 \tag{E7.7.7}$$

$$W''_x(l, s) = 0 \tag{E7.7.8}$$

The applied concentrated force can be expressed as

$$f(x, t) = f_0 \delta(x - \xi) h(t) \equiv f_0 \sin \omega t \, \delta(x - \xi) \tag{E7.7.9}$$

The Laplace transform of Eq. (E7.7.9) gives

$$F(x, s) = f_0 \delta(x - \xi) H(s) \qquad \text{(E7.7.10)}$$

where

$$H(s) = L[h(t)] = L[\sin \omega t] = \frac{\omega}{s^2 + \omega^2} \qquad \text{(E7.7.11)}$$

Since the initial conditions are zero,

$$w_0(x) = \dot{w}_0(x) = 0 \qquad \text{(E7.7.12)}$$

and hence Eq. (7.165) yields

$$G(x, s) = \frac{F(x, s)}{EI} = \frac{f_0}{EI} \delta(x - \xi) H(s) \qquad \text{(E7.7.13)}$$

Using the boundary conditions at $x = 0$ [Eqs. (E7.7.5) and (E7.7.6)] and Eq. (E7.7.13), Eq. (7.167) can be expressed as

$$W(x, s) = \frac{1}{2\beta^3} \int_0^x \frac{f_0}{EI} \delta(\eta - \xi) H(s) [\sinh \beta(x - \eta) - \sin \beta(x - \eta)] \, d\eta$$
$$+ c_1 \sinh \beta x + c_2 \sin \beta x \qquad \text{(E7.7.14)}$$

where

$$c_1 = \frac{1}{2\beta} W_x'(0, s) + \frac{1}{2\beta^3} W_x'''(0, s) \qquad \text{(E7.7.15)}$$

$$c_2 = \frac{1}{2\beta} W_x'(0, s) - \frac{1}{2\beta^3} W_x'''(0, s) \qquad \text{(E7.7.16)}$$

By differentiating $W(x, s)$ given by Eq. (E7.7.14) with respect to x and using the conditions of Eqs. (E7.7.7) and (E7.7.8), we obtain

$$W(l, s) = 0 = \frac{f_0 H(s)}{2\beta^3 EI} \int_0^l \delta(\eta - \xi) [\sinh \beta(l - \eta) - \sin \beta(l - \eta)] \, d\eta$$
$$+ c_1 \sinh \beta l + c_2 \sin \beta l \qquad \text{(E7.7.17)}$$

$$W_x''(l, s) = 0 = \frac{f_0 H(s)}{2\beta^3 EI} \int_0^l \delta(\eta - \xi) [\sinh \beta(l - \eta) + \sin \beta(l - \eta)] \, d\eta$$
$$+ c_1 \sinh \beta l - c_2 \sin \beta l \qquad \text{(E7.7.18)}$$

The solution of Eqs. (E7.7.17) and (E7.7.18) gives

$$c_1 = -\frac{f_0 H(s) \sinh \beta(l - \xi)}{2\beta^3 EI \sinh \beta l} \qquad \text{(E7.7.19)}$$

$$c_2 = \frac{f_0 H(s) \sin \beta(l - \xi)}{2\beta^3 EI \sin \beta l} \qquad \text{(E7.7.20)}$$

Thus, Eq. (E7.7.18) becomes

$$W(x, s) = \frac{f_0 H(s)}{2\beta^3 EI} \int_0^x \delta(\eta - \xi)[\sinh \beta(x - \eta) - \sin \beta(x - \eta)]\, d\eta$$

$$- \frac{f_0 H(s)}{2\beta^3 EI} \frac{\sinh \beta(l - \xi) \sinh \beta x \sin \beta l - \sin \beta(l - \xi) \sin \beta x \sinh \beta l}{(\sinh \beta l \sin \beta l)}$$

$$= \frac{f_0 H(s)}{2\beta^3 EI}[\sinh \beta(x - \xi) - \sin \beta(x - \xi)]\delta(x - \xi)$$

$$+ \frac{\sinh \beta(l - \xi) \sinh \beta x \sin \beta l - \sin \beta(l - \xi) \sin \beta x \sinh \beta l}{\sinh \beta l \sin \beta l} \qquad \text{(E7.7.21)}$$

which results in the solution

$$w(x, t) = \frac{f_0 \sin \omega t}{2\beta^3 EI} \left\{ [\sin \beta(x - \xi) - \sinh \beta(x - \xi)]\, \delta(x - \xi) \right.$$

$$\left. + \frac{\sinh \beta(l - \xi) \sinh \beta x \sin \beta l - \sin \beta(l - \xi) \sin \beta x \sinh \beta l}{\sinh \beta l \sin \beta l} \right\} \qquad \text{(E7.7.22)}$$

The value of β^4 can be obtained from Eq. (7.164), by substituting $s = i\omega$ as

$$\beta^4 = \frac{\rho A \omega^2}{EI} \qquad \text{(E7.7.23)}$$

7.10   RECENT CONTRIBUTIONS

Fast Fourier Transforms   The fast Fourier transform algorithm and the associated programming considerations in the calculation of sine, cosine, and Laplace transforms was presented by Cooley et al. [13]. The problem of establishing the correspondence between discrete and continuous functions is described.

Beams   Cobble and Fang [14] considered the finite transform solution of the damped cantilever beam equation with distributed load, elastic support, and the wall edge elastically restrained against rotation. The solution is based on the properties of a Hermitian operator and its orthogonal basis vectors.

Membranes   The general solution of the vibrating annular membrane with arbitrary loading, initial conditions, and time-dependent boundary conditions was given by Sharp [15].

Hankel Transform   The solution of the scalar wave equation of an annular membrane, in which the motion is symmetrical about the origin, for arbitrary initial and boundary conditions was given in Ref. [16]. The solution is obtained by using a finite Hankel transform. An example is given to illustrate the procedure and the solution is compared to the one given by the method of separation of variables.

Plates   A method of determining a finite integral transform that will remove the presence of one of the independent variables in a fourth-order partial differential equation is applied to the equation of motion of classical plate theory for complete and annular circular plates subjected to various boundary conditions by Anderson [17]. The method is expected to be particularly useful for the solution of plate vibration problems with time-dependent boundary conditions. Forced torsional vibration of thin, elastic, spherical, and hemispherical shells subjected to either a free or a restrained edge was considered by Anderson in Ref. [18].

z Transform   Application of the *z*-transform method to the solution of the wave equation was presented by Tsai et al. [19]. In the conventional method of solution using the Laplace transformation, the conversion, directly from the *s* domain to the *t* domain to find the time function, sometimes proves to be very difficult and yields a solution in the form of an infinite series. However, if the *s* domain solution is first transformed to the *z* domain and then converted to the time domain, the process of inverse transformation is simplified and a closed-form solution may be obtained.

REFERENCES

1. O. Heaviside, *Electromagnetic Theory*, 1899; reprint by Dover, New York, 1950.

2. T. Bromwich, Normal coordinates in dynamical systems, *Proceedings of the London Mathematical Society*, Ser. 2, Vol. 15, pp. 401–448, 1916.

3. K. W. Wagner, Über eine Formel Von Heaviside zur Berechnung von Einschaltvorgangen, *Archiv fuver Elektrotechnik*, Vol. 4, pp. 159–193, 1916.

4. J. R. Carson, On a general expansion theorem for the transient oscillations of a connected system, *Physics Review*, Ser. 2, Vol. 10, pp. 217–225, 1917.

5. B. Van der Pol, A simple proof and extension of Heaviside's operational calculus for invariable systems, *Philosophical Magazine*, Ser. 7, pp. 1153–1162, 1929.

6. L. Debnath, *Integral Transforms and Their Applications*, CRC Press, Boca Raton, FL, 1995.

7. I. N. Sneddon, *Fourier Transforms*, McGraw-Hill, New York, 1951.

8. M. R. Spiegel, *Theory and Problems of Laplace Transforms*, Schaum, New York, 1965.

9. W. T. Thomson, *Laplace Transformation*, 2nd ed., Prentice-Hall, Englewood Cliffs, NJ, 1960.

10. C. J. Tranter, *Integral Transforms in Mathematical Physics*, Methuen, London, 1959.

11. E. C. Titchmarsh, *Introduction to the Theory of Fourier Integrals*, Oxford University Press, New York, 1948.

12. W. Nowacki, *Dynamics of Elastic Systems*, translated by H. Zorski, Wiley, New York, 1963.

13. J. W. Cooley, P. A. W. Lewis, and P. D. Welch, The fast Fourier transform algorithm: programming considerations in the calculation of sine, cosine and Laplace transforms, *Journal of Sound and Vibration*, Vol. 12, No. 3, pp. 315–337, 1970.

14. M. H. Cobble and P. C. Fang, Finite transform solution of the damped cantilever beam equation having distributed load, elastic support, and the wall edge elastically restrained against rotation, *Journal of Sound and Vibration*, Vol. 6, No. 2, pp. 187–198, 1967.

15. G. R. Sharp, Finite transform solution of the vibrating annular membrane, *Journal of Sound and Vibration*, Vol. 6, No. 1, pp. 117–128, 1967.

16. G. R. Sharp, Finite transform solution of the symmetrically vibrating annular membrane, *Journal of Sound and Vibration*, Vol. 5, No. 1, pp. 1–8, 1967.

17. G. L. Anderson, On the determination of finite integral transforms for forced vibrations of circular plates, *Journal of Sound and Vibration*, Vol. 9, No. 1, pp. 126–144, 1969.

18. G. L. Anderson, On Gegenbauer transforms and forced torsional vibrations of thin spherical shells, *Journal of Sound and Vibration*, Vol. 12, No. 3, pp. 265–275, 1970.

19. S. C. Tsai, E. C. Ong, B. P. Tan, and P. H. Wong, Application of the z-transform method to the solution of the wave equation, *Journal of Sound and Vibration*, Vol. 19, No. 1, pp. 17–20, 1971.

PROBLEMS

7.1 Find the Fourier transforms of the following functions:

(a) $f(x) = \frac{1}{x^2 + a^2}$

(b) $f(x) = \delta(x - a)e^{-c^2 x^2}$

7.2 Find the Fourier cosine transforms of the following functions:

(a) $f(x) = e^{-x}$

(b) $f(x) = \begin{cases} a, & 0 < x < a \\ 0, & x > a \end{cases}$

7.3 Find the Fourier sine transforms of the following functions:

(a) $f(x) = e^{-x}$

(b) $f(x) = \begin{cases} a, & 0 < x < a \\ 0, & x > a \end{cases}$

7.4 Find the Fourier cosine transforms of the following functions:

(a) $f(x) = \begin{cases} \sin x, & 0 \le x \le \pi \\ 0, & \text{otherwise} \end{cases}$

(b) $f(x) = \begin{cases} 1 - x^2, & 0 \le x \le 1 \\ 0, & \text{otherwise} \end{cases}$

7.5 Find the Fourier sine transforms of the following functions:

(a) $f(x) = \begin{cases} \sin x, & 0 \le x \le \pi \\ 0, & \text{otherwise} \end{cases}$

(b) $f(x) = \begin{cases} 1 - x^2, & 0 \le x \le 1 \\ 0, & \text{otherwise} \end{cases}$

7.6 Find the Laplace transforms of the following functions:

(a) $f(t) = t$

(b) $f(t) = e^{\alpha t}$

(c) $f(t) = \sin \alpha t$

(d) $f(t) = \cos \alpha t$

7.7 Find the Laplace transforms of the following functions:

(a) $f(t) = \begin{cases} 6, & 0 < t < 2 \\ 0, & t > 2 \end{cases}$

(b) $f(t) = \begin{cases} \sin t, & 0 < t < \pi \\ 0, & \pi < t < 2\pi \end{cases}$

7.8 Find the Laplace transforms of the following functions:

(a) $f(t) = \begin{cases} \cos(t - \pi/4), & t > \pi/4 \\ 0, & t < \pi/4 \end{cases}$

(b) Heaviside's unit step function:

$$f(t) = U(t - a) = \begin{cases} 1, & t > a \\ 0, & t < a \end{cases}$$

7.9 A single-degree-of-freedom spring–mass–damper system is subjected to a displacement x_0 and velocity $\dot{x}_0$ at time $t = 0$. Determine the resulting motion of the mass (m) using Laplace transforms. Assume the spring and damping forces to be kx and $c\dot{x}$, where k is the spring constant and $\dot{x} = dx/dt$ is the velocity of the mass.

7.10 Derive an expression for the response of a uniform beam of length l fixed at both ends when subjected to a concentrated force $f_0(t)$ at $x = \xi, 0 < \xi < l$. Assume the initial conditions of the beam to be zero. Use Fourier transforms.

7.11 Find the response of a uniform beam of length l fixed at both the ends when subjected to an impulse $\hat{G}$ at $x = \xi, 0 < \xi < l$. Assume that the beam is in equilibrium before the impulse is applied. Use Fourier transforms.

7.12 Find the free transverse displacement of a semi-infinite string using Fourier transforms. The governing equation is

$$c^2 \frac{\partial^2 w}{\partial x^2} = \frac{\partial^2 w}{\partial t^2}$$

and the initial conditions are

$$w(x, 0) = w_0(x), \qquad \frac{\partial w}{\partial t}(x, 0) = \dot{w}_0$$

Assume that the string is fixed at $x = 0$ and stretched along the positive x axis under tension P.

7.13 Consider a finite string of length l fixed at $x = 0$ and $x = l$, subjected to tension P. Find the transverse displacement of a string that is initially at rest and subjected to an impulse $\hat{F}$ at point $x = a, 0 < a < l$ using the Fourier transform method.

7.14 Find the steady-state transverse vibration response of a string of length l fixed at both ends, subjected to the force

$$f(x, t) = f_0 \sin \Omega t$$

using the Fourier transform method.

7.15 Find the Laplace transforms of the following functions:

(a) $f(t) = \begin{cases} 0, & t < 0 \\ at, & t \geq 0 \end{cases}$

(b) $f(t) = e^{\alpha t}, \alpha$ is real

(c) $f(t) = 2e^{-2t} \sin 3t$

7.16 Find the solution of the following differential equation using Laplace transforms:

$$\frac{d^2 w}{dx^2} + 4 \frac{dw}{dx} + 3w = \sin 2t$$

with $w(0) = 1$ and $dw/dt(0) = -1$.

7.17 The longitudinal vibration of a uniform bar of length l is governed by the equation

$$c^2 \frac{\partial^2 u(x, t)}{\partial x^2} = \frac{\partial^2 u(x, t)}{\partial t^2}$$

where $c^2 = E/\rho$ with E and ρ denoting Young's modulus and the mass density of the bar respectively. The bar is fixed at $x = 0$ and free at $x = l$. Find the free vibration response of the bar subject to the initial conditions

$$u(x, 0) = u_0(x), \qquad \frac{\partial u}{\partial t}(x, 0) = \dot{u}_0(x)$$

using Laplace transforms.

7.18 Find the longitudinal vibration response of a uniform bar fixed at $x = 0$ and subjected to an axial force $f(t)$ at $x = l$, using Laplace transforms. The equation of motion is given in Problem 7.17.

7.19 A uniform bar fixed at $x = 0$, is subjected to a sudden displacement of magnitude u_0 at $x = l$. Find the ensuing axial motion of the bar using Laplace transforms. The governing equation of the bar is given in Problem 7.17.

7.20 A taut string of length 1, fixed at $x = 0$ and $x = 1$ is subjected to tension P. If the string is given an initial displacement

$$w(x, 0) = w_0(x) = \begin{cases} 2x, & 0 < x < 0.5 \\ 2(1 - x), & 0.5 < x < 1 \end{cases}$$

and released with zero velocity, determine the ensuing motion of the string.

8

Transverse Vibration of Strings

8.1 INTRODUCTION

It is well known that most important musical instruments, including the violin and the guitar, involve strings whose natural frequencies and mode shapes play a significant role in their performance. The characteristics of many engineering systems, such as guy wires, electric transmission lines, ropes and belts used in machinery, and thread manufacture, can be derived from a study of the dynamics of taut strings. The free and forced transverse vibration of strings is considered in this chapter. As will be seen in subsequent chapters, the equation governing the transverse vibration of strings will have the same form as the equations of motion of longitudinal vibration of bars and torsional vibration of shafts.

8.2 EQUATION OF MOTION

8.2.1 Equilibrium Approach

Figure 8.1 shows a tightly stretched elastic string or cable of length l subjected to a distributed transverse force $f(x,t)$ per unit length. The string is assumed to be supported at the ends on elastic springs of stiffness k_1 and k_2. By assuming the transverse displacement of the string $w(x,t)$ to be small, Newton's second law of motion can be applied for the motion of an element of the string in the z direction as

$$\text{net force acting on an element} = \text{inertia force acting on the element} \qquad (8.1)$$

If P is the tension, ρ is the mass per unit length, and θ is the angle made by the deflected string with the x axis, Eq. (8.1) can be rewritten, for an element of length dx, as

$$(P + dP)\sin(\theta + d\theta) + f\,dx - P\sin\theta = \rho\,dx\frac{\partial^2 w}{\partial t^2} \qquad (8.2)$$

Noting that

$$dP = \frac{\partial P}{\partial x}\,dx \qquad (8.3)$$

$$\sin\theta \approx \tan\theta = \frac{\partial w}{\partial x} \qquad (8.4)$$

$$\sin(\theta + d\theta) \approx \tan(\theta + d\theta) = \frac{\partial w}{\partial x} + \frac{\partial^2 w}{\partial x^2}\,dx \qquad (8.5)$$

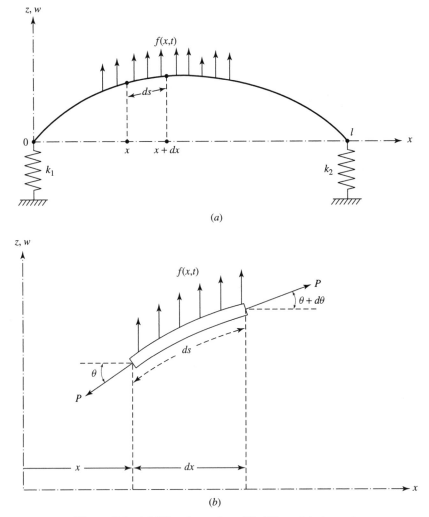

Figure 8.1 (*a*) Vibrating string; (*b*) differential element.

Eq. (8.2) can be expressed as

$$\frac{\partial}{\partial x}\left[P\frac{\partial w(x,t)}{\partial x}\right] + f(x,t) = \rho(x)\frac{\partial^2 w(x,t)}{\partial t^2}$$ (8.6)

If the string is uniform and the tension is constant, Eq. (8.6) takes the form

$$P\frac{\partial^2 w(x,t)}{\partial x^2} + f(x,t) = \rho\frac{\partial^2 w(x,t)}{\partial t^2}$$ (8.7)

For free vibration, $f(x,t) = 0$ and Eq. (8.7) reduces to

$$P\frac{\partial^2 w(x,t)}{\partial x^2} = \rho\frac{\partial^2 w(x,t)}{\partial t^2}$$ (8.8)

which can be rewritten as

$$c^2 \frac{\partial^2 w}{\partial x^2} = \frac{\partial^2 w}{\partial t^2} \tag{8.9}$$

where

$$c = \left(\frac{P}{\rho}\right)^{1/2} \tag{8.10}$$

Equation (8.9) is called the *one-dimensional wave equation*.

8.2.2 Variational Approach

Strain Energy   There are three sources of strain energy for the taut string shown in Fig. 8.1. The first is due to the deformation of the string over $0 \leq x \leq l$, where the tension $P(x)$ tries to restore the deflected string to the equilibrium position; the second is due to the deformation of the spring at $x = 0$; and the third is due to the deformation of the spring at $x = l$. The length of a differential element dx in the deformed position, ds, can be expressed as

$$ds = \left[(dx)^2 + \left(\frac{\partial w}{\partial x} dx\right)^2\right]^{1/2} = \left[1 + \left(\frac{\partial w}{\partial x}\right)^2\right]^{1/2} dx \approx \left[1 + \frac{1}{2}\left(\frac{\partial w}{\partial x}\right)^2\right] dx \tag{8.11}$$

by assuming the slope of the deflected string, $\partial w / \partial x$, to be small. The strain energy due to the deformations of the springs at $x = 0$ and $x = l$ is given by $\frac{1}{2}k_1 w^2(0,t)$ and $\frac{1}{2}k_2 w^2(l,t)$, and the strain energy associated with the deformation of the string is given by the work done by the tensile force $P(x)$ while moving through the distance $ds - dx$:

$$\int_0^l P(x)[ds(x,t) - dx] = \frac{1}{2}\int_0^l P(x)\left[\frac{\partial w(x,t)}{\partial x}\right]^2 dx \tag{8.12}$$

Thus, the total strain energy, π, is given by

$$\pi = \frac{1}{2}\int_0^l P(x)\left[\frac{\partial w(x,t)}{\partial x}\right]^2 dx + \frac{1}{2}k_1 w^2(0,t) + \frac{1}{2}k_2 w^2(l,t) \tag{8.13}$$

Kinetic Energy   The kinetic energy of the string is given by

$$T = \frac{1}{2}\int_0^l \rho(x)\left[\frac{\partial w(x,t)}{\partial t}\right]^2 dx \tag{8.14}$$

where $\rho(x)$ is the mass of the string per unit length.

Work Done by External Forces   The work done by the nonconservative distributed load acting on the string, $f(x,t)$, can be expressed as

$$W = \int_0^l f(x,t)w(x,t)\,dx \tag{8.15}$$

Hamilton's principle gives

$$\delta \int_{t_1}^{t_2} (T - \pi + W)\, dt = 0 \tag{8.16}$$

or

$$\delta \int_{t_1}^{t_2} \left[\frac{1}{2} \int_0^l \rho(x) \left(\frac{\partial w}{\partial t} \right)^2 dx - \frac{1}{2} \int_0^l P(x) \left(\frac{\partial w}{\partial x} \right)^2 dx - \frac{1}{2} k_1 w^2(0,t) \right.$$
$$\left. - \frac{1}{2} k_2 w^2(l,t) + \int_0^l f(x,t) w\, dx \right] dt = 0 \tag{8.17}$$

The variations of the individual terms appearing in Eq. (8.17) can be carried out as follows:

$$\int_{t_1}^{t_2} \delta T\, dt = \int_{t_1}^{t_2} dt \int_0^l \rho(x) \frac{\partial w}{\partial t} \delta \frac{\partial w}{\partial t}\, dx = \int_{t_1}^{t_2} dt \int_0^l \rho \frac{\partial w}{\partial t} \frac{\partial (\delta w)}{\partial t}\, dx \tag{8.18}$$

using the interchangeability of the variation and differentiation processes. Equation (8.18) can be evaluated by using integration by parts with respect to time:

$$\int_0^l \left[\int_{t_1}^{t_2} \rho \frac{\partial w}{\partial t} \frac{\partial (\delta w)}{\partial t}\, dt \right] dx = \int_0^l \left\{ \left(\rho \frac{\partial w}{\partial t} \delta w \Big|_{t_1}^{t_2} \right) \right.$$
$$\left. - \int_0^l \left[\int_{t_1}^{t_2} \frac{\partial}{\partial t} \left(\rho \frac{\partial w}{\partial t} \right) \delta w\, dt \right] \right\} dx \tag{8.19}$$

Using the fact that $\delta w = 0$ at $t = t_1$ and $t = t_2$ and assuming $\rho(x)$ to be constant, Eq. (8.19) yields

$$\int_{t_1}^{t_2} \delta T\, dt = - \int_{t_1}^{t_2} \left(\int_0^l \rho \frac{\partial^2 w}{\partial t^2} \delta w\, dx \right) dt \tag{8.20}$$

The second term of Eq. (8.16) can be written as

$$\int_{t_1}^{t_2} \delta \pi\, dt = \int_{t_1}^{t_2} \left[\int_0^l P(x) \frac{\partial w}{\partial x} \frac{\partial}{\partial x} (\delta w)\, dx + k_1 w(0,t) \delta w(0,t) \right.$$
$$\left. + k_2 w(l,t) \delta w(l,t) \right] dt \tag{8.21}$$

By using integration by parts with respect to x, Eq. (8.21) can be expressed as

$$\int_{t_1}^{t_2} \delta \pi\, dt = \int_{t_1}^{t_2} \left[P \frac{\partial w}{\partial x} \delta w \Big|_0^l - \int_0^l \frac{\partial}{\partial x} \left(P \frac{\partial w}{\partial x} \right) \delta w\, dx + k_1 w(0,t) \delta w(0,t) \right.$$
$$\left. + k_2 w(l,t) \delta w(l,t) \right] dt \tag{8.22}$$

The third term of Eq. (8.16) can be written as

$$\int_{t_1}^{t_2} \delta W\, dt = \int_{t_1}^{t_2} \left[\int_0^l f(x,t) \delta w(x,t)\, dx \right] dt \tag{8.23}$$

By inserting Eqs. (8.20), (8.22), and (8.23) into Eq. (8.16) and collecting the terms, we obtain

$$\int_{t_1}^{t_2} \left\{ \int_0^l \left[-\rho \frac{\partial^2 w}{\partial t^2} + \frac{\partial}{\partial x}\left(P\frac{\partial w}{\partial x}\right) + f \right] \delta w \, dx + \left(P\frac{\partial w}{\partial x} - k_1 w \right) \delta w \Big|_{x=0} \right.$$
$$\left. - \left(P\frac{\partial w}{\partial x} + k_2 w \right) \delta w|_{x=l} \right\} dt = 0 \tag{8.24}$$

Since the variation δw over the interval $0 < x < l$ is arbitrary, Eq. (8.24) can be satisfied only when the individual terms of Eq. (8.24) are equal to zero:

$$\frac{\partial}{\partial x}\left(P\frac{\partial w}{\partial x}\right) + f = \rho \frac{\partial^2 w}{\partial t^2}, \qquad 0 < x < l \tag{8.25}$$

$$\left(P\frac{\partial w}{\partial x} - k_1 w \right) \delta w = 0, \qquad x = 0 \tag{8.26}$$

$$\left(P\frac{\partial w}{\partial x} + k_2 w \right) \delta w = 0, \qquad x = l \tag{8.27}$$

Equation (8.25) denotes the equation of motion while Eqs. (8.26) and (8.27) represent the boundary conditions. Equation (8.26) can be satisfied when $w(0,t) = 0$ or when $P[\partial w/\partial x](0,t) - k_1 w(0,t) = 0$. Since the displacement w cannot be zero for all time at $x = 0$, Eq. (8.26) can only be satisfied by setting

$$P\frac{\partial w}{\partial x} - k_1 w = 0 \qquad \text{at } x = 0 \tag{8.28}$$

Similarly, Eq. (8.27) leads to the condition

$$P\frac{\partial w}{\partial x} + k_2 w = 0 \qquad \text{at } x = l \tag{8.29}$$

Thus, the differential equation of motion of the string is given by Eq. (8.25) and the corresponding boundary conditions by Eqs. (8.28) and (8.29).

8.3 INITIAL AND BOUNDARY CONDITIONS

The equation of motion, Eq. (8.6), or its special forms (8.7) and (8.8) or (8.9), is a partial differential equation of order 2 in x as well as t. Thus, two boundary conditions and two initial conditions are required to find the solution, $w(x,t)$. If the string is given an initial deflection $w_0(x)$ and an initial velocity $\dot{w}_0(x)$, the initial conditions can be stated as

$$w(x, t = 0) = w_0(x)$$
$$\frac{\partial w}{\partial t}(x, t = 0) = \dot{w}_0(x) \tag{8.30}$$

If the string is fixed at $x = 0$, the displacement is zero and hence the boundary conditions will be

$$w(x = 0, t) = 0, \qquad t \geq 0 \tag{8.31}$$

If the string is connected to a pin that is free to move in a transverse direction, the end will not be able to support any transverse force, and hence the boundary condition will be

$$P(x)\frac{\partial w}{\partial x}(x,t) = 0 \tag{8.32}$$

If the axial force is constant and the end $x = 0$ is free, Eq. (8.32) becomes

$$\frac{\partial w}{\partial x}(0,t) = 0, \qquad t \geq 0 \tag{8.33}$$

If the end $x = 0$ of the string is connected to an elastic spring of stiffness k, the boundary condition will be

$$P(x)\left.\frac{\partial w}{\partial x}(x,t)\right|_{x=0} = kw(x,t)\Big|_{x=0}, \qquad t \geq 0 \tag{8.34}$$

Some of the possible boundary conditions of a string are summarized in Table 8.1.

8.4 FREE VIBRATION OF AN INFINITE STRING

Consider a string of infinite length. The free vibration equation of the string, Eq. (8.9), is solved using three different approaches in this section.

8.4.1 Traveling-Wave Solution

The solution of the wave equation (8.9) can be expressed as

$$w(x,t) = F_1(x - ct) + F_2(x + ct) \tag{8.35}$$

where F_1 and F_2 are arbitrary functions of $(x - ct)$ and $(x + ct)$, respectively. The solution given by Eq. (8.35) is known as *D'Alembert's solution*. The validity of Eq. (8.35) can be established by differentiating Eq. (8.35) as

$$\frac{\partial^2 w}{\partial x^2}(x,t) = F_1''(x - ct) + F_2''(x + ct) \tag{8.36}$$

$$\frac{\partial^2 w}{\partial t^2}(x,t) = c^2 F_1''(x - ct) + c^2 F_2''(x + ct) \tag{8.37}$$

where a prime indicates a derivative with respect to the respective argument. By substituting Eqs. (8.36) and (8.37) into Eq. (8.9), we find that Eq. (8.9) is satisfied. The functions $F_1(x - ct)$ and $F_2(x + ct)$ denote waves that propagate in the positive and negative directions of the x axis, respectively, with a velocity c. The functions F_1 and F_2 can be determined from the known initial conditions of the string. Using the initial conditions of Eq. (8.30), Eq. (8.35) yields

$$F_1(x) + F_2(x) = w_0(x) \tag{8.38}$$

$$-cF_1'(x) + cF_2'(x) = \dot{w}_0(x) \tag{8.39}$$

Table 8.1   Boundary Conditions of a String

Support conditions of the string	Boundary conditions to be satisfied
1. Both ends fixed	

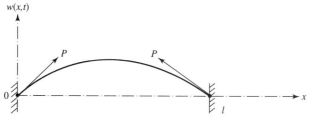

$$w(0,t) = 0$$
$$w(l,t) = 0$$

2. Both ends free

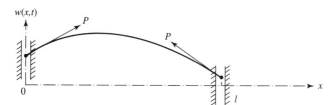

$$\frac{\partial w}{\partial x}(0,t) = 0$$

$$\frac{\partial w}{\partial x}(l,t) = 0$$

3. Both ends attached with masses

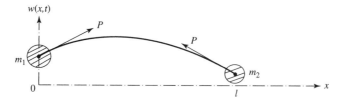

$$m_1\frac{\partial^2 w}{\partial t^2}(0,t) = P\frac{\partial w}{\partial x}(0,t)$$

$$-m_2\frac{\partial^2 w}{\partial t^2}(l,t) = P\frac{\partial w}{\partial x}(l,t)$$

4. Both ends attached with springs

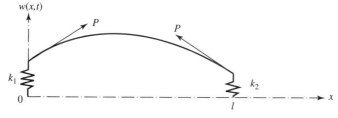

$$k_1 w(0,t) = P\frac{\partial w}{\partial x}(0,t)$$

$$-k_2 w(l,t) = P\frac{\partial w}{\partial x}(l,t)$$

5. Both ends attached with dampers

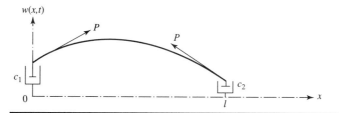

$$c_1\frac{\partial w}{\partial t}(0,t) = P\frac{\partial w}{\partial x}(0,t)$$

$$-c_2\frac{\partial w}{\partial t}(l,t) = P\frac{\partial w}{\partial x}(l,t)$$

where a prime in Eq. (8.39) denotes a derivative with respect to the respective argument at $t = 0$ (i.e., with respect to x). By integrating Eq. (8.39) with respect to x, we obtain

$$-F_1(x) + F_2(x) = \frac{1}{c} \int_{x_0}^{x} \dot{w}_0(\overline{x}) \, d\overline{x} \qquad (8.40)$$

where x_0 is a constant. Equations (8.38) and (8.40) can be solved to find $F_1(x)$ and $F_2(x)$ as

$$F_1(x) = \frac{1}{2} \left[w_0(x) - \frac{1}{c} \int_{x_0}^{x} \dot{w}_0(\overline{x}) \, d\overline{x} \right] \qquad (8.41)$$

$$F_2(x) = \frac{1}{2} \left[w_0(x) + \frac{1}{c} \int_{x_0}^{x} \dot{w}_0(\overline{x}) \, d\overline{x} \right] \qquad (8.42)$$

By replacing x by $x - ct$ and $x + ct$, respectively, in Eqs. (8.41) and (8.42), we can express the wave solution of the string, $w(x,t)$, as

$$w(x,t) = F_1(x - ct) + F_2(x + ct)$$
$$= \frac{1}{2} \left[w_0(x - ct) + w_0(x + ct) + \frac{1}{2c} \int_{x-ct}^{x+ct} \dot{w}_0(\overline{x}) \, d\overline{x} \right] \qquad (8.43)$$

The solution given by Eq. (8.43) can be rewritten as

$$w(x,t) = w_d(x,t) + w_v(x,t) \qquad (8.44)$$

where $w_d(x,t)$ represents a wave propagating due to a known initial displacement $w_0(x)$ with zero initial velocity, and $w_v(x,t)$ indicates a wave moving due to the initial velocity $\dot{w}_0(x)$ with zero initial displacement. A typical wave traveling due to initial displacement (introduced by pulling the string slightly in the transverse direction with zero velocity) is shown in Fig. 8.2.

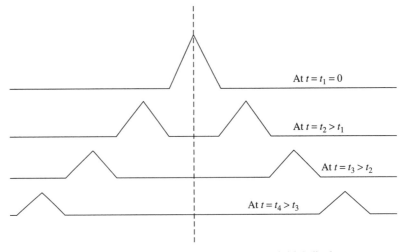

Figure 8.2   Propagation of a wave caused by an initial displacement.

8.4.2   Fourier Transform–Based Solution

To find the free vibration response of an infinite string $(-\infty < x < \infty)$ under the initial conditions of Eq. (8.30), we take the Fourier transform of Eq. (8.9). For this, we multiply Eq. (8.9) by e^{iax} and integrate from $x = -\infty$ to $x = \infty$:

$$\int_{-\infty}^{\infty} \frac{\partial^2 w(x,t)}{\partial x^2} e^{iax}\, dx = \frac{1}{c^2} \frac{\partial^2}{\partial t^2} \int_{-\infty}^{\infty} w(x,t) e^{iax}\, dx \qquad (8.45)$$

Integration of the left-hand side of Eq. (8.45) by parts results in

$$\int_{-\infty}^{\infty} \frac{\partial^2 w(x,t)}{\partial x^2} e^{iax}\, dx = \frac{\partial w}{\partial x} e^{iax} \Big|_{-\infty}^{\infty} - ia \int_{-\infty}^{\infty} \frac{\partial w}{\partial x} e^{iax}\, dx$$

$$= \frac{\partial w}{\partial x} e^{iax} \Big|_{-\infty}^{\infty} - \left[iaw e^{iax} \Big|_{-\infty}^{\infty} - (ia)^2 \int_{-\infty}^{\infty} w e^{iax}\, dx \right]$$

$$= \left(\frac{\partial w}{\partial x} - iaw \right) e^{iax} \Big|_{-\infty}^{\infty} - a^2 \int_{-\infty}^{\infty} w e^{iax}\, dx \qquad (8.46)$$

Assuming that both w and $\partial w / \partial x$ tend to zero as $|x| \to \infty$, the first term on the right-hand side of Eq. (8.46) vanishes. Using Eq. (7.16), the Fourier transform of $w(x,t)$ is defined as

$$W(a,t) = \frac{1}{\sqrt{2\pi}} \int_{-\infty}^{\infty} w(x,t) e^{iax}\, dx \qquad (8.47)$$

and Eq. (8.45) can be rewritten in the form

$$\frac{1}{c^2} \frac{d^2 W}{dt^2} + a^2 W = 0 \qquad (8.48)$$

Note that the use of the Fourier transform reduced the partial differential equation (8.9) into the ordinary differentiation equation (8.48). The solution of Eq. (8.48) can be expressed as

$$W(a,t) = C_1 e^{iact} + C_2 e^{-iact} \qquad (8.49)$$

where the constants C_1 and C_2 can be evaluated using the initial conditions, Eqs. (8.30). By taking the Fourier transforms of the initial displacement $[w = w_0(x)]$ and initial velocity $[\partial w / \partial t = \dot{w}_0(x)]$, we obtain

$$W(a, t = 0) = \frac{1}{\sqrt{2\pi}} \int_{-\infty}^{\infty} w_0(x) e^{iax}\, dx = W_0(a) \qquad (8.50)$$

$$\frac{dW}{dt}(a, t = 0) = \frac{1}{\sqrt{2\pi}} \int_{-\infty}^{\infty} \dot{w}_0(x) e^{iax}\, dx = \dot{W}_0(a) \qquad (8.51)$$

The use of Eqs. (8.50) and (8.51) in (8.49) leads to

$$W_0(a) = C_1 + C_2 \tag{8.52}$$

$$\dot{W}_0(a) = iac(C_1 - C_2) \tag{8.53}$$

whose solution gives

$$C_1 = \frac{W_0}{2} + \frac{1}{2iac}\dot{W}_0 \tag{8.54}$$

$$C_2 = \frac{W_0}{2} - \frac{1}{2iac}\dot{W}_0 \tag{8.55}$$

Thus, Eq. (8.49) can be expressed as

$$W(a,t) = \frac{1}{2}W_0(a)(e^{iact} + e^{-iact}) + \frac{1}{2iac}\dot{W}_0(a)(e^{iact} - e^{-iact}) \tag{8.56}$$

By using the inverse Fourier transform of Eq. (8.47), we obtain

$$w(x,t) = \frac{1}{\sqrt{2\pi}} \int_{-\infty}^{\infty} W(a,t)e^{-iax}\, da \tag{8.57}$$

which, in view of Eq. (8.56), becomes

$$w(x,t) = \frac{1}{2}\left\{ \frac{1}{\sqrt{2\pi}} \int_{-\infty}^{\infty} W_0(a)[e^{-ia(x-ct)} + e^{-ia(x+ct)}]\, da \right\}$$
$$+ \frac{1}{2c}\left\{ \frac{1}{\sqrt{2\pi}} \int_{-\infty}^{\infty} \frac{\dot{W}_0(a)}{ia}[e^{-ia(x-ct)} - e^{-ia(x+ct)}]\, da \right\} \tag{8.58}$$

Note that the inverse Fourier transforms of $W_0(a)$ and $\dot{W}_0(a)$, Eqs. (8.50) and (8.51), can be obtained as

$$w_0(x) = \frac{1}{\sqrt{2\pi}} \int_{-\infty}^{\infty} W_0(a)e^{-iax}\, da \tag{8.59}$$

$$\dot{w}_0(\xi) = \frac{1}{\sqrt{2\pi}} \int_{-\infty}^{\infty} \dot{W}_0(a)e^{-ia\xi}\, da \tag{8.60}$$

so that

$$w_0(x \mp ct) = \frac{1}{\sqrt{2\pi}} \int_{-\infty}^{\infty} W_0(a)e^{-ia(x\mp ct)}\, da \tag{8.61}$$

By integrating Eq. (8.60) with respect to ξ from $x - ct$ to $x + ct$, we obtain

$$\int_{x-ct}^{x+ct} \dot{w}_0(\xi)\, d\xi = \frac{1}{\sqrt{2\pi}} \int_{-\infty}^{\infty} \frac{\dot{W}_0(a)}{ia}[e^{-ia(x-ct)} - e^{-ia(x+ct)}]\, da \tag{8.62}$$

When Eqs. (8.61) and (8.62) are substituted into Eq. (8.58), we obtain

$$w(x,t) = \frac{1}{2}[w_0(x + ct) + w_0(x - ct)] + \frac{1}{2c}\int_{x-ct}^{x+ct} \dot{w}_0(\xi)\,d\xi \qquad (8.63)$$

which can be seen to be identical to Eq. (8.43).

8.4.3   Laplace Transform–Based Solution

The Laplace transforms of the terms in the governing equation (8.9) lead to

$$L\left[\frac{\partial^2 w}{\partial x^2}\right] = \frac{d^2 W(x, s)}{dx^2} \qquad (8.64)$$

$$L\left[\frac{\partial^2 w}{\partial t^2}\right] = s^2 W(x, s) - sw(x, 0) - \frac{\partial w}{\partial t}(x, 0) \qquad (8.65)$$

where

$$W(x, s) = \int_0^\infty e^{-st} w(x,t)\,dt \qquad (8.66)$$

Using Eqs. (8.64) and (8.65) along with the initial conditions of Eq. (8.30), Eq. (8.9) can be expressed as

$$c^2 \frac{d^2 W}{dx^2} = s^2 W - sw_0(x) - \dot{w}_0(x) \qquad (8.67)$$

Now, we take the Fourier transform of Eq. (8.67). For this, we multiply Eq. (8.67) by e^{ipx} and integrate with respect to x from $-\infty$ to $+\infty$, to obtain

$$c^2 \int_{-\infty}^{\infty} \frac{d^2 W}{dx^2} e^{ipx}\,dx = s^2 \int_{-\infty}^{\infty} W e^{ipx}\,dx - s\int_{-\infty}^{\infty} w_0(x)e^{ipx}\,dx - \int_{-\infty}^{\infty} \dot{w}_0(x)e^{ipx}\,dx \qquad (8.68)$$

The integral on the left-hand side of Eq. (8.68) can be evaluated by parts:

$$\int_{-\infty}^{\infty} \frac{d^2 W}{dx^2} e^{ipx}\,dx = \frac{dW}{dx} e^{ipx}\Big|_{-\infty}^{+\infty} - \int_{-\infty}^{\infty} \frac{dW}{dx} ipe^{ipx}\,dx$$

$$= \frac{dW}{dx} e^{ipx}\Big|_{-\infty}^{+\infty} - ipW e^{ipx}\Big|_{-\infty}^{+\infty} - p^2 \int_{-\infty}^{+\infty} W e^{ipx}\,dx \qquad (8.69)$$

Assuming that the deflection, $W(x, s)$, and the slope, $dW(x, s)/dx$, tend to be zero as $x \rightarrow \pm\infty$, Eq. (8.69) reduces to

$$\int_{-\infty}^{+\infty} \frac{d^2 W}{dx^2} e^{ipx}\,dx = -p^2 \int_{-\infty}^{+\infty} W e^{ipx}\,dx \qquad (8.70)$$

and hence Eq. (8.68) can be rewritten as

$$(c^2 p^2 + s^2) \int_{-\infty}^{+\infty} W(x,s)e^{ipx}\,dx = s \int_{-\infty}^{+\infty} w_0(x)e^{ipx}\,dx + \int_{-\infty}^{+\infty} \dot{w}_0(x)e^{ipx}\,dx$$

or

$$\overline{W}(c^2 p^2 + s^2) = s\overline{W}_0 + \overline{\dot{W}}_0$$

or

$$\overline{W} = \frac{s\overline{W}_0 + \overline{\dot{W}}_0}{c^2 p^2 + s^2} \tag{8.71}$$

where

$$\overline{W}(p,s) = \frac{1}{\sqrt{2\pi}} \int_{-\infty}^{+\infty} W(x,s)e^{ipx}\,dx \tag{8.72}$$

$$\overline{W}_0(p) = \frac{1}{\sqrt{2\pi}} \int_{-\infty}^{+\infty} w_0(x)e^{ipx}\,dx \tag{8.73}$$

$$\overline{\dot{W}}_0(p) = \frac{1}{\sqrt{2\pi}} \int_{-\infty}^{+\infty} \dot{w}_0(x)e^{ipx}\,dx \tag{8.74}$$

Now we first take the inverse Fourier transform of $\overline{W}(p,s)$ to obtain

$$W(x,s) = \frac{1}{\sqrt{2\pi}} \int_{-\infty}^{\infty} \frac{s\overline{W}_0(p) + \overline{\dot{W}}_0(p)}{c^2 p^2 + s^2}e^{-ipx}\,dx \tag{8.75}$$

and next we take the inverse Laplace transform of $W(x,s)$ to obtain

$$w(x,t) = L^{-1}[W(x,s)] \tag{8.76}$$

Noting that

$$L^{-1}\left[\frac{s}{c^2 p^2 + s^2}\right] = \cos pct \tag{8.77}$$

and

$$L^{-1}\left[\frac{1}{c^2 p^2 + s^2}\right] = \frac{1}{pc}\sin pct \tag{8.78}$$

Eqs. (8.76) and (8.75) yield

$$w(x,t) = \frac{1}{\sqrt{2\pi}} \int_{-\infty}^{\infty} \left[\overline{W}_0(p)\cos pct + \frac{1}{pc}\overline{\dot{W}}_0(p)\sin pct\right]e^{-ipx}\,dp \tag{8.79}$$

where

$$w_0(x) = \frac{1}{\sqrt{2\pi}} \int_{-\infty}^{\infty} \overline{W}_0(p)e^{-ipx}\,dp \tag{8.80}$$

$$\dot{w}_0(x) = \frac{1}{\sqrt{2\pi}} \int_{-\infty}^{\infty} \overline{\dot{W}}_0(p)e^{-ipx}\,dp \tag{8.81}$$

From Eqs. (8.80) and (8.81), we can write

$$w_0(x \pm ct) = \frac{1}{\sqrt{2\pi}} \int_{-\infty}^{\infty} \overline{W}_0(p) e^{-ip(x \pm ct)} \, dp \tag{8.82}$$

$$\int_{x-ct}^{x+ct} \dot{w}_0(\xi) \, d\xi = \frac{1}{\sqrt{2\pi}} \int_{-\infty}^{\infty} \frac{\overline{\dot{W}}_0(p)}{ip} [e^{-ip(x-ct)} - e^{ip(x+ct)}] \, dp \tag{8.83}$$

In addition, the following identities are valid:

$$\cos pct = \frac{1}{2}(e^{ipct} + e^{-ipct}) \tag{8.84}$$

$$\sin pct = \frac{1}{2i}(e^{ipct} - e^{-ipct}) \tag{8.85}$$

Thus Eq. (8.79) can be rewritten as

$$w(x,t) = \frac{1}{2}[w_0(x + ct) + w_0(x - ct)] + \frac{1}{2c} \int_{x-ct}^{x+ct} \dot{w}_0(\xi) \, d\xi \tag{8.86}$$

which can be seen to be the same as the solution given by Eqs. (8.43) and (8.63). Note that Fourier transforms were used in addition to Laplace transforms in the current approach.

8.5   FREE VIBRATION OF A STRING OF FINITE LENGTH

The solution of the free vibration equation, Eq. (8.9), can be found using the method of separation of variables. In this method, the solution is written as

$$w(x,t) = W(x)T(t) \tag{8.87}$$

where $W(x)$ is a function of x only and $T(t)$ is a function of t only. By substituting Eq. (8.87) into Eq. (8.9), we obtain

$$\frac{c^2}{W} \frac{d^2 W}{dx^2} = \frac{1}{T} \frac{d^2 T}{dt^2} \tag{8.88}$$

Noting that the left-hand side of Eq. (8.88) depends only on x while the right-hand side depends only on t, their common value must be a constant, a, and hence

$$\frac{c^2}{W} \frac{d^2 W}{dx^2} = \frac{1}{T} \frac{d^2 T}{dt^2} = a \tag{8.89}$$

Equation (8.89) can be written as two separate equations:

$$\frac{d^2 W}{dx^2} - \frac{a}{c^2} W = 0 \tag{8.90}$$

$$\frac{d^2 T}{dx^2} - aT = 0 \tag{8.91}$$

The constant a is usually negative[1] and hence, by setting $a = -\omega^2$, Eqs. (8.90) and (8.91) can be rewritten as

$$\frac{d^2 W}{dx^2} + \frac{\omega^2}{c^2} W = 0 \tag{8.92}$$

$$\frac{d^2 T}{dt^2} + \omega^2 T = 0 \tag{8.93}$$

The solution of Eqs. (8.92) and (8.93) can be expressed as

$$W(x) = A \cos \frac{\omega x}{c} + B \sin \frac{\omega x}{c} \tag{8.94}$$

$$T(t) = C \cos \omega t + D \sin \omega t \tag{8.95}$$

where ω is the frequency of vibration, the constants A and B can be evaluated from the boundary conditions, and the constants C and D can be determined from the initial conditions of the string.

8.5.1 Free Vibration of a String with Both Ends Fixed

If the string is fixed at both ends, the boundary conditions are given by

$$w(0,t) = w(l,t) = 0, \qquad t \geq 0 \tag{8.96}$$

Equations (8.96) and (8.94) yield

$$W(0) = 0 \tag{8.97}$$

$$W(l) = 0 \tag{8.98}$$

The condition of Eq. (8.97) requires that

$$A = 0 \tag{8.99}$$

in Eq. (8.94). Using Eqs. (8.98) and (8.99) in Eq. (8.94), we obtain

$$B \sin \frac{\omega l}{c} = 0 \tag{8.100}$$

---

[1]To show that a is usually a negative quantity, multiply Eq. (8.90) by $W(x)$ and integrate with respect to x from 0 to l to obtain

$$\int_0^l W(x) \frac{d^2 W(x)}{dx^2} \, dx = \frac{a}{c^2} \int_0^l W^2(x) \, dx \tag{a}$$

Equation (a) indicates that the sign of a will be same as the sign of the integral on the left-hand side. The left-hand side of Eq. (a) can be integrated by parts to obtain

$$\int_0^l W(x) \frac{d^2 W(x)}{dx^2} \, dx = W(x) \frac{dW(x)}{dx} \Big|_0^l - \int_0^l \left[\frac{dW(x)}{dx} \right]^2 dx \tag{b}$$

The first term on the right-hand side of Eq. (b) can be seen to be zero or negative for a string with any combination of fixed end ($W = 0$), free end ($dW/dx = 0$), or elastically supported end ($P\,dW/dx = -kW$), where k is the spring constant of the elastic support. Thus, the integral on the left-hand side of Eq. (a), and hence the sign of a is negative.

For a nontrivial solution, B cannot be zero and hence

$$\sin \frac{\omega l}{c} = 0 \tag{8.101}$$

Equation (8.101) is called the *frequency* or *characteristic equation*, and the values of ω that satisfy Eq. (8.101) are called the *eigenvalues* (or *characteristic values* or *natural frequencies*) of the string. The nth root of Eq. (8.101) is given by

$$\frac{\omega_n l}{c} = n\pi, \qquad n = 1, 2, \ldots \tag{8.102}$$

and hence the nth natural frequency of vibration of the string is given by

$$\omega_n = \frac{nc\pi}{l}, \qquad n = 1, 2, \ldots \tag{8.103}$$

The transverse displacement of the string, corresponding to ω_n, known as the nth *mode of vibration* or nth *harmonic* or nth *normal mode* of the string is given by

$$w_n(x,t) = W_n(x)T_n(t) = \sin \frac{n\pi x}{l} \left(C_n \cos \frac{nc\pi t}{l} + D_n \sin \frac{nc\pi t}{l} \right) \tag{8.104}$$

In the nth mode, each point of the string vibrates with an amplitude proportional to the value of W_n at that point with a circular frequency ω_n. The first four modes of vibration are shown in Fig. 8.3. The mode corresponding to $n = 1$ is called the *fundamental mode*, ω_1 is called the *fundamental frequency*, and

$$\tau_1 = \frac{2\pi}{\omega_1} = \frac{2l}{c} \tag{8.105}$$

is called the *fundamental period*. The points at which $w_n = 0$ for $t \geq 0$ are called *nodes*. It can be seen that the fundamental mode has two nodes (at $x = 0$ and $x = l$), the second mode has three nodes (at $x = 0$, $x = l/2$, and $x = l$), and so on.

The free vibration of the string, which satisfies the boundary conditions of Eqs. (8.97) and (8.98), can be found by superposing all the natural modes $w_n(x)$ as

$$w(x,t) = \sum_{n=1}^{\infty} w_n(x,t) = \sum_{n=1}^{\infty} \sin \frac{n\pi x}{l} \left(C_n \cos \frac{nc\pi t}{l} + D_n \sin \frac{nc\pi t}{l} \right) \tag{8.106}$$

This equation represents the general solution of Eq. (8.9) and includes all possible vibrations of the string. The particular vibration that occurs is uniquely determined by the initial conditions specified. The initial conditions provide unique values of the constants C_n and D_n in Eq. (8.106). For the initial conditions stated in Eq. (8.30), Eq. (8.106) yields

$$\sum_{n=1}^{\infty} C_n \sin \frac{n\pi x}{l} = w_0(x) \tag{8.107}$$

$$\sum_{n=1}^{\infty} \frac{nc\pi}{l} D_n \sin \frac{n\pi x}{l} = \dot{w}_0(x) \tag{8.108}$$

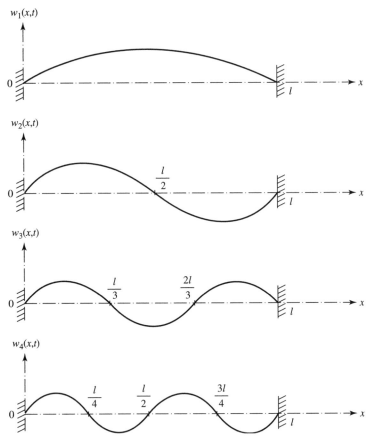

Figure 8.3   Mode shapes of a string.

Noting that Eqs. (8.107) and (8.108) denote Fourier sine series expansions of $w_0(x)$ and $\dot{w}_0(x)$ in the interval $0 \leq x \leq l$, the values of C_n and D_n can be determined by multiplying Eqs. (8.107) and (8.108) by $\sin \frac{n\pi x}{l}$ and integrating with respect to x from 0 to l. This gives the constants C_n and D_n as

$$C_n = \frac{2}{l} \int_0^l w_0(x) \sin \frac{n\pi x}{l}\, dx \tag{8.109}$$

$$D_n = \frac{2}{nc\pi} \int_0^l \dot{w}_0(x) \sin \frac{n\pi x}{l}\, dx \tag{8.110}$$

Note that the solution given by Eq. (8.106) represents the method of mode superposition since the response is expressed as a superposition of the normal modes. As indicated earlier, the procedure is applicable in finding not only the free vibration response but also the forced vibration response of any continuous system.

Example 8.1   Find the free vibration response of a fixed–fixed string whose middle point is pulled out by a distance h and then let it go at time $t = 0$ as shown in Fig. 8.4.

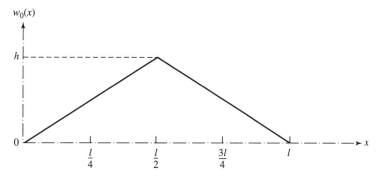

Figure 8.4 Initial deflection of the string.

SOLUTION   The free vibration solution if the string is given by Eq. (8.106) with C_n and D_n given by Eqs. (8.109) and (8.110), respectively. In the present case, the initial displacement can be represented as

$$w_0(x) = \begin{cases} \dfrac{2\,hx}{l} & \text{for } 0 \leq x \leq \dfrac{l}{2} \\[2mm] \dfrac{2\,h(l-x)}{l} & \text{for } \dfrac{l}{2} \leq x \leq l \end{cases} \tag{E8.1.1}$$

and the initial velocity is zero:

$$\dot{w}_0(x) = 0 \tag{E8.1.2}$$

Equations (E8.1.2) and (8.106) yield

$$D_n = 0 \tag{E8.1.3}$$

Thus, the free vibration response becomes

$$w(x,t) = \sum_{n=1}^{\infty} C_n \sin \frac{n\pi x}{l} \cos \frac{nc\pi t}{l} \tag{E8.1.4}$$

The constant C_n can be evaluated using Eq. (8.109) as

$$C_n = \frac{2}{l} \int_0^l w_0(x) \sin \frac{n\pi x}{l}\, dx$$

$$= \frac{2}{l} \left[\int_0^{l/2} \frac{2hx}{l} \sin \frac{n\pi x}{l}\, dx + \int_{l/2}^{l} \frac{2h}{l}(l-x) \sin \frac{n\pi x}{l}\, dx \right]$$

$$= \begin{cases} \dfrac{8h}{\pi^2 n^2} \sin \dfrac{n\pi}{2} & \text{for } n = 1, 3, 5, \ldots \\[2mm] 0 & \text{for } n = 2, 4, 6, \ldots \end{cases} \tag{E8.1.5}$$

Noting the relation

$$\sin \frac{n\pi}{2} = (-1)^{(n-1)/2}, \qquad n = 1, 3, 5, \dots \qquad \text{(E8.1.6)}$$

the free vibration response of the string can be expressed as

$$w(x,t) = \frac{8h}{\pi^2} \left(\sin \frac{\pi x}{l} \cos \frac{\pi ct}{l} - \frac{1}{9} \sin \frac{3\pi x}{l} \cos \frac{3\pi ct}{l} + \cdots \right) \qquad \text{(E8.1.7)}$$

The solution given by Eq. (E8.1.7), using different number of terms, is shown in Fig. 8.5. The fast convergence of the series of Eq. (E8.1.7) can be seen from the figure.

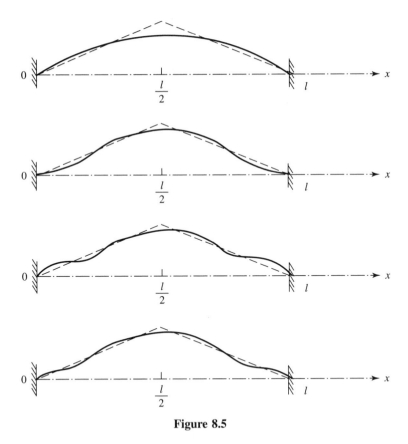

Figure 8.5

Example 8.2 A string of length l fixed at both ends is struck at $t = 0$ such that the initial displacement distribution is zero and the initial velocity distribution is given by

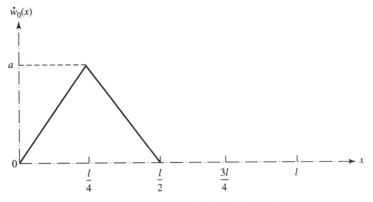

Figure 8.6   Initial velocity of the string

(Fig. 8.6):

$$\dot{w}_0(x) = \begin{cases} \dfrac{4ax}{l}, & 0 \le x \le \dfrac{l}{4} \\[3mm] \dfrac{4a}{l}\left(\dfrac{l}{2} - x\right), & \dfrac{l}{4} \le x \le \dfrac{l}{2} \\[3mm] 0, & \dfrac{l}{2} \le x \le l \end{cases} \qquad (E8.2.1)$$

Find the resulting free vibration response of the string.

SOLUTION   Since the initial displacement of the string is zero,

$$w_0(x) = 0 \qquad (E8.2.2)$$

and hence Eq. (8.106) gives

$$C_n = 0 \qquad (E8.2.3)$$

Thus, the free vibration solution becomes

$$w(x,t) = \sum_{n=1}^{\infty} D_n \sin\frac{n\pi x}{l} \sin\frac{nc\pi t}{l} \qquad (E8.2.4)$$

The constant D_n can be evaluated using Eqs. (8.110) and (E8.2.1) as

$$\begin{aligned} D_n &= \frac{2}{nc\pi} \int_0^l \dot{w}_0(x) \sin\frac{n\pi x}{l}\, dx \\[2mm] &= \frac{2}{nc\pi}\left[\int_0^{l/4} \frac{4ax}{l}\sin\frac{n\pi x}{l}\, dx + \int_{l/4}^{l/2} \frac{4a}{l}\left(\frac{l}{2} - x\right)\sin\frac{n\pi x}{l}\, dx\right] \\[2mm] &= \frac{8a}{\pi^2 n^2}\left(2\sin\frac{\pi n}{4} - \sin\frac{\pi n}{2}\right) \qquad (E8.2.5) \end{aligned}$$

Thus, the free vibration response of the string is given by

$$
w(x,t) = \frac{8al}{\pi^3 c} \left[(\sqrt{2} - 1) \sin \frac{\pi x}{l} \sin \frac{\pi ct}{l} + \frac{1}{4} \sin \frac{2\pi x}{l} \sin \frac{2\pi ct}{l} \right.
$$

$$
\left. + \frac{\sqrt{2} + 1}{27} \sin \frac{3\pi x}{l} \sin \frac{3\pi ct}{l} - \frac{\sqrt{2} + 1}{125} \sin \frac{5\pi x}{l} \sin \frac{5\pi ct}{l} + \cdots \right] \quad \text{(E8.2.6)}
$$

It is to be noted that the modes involving $n = 4, 8, 12, \ldots$ are absent in Eq. (E8.2.6). The solution given by Eq. (E8.2.6), using a different number of terms is shown in Fig. 8.7. The fast convergence of the series of Eq. (E8.2.6) can be seen from the figure.

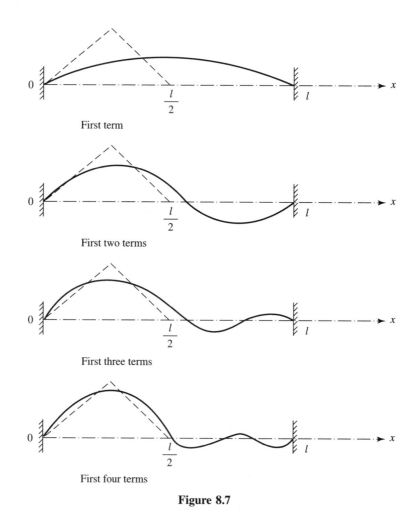

Figure 8.7

Example 8.3   Find the natural frequencies and mode shapes of a taut wire supported at the ends by springs as shown in Fig. 8.8.

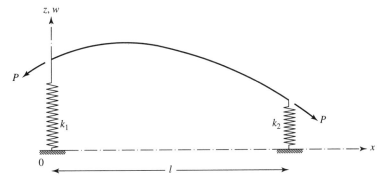

Figure 8.8   Taut wire supported by springs at the two ends.

SOLUTION   The nth mode of vibration of the wire can be represented as

$$w_n(x,t) = W_n(x)T_n(t)$$

$$= \left(A_n \cos \frac{\omega_n x}{c} + B_n \sin \frac{\omega_n x}{c} \right) (C_n \cos \omega_n t + D_n \sin \omega_n t). \qquad (\text{E8.3.1})$$

At the two ends of the wire, the spring force must be in equilibrium with the z component of the tensile force P in the wire. Thus, the boundary conditions can be expressed as

$$P \frac{\partial w}{\partial x}(0,t) = k_1 w(0,t) \qquad (\text{E8.3.2})$$

$$P \frac{\partial w}{\partial x}(l,t) = -k_2 w(l,t) \qquad (\text{E8.3.3})$$

Substituting Eq. (E8.3.1) into Eqs. (E8.3.2) and (E8.3.3), we obtain

$$P \frac{dW_n(0)}{dx} = k_1 W_n(0) \qquad (\text{E8.3.4})$$

or

$$PB_n \frac{\omega_n}{c} = k_1 A_n \qquad (\text{E8.3.5})$$

$$P \frac{dW_n(l)}{dx} = -k_2 W_n(l) \qquad (\text{E8.3.6})$$

or

$$P \left(-\frac{A_n \omega_n}{c} \sin \frac{\omega_n l}{c} + \frac{B_n \omega_n}{c} \cos \frac{\omega_n l}{c} \right) = -k_2 \left(A_n \cos \frac{\omega_n l}{c} + B_n \sin \frac{\omega_n l}{c} \right) \qquad (\text{E8.3.7})$$

Equations (E8.3.5) and (E8.3.7) can be rewritten as

$$A_n(k_1) - B_n \frac{P\omega_n}{c} = 0 \qquad (\text{E8.3.8})$$

$$A_n \left(k_2 \cos \frac{\omega_n l}{c} - \frac{P\omega_n}{c} \sin \frac{\omega_n l}{c} \right) + B_n \left(k_2 \sin \frac{\omega_n l}{c} + \frac{P\omega_n}{c} \cos \frac{\omega_n l}{c} \right) = 0 \qquad (\text{E8.3.9})$$

For a nontrivial solution of the constants A_n and B_n, the determinant of their coefficient matrix must be zero:

$$k_1 \left(k_2 \sin \frac{\omega_n l}{c} + \frac{P \omega_n}{c} \cos \frac{\omega_n l}{c} \right) + \frac{P \omega_n}{c} \left(k_2 \cos \frac{\omega_n l}{c} - \frac{P \omega_n}{c} \sin \frac{\omega_n l}{c} \right) = 0$$

$$\text{(E8.3.10)}$$

Defining

$$\alpha_n = \frac{\omega_n l}{c} \tag{E8.3.11}$$

$$\beta_n = \frac{\rho l \omega_n^2}{k_1} \tag{E8.3.12}$$

$$\gamma_n = \frac{\rho l \omega_n^2}{k_2} \tag{E8.3.13}$$

the frequency equation, Eq. (E8.3.10), can be expressed as

$$\left(1 - \frac{\alpha_n^2}{\beta_n \gamma_n} \right) \tan \alpha_n - \left(\frac{1}{\beta_n} + \frac{1}{\gamma_n} \right) \alpha_n = 0 \tag{E8.3.14}$$

Using the relation

$$B_n = A_n \frac{k_1 c}{P \omega_n} = A_n \frac{\alpha_n}{\beta_n} \tag{E8.3.15}$$

From Eq. (E8.3.8), the modal function $W_n(x)$ can be written as

$$W_n(x) = C_n \left(\cos \frac{\omega_n x}{c} + \frac{\alpha_n}{\beta_n} \sin \frac{\omega_n x}{c} \right) = C_n \left(\cos \frac{\alpha_n x}{l} + \frac{\alpha_n}{\beta_n} \sin \frac{\alpha_n x}{l} \right) \tag{E8.3.16}$$

Notes

1. If k_1 and k_2 are both large, $k_1 \to \infty$ and $k_2 \to \infty$ and the frequency equation, Eq. (E8.3.10), reduces to

$$\sin \frac{\omega_n l}{c} = 0 \tag{E8.3.17}$$

Equation (E8.3.17) corresponds to the frequency equation of a wire with both ends fixed.

2. If k_1 and k_2 are both small, $1/\beta_n \to 0$ and $1/\gamma_n \to 0$ and Eq. (E8.3.14) gives the frequencies as

$$\tan \alpha_n = 0 \quad \text{or} \quad \alpha_n = n\pi \quad \text{or} \quad \omega_n = \frac{n\pi c}{l} \tag{E8.3.18}$$

and Eq. (E8.3.16) yields the modal functions as

$$W_n(x) = C_n \cos \frac{\omega_n x}{l} \tag{E8.3.19}$$

It can be observed that this solution corresponds to that of a wire with both ends free.

3. If k_1 is large and k_2 is small, $k_1 \to \infty$ and $k_2 \to 0$, and Eq. (E8.3.10) yields

$$\cos \frac{\omega_n l}{c} = 0 \quad \text{or} \quad \frac{\omega_n l}{c} = \frac{(2n-1)\pi}{2}$$

or

$$\omega_n = \frac{(2n-1)\pi c}{2l} \tag{E8.3.20}$$

and Eq. (E8.3.16) gives the modal functions as

$$W_n(x) = C_n \sin \frac{\alpha_n x}{l} = C_n \sin \frac{\omega_n x}{c} \tag{E8.3.21}$$

This solution corresponds to that of a wire which is fixed at $x = 0$ and free at $x = l$.

8.6   FORCED VIBRATION

The equation of motion governing the forced vibration of a uniform string subjected to a distributed load $f(x,t)$ per unit length is given by

$$\rho \frac{\partial^2 w(x,t)}{\partial t^2} - P \frac{\partial^2 w(x,t)}{\partial x^2} = f(x,t) \tag{8.111}$$

Let the string be fixed at both ends so that the boundary conditions become

$$w(0,t) = 0 \tag{8.112}$$

$$w(l,t) = 0 \tag{8.113}$$

The solution of the homogeneous equation [with $f(x,t) = 0$ in Eq. (8.111)], which represents free vibration, can be expressed as [see Eq. (8.106)]

$$w(x,t) = \sum_{n=1}^{\infty} \sin \frac{n\pi x}{l} \left(C_n \cos \frac{nc\pi t}{l} + D_n \sin \frac{nc\pi t}{l} \right) \tag{8.114}$$

The solution of the nonhomogeneous equation [with $f(x,t)$ in Eq. (8.111)], which also satisfies the boundary conditions of Eqs. (8.112) and (8.113), can be assumed to be of the form

$$w(x,t) = \sum_{n=1}^{\infty} \sin \frac{n\pi x}{l} \eta_n(t) \tag{8.115}$$

where $\eta_n(t)$ denotes the generalized coordinates. By substituting Eq. (8.115) into Eq. (8.111), we obtain

$$\rho \sum_{n=1}^{\infty} \sin \frac{n\pi x}{l} \frac{d^2 \eta_n(t)}{dt^2} + P \sum_{n=1}^{\infty} \left(\frac{n\pi}{l} \right)^2 \sin \frac{n\pi x}{l} \eta_n(t) = f(x,t) \tag{8.116}$$

Multiplication of Eq. (8.116) by $\sin(n\pi x/l)$ and integration from 0 to l, along with the use of the orthogonality of the functions $\sin(i\pi x/l)$, leads to

$$\frac{d^2\eta_n(t)}{dt^2} + \frac{n^2 c^2 \pi^2}{l^2}\eta_n(t) = \frac{2}{\rho l}Q_n(t) \tag{8.117}$$

where

$$Q_n(t) = \int_0^l f(x,t)\sin\frac{n\pi x}{l}\,dx \tag{8.118}$$

The solution of Eq. (8.117), including the homogeneous solution and the particular integral, can be expressed as

$$\eta_n(t) = C_n \cos\frac{nc\pi t}{l} + D_n \sin\frac{nc\pi t}{l} + \frac{2}{nc\pi\rho}\int_0^t Q_n(\tau)\sin\frac{nc\pi(t-\tau)}{l}\,d\tau \tag{8.119}$$

Thus, in view of Eq. (8.115), the forced vibration response of the string is given by

$$w(x,t) = \sum_{n=1}^{\infty}\left(C_n \cos\frac{nc\pi t}{l} + D_n \sin\frac{nc\pi t}{l}\right)\sin\frac{n\pi x}{l}$$

$$+\frac{2}{c\pi\rho}\sum_{n=1}^{\infty}\frac{1}{n}\sin\frac{n\pi x}{l}\int_0^t Q_n(\tau)\sin\frac{nc\pi(t-\tau)}{l}\,d\tau \tag{8.120}$$

where the constants C_n and D_n are determined from the initial conditions of the string.

Example 8.4   Find the forced vibration response of a uniform taut string fixed at both ends when a uniformly distributed force f_0 per unit length is applied. Assume the initial displacement and the initial velocity of the string to be zero.

SOLUTION   For a uniformly distributed force $f(x,t) = f_0$, Eq. (8.118) gives

$$Q_n(t) = \int_0^l f_0 \sin\frac{n\pi x}{l}\,dx = \frac{2l f_0}{n\pi}, \qquad n = 1, 3, 5, \ldots \tag{E8.4.1}$$

and hence

$$\int_0^t Q_n(\tau)\sin\frac{n\pi c(t-\tau)}{l}\,d\tau = \frac{2l f_0}{n\pi}\int_0^t \sin\frac{n\pi c(t-\tau)}{l}\,d\tau$$

$$= -\frac{2l f_0}{n\pi}\frac{l}{n\pi c}\int_{n\pi ct/l}^0 \sin y\,dy$$

$$= \frac{2l^2 f_0}{n\pi^2 c}\left(1 - \cos\frac{n\pi ct}{l}\right), \qquad n = 1, 3, 5, \ldots \tag{E8.4.2}$$

Thus, the forced response of the string becomes [Eq. (8.120)]:

$$w(x,t) = \sum_{n=1}^{\infty} \left(C_n \cos \frac{nc\pi t}{l} + D_n \sin \frac{nc\pi t}{l} \right) \sin \frac{n\pi x}{l}$$

$$+ \frac{4l^2 f_0}{c^2 \pi^3 \rho} \sum_{n=1,3,5,...}^{\infty} \frac{1}{n^3} \sin \frac{n\pi x}{l} \left(1 - \cos \frac{nc\pi t}{l} \right) \qquad \text{(E8.4.3)}$$

Use of the initial conditions

$$w(x, 0) = 0 \qquad \text{(E8.4.4)}$$

$$\frac{\partial w(x, 0)}{\partial t} = 0 \qquad \text{(E8.4.5)}$$

in Eq. (E8.4.3) yields

$$\sum_{n=1}^{\infty} C_n \sin \frac{n\pi x}{l} = 0 \qquad \text{(E8.4.6)}$$

$$\sum_{n=1}^{\infty} \frac{nc\pi}{l} D_n \sin \frac{n\pi x}{l} = 0 \qquad \text{(E8.4.7)}$$

Equations (E8.4.6) and (E8.4.7) result in

$$C_n = D_n = 0 \qquad \text{(E8.4.8)}$$

and hence the forced vibration response of the string is given by [see Eq. (E8.4.3)]:

$$w(x,t) = \frac{4l^2 f_0}{c^2 \pi^3 \rho} \sum_{n=1,3,5,...}^{\infty} \frac{1}{n^3} \sin \frac{n\pi x}{l} \left(1 - \cos \frac{nc\pi t}{l} \right) \qquad \text{(E8.4.9)}$$

Example 8.5   Find the steady-state forced vibration response of a fixed–fixed string subjected to a concentrated force $F(t) = F_0$ at $x = x_0$.

SOLUTION   The applied force can be represented as

$$f(x,t) = F(t)\delta(x - x_0) \qquad \text{(E8.5.1)}$$

The steady-state forced vibration response of the string can be expressed, using Eq. (8.120), as

$$w(x,t) = \frac{2}{c\pi\rho} \sum_{n=1}^{\infty} \frac{1}{n} \sin \frac{n\pi x}{l} \int_0^l Q_n(\tau) \sin \frac{nc\pi(t-\tau)}{l} d\tau \qquad \text{(E8.5.2)}$$

where $Q_n(t)$ is given by Eq. (8.118):

$$Q_n(t) = \int_0^l f(x,t) \sin \frac{n\pi x}{l} dx = \int_0^l F(t)\delta(x - x_0) \sin \frac{n\pi x}{l} dx$$

$$= F(t) \sin \frac{n\pi x_0}{l} \qquad \text{(E8.5.3)}$$

The function $F(t) = F_0$ can be denoted as

$$F(t) = F_0 H(t) \tag{E8.5.4}$$

where $H(t)$ is the Heaviside function, defined by

$$H(t) = \begin{cases} 0, & t \leq 0 \\ 1, & t \geq 0 \end{cases} \tag{E8.5.5}$$

Substitution of Eqs. (E8.5.4) and (E8.5.3) into (E8.5.2) results in

$$w(x,t) = \frac{2F_0}{c\pi\rho} \sum_{n=1}^{\infty} \frac{1}{n} \sin \frac{n\pi x}{l} \int_0^l H(\tau) \sin \frac{n\pi x_0}{l} \sin \frac{n\pi c(t-\tau)}{l} d\tau$$

$$= \frac{2F_0 l}{\pi^2 c^2 \rho} \sum_{n=1}^{\infty} \frac{1}{n^2} \sin \frac{n\pi x_0}{l} \sin \frac{n\pi x}{l} \left(1 - \cos \frac{n\pi ct}{l}\right) \tag{E8.5.6}$$

Example 8.6   Find the steady-state response of a fixed–fixed string subjected to a load moving at a constant velocity v given by

$$f(x,t) = \begin{cases} F(t)\delta(x - vt), & 0 \leq vt \leq l \\ 0, & vt > l \end{cases} \tag{E8.6.1}$$

where $F(t)$ is a suddenly applied force F_0.

SOLUTION   The steady-state response of the string is given by Eq. (8.120) as

$$w(x,t) = \frac{2}{c\pi\rho} \sum_{n=1}^{\infty} \frac{1}{n} \sin \frac{n\pi x}{l} \int_0^l Q_n(\tau) \sin \frac{n\pi c(t-\tau)}{l} d\tau \tag{E8.6.2}$$

where $Q_n(t)$ is given by Eq. (8.118):

$$Q_n(t) = \int_0^l f(x,t) \sin \frac{n\pi x}{l} dx \tag{E8.6.3}$$

Using Eq. (E8.6.1), $Q_n(t)$ can be evaluated as

$$Q_n(t) = \int_0^l F(t)\delta(x - vt) \sin \frac{n\pi x}{l} dx = F(t) \sin \frac{n\pi vt}{l} \tag{E8.6.4}$$

Thus, Eq. (E8.6.2) becomes

$$w(x,t) = \frac{2}{c\pi\rho} \sum_{n=1}^{\infty} \frac{1}{n} \sin \frac{n\pi x}{l} \int_0^l F(\tau) \sin \frac{n\pi vt}{l} \sin \frac{n\pi c(t-\tau)}{l} d\tau \tag{E8.6.5}$$

Using

$$F(t) = F_0 H(t) \tag{E8.6.6}$$

Eq. (E8.6.5) can be evaluated as

$$w(x,t) = \frac{2F_0}{c\pi\rho} \sum_{n=1}^{\infty} \frac{1}{n} \sin\frac{n\pi x}{l} \int_0^l H(\tau) \sin\frac{n\pi vt}{l} \sin\frac{n\pi c(t-\tau)}{l} \, d\tau$$

$$= \frac{2F_0}{c\pi\rho} \sum_{n=1}^{\infty} \frac{\sin(n\pi x/l)}{n\left(n^2\pi^2v^2/l^2 - n^2\pi^2c^2/l^2\right)} \left(\frac{n\pi v}{l} \sin\frac{n\pi c}{l}t - \frac{n\pi c}{l} \sin\frac{n\pi v}{l}t\right)$$

$$(E8.6.7)$$

8.7 RECENT CONTRIBUTIONS

The D'Alembert's solution of Eq. (8.8), as given by Eq. (8.35), is obtained by assuming that the increase in tension due to stretching is negligible. If this assumption is not made, Eq. (8.9) becomes [4]

$$\frac{\partial^2 w}{\partial t^2} = \left[c^2 + \frac{1}{2}c_1^2 \left(\frac{\partial w}{\partial x}\right)^2\right] \frac{\partial^2 w}{\partial x^2} \tag{8.121}$$

where

$$c_1 = \sqrt{\frac{E}{\rho_0}} \tag{8.122}$$

with ρ_0 denoting the density of the string. Here c_1 denotes the speed of compressional longitudinal wave through the string. An approximate solution of Eq. (8.121) was presented by Bolwell [4]. The dynamics of cables, chains, taut inclined cables, and hanging cables was considered by Triantafyllou [5, 6]. In particular, the problem of linear transverse vibration of an elastic string hanging freely under its own weight presents a paradox, in that a solution can be obtained only when the lower end is free. An explanation of the paradox was given by Triantafyllou [6], who also showed that the paradox can be removed by including bending stiffness using singular perturbations.

A mathematical model of the excitation of a vibrating system by a plucking action was studied by Griffel [7]. The mechanism is of the type used in musical instruments [8]. The effectiveness of the mechanism is computed over a range of the relevant parameters. In Ref. [9], Simpson derived the equations of in-plane motion of an elastic catenary translating uniformly between its end supports in an Eulerian frame of reference. The approximate analytical solution of these equations is given for a shallow catenary in which the tension is dominated by the cable section modulus. Although the mathematical description of a vibrating string is given by the wave equation, a quantum model of information theory was used by Barrett to obtain a one-degree-of-freedom mechanical system governed by a second-order differential equation [10].

The vibration of a sectionally uniform string from an initial state was considered by Beddoe [11]. The problem was formulated in terms of reflections and transmissions of progressive waves and solved using the Laplace transform method without incorporating the orthogonality relationships. The exact equations of motion of a string were formulated by Narasimha [12], and a systematic procedure was described for obtaining approximations to the equations to any order, making only the assumption that the strain in the material of the string is small. It was shown that the lowest-order equations in the scheme, which were nonlinear, were used to describe the response of the string near resonance.

Electrodischarge machining (EDM) is a noncontact process of electrically removing (cutting) material from conductive workpieces. In this process, a high potential difference is generated between a wire and a workpiece by charging them positively and negatively, respectively. The potential difference causes sparks between the wires and the workpiece. By moving the wire forward and sideways, the contour desired can be cut on the workpiece. In Ref. [13], Shahruz developed a mathematical model for the transverse vibration of the moving wire used in the EDM process in the form of a nonlinear partial differential equation. The equation was solved, and it was shown that the transverse vibration of the wire is stable and decays to zero for wire axial speeds below a critical value.

A comprehensive view of cable structures was presented by Irvine [14]. The natural frequencies and mode shapes of cables with attached masses have been determined by Sergev and Iwan [15]. The linear theory of free vibrations of a suspended cable has been outlined by Irvine and Caughey [16]. Yu presented explicit vibration solutions of a cable under complicated loads [17]. A theoretical and experimental analysis of free and forced vibration of sagged cable/mass suspension has been presented by Cheng and Perkins [18]. The linear dynamics of a translating string on an elastic foundation was considered by Perkins [19].

REFERENCES

1. W. Nowacki, *Dynamics of Elastic Systems*, translated by H. Zorski, Wiley, New York, 1963.

2. N. W. McLachlan, *Theory of Vibrations*, Dover, New York, 1951.

3. S. Timoshenko, D. H. Young, and W. Weaver, Jr., *Vibration Problems in Engineering*, 4th ed., Wiley, New York, 1974.

4. J. E. Bolwell, The flexible string's neglected term, *Journal of Sound and Vibration*, Vol. 206, No. 4, pp. 618–623, 1997.

5. M. S. Triantafyllou, Dynamics of cables and chains, *Shock and Vibration Digest*, Vol. 19, pp. 3–5 1987.

6. M. S. Triantafyllou and G. S. Triantafyllou, The paradox of the hanging string: an explanation using singular perturbations, *Journal of Sound and Vibration*, Vol. 148, No. 2, pp. 343–351, 1991.

7. D. H. Griffel, The dynamics of plucking, *Journal of Sound and Vibration*, Vol. 175, No. 3, pp. 289–297, 1994.

8. N. H. Fletcher and T. D. Rossing, *The Physics of Musical Instruments*, Springer-Verlag, New York, 1991.

9. A. Simpson, On the oscillatory motions of translating elastic cables, *Journal of Sound and Vibration*, Vol. 20, No. 2, pp. 177–189, 1972.

10. T. W. Barrett, On vibrating strings and information theory, *Journal of Sound and Vibration*, Vol. 20, No. 3, pp. 407–412, 1972.

11. B. Beddoe, Vibration of a sectionally uniform string from an initial state, *Journal of Sound and Vibration*, Vol. 4, No. 2, pp. 215–223, 1966.

12. R. Narasimha, Non-linear vibration of an elastic string, *Journal of Sound and Vibration*, Vol. 8, No. 1, pp. 134–146, 1968.

13. S. M. Shahruz, Vibration of wires used in electro-discharge machining, *Journal of Sound and Vibration*, Vol. 266, No. 5, pp. 1109–1116, 2003.

14. H. M. Irvine, *Cable Structures*, MIT Press, Cambridge, MA, 1981.

15. S. S. Sergev and W. D. Iwan, The natural frequencies and mode shapes of cables with attached masses, *Journal of Energy Resources Technology*, Vol. 103, pp. 237–242, 1981.

16. H. M. Irvine and T. K. Caughey, The linear theory of free vibrations of a suspended cable, *Proceedings of the Royal Society, London*, Vol. A-341, pp. 299–315, 1974.

17. P. Yu, Explicit vibration solutions of a cable under complicated loads, *Journal of Applied Mechanics*, Vol. 64, No. 4, pp. 957–964, 1997.

18. S.-P. Cheng and N. C. Perkins, Theoretical and experimental analysis of the forced response of sagged cable/mass suspension, *Journal of Applied Mechanics*, Vol. 61, No. 4, pp. 944–948, 1994.

19. N. C. Perkins, Linear dynamics of a translating string on an elastic foundation, *Journal of Vibration and Acoustics*, Vol. 112, No. 1, pp. 2–7, 1990.

PROBLEMS

8.1 Find the free vibration response of a fixed–fixed string of length l which is given an initial displacement

$$w_0(x) = h \sin \frac{2\pi x}{l}$$

and initial velocity $\dot{w}_0(x) = 0$.

8.2 A steel wire of diameter $\frac{1}{32}$ in. and length 3 ft is fixed at both ends and is subjected to a tension of 200 lb. Find the first four natural frequencies and the corresponding mode shapes of the wire.

8.3 Determine the stress that needs to be applied to the wire of Problem 8.2 to reduce its fundamental natural frequency of vibration by 50 % of the value found in Problem 8.2.

8.4 A string of length l is fixed at $x = 0$ and subjected to a transverse force $f(t) = f_0 \cos \omega t$ at $x = l$. Find the resulting vibration of the string.

8.5 Find the forced vibration response of a fixed–fixed string of length l that is subjected to the distributed transverse force $f(x,t) = F(x)e^{i\omega t}$.

8.6 A uniform string of length l is fixed at both ends and is subjected to the following initial conditions:

$$w(x, 0) = x_0 \sin \frac{2\pi x}{l}, \qquad \dot{w}(x, 0) = -y_0 \sin \frac{2\pi x}{l}$$

8.7 Derive the boundary conditions corresponding to support conditions 3, 4, and 5 of Table 8.1 from equilibrium considerations.

8.8 The transverse vibration of a string of length $l = 2$ is governed by the equation

$$16 \frac{\partial^2 w}{\partial x^2} = \frac{\partial^2 w}{\partial t^2}$$

The boundary and initial conditions of the string are given by

$$w(0,t) = 0, \qquad w(2,t) = 0$$

$$w(x, 0) = 0.1x(2 - x), \qquad \frac{\partial w}{\partial t}(x, 0) = 0$$

Find the displacement of the string, $w(x,t)$.

8.9 A semi-infinite string has one end at $x = 0$ and the other end at $x = \infty$. It is initially at rest on the x axis and the end $x = 0$ is made to oscillate with a transverse displacement of $w(0,t) = c \sin \Omega t$. Find the transverse displacement of the string, $w(x,t)$.

8.10 Find the natural frequencies of transverse vibration of a taut string of length l resting on linear springs of stiffnesses k_1 and k_2 at the ends $x = 0$ and $x = l$, respectively. Assume the following data: $P = 1000$ N, $\rho = 0.1$ kg/m, and $k_1 = k_2 = 5000$ N/m.

9

Longitudinal Vibration of Bars

9.1 INTRODUCTION

A straight elastic bar can undergo longitudinal, torsional, and lateral vibration. Among these, the longitudinal vibration is the simplest to analyze. If x denotes the longitudinal (centroidal) axis and y and z represent the principal directions of the cross section, the longitudinal vibrations take place in the x direction, torsional vibrations occur about the x axis, and lateral vibrations involve motion in either the xy plane or the xz plane. These vibrations may be coupled in some cases. For example, if the cross section is not symmetric about the y or z axis, the torsional and lateral vibrations are coupled. If the bar is pretwisted along the x direction, the lateral vibrations in the xy and xz planes are coupled. We consider first the longitudinal vibration of a bar using a simple theory.

9.2 EQUATION OF MOTION USING SIMPLE THEORY

We consider a simple theory for the longitudinal vibration of bars based on the following assumptions:

1. The cross sections of the bar originally plane remain plane during deformation.
2. The displacement components in the bar (except for the component parallel to the bar's longitudinal axis) are negligible.

These assumptions permit the specification of the displacement as a function of the single space coordinate denoting location along the length of the bar. Although lateral displacement components exist in any cross section, the second assumption can be shown to be valid for vibrations involving wavelengths that are long compared to the cross-sectional dimensions of the bar. We shall derive the equation of motion using two different approaches: by applying Newton's second law of motion and from Hamilton's principle.

9.2.1 Using Newton's Second Law of Motion

For an elastic bar of length l, Young's modulus E, and mass density ρ with varying cross-sectional area $A(x)$ as shown in Fig. 9.1(a), the equation of motion has been derived, using Newton's second law of motion, in Section 3.4 as

$$\frac{\partial}{\partial x}\left[E(x)A(x)\frac{\partial u(x,t)}{\partial x}\right] + f(x,t) = \rho(x)A(x)\frac{\partial^2 u(x,t)}{\partial t^2} \qquad (9.1)$$

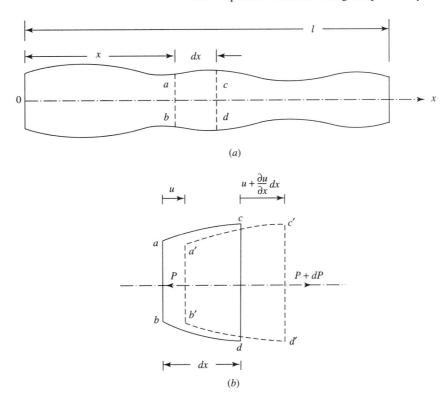

Figure 9.1   Longitudinal vibration of a bar.

For a uniform bar, Eq. (9.1) reduces to

$$EA\frac{\partial^2 u(x,t)}{\partial x^2} + f(x,t) = \rho A\frac{\partial^2 u(x,t)}{\partial t^2} \tag{9.2}$$

9.2.2   Using Hamilton's Principle

During longitudinal vibration, the cross section of the bar located at a distance x from the origin undergoes an axial displacement of u while the cross section located at a distance $x + dx$ undergoes an axial displacement of $u + du = u + (\partial u/\partial x)\,dx$, as shown in Fig. 9.1($b$). Since the deformation of the cross section in the y and z directions (v and w) is assumed to be negligible, the displacement field can be expressed as

$$u = u(x,t), \qquad v = 0, \qquad w = 0 \tag{9.3}$$

The strains in the cross section at x are given by

$$\varepsilon_{xx} = \frac{\partial u}{\partial x}, \qquad \varepsilon_{yy} = \varepsilon_{zz} = 0, \qquad \varepsilon_{xy} = \varepsilon_{yz} = \varepsilon_{zx} = 0 \tag{9.4}$$

Note that the displacements v and w, and the strains ε_{yy} and ε_{zz}, will not be zero, due to Poisson's effect in a real bar; they are assumed to be zero in the simple theory. The stresses acting in the cross section at x, corresponding to the strains given by Eq. (9.4), are

$$\sigma_{xx} = E\frac{\partial u}{\partial x}, \qquad \sigma_{yy} = \sigma_{zz} = 0, \qquad \sigma_{xy} = \sigma_{yz} = \sigma_{zx} = 0 \tag{9.5}$$

The strain and kinetic energies of the bar can be found as

$$\pi = \frac{1}{2}\int_0^l \sigma_{xx}\varepsilon_{xx}A\,dx = \frac{1}{2}\int_0^l EA\left(\frac{\partial u}{\partial x}\right)^2 dx \tag{9.6}$$

$$T = \frac{1}{2}\int_0^l \rho A\left(\frac{\partial u}{\partial t}\right)^2 dx \tag{9.7}$$

The work done by the external force $f(x, t)$ is given by

$$W = \int_0^l f(x, t)u\,dx \tag{9.8}$$

The generalized Hamilton's principle can be stated as

$$\delta\int_{t_1}^{t_2}(T - \pi + W)\,dt = 0 \tag{9.9}$$

The substitution of Eqs. (9.6)–(9.8) into Eq. (9.9) yields the equation of motion and the associated boundary conditions as (see Problem 9.8)

$$\frac{\partial}{\partial x}\left(EA\frac{\partial u}{\partial x}\right) + f = \rho A\frac{\partial^2 u}{\partial t^2} \tag{9.10}$$

$$EA\frac{\partial u}{\partial x}\delta u\Big|_0^l = 0 \tag{9.11}$$

Note that Eq. (9.11) will be satisfied for a free boundary where

$$\sigma_{xx} = EA\frac{\partial u}{\partial x} = 0 \tag{9.12}$$

or when the displacement is specified at the boundary with $\delta u = 0$; for a fixed end, the boundary condition is

$$u = 0 \tag{9.13}$$

9.3 FREE VIBRATION SOLUTION AND NATURAL FREQUENCIES

The equation governing the free vibration of bars can be obtained by setting $f = 0$ in Eqs. (9.1) and (9.2). For nonuniform bars:

$$\frac{\partial}{\partial x}\left[EA(x)\frac{\partial u(x, t)}{\partial x}\right] = \rho A(x)\frac{\partial^2 u(x, t)}{\partial t^2} \tag{9.14}$$

For uniform bars:

$$EA\frac{\partial^2 u(x, t)}{\partial x^2} = \rho A\frac{\partial^2 u(x, t)}{\partial t^2}$$

or

$$c^2\frac{\partial^2 u(x, t)}{\partial x^2} = \frac{\partial^2 u(x, t)}{\partial t^2} \tag{9.15}$$

where

$$c = \sqrt{\frac{E}{\rho}} \tag{9.16}$$

The solution of Eq. (9.15) can be obtained using either the wave solution approach or the method of separation of variables. The wave solution of Eq. (9.15) can be expressed, as in the case of vibration of strings, as

$$u(x, t) = f_1(x - ct) + f_2(x + ct) \tag{9.17}$$

Although this solution [Eq. (9.17)] is useful in the study of certain impact and wave propagation problems involving impulses of very short duration, it is not very useful in the study of vibration problems. The method of separation of variables followed by the eigenvalue and modal analyses is more useful in the study of vibrations.

9.3.1   Solution Using Separation of Variables

To develop the solution using the method of separation of variables, the solution of Eq. (9.15) is written as

$$U(x, t) = U(x)T(t) \tag{9.18}$$

where U and T depend on only x and t, respectively. Substitution of Eq. (9.18) into Eq. (9.15) leads to

$$\frac{c^2}{U}\frac{d^2 U}{dx^2} = \frac{1}{T}\frac{d^2 T}{dt^2} \tag{9.19}$$

Since the left-hand side of Eq. (9.19) depends only on x and the right-hand side depends only on t, their common value must be a constant, which can be shown to be a negative number (see Problem 9.7), denoted as $-\omega^2$. Thus, Eq. (9.19) can be written as two separate equations:

$$\frac{d^2 U(x)}{dx^2} + \frac{\omega^2}{c^2}U(x) = 0 \tag{9.20}$$

$$\frac{d^2 T(t)}{dt^2} + \omega^2 T(t) = 0 \tag{9.21}$$

The solution of Eqs. (9.20) and (9.21) can be represented as

$$U(x) = A \cos \frac{\omega x}{c} + B \sin \frac{\omega x}{c} \tag{9.22}$$

$$T(t) = C \cos \omega t + D \sin \omega t \tag{9.23}$$

where ω denotes the frequency of vibration, the function $U(x)$ represents the normal mode, the constants A and B can be evaluated from the boundary conditions, the function $T(t)$ indicates harmonic motion, and the constants C and D can be determined from the initial conditions of the bar. The complete solution of Eq. (9.15) becomes

$$u(x, t) = U(x)T(t) = \left(A \cos \frac{\omega x}{c} + B \sin \frac{\omega x}{c} \right) (C \cos \omega t + D \sin \omega t) \tag{9.24}$$

The common boundary conditions of the bar are as follows. For the fixed end:

$$u = 0 \tag{9.25}$$

For the free end:

$$\frac{\partial u}{\partial x} = 0 \tag{9.26}$$

Some possible boundary conditions of a bar are shown in Table 9.1. The application of the boundary conditions in Eq. (9.22) leads to the frequency equation whose solution yields the eigenvalues. The substitution of any specific eigenvalue in Eq. (9.22) gives the corresponding eigenfunction.

If the axial displacement and the axial velocity of the bar are specified as $u_0(x)$ and $\dot{u}_0(x)$, respectively, at time $t = 0$, the initial conditions can be stated as

$$u(x, t = 0) = u_0(x) \tag{9.27}$$

$$\frac{\partial u}{\partial t}(x, t = 0) = \dot{u}_0(x) \tag{9.28}$$

The following examples illustrate the formulation of boundary conditions, the determination of natural frequencies for specified boundary conditions of the bar, and the method of finding the free vibration solution of the bar in longitudinal vibration.

Example 9.1   The ends of a uniform bar are connected to masses, springs, and viscous dampers as shown in Fig. 9.2(a). State the boundary conditions of the bar in axial vibration.

SOLUTION   If the axial displacement, velocity, and acceleration of the bar at $x = 0$ are assumed to be positive with values $u(0, t)$, $\partial u/\partial t(0, t)$, and $\partial^2 u/\partial t^2(0, t)$, respectively, the spring force $k_1 u(0, t)$, the damping force $c_1[\partial u/\partial t](0, t)$, and the inertia force $m_1[\partial^2 u/\partial t^2](0, t)$ act toward the left as shown in the free-body diagram of Fig. 9.2(b). The boundary condition at $x = 0$ can be expressed as

[force (reaction) in bar at $x = 0$]

$\quad$ = (sum of spring, damper, and inertia forces at $x = 0$)

Table 9.1 Boundary Conditions of a Bar in Longitudinal Vibration

End conditions of the bar	Boundary conditions	Frequency equation	Mode shapes (normal functions)	Natural frequencies of vibration
1. Fixed–free	$u(0,t)=0$ $\dfrac{\partial u}{\partial x}(l,t)=0$	$\cos\dfrac{\omega l}{c}=0$	$U_n(x)=$ $C_n \sin\dfrac{(2n+1)\pi x}{2l}$	$\omega_n=\dfrac{(2n+1)\pi c}{2l},$ $n=0,1,2,\dots$
2. Fixed–fixed	$u(0,t)=0$ $u(l,t)=0$	$\sin\dfrac{\omega l}{c}=0$	$U_n(x)=$ $C_n \sin\dfrac{n\pi x}{l}$	$\omega_n=\dfrac{n\pi c}{l},$ $n=1,2,3,\dots$
3. Fixed–attached mass	$u(0,t)=0$ $EA\dfrac{\partial u}{\partial x}(l,t)=$ $-M\dfrac{\partial^2 u}{\partial t^2}(l,t)$	$\alpha\tan\alpha=\beta$ $\alpha=\dfrac{\omega l}{c}$ $\beta=\dfrac{m}{M}$	$U_n(x)=$ $C_n \sin\dfrac{\omega_n x}{c}$	$\omega_n=\dfrac{\alpha_n c}{l},$ $n=1,2,3,\dots$
4. Fixed–attached spring	$u(0,t)=0$ $EA\dfrac{\partial u}{\partial x}(l,t)=$ $-ku(l,t)$	$\alpha\tan\alpha=-\gamma$ $\alpha=\dfrac{\omega l}{c}$ $\gamma=\dfrac{m\omega^2}{k}$	$U_n(x)=$ $C_n \sin\dfrac{\omega_n x}{c}$	$\omega_n=\dfrac{\alpha_n c}{l},$ $n=1,2,3,\dots$

(continued overleaf)

Table 9.1 (*continued*)

End conditions of the bar	Boundary conditions	Frequency equation	Mode shapes (normal functions)	Natural frequencies of vibration
5. Fixed–attached spring and mass	$u(0,t)=0$ $EA\dfrac{\partial u}{\partial x}(l,t) = -M\dfrac{\partial^2 u}{\partial t^2}(l,t) - ku(l,t)$	$\alpha\cot\alpha = \dfrac{\alpha^2}{\beta} - \dfrac{k}{k_0}$ $\beta = \dfrac{M}{m}$ $k_0 = \dfrac{AE}{l}$	$U_n(x) = C_n\sin\dfrac{\omega_n x}{c}$	$\omega_n = \dfrac{\alpha_n c}{l}$, $n = 1,2,3,\ldots$
6. Free–free	$\dfrac{\partial u}{\partial x}(0,t)=0$ $\dfrac{\partial u}{\partial x}(l,t)=0$	$\sin\dfrac{\omega l}{c} = 0$	$U_n(x) = C_n\cos\dfrac{n\pi x}{l}$	$\omega_n = \dfrac{n\pi c}{l}$, $n = 0,1,2,\ldots$
7. Free–attached mass	$\dfrac{\partial u}{\partial x}(0,t)=0$ $EA\dfrac{\partial u}{\partial x}(l,t) = -M\dfrac{\partial^2 u}{\partial t^2}(l,t)$	$\tan\alpha = -\alpha\beta$	$U_n(x) = C_n\cos\dfrac{\omega_n x}{c}$	$\omega_n = \dfrac{\alpha_n c}{l}$, $n = 1,2,3,\ldots$
8. Free–attached spring	$\dfrac{\partial u}{\partial x}(0,t)=0$ $EA\dfrac{\partial u}{\partial x}(l,t) = -ku(l,t)$	$\alpha\cot\alpha = \delta$ $\delta = \dfrac{AE}{lk}$	$U_n(x) = C_n\cos\dfrac{\omega_n x}{c}$	$\omega_n = \dfrac{\alpha_n c}{l}$, $n = 1,2,3,\ldots$

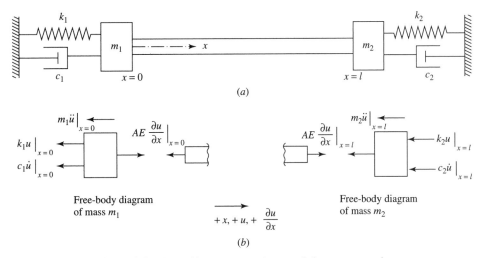

Figure 9.2   Bar with masses, springs, and dampers at ends.

or

$$AE\frac{\partial u}{\partial x}(0, t) = k_1 u(0, t) + c_1\frac{\partial u}{\partial t}(0, t) + m_1\frac{\partial^2 u}{\partial t^2}(0, t) \qquad (E9.1.1)$$

In a similar manner, the boundary condition at $x = l$ can be expressed as

$$AE\frac{\partial u}{\partial x}(l, t) = -k_2 u(l, t) - c_2\frac{\partial u}{\partial t}(l, t) - m_2\frac{\partial^2 u}{\partial t^2}(l, t) \qquad (E9.1.2)$$

Example 9.2   A uniform bar of length l, cross-sectional area A, density ρ, and Young's modulus E, is fixed at $x = 0$ and a rigid mass M is attached at $x = l$ [Fig. 9.3(a)]. Determine the natural frequencies and mode shapes of longitudinal vibration of the bar.

SOLUTION   The solution for the free longitudinal vibration of a bar is given by Eq. (9.24):

$$u(x, t) = \left(\underset{\sim}{A}\cos\frac{\omega x}{c} + \underset{\sim}{B}\sin\frac{\omega x}{c}\right)(C\cos\omega t + D\sin\omega t) \qquad (E9.2.1)$$

The boundary condition at the fixed end, $x = 0$, is given by

$$u(0, t) = 0 \qquad (E9.2.2)$$

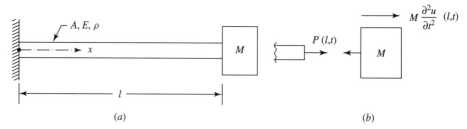

Figure 9.3 Longitudinal vibration of a bar, fixed at $x = 0$ and mass attached at $x = l$: (a) bar with end mass M; (b) free body diagram of mass M.

The boundary condition at $x = l$ can be expressed from the free-body diagram of the mass shown in Fig. 9.3(b) as

$$\text{reaction force} = P(l, t) = A\sigma(l, t) = AE\frac{\partial u}{\partial x}(l, t)$$

$$= -\text{inertia force} = -M\frac{\partial^2 u}{\partial t^2}(l, t) \tag{E9.2.3}$$

Equations (E9.2.2) and (E9.2.1) give

$$\underset{\sim}{A} = 0 \tag{E9.2.4}$$

and hence Eq. (E9.2.1) becomes

$$u(x, t) = \underset{\sim}{B} \sin \frac{\omega x}{c}(C \cos \omega t + D \sin \omega t) \tag{E9.2.5}$$

Equation (E9.2.5) gives

$$\frac{\partial u}{\partial x} = \underset{\sim}{B}\frac{\omega}{c} \cos \frac{\omega x}{c}(C \cos \omega t + D \sin \omega t) \tag{E9.2.6}$$

$$\frac{\partial^2 u}{\partial t^2} = -\underset{\sim}{B}\omega^2 \sin \frac{\omega x}{c}(C \cos \omega t + D \sin \omega t) \tag{E9.2.7}$$

Using Eqs. (E9.2.6) and (E9.2.7), Eq. (E9.2.3) can be expressed as

$$AE\frac{\omega}{c}\underset{\sim}{B} \cos \frac{\omega l}{c}(C \cos \omega t + D \sin \omega t) = M\omega^2\underset{\sim}{B} \sin \frac{\omega l}{c}(C \cos \omega t + D \sin \omega t)$$

or

$$\tan \frac{\omega l}{c} = \frac{AE}{M\omega c} \tag{E9.2.8}$$

By introducing the mass of the bar, m, as

$$m = \rho A l \tag{E9.2.9}$$

Eq. (E9.2.8) can be rewritten as

$$\alpha \tan \alpha = \beta \tag{E9.2.10}$$

where

$$\alpha = \frac{\omega l}{c} \qquad (E9.2.11)$$

$$\beta = \frac{\rho A l}{M} = \frac{m}{M} \qquad (E9.2.12)$$

Equation (E9.2.10) is the frequency equation in the form of a transcendental equation which has an infinite number of roots. For the nth root, Eq. (E9.2.10) can be written as

$$\alpha_n \tan \alpha_n = \beta, \qquad n = 1, 2, \ldots \qquad (E9.2.13)$$

with

$$\alpha_n = \frac{\omega_n l}{c} \quad \text{or} \quad \omega_n = \frac{\alpha_n c}{l} \qquad (E9.2.14)$$

The mode shapes corresponding to the natural frequency ω_n can be expressed as

$$U_n(x) = \underset{\sim}{B}_n \sin \frac{\omega_n x}{c}, \qquad n = 1, 2, \ldots \qquad (E9.2.15)$$

The first 10 roots of Eq. (E9.2.13) for different values of the mass ratio β are given in Table 9.2.

Table 9.2   Roots of Eq. (E9.2.13)

	Value of α_n for:				
n	$\beta = 0$	$\beta = 10$	$\beta = 1$	$\beta = \frac{1}{10}$	$\beta = \frac{1}{100}$
1	0	1.4289	0.8603	0.3111	0.0998
2	3.1416	4.3058	3.4256	3.1731	3.1448
3	6.2832	7.2281	6.4373	6.2991	6.2848
4	9.4248	10.2003	9.5293	9.4354	9.4258
5	12.5664	13.2142	12.6453	12.5743	12.5672
6	15.7080	16.2594	15.7713	15.7143	15.7086
7	18.8496	19.3270	18.9024	18.8549	18.8501
8	21.9911	22.4108	22.0365	21.9957	21.9916
9	25.1327	25.5064	25.1724	25.1367	25.1331
10	28.2743	28.6106	28.3096	28.2779	28.2747

Example 9.3   A uniform bar of length l, cross-sectional area A, density ρ, and Young's modulus E is free at $x = 0$ and attached to a spring of stiffness K at $x = l$, as shown in Fig. 9.4(a). Determine the natural frequencies and the mode shapes of longitudinal vibration of the bar.

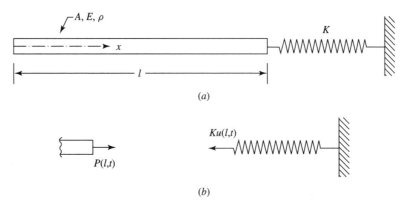

Figure 9.4   Bar free at $x = 0$ and attached to a spring at $x = l$.

SOLUTION   The solution for the free longitudinal vibration of a bar is given by Eq. (9.24):

$$u(x, t) = \left(\underset{\sim}{A} \cos \frac{\omega x}{c} + \underset{\sim}{B} \sin \frac{\omega x}{c} \right) (C \cos \omega t + D \sin \omega t) \tag{E9.3.1}$$

Since the end $x = 0$ is free, we have

$$AE \frac{\partial u}{\partial x}(0, t) = 0 \quad \text{or} \quad \frac{\partial u}{\partial x}(0, t) = 0 \tag{E9.3.2}$$

Equations (E9.3.1) and (E9.3.2) yield

$$\underset{\sim}{B} = 0 \tag{E9.3.3}$$

Thus, Eq. (E9.3.1) reduces to

$$u(x, t) = \underset{\sim}{A} \cos \frac{\omega x}{c} (C \cos \omega t + D \sin \omega t) \tag{E9.3.4}$$

The boundary condition at $x = l$ can be expressed as [Fig. 9.4(b)]

$$\text{reaction force} = -\text{spring force}$$

that is,

$$AE \frac{\partial u}{\partial x}(l, t) = -Ku(l, t) \tag{E9.3.5}$$

Equations (E9.3.4) and (E9.3.5) lead to

$$-\underset{\sim}{A} \frac{\omega AE}{c} \sin \frac{\omega l}{c} (C \cos \omega t + D \sin \omega t) = -K \underset{\sim}{A} \cos \frac{\omega l}{c} (C \cos \omega t + D \sin \omega t)$$

or

$$\frac{AE\omega}{cK} = \cot \frac{\omega l}{c} \tag{E9.3.6}$$

Introducing the mass of the bar, m, as

$$m = \rho A l \qquad (E9.3.7)$$

Eq. (E9.3.6) can be rewritten as

$$\alpha \cot \alpha = \beta \qquad (E9.3.8)$$

where

$$\alpha = \frac{\omega l}{c} \qquad (E9.3.9)$$

$$\beta = \frac{k}{K} \qquad (E9.3.10)$$

$$k = \frac{AE}{l} \qquad (E9.3.11)$$

denotes the stiffness of the bar. Equation (E9.3.8) denotes the frequency equation in the form of a transcendental equation with an infinite number of roots. For the nth root, Eq. (E9.3.8) can be expressed as

$$\alpha_n \cot \alpha_n = \beta_n, \qquad n = 1, 2, \ldots \qquad (E9.3.12)$$

with

$$\alpha_n = \frac{\omega_n l}{c} \quad \text{or} \quad \omega_n = \frac{\alpha_n c}{l} \qquad (E9.3.13)$$

The mode shape corresponding to the natural frequency ω_n can be expressed as

$$U_n(x) = \underset{\sim}{A}_n \cos \frac{\omega_n x}{c}, \qquad n = 1, 2, \ldots \qquad (E9.3.14)$$

The first 10 roots of Eq. (E9.3.12) for different values of the stiffness ratio $\beta = k/K$ are given in Table 9.3.

Table 9.3   Roots of Eq. (E9.3.12)

	Value of α_n for:				
n	$\beta = 1$	$\beta = \frac{1}{5}$	$\beta = \frac{1}{10}$	$\beta = \frac{1}{50}$	$\beta = \frac{1}{100}$
1	3.145	1.435	1.505	1.555	1.565
2	4.495	3.145	3.145	3.145	3.145
3	6.285	4.665	4.695	4.705	4.715
4	7.725	6.285	6.285	6.285	6.285
5	9.425	7.825	7.845	7.855	7.855
6	10.905	9.425	9.425	9.425	9.425
7	12.565	10.975	10.985	10.995	10.995
8	14.065	12.565	12.565	12.565	12.565
9	15.705	14.125	14.135	14.135	14.135
10	17.225	15.705	15.705	15.705	15.705

Example 9.4 Find the natural frequencies of vibration and the mode shapes of a bar with free ends.

SOLUTION The boundary conditions of a free–free bar can be expressed as

$$\frac{\partial u}{\partial x}(0, t) = \frac{dU}{dx}(0) = 0, \qquad t \geq 0 \tag{E9.4.1}$$

$$\frac{\partial u}{\partial x}(l, t) = \frac{dU}{dx}(l) = 0, \qquad t \geq 0 \tag{E9.4.2}$$

In the solution

$$U(x) = \underset{\sim}{A} \cos \frac{\omega x}{c} + \underset{\sim}{B} \sin \frac{\omega x}{c} \tag{E9.4.3}$$

$$\frac{dU}{dx}(x) = -\underset{\sim}{A} \frac{\omega}{c} \sin \frac{\omega x}{c} + \underset{\sim}{B} \frac{\omega}{c} \cos \frac{\omega x}{c} \tag{E9.4.4}$$

use of the condition, Eq. (E9.4.1), gives

$$\underset{\sim}{B} = 0 \tag{E9.4.5}$$

The condition of Eq. (E9.4.2) leads to the frequency equation (noting that $\underset{\sim}{A}$ cannot be equal to zero for a nontrivial solution):

$$\sin \frac{\omega l}{c} = 0 \tag{E9.4.6}$$

which yields

$$\frac{\omega l}{c} = n\pi, \qquad n = 1, 2, \ldots \tag{E9.4.7}$$

As different values of n give different frequencies of the various modes of vibration, the nth frequency and the corresponding mode shape can be expressed as

$$\omega_n = \frac{n\pi c}{l}, \qquad n = 1, 2, \ldots \tag{E9.4.8}$$

$$U_n(x) = \underset{\sim}{A} \cos \frac{\omega_n x}{c} = \underset{\sim}{A} \cos \frac{n\pi x}{l}, \qquad n = 1, 2, \ldots \tag{E9.4.9}$$

The first three frequencies and the corresponding mode shapes, given by Eqs. (E9.4.8) and (E9.4.9), are shown in Table 9.4.

9.3.2 Orthogonality of Eigenfunctions

The differential equation governing the free longitudinal vibration of a prismatic bar, Eq. (9.1) with $f = 0$, can be written in general form as

$$L[u(x, t)] + M \frac{\partial^2 u}{\partial t^2}(x, t) = 0 \tag{9.29}$$

Table 9.4   First Three Mode Shapes of a Free–Free bar

Mode number, n	Natural frequency, ω_n	Mode shape, $U_n(X)$
1	$\dfrac{\pi c}{l}$	
2	$\dfrac{2\pi c}{l}$	
3	$\dfrac{3\pi c}{l}$	

where

$$L = -R\frac{\partial^2}{\partial x^2} \equiv -EA\frac{\partial^2}{\partial x^2}, \qquad M = \rho A \tag{9.30}$$

$R = EA$ denotes the axial rigidity and ρA indicates the mass per unit length of the bar. For free vibration (with harmonic motion) in the ith natural mode, we can write

$$u_i(x, t) = U_i(x)(C_i \cos \omega_i t + D_i \sin \omega_i t) \tag{9.31}$$

Substituting Eq. (9.31) into Eq. (9.29), we obtain

$$RU_i''(x) + M\omega_i^2 U_i(x) = 0 \tag{9.32}$$

where a prime denotes a derivative with respect to x. Equation (9.32) can be rewritten as an eigenvalue problem

$$U_i''(x) = \lambda_i U_i(x) \tag{9.33}$$

where $U_i(x)$ is the eigenfunction or normal function determined from the boundary conditions and

$$\lambda_i = -\frac{M\omega_i^2}{R} = -\left(\frac{\omega_i}{c}\right)^2 \tag{9.34}$$

is the eigenvalue with

$$c = \sqrt{\frac{R}{M}} = \sqrt{\frac{E}{\rho}} \tag{9.35}$$

Let $U_i(x)$ and $U_j(x)$ denote the eigenfunctions corresponding to the natural frequencies ω_i and ω_j, respectively, so that

$$U_i'' = \lambda_i U_i \tag{9.36}$$

$$U_j'' = \lambda_j U_j \tag{9.37}$$

Multiply Eq. (9.36) by U_j and Eq. (9.37) by U_i and integrate the resulting equations from 0 to l to obtain

$$\int_0^l U_i'' U_j \, dx = \lambda_i \int_0^l U_i U_j \, dx \tag{9.38}$$

$$\int_0^l U_j'' U_i \, dx = \lambda_j \int_0^l U_i U_j \, dx \tag{9.39}$$

Integrate the left-hand sides of Eqs. (9.38) and (9.39) by parts:

$$U_i' U_j \big|_0^l - \int_0^l U_i' U_j' \, dx = \lambda_i \int_0^l U_i U_j \, dx \tag{9.40}$$

$$U_j' U_i \big|_0^l - \int_0^l U_i' U_j' \, dx = \lambda_j \int_0^l U_i U_j \, dx \tag{9.41}$$

The first terms on the left-hand sides of Eqs. (9.40) and (9.41) are zero if the end of the bar is either fixed or free. Subtract Eq. (9.41) from (9.40) to obtain

$$(\lambda_i - \lambda_j) \int_0^l U_i U_j \, dx = 0 \tag{9.42}$$

When the eigenvalues are distinct $\lambda_i \neq \lambda_j$, Eq. (9.42) gives the orthogonality principle for normal functions:

$$\int_0^l U_i U_j \, dx = 0, \qquad i \neq j \tag{9.43}$$

In view of Eq. (9.43). Eqs. (9.40) and (9.39) yield

$$\int_0^l U_i' U_j' \, dx = 0, \qquad i \neq j \tag{9.44}$$

$$\int_0^l U_i'' U_j \, dx = 0, \qquad i \neq j \tag{9.45}$$

Equations (9.43)–(9.45) indicate that the orthogonality is valid not only among the eigenfunctions, but also among the derivatives of the eigenfunctions.

Note   The orthogonality relationships for a bar with a mass M attached at the end $x = l$ [as in Fig. 9.3(a)] can be developed as follows. Rewrite Eq. (9.32) corresponding to two distinct eigenvalues i and j as

$$RU_i'' = -m\omega_i^2 U_i, \qquad RU_j'' = -m\omega_j^2 U_j \tag{9.46}$$

with $m \equiv M = \rho A$. To include the boundary condition at $x = l$ in the orthogonality relation, we write the boundary condition for eigenvalues i and j as

$$RU_i'(l) = \underline{M}\omega_i^2 U_i(l), \qquad RU_j'(l) = \underline{M}\omega_j^2 U_j(l) \tag{9.47}$$

Using a procedure similar to the one used in deriving Eq. (9.43), we obtain the orthogonality condition as (see Problem 9.21)

$$m \int_0^l U_i U_j \, dx + \underline{M} U_i(l) U_j(l) = 0, \qquad i \neq j \tag{9.48}$$

9.3.3   Free Vibration Response due to Initial Excitation

The response or displacement of the bar during longitudinal vibration can be expressed in terms of the normal functions $U_i(x)$, using the expansion theorem, as

$$u(x, t) = \sum_{i=1}^{\infty} U_i(x) \eta_i(t) \tag{9.49}$$

Substitution of Eq. (9.49) into Eq. (9.29) results in

$$\sum_{i=1}^{\infty} [R\eta_i(t) U_i''(x) + M\ddot{\eta}_i(t) U_i(x)] = 0 \tag{9.50}$$

Multiplication of Eq. (9.50) by $U_j(x)$ and integration from 0 to l yields

$$\sum_{i=1}^{\infty} \left[R\eta_i(t) \int_0^l U_i'' U_j \, dx + M\ddot{\eta}_i \int_0^l U_i U_j \, dx \right] = 0 \tag{9.51}$$

In view of the orthogonality relationships, Eqs. (9.43) and (9.45), Eq. (9.51) reduces to

$$M_i \ddot{\eta}_i(t) + R_i \eta_i(t) = 0, \qquad i = 1, 2, \ldots \tag{9.52}$$

where M_i and R_i denote the generalized mass and generalized stiffness (or rigidity), respectively, in mode i:

$$M_i = M \int_0^l U_i^2 \, dx \tag{9.53}$$

$$R_i = R \int_0^l U_i'' U_i \, dx = -R \int_0^l (U_i')^2 \, dx = \omega_i^2 M_i \tag{9.54}$$

If the eigenfunctions are normalized with respect to the mass distribution as

$$M_i = M \int_0^l U_i^2 \, dx = 1 \tag{9.55}$$

Eq. (9.54) gives $R_i = \omega_i^2$ and Eq. (9.52) becomes

$$\ddot{\eta}_i(t) + \omega_i^2 \eta_i(t) = 0, \qquad i = 1, 2, \ldots \tag{9.56}$$

The solution of Eq. (9.56) is given by

$$\eta_i(t) = \eta_i(0) \cos \omega_i t + \frac{\dot{\eta}_i(0)}{\omega_i} \sin \omega_i t \tag{9.57}$$

where $\eta_i(0) = \eta_{i0}$ and $\dot{\eta}_i(0) = \dot{\eta}_{i0}$ are the initial values of $\eta_i(t)$ and $\dot{\eta}_i(t)$, which can be determined from the initial values of the displacement and velocity given by Eqs. (9.27) and (9.28). For this, first we express $u_0(x)$ and $\dot{u}_0(x)$ using Eq. (9.49) as

$$u_0(x) = \sum_{i=1}^{\infty} U_i(x)\eta_{i0} \tag{9.58}$$

$$\dot{u}_0(x) = \sum_{i=1}^{\infty} U_i(x)\dot{\eta}_{i0} \tag{9.59}$$

Multiplication of Eqs. (9.58) and (9.59) by $U_j(x)$ and integration from 0 to l result in

$$\int_0^l u_0(x)U_j(x) \, dx = \sum_{i=1}^{\infty} \eta_{i0} \int_0^l U_i(x)U_j(x) \, dx = \eta_{j0} \tag{9.60}$$

$$\int_0^l \dot{u}_0(x)U_j(x) \, dx = \sum_{i=1}^{\infty} \dot{\eta}_{i0} \int_0^l U_i(x)U_j(x) \, dx = \dot{\eta}_{j0} \tag{9.61}$$

in view of the orthogonality of the normal modes. Thus, the jth generalized coordinate can be determined from Eq. (9.57). The total response of the bar can be expressed as [Eq. (9.49)]

$$u(x, t) = \sum_{i=1}^{\infty} u_i(x, t) = \sum_{i=1}^{\infty} U_i(x) \left(\eta_{i0} \cos \omega_i t + \frac{\dot{\eta}_{i0}}{\omega_i} \sin \omega_i t \right) \tag{9.62}$$

Example 9.5   Find the free vibration response of a uniform bar with free ends due to initial displacement and velocity.

SOLUTION   The free vibratory motion of the free–free bar in the nth mode can be expressed, using Eq. (9.24), as

$$u_n(x, t) = U_n(x)T_n(t) = \cos \frac{n\pi x}{l} \left(C_n \cos \frac{n\pi c}{l} t + D_n \sin \frac{n\pi c}{l} t \right) \tag{E9.5.1}$$

where C_n and D_n are constants. By superposing the solutions given by Eq. (E9.5.1), we can represent any longitudinal vibration of the bar in the form

$$u(x, t) = \sum_{n=1}^{\infty} \cos \frac{n\pi x}{l} \left(C_n \cos \frac{n\pi c}{l} t + D_n \sin \frac{n\pi c}{l} t \right) \tag{E9.5.2}$$

where the constants C_n and D_n can be determined from the initial conditions specified. If the initial displacement and initial velocity of the bar are specified as

$$u(x, 0) = u_0(x), \qquad \frac{\partial u}{\partial t}(x, 0) = \dot{u}_0(x) \tag{E9.5.3}$$

then Eq. (E9.5.2) gives

$$u(x, 0) = u_0(x) = \sum_{n=1}^{\infty} C_n \cos \frac{n\pi x}{l} \tag{E9.5.4}$$

$$\frac{\partial u}{\partial t}(x, 0) = \dot{u}_0(x) = \sum_{n=1}^{\infty} \frac{n\pi c}{l} D_n \cos \frac{n\pi x}{l} \tag{E9.5.5}$$

To determine the constant C_n in Eq. (E9.5.4), we multiply both sides of Eq. (E9.5.4) by the mth mode shape, $\cos(m\pi x/l)$, and integrate from 0 to l:

$$\int_0^l u_0(x) \cos \frac{m\pi x}{l} \, dx = \int_0^l \sum_{n=1}^{\infty} C_n \cos \frac{n\pi x}{l} \cos \frac{m\pi x}{l} \, dx \tag{E9.5.6}$$

Noting that

$$\int_0^l \cos \frac{n\pi x}{l} \cos \frac{m\pi x}{l} = \begin{cases} 0, & m \neq n \\ \dfrac{l}{2}, & m = n \end{cases} \tag{E9.5.7}$$

Eq. (E9.5.6) can be simplified to obtain

$$C_n = \frac{2}{l} \int_0^l u_0(x) \cos \frac{n\pi x}{l} \, dx \tag{E9.5.8}$$

Using a similar procedure, the constant D_n in Eq. (E9.5.5) can be determined as

$$D_n = \frac{2}{n\pi c} \int_0^l \dot{u}_0(x) \cos \frac{n\pi x}{l} \, dx \tag{E9.5.9}$$

Example 9.6 Consider a free–free bar of uniform cross-sectional area. It is subjected to an axial compressive force at each end. Find the free vibration response of the bar when the forces are suddenly removed.

SOLUTION   We assume that the middle of the bar remains stationary. The displacement of the bar just before the forces are removed (one-half of the initial displacement at each end) is given by

$$u_0 = u(x, 0) = \frac{\varepsilon_0 l}{2} - \varepsilon_0 x \tag{E9.6.1}$$

and the initial velocity by

$$\dot{u}_0 = \frac{\partial u}{\partial t}(x, 0) = 0 \tag{E9.6.2}$$

where ε_0 denotes the compressive strain at the ends at time $t = 0$.

Using Eqs. (E9.5.8) and (E9.5.9) and Eqs. (E9.6.1) and (E9.6.2), the constants C_n and D_n can be evaluated as

$$
\begin{aligned}
C_n &= \frac{2}{l} \int_0^l \left(\frac{\varepsilon_0 l}{2} - \varepsilon_0 x \right) \cos \frac{n\pi x}{l} \, dx \\
&= \frac{\varepsilon_0 l}{n\pi} \int_0^l \cos \frac{n\pi x}{l} \, d\left(\frac{n\pi x}{l} \right) - 2\varepsilon_0 \frac{l}{n^2\pi^2} \int_0^l \frac{n\pi x}{l} \cos \frac{n\pi x}{l} \, d\left(\frac{n\pi x}{l} \right) \\
&= \begin{cases} 0, & n \text{ is even} \\ \dfrac{4\varepsilon_0 l}{n^2\pi^2}, & n \text{ is odd} \end{cases} \tag{E9.6.3}
\end{aligned}
$$

$$D_n = 0 \tag{E9.6.4}$$

Thus, the general solution for the longitudinal vibration of the free–free bar can be expressed as [see Eq. (E9.5.2)]

$$u(x, t) = \frac{4\epsilon_0 l}{\pi^2} \sum_{n=1,3,5,\cdots}^{\infty} \frac{1}{n^2} \cos \frac{n\pi x}{l} \cos \frac{n\pi cx}{l} \tag{E9.6.5}$$

Example 9.7 A bar of uniform cross-sectional area A, length l, modulus of elasticity E, and density ρ is fixed at both ends. It is subjected to an axial force F_0 at the middle [Fig. 9.5(a)] and is suddenly removed at $t = 0$. Find the resulting vibration of the bar.

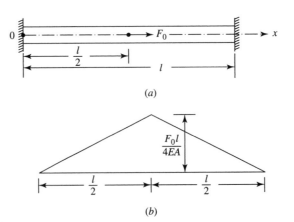

Figure 9.5 (a) Bar subjected to axial force F_0 at the middle; (b) initial displacement distribution, $u_0(x)$.

SOLUTION   The tensile strain induced in the left half of the bar is given by

$$\varepsilon = \frac{F_0}{2EA} \tag{E9.7.1}$$

which is equal in magnitude to the compressive strain in the right half of the bar. Thus, the initial displacement of the bar can be expressed as [see Fig. 9.5(b)]

$$u(x, 0) = u_0(x) = \begin{cases} \varepsilon x = \dfrac{F_0 x}{2EA}, & 0 \le x \le \dfrac{l}{2} \\[2mm] \varepsilon(l - x) = \dfrac{F_0(l - x)}{2EA}, & \dfrac{l}{2} \le x \le l \end{cases} \tag{E9.7.2}$$

Since the initial velocity is zero, we have

$$\frac{\partial u}{\partial t}(x, 0) = \dot{u}_0(x) = 0, \qquad 0 \le x \le l \tag{E9.7.3}$$

To find the general solution of the bar, we note the boundary conditions

$$u(0, t) = 0, \qquad t \ge 0 \tag{E9.7.4}$$

$$u(l, t) = 0, \qquad t \ge 0 \tag{E9.7.5}$$

The use of Eq. (E9.7.4) in Eq. (9.24) gives $\underset{\sim}{A} = 0$, and the use of Eq. (E9.7.5) gives the frequency equation:

$$\underset{\sim}{B} \sin \frac{\omega l}{c} = 0 \quad \text{or} \quad \sin \frac{\omega l}{c} = 0 \tag{E9.7.6}$$

The natural frequencies are given by

$$\frac{\omega_n l}{c} = n\pi, n = 1, 2, \ldots$$

or

$$\omega_n = \frac{n\pi c}{l}, \qquad n = 1, 2, \ldots \tag{E9.7.7}$$

and the corresponding mode shapes by

$$U_n(x) = \underset{\sim}{B}_n \sin \frac{n\pi x}{l}, \qquad n = 1, 2, \ldots \tag{E9.7.8}$$

The general free vibration solution of the bar can be expressed using the mode superposition approach as

$$u(x, t) = \sum_{n=1}^{\infty} u_n(x, t) = \sum_{n=1}^{\infty} \sin \frac{n\pi x}{l} \left(C_n \cos \frac{n\pi ct}{l} + D_n \sin \frac{n\pi ct}{l} \right) \tag{E9.7.9}$$

Using the initial velocity condition, Eq. (E9.7.3), in Eq. (E9.7.9) gives $D_n = 0$. The use of the initial displacement condition, Eq. (E9.7.2), in Eq. (E9.7.9) yields

$$
\begin{aligned}
C_n &= \frac{2}{l} \int_0^{l/2} \frac{F_0 x}{2EA} \sin \frac{n\pi x}{l} \, dx + \frac{2}{l} \int_{l/2}^{l} \frac{F_0(l-x)}{EA} \sin \frac{n\pi x}{l} \, dx \\
&= \frac{F_0}{EAl} \frac{l^2}{n^2\pi^2} \int_0^{l/2} \frac{n\pi x}{l} \sin \frac{n\pi x}{l} \, d\left(\frac{n\pi x}{l}\right) \\
&\quad + \frac{F_0 l}{EA} \frac{l}{n\pi} \int_{l/2}^{l} \sin \frac{n\pi x}{l} \, d\left(\frac{n\pi x}{l}\right) \\
&\quad - \frac{F_0}{EAl} \frac{l^2}{n^2\pi^2} \int_{l/2}^{l} \frac{n\pi x}{l} \sin \frac{n\pi x}{l} \, d\left(\frac{n\pi x}{l}\right) \\
&= \begin{cases} \dfrac{2F_0 l}{AE\pi^2} \dfrac{(-1)^{(n-1)/2}}{n^2} & \text{if } n \text{ is odd} \\[2mm] 0 & \text{if } n \text{ is even} \end{cases}
\end{aligned} \tag{E9.7.10}
$$

Thus, the free vibration solution of the bar becomes

$$
u(x,t) = \frac{2F_0 l}{AE\pi^2} \sum_{n=1,3,5,\ldots}^{\infty} \frac{(-1)^{(n-1)/2}}{n^2} \sin \frac{n\pi x}{l} \cos \frac{n\pi ct}{l} \tag{E9.7.11}
$$

9.4 FORCED VIBRATION

The equation of motion for the longitudinal vibration of a prismatic bar subjected to a distributed force $f(x,t)$ per unit length can be expressed in a general form as

$$
-Ru''(x,t) + M\ddot{u}(x,t) = f(x,t) \tag{9.63}
$$

or

$$
-c^2 u''(x,t) + \ddot{u}(x,t) = \underset{\sim}{f}(x,t) \tag{9.64}
$$

where R and M are given by Eq. (9.30), c by Eq. (9.35), and

$$
\underset{\sim}{f} = \frac{f}{M} = \frac{f}{\rho A} \tag{9.65}
$$

In modal analysis, the forced vibration response is assumed to be given by the sum of products of normal modes and generalized coordinates as indicated by Eq. (9.49). By substituting Eq. (9.49) in Eq. (9.64) for $u(x,t)$, multiplying by $U_j(x)$, and integrating from 0 to l, we obtain

$$
\sum_{i=1}^{\infty} \left[-c^2 \eta_i \int_0^l U_i''(x)U_j(x) \, dx + \ddot{\eta}_i \int_0^l U_i(x)U_j(x) \, dx \right]
$$

$$
= \int_0^l U_j(x) \underset{\sim}{f}(x,t) \, dx \tag{9.66}
$$

In view of the orthogonality relations, Eqs. (9.43) and (9.45), Eq. (9.66) becomes (for $i = j$)

$$\ddot{\eta}_i + \omega_i^2 \eta_i = \int_0^l U_i(x) \underset{\sim}{f}(x, t)\, dx \tag{9.67}$$

Equation (9.67) represents a second-order ordinary differential equation for the generalized coordinate $\eta_i(t)$. The solution of Eq. (9.67) can be obtained, using a Duhamel integral, as

$$\eta_i(t) = \frac{1}{\omega_i} \int_0^l U_i(x) \int_0^t \underset{\sim}{f}(x, \tau) \sin \omega_i(t - \tau)\, d\tau\, dx \tag{9.68}$$

Thus, the total steady-state forced longitudinal vibration response of the bar is given by (ignoring the effect of initial conditions)

$$u(x, t) = \sum_{i=1}^{\infty} \frac{U_i(x)}{\omega_i} \int_0^l U_i(x) \int_0^t \underset{\sim}{f}(x, \tau) \sin \omega_i(t - \tau)\, d\tau\, dx \tag{9.69}$$

Note   If the bar is subjected to an axial concentrated force $F_m(t)$ at $x = x_m$, there is no need for integration over the length of the bar, and Eq. (9.69) takes the form

$$u(x, t) = \sum_{i=1}^{\infty} \frac{U_i(x) U_i(x = x_m)}{\omega_i} \int_0^t \frac{F_m(\tau)}{\rho A} \sin \omega_i(t - \tau)\, d\tau \tag{9.70}$$

Example 9.8   Consider a prismatic bar fixed at both ends. Find the steady-state response of the bar if the following loads are applied suddenly at the same time (see Fig. 9.6): a uniformly distributed longitudinal force of magnitude f_0 per unit length, and an axial concentrated force F_0 at the middle point of the bar, $x = l/2$.

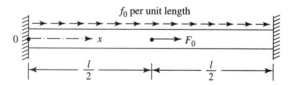

Figure 9.6   Bar subjected to distributed and concentrated loads.

SOLUTION   We can find the steady-state response of the bar by superposing the responses due to the two loads. To find the response due to the uniformly distributed load, we use

$$\underset{\sim}{f}(x, \tau) = \frac{f_0}{\rho A} \tag{E9.8.1}$$

in Eq. (9.69) to obtain

$$u(x, t) = \sum_{i=1}^{\infty} \frac{U_i(x)}{\omega_i} \int_0^l U_i(x) \frac{f_0}{\rho A} \int_0^t \sin \omega_i(t - \tau)\, d\tau\, dx \tag{E9.8.2}$$

where from the free vibration analysis [Eq. (E9.7.8)] we have

$$U_i(x) = B_i \sin \frac{i\pi x}{l} \tag{E9.8.3}$$

$$\omega_i = \frac{i\pi c}{l} \tag{E9.8.4}$$

and B_i is a constant $(i = 1, 2, \ldots)$. When $U_i(x)$ is normalized as

$$\int_0^l U_i^2(x)\, dx = 1 \tag{E9.8.5}$$

or

$$B_i^2 \int_0^l \sin^2 \frac{i\pi x}{l}\, dx = 1 \tag{E9.8.6}$$

we obtain

$$B_i = \sqrt{\frac{2}{l}} \tag{E9.8.7}$$

and hence

$$U_i(x) = \sqrt{\frac{2}{l}} \sin \frac{i\pi x}{l} \tag{E9.8.8}$$

Thus, Eq. (E9.8.2) becomes

$$
\begin{aligned}
u(x, t) &= \sum_{n=1}^{\infty} \sqrt{\frac{2}{l}}\frac{l}{n\pi c} \sin \frac{n\pi x}{l} \int_0^l \sqrt{\frac{2}{l}} \sin \frac{n\pi x}{l} \frac{f_0}{\rho A} \int_{\tau=0}^t \sin \omega_n(t - \tau)\, d\tau\, dx \\
&= \sum_{n=1,3,5,\ldots} \frac{4 f_0 l^2}{\pi^3 c^2 \rho A} \frac{1}{n^3} \sin \frac{n\pi x}{l} \left(1 - \cos \frac{n\pi ct}{l} \right)
\end{aligned}
\tag{E9.8.9}
$$

To find the response of the bar due to the concentrated load, we use $F_m(\tau) = F_0$ in Eq. (9.70), so that

$$
\begin{aligned}
u(x, t) &= \sum_{n=1}^{\infty} \sqrt{\frac{2}{l}} \frac{\sin(n\pi x/l)\sqrt{2/l}\sin(n\pi/2)}{n\pi c/l} \frac{F_0}{\rho A} \int_0^t \sin \omega_n(t - \tau)\, d\tau \\
&= \sum_{n=1,3,5,\ldots} \frac{2 F_0 l}{\pi^2 c^2 \rho A} \frac{1}{n^2} \sin \frac{n\pi x}{l} (-1)^{(n-1)/2} \left(1 - \cos \frac{n\pi ct}{l} \right)
\end{aligned}
\tag{E9.8.10}
$$

Thus, the total response of the bar is given by the sum of the two responses given by Eqs. (E9.8.9) and (E9.8.10):

$$u(x, t) = \frac{2l}{\pi^2 c^2 \rho A} \sum_{n=1,3,5,\ldots} \sin \frac{n\pi x}{l} \left(1 - \cos \frac{n\pi ct}{l} \right) \left[\frac{2 f_0 l}{\pi n^3} + \frac{F_0}{n^2} (-1)^{(n-1)/2} \right] \tag{E9.8.11}$$

9.5   RESPONSE OF A BAR SUBJECTED TO LONGITUDINAL SUPPORT MOTION

Let a prismatic bar be subjected to a support or base motion, $u_b(t) = p(t)$ in the axial direction as shown in Fig. 9.7. The equation of motion for the longitudinal vibration of the bar can be obtained as

$$\rho A \frac{\partial^2 u(x, t)}{\partial t^2} - EA \frac{\partial^2}{\partial x^2}[u(x, t) - u_b(t)] = 0 \tag{9.71}$$

By defining a new variable $v(x, t)$ that denotes the displacement of any point in the bar relative to the base as

$$v(x, t) = u(x, t) - u_b(t) \tag{9.72}$$

we can write

$$\frac{\partial^2 u}{\partial t^2} = \frac{\partial^2 v}{\partial t^2} + \frac{\partial^2 p}{\partial t^2} \tag{9.73}$$

Using Eqs. (9.72) and (9.73), Eq. (9.71) can be rewritten as

$$\rho A \frac{\partial^2 v(x, t)}{\partial t^2} - EA \frac{\partial^2 v(x, t)}{\partial x^2} = -\rho A \frac{\partial^2 p(t)}{\partial t^2} \tag{9.74}$$

A comparison of Eq. (9.74) with Eq. (9.10) shows that the term on the right-hand side of Eq. (9.74) denotes equivalent distributed loading induced by the base motion. By dividing Eq. (9.74) by ρA, we obtain

$$\frac{\partial^2 v}{\partial t^2} - c^2 \frac{\partial^2 v}{\partial x^2} = -\frac{\partial^2 p}{\partial t^2} \tag{9.75}$$

Since Eq. (9.75) is similar to Eq. (9.64), we can find the equation for the ith generalized coordinate $\eta_i(t)$ in Eq. (9.49) as

$$\frac{\partial^2 \eta_i}{\partial t^2} + \omega_i^2 \eta_i = -\frac{\partial^2 p}{\partial t^2} \int_0^l U_i(x)\, dx, \qquad i = 1, 2, \ldots \tag{9.76}$$

The solution of Eq. (9.76) can be expressed, using a Duhamel integral, as

$$\eta_i(t) = -\frac{1}{\omega_i} \int_0^l U_i(x)\, dx \int_0^t \frac{\partial^2 p}{\partial t^2}(\tau) \sin \omega_i (t - \tau)\, d\tau \tag{9.77}$$

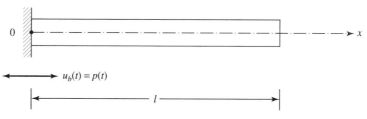

Figure 9.7   Bar with support motion.

The total solution for $v(x, t)$ can be obtained by superposing all the normal-mode responses as

$$v(x, t) = -\sum_{i=1}^{\infty} \frac{U_i(x)}{\omega_i} \int_0^l U_i(x)\, dx \int_0^t \frac{\partial^2 p}{\partial t^2}(\tau) \sin \omega_i(t - \tau)\, d\tau \qquad (9.78)$$

Finally, the longitudinal vibrational motion of the bar can be found from Eq. (9.72) as

$$u(x, t) = u_b(t) + v(x, t) \qquad (9.79)$$

9.6 RAYLEIGH THEORY

9.6.1 Equation of Motion

In this theory, the inertia of the lateral motions by which the cross sections are extended or contracted in their own planes is considered. But the contribution of shear stiffness to the strain energy is neglected. An element in the cross section of the bar, located at the coordinates y and z, undergoes the lateral displacements $-vy(\partial u/\partial x)$ and $-vz(\partial u/\partial x)$, respectively, along the y and z directions, with v denoting Poisson's ratio [2, 3, 6]. Thus, the displacement field is given by

$$u = u(x, t), \qquad v = -vy\frac{\partial u(x, t)}{\partial x}, \qquad w = -vz\frac{\partial u(x, t)}{\partial x} \qquad (9.80)$$

The strain energy of the bar and the work done by the external forces are given by Eqs. (9.6) and (9.8), while the kinetic energy of the bar can be obtained as

$$\begin{aligned}
T &= \frac{1}{2}\int_0^l dx \int_0^A \rho\, dA \left(\frac{\partial u}{\partial t}\right)^2 + \frac{1}{2}\int_0^l dx \int_0^A \rho\, dA \left[\left(\frac{\partial v}{\partial t}\right)^2 + \left(\frac{\partial w}{\partial t}\right)^2\right] \\
&= \frac{1}{2}\int_0^l \rho A \left(\frac{\partial u}{\partial t}\right)^2 dx + \frac{1}{2}\int_0^l dx \int_0^A \rho\, dA \left[\left(-vy\frac{\partial^2 u}{\partial x\partial t}\right)^2 + \left(-vz\frac{\partial^2 u}{\partial x\partial t}\right)^2\right] \\
&= \frac{1}{2}\int_0^l \rho A \left(\frac{\partial u}{\partial t}\right)^2 dx + \frac{1}{2}\int_0^l \rho v^2 I_p \left(\frac{\partial^2 u}{\partial x\partial t}\right)^2 dx \qquad (9.81)
\end{aligned}$$

where I_p is the polar moment of the inertia of the cross section, defined by

$$I_p = \int_A (y^2 + z^2)\, dA \qquad (9.82)$$

The application of extended Hamilton's principle gives

$$\delta \int_{t_1}^{t_2} dt \int_0^l \left[\frac{1}{2}\rho A \left(\frac{\partial u}{\partial t}\right)^2 + \frac{1}{2}\rho v^2 I_p \left(\frac{\partial^2 u}{\partial x\partial t}\right)^2 - \frac{1}{2}EA \left(\frac{\partial u}{\partial x}\right)^2 + fu\right] dx = 0$$

$$(9.83)$$

yielding the equation of motion and the boundary conditions as

$$-\frac{\partial}{\partial x}\left(\rho v^2 I_p \frac{\partial^3 u}{\partial x \partial t^2}\right) - \frac{\partial}{\partial x}\left(EA\frac{\partial u}{\partial x}\right) + \rho A \frac{\partial^2 u}{\partial t^2} = f \tag{9.84}$$

$$\left(EA\frac{\partial u}{\partial x} + \rho v^2 I_p \frac{\partial^3 u}{\partial x \partial t^2}\right)\delta u \bigg|_0^l = 0 \tag{9.85}$$

Note that Eq. (9.85) is satisfied if the bar is either fixed or free at the ends $x = 0$ and $x = l$. At a fixed end, $u = 0$ and hence $\delta u = 0$, while

$$EI\frac{\partial u}{\partial x} + \rho v^2 I_p \frac{\partial^3 u}{\partial x \partial t^2} = 0 \tag{9.86}$$

at a free end.

9.6.2 Natural Frequencies and Mode Shapes

For the free axial vibration of a uniform bar, we set $f = 0$ and Eqs. (9.84) and (9.85) reduce to

$$\rho v^2 I_p \frac{\partial^4 u}{\partial x^2 \partial t^2} + EA\frac{\partial^2 u}{\partial x^2} - \rho A \frac{\partial^2 u}{\partial t^2} = 0 \tag{9.87}$$

$$\left(EA\frac{\partial u}{\partial x} + \rho v^2 I_p \frac{\partial^3 u}{\partial x \partial t^2}\right)\delta u \bigg|_0^l = 0 \tag{9.88}$$

The natural frequencies of the bar can be determined using a harmonic solution

$$u(x, t) = U(x) \cos \omega t \tag{9.89}$$

Using Eq. (9.89), Eq. (9.87) can be expressed as

$$(-\rho v^2 I_p \omega^2 + EA)\frac{d^2 U}{dx^2} + \rho A \omega^2 U = 0 \tag{9.90}$$

The solution of the second-order ordinary differential equation, Eq. (9.90) can be written as

$$U(x) = C_1 \cos px + C_2 \sin px \tag{9.91}$$

where

$$p = \sqrt{\frac{\rho A \omega^2}{EA - \rho v^2 I_p \omega^2}} \tag{9.92}$$

and C_1 and C_2 are constants to be determined from the boundary conditions of the bar.

Bar with Both Ends Fixed   For a bar fixed at both ends,

$$U(x = 0) = 0 \quad \text{and} \quad U(x = l) = 0 \tag{9.93}$$

Equations (9.91) and (9.93) lead to

$$C_1 = 0 \tag{9.94}$$

$$\sin pl = 0 \tag{9.95}$$

Equation (9.95) gives the frequencies of vibration:

$$pl = n\pi, \qquad n = 1, 2, \ldots$$

or

$$\omega_n^2 = \frac{n^2\pi^2}{(1 + \nu^2 I_p n^2 \pi^2 / Al^2)} \frac{E}{\rho l^2}, \qquad n = 1, 2, \ldots \tag{9.96}$$

The mode shape corresponding to the frequency ω_n is given by

$$U_n(x) = \sin n\pi x, \qquad n = 1, 2, \ldots \tag{9.97}$$

It can be seen that the mode shapes [Eq. (9.97)] are identical to those given by the simple theory, whereas the natural frequencies [Eq. (9.96)] are reduced by the factor

$$\left(1 + \frac{\nu^2 I_p n^2 \pi^2}{Al^2}\right)^{1/2}$$

compared to those given by the simple theory.

9.7 BISHOP'S THEORY

9.7.1 Equation of Motion

This theory considers the effect not only of the inertia of the lateral motions but also of the shear stiffness [1, 3, 6]. The displacement field is given by Eq. (9.80). The strains in the cross section can be obtained as

$$\varepsilon_{xx} = \frac{\partial u}{\partial x}, \qquad \varepsilon_{yy} = \frac{\partial v}{\partial y} = -\nu\frac{\partial u}{\partial x}, \qquad \varepsilon_{zz} = \frac{\partial w}{\partial z} = -\nu\frac{\partial u}{\partial x},$$

$$\varepsilon_{xy} = \left(\frac{\partial u}{\partial y} + \frac{\partial v}{\partial x}\right) = -\nu y\frac{\partial^2 u}{\partial x^2}, \qquad \varepsilon_{yz} = \left(\frac{\partial v}{\partial z} + \frac{\partial w}{\partial y}\right) = 0,$$

$$\varepsilon_{zx} = \left(\frac{\partial u}{\partial z} + \frac{\partial w}{\partial x}\right) = -\nu z\frac{\partial^2 u}{\partial x^2} \tag{9.98}$$

The stresses induced in the cross section of the bar can be determined, using the three-dimensional Hooke's law, as

$$
\begin{Bmatrix} \sigma_{xx} \\ \sigma_{yy} \\ \sigma_{zz} \\ \sigma_{xy} \\ \sigma_{yz} \\ \sigma_{zx} \end{Bmatrix} = \frac{E}{(1+v)(1-2v)}
\begin{bmatrix}
1-v & v & v & 0 & 0 & 0 \\
v & 1-v & v & 0 & 0 & 0 \\
v & v & 1-v & 0 & 0 & 0 \\
0 & 0 & 0 & \frac{1-2v}{2} & 0 & 0 \\
0 & 0 & 0 & 0 & \frac{1-2v}{2} & 0 \\
0 & 0 & 0 & 0 & 0 & \frac{1-2v}{2}
\end{bmatrix}
\begin{Bmatrix} \varepsilon_{xx} \\ \varepsilon_{yy} \\ \varepsilon_{zz} \\ \varepsilon_{xy} \\ \varepsilon_{yz} \\ \varepsilon_{zx} \end{Bmatrix}
\tag{9.99}
$$

Substitution of Eq. (9.98) in Eq. (9.99) results in

$$
\begin{Bmatrix} \sigma_{xx} \\ \sigma_{yy} \\ \sigma_{zz} \\ \sigma_{xy} \\ \sigma_{yz} \\ \sigma_{zx} \end{Bmatrix} =
\begin{Bmatrix} E\frac{\partial u}{\partial x} \\ 0 \\ 0 \\ -vGy\frac{\partial^2 u}{\partial x^2} \\ 0 \\ -vGz\frac{\partial^2 u}{\partial x^2} \end{Bmatrix}
\tag{9.100}
$$

The strain energy of the bar can be computed as

$$
\pi = \frac{1}{2} \iiint_V (\sigma_{xx}\varepsilon_{xx} + \sigma_{yy}\varepsilon_{yy} + \sigma_{zz}\varepsilon_{zz} + \sigma_{xy}\varepsilon_{xy} + \sigma_{yz}\varepsilon_{yz} + \sigma_{zx}\varepsilon_{zx})\, dV
$$

$$
= \frac{1}{2} \int_0^l dx \iint_0^A dA \left[E\left(\frac{\partial u}{\partial x}\right)^2 + 0 + 0 + v^2 Gy^2 \left(\frac{\partial^2 u}{\partial x^2}\right)^2 + 0 + v^2 Gz^2 \left(\frac{\partial^2 u}{\partial x^2}\right)^2 \right]
$$

$$
= \frac{1}{2} \int_0^l \left[EA\left(\frac{\partial u}{\partial x}\right)^2 + v^2 GI_p \left(\frac{\partial^2 u}{\partial x^2}\right)^2 \right] dx
\tag{9.101}
$$

The kinetic energy of the bar and the work done by the external forces are given by Eqs. (9.81) and (9.8), respectively. The extended Hamilton's principle can be expressed as

$$
\delta \int_{t_1}^{t_2} (T - \pi + W)\, dt = 0
\tag{9.102}
$$

By substituting Eqs. (9.81), (9.101), and (9.8) for T, π, and W, respectively, in Eq. (9.102) and simplifying results in the following equation of motion and the associated boundary conditions,

$$
\frac{\partial^2}{\partial x^2}\left(v^2 GI_p \frac{\partial^2 u}{\partial x^2}\right) - \frac{\partial}{\partial x}\left(v^2 \rho I_p \frac{\partial^3 u}{\partial x \partial t^2}\right) - \frac{\partial}{\partial x}\left(EA\frac{\partial u}{\partial x}\right) + \rho A \frac{\partial^2 u}{\partial t^2} = f
\tag{9.103}
$$

$$
\left[EA\frac{\partial u}{\partial x} + v^2 \rho I_p \frac{\partial^3 u}{\partial x \partial t^2} - v^2 \frac{\partial}{\partial x}\left(GI_p \frac{\partial^2 u}{\partial x^2}\right) \right] \delta u \Big|_0^l + \left(v^2 GI_p \frac{\partial^2 u}{\partial x^2}\right) \delta\left(\frac{\partial u}{\partial x}\right)\Big|_0^l = 0
\tag{9.104}
$$

it can be seen that if an end is rigidly fixed, $u = \partial u/\partial x = 0$ and hence $\delta u = \delta(\partial u/\partial x) = 0$ in Eq. (9.104).

9.7.2 Natural Frequencies and Mode Shapes

For a uniform bar undergoing free vibration ($f = 0$), Eqs. (9.103) and (9.104) can be written as

$$v^2 G I_p \frac{\partial^4 u}{\partial x^4} - \rho v^2 I_p \frac{\partial^4 u}{\partial x^2 \partial t^2} - EA \frac{\partial^2 u}{\partial x^2} + \rho A \frac{\partial^2 u}{\partial t^2} = 0 \qquad (9.105)$$

$$\left(EA \frac{\partial u}{\partial x} + v^2 \rho I_p \frac{\partial^3 u}{\partial x \partial t^2} - v^2 G I_p \frac{\partial^3 u}{\partial x^3} \right) \delta u \bigg|_0^l + \left(v^2 G I_p \frac{\partial^2 u}{\partial x^2} \right) \delta \left(\frac{\partial u}{\partial x} \right) \bigg|_0^l = 0 \quad (9.106)$$

The natural frequencies of the bar can be found using a harmonic solution:

$$u(x, t) = U(x) \cos \omega t \qquad (9.107)$$

Substitution of Eq. (9.107) into (9.105) leads to

$$v^2 G I_p \frac{d^4 U}{dx^4} + (\rho v^2 I_p \omega^2 - EA) \frac{d^2 U}{dx^2} - \rho A \omega^2 U = 0 \qquad (9.108)$$

By assuming the solution of Eq. (9.108) as

$$U(x) = C e^{px} \qquad (9.109)$$

where C and p are constants, the auxiliary equation can be obtained as

$$v^2 G I_p p^4 + (\rho v^2 I_p \omega^2 - EA) p^2 - \rho A \omega^2 = 0 \qquad (9.110)$$

Equation (9.110) is a quadratic equation in p^2 whose roots are given by

$$p^2 = \frac{(EA - \rho v^2 I_p \omega^2) \pm \sqrt{(EA - \rho v^2 I_p \omega^2)^2 + 4v^2 G I_p \rho A \omega^2}}{2v^2 G I_p} = a \pm b \quad (9.111)$$

where

$$a = \frac{EA - \rho v^2 I_p \omega^2}{2v^2 G I_p} \qquad (9.112)$$

$$b = \frac{\sqrt{(EA - \rho v^2 I_p \omega^2)^2 + 4v^2 G I_p \rho A \omega^2}}{2v^2 G I_p} \qquad (9.113)$$

Since $b > a$, the roots can be expressed as

$$p_1 = -p_2 = s_1 = \sqrt{a + b}, \qquad p_3 = -p_4 = i s_2 = i \sqrt{b - a} \qquad (9.114)$$

Thus, the general solution of Eq. (9.110) can be written as

$$U(x) = \overline{C}_1 e^{s_1 x} + \overline{C}_2 e^{-s_1 x} + \overline{C}_3 e^{i s_2} + \overline{C}_4 e^{-i s_2} \qquad (9.115)$$

where the constants $\overline{C}_1$, $\overline{C}_2$, $\overline{C}_3$, and $\overline{C}_4$ are to be determined from the boundary conditions of the bar. Noting that $\sinh x = \frac{1}{2}(e^x - e^{-x})$, $\cosh x = \frac{1}{2}(e^x + e^{-x})$, $\sin x = (1/2i)(e^{ix} - e^{-ix})$, and $\cos x = \frac{1}{2}(e^{ix} + e^{-ix})$, Eq. (9.115) can be rewritten as

$$U(x) = C_1 \cosh s_1 x + C_2 \sinh s_1 x + C_3 \cos s_2 x + C_4 \sin s_2 x \qquad (9.116)$$

where C_1, C_2, C_3, and C_4 are constants.

Bar Fixed Loosely at Both Ends   If the bar is fixed loosely at both ends, the axial displacement and shear strain will be zero at each end, so that

$$U(0) = 0 \tag{9.117}$$

$$U(l) = 0 \tag{9.118}$$

$$\frac{d^2U}{dx^2}(0) = 0 \tag{9.119}$$

$$\frac{d^2U}{dx^2}(l) = 0 \tag{9.120}$$

Equations (9.116)–(9.120) lead to

$$C_1 + C_3 = 0 \tag{9.121}$$

$$C_1 \cosh s_1 l + C_2 \sinh s_1 l + C_3 \cos s_2 l + C_4 \sin s_2 l = 0 \tag{9.122}$$

$$C_1 s_1^2 - C_3 s_2^2 = 0 \tag{9.123}$$

$$C_1 s_1^2 \cosh s_1 l + C_2 s_1^2 \sinh s_1 l - C_3 s_2^2 \cos s_2 l - C_4 s_2^2 \sin s_2 l = 0 \tag{9.124}$$

Equations (9.121) and (9.123) give

$$C_1 = C_3 = 0 \tag{9.125}$$

and Eqs. (9.122) and (9.124) reduce to

$$C_2 \sinh s_1 l + C_4 \sin s_2 l = 0 \tag{9.126}$$

$$C_2 s_1^2 \sinh s_1 l - C_4 s_2^2 \sin s_2 l = 0 \tag{9.127}$$

The condition for a nontrivial solution of C_2 and C_4 in Eqs. (9.126) and (9.127) is

$$\begin{vmatrix} \sinh s_1 l & \sin s_2 l \\ s_1^2 \sinh s_1 l & -s_2^2 \sin s_2 l \end{vmatrix} = 0$$

or

$$\sinh s_1 l \sin s_2 l = 0 \tag{9.128}$$

Since $\sinh s_1 l \neq 0$ for nonzero values of $s_1 l$, Eq. (9.128) leads to the frequency equation

$$\sin s_2 l = 0 \tag{9.129}$$

The natural frequencies are given by

$$s_2 l = n\pi, \quad n = 1, 2, \ldots \tag{9.130}$$

Using Eqs. (9.114), (9.112), and (9.113) in (9.130), we can express the natural frequencies as (see Problem 9.6)

$$\omega_n^2 = \frac{n^2 \pi^2 E}{\rho l^2} \left(\frac{AEl^2 + v^2 G I_p n^2 \pi^2}{AEl^2 + v^2 EI_p n^2 \pi^2} \right) \tag{9.131}$$

The mode shape corresponding to the frequency ω_n is given by

$$U_n(x) = \sin n\pi x, \qquad n = 1, 2, \ldots \tag{9.132}$$

It can be observed that the mode shapes [Eq. (9.132)] are identical to those given by the simple theory, whereas the natural frequencies [Eq. (9.131)] are reduced by the factor

$$\left(\frac{AEl^2 + v^2 G I_p n^2 \pi^2}{AEl^2 + v^2 E I_p n^2 \pi^2}\right)^{1/2}$$

compared to those given by the simple theory.

9.7.3   Forced Vibration Using Modal Analysis

The equation of motion of a prismatic bar in longitudinal vibration, Eq. (9.105), can be expressed as

$$M\ddot{u} + Lu = f \tag{9.133}$$

where

$$M = \rho A - v^2 \rho I_p \frac{\partial^2}{\partial x^2} \tag{9.134}$$

$$L = v^2 G I_p \frac{\partial^4}{\partial x^4} - EI \frac{\partial^2}{\partial x^2} \tag{9.135}$$

In modal analysis, the solution is expressed as the sum of natural modes as

$$u(x, t) = \sum_{i=1}^{\infty} U_i(x)\eta_i(t) \tag{9.136}$$

so that the equation of motion for the ith normal mode becomes

$$(M[U_i(x)])\ddot{\eta}_i(t) + (L[U_i(x)])\eta_i(t) = f(x, t) \tag{9.137}$$

By multiplying Eq. (9.137) by $U_j(x)$ and integrating from 0 to l, we obtain

$$\int_0^l (M[U_i(x)])U_j(x)\ddot{\eta}_i(t)\, dx + \int_0^l (L[U_i(x)])U_j(x)\eta_i(t)\, dx = \int_0^l f(x, t)U_j(x)\, dx \tag{9.138}$$

In view of the orthogonality relations among natural modes, Eq. (9.138) reduces to

$$M_i\ddot{\eta}_i + K_i\eta_i = f_i, \qquad i = 1, 2, \ldots \tag{9.139}$$

where M_i is the generalized mass, K_i is the generalized stiffness, and f_i is the generalized force in the ith mode, given by

$$M_i = \int_0^l (M[U_i(x)])U_i(x)\,dx \tag{9.140}$$

$$K_i = \int_0^l (L[U_i(x)])U_i(x)\,dx \tag{9.141}$$

$$f_i = \int_0^l f(x,t)U_i(x)\,dx \tag{9.142}$$

The solution of Eq. (9.139) can be expressed, using a Duhamel integral, as

$$\eta_i(t) = \eta_0 \cos\omega_i t + \frac{\dot\eta_i(0)}{\omega_i}\sin\omega_i t + \frac{1}{M_i\omega_i}\int_0^t f_i(\tau)\sin\omega_i(t-\tau)\,d\tau, \qquad i = 1, 2, \ldots \tag{9.143}$$

where ω_i is the ith natural frequency given by

$$\omega_i = \sqrt{\frac{K_i}{M_i}}, \qquad i = 1, 2, \ldots \tag{9.144}$$

and $\eta_i(0)$ and $\dot\eta_i(0)$ are the initial values of the generalized displacement $\eta_i(t)$ and generalized velocity $\dot\eta_i(t)$. If $u_0(x) = u(x, 0)$ and $\dot u_0(x) = \dot u(x, 0)$ are the initial values specified for longitudinal displacement and velocity, we can express

$$u_0(x) = \sum_{i=1}^\infty U_i(x)\eta_i(0) \tag{9.145}$$

$$\dot u_0(x) = \sum_{i=1}^\infty U_i(x)\dot\eta_i(0) \tag{9.146}$$

Multiplying Eqs. (9.145) and (9.146) by $M[U_j(x)]$ and integrating from 0 to l results in

$$\int_0^l u_0(x)M[U_j(x)]\,dx = \sum_{i=1}^\infty \int_0^l \eta_i(0)U_i(x)M[U_j(x)]\,dx \tag{9.147}$$

$$\int_0^l \dot u_0(x)M[U_j(x)]\,dx = \sum_{i=1}^\infty \int_0^l \dot\eta_i(0)U_i(x)M[U_j(x)]\,dx \tag{9.148}$$

When the property of orthogonality of normal modes is used, Eqs. (9.147) and (9.148) yield

$$\eta_i(0) = \frac{1}{M_i}\int_0^l u_0(x)M[U_i(x)]\,dx, \qquad i = 1, 2, \ldots \tag{9.149}$$

$$\dot\eta_i(0) = \frac{1}{M_i}\int_0^l \dot u_0(x)M[U_i(x)]\,dx, \qquad i = 1, 2, \ldots \tag{9.150}$$

Finally, the total axial motion (displacement) of the bar can be expressed as

$$u(x, t) = \sum_{i=1}^{\infty} \left[\eta_i(0) \cos \omega_i t + \frac{\dot{\eta}_i(0)}{\omega_i} \sin \omega_i t + \frac{1}{M_i \omega_i} \int_0^t f_i(\tau) \sin \omega_i (t - \tau) \, d\tau \right] U_i(x)$$

(9.151)

Example 9.9   Determine the steady-state response of a prismatic bar fixed loosely at both ends when an axial force F_0 is suddenly applied at the middle as shown in Fig. 9.8.

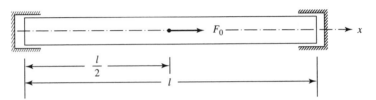

Figure 9.8   Bar supported loosely at ends.

SOLUTION   The natural frequencies and normal modes of the bar are given by Eqs. (9.131) and (9.132):

$$\omega_i^2 = \frac{i^2 \pi^2 E}{\rho l^2} \frac{AEl^2 + v^2 G I_p i^2 \pi^2}{AEl^2 + v^2 E I_p i^2 n^2}$$

(E9.9.1)

$$U_i(x) = \sin i\pi x$$

(E9.9.2)

The generalized mass M_i and the generalized stiffness K_i in mode i can be determined as

$$M_i = \int_0^l (M[U_i(x)]) U_i(x) \, dx = \int_0^l \left[\left(\rho A - v^2 \rho I_p \frac{\partial^2}{\partial x^2} \right) \sin i\pi x \right] \sin i\pi x \, dx$$

$$= (\rho A + v^2 \rho I_p i^2 \pi^2) \frac{l}{2}$$

(E9.9.3)

$$K_i = \int_0^l (L[U_i(x)]) U_i(x) \, dx = \int_0^l \left[\left(v^2 G I_p \frac{\partial^4}{\partial x^4} - EI \frac{\partial^2}{\partial x^2} \right) \sin i\pi x \right] \sin i\pi x \, dx$$

$$= (v^2 G I_p i^4 \pi^4 + EI i^2 \pi^2) \frac{l}{2}$$

(E9.9.4)

The applied axial force can be represented as

$$f(x, t) = F_0 H(t) \delta \left(x - \frac{l}{2} \right)$$

(E9.9.5)

where $H(t)$ is the Heaviside unit step function and δ is the Dirac delta function. The generalized force in the ith normal mode can be computed as

$$f_i(t) = \int_0^l f(x, t)U_i(x)\, dx = \int_0^l F_0 H(t)\delta\left(x - \frac{l}{2}\right)U_i(x)\, dx$$

$$= F_0 H(t)U_i\left(x = \frac{l}{2}\right) = F_0 H(t)\sin\frac{i\pi l}{2} \qquad \text{(E9.9.6)}$$

Thus, the steady-state solution of the ith generalized coordinate is given by the solution of Eq. (9.139) as

$$\eta_i(t) = \frac{1}{M_i\omega_i}\int_0^t f_i(\tau)\sin\omega_i(t - \tau)\, d\tau, \qquad i = 1, 2, \ldots \qquad \text{(E9.9.7)}$$

which can be written as

$$\eta_i(t) = \frac{1}{M_i\omega_i}\int_0^t F_0 H(\tau)\sin\frac{i\pi l}{2}\sin\omega_i(t - \tau)\, d\tau$$

$$= \frac{F_0 \sin(i\pi l/2)}{M_i\omega_i}\int_0^t H(\tau)\sin\omega_i(t - \tau)\, d\tau = \frac{F_0 \sin(i\pi l/2)}{M_i\omega_i^2}(1 - \cos\omega_i t)$$

$$\text{(E9.9.8)}$$

Thus, the total steady-state response of the bar is given by

$$u(x, t) = \sum_{i=1}^{\infty} \frac{F_0 \sin(i\pi l/2)}{M_i\omega_i^2}(1 - \cos\omega_i t) \qquad \text{(E9.9.9)}$$

9.8 RECENT CONTRIBUTIONS

Additional problems of longitudinal vibration, including the determination of the natural frequencies of nonuniform bars, and free and forced vibration of uniform viscoelastic and viscoelastically coated bars, are discussed in detail by Rao [3].

A comparative evaluation of the approximate solutions given by discretization methods such as the finite element and finite difference methods for the free axial vibration of uniform rods was made by Ramesh and Itku [14]. The solution of the wave equation, which describes the axial free vibration of uniform rods in terms of eigenvalues and eigenfunctions, was used as a basis for comparison of the approximate solutions. It was observed that the frequencies given by the discretization methods were influenced significantly and the mode shapes were relatively insensitive to the choice of mass lumping scheme.

Kukla et al. [15] considered the problem of longitudinal vibration of two rods coupled by many translational springs using the Green's function method. The frequencies of longitudinal vibration of a uniform rod with a tip mass or spring was considered by Kohoutek [8]. Raj and Sujith [9] developed closed-form solutions for the free longitudinal vibration of inhomogeneous rods. The longitudinal impulsive response analysis of variable-cross-section bars was presented by Matsuda et al. [10].

Exact analytical solutions for the longitudinal vibration of bars with a nonuniform cross section were presented by Li [11] and Kumar and Sujith [12].

The solutions are found in terms of special functions such as the Bessel and Neumann as well as trigonometric functions. Simple expressions are given to predict the natural frequencies of nonuniform bars with various boundary conditions. The equation of motion of a vibrating Timoshenko column is discussed by Kounadis in Ref. [13].

REFERENCES

1. R. E. D. Bishop, Longitudinal waves in beams, *Aeronautical Quarterly*, Vol. 3, No. 2, pp. 280–293, 1952.

2. J. W. S. Rayleigh, *The Theory of Sound*, Dover, New York, 1945.

3. J. S. Rao, *Advanced Theory of Vibration*, Wiley Eastern, New Delhi, India, 1992.

4. A. E. H. Love, *A Treatize on the Mathematical Theory of Elasticity*, 4th ed., Dover, New York, 1944.

5. S. Timoshenko, D. H. Young, and W. Weaven, Jr., *Vibration Problems in Engineering*, 4th ed., Wiley, New York, 1974.

6. E. Volterra and E. C. Zachmanoglou, *Dynamics of Vibrations*, Charles E. Merrill, Columbus, OH, 1965.

7. S. K. Clark, *Dynamics of Continuous Elements*, Prentice-Hall, Englewood Cliffs, NJ, 1972.

8. R. Kohoutek, Natural longitudinal frequencies of a uniform rod with a tip mass or spring, *Journal of Sound and Vibration*, Vol. 77, No. 1, pp. 147–148, 1981.

9. A. Raj and R. I. Sujith, Closed-form solutions for the free longitudinal vibration of inhomogeneous rods, *Journal of Sound and Vibration*, Vol. 283, No. 3–5, pp. 1015–1030, 2005.

10. H. Matsuda, T. Sakiyama, C. Morita, and M. Kawakami, Longitudinal impulsive response analysis of variable cross-section bars, *Journal of Sound and Vibration*, Vol. 181, No. 3, pp. 541–551, 1995.

11. Q. S. Li, Exact solutions for free longitudinal vibration of non-uniform rods, *Journal of Sound and Vibration*, Vol. 234, No. 1, pp. 1–19, 2000.

12. B. M. Kumar and R. I. Sujith, Exact solutions for the longitudinal vibration of non-uniform rods, *Journal of Sound and Vibration*, Vol. 207, No. 5, pp. 721–729, 1997.

13. N. Kounadis, On the derivation of equation of motion for a vibrating Timoshenko column, *Journal of Sound and Vibration*, Vol. 73, No. 2, pp. 174–184, 1980.

14. A. V. Ramesh and S. Utku, A comparison of approximate solutions for the axial free vibrations of uniform rods, *Journal of Sound and Vibration*, Vol. 139, No. 3, pp. 407–424, 1990.

15. S. Kukla, J. Przybylski, and L. Tomski, Longitudinal vibration of rods coupled by translational springs, *Journal of Sound and Vibration*, Vol. 185, No. 4, pp. 717–722, 1995.

PROBLEMS

9.1   Derive the frequency equation for the longitudinal vibration of the bar shown in Fig. 9.9.

9.2   Derive the equation of motion for the longitudinal vibration of a bar by including the damping force that is proportional to the longitudinal velocity.

9.3   A uniform bar is fixed at one end and free at the other end. Find the longitudinal vibration response of the bar subject to the initial conditions $u(x, 0) = U_0 x^2$ and $\dot{u}(x, 0) = 0$.

9.4   Consider a uniform bar fixed at one end and carrying a mass M at the other end. Find the longitudinal vibration response of the bar when its fixed end is subjected to a harmonic axial motion as shown in Fig. 9.10.

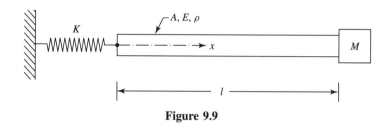

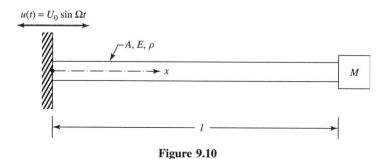

Figure 9.9

Figure 9.10

9.5 Specialize Eq. (E9.2.10) to the case where the mass of the bar is negligible compared to the mass attached. Solve the resulting equation to find the fundamental frequency of vibration of the bar.

9.6 Derive Eq. (9.131).

9.7 Show that the expressions on either side of the equality sign in Eq. (9.19) is equal to a negative quantity.

9.8 Derive Eqs. (9.10) and (9.11) from Eq. (9.9).

9.9 Derive the frequency equation for the longitudinal vibration of a uniform bar fixed at $x = 0$ and attached to a mass M and spring of stiffness k at $x = l$ (case 5 of Table 9.1).

9.10 Derive the frequency equation for the longitudinal vibration of a uniform bar free at $x = 0$ and attached to a mass M at $x = l$ (case 7 of Table 9.1).

9.11 Derive Eqs. (9.84) and (9.85) from Hamilton's principle.

9.12 Derive Eqs. (9.103) and (9.104) from Hamilton's principle.

9.13 Consider a uniform free–free bar. If the ends $x = 0$ and $x = l$ are subjected to the displacements $u(0, t) = U_1 e^{i\Omega t}$ and $u(l, t) = U_2 e^{i\Omega t}$, determine the axial motion of the bar, $u(x, t), 0 < x < l, t > 0$.

9.14 A uniform bar fixed at $x = 0$ and free at $x = l$ is subjected to a distributed axial force $f(x, t) = x^2 \sin 2t$. Determine the resulting axial motion of the bar.

9.15 The ends of a uniform bar are connected to two springs as shown in Fig. 9.11. Derive the frequency equation corresponding to the axial vibration of the bar.

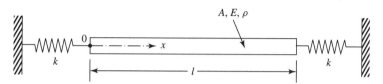

Figure 9.11

9.16 A uniform bar is fixed at $x = 0$ and is subjected to a sudden axial force f_0 (shown in Fig. 9.12) at $x = l$. Find the ensuing axial motion of the bar at $x = l$.

9.17 A uniform bar of length l, cross-sectional area A, Young's modulus E, and mass density ρ strikes a spring of stiffness k with a velocity V as shown in Fig. 9.13. Find the resulting axial motion of the bar, $u(x, t)$, measured from the instant the bar strikes the spring.

9.18 A uniform bar of length l, cross-sectional area A, Young's modulus E, and mass density ρ is fixed at $x = 0$ and carries a mass M at $x = l$. The end $x = l$ is subjected to an axial force $F(t) = F_0 \sin \Omega t$ as shown in Fig. 9.14. Determine the steady-state response, $u(x, t)$, of the bar.

9.19 Find the longitudinal vibration response of a uniform bar of length l, fixed at $x = 0$ and free at $x = l$, when the end $x = 0$ is subjected to an axial harmonic displacement, $u_b(t) = c \sin \omega t$ where c is a constant and ω is the frequency.

9.20 Find the steady state axial motion of a prismatic bar of length l, fixed at $x = 0$, when an axial force $F(t) = F_0$ acts at the end $x = l$ using the Laplace transform approach.

9.21 Derive the orthogonality relationships for a bar, fixed at $x = 0$, carrying a mass $\underset{\sim}{M}$ at $x = l$.

Figure 9.12

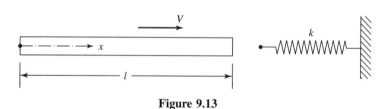

Figure 9.13

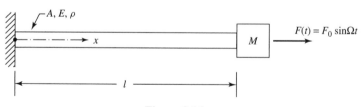

Figure 9.14

10

Torsional Vibration of Shafts

10.1 INTRODUCTION

Many rotating shafts and axles used for power transmission experience torsional vibration, particularly when the prime mover is a reciprocating engine. The shafts used in high-speed machinery, especially those carrying heavy wheels, are subjected to dynamic torsional forces and vibration. A solid or hollow cylindrical rod of circular section undergoes torsional displacement or twisting such that each transverse section remains in its own plane when a torsional moment is applied. In this case the cross sections of the rod do not experience any motion parallel to the axis of the rod. However, if the cross section of the rod is not circular, the effect of a twist will be more involved. In this case the twist will be accompanied by a warping of normal cross sections. The torsional vibrations of uniform and nonuniform rods with circular cross section and rods with noncircular section are considered in this chapter. For noncircular sections, the equations of motion are derived using both the Saint-Venant and Timoshenko–Gere theories. The methods of determining the torsional rigidity of noncircular rods is presented using the Prandtl stress function and the Prandtl membrane analogy.

10.2 ELEMENTARY THEORY: EQUATION OF MOTION

10.2.1 Equilibrium Approach

Consider an element of a nonuniform circular shaft between two cross sections at x and $x + dx$, as shown in Fig. 10.1(a). Let $M_t(x, t)$ denote the torque induced in the shaft at x and time t and $M_t(x, t) + dM_t(x, t)$ the torque induced in the shaft at $x + dx$ and at the same time t. If the angular displacement of the cross section at x is denoted as $\theta(x, t)$, the angular displacement of the cross section at $x + dx$ can be represented as $\theta(x, t) + d\theta(x, t)$. Let the external torque acting on the shaft per unit length be denoted $m_t(x, t)$. The inertia torque acting on the element of the shaft is given by $I_0 dx(\partial^2\theta/\partial t^2)$, where I_0 is the mass polar moment of inertia of the shaft per unit length. Noting that $dM_t = (\partial M_t/\partial x)\, dx$ and $d\theta = (\partial\theta/\partial x)dx$, Newton's second law of motion can be applied to the element of the shaft to obtain the equation of motion as

$$\left(M_t + \frac{\partial M_t}{\partial x}\, dx\right) - M_t + m_t\, dx = I_0\, dx\, \frac{\partial^2\theta}{\partial t^2} \tag{10.1}$$

From strength of materials, the relationship between the torque in the shaft and the angular displacement is given by [1]

$$M_t = GI_p \frac{\partial \theta}{\partial x} \tag{10.2}$$

where G is the shear modulus and $I_p = J$ is the polar moment of inertia of the cross section of the shaft. Using Eq. (10.2), the equation of motion, Eq. (10.1), can be expressed as

$$\frac{\partial}{\partial x}\left(GI_p \frac{\partial \theta(x,t)}{\partial x}\right) + m_t(x,t) = I_0 \frac{\partial^2 \theta(x,t)}{\partial t^2} \tag{10.3}$$

10.2.2 Variational Approach

The equation of motion of a nonuniform shaft, using the variational approach, has been derived in Section 4.11.1. In this section the variational approach is used to derive the equation of motion and the boundary conditions for a nonuniform shaft with torsional springs (with stiffnesses k_{t1} and k_{t2}) and masses (with mass moments of inertia I_{10} and I_{20}) attached at each end as shown in Fig. 10.1.

The cross sections of the shaft are assumed to remain plane before and after angular deformation. Since the cross section of the shaft at x undergoes an angular displacement

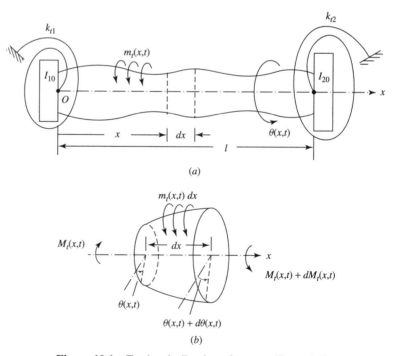

Figure 10.1 Torsional vibration of a nonuniform shaft.

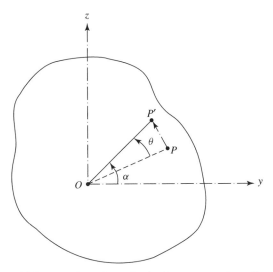

Figure 10.2   Rotation of a point in the cross section of a shaft.

$\theta(x, t)$ about the center of twist, the shape of the cross section does not change. The cross section simply rotates about the x axis. A typical point P rotates around the x axis by a small angle θ as shown in Fig. 10.2. The displacements of point P parallel to the y and z axes are given by the projections of the displacement PP' on oy and oz:

$$v(y, z) = OP' \cos \alpha - OP \cos(\alpha - \theta)$$

$$= OP' \cos \alpha - OP \cos \alpha \cos \theta - OP \sin \alpha \sin \theta \qquad (10.4)$$

$$w(y, z) = OP' \sin \alpha - OP \sin(\alpha - \theta)$$

$$= OP' \sin \alpha - OP \sin \alpha \cos \theta + OP \cos \alpha \sin \theta$$

Since θ is small, we can write

$$\sin \theta \approx \theta, \qquad \cos \theta \approx 1$$

$$OP \cos \alpha \simeq OP' \cos \alpha = y$$

$$OP \sin \alpha \simeq OP' \sin \alpha = z \qquad (10.5)$$

so that

$$v(y, z) = -z\theta, \qquad w(y, z) = y\theta \qquad (10.6)$$

Thus, the displacement components of the shaft parallel to the three coordinate axes can be expressed as

$$u(x, t) = 0$$

$$v(x, t) = -z\theta(x, t)$$

$$w(x, t) = y\theta(x, t) \qquad (10.7)$$

The strains in the shaft are assumed to be

$$\varepsilon_{xy} = \frac{\partial u}{\partial y} + \frac{\partial v}{\partial x} = -z\frac{\partial \theta}{\partial x}$$

$$\varepsilon_{xz} = \frac{\partial u}{\partial z} + \frac{\partial w}{\partial x} = y\frac{\partial \theta}{\partial x} \tag{10.8}$$

$$\varepsilon_{xx} = \varepsilon_{yy} = \varepsilon_{zz} = \varepsilon_{yz} = 0$$

and the corresponding stresses are given by

$$\sigma_{xy} = -Gz\frac{\partial \theta}{\partial x}$$

$$\sigma_{xz} = Gy\frac{\partial \theta}{\partial x} \tag{10.9}$$

$$\sigma_{xx} = \sigma_{yy} = \sigma_{zz} = \sigma_{yz} = 0$$

The strain energy of the shaft and the torsional springs is given by

$$\pi = \frac{1}{2}\iiint\limits_{V}(\sigma_{xx}\varepsilon_{xx} + \sigma_{yy}\varepsilon_{yy} + \sigma_{zz}\varepsilon_{zz} + \sigma_{xy}\varepsilon_{xy} + \sigma_{xz}\varepsilon_{xz} + \sigma_{yz}\varepsilon_{yz})\,dV$$

$$+ \left[\tfrac{1}{2}k_{t1}\theta^2(0, t) + \tfrac{1}{2}k_{t2}\theta^2(l, t)\right]$$

$$= \frac{1}{2}\int_{x=0}^{l}\left\{\iint\limits_{A}\left[G\left(\frac{\partial \theta}{\partial x}\right)^2(y^2 + z^2)\right]dA\right\}dx + \left[\frac{1}{2}k_{t1}\theta^2(0, t) + \frac{1}{2}k_{t2}\theta^2(l, t)\right]$$

$$= \frac{1}{2}\int_0^l GI_p\left(\frac{\partial \theta}{\partial x}\right)^2 dx + \left[\frac{1}{2}k_{t1}\theta^2(0, t) + \frac{1}{2}k_{t2}\theta^2(l, t)\right] \tag{10.10}$$

where $I_p = \iint\limits_{A}(y^2 + z^2)dA$. The kinetic energy of the shaft can be expressed as

$$T = \frac{1}{2}\iiint\limits_{V}\rho\left[\left(\frac{\partial u}{\partial t}\right)^2 + \left(\frac{\partial v}{\partial t}\right)^2 + \left(\frac{\partial w}{\partial t}\right)^2\right]dV$$

$$+ \left[\frac{1}{2}I_{10}\left(\frac{\partial \theta}{\partial t}(0, t)\right)^2 + \frac{1}{2}I_{20}\left(\frac{\partial \theta}{\partial t}(l, t)\right)^2\right]$$

$$= \frac{1}{2}\int_0^l \rho I_p\left(\frac{\partial \theta}{\partial t}\right)^2 dx + \left[\frac{1}{2}I_{10}\left(\frac{\partial \theta}{\partial t}(0, t)\right)^2 + \frac{1}{2}I_{20}\left(\frac{\partial \theta}{\partial t}(l, t)\right)^2\right] \tag{10.11}$$

The work done by the external torque $m_t(x, t)$ can be represented as

$$W = \int_0^l m_t\theta\,dx \tag{10.12}$$

The application of the generalized Hamilton's principle yields

$$\delta \int_{t_1}^{t_2} (\pi - T - W)\, dt = 0$$

or

$$
\delta \int_{t_1}^{t_2} \left\{ \frac{1}{2} \int_0^l GI_p \left(\frac{\partial \theta}{\partial x} \right)^2 dx + \left[\frac{1}{2} k_{t1} \theta^2(0,t) + \frac{1}{2} k_{t2} \theta^2(l,t) \right] \right.
$$
$$
- \frac{1}{2} \int_0^l \rho I_p \left(\frac{\partial \theta}{\partial t} \right)^2 dx - \left[\frac{1}{2} I_{10} \left(\frac{\partial \theta}{\partial t}(0,t) \right)^2 + \frac{1}{2} I_{20} \left(\frac{\partial \theta}{\partial t}(l,t) \right)^2 \right]
$$
$$
\left. - \int_0^l m_t \theta\, dx \right\} dt = 0 \tag{10.13}
$$

The variations in Eq. (10.13) can be evaluated using integration by parts to obtain

$$
\delta \int_0^l \frac{1}{2} GI_p \left(\frac{\partial \theta}{\partial x} \right)^2 dx = \int_0^l GI_p \left(\frac{\partial \theta}{\partial x} \right) \frac{\partial(\delta\theta)}{\partial x}\, dx
$$
$$
= GI_p \frac{\partial \theta}{\partial x} \delta\theta \Big|_0^l - \int_0^l \frac{\partial}{\partial x} \left(GI_p \frac{\partial \theta}{\partial x} \right) \delta\theta\, dx \tag{10.14}
$$

$$
\delta \int_{t_1}^{t_2} \left[\frac{1}{2} k_{t1} \theta^2(0,t) + \frac{1}{2} k_{t2} \theta^2(l,t) \right] dt
$$
$$
= \int_{t_1}^{t_2} \left[k_{t1}\theta(0,t)\delta\theta(0,t) + k_{t2}\theta(l,t)\delta\theta(l,t) \right] dt \tag{10.15}
$$

$$
\delta \int_{t_1}^{t_2} \left[\frac{1}{2} \int_0^l \rho I_p \left(\frac{\partial \theta}{\partial t} \right)^2 dx \right] dt
$$
$$
= \int_0^l \left(\rho I_p \frac{\partial \theta}{\partial t} \delta\theta \Big|_{t_1}^{t_2} \right) dx - \int_0^l \left(\int_{t_1}^{t_2} \rho I_p \frac{\partial^2 \theta}{\partial t^2} \delta\theta\, dt \right) dx
$$
$$
= - \int_{t_1}^{t_2} \left(\int_0^l \rho I_p \frac{\partial^2 \theta}{\partial t^2} \delta\theta\, dx \right) dt \tag{10.16}
$$

$$
\delta \int_{t_1}^{t_2} \left[\frac{1}{2} I_{10} \left(\frac{\partial \theta}{\partial t}(0,t) \right)^2 + \frac{1}{2} I_{20} \left(\frac{\partial \theta}{\partial t}(l,t) \right)^2 \right] dt
$$
$$
= - \int_{t_1}^{t_2} \left[I_{10} \frac{\partial^2 \theta(0,t)}{\partial t^2} \delta\theta(0,t) + I_{20} \frac{\partial^2 \theta(l,t)}{\partial t^2} \delta\theta(l,t) \right] dt \tag{10.17}
$$

Note that integration by parts with respect to time, along with the fact that $\delta\theta = 0$ at $t = t_1$ and $t = t_2$, has been used in deriving Eqs. (10.16) and (10.17). By using

Eqs. (10.14)–(10.17) in Eq. (10.13), we obtain

$$\int_{t_1}^{t_2} \left\{ GI_p \frac{\partial \theta}{\partial x} \delta\theta \Big|_0^l + k_{t1}\theta\delta\theta|_0 \right.$$

$$+ I_{10}\frac{\partial^2 \theta}{\partial t^2}\delta\theta \Big|_0 + k_{t2}\theta\delta\theta|^l + I_{20}\frac{\partial^2 \theta}{\partial t^2}\delta\theta\Big|^l \right\} dt$$

$$+ \int_{t_1}^{t_2} \left\{ \int_0^l \left[-\frac{\partial}{\partial x}\left(GI_p \frac{\partial \theta}{\partial x}\right) + \rho I_p \frac{\partial^2 \theta}{\partial t^2} - m_t \right] \delta\theta \, dx \right\} dt = 0 \quad (10.18)$$

By setting the two expressions under the braces in each term of Eq. (10.18) equal to zero, we obtain the equation of motion for the torsional vibration of the shaft as

$$I_0 \frac{\partial^2 \theta}{\partial t^2} = \frac{\partial}{\partial x}\left(GI_p \frac{\partial \theta}{\partial x}\right) + m_t(x, t) \quad (10.19)$$

where $I_0 = \rho I_p$ is the mass moment of inertia of the shaft per unit length, and the boundary conditions as

$$\left(-GI_p \frac{\partial \theta}{\partial x} + k_{t1}\theta + I_{10}\frac{\partial^2 \theta}{\partial t^2}\right)\delta\theta = 0 \qquad \text{at } x = 0$$

$$\left(GI_p \frac{\partial \theta}{\partial x} + k_{t2}\theta + I_{20}\frac{\partial^2 \theta}{\partial t^2}\right)\delta\theta = 0 \qquad \text{at } x = l \quad (10.20)$$

Each of the equations in (10.20) can be satisfied in two ways but will be satisfied only one way for any specific end conditions of the shaft. The boundary conditions implied by Eqs. (10.20) are as follows. At $x = 0$, either θ is specified (so that $\delta\theta = 0$) or

$$\left(GI_p \frac{\partial \theta}{\partial x} - k_{t1}\theta - I_{10}\frac{\partial^2 \theta}{\partial t^2}\right) = 0 \quad (10.21)$$

At $x = l$, either θ is specified (so that $\delta\theta = 0$) or

$$\left(GI_p \frac{\partial \theta}{\partial x} + k_{t1}\theta + I_{10}\frac{\partial^2 \theta}{\partial t^2}\right) = 0 \quad (10.22)$$

In the present case, the second conditions stated in each of Eqs. (10.21) and (10.22) are valid.

10.3   FREE VIBRATION OF UNIFORM SHAFTS

For a uniform shaft, Eq. (10.19) reduces to

$$GI_p \frac{\partial^2 \theta}{\partial x^2}(x, t) + m_t(x, t) = I_0 \frac{\partial^2 \theta}{\partial t^2}(x, t) \quad (10.23)$$

By setting $m_t(x, t) = 0$, we obtain the free vibration equation

$$c^2 \frac{\partial^2 \theta}{\partial x^2}(x, t) = \frac{\partial^2 \theta}{\partial t^2}(x, t) \quad (10.24)$$

where

$$c = \sqrt{\frac{GI_p}{I_0}} \qquad (10.25)$$

It can be observed that Eqs. (10.23)–(10.25) are similar to the equations derived in the cases of transverse vibration of a string and longitudinal vibration of a bar. For a uniform shaft, $I_0 = \rho I_p$ and Eq. (10.25) takes the form

$$c = \sqrt{\frac{G}{\rho}} \qquad (10.26)$$

By assuming the solution as

$$\theta(x, t) = \Theta(x)T(t) \qquad (10.27)$$

Eq. (10.24) can be written as two separate equations:

$$\frac{d^2\Theta(x)}{\partial x^2} + \frac{\omega^2}{c^2}\Theta(x) = 0 \qquad (10.28)$$

$$\frac{d^2 T}{\partial t^2} + \omega^2 T(t) = 0 \qquad (10.29)$$

The solutions of Eqs. (10.28) and (10.29) can be expressed as

$$\Theta(x) = A \cos \frac{\omega x}{c} + B \sin \frac{\omega x}{c} \qquad (10.30)$$

$$T(t) = C \cos \omega t + D \sin \omega t \qquad (10.31)$$

where A, B, C, and D are constants. If ω_n denotes the nth frequency of vibration and $\Theta_n(\theta)$ the corresponding mode shape, the general free vibration solution of Eq. (10.24) is given by

$$\theta(x, t) = \sum_{n=1}^{\infty} \Theta_n(\theta)T_n(t)$$

$$= \sum_{n=1}^{\infty} \left(A_n \cos \frac{\omega_n x}{c} + B_n \sin \frac{\omega_n x}{c} \right) (C_n \cos \omega_n t + D_n \sin \omega_n t) \qquad (10.32)$$

The constraints C_n and D_n can be evaluated from the initial conditions, and the constraints A_n and B_n can be determined (not the absolute values, only their relative values) from the boundary conditions of the shaft. The initial conditions are usually stated in terms of the initial angular displacement and angular velocity distributions of the shaft.

10.3.1   Natural Frequencies of a Shaft with Both Ends Fixed

For a uniform circular shaft of length l fixed at both ends, the boundary conditions are given by

$$\theta(0, t) = 0 \qquad (10.33)$$

$$\theta(l, t) = 0 \qquad (10.34)$$

The free vibration solution is given by Eq. (10.27):

$$\theta(x, t) = \Theta(\theta)T(t) \equiv \left(A \cos \frac{\omega x}{c} + B \sin \frac{\omega x}{c} \right)(C \cos \omega t + D \sin \omega t) \tag{10.35}$$

Equations (10.33) and (10.35) yield

$$A = 0 \tag{10.36}$$

and the solution can be expressed as

$$\theta(x, t) = \sin \frac{\omega x}{c}(C' \cos \omega t + D' \sin \omega t) \tag{10.37}$$

where C' and D' are new constants. The use of Eq. (10.34) in (10.37) gives the frequency equation

$$\sin \frac{\omega l}{c} = 0 \tag{10.38}$$

The natural frequencies of vibration are given by the roots of Eq. (10.38) as

$$\frac{\omega l}{c} = n\pi, \qquad n = 1, 2, \ldots$$

or

$$\omega_n = \frac{n\pi c}{l}, \qquad n = 1, 2, \ldots \tag{10.39}$$

The mode shape corresponding to the natural frequency ω_n can be expressed as

$$\Theta_n(x) = B_n \sin \frac{\omega_n x}{c}, \qquad n = 1, 2, \ldots \tag{10.40}$$

The free vibration solution of the fixed–fixed shaft is given by a linear combination of its normal modes:

$$\theta(x, t) = \sum_{n=1}^{\infty} \sin \frac{\omega_n x}{c}(C'_n \cos \omega_n t + D'_n \sin \omega_n t) \tag{10.41}$$

10.3.2   Natural Frequencies of a Shaft with Both Ends Free

Since the torque, $M_t = GI_p(\partial\theta/\partial x)$, is zero at a free end, the boundary conditions of a free–free shaft are given by

$$\frac{\partial\theta}{\partial x}(0, t) = 0 \tag{10.42}$$

$$\frac{\partial\theta}{\partial x}(l, t) = 0 \tag{10.43}$$

In view of Eq. (10.27), Eqs. (10.42) and (10.43) can be expressed as

$$\frac{d\Theta}{dx}(0) = 0 \tag{10.44}$$

$$\frac{d\Theta}{dx}(l) = 0 \tag{10.45}$$

Equation (10.30) gives

$$\Theta(x) = A \cos \frac{\omega x}{c} + B \sin \frac{\omega x}{c} \tag{10.46}$$

$$\frac{d\Theta}{dx}(x) = -\frac{A\omega}{c} \sin \frac{\omega x}{c} + \frac{B\omega}{c} \cos \frac{\omega x}{c} \tag{10.47}$$

Equations (10.44) and (10.47) yield

$$B = 0 \tag{10.48}$$

and Eqs. (10.45) and (10.47) result in

$$\sin \frac{\omega l}{c} = 0 \tag{10.49}$$

The roots of Eq. (10.49) are given by

$$\omega_n = \frac{n\pi c}{l} = \frac{n\pi}{l}\sqrt{\frac{G}{\rho}}, \qquad n = 1, 2, \ldots \tag{10.50}$$

The nth normal mode is given by

$$\Theta_n(x) = A_n \cos \frac{\omega_n x}{c}, \qquad n = 1, 2, \ldots \tag{10.51}$$

The free vibration solution of the shaft can be expressed as [see Eq. (10.32)]

$$\theta(x, t) = \sum_{n=1}^{\infty} \Theta_n(x) T_n(t) = \sum_{n=1}^{\infty} \cos \frac{\omega_n x}{c} (C_n \cos \omega_n t + D_n \sin \omega_n t) \tag{10.52}$$

where the constants C_n and D_n can be determined from the initial conditions of the shaft.

10.3.3   Natural Frequencies of a Shaft Fixed at One End and Attached to a Torsional Spring at the Other

For a uniform circular shaft fixed at $x = 0$ and attached to a torsional spring of stiffness k_t at $x = l$ as shown in Fig. 10.3, the boundary conditions are given by

$$\theta(0, t) = 0 \tag{10.53}$$

$$M_t(l, t) = GI_p \frac{\partial \theta}{\partial x}(l, t) = -k_t \theta(l, t) \tag{10.54}$$

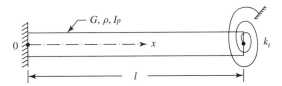

Figure 10.3   Shaft fixed at $x = 0$ and a torsional spring attached at $x = l$.

The free vibration solution of a shaft is given by Eq. (10.27):

$$\theta(x, t) = \Theta(\theta)T(t) \equiv \left(A \cos \frac{\omega x}{c} + B \sin \frac{\omega x}{c}\right)(C \cos \omega t + D \sin \omega t) \qquad (10.55)$$

The use of the boundary condition of Eq. (10.53) in Eq. (10.55) gives

$$A = 0 \qquad (10.56)$$

and the solution can be expressed as

$$\theta(x, t) = \sin \frac{\omega x}{c}(\overline{C} \cos \omega t + \overline{D} \sin \omega t) \qquad (10.57)$$

The use of the boundary condition of Eq. (10.54) in Eq. (10.57) yields the frequency equation

$$\frac{\omega G I_p}{c} \cos \frac{\omega l}{c} = -k_t \sin \frac{\omega l}{c} \qquad (10.58)$$

Using Eq. (10.26), Eq. (10.58) can be rewritten as

$$\alpha \tan \alpha = -\beta \qquad (10.59)$$

where

$$\alpha = \frac{\omega l}{c}, \qquad \beta = \frac{\omega^2 \rho l I_p}{k_t} \qquad (10.60)$$

The roots of the frequency equation (10.60) give the natural frequencies of vibration of the shaft as

$$\omega_n = \frac{\alpha_n c}{l}, \qquad n = 1, 2, \ldots \qquad (10.61)$$

and the corresponding mode shapes as

$$\Theta_n(x) = B_n \sin \frac{\omega_n x}{c}, \qquad n = 1, 2, \ldots \qquad (10.62)$$

Finally, the free vibration solution of the shaft can be expressed as

$$\theta(x, t) = \sum_{n=1}^{\infty} \sin \frac{\omega_n x}{c}(C_n \cos \omega_n t + D_n \sin \omega_n t) \qquad (10.63)$$

Several possible boundary conditions for the torsional vibration of a uniform shaft are given in Table 10.1 along with the corresponding frequency equations and the mode shapes.

Example 10.1 Determine the free torsional vibration solution of a uniform shaft carrying disks at both the ends as shown in Fig. 10.4.

Table 10.1 Boundary Conditions of a Uniform Shaft in Torsional Vibration

End conditions of shaft	Boundary conditions	Frequency equation	Mode shape (normal function)	Natural frequencies
1. Fixed–free	$\theta(0, t) = 0$ $\dfrac{\partial \theta}{\partial x}(l, t) = 0$	$\cos \dfrac{\omega l}{c} = 0$	$\Theta_n(x) = C_n \sin \dfrac{(2n+1)\pi x}{2l}$	$\omega_n = \dfrac{(2n+1)\pi c}{2l}$, $n = 0, 1, 2, \ldots$
2. Free–free	$\dfrac{\partial \theta}{\partial x}(0, t) = 0$ $\dfrac{\partial \theta}{\partial x}(l, t) = 0$	$\sin \dfrac{\omega l}{c} = 0$	$\Theta_n(x) = C_n \cos \dfrac{n\pi x}{l}$	$\omega_n = \dfrac{n\pi c}{l}$, $n = 0, 1, 2, \ldots$
3. Fixed–fixed	$\theta(0, t) = 0$ $\theta(l, t) = 0$	$\sin \dfrac{\omega l}{c} = 0$	$\Theta_n(x) = C_n \sin \dfrac{n\pi x}{l}$	$\omega_n = \dfrac{n\pi c}{l}$, $n = 1, 2, 3, \ldots$
4. Fixed–disk	$\theta(0, t) = 0$ $GJ\dfrac{\partial \theta}{\partial x}(l, t) = -I_d \dfrac{\partial^2 \theta}{\partial t^2}(l, t)$	$\alpha \tan \alpha = \beta$ $\alpha = \dfrac{\omega l}{c}$ $\beta = \dfrac{\rho J l}{I_d}$	$\Theta_n(x) = C_n \sin \dfrac{\omega_n x}{c}$	$\omega_n = \dfrac{\alpha_n c}{l}$, $n = 1, 2, 3, \ldots$

(continued overleaf)

Table 10.1 (*continued*)

End conditions of shaft	Boundary conditions	Frequency equation	Mode shape (normal function)	Natural frequencies
5. Fixed–torsional spring	$\theta(0, t) = 0$ $GJ\dfrac{\partial \theta}{\partial x}(l, t) = -k_t\theta(l, t)$	$\alpha \tan \alpha = -\beta$ $\alpha = \dfrac{\omega l}{c}$ $\beta = \dfrac{\omega^2 \rho J l}{k_t}$	$\Theta_n(x) = C_n \sin \dfrac{\omega_n x}{c}$	$\omega_n = \dfrac{\alpha_n c}{l},$ $n = 1, 2, 3, \cdots$
6. Free–disk	$\dfrac{\partial \theta}{\partial x}(0, t) = 0$ $GJ\dfrac{\partial \theta}{\partial x}(l, t) = -I_d\dfrac{\partial^2\theta}{\partial t^2}(l, t)$	$\alpha \cot \alpha = -\beta$ $\alpha = \dfrac{\omega l}{c}$ $\beta = \dfrac{\rho J l}{I_d}$	$\Theta_n(x) = C_n \cos \dfrac{\omega_n x}{c}$	$\omega_n = \dfrac{\alpha_n c}{l},$ $n = 1, 2, 3, \cdots$
7. Free–torsional spring	$\dfrac{\partial \theta}{\partial x}(0, t) = 0$ $GJ\dfrac{\partial \theta}{\partial x}(l, t) = -k_t\theta(l, t)$	$\alpha \cot \alpha = \beta$ $\alpha = \dfrac{\omega l}{c}$ $\beta = \dfrac{\omega^2 \rho J l}{k_t}$	$\Theta_n(x) = C_n \cos \dfrac{\omega_n x}{c}$	$\omega_n = \dfrac{\alpha_n c}{l},$ $n = 1, 2, 3, \cdots$
8. Disk–disk	$GJ\dfrac{\partial \theta}{\partial x}(0, t) = I_1\dfrac{\partial^2\theta}{\partial t^2}(0, t)$ $GJ\dfrac{\partial \theta}{\partial x}(l, t) = -I_2\dfrac{\partial^2\theta}{\partial t^2}(l, t)$	$\tan \alpha = \dfrac{\alpha(\beta_1 + \beta_2)}{(\alpha^2 - \beta_1\beta_2)}$ $\alpha = \dfrac{\omega l}{c}$ $\beta_1 = \dfrac{I_0}{I_1} = \dfrac{\rho J l}{I_1}$ $\beta_2 = \dfrac{I_0}{I_2} = \dfrac{\rho J l}{I_2}$	$\Theta_n(x) = $ $C_n\left(\cos \dfrac{\alpha_n x}{l} - \dfrac{\alpha_n}{\beta_1}\sin \dfrac{\alpha_n x}{l}\right)$	$\omega_n = \dfrac{\alpha_n c}{l},$ $n = 1, 2, 3, \cdots$

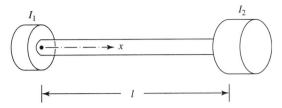

Figure 10.4 Shaft with disks at both ends, under torsional vibration.

SOLUTION The boundary conditions, with the inertial torques exerted by the disks, can be expressed as

$$GJ \frac{\partial \theta}{\partial x}(0, t) = I_1 \frac{\partial^2 \theta}{\partial t^2}(0, t) \tag{E10.1.1}$$

$$GJ \frac{\partial \theta}{\partial x}(l, t) = -I_2 \frac{\partial^2 \theta}{\partial t^2}(l, t) \tag{E10.1.2}$$

Assuming the solution in the nth mode of vibration as

$$\theta_n(x, t) = \Theta_n(x)(C_n \cos \omega_n t + D_n \sin \omega_n t) \tag{E10.1.3}$$

where

$$\Theta_n(x) = A_n \cos \frac{\omega_n x}{c} + B_n \sin \frac{\omega_n x}{c} \tag{E10.1.4}$$

the boundary conditions of Eqs. (E10.1.1) and (E10.1.2) can be rewritten as

$$GJ \frac{d\Theta_n}{dx}(0) = -I_1 \omega_n^2 \Theta_n(0)$$

or

$$GJ \frac{\omega_n}{c} B_n = -I_1 \omega_n^2 A_n \tag{E10.1.5}$$

and

$$GJ \frac{d\Theta_n}{dx}(l) = I_2 \omega_n^2 \Theta_n(l)$$

or

$$GJ \frac{\omega_n}{c} \left(-A_n \sin \frac{\omega_n l}{c} + B_n \cos \frac{\omega_n l}{c} \right) = I_2 \omega_n^2 \left(A_n \cos \frac{\omega_n l}{c} + B_n \sin \frac{\omega_n l}{c} \right) \tag{E10.1.6}$$

Equations (E10.1.5) and (E10.1.6) represent a system of two homogeneous algebraic equations in the two unknown constants A_n and B_n, which can be rewritten in matrix form as

$$\begin{bmatrix} I_1 \omega_n^2 & GJ \frac{\omega_n}{c} \\ GJ \frac{\omega_n}{c} \sin \frac{\omega_n l}{c} + I_2 \omega_n^2 \cos \frac{\omega_n l}{c} & -GJ \frac{\omega_n}{c} \cos \frac{\omega_n l}{c} + I_2 \omega_n^2 \sin \frac{\omega_n l}{c} \end{bmatrix}$$

$$\cdot \begin{Bmatrix} A_n \\ B_n \end{Bmatrix} = \begin{Bmatrix} 0 \\ 0 \end{Bmatrix} \tag{E10.1.7}$$

The determinant of the coefficient matrix in Eq. (E10.1.7) is set equal to zero for a nontrivial solution of A_n and B_n to obtain the frequency equation as

$$-I_1 \omega_n^3 \frac{GJ}{c} \cos \frac{\omega_n l}{c} + I_1 I_2 \omega_n^4 \sin \frac{\omega_n l}{c} - G^2 J^2 \frac{\omega_n^2}{c^2} \sin \frac{\omega_n l}{c} - GJ \frac{\omega_n^3}{c} I_2 \cos \frac{\omega_n l}{c} = 0$$

$$(E10.1.8)$$

Rearranging the terms, Eq.(E10.1.8) can be expressed as

$$\left(\frac{\alpha_n^2}{\beta_1 \beta_2} - 1 \right) \tan \alpha_n = \alpha_n \left(\frac{1}{\beta_1} + \frac{1}{\beta_2} \right) \qquad (E10.1.9)$$

where

$$\alpha_n = \frac{\omega_n l}{c} \qquad (E10.1.10)$$

$$\beta_1 = \frac{\rho J l}{I_1} = \frac{I_0}{I_1} \qquad (E10.1.11)$$

$$\beta_2 = \frac{\rho J l}{I_2} = \frac{I_0}{I_2} \qquad (E10.1.12)$$

Thus, the mode shapes or normal functions can be expressed, using Eq. (E10.1.5) in (E10.1.3), as

$$\Theta_n(x) = A_n \left(\cos \frac{\alpha_n x}{l} - \frac{\alpha_n}{\beta_1} \sin \frac{\alpha_n x}{l} \right) \qquad (E10.1.13)$$

Thus, the complete free vibration solution of the shaft with disks is given by

$$\theta(x, t) = \sum_{n=1}^{\infty} \Theta_n(x)(C_n \cos \omega_n t + D_n \sin \omega_n t)$$

$$= \sum_{n=1}^{\infty} \left(\cos \frac{\alpha_n x}{l} - \frac{\alpha_n}{\beta_1} \sin \frac{\alpha_n x}{l} \right) (\underset{\sim}{C_n} \cos \omega_n t + \underset{\sim}{D_n} \sin \omega_n t) \qquad (E10.1.14)$$

where $\underset{\sim}{C_n}$ and $\underset{\sim}{D_n}$ denote new constants whose values can be determined from the initial conditions specified.

Notes

1. If the mass moments of inertia of the disks I_1 and I_2 are large compared to the mass moment of inertia of the shaft I_0, β_1 and β_2 will be small and the frequency equation, Eq. (E10.1.9), can be written as

$$\alpha_n \tan \alpha_n \approx \beta_1 + \beta_2 \qquad (E10.1.15)$$

2. If the mass moments of inertia of the disks I_1 and I_2 are small compared to the mass moment of inertia of the shaft I_0, β_1 and β_2 will be large and the frequency equation, Eq. (E10.1.9), can be written as

$$\tan \alpha_n \approx 0 \qquad (E10.1.16)$$

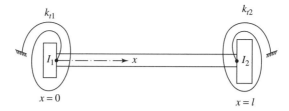

Figure 10.5   Shaft attached to a heavy disk and a torsional spring at each end.

Example 10.2   Derive the frequency equation of a uniform shaft attached to a heavy disk and a torsional spring at each end as shown in Fig. 10.5.

SOLUTION   The free vibration of the shaft in the ith mode is given by

$$\theta_i(x, t) = \Theta_i(x)(C_i \cos \omega_i t + D_i \sin \omega_i t) \tag{E10.2.1}$$

where

$$\Theta_i(x) = A_i \cos \frac{\omega_i x}{c} + B_i \sin \frac{\omega_i x}{c} \tag{E10.2.2}$$

The boundary conditions, considering the resulting torques of the torsional springs and the inertial torques of the heavy disks, can be stated as

$$GI_p \frac{\partial \theta}{\partial x}(0, t) = I_1 \frac{\partial^2 \theta}{\partial t^2}(0, t) + k_{t_1}\theta(0, t) \tag{E10.2.3}$$

$$GI_p \frac{\partial \theta}{\partial x}(l, t) = -I_2 \frac{\partial^2 \theta}{\partial t^2}(l, t) - k_{t_2}\theta(l, t) \tag{E10.2.4}$$

Using Eq. (E10.2.1), Eqs. (E10.2.3) and (E10.2.4) can be expressed as

$$GI_p \frac{d\Theta_i(0)}{dx} = -I_1 \omega_i^2 \Theta_i(0) + k_{t_1}\Theta_i(0) \tag{E10.2.5}$$

$$GI_p \frac{d\Theta_i(l)}{dx} = I_2 \omega_i^2 \Theta_i(l) - k_{t_2}\Theta_i(l) \tag{E10.2.6}$$

Equation (E10.2.2) gives

$$\frac{d\Theta_i(x)}{dx} = -\frac{A_i \omega_i}{c} \sin \frac{\omega_i x}{c} + \frac{B_i \omega_i}{c} \cos \frac{\omega_i x}{c} \tag{E10.2.7}$$

In view of Eqs. (E10.2.2) and (E10.2.7), Eqs. (E10.2.5) and (E10.2.6) yield

$$A_i(I_1 \omega_i^2 - k_{t_1}) + B_i \frac{GI_p \omega_i}{c} = 0 \tag{E10.2.8}$$

$$A_i \left[\frac{GI_p \omega_i}{c} \sin \frac{\omega_i l}{c} + (I_2 \omega_i^2 - k_{t_2}) \cos \frac{\omega_i l}{c} \right]$$
$$+ B_i \left[-\frac{GI_p \omega_i}{c} \cos \frac{\omega_i l}{c} + (I_2 \omega_i^2 - k_{t_2}) \sin \frac{\omega_i l}{c} \right] = 0 \tag{E10.2.9}$$

For a nontrivial solution of A_i and B_i, the determinant of the coefficient matrix must be equal to zero in Eqs. (E10.2.8) and (E10.2.9). This gives the desired frequency equation as

$$\begin{vmatrix} I_1\omega_i^2 - k_{t_1} & \dfrac{GI_p\omega_i}{c} \\ \dfrac{GI_p\omega_i}{c}\sin\dfrac{\omega_i l}{c} + (I_2\omega_i^2 - k_{t_2})\cos\dfrac{\omega_i l}{c} & -\dfrac{GI_p\omega_i}{c}\cos\dfrac{\omega_i l}{c} + (I_2\omega_i^2 - k_{t_2})\sin\dfrac{\omega_i l}{c} \end{vmatrix} = 0$$

(E10.2.10)

Example 10.3   Derive the orthogonality relationships for a shaft in torsional vibration.

SOLUTION
Case (i): Shaft with simple boundary conditions   The eigenvalue problem of the shaft, corresponding to two distinct natural frequencies of vibration ω_i and ω_j, can be expressed as [from Eq. (10.28)]

$$\Theta_i''(x) + \frac{\omega_i^2}{c^2}\Theta_i(x) = 0 \tag{E10.3.1}$$

$$\Theta_j''(x) + \frac{\omega_j^2}{c^2}\Theta_j(x) = 0 \tag{E10.3.2}$$

where a prime denotes a derivative with respect to x. Multiply Eq. (E10.3.1) by $\Theta_j(x)$ and Eq. (E10.3.2) by $\Theta_i(x)$ and integrate the resulting equations from 0 to l to obtain

$$\int_0^l \Theta_i''\Theta_j\,dx + \frac{\omega_i^2}{c^2}\int_0^l \Theta_i\Theta_j\,dx = 0 \tag{E10.3.3}$$

$$\int_0^l \Theta_j''\Theta_i\,dx + \frac{\omega_j^2}{c^2}\int_0^l \Theta_i\Theta_j\,dx = 0 \tag{E10.3.4}$$

Integrating the left-hand sides of Eqs. (E10.3.3) and (E10.3.4) by parts results in

$$\Theta_i'\Theta_j\Big|_0^l - \int_0^l \Theta_i'\Theta_j'\,dx + \frac{\omega_i^2}{c^2}\int_0^l \Theta_i\Theta_j\,dx = 0 \tag{E10.3.5}$$

$$\Theta_j'\Theta_i\Big|_0^l - \int_0^l \Theta_i'\Theta_j'\,dx + \frac{\omega_i^2}{c^2}\int_0^l \Theta_i\Theta_j\,dx = 0 \tag{E10.3.6}$$

If the ends of the shaft are either fixed ($\Theta_i = \Theta_j = 0$) or free ($\Theta_i' = \Theta_j' = 0$), the first terms of Eqs. (E10.3.5) and (E10.3.6) will be zero. By subtracting the resulting Equation (E10.3.6) from (E10.3.5), we obtain

$$(\omega_i^2 - \omega_j^2)\int_0^l \Theta_i\Theta_j\,dx = 0 \tag{E10.3.7}$$

For distinct eigenvalues, $\omega_i \neq \omega_j$, and Eq. (E10.3.7) gives the orthogonality relation for normal modes of the shaft as

$$\int_0^l \Theta_i\Theta_j\,dx = 0,\, i \neq j \tag{E10.3.8}$$

In view of Eq. (E10.3.8), Eqs. (E10.3.5) and (E10.3.3) give

$$\int_0^l \Theta_i' \Theta_j' \, dx = 0, \qquad i \neq j \tag{E10.3.9}$$

$$\int_0^l \Theta_i'' \Theta_j \, dx = 0, \qquad i \neq j \tag{E10.3.10}$$

Equations (E10.3.8)–(E10.3.10) denote the desired orthogonality relationships for the torsional vibration of a shaft.

Case (ii): Shaft with disks at both ends The eigenvalue problem of the shaft corresponding to two distinct natural frequencies of vibration can be expressed as [from Eq. (10.28)]

$$GI_p \Theta_i''(x) + \omega_i^2 I_0 \Theta_i(x) = 0 \tag{E10.3.11}$$

$$GI_p \Theta_j''(x) + \omega_j^2 I_0 \Theta_j(x) = 0 \tag{E10.3.12}$$

Multiplying Eq. (E10.3.11) by $\Theta_j(x)$ and Eq. (E10.3.12) by $\Theta_i(x)$ and integrating over the length of the shaft, we have

$$GI_p \int_0^l \Theta_i'' \Theta_j \, dx = -I_0 \omega_i^2 \int_0^l \Theta_i \Theta_j \, dx \tag{E10.3.13}$$

$$GI_p \int_0^l \Theta_j'' \Theta_i \, dx = -I_0 \omega_j^2 \int_0^l \Theta_i \Theta_j \, dx \tag{E10.3.14}$$

The disk located at $x = 0$ and $x = l$ must also be considered in developing the orthogonality relationship and hence the boundary conditions given by Eqs. (E10.2.5) and (E10.2.6) (without the torsional springs) are written for modes i and j as

$$GI_p \Theta_{i_0}' = -I_1 \omega_i^2 \Theta_{i_0} \tag{E10.3.15}$$

$$GI_p \Theta_{j_0}' = -I_1 \omega_j^2 \Theta_{j_0} \tag{E10.3.16}$$

$$GI_p \Theta_{i_l}' = I_2 \omega_i^2 \Theta_{i_l} \tag{E10.3.17}$$

$$GI_p \Theta_{j_l}' = I_2 \omega_j^2 \Theta_{j_l} \tag{E10.3.18}$$

where

$$\Theta_{i_0}' = \frac{d\Theta_i}{dx}(x = 0), \qquad \Theta_{i_l}' = \frac{d\Theta_i}{dx}(x = l) \tag{E10.3.19}$$

Multiply Eqs. (E10.3.15)–(E10.3.19), respectively by $\Theta_{j_0}, \Theta_{i_0}, \Theta_{j_l}$ and Θ_{i_l} to obtain

$$GI_p \Theta_{i_0}' \Theta_{j_0} = -I_1 \omega_i^2 \Theta_{i_0} \Theta_{j_0} \tag{E10.3.20}$$

$$GI_p \Theta_{j_0}' \Theta_{i_0} = -I_1 \omega_j^2 \Theta_{i_0} \Theta_{j_0} \tag{E10.3.21}$$

$$GI_p \Theta_{i_l}' \Theta_{j_l} = I_2 \omega_i^2 \Theta_{i_l} \Theta_{j_l} \tag{E10.3.22}$$

$$GI_p \Theta_{j_l}' \Theta_{i_l} = I_2 \omega_j^2 \Theta_{i_l} \Theta_{j_l} \tag{E10.3.23}$$

Add Eqs. (E10.3.20) and (E10.3.21) to (E10.3.13) and subtract Eqs. (E10.3.22) and (E10.3.23) from (E10.3.14) to produce the combined relationships

$$GI_p \int_0^l \Theta_i'' \Theta_j \, dx + GI_p \Theta_{i_0}' \Theta_{j_0} + GI_p \Theta_{j_0}' \Theta_{i_0}$$

$$= -I_0 \omega_i^2 \int_0^l \Theta_i \Theta_j \, dx - I_1 \omega_i^2 \Theta_{i_0} \Theta_{j_0} - I_1 \omega_j^2 \Theta_{i_0} \Theta_{j_0} \qquad \text{(E10.3.24)}$$

$$GI_p \int_0^l \Theta_j'' \Theta_i \, dx - GI_p \Theta_{i_l}' \Theta_{j_l} - GI_p \Theta_{j_l}' \Theta_{i_l}$$

$$= -I_0 \omega_i^2 \int_0^l \Theta_i \Theta_j \, dx - I_2 \omega_i^2 \Theta_{i_l} \Theta_{j_l} - I_2 \omega_j^2 \Theta_{i_l} \Theta_{j_l} \qquad \text{(E10.3.25)}$$

Carrying out the integrations on the left-hand sides of Eqs. (E10.3.24) and (E10.3.25) by parts, we obtain

$$GI_p \Theta_{i_l}' \Theta_{j_l} - GI_p \int_0^l \Theta_i' \Theta_j' \, dx + GI_p \Theta_{j_0}' \Theta_{i_0}$$

$$= -I_0 \omega_i^2 \int_0^l \Theta_i \Theta_j \, dx - I_1 \omega_i^2 \Theta_{i_0} \Theta_{j_0} - I_1 \omega_j^2 \Theta_{i_0} \Theta_{j_0} \qquad \text{(E10.3.26)}$$

$$- GI_p \Theta_{j_0}' \Theta_{i_0} - GI_p \int_0^l \Theta_j' \Theta_i' \, dx - GI_p \Theta_{i_l}' \Theta_{j_l}$$

$$= -I_0 \omega_j^2 \int_0^l \Theta_i \Theta_j \, dx - I_2 \omega_i^2 \Theta_{i_l} \Theta_{j_l} - I_2 \omega_j^2 \Theta_{i_l} \Theta_{j_l} \qquad \text{(E10.3.27)}$$

Subtract Eqs. (E10.3.20) and (E10.3.21) from (E10.3.13) to obtain

$$GI_p \int_0^l \Theta_i'' \Theta_j \, dx - GI_p \Theta_{i_0}' \Theta_{j_0} - GI_p \Theta_{j_0}' \Theta_{i_0}$$

$$= -I_0 \omega_i^2 \int_0^l \Theta_i \Theta_j \, dx + I_1 \omega_i^2 \Theta_{i_0} \Theta_{j_0} + I_1 \omega_j^2 \Theta_{i_0} \Theta_{j_0} \qquad \text{(E10.3.28)}$$

Integration on the left-hand side of Eq. (E10.3.28) by parts results in

$$GI_p \Theta_{i_l}' \Theta_{j_l} - GI_p \Theta_{i_0}' \Theta_{j_0} - GI_p \int_0^l \Theta_i' \Theta_j' \, dx - GI_p \Theta_{i_0}' \Theta_{j_0} - GI_p \Theta_{j_0}' \Theta_{i_0}$$

$$= -I_0 \omega_i^2 \int_0^l \Theta_i \Theta_j \, dx + I_1 \omega_i^2 \Theta_{i_0} \Theta_{j_0} + I_1 \omega_j^2 \Theta_{i_0} \Theta_{j_0} \qquad \text{(E10.3.29)}$$

Subtraction of Eq. (E10.3.29) from Eq. (E10.3.27) results in

$$2GI_p (\Theta_{i_0}' \Theta_{j_0} - \Theta_{i_l}' \Theta_{j_l})$$

$$= (\omega_i^2 - \omega_j^2) I_0 \int_0^l \Theta_i \Theta_j \, dx - I_1 (\omega_j^2 + \omega_i^2) \Theta_{i_0} \Theta_{j_0} - I_2 (\omega_i^2 + \omega_j^2) \Theta_{i_l} \Theta_{j_l}$$

$$\text{(E10.3.30)}$$

By using two times the result obtained by subtracting Eq. (E10.3.22) from (E10.3.20) on the left-hand side of Eq. (E10.3.30), we obtain, after simplification,

$$(\omega_i^2 - \omega_j^2)\left(I_0 \int_0^l \Theta_i \Theta_j \, dx + I_1 \Theta_{i_0} \Theta_{j_0} + I_2 \Theta_{i_l} \Theta_{j_l}\right) = 0 \qquad \text{(E10.3.31)}$$

For $\omega_i \neq \omega_j$, Eq. (E10.3.31) gives

$$I_0 \int_0^l \Theta_i \Theta_j \, dx + I_1 \Theta_{i_0} \Theta_{j_0} + I_2 \Theta_{i_l} \Theta_{j_l} = 0 \qquad \text{(E10.3.32)}$$

Addition of Eqs. (E10.3.13) and (E10.3.20) and subtraction of Eq. (E10.3.22) from the result yields

$$GI_p\left(\int_0^l \Theta_i'' \Theta_j \, dx + \Theta_{i_0}' \Theta_{j_0} - \Theta_{i_l}' \Theta_{j_l}\right)$$
$$= -\omega_i^2\left(I_0 \int_0^l \Theta_i \Theta_j \, dx + I_1 \Theta_{i_0} \Theta_{j_0} + I_2 \Theta_{i_l} \Theta_{j_l}\right) \qquad \text{(E10.3.33)}$$

In view of Eq. (E10.3.32), Eq. (E10.3.33) gives

$$GI_p\left(\int_0^l \Theta_i'' \Theta_j \, dx + \Theta_{i_0}' \Theta_{j_0} - \Theta_{i_l}' \Theta_{j_l}\right) = 0 \qquad \text{(E10.3.34)}$$

Finally, the addition of Eqs. (E10.3.26) and (E10.3.27) gives

$$-2GI_p \int_0^l \Theta_i' \Theta_j' \, dx = -\left(\omega_i^2 + \omega_j^2\right)\left(I_0 \int_0^l \Theta_i \Theta_j \, dx + I_1 \Theta_{i_0} \Theta_{j_0} + I_2 \Theta_{i_l} \Theta_{j_l}\right)$$
$$\text{(E10.3.35)}$$

In view of Eq. (E10.3.32), Eq. (E10.3.35) reduces to

$$GI_p \int_0^l \Theta_i' \Theta_j' \, dx = 0 \qquad \text{(E10.3.36)}$$

Equations (E10.3.32), (E10.3.34) and (E10.3.36) denote the orthogonality relations for torsional vibration of a uniform shaft with heavy disks at both ends.

10.4   FREE VIBRATION RESPONSE DUE TO INITIAL CONDITIONS: MODAL ANALYSIS

The angular displacement of a shaft in torsional vibration can be expressed in terms of normal modes $\Theta_i(x)$ using the expansion theorem, as

$$\theta(x, t) = \sum_{i=1}^{\infty} \Theta_i(x)\eta_i(t) \qquad \text{(10.64)}$$

where $\eta_i(t)$ is the ith generalized coordinate. Substituting Eq. (10.64) into Eq. (10.24), we obtain

$$c^2 \sum_{i=1}^{\infty} \Theta_i''(x)\eta_i(t) = \sum_{i=1}^{\infty} \Theta_i(x)\ddot{\eta}_i(t) \tag{10.65}$$

where $\Theta_i''(x) = d^2\Theta_i(x)/dx^2$ and $\ddot{\eta}_i(t) = d^2\eta_i(t)/dt^2$. Multiplication of Eq. (10.65) by $\Theta_j(x)$ and integration from 0 to l yields

$$c^2 \sum_{i=1}^{\infty} \int_0^l \Theta_i''(x)\Theta_j(x)\,dx\,\eta_i(t) = \sum_{i=1}^{\infty} \int_0^{\infty} \Theta_i(x)\Theta_j(x)\,dx\,\ddot{\eta}_i(t) \tag{10.66}$$

In view of the orthogonality relationships, Eqs. (E10.3.8) and (E10.3.10), Eq. (10.66) reduces to

$$c^2 \int_0^l \Theta_i''(x)\Theta_i(x)\,dx\,\eta_i(t) = \int_0^{\infty} \Theta_i^2(x)\,dx\,\ddot{\eta}_i(t)$$

or

$$-\omega_i^2 \left(\int_0^l \Theta_i^2(x)\,dx \right) \eta_i(t) = \left(\int_0^l \Theta_i^2(x)\,dx \right) \ddot{\eta}_i(t) \tag{10.67}$$

Equation (10.67) yields

$$\ddot{\eta}_i(t) + \omega_i^2 \eta_i(t) = 0, \qquad i = 1, 2, \ldots \tag{10.68}$$

The solution of Eq. (10.68) is given by

$$\eta_i(t) = \eta_{i_0} \cos \omega_i t + \frac{\dot{\eta}_{i_0}}{\omega_i} \sin \omega_i t \tag{10.69}$$

where $\eta_{i_0} = \eta_i(t = 0)$ and $\dot{\eta}_{i_0} = \dot{\eta}_i(t = 0)$ denote the initial values of the generalized coordinate $\eta_i(t)$ and the generalized velocity $\dot{\eta}_i(t)$, respectively.

Initial Conditions   If the initial conditions of the shaft are given by

$$\theta(x, 0) = \theta_0(x) \tag{10.70}$$

$$\frac{\partial \theta}{\partial t}(x, 0) = \dot{\theta}_0(x) \tag{10.71}$$

Eq. (10.64) gives

$$\theta_0(x) = \sum_{i=1}^{\infty} \Theta_i(x)\eta_{i_0} \tag{10.72}$$

$$\dot{\theta}_0(x) = \sum_{i=1}^{\infty} \Theta_i(x)\dot{\eta}_{i_0} \tag{10.73}$$

By multiplying Eqs. (10.72) and (10.73) by $\Theta_j(x)$ and integrating from 0 to l, we obtain

$$\int_0^l \theta_0(x)\Theta_j(x)\,dx = \sum_{i=1}^{\infty} \eta_{i0} \int_0^l \Theta_i(x)\Theta_j(x)\,dx = \eta_{j0}, \qquad j = 1, 2, \ldots \quad (10.74)$$

$$\int_0^l \dot{\theta}_0(x)\Theta_j(x)\,dx = \sum_{i=1}^{\infty} \dot{\eta}_{i0} \int_0^l \Theta_i(x)\Theta_j(x)\,dx = \dot{\eta}_{j0}, \qquad j = 1, 2, \ldots \quad (10.75)$$

in view of the orthogonality of normal modes [Eq. (E10.3.8)]. Using the initial values of $\eta_j(t)$ and $\dot{\eta}_j(t)$, Eqs. (10.74) and (10.75), the free vibration response of the shaft can be determined from Eqs. (10.69) and (10.64):

$$\theta(x, t) = \sum_{i=1}^{\infty} \Theta_i(x) \left(\eta_{i0} \cos \omega_i t + \frac{\dot{\eta}_{i0}}{\omega_i} \sin \omega_i t \right) \quad (10.76)$$

Example 10.4 Find the free vibration response of an unrestrained uniform shaft shown in Fig. 10.6 when it is twisted by an equal and apposite angle a_0 at the two ends at $t = 0$ and then released.

SOLUTION   The initial displacement of the shaft can be expressed as

$$\theta(x, 0) = \theta_0(x) = a_0 \left(2\frac{x}{l} - 1 \right) \quad (E10.4.1)$$

which gives the angular deflections as $-a_0$ at $x = 0$ and a_0 at $x = l$. The initial velocity can be assumed to be zero:

$$\dot{\theta}(x, 0) = \dot{\theta}_0 = 0 \quad (E10.4.2)$$

The natural frequencies and the mode shapes of the shaft are given by Eqs. (10.50) and (10.51):

$$\omega_i = \frac{i\pi}{l} \sqrt{\frac{G}{\rho}}, \qquad i = 1, 2, \ldots \quad (E10.4.3)$$

$$\Theta_i(x) = A_i \cos \frac{\omega_i x}{c}, \qquad i = 1, 2, \ldots \quad (E10.4.4)$$

The mode shapes are normalized as

$$\int_0^l \Theta_i^2(x)\,dx = A_i^2 \int_0^l \cos^2 \frac{\omega_i x}{c}\,dx = 1 \quad (E10.4.5)$$

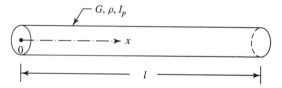

Figure 10.6   Unrestrained (free–free) shaft.

which gives

$$A_i = \sqrt{\frac{2}{l}} \qquad \text{(E10.4.6)}$$

and hence

$$\Theta_i(x) = \sqrt{\frac{2}{l}} \cos \frac{\omega_i x}{c}, \qquad i = 1, 2, \ldots \qquad \text{(E10.4.7)}$$

The initial values of the generalized displacement and generalized velocity can be determined using Eqs. (10.74) and (10.75) as:

$$\eta_{i_0} = \int_0^l \theta_0(x)\Theta_i(x)\, dx \qquad \text{(E10.4.8)}$$

$$= \int_0^l a_0 \left(2\frac{x}{l} - 1\right)\sqrt{\frac{2}{l}}\cos\frac{\omega_i x}{c}\, dx = a_0 \sqrt{\frac{2}{l}} \int_0^l \left(2\frac{x}{l} - 1\right)\cos\frac{\omega_i x}{c}\, dx \quad \text{(E10.4.9)}$$

$$= \begin{cases} -\dfrac{4\sqrt{2l}a_0}{i^2\pi^2}, & i = 1, 3, 5, \ldots \\ 0, & i = 2, 4, 6, \ldots \end{cases} \qquad \text{(E10.4.10)}$$

$$\dot{\eta}_{i_0} = \int_0^l \dot{\theta}_0(x)\Theta_i(x)\, dx = 0 \qquad \text{(E10.4.11)}$$

Thus, the free vibration response of the shaft is given by Eq. (10.76)

$$\theta(x, t) = -\frac{8a_0}{\pi^2} \sum_{i=1,3,5,\ldots}^{\infty} \cos\frac{i\pi x}{l} \cos\frac{i\pi ct}{l} \qquad \text{(E10.4.12)}$$

10.5 FORCED VIBRATION OF A UNIFORM SHAFT: MODAL ANALYSIS

The equation of motion of a uniform shaft subjected to distributed external torque, $m_t(x, t)$, is given by Eq. (10.23):

$$GI_p \frac{\partial^2\theta}{\partial x^2}(x, t) + m_t(x, t) = I_0 \frac{\partial^2\theta}{\partial t^2}(x, t) \qquad \text{(10.77)}$$

The solution of Eq. (10.77) using modal analysis is expressed as

$$\theta(x, t) = \sum_{n=1}^{\infty} \Theta_n(x)\eta_n(t) \qquad \text{(10.78)}$$

where $\Theta_n(x)$ is the nth normalized normal mode and $\eta_n(t)$ is the nth generalized coordinate. The normal modes $\Theta_n(x)$ are determined by solving the eigenvalue problem

$$GI_p \frac{d^2\Theta_n(x)}{dx^2} + I_0\omega_n^2\Theta_n(x) = 0 \qquad \text{(10.79)}$$

by applying the boundary conditions of the shaft. By substituting Eq. (10.78) into (10.77), we obtain

$$\sum_{n=1}^{\infty} GI_p \Theta_n''(x)\eta_n(t) + m_t(x, t) = \sum_{n=1}^{\infty} I_0 \Theta_n(x)\ddot{\eta}_n(t) \tag{10.80}$$

where

$$\Theta_n''(x) = \frac{d^2\Theta_n(x)}{dx^2}, \qquad \ddot{\eta}_n(t) = \frac{d^2\eta_n(t)}{dt^2} \tag{10.81}$$

Using Eq. (10.79), Eq. (10.80) can be rewritten as

$$-\sum_{n=1}^{\infty} I_0\omega_n^2 \Theta_n(x)\eta_n(t) + m_t(x, t) = \sum_{n=1}^{\infty} I_0 \Theta_n(x)\ddot{\eta}_n(t) \tag{10.82}$$

Multiplication of Eq. (10.82) by $\Theta_m(x)$ and integration from 0 to l result in

$$-I_0\omega_n^2\eta_n(t)\sum_{n=1}^{\infty} \int_0^l \Theta_n(x)\Theta_m(x)\,dx + \int_0^l m_t(x, t)\Theta_m(x)\,dx$$

$$= I_0\ddot{\eta}_n(t)\int_0^l \Theta_n(x)\Theta_m(x)\,dx \tag{10.83}$$

In view of the orthogonality relationships, Eq. (E10.3.8), Eq. (10.83) reduces to

$$\ddot{\eta}_n(t) + \omega_n^2\eta_n(t) = Q_n(t), \qquad n = 1, 2, \ldots \tag{10.84}$$

where the normal modes are assumed to satisfy the normalization condition

$$\int_0^l \Theta_n^2(x)\,dx = 1, \qquad n = 1, 2, \ldots \tag{10.85}$$

and $Q_n(t)$, called the *generalized force* in nth mode, is given by

$$Q_n(t) = \frac{1}{I_0}\int_0^l m_t(x, t)\Theta_n(x)\,dx \tag{10.86}$$

The complete solution of Eq. (10.84) can be expressed as

$$\eta_n(t) = A_n \cos\omega_n t + B_n \sin\omega_n t + \frac{1}{\omega_n}\int_0^t Q_n(\tau)\sin\omega_n(t - \tau)\,d\tau \tag{10.87}$$

where the constants A_n and B_n can be determined from the initial conditions of the shaft. Thus, the forced vibration response of the shaft [i.e., the solution of Eq. (10.77)], is given by

$$\theta(x, t) = \sum_{n=1}^{\infty} \left[A_n \cos\omega_n t + B_n \sin\omega_n t + \frac{1}{\omega_n}\int_0^t Q_n(\tau)\sin\omega_n(t - \tau)\,d\tau \right]\Theta_n(x) \tag{10.88}$$

The steady-state response of the shaft, without considering the effect of initial conditions, can be obtained from Eq. (10.88), as

$$\theta(x, t) = \sum_{n=1}^{\infty} \frac{\Theta_n(x)}{\omega_n} \int_0^t Q_n(\tau) \sin \omega_n(t - \tau) \, d\tau \tag{10.89}$$

Note that if the shaft is unrestrained (free at both ends), the rigid-body displacement, $\bar{\theta}(t)$, is to be added to the solution given by Eq. (10.89). If $M_t(t)$ denotes the torque applied to the shaft, the rigid-body motion of the shaft, $\bar{\theta}(t)$, can be determined from the relation

$$I_0 \frac{d^2 \bar{\theta}(t)}{dt^2} = M_t(t) \tag{10.90}$$

where $I_0 = \rho l I_p$ denotes the mass moment of inertia of the shaft and $d^2 \bar{\theta}(t)/dt^2$ indicates the acceleration of rigid-body motion.

Example 10.5   Find the steady-state response of a shaft, free at both ends, when subjected to a torque $M_t = a_0 t$, where a_0 is a constant at $x = l$.

SOLUTION   The steady-state response of the shaft is given by Eq. (10.89):

$$\theta(x, t) = \sum_{n=1}^{\infty} \frac{\Theta_n(x)}{\omega_n} \int_0^t Q_n(\tau) \sin \omega_n(t - \tau) \, d\tau \tag{E10.5.1}$$

where the generalized force $Q_n(t)$ is given by Eq. (10.86):

$$Q_n(t) = \frac{1}{I_0} \int_0^l m_t(x, t) \Theta_n(x) \, dx \tag{E10.5.2}$$

Since the applied torque is concentrated at $x = l$, $m_t(x, t)$ can be represented as

$$m_t(x, t) = M_t \delta(x - l) = a_0 t \delta(x - l) \tag{E10.5.3}$$

where δ is the Dirac delta function. Thus,

$$Q_n(t) = \frac{1}{I_0} \int_0^l a_0 t \delta(x - l) \Theta_n(x) \, dx = \frac{a_0}{I_0} \Theta_n(l) t \tag{E10.5.4}$$

For a free–free shaft, the natural frequencies and mode shapes are given by Eqs. (10.50) and (10.51):

$$\omega_n = \frac{n \pi c}{l} = \frac{n \pi}{l} \sqrt{\frac{G}{\rho}}, \qquad n = 1, 2, \ldots \tag{E10.5.5}$$

$$\Theta_n(x) = A_n \cos \frac{\omega_n x}{c}, \qquad n = 1, 2, \ldots \tag{E10.5.6}$$

When $\Theta_n(x)$ is normalized as

$$\int_0^l \Theta_n^2(x) \, dx = 1 \tag{E10.5.7}$$

we obtain

$$A_n = \sqrt{\frac{2}{l}}, \qquad n = 1, 2, \dots \tag{E10.5.8}$$

$$\Theta_n(x) = \sqrt{\frac{2}{l}} \cos \frac{n\pi x}{l}, \qquad n = 1, 2, \dots \tag{E10.5.9}$$

The integral in Eq. (E10.5.1) can be evaluated as

$$\int_0^t Q_n(\tau) \sin \omega_n(t - \tau)\, d\tau = \frac{a_0}{I_0} \sqrt{\frac{2}{l}} \cos n\pi \int_0^t \tau \sin \omega_n(t - \tau)\, d\tau \tag{E10.5.10}$$

$$= \frac{a_0}{I_0 \omega_n} \sqrt{\frac{2}{l}} \cos n\pi \left(t - \frac{1}{\omega_n} \sin \omega_n t \right) \tag{E10.5.11}$$

Thus, the steady-state response, given by Eq. (E10.5.1), becomes

$$\theta(x, t) = \sum_{n=1}^{\infty} \frac{1}{\omega_n} \sqrt{\frac{2}{l}} \cos \frac{n\pi x}{l} \left(\frac{a_0}{I_0 \omega_n} \sqrt{\frac{2}{l}} \cos n\pi \right) \left(t - \frac{1}{\omega_n} \sin \omega_n t \right) \tag{E10.5.12}$$

$$= \sum_{n=2,4,\dots}^{\infty} \frac{2a_0}{l I_0 \omega_n^2} \cos \frac{n\pi x}{l} \left(t - \frac{1}{\omega_n} \sin \omega_n t \right)$$

$$- \sum_{n=1,3,\dots}^{\infty} \frac{2a_0}{l I_0 \omega_n^2} \cos \frac{n\pi x}{l} \left(t - \frac{1}{\omega_n} \sin \omega_n t \right) \tag{E10.5.13}$$

Since the shaft is unrestrained, the rigid-body motion is to be added to Eq. (E10.5.13).

Using $M_t = a_0 t$ in Eq. (10.90) and integrating it twice with respect to t, we obtain the rigid-body rotation of the shaft as

$$\bar{\theta}(t) = \frac{a_0}{I_0} \frac{t^3}{6} \tag{E10.5.14}$$

Thus, the complete torsional vibration response of the shaft becomes

$$\theta(t) = \frac{a_0}{6 I_0} t^3 + \frac{2a_0}{l I_0} \left[\sum_{n=2,4,\dots}^{\infty} \frac{1}{\omega_n^2} \cos \frac{n\pi x}{l} \left(t - \frac{1}{\omega_n} \sin \omega_n t \right) \right.$$

$$\left. - \sum_{n=1,3,\dots}^{\infty} \frac{1}{\omega_n^2} \cos \frac{n\pi x}{l} \left(t - \frac{1}{\omega_n} \sin \omega_n t \right) \right] \tag{E10.5.15}$$

10.6   TORSIONAL VIBRATION OF NONCIRCULAR SHAFTS: SAINT-VENANT'S THEORY

For a shaft or bar of noncircular cross section subjected to torsion, the cross sections do not simply rotate with respect to one another as in the case of a circular shaft, but they are deformed, too. The originally plane cross sections of the shaft do not remain plane but warp out of their own planes after twisting as shown in Fig. 10.7. Thus, the

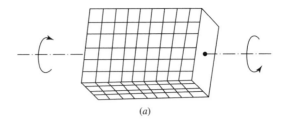

(a)

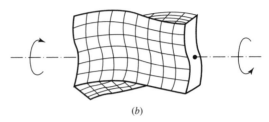

(b)

Figure 10.7 Shaft with rectangular cross section under torsion: (a) before deformation; (b) after deformation.

points in the cross section undergo an axial displacement. A function $\psi(y, z)$, known as the *warping function*, is used to denote the axial displacement as

$$u = \psi(y, z)\frac{\partial \theta}{\partial x} \tag{10.91}$$

where $\partial \theta/\partial x$ denotes the rate of twist along the shaft, assumed to be a constant. The other components of displacement in the shaft are given by

$$v = -z\theta(x, t) \tag{10.92}$$

$$w = y\theta(x, t) \tag{10.93}$$

The strains are given by

$$\varepsilon_{xx} = \frac{\partial u}{\partial x} = 0 \qquad (\text{since} \partial \theta/\partial x \text{ is a constant})$$

$$\varepsilon_{yy} = \frac{\partial v}{\partial y} = 0$$

$$\varepsilon_{zz} = \frac{\partial w}{\partial z} = 0$$

$$\varepsilon_{xy} = \frac{\partial u}{\partial y} + \frac{\partial v}{\partial x} = \left(\frac{\partial \psi}{\partial y} - z\right)\frac{\partial \theta}{\partial x} \tag{10.94}$$

$$\varepsilon_{xz} = \frac{\partial u}{\partial z} + \frac{\partial w}{\partial x} = \left(\frac{\partial \psi}{\partial z} + y\right)\frac{\partial \theta}{\partial x}$$

$$\varepsilon_{yz} = \frac{\partial v}{\partial z} + \frac{\partial w}{\partial y} = -\theta + \theta = 0$$

The corresponding stresses can be determined as

$$\sigma_{xx} = \sigma_{yy} = \sigma_{zz} = \sigma_{yz} = 0, \qquad \sigma_{xy} = G\left(\frac{\partial \psi}{\partial y} - z\right)\frac{\partial \theta}{\partial x},$$

$$\sigma_{xz} = G\left(\frac{\partial \psi}{\partial z} + y\right)\frac{\partial \theta}{\partial x} \tag{10.95}$$

The strain energy of the shaft is given by

$$\pi = \frac{1}{2}\iiint_V (\sigma_{xx}\varepsilon_{xx} + \sigma_{yy}\varepsilon_{yy} + \sigma_{zz}\varepsilon_{zz} + \sigma_{xy}\varepsilon_{xy} + \sigma_{xz}\varepsilon_{xz} + \sigma_{yz}\varepsilon_{yz})\,dV$$

$$= \frac{1}{2}\int_{x=0}^{l}\iint_A G\left[\left(\frac{\partial \psi}{\partial y} - z\right)^2 + \left(\frac{\partial \psi}{\partial z} + y\right)^2\right]\left(\frac{\partial \theta}{\partial x}\right)^2 dA\,dx \tag{10.96}$$

Defining the torsional rigidity (C) of its noncircular section of the shaft as

$$C = \iint_A G\left[\left(\frac{\partial \psi}{\partial y} - z\right)^2 + \left(\frac{\partial \psi}{\partial z} + y\right)^2\right]dA \tag{10.97}$$

the strain energy of the shaft can be expressed as

$$\pi = \frac{1}{2}\int_0^l C\left(\frac{\partial \theta}{\partial x}\right)^2 dx \tag{10.98}$$

Neglecting the inertia due to axial motion, the kinetic energy of the shaft can be written as in Eq. (10.11). The work done by the applied torque is given by Eq. (10.12). Hamilton's principle can be written as

$$\delta\int_{t_1}^{t_2}\int_0^l \left\{\iint_A \frac{1}{2}G\left[\left(\frac{\partial \psi}{\partial y} - z\right)^2 + \left(\frac{\partial \psi}{\partial z} + y\right)^2\right]dA\right\}\left(\frac{\partial \theta}{\partial x}\right)^2 dx\,dt$$

$$-\delta\int_{t_1}^{t_2}\int_0^l \frac{1}{2}\rho I_p\left(\frac{\partial \theta}{\partial t}\right)^2 dx\,dt - \delta\int_{t_1}^{t_2}\int_0^l m_t\theta\,dx\,dt = 0 \tag{10.99}$$

The first integral of Eq. (10.99) can be evaluated as

$$\delta\int_{t_1}^{t_2}\int_0^l \iint_A \frac{1}{2}G\left[\left(\frac{\partial \psi}{\partial y} - z\right)^2 + \left(\frac{\partial \psi}{\partial z} + y\right)^2\right]\left(\frac{\partial \theta}{\partial x}\right)^2 dA\,dx\,dt$$

$$= \int_{t_1}^{t_2}\int_0^l \iint_A G\left[\left(\frac{\partial \psi}{\partial y} - z\right)^2 + \left(\frac{\partial \psi}{\partial z} + y\right)^2\right]\frac{\partial \theta}{\partial x}\frac{\partial (\delta\theta)}{\partial x}\,dA\,dx\,dt$$

$$+ \int_{t_1}^{t_2}\int_0^l \iint_A G\left(\frac{\partial \theta}{\partial x}\right)^2\left[\left(\frac{\partial \psi}{\partial y} - z\right)\frac{\partial (\delta\psi)}{\partial y} + \left(\frac{\partial \psi}{\partial z} + y\right)\frac{\partial (\delta\psi)}{\partial z}\right]dA\,dx\,dt$$

$$\tag{10.100}$$

The first integral term on the right-hand side of Eq. (10.100) can be expressed as [see Eq. (10.97)]

$$\int_{t_1}^{t_2} \int_0^l GC\frac{\partial \theta}{\partial x}\frac{\partial(\delta\theta)}{\partial x}\,dx\,dt = \int_{t_1}^{t_2}\left[\int_0^l -\frac{\partial}{\partial x}\left(GC\frac{\partial \theta}{\partial x}\right)\delta\theta + GC\frac{\partial \theta}{\partial x}\delta\theta\Big|_0^l\right]dt$$

(10.101)

The second integral term on the right-hand side of Eq. (10.100) is set equal to zero independently:

$$\int_{t_1}^{t_2}\int_0^l \iint_A G\left(\frac{\partial \theta}{\partial x}\right)^2\left[\left(\frac{\partial \psi}{\partial y} - z\right)\frac{\partial(\delta\psi)}{\partial y} + \left(\frac{\partial \psi}{\partial z} + y\right)\frac{\partial(\delta\psi)}{\partial z}\right]dy\,dz\,dx\,dt = 0$$

(10.102)

Integrating Eq. (10.102) by parts, we obtain

$$\int_{t_1}^{t_2}\left[\int_0^l G\left(\frac{\partial \theta}{\partial x}\right)^2 dx\right]\left[-\iint_A \frac{\partial}{\partial y}\left(\frac{\partial \psi}{\partial y} - z\right)\delta\psi\,dy\,dz + \int_\xi \left(\frac{\partial \psi}{\partial y} - z\right)l_y\delta\psi\,d\xi\right.$$
$$\left.-\iint_A \frac{\partial}{\partial z}\left(\frac{\partial \psi}{\partial z} + y\right)\delta\psi\,dy\,dz + \int_\xi \left(\frac{\partial \psi}{\partial z} + y\right)l_z\delta\psi\,d\xi\right]dt = 0$$

(10.103)

where ξ is the bounding curve of the cross section and $l_y(l_z)$ is the cosine of the angle between the normal to the bounding curve and the $y(z)$ direction. Equation (10.103) yields the differential equation for the warping function ψ as

$$\left(\frac{\partial^2 \psi}{\partial y^2} + \frac{\partial^2 \psi}{\partial z^2}\right) = \nabla^2\psi = 0$$

(10.104)

and the boundary condition on ψ as

$$\left(\frac{\partial \psi}{\partial y} - z\right)l_y + \left(\frac{\partial \psi}{\partial z} + y\right)l_z = 0$$

(10.105)

Physically, Eq. (10.105) represents that the shear stress normal to the boundary must be zero at every point on the boundary of the cross section of the shaft. When Eq. (10.101) is combined with the second and third integrals of Eq. (10.99), it leads to the equation of motion as

$$\rho I_p \frac{\partial^2 \theta(x, t)}{\partial t^2} = \frac{\partial}{\partial x}\left(C\frac{\partial \theta(x, t)}{\partial x}\right) + m_t(x, t)$$

(10.106)

and the boundary conditions on θ as

$$C\frac{\partial \theta}{\partial x}\delta\theta\Big|_0^l = 0$$

(10.107)

10.7   TORSIONAL VIBRATION OF NONCIRCULAR SHAFTS, INCLUDING AXIAL INERTIA

Love included the inertia due to the axial motion caused by the warping of the cross section in deriving the equation of motion of a shaft in torsional vibration [4, 10]. In this case, the kinetic energy of the shaft is given by

$$T = \frac{1}{2} \int_0^l \iint_A \rho \left[\psi^2(y, z) \left(\frac{\partial^2 \theta}{\partial t \partial x} \right)^2 + z^2 \left(\frac{\partial \theta}{\partial t} \right)^2 + y^2 \left(\frac{\partial \theta}{\partial t} \right)^2 \right] dA\, dx$$

$$= I_1 + I_2 \tag{10.108}$$

where

$$I_1 = \frac{1}{2} \int_0^l \iint_A \rho \psi^2 \left(\frac{\partial^2 \theta}{\partial t \partial x} \right)^2 dA\, dx \tag{10.109}$$

$$I_2 = \frac{1}{2} \int_0^l \rho I_p \left(\frac{\partial \theta}{\partial t} \right)^2 dx \tag{10.110}$$

Note that I_1 denotes the axial inertia term. The variation associated with I_1 in Hamilton's principle leads to

$$\delta \int_{t_1}^{t_2} I_1\, dt = \int_{t_1}^{t_2} \delta I_1\, dt$$

$$= \int_{t_1}^{t_2} \int_0^l \rho \left(\iint_A \psi^2\, dA \right) \frac{\partial^2 \theta}{\partial t \partial x} \frac{\partial^2 (\delta \theta)}{\partial t \partial x} dx\, dt$$

$$+ \int_{t_1}^{t_2} \int_0^l \rho \left(\frac{\partial^2 \theta}{\partial t \partial x} \right)^2 \left(\iint_A \psi \delta \psi\, dA \right) dx\, dt \tag{10.111}$$

Denoting

$$I_\psi = \iint_A \psi^2\, dA \tag{10.112}$$

$$I_\theta = \int_0^l \rho \left(\frac{\partial^2 \theta}{\partial t \partial x} \right)^2 dx \tag{10.113}$$

the integrals in Eq. (10.111) can be evaluated to obtain

$$\delta \int_{t_1}^{t_2} I_1\, dt = -\int_{t_1}^{t_2} \int_0^l \frac{\partial}{\partial t} \left(\rho I_\psi \frac{\partial^2 \theta}{\partial t \partial x} \right) \frac{\partial (\delta \theta)}{\partial x} dx\, dt + \int_{t_1}^{t_2} I_\theta \iint_A \psi \delta \psi\, dA\, dt$$

$$= \int_{t_1}^{t_2} \int_0^l \frac{\partial^2}{\partial x \partial t} \left(\rho I_\psi \frac{\partial^2 \theta}{\partial t \partial x} \right) \delta \theta\, dx\, dt - \int_{t_1}^{t_2} \frac{\partial}{\partial t} \left(\rho I_\psi \frac{\partial^2 \theta}{\partial t \partial x} \right) \delta \theta \Big|_0^l\, dt$$

$$+ \int_{t_1}^{t_2} I_\theta \iint_A \psi \delta \psi\, dA\, dt \tag{10.114}$$

The first, second, and third terms on the right-hand side of Eq. (10.114) contribute to the equation of motion, Eq. (10.106), the boundary conditions on θ, Eq. (10.107), and to the differential equation for ψ, Eq. (10.104), respectively. The new equations are given by

$$\rho I_p \frac{\partial^2 \theta}{\partial t^2} = \frac{\partial^2}{\partial t \, \partial x}\left(\rho I_\psi \frac{\partial^2 \theta}{\partial x \, \partial t}\right) + \frac{\partial}{\partial x}\left(C\frac{\partial \theta}{\partial x}\right) + m_t \tag{10.115}$$

$$\left(C\frac{\partial \theta}{\partial x} + \rho I_\psi \frac{\partial^3 \theta}{\partial x \, \partial t^2}\right)\delta\theta\Big|_0^l = 0 \tag{10.116}$$

$$\frac{\partial^2 \psi}{\partial y^2} + \frac{\partial^2 \psi}{\partial z^2} + \frac{I_\theta}{I_g}\psi = 0 \tag{10.117}$$

$$\left(\frac{\partial \psi}{\partial y} - z\right)l_y + \left(\frac{\partial \psi}{\partial z} + y\right)l_z = 0 \tag{10.118}$$

where

$$I_g = \int_0^l G\left(\frac{\partial \theta}{\partial x}\right)^2 dx \tag{10.119}$$

10.8 TORSIONAL VIBRATION OF NONCIRCULAR SHAFTS: TIMOSHENKO–GERE THEORY

In this theory also, the displacement components of a point in the cross section are assumed to be [4, 11, 17]

$$u = \psi(y, z)\frac{\partial \theta}{\partial x}(x, t) \tag{10.120}$$

$$v = -z\theta(x, t) \tag{10.121}$$

$$w = y\theta(x, t) \tag{10.122}$$

where $\partial\theta/\partial x$ is not assumed to be a constant. The components of strains can be obtained as

$$\varepsilon_{xx} = \frac{\partial u}{\partial x} = \psi(y, z)\frac{\partial^2 \theta}{\partial x^2} \tag{10.123}$$

$$\varepsilon_{yy} = \frac{\partial v}{\partial y} = 0 \tag{10.124}$$

$$\varepsilon_{zz} = \frac{\partial w}{\partial z} = 0 \tag{10.125}$$

$$\varepsilon_{xy} = \frac{\partial v}{\partial x} + \frac{\partial u}{\partial y} = \left(\frac{\partial \psi}{\partial y} - z\right)\frac{\partial \theta}{\partial x} \tag{10.126}$$

$$\varepsilon_{xz} = \frac{\partial u}{\partial z} + \frac{\partial w}{\partial x} = \left(\frac{\partial \psi}{\partial z} + y\right)\frac{\partial \theta}{\partial x} \tag{10.127}$$

$$\varepsilon_{yz} = \frac{\partial v}{\partial z} + \frac{\partial w}{\partial y} = -\theta + \theta = 0 \tag{10.128}$$

The component of stress are given by

$$\begin{Bmatrix} \sigma_{xx} \\ \sigma_{yy} \\ \sigma_{zz} \\ \sigma_{xy} \\ \sigma_{yz} \\ \sigma_{zx} \end{Bmatrix} = \frac{E}{(1+v)(1-2v)} \begin{bmatrix} 1-v & v & v & 0 & 0 & 0 \\ v & 1-v & v & 0 & 0 & 0 \\ v & v & 1-v & 0 & 0 & 0 \\ 0 & 0 & 0 & \frac{1-2v}{2} & 0 & 0 \\ 0 & 0 & 0 & 0 & \frac{1-2v}{2} & 0 \\ 0 & 0 & 0 & 0 & 0 & \frac{1-2v}{2} \end{bmatrix} \begin{Bmatrix} \varepsilon_{xx} \\ \varepsilon_{yy} \\ \varepsilon_{zz} \\ \varepsilon_{xy} \\ \varepsilon_{yz} \\ \varepsilon_{zx} \end{Bmatrix}$$

$$(10.129)$$

that is,

$$\sigma_{xx} = \frac{E(1-v)\psi}{(1+v)(1-2v)} \frac{\partial^2 \theta}{\partial x^2} \approx E\psi \frac{\partial^2 \theta}{\partial x^2} \tag{10.130}$$

$$\sigma_{yy} = \sigma_{zz} = \frac{Ev\psi}{(1+v)(1-2v)} \frac{\partial^2 \theta}{\partial x^2} \approx 0 \tag{10.131}$$

$$\sigma_{xy} = \frac{E}{(1+v)(1-2v)} \frac{1-2v}{2} \left(\frac{\partial \psi}{\partial y} - z \right) \frac{\partial \theta}{\partial x} = G \left(\frac{\partial \psi}{\partial y} - z \right) \frac{\partial \theta}{\partial x} \tag{10.132}$$

$$\sigma_{yz} = 0 \tag{10.133}$$

$$\sigma_{zx} = \frac{E}{(1+v)(1-2v)} \frac{1-2v}{2} \left(\frac{\partial \psi}{\partial z} + y \right) \frac{\partial \theta}{\partial x} = G \left(\frac{\partial \psi}{\partial z} + y \right) \frac{\partial \theta}{\partial x} \tag{10.134}$$

Note that the effect of Poisson's ratio is neglected in Eqs. (10.130) and (10.131). The strain energy of the shaft can be determined as

$$\pi = \frac{1}{2} \iiint\limits_V (\sigma_{xx}\varepsilon_{xx} + \sigma_{yy}\varepsilon_{yy} + \sigma_{zz}\varepsilon_{zz} + \sigma_{xy}\varepsilon_{xy} + \sigma_{xz}\varepsilon_{xz} + \sigma_{yz}\varepsilon_{yz}) \, dV$$

$$= I_1 + I_2 \tag{10.135}$$

where

$$I_1 = \frac{1}{2} \int_{x=0}^{l} \iint_A E \left(\psi \frac{\partial^2 \theta}{\partial x^2} \right)^2 dA \, dx \tag{10.136}$$

$$I_2 = \frac{1}{2} \int_{x=0}^{l} \iint_A \left\{ G \left[\left(\frac{\partial \psi}{\partial y} - z \right) \frac{\partial \theta}{\partial x} \right]^2 + G \left[\left(\frac{\partial \psi}{\partial z} + y \right) \frac{\partial \theta}{\partial x} \right]^2 \right\} dA \, dx \tag{10.137}$$

The variation of the integral I_1 can be evaluated as

$$\delta I_1 = \delta \left[\frac{1}{2} \int_{t_1}^{t_2} \int_{x=0}^{l} \iint_A E\psi^2 \left(\frac{\partial^2 \theta}{\partial x^2} \right)^2 dA \, dx \right]$$

$$= \int_{t_1}^{t_2} \int_{x=0}^{l} \iint_A E\psi \, \delta\psi \left(\frac{\partial^2 \theta}{\partial x^2} \right)^2 dA \, dx \, dt$$

$$+ \int_{t_1}^{t_2} \int_{x=0}^{l} \iint_A E\psi^2 \frac{\partial^2 \theta}{\partial x^2} \frac{\partial^2 (\delta\theta)}{\partial x^2} \, dA \, dx \, dt$$

$$= \int_{t_1}^{t_2} \int_{x=0}^{l} \iint_A E\psi \delta\psi \left(\frac{\partial^2 \theta}{\partial x^2}\right)^2 dA \, dx \, dt + \int_{t_1}^{t_2} \int_{x=0}^{l} E I_\psi \frac{\partial^2 \theta}{\partial x^2} \frac{\partial^2 (\delta\theta)}{\partial x^2} \, dx \, dt$$

$$(10.138)$$

where

$$I_\psi = \iint_A \psi^2 \, dA \tag{10.139}$$

When the second term on the right-hand side of Eq. (10.138) is integrated by parts, we obtain

$$\delta I_1 = \int_{t_1}^{t_2} \int_{x=0}^{l} \iint_A E\psi \delta\psi \left(\frac{\partial^2 \theta}{\partial x^2}\right)^2 dA \, dx \, dt + \int_{t_1}^{t_2} \left[E I_\psi \frac{\partial^2 \theta}{\partial x^2} \delta\left(\frac{\partial \theta}{\partial x}\right)\Big|_0^l \right.$$

$$\left. - \frac{\partial}{\partial x}\left(E I_\psi \frac{\partial^2 \theta}{\partial x^2}\right) \delta\theta \Big|_0^l + \int_0^l \frac{\partial^2}{\partial x^2}\left(E I_\psi \frac{\partial^2 \theta}{\partial x^2}\right) \delta\theta \, dx \right] dt \tag{10.140}$$

The variation of the integral I_2 can be evaluated as indicated in Eqs. (10.100), (10.101) and (10.103). The expressions for the kinetic energy and the work done by the applied torque are given by Eqs. (10.108) and (10.12), respectively, and hence their variations can be evaluated as indicated earlier. The application of Hamilton's principle leads to the equation of motion for $\theta(x, t)$:

$$\rho I_p \frac{\partial^2 \theta}{\partial t^2} - \frac{\partial^2}{\partial t \, \partial x}\left(\rho I_\psi \frac{\partial^2 \theta}{\partial x \, \partial t}\right) - \frac{\partial}{\partial x}\left(C \frac{\partial \theta}{\partial x}\right) + \frac{\partial^2}{\partial x^2}\left(E I_\psi \frac{\partial^2 \theta}{\partial x^2}\right) = m_t(x, t)$$

$$(10.141)$$

and the boundary conditions

$$\left[C \frac{\partial \theta}{\partial x} + \rho I_\psi \frac{\partial^3 \theta}{\partial t^2 \, \partial x} - \frac{\partial}{\partial x}\left(E I_\psi \frac{\partial^2 \theta}{\partial x^2}\right) \right] \delta\theta \Big|_0^l = 0 \tag{10.142}$$

$$E I_\psi \frac{\partial^2 \theta}{\partial x^2} \delta\left(\frac{\partial \theta}{\partial x}\right)\Big|_0^l = 0 \tag{10.143}$$

The differential equation for the warping function ψ becomes

$$\int_0^l G\left(\frac{\partial \theta}{\partial x}\right)^2 dx \left(\frac{\partial^2 \psi}{\partial y^2} + \frac{\partial^2 \psi}{\partial z^2}\right) + \left[\int_0^l \rho\left(\frac{\partial^2 \theta}{\partial x \, \partial t}\right)^2 dx - \int_0^l E\left(\frac{\partial^2 \theta}{\partial x^2}\right)^2 dx\right] \psi = 0$$

$$(10.144)$$

with the boundary condition on ψ given by

$$\left(\frac{\partial \psi}{\partial y} - z\right) l_y + \left(\frac{\partial \psi}{\partial z} + y\right) l_z = 0 \tag{10.145}$$

10.9   TORSIONAL RIGIDITY OF NONCIRCULAR SHAFTS

It is necessary to find the torsional rigidity C of the shaft in order to find the solution of the torsional vibration problem, Eq. (10.106). The torsional rigidity can be determined by solving the Laplace equation, Eq. (10.104):

$$\nabla^2 \psi = \frac{\partial^2 \psi}{\partial y^2} + \frac{\partial^2 \psi}{\partial z^2} = 0 \tag{10.146}$$

subject to the boundary condition, Eq. (10.105):

$$\left(\frac{\partial \psi}{\partial y} - z \right) l_y + \left(\frac{\partial \psi}{\partial z} + y \right) l_z = 0 \tag{10.147}$$

which is equivalent to

$$\sigma_{xy} l_y + \sigma_{xz} l_z = 0 \tag{10.148}$$

Since the solution of Eq. (10.146), for the warping function ψ, subject to the boundary condition of Eq. (10.147) or (10.148) is relatively more difficult, we use an alternative procedure which leads to a differential equation similar to Eq. (10.146), and a boundary condition that is much simpler in form than Eq. (10.147) or (10.148). For this we express the stresses σ_{xy} and σ_{xz} in terms of a function $\Phi(y, z)$, known as the *Prandtl stress function*, as [3, 7]

$$\sigma_{xy} = \frac{\partial \Phi}{\partial z}, \qquad \sigma_{xz} = -\frac{\partial \Phi}{\partial y} \tag{10.149}$$

The stress field corresponding to Saint-Venant's theory, Eq. (10.95), along with Eq. (10.149), satisfies the equilibrium equations:

$$\frac{\partial \sigma_{xx}}{\partial x} + \frac{\partial \sigma_{xy}}{\partial y} + \frac{\partial \sigma_{xz}}{\partial z} = 0$$

$$\frac{\partial \sigma_{xy}}{\partial x} + \frac{\partial \sigma_{yy}}{\partial y} + \frac{\partial \sigma_{yz}}{\partial z} = 0$$

$$\frac{\partial \sigma_{xz}}{\partial x} + \frac{\partial \sigma_{yz}}{\partial y} + \frac{\partial \sigma_{zz}}{\partial z} = 0 \tag{10.150}$$

By equating the corresponding expressions of σ_{xy} and σ_{xz} given by Eqs. (10.95) and (10.149), we obtain

$$G \frac{\partial \theta}{\partial x} \left(\frac{\partial \psi}{\partial y} - z \right) = \frac{\partial \Phi}{\partial z} \tag{10.151}$$

$$G \frac{\partial \theta}{\partial x} \left(\frac{\partial \psi}{\partial z} + y \right) = -\frac{\partial \Phi}{\partial y} \tag{10.152}$$

Differentiating Eq. (10.151) with respect to z and Eq. (10.152) with respect to y and subtracting the resulting equations one from the other leads to the Poisson equation:

$$\nabla^2 \Phi = \frac{\partial^2 \Phi}{\partial y^2} + \frac{\partial^2 \Phi}{\partial z^2} = -2G\beta \tag{10.153}$$

where

$$\frac{\partial \theta}{\partial x} = \beta \tag{10.154}$$

is assumed to be a constant. The condition to be satisfied by the stress function Φ on the boundary can be derived by considering a small element of the rod at the boundary as shown in Fig. 10.8. The component of shear stress along the normal direction n can be expressed as

$$\sigma_{xn} = \sigma_{xy}l_y + \sigma_{xz}l_z = 0 \tag{10.155}$$

since the boundary is stress-free. In Eq. (10.155), the direction cosines are given by

$$l_y = \cos \alpha = \frac{dz}{dt}, \qquad l_z = \sin \alpha = -\frac{dy}{dt} \tag{10.156}$$

where t denotes the tangential direction. Using Eq. (10.149), the boundary condition, Eq. (10.155), can be written as

$$\frac{\partial \Phi}{\partial z}l_y - \frac{\partial \Phi}{\partial y}l_z = 0 \tag{10.157}$$

The rate of change of Φ along the tangential direction at the boundary (t) can be expressed as

$$\frac{d\Phi}{dt} = \frac{\partial \Phi}{\partial y}\frac{dy}{dt} + \frac{\partial \Phi}{\partial z}\frac{dz}{dt} = -\frac{\partial \Phi}{\partial y}\sin \alpha + \frac{\partial \Phi}{\partial z}\cos \alpha = 0 \tag{10.158}$$

using Eqs. (10.156) and (10.157). Equation (10.158) indicates that the stress function Φ is a constant on the boundary of the cross section of the rod. Since the magnitude

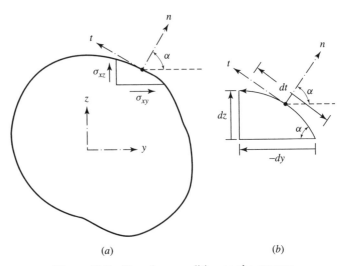

(a)   (b)

Figure 10.8 Boundary condition on the stresses.

of this constant does not affect the stress, which contains only derivatives of Φ, we choose, for convenience,

$$\Phi = 0 \qquad (10.159)$$

to be the boundary condition.

Next we derive a relation between the unknown angle β(angle of twist per unit length) and the torque (M_t) acting on the rod. For this, consider the cross section of the twisted rod as shown in Fig. 10.9. The moment about the x axis of all the forces acting on a small elemental area dA located at the point (y,z) is given by

$$(\sigma_{xz}\,dA)y - (\sigma_{xy}\,dA)z \qquad (10.160)$$

The resulting moment can be found by integrating the expression in Eq. (10.160) over the entire area of cross section of the bar as

$$M_t = \iint\limits_{A} (\sigma_{xz}y - \sigma_{xy}z)\,dA = -\iint\limits_{A} \left(\frac{\partial \Phi}{\partial y}y + \frac{\partial \Phi}{\partial z}z\right)dA \qquad (10.161)$$

Each term under the integral sign in Eq. (10.161) can be integrated by parts to obtain (see Fig. 10.9):

$$\iint\limits_{A} \frac{\partial \Phi}{\partial y}y\,dA = \iint\limits_{A} \frac{\partial \Phi}{\partial y}y\,dy\,dz = \int dz \int_{P_1}^{P_2} \frac{\partial \Phi}{\partial y}\,dy$$

$$= \int dz \left(\Phi y \Big|_{P_3}^{P_4} - \int_{P_1}^{P_2} \Phi\,dy\right) = -\int dz \int_{P_1}^{P_2} \Phi\,dy = -\iint\limits_{A} \Phi\,dA$$

$$(10.162)$$

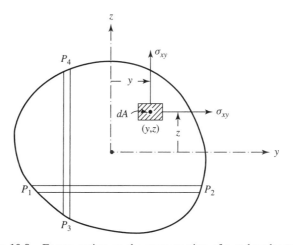

Figure 10.9 Forces acting on the cross section of a rod under torsion.

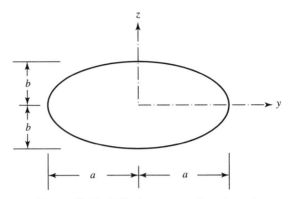

Figure 10.10   Elliptic cross section of a rod.

since $\Phi = 0$ at the points P_1 and P_2. Similarly,

$$\iint_A \frac{\partial \Phi}{\partial z} z \, dA = \iint_A \frac{\partial \Phi}{\partial z} z \, dy \, dz = \int dy \int_{P_3}^{P_4} \frac{\partial \Phi}{\partial z} z \, dz$$

$$= \int dy \left(\Phi z \Big|_{P_3}^{P_4} - \int_{P_3}^{P_4} \Phi \, dz \right) = -\iint_A \Phi \, dA \qquad (10.163)$$

Thus, the torque on the cross section (M_t) is given by

$$M_t = 2 \iint_A \Phi \, dA \qquad (10.164)$$

The function Φ satisfies the linear differential (Poisson) equation given by Eq. (10.153) and depends linearly on $G\beta$, so that Eq. (10.164) produces an equation of the form $M_t = GJ\beta = C\beta$, where J is called the *torsional constant* (J is the polar moment of inertia of the cross section for a circular section) and C is called the *torsional rigidity*. Thus, Eq. (10.164) can be used to find the torsional rigidity (C).

Note   There are very few cross-sectional shapes for which Eq. (10.164) can be evaluated in closed form to find an exact solution of the torsion problem. The following example indicates the procedure of finding an exact closed-form solution for the torsion problem for an elliptic cross section.

Example 10.6   Find the torsional rigidity of a rod with an elliptic cross section (Fig. 10.10).

SOLUTION   The equation of the boundary of the ellipse can be expressed as

$$f(y, z) = 1 - \frac{y^2}{a^2} - \frac{z^2}{b^2} = 0 \qquad (E10.6.1)$$

Noting that $\nabla^2 f$ is a constant, we take the stress function $\Phi(y, z)$ as

$$\Phi(y, z) = c \left(1 - \frac{y^2}{a^2} - \frac{z^2}{b^2} \right) \qquad (E10.6.2)$$

where c is a constant. Using Eq. (E10.6.2) in Eq. (10.153), we obtain

$$\nabla^2 \Phi = \frac{\partial^2 \Phi}{\partial y^2} + \frac{\partial^2 \Phi}{\partial z^2} = -2c \left(\frac{1}{a^2} + \frac{1}{b^2} \right) = -2G\beta \qquad (E10.6.3)$$

If we choose

$$c = \frac{G\beta a^2 b^2}{a^2 + b^2} \qquad (E10.6.4)$$

the function Φ satisfies not only the differential equation, Eq. (10.153), but also the boundary condition, Eq. (10.159). The stresses σ_{xy} and σ_{xz} become

$$\sigma_{xy} = \frac{\partial \Phi}{\partial z} = -\frac{2G\beta a^2}{(a^2 + b^2)} z \qquad (E10.6.5)$$

$$\sigma_{xz} = -\frac{\partial \Phi}{\partial y} = \frac{2G\beta b^2}{(a^2 + b^2)} y \qquad (E10.6.6)$$

The torque (M_t) can be obtained as

$$M_t = \iint_A (\sigma_{xz} y - \sigma_{xy} z) \, dA = \frac{2G\beta}{a^2 + b^2} \iint_A (y^2 b^2 + z^2 a^2) \, dA \qquad (E10.6.7)$$

Noting that

$$\iint_A y^2 \, dA = \iint_A y^2 \, dy \, dz = I_z = \frac{1}{4} \pi b a^3 \qquad (E10.6.8)$$

$$\iint_A z^2 \, dA = \iint_A z^2 \, dz \, dy = I_y = \frac{1}{4} \pi a b^3 \qquad (E10.6.9)$$

Eq. (E10.6.7) yields

$$M_t = \frac{\pi a^3 b^3}{a^2 + b^2} G\beta \qquad (E10.6.10)$$

When M_t is expressed as

$$M_t = GJ\beta = C\beta \qquad (E10.6.11)$$

Eq. (E10.6.10) gives the torsional constant J as

$$J = \frac{\pi a^3 b^3}{a^2 + b^2} \qquad (E10.6.12)$$

and the torsional rigidity C as

$$C = \frac{\pi a^3 b^3}{a^2 + b^2} G = \frac{M_t}{\beta} \qquad (E10.6.13)$$

The rate of twist can be expressed in terms of the torque as

$$\beta = \frac{M_t}{C} = \frac{M_t (a^2 + b^2)}{G \pi a^3 b^3} \qquad (E10.6.14)$$

10.10 PRANDTL'S MEMBRANE ANALOGY

Prandtl observed that the differential equation for the stress function, Eq. (10.153), is of the same form as the equation that describes the deflection of a membrane or soap film under transverse pressure [see Eq. (13.1) without the right-hand-side inertia term]. This analogy between the torsion and membrane problems has been used in determining the torsional rigidity of rods with noncircular cross sections experimentally [3, 4]. An actual experiment with a soap bubble would consist of an airtight box with a hole cut on one side (Fig. 10.11). The shape of the hole is the same as the cross section of the rod in torsion. First, a soap film is created over the hole. Then air under pressure (p) is pumped into the box. This causes the soap film to deflect transversely as shown in Fig. 10.11. If P denotes the uniform tension in the soap film, the small transverse deflection of the soap film (w) is governed by the equation [see Eq. (13.1) without the right-hand-side inertia term]

$$P \left(\frac{\partial^2 w}{\partial y^2} + \frac{\partial^2 w}{\partial z^2} \right) + p = 0 \tag{10.165}$$

or

$$\nabla^2 w = -\frac{p}{P} \tag{10.166}$$

in the hole region (cross section) and

$$w = 0 \tag{10.167}$$

on the boundary of the hole (cross section). Note that the differential equation and the boundary condition, Eqs. (10.166) and (10.167), are of precisely the same form as for the stress function Φ, namely, Eqs. (10.153) and (10.159):

$$\nabla^2 \Phi = -2G\beta \tag{10.168}$$

in the interior, and

$$\Phi = 0 \tag{10.169}$$

on the boundary. Thus, the soap bubble represents the surface of the stress function with

$$\frac{w}{p/P} = \frac{\Phi}{2G\beta} \tag{10.170}$$

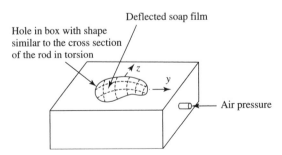

Figure 10.11 Soap film for the membrane analogy.

or

$$\Phi = \underset{\sim}{C} w \qquad (10.171)$$

where $\underset{\sim}{C}$ denotes a proportionality constant:

$$\underset{\sim}{C} = \frac{2G\beta P}{p} \qquad (10.172)$$

The analogous quantities in the two cases are given in Table 10.2.

The membrane analogy provides more than an experimental technique for the solution of torsion problem. It also serves as the basis for obtaining approximate analytical solutions for rods with narrow cross sections and open thin-walled cross sections. Table 10.3 gives the values of the maximum shear stress and the angle of twist per unit length for some commonly encountered cross-sectional shapes of rods.

Example 10.7   Determine the torsional rigidity of a rod with a rectangular cross section as shown in Fig. 10.12.

SOLUTION   We seek a solution of the membrane equation, (10.166), and use it for the stress function, Eq. (10.171). The governing equation for the deflection of a membrane is

$$\frac{\partial^2 w}{\partial y^2} + \frac{\partial^2 w}{\partial z^2} = -\frac{p}{P}, \qquad -a \leq y \leq a, -b \leq z \leq b \qquad (E10.7.1)$$

Table 10.2   Prandtl's Membrane Analogy

Soap bubble (membrane) problem	Torsion problem
w	Φ
$\frac{1}{P}$	G
p	2β
$-\frac{\partial w}{\partial z}, \frac{\partial w}{\partial y}$	σ_{xy}, σ_{xz}
2 (volume under bubble)	M_t

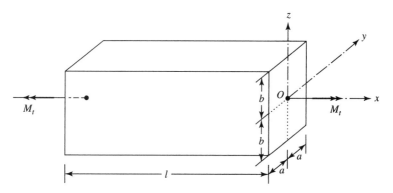

Figure 10.12   Rod with a rectangular cross section.

Table 10.3   Torsional Properties of Shafts with Various Cross Sections

Cross section	Angle of twist per unit length, θ	Maximum shear stress, τ_{max}
1. Solid circular shaft		
	$\dfrac{2T}{\pi G R^4}$	$\dfrac{2T}{\pi R^3}$
2. Thick-walled tube	$T =$ torque $G =$ shear modulus $\dfrac{2T}{\pi G(R_0^4 - R_i^4)}$	$\dfrac{2T R_0}{\pi(R_0^4 - R_i^4)}$
3. Thin-walled tube		
	$\dfrac{T}{2\pi G R^3 t}$	$\dfrac{T}{2\pi R^2 t}$
4. Solid elliptic shaft		
$\left(\dfrac{x}{a}\right)^2 + \left(\dfrac{y}{b}\right)^2 - 1 = 0$	$\dfrac{(a^2 + b^2)T}{\pi G a^3 b^3}$	$\dfrac{2T}{\pi ab^2}$
5. Hollow elliptic tube		
	$\dfrac{\sqrt{2(a^2 + b^2)}T}{4\pi G a^2 b^2 t}$	$\dfrac{T}{2\pi abt}$

(*continued on next page*)

Table 10.3   (*continued*)

Cross section	Angle of twist per unit length, θ	Maximum shear stress, τ_{max}
6. Solid square shaft		

$$\frac{7.092T}{Ga^4}$$

$$\frac{4.808T}{a^3}$$

7. Solid rectangular shaft

$$\frac{T}{\alpha Gab^3}$$

$\dfrac{a}{b}$	α	β
1.0	0.141	0.208
2.0	0.229	0.246
3.0	0.263	0.267
5.0	0.291	0.292
10.0	0.312	0.312
∞	0.333	0.333

$$\frac{T}{\beta ab^2}$$

8. Hollow rectangular shaft

$$\frac{(at_2 + bt_1)T}{2Gt_1t_2a^2b^2}$$

$$\tau_{max\,1} = \frac{T}{2abt_1}$$

$$\tau_{max\,2} = \frac{T}{2abt_2}$$

9. Solid equilateral triangular shaft

$$\frac{26T}{Ga^4}$$

$$\frac{13T}{a^3}$$

10. Thin-walled tube

$$\frac{TS}{4G\tilde{A}^2t}$$

S = circumference of the centerline of the tube (midwall perimeters)

$\tilde{A}$ = area enclosed by the midwall perimeters

$$\frac{T}{2\tilde{A}t}$$

and the boundary conditions are

$$w(-a, z) = w(a, z) = 0, \qquad -b \le z \le b \qquad \text{(E10.7.2)}$$

$$w(y, -b) = w(y, b) = 0, \qquad -a \le y \le a \qquad \text{(E10.7.3)}$$

The deflection shape, $w(y, z)$, that satisfies the boundary conditions at $y = \mp a$, Eq. (E10.7.2), can be expressed as

$$w(y, z) = \sum_{i=1,3,5,\ldots}^{\infty} a_i \cos \frac{i\pi y}{2a} Z_i(z) \qquad \text{(E10.7.4)}$$

where a_i is a constant and $Z_i(z)$ is a function of z to be determined. Substituting Eq. (E10.7.4) into Eq. (E10.7.1), we obtain

$$\sum_{i=1,3,5,\ldots}^{\infty} a_i \cos \frac{i\pi y}{2a} \left(-\frac{i^2\pi^2}{4a^2} Z_i + \frac{d^2 Z_i}{dz^2} \right) = -\frac{p}{P} \qquad \text{(E10.7.5)}$$

When the relation

$$\sum_{i=1,3,5,\ldots}^{\infty} \frac{4}{i\pi}(-1)^{(i-1)/2} \cos \frac{i\pi y}{2a} = 1 \qquad \text{(E10.7.6)}$$

is introduced on the right-hand side of Eq. (E10.7.5), the equation yields

$$\frac{d^2 Z_i}{dz^2} - \frac{i^2\pi^2}{4a^2} Z_i = -\frac{p}{P} \frac{4}{i\pi a_i}(-1)^{(i-1)/2} \qquad \text{(E10.7.7)}$$

The solution of this second-order differential equation can be expressed as

$$Z_i(z) = A_i \cosh \frac{i\pi z}{2a} + B_i \sinh \frac{i\pi z}{2a} + \frac{16pa^2}{Pi^2\pi^2 a_i}(-1)^{(i-1)/2} \qquad \text{(E10.7.8)}$$

where A_i and B_i are constants to be determined from the boundary conditions at $z = \mp b$. Using the condition $Z_i(z = -b) = 0$ in Eq. (E10.7.8) yields $B_i = 0$, and the condition $Z_i(z = b) = 0$ leads to

$$A_i = \frac{16pa^2}{Pi^2\pi^2 a_i} \frac{1}{\cosh(i\pi b/2a)} \qquad \text{(E10.7.9)}$$

Thus, the function $Z_i(z)$ and the deflection of the rectangular membrane can be expressed as

$$Z(z) = \frac{16pa^2}{Pi^2\pi^2 a_i}(-1)^{(i-1)/2}\left[1 - \frac{\cosh(i\pi z/2a)}{\cosh(i\pi b/2a)} \right] \qquad \text{(E10.7.10)}$$

$$w(y, z) = \frac{16pa^2}{P\pi^2} \sum_{i=1,3,5,\ldots}^{\infty} \frac{1}{i^2}(-1)^{(i-1)/2}\left[1 - \frac{\cosh(i\pi z/2a)}{\cosh(i\pi b/2a)} \right] \cos \frac{i\pi y}{2a}$$

$$\text{(E10.7.11)}$$

Equations (E10.7.11), (10.171), and (10.172) yield the stress function $\Phi(y, z)$ as

$$\Phi(y, z) = \frac{32G\beta a^2}{\pi^2} \sum_{i=1,3,5,\ldots}^{\infty} \frac{1}{i^2}(-1)^{(i-1)/2} \left[1 - \frac{\cosh(i\pi z/2a)}{\cosh(i\pi b/2a)}\right] \cos\frac{i\pi y}{2a} \quad (E10.7.12)$$

The torque on the rod, M_t, can be determined as [see Eq. (10.164)]

$$M_t = 2 \iint_A \Phi \, dA$$

$$= \frac{64G\beta a^2}{\pi^2} \int_{-a}^{a} \int_{-b}^{b} \sum_{i=1,3,5,\ldots}^{\infty} \frac{1}{i^2}(-1)^{(i-1)/2} \left[1 - \frac{\cosh(i\pi z/2a)}{\cosh(i\pi b/2a)}\right] \cos\frac{i\pi y}{2a} \, dy \, dz$$

$$= \frac{32G\beta(2a)^3(2b)}{\pi^4} \sum_{i=1,3,5,\ldots}^{\infty} \frac{1}{i^4} - \frac{64G\beta(2a)^4}{\pi^5} \sum_{i=1,3,5,\ldots}^{\infty} \frac{1}{i^5} \tanh\frac{i\pi b}{2a} \quad (E10.7.13)$$

Using the identity

$$\sum_{i=1,3,5,\ldots}^{\infty} \frac{1}{i^4} = \frac{\pi^4}{96} \quad (E10.7.14)$$

Eq. (E10.7.13) can be rewritten as

$$M_t = \frac{1}{3}G\beta(2a)^3(2b)\left(1 - \frac{192a}{\pi^5 b} \sum_{i=1,3,5,\ldots}^{\infty} \frac{1}{i^5} \tanh\frac{i\pi b}{2a}\right) \quad (E10.7.15)$$

The torsional rigidity of the rectangular cross section (C) can be found as

$$C = \frac{M_t}{\beta} = kG(2a)^3(2b) \quad (E10.7.16)$$

where

$$k = \frac{1}{3}\left(1 - \frac{192}{\pi^5}\frac{a}{b} \sum_{i=1,3,5,\ldots}^{\infty} \frac{1}{i^5} \tanh i\pi\frac{b}{a}\right) \quad (E10.7.17)$$

For any given rectangular cross section, the ratio b/a is known and hence the series in Eq. (E10.7.17) can be evaluated to find the value of k to any desired accuracy. The values of k for a range of b/a are given in Table 10.4.

10.11 RECENT CONTRIBUTIONS

The torsional vibration of tapered rods with rectangular cross section, pre-twisted uniform rods, and pre-twisted tapered rods is presented by Rao [4]. In addition, several refined theories of torsional vibration of rods are also presented in Ref. [4].

Table 10.4   Values of k in
Eq. (E10.7.16)

b/a	k	b/a	k
1.0	0.141	4.0	0.281
1.5	0.196	5.0	0.291
2.0	0.229	6.0	0.299
2.5	0.249	10.0	0.312
3.0	0.263	∞	0.333

Torsional Vibration of Bars   The torsional vibration of beams with a rectangular cross section is presented by Vet [8]. An overview of the vibration problems associated with turbomachinery is given by Vance [9]. The free vibration coupling of bending and torsion of a uniform spinning beam was studied by Filipich and Rosales [16]. The exact solution was presented and a numerical example was presented to point out the influence of whole coupling.

Torsional Vibration of Thin-Walled Beams   The torsional vibration of beams of thin-walled open section has been studied by Gere [11]. The behavior of torsion of bars with warping restraint is studied using Hamilton's principle by Lo and Goulard [12].

Vibration of a Cracked Rotor   The coupling between longitudinal, lateral, and torsional vibrations of a cracked rotor was studied by Darpe et al. [13]. In this work, the stiffness matrix of a Timoshenko beam element was modified to account for the effect of a crack and all six degrees of freedom per node were considered.

Torsional Vibration Control   The torsional vibration control of a shaft through active constrained layer damping treatments has been studied by Shen et al. [14]. The equation of motion of the arrangement, consisting of piezoelectric and viscoelastic layers, is derived and its stability and controllability are discussed.

Torsional Vibration of Machinery Drives   The startup torque in an electrical induction motor can create problems when the motor is connected to mechanical loads such as fans and pumps through shafts. The interrelationship between the electric motor and the mechanical system, which is effectively a multimass oscillatory system, has been examined by Ran et al. [15].

REFERENCES

1. W. F. Riley, L. D. Sturges, and D. H. Morris, *Mechanics of Materials*, 5th ed., Wiley, New York, 1999.
2. S. Timoshenko, D. H. Young, and W. Weaver, Jr., *Vibration Problems in Engineering*, 4th ed., Wiley, New York, 1974.
3. W. B. Bickford, *Advanced Mechanics of Materials*, Addison-Wesley, Reading, MA, 1998.
4. J. S. Rao, *Advanced Theory of Vibration*, Wiley, New York, 1992.

5. S. K. Clark, *Dynamics of Continuous Elements*, Prentice-Hall, Englewood Cliffs, NJ, 1972.

6. S. S. Rao, *Mechanical Vibrations*, 4th ed., Prentice Hall, Upper Saddle River, NJ, 2004.

7. J. H. Faupel and F. E. Fisher, *Engineering Design*, 2nd ed., Wiley-Interscience, New York, 1981.

8. M. Vet, Torsional vibration of beams having rectangular cross sections, *Journal of the Acoustical Society of America*, Vol. 34, p. 1570, 1962.

9. J. M. Vance, *Rotordynamics of Turbomachinery*, Wiley, New York, 1988.

10. A. E. H. Love, *A Treatise on the Mathematical Theory of Elasticity*, 4th Ed., Dover, New York, 1944.

11. J. M. Gere, Torsional vibrations of beams of thin walled open section, *Journal of Applied Mechanics*, Vol. 21, p. 381, 1954.

12. H. Lo and M. Goulard, Torsion with warping restraint from Hamilton's principle, *Proceedings of the 2nd Midwestern Conference on Solid Mechanics*, 1955, p. 68.

13. A. K. Darpe, K. Gupta, and A. Chawla, Coupled bending, longitudinal and torsional vibration of a cracked rotor, *Journal of Sound and Vibration*, Vol. 269, No. 1–2, pp. 33–60, 2004.

14. I. Y. Shen, W. Guo, and Y. C. Pao, Torsional vibration control of a shaft through active constrained layer damping treatments, *Journal of Vibration and Acoustics*, Vol. 119, No. 4, pp. 504–511, 1997.

15. L. Ran, R. Yacamini, and K. S. Smith, Torsional vibrations in electrical induction motor drives during start up, *Journal of Vibration and Acoustics*, Vol. 118, No. 2, pp. 242–251, 1996.

16. C. P. Filipich and M. B. Rosales, Free flexural–torsional vibrations of a uniform spinning beam, *Journal of Sound and Vibration*, Vol. 141, No. 3, pp. 375–387, 1990.

17. S. P. Timoshenko, Theory of bending, torsion and buckling of thin-walled member of open cross-section, *Journal of the Franklin Institute*, Vol. 239, pp. 201, 249, and 343, 1945.

PROBLEMS

10.1 A shaft with a uniform circular cross section of diameter d and length l carries a heavy disk of mass moment of inertia I_1 at the center. Find the first three natural frequencies and the corresponding modes of the shaft in torsional vibration. Assume that the shaft is fixed at both the ends.

10.2 A shaft with a uniform circular cross section of diameter d and length l carries a heavy disk of mass moment of inertia I_1 at the center. If both ends of the shaft are fixed, determine the free vibration response of the system when the disk is given an initial angular displacement of θ_0 and a zero initial angular velocity.

10.3 A uniform shaft supported at $x = 0$ and rotating at an angular velocity Ω is suddenly stopped at the end $x = 0$. If the end $x = l$ is free and the cross section of the shaft is tubular with inner and outer radii r_i and r_o,

respectively, find the subsequent time variation of the angular displacement of the shaft.

10.4 A uniform shaft of length l is fixed at $x = 0$ and free at $x = l$. Find the forced vibration response of the shaft if a torque $M_t(t) = M_{t0} \cos \Omega t$ is applied at the free end. Assume the initial conditions of the shaft to be zero.

10.5 Find the first three natural frequencies of torsional vibration of a shaft of length 1 m and diameter 20 mm for the following end conditions:

(a) Both ends are fixed.

(b) One end is fixed and the other end is free.

(c) Both ends are free.

Material of the shaft: steel with $\rho = 7800$ kg/m^3 and $G = 0.8 \times 10^{11}$ N/m^2.

10.6 Solve Problem 10.5 by assuming the material of the shaft to be aluminum with $\rho = 2700$ kg/m^3 and $G = 0.26 \times 10^{11}$ N/m^2.

10.7 Consider two shafts each of length l with thin-walled tubular sections, one in the form of a circle and the other in the form of a square, as shown in Fig. 10.13.

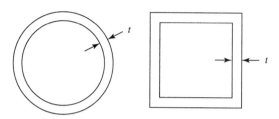

Figure 10.13

Assuming the same wall thickness of the tubes and the same total area of the region occupied by the material (material area), compare the fundamental natural frequencies of torsional vibration of the shafts.

10.8 Solve Problem 10.7 by assuming the tube wall thickness and the interior cavity areas of the tubes to be the same.

10.9 Determine the velocity of propagation of torsional waves in the drive shaft of an automobile for the following data:

(a) Cross section: circular with diameter 100 mm; material: steel with $\rho = 7800$ kg/m^3 and $G = 0.8 \times 10^{11}$ N/m^2.

(b) Cross section: hollow with inner diameter 80 mm and outer diameter 120 mm; material: aluminum with $\rho = 2700$ kg/m^3 and $G = 0.26 \times 10^{11}$ N/m^2.

10.10 Find the first three natural frequencies of torsional vibration of a shaft fixed at $x = 0$ and a disk of mass moment of inertia $I_1 = 20$ kg·m^2/rad attached at $x = l$. Shaft: uniform circular cross section of diameter 20 mm and length 1 m; material of shaft: steel with $\rho = 7800$ kg/m^3 and $G = 0.8 \times 10^{11}$ N/m^2.

10.11 Find the fundamental natural frequency of torsional vibration of the shaft described in Problem 10.10 using a single-degree-of-freedom model.

10.12 Find the free torsional vibration response of a uniform shaft of length l subjected to an initial angular displacement $\theta(x, 0) = \theta_0(x/l)$ and an initial angular velocity $\dot{\theta}(x, 0) = 0$ using modal analysis. Assume the shaft to be fixed at $x = 0$ and free at $x = l$.

10.13 Find the free torsional vibration response of a uniform shaft of length l subjected to an initial angular displacement $\theta(x, 0) = 0$ and an initial angular velocity $\dot{\theta}(x, 0) = V_0\delta(x - l)$ using modal analysis. Assume the shaft to be fixed at $x = 0$ and free at $x = l$.

10.14 Derive the frequency equation for the torsional vibration of a uniform shaft with a torsional spring of stiffness k_t attached to each end.

10.15 Find the steady-state response of a shaft fixed at both ends when subjected to a torque $M_t(x, t) = M_{t0} \sin \Omega t$ at $x = l/4$ using modal analysis.

11

Transverse Vibration of Beams

11.1 INTRODUCTION

The free and forced transverse vibration of beams is considered in this chapter. The equations of motion of a beam are derived according to the Euler–Bernoulli, Rayleigh, and Timoshenko theories. The Euler–Bernoulli theory neglects the effects of rotary inertia and shear deformation and is applicable to an analysis of thin beams. The Rayleigh theory considers the effect of rotary inertia, and the Timoshenko theory considers the effects of both rotary inertia and shear deformation. The Timoshenko theory can be used for thick beams. The equations of motion for the transverse vibration of beams are in the form of fourth-order partial differential equations with two boundary conditions at each end. The different possible boundary conditions of the beam can involve spatial derivatives up to third order. The responses of beams under moving loads, beams subjected to axial force, rotating beams, continuous beams, and beams on elastic foundation are considered using thin beam (Euler–Bernoulli) theory. The free vibration solution, including the determination of natural frequencies and mode shapes, is considered according to the three theories.

11.2 EQUATION OF MOTION: EULER–BERNOULLI THEORY

The governing equation of motion and boundary conditions of a thin beam have been derived by considering an element of the beam shown in Fig. 11.1(b), using Newton's second law of motion (equilibrium approach), in Section 3.5. The equation of motion was derived in Section 4.11.2 using the extended Hamilton's principle (variational approach). We now derive the equation of motion and boundary conditions corresponding to the transverse vibration of a thin beam connected to a mass, a linear spring, and a torsional spring at each end [Fig. 11.1(a), (b)] using the generalized Hamilton's principle. In the Euler–Bernoulli or thin beam theory, the rotation of cross sections of the beam is neglected compared to the translation. In addition, the angular distortion due to shear is considered negligible compared to the bending deformation. The thin beam theory is applicable to beams for which the length is much larger than the depth (at least 10 times) and the deflections are small compared to the depth. When the transverse displacement of the centerline of the beam is w, the displacement components of any point in the cross section, when plane sections remain plane and normal

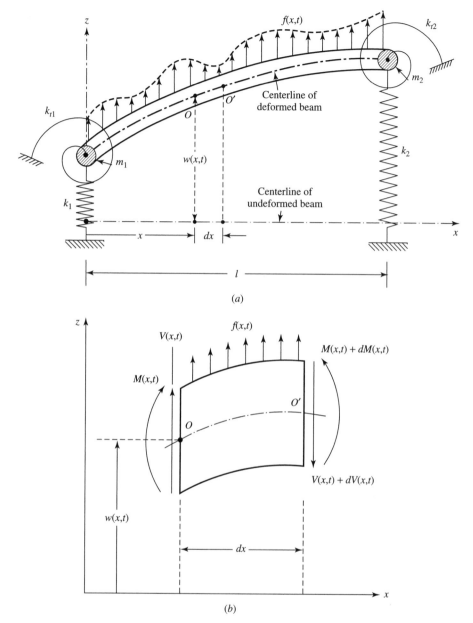

Figure 11.1(a), (b) Beam in bending.

to the centerline, are given by [Fig. 11.1(c)]

$$u = -z\frac{\partial w(x, t)}{\partial x}, \qquad v = 0, \quad w = w(x, t) \tag{11.1}$$

where u, v, and w denote the components of displacement parallel to x, y, and z directions, respectively. The components of strain and stress corresponding to this

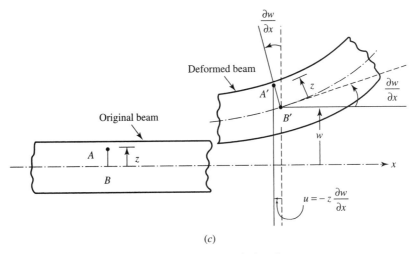

(c)

Figure 11.1(c)   Beam in bending.

displacement field are given by

$$\varepsilon_{xx} = \frac{\partial u}{\partial x} = -z\frac{\partial^2 w}{\partial x^2}, \qquad \varepsilon_{yy} = \varepsilon_{zz} = \varepsilon_{xy} = \varepsilon_{yz} = \varepsilon_{zx} = 0$$

$$\sigma_{xx} = -Ez\frac{\partial^2 w}{\partial x^2}, \qquad \sigma_{yy} = \sigma_{zz} = \sigma_{xy} = \sigma_{yz} = \sigma_{zx} = 0$$

(11.2)

The strain energy of the system (π) can be expressed as

$$\begin{aligned}
\pi &= \frac{1}{2}\iiint_V (\sigma_{xx}\varepsilon_{xx} + \sigma_{yy}\varepsilon_{yy} + \sigma_{zz}\varepsilon_{zz} + \sigma_{xy}\varepsilon_{xy} + \sigma_{yz}\varepsilon_{yz} + \sigma_{zx}\varepsilon_{zx})\,dV \\
&\quad + \left[\frac{1}{2}k_1 w^2(0,t) + \frac{1}{2}k_{t1}\left(\frac{\partial w}{\partial x}(0,t)\right)^2 + \frac{1}{2}k_2 w^2(l,t) + \frac{1}{2}k_{t2}\left(\frac{\partial w}{\partial x}(l,t)\right)^2\right] \\
&= \frac{1}{2}\int_0^l EI\left(\frac{\partial^2 w}{\partial x^2}\right)^2 dx \\
&\quad + \left[\frac{1}{2}k_1 w^2(0,t) + \frac{1}{2}k_{t1}\left(\frac{\partial w}{\partial x}(0,t)\right)^2 + \frac{1}{2}k_2 w^2(l,t) + \frac{1}{2}k_{t2}\left(\frac{\partial w}{\partial x}(l,t)\right)^2\right]
\end{aligned}$$

(11.3)

where the first term on the right-hand side of Eq. (11.3) denotes the strain energy of the beam, the second term, in brackets, indicates the strain energy of the springs, and I denotes the area moment of inertia of the cross section of the beam about the y axis:

$$I = I_y = \iint_A z^2\,dA$$

(11.4)

The kinetic energy of the system (T) is given by

$$T = \frac{1}{2} \int_0^l \int\int_A \rho \left(\frac{\partial w}{\partial t}\right)^2 dA\, dx + \left[\frac{1}{2} m_1 \left(\frac{\partial w}{\partial t}(0, t)\right)^2 + \frac{1}{2} m_2 \left(\frac{\partial w}{\partial t}(l, t)\right)^2\right]$$

$$= \frac{1}{2} \int_0^l \rho A \left(\frac{\partial w}{\partial t}\right)^2 dx + \left[\frac{1}{2} m_1 \left(\frac{\partial w}{\partial t}(0, t)\right)^2 + \frac{1}{2} m_2 \left(\frac{\partial w}{\partial t}(l, t)\right)^2\right] \quad (11.5)$$

where the first term on the right-hand side of Eq. (11.5) represents the kinetic energy of the beam, and the second term, in brackets, indicates the kinetic energy of the attached masses. The work done by the distributed transverse load $f(x, t)$ is given by

$$W = \int_0^l f w\, dx \quad (11.6)$$

The application of the generalized Hamilton's principle gives

$$\delta \int_{t_1}^{t_2} (\pi - T - W)\, dt = 0$$

or

$$\delta \int_{t_1}^{t_2} \left\{ \frac{1}{2} \int_0^l EI \left(\frac{\partial^2 w}{\partial x^2}\right)^2 dx \right.$$

$$+ \left[\frac{1}{2} k_1 w^2(0, t) + \frac{1}{2} k_{t1} \left(\frac{\partial w}{\partial x}(0, t)\right)^2 + \frac{1}{2} k_2 w^2(l, t) + \frac{1}{2} k_{t2} \left(\frac{\partial w}{\partial x}(l, t)\right)^2\right]$$

$$- \frac{1}{2} \int_0^l \rho A \left(\frac{\partial w}{\partial t}\right)^2 dx - \left[\frac{1}{2} m_1 \left(\frac{\partial w}{\partial t}(0, t)\right)^2\right.$$

$$+ \left. \frac{1}{2} m_2 \left(\frac{\partial w}{\partial t}(l, t)\right)^2\right] - \int_0^l f w\, dx \left. \right\}\, dt = 0 \quad (11.7)$$

The variations in Eq. (11.7) can be evaluated using integration by parts, similar to those used in Eqs. (4.181), (4.184) and (4.185), to obtain

$$\delta \int_{t_1}^{t_2} \frac{1}{2} \int_0^l EI \left(\frac{\partial^2 w}{\partial x^2}\right)^2 dx\, dt = \int_{t_1}^{t_2} \left[EI \frac{\partial^2 w}{\partial x^2} \delta \left(\frac{\partial w}{\partial x}\right) \Big|_0^l - \frac{\partial}{\partial x} \left(EI \frac{\partial^2 w}{\partial x^2}\right) \delta w \Big|_0^l \right.$$

$$+ \left. \int_0^l \frac{\partial^2}{\partial x^2} \left(EI \frac{\partial^2 w}{\partial x^2}\right) \delta w\, dx \right] dt$$

$$\delta \int_{t_1}^{t_2} \left[\frac{1}{2} k_1 w^2(0, t) + \frac{1}{2} k_{t1} \left(\frac{\partial w}{\partial x}(0, t)\right)^2 + \frac{1}{2} k_2 w^2(l, t) + \frac{1}{2} k_{t2} \left(\frac{\partial w}{\partial x}(l, t)\right)^2\right] dt$$

$$= \int_{t_1}^{t_2} \left[k_1 w(0, t) \delta w(0, t) + k_{t1} \frac{\partial w(0, t)}{\partial x} \delta \left(\frac{\partial w(0, t)}{\partial x}\right) \right.$$

$$+ \left. k_2 w(l, t) \delta w(l, t) + k_{t2} \frac{\partial w(l, t)}{\partial x} \delta \left(\frac{\partial w(l, t)}{\partial x}\right) \right] dt \quad (11.8)$$

$$\delta \int_{t_1}^{t_2} \left[\frac{1}{2} \int_0^l \rho A \left(\frac{\partial w}{\partial t} \right)^2 dx \right] dt$$

$$= \int_0^l \left(\rho A \frac{\partial w}{\partial t} \delta w \Big|_{t_1}^{t_2} \right) dx - \int_0^l \left(\int_{t_1}^{t_2} \rho A \frac{\partial^2 w}{\partial t^2} \delta w\, dt \right) dx$$

$$= -\int_{t_1}^{t_2} \left(\int_0^l \rho A \frac{\partial^2 w}{\partial t^2} \delta w\, dx \right) dt$$

$$\delta \int_{t_1}^{t_2} \left[\frac{1}{2} m_1 \left(\frac{\partial w}{\partial t}(0,t) \right)^2 + \frac{1}{2} m_2 \left(\frac{\partial w}{\partial t}(l,t) \right)^2 \right] dt$$

$$= -\int_{t_1}^{t_2} \left[m_1 \frac{\partial^2 w(0,t)}{\partial t^2} \delta w(0,t) + m_2 \frac{\partial^2 w(l,t)}{\partial t^2} \delta w(l,t) \right] dt \qquad (11.9)$$

Note that integration by parts with respect to time, along with the fact that $\delta w = 0$ at $t = t_1$ and $t = t_2$, is used in deriving Eqs. (11.9).

$$\delta \int_{t_1}^{t_2} \left(\int_0^l f w\, dx \right) dt = \int_{t_1}^{t_2} \int_0^l f \delta w\, dx\, dt \qquad (11.10)$$

In view of Eqs. (11.8)–(11.10), Eq. (11.7) becomes

$$\int_{t_1}^{t_2} \left\{ EI \frac{\partial^2 w}{\partial x^2} \delta \left(\frac{\partial w}{\partial x} \right) \Big|_0^l - \frac{\partial}{\partial x} \left(EI \frac{\partial^2 w}{\partial x^2} \right) \delta w \Big|_0^l + k_1 w \delta w \Big|_0 + k_{t1} \frac{\partial w}{\partial x} \delta \left(\frac{\partial w}{\partial x} \right) \Big|_0 \right.$$

$$\left. + m_1 \frac{\partial^2 w}{\partial t^2} \delta w \Big|_0 + k_2 w \delta w |^l + k_{t2} \frac{\partial w}{\partial x} \delta \left(\frac{\partial w}{\partial x} \right) \Big|^l + m_2 \frac{\partial^2 w}{\partial t^2} \delta w \Big|^l \right\} dt$$

$$+ \int_{t_1}^{t_2} \left\{ \int_0^l \left[\frac{\partial^2}{\partial x^2} \left(EI \frac{\partial^2 w}{\partial x^2} \right) + \rho A \frac{\partial^2 w}{\partial t^2} - f \right] \delta w\, dx \right\} dt = 0 \qquad (11.11)$$

By setting the two expressions within the braces in each term of Eq. (11.11) equal to zero, we obtain the differential equation of motion for the transverse vibration of the beam as

$$\frac{\partial^2}{\partial x^2} \left(EI \frac{\partial^2 w}{\partial x^2} \right) + \rho A \frac{\partial^2 w}{\partial t^2} = f(x,t) \qquad (11.12)$$

and the boundary conditions as

$$EI \frac{\partial^2 w}{\partial x^2} \delta \left(\frac{\partial w}{\partial x} \right) \Big|_0^l - \frac{\partial}{\partial x} \left(EI \frac{\partial^2 w}{\partial x^2} \right) \delta w \Big|_0^l + k_1 w \delta w |_0 + k_{t1} \frac{\partial w}{\partial x} \delta \left(\frac{\partial w}{\partial x} \right) \Big|_0$$

$$+ m_1 \frac{\partial^2 w}{\partial t^2} \delta w |_0 + k_2 w \delta w |^l + k_{t2} \frac{\partial w}{\partial x} \delta \left(\frac{\partial w}{\partial x} \right) |^l + m_2 \frac{\partial^2 w}{\partial t^2} \delta w |^l = 0 \qquad (11.13)$$

To satisfy Eq. (11.13), the following conditions are to be satisfied:

$$\left(-EI\frac{\partial^2 w}{\partial x^2} + k_{t1}\frac{\partial w}{\partial x}\right)\delta\left(\frac{\partial w}{\partial x}\right)\bigg|_{x=0} = 0 \tag{11.14}$$

$$\left(EI\frac{\partial^2 w}{\partial x^2} + k_{t2}\frac{\partial w}{\partial x}\right)\delta\left(\frac{\partial w}{\partial x}\right)\bigg|_{x=l} = 0 \tag{11.15}$$

$$\left[\frac{\partial}{\partial x}\left(EI\frac{\partial^2 w}{\partial x^2}\right) + k_1 w + m_1\frac{\partial^2 w}{\partial t^2}\right]\delta w\bigg|_{x=0} = 0 \tag{11.16}$$

$$\left[-\frac{\partial}{\partial x}\left(EI\frac{\partial^2 w}{\partial x^2}\right) + k_2 w + m_2\frac{\partial^2 w}{\partial t^2}\right]\delta w\bigg|_{x=l} = 0 \tag{11.17}$$

Each of the equations in (11.14)–(11.17) can be satisfied in two ways but will be satisfied in only one way for any specific support conditions of the beam. The boundary conditions implied by Eqs. (11.14)–(11.17) are as follows, At $x = 0$,

$$\frac{\partial w}{\partial x} = \text{constant}\left[\text{so that } \delta\left(\frac{\partial w}{\partial x}\right) = 0\right] \quad \text{or} \quad \left(-EI\frac{\partial^2 w}{\partial x^2} + k_{t1}\frac{\partial w}{\partial x}\right) = 0 \tag{11.18}$$

and

$$w = \text{constant(so that } \delta w = 0) \quad \text{or} \quad \left[\frac{\partial}{\partial x}\left(EI\frac{\partial^2 w}{\partial x^2}\right) + k_1 w + m_1\frac{\partial^2 w}{\partial t^2}\right] = 0 \tag{11.19}$$

At $x = l$,

$$\frac{\partial w}{\partial x} = \text{constant}\left[\text{so that } \delta\left(\frac{\partial w}{\partial x}\right) = 0\right] \quad \text{or} \quad \left[EI\frac{\partial^2 w}{\partial x^2} + k_{t2}\frac{\partial w}{\partial x}\right] = 0 \tag{11.20}$$

and

$$w = \text{constant(so that } \delta w = 0) \quad \text{or} \quad \left[-\frac{\partial}{\partial x}\left(EI\frac{\partial^2 w}{\partial x^2}\right) + k_2 w + m_2\frac{\partial^2 w}{\partial t^2}\right] = 0 \tag{11.21}$$

In the present case, the second conditions stated in each of Eqs. (11.18)–(11.21)are valid.

Note   The boundary conditions of a beam corresponding to different types of end supports are given in Table 11.1. The boundary conditions for supports, other than the supports shown in the table, can be obtained from Hamilton's principle by including the appropriate energy terms in the formulation of equations as illustrated for the case of attached masses and springs in this section.

11.3   FREE VIBRATION EQUATIONS

For free vibration, the external excitation is assumed to be zero:

$$f(x, t) = 0 \tag{11.22}$$

Table 11.1 Boundary Conditions of a Beams[†]

Boundary condition	At left end ($x = 0$)	At right end ($x = l$)
1. Free end (bending moment = 0, shear force = 0)	$x = 0$ $$EI\frac{\partial^2 w}{\partial x^2}(0,t) = 0$$ $$\frac{\partial}{\partial x}\left(EI\frac{\partial^2 w}{\partial x^2}\right)\Big|_{(0,t)} = 0$$	$x = l$ $$EI\frac{\partial^2 w}{\partial x^2}(l,t) = 0$$ $$\frac{\partial}{\partial x}\left(EI\frac{\partial^2 w}{\partial x^2}\right)\Big|_{(l,t)} = 0$$
2. Fixed end (deflection = 0, slope = 0)	$x = 0$ $$w(0,t) = 0$$ $$\frac{\partial w}{\partial x}(0,t) = 0$$	$x = l$ $$w(l,t) = 0$$ $$\frac{\partial w}{\partial x}(l,t) = 0$$
3. Simply supported end (deflection = 0, bending moment = 0)	$x = 0$ $$w(0,t) = 0$$ $$EI\frac{\partial^2 w}{\partial x^2}(0,t) = 0$$	$x = l$ $$w(l,t) = 0$$ $$EI\frac{\partial^2 w}{\partial x^2}(l,t) = 0$$
4. Sliding end (slope = 0, shear force = 0)	$x = 0$ $$\frac{\partial w}{\partial x}(0,t) = 0$$ $$\frac{\partial}{\partial x}\left(EI\frac{\partial^2 w}{\partial x^2}\right)\Big|_{(0,t)} = 0$$	$x = l$ $$\frac{\partial w}{\partial x}(l,t) = 0$$ $$\frac{\partial}{\partial x}\left(EI\frac{\partial^2 w}{\partial x^2}\right)\Big|_{(l,t)} = 0$$
5. End spring (spring constant $= \underline{k}$)	$x = 0$ $$\frac{\partial}{\partial x}\left(EI\frac{\partial^2 w}{\partial x^2}\right)\Big|_{(0,t)} = -\underline{k}w(0,t)$$ $$EI\frac{\partial^2 w}{\partial x^2}(0,t) = 0$$	$x = l$ $$\frac{\partial}{\partial x}\left(EI\frac{\partial^2 w}{\partial x^2}\right)\Big|_{(l,t)} = \underline{k}w(l,t)$$ $$EI\frac{\partial^2 w}{\partial x^2}(l,t) = 0$$

(continued overleaf)

Table 11.1 *(continued)*

Boundary condition	At left end ($x = 0$)	At right end ($x = l$)
6. End damper (damping constant $= \underset{\sim}{c}$)	$x = 0$ $$\frac{\partial}{\partial x}\left(EI\frac{\partial^2 w}{\partial x^2}\right)\bigg|_{(0,t)} = -\underset{\sim}{c}\frac{\partial w}{\partial t}(0,t)$$ $$EI\frac{\partial^2 w}{\partial x^2}(0,t)=0$$	$x = l$ $$\frac{\partial}{\partial x}\left(EI\frac{\partial^2 w}{\partial x^2}\right)\bigg|_{(l,t)} = \underset{\sim}{c}\frac{\partial w}{\partial t}(l,t)$$ $$EI\frac{\partial^2 w}{\partial x^2}(l,t)=0$$
7. End mass (mass $= \underset{\sim}{m}$ with negligible moment of inertia)	$x = 0$ $$\frac{\partial}{\partial x}\left(EI\frac{\partial^2 w}{\partial x^2}\right)\bigg|_{(0,t)} = -\underset{\sim}{m}\frac{\partial^2 w}{\partial t^2}(0,t)$$ $$EI\frac{\partial^2 w}{\partial x^2}(0,t)=0$$	$x = l$ $$\frac{\partial}{\partial x}\left(EI\frac{\partial^2 w}{\partial x^2}\right)\bigg|_{(l,t)} = \underset{\sim}{m}\frac{\partial^2 w}{\partial t^2}(l,t)$$ $$EI\frac{\partial^2 w}{\partial x^2}(l,t)=0$$
8. End mass with moment of inertia (mass $= \underset{\sim}{m}$, moment of inertia $= \underset{\sim}{J}_0$)	$x = 0$ $$EI\frac{\partial^2 w}{\partial x^2}(0,t) = -\underset{\sim}{J}_0\frac{\partial^3 w}{\partial x\,\partial t^2}(0,t)$$ $$\frac{\partial}{\partial x}\left(EI\frac{\partial^2 w}{\partial x^2}\right)\bigg|_{(0,t)} = -\underset{\sim}{m}\frac{\partial^2 w}{\partial t^2}(0,t)$$	$x = l$ $$EI\frac{\partial^2 w}{\partial x^2}(l,t) = \underset{\sim}{J}_0\frac{\partial^3 w}{\partial x\,\partial t^2}(l,t)$$ $$\frac{\partial}{\partial x}\left(EI\frac{\partial^2 w}{\partial x^2}\right)\bigg|_{(l,t)} = \underset{\sim}{m}\frac{\partial^2 w}{\partial t^2}(l,t)$$

†Source: Ref. [23].

and hence the equation of motion, Eq. (11.13), reduces to

$$\frac{\partial^2}{\partial x^2}\left[EI(x)\frac{\partial^2 w(x,t)}{\partial x^2}\right] + \rho A(x)\frac{\partial^2 w(x,t)}{\partial t^2} = 0 \qquad (11.23)$$

For a uniform beam, Eq. (11.23) can be expressed as

$$c^2\frac{\partial^4 w}{\partial x^4}(x,t) + \frac{\partial^2 w}{\partial t^2}(x,t) = 0 \qquad (11.24)$$

where

$$c = \sqrt{\frac{EI}{\rho A}} \qquad (11.25)$$

11.4   FREE VIBRATION SOLUTION

The free vibration solution can be found using the method of separation of variables as

$$w(x,t) = W(x)T(t) \qquad (11.26)$$

Using Eq. (11.26) in Eq. (11.24) and rearranging yields

$$\frac{c^2}{W(x)}\frac{d^4 W(x)}{dx^4} = -\frac{1}{T(t)}\frac{d^2 T(t)}{dt^2} = a = \omega^2 \qquad (11.27)$$

where $a = \omega^2$ can be shown to be a positive constant (see Problem 11.20). Equation (11.27) can be rewritten as two equations:

$$\frac{d^4 W(x)}{dx^4} - \beta^4 W(x) = 0 \qquad (11.28)$$

$$\frac{d^2 T(t)}{dt^2} + \omega^2 T(t) = 0 \qquad (11.29)$$

where

$$\beta^4 = \frac{\omega^2}{c^2} = \frac{\rho A\omega^2}{EI} \qquad (11.30)$$

The solution of Eq. (11.29) is given by

$$T(t) = A\cos\omega t + B\sin\omega t \qquad (11.31)$$

where A and B are constants that can be found from the initial conditions. The solution of Eq. (11.28) is assumed to be of exponential form as

$$W(x) = Ce^{sx} \qquad (11.32)$$

where C and s are constants. Substitution of Eq. (11.32) into Eq. (11.28) results in the auxiliary equation

$$s^4 - \beta^4 = 0 \qquad (11.33)$$

The roots of this equation are given by

$$s_{1,2} = \pm\beta, \qquad s_{3,4} = \pm i\beta \qquad (11.34)$$

Thus, the solution of Eq. (13.28) can be expressed as

$$W(x) = C_1 e^{\beta x} + C_2 e^{-\beta x} + C_3 e^{i\beta x} + C_4 e^{-i\beta x} \tag{11.35}$$

where C_1, C_2, C_3, and C_4 are constants. Equation (11.35) can be expressed more conveniently as

$$W(x) = C_1 \cos \beta x + C_2 \sin \beta x + C_3 \cosh \beta x + C_4 \sinh \beta x \tag{11.36}$$

or

$$W(x) = C_1(\cos \beta x + \cosh \beta x) + C_2(\cos \beta x - \cosh \beta x)$$
$$+ C_3(\sin \beta x + \sinh \beta x) + C_4(\sin \beta x - \sinh \beta x) \tag{11.37}$$

where C_1, C_2, C_3, and C_4, are different constants in each case. The natural frequencies of the beam can be determined from Eq. (11.30) as

$$\omega = \beta^2 \sqrt{\frac{EI}{\rho A}} = (\beta l)^2 \sqrt{\frac{EI}{\rho A l^4}} \tag{11.38}$$

The function $W(x)$ is known as the *normal mode* or *characteristic function* of the beam and ω is called the *natural frequency* of vibration. For any beam, there will be an infinite number of normal modes with one natural frequency associated with each normal mode. The unknown constants C_1 to C_4 in Eq. (11.36) or (11.37) and the value of β in Eq. (11.38) can be determined from the known boundary conditions of the beam.

If the ith natural frequency is denoted as ω_i and the corresponding normal mode as $W_i(x)$, the total free vibration response of the beam can be found by superposing the normal modes as

$$w(x, t) = \sum_{i=1}^{\infty} W_i(x)(A_i \cos \omega_i t + B_i \sin \omega_i t) \tag{11.39}$$

where the constants A_i and B_i can be determined from the initial conditions of the beam.

11.5 FREQUENCIES AND MODE SHAPES OF UNIFORM BEAMS

The natural frequencies and mode shapes of beams with a uniform cross section with different boundary conditions are considered in this section.

11.5.1 Beam Simply Supported at Both Ends

The transverse displacement and the bending moment are zero at a simply supported (or pinned or hinged) end. Hence, the boundary conditions can be stated as

$$W(0) = 0 \tag{11.40}$$

$$EI \frac{d^2 W}{dx^2}(0) = 0 \quad \text{or} \quad \frac{d^2 W}{dx^2}(0) = 0 \tag{11.41}$$

$$W(l) = 0 \tag{11.42}$$

$$EI\frac{d^2W}{dx^2}(l) = 0 \quad \text{or} \quad \frac{d^2W}{dx^2}(l) = 0 \tag{11.43}$$

When used in the solution of Eq. (11.37), Eqs. (11.40) and (11.41) yield

$$C_1 = C_2 = 0 \tag{11.44}$$

When used with Eq. (11.37), Eqs. (11.42) and (11.43) result in

$$C_3(\sin \beta l + \sinh \beta l) + C_4(\sin \beta l - \sinh \beta l) = 0 \tag{11.45}$$

$$-C_3(\sin \beta l - \sinh \beta l) - C_4(\sin \beta l + \sinh \beta l) = 0 \tag{11.46}$$

Equations (11.45) and (11.46) denote a system of two equations in the two unknowns C_3 and C_4. For a nontrivial solution of C_3 and C_4, the determinant of the coefficients must be equal to zero. This leads to

$$-(\sin \beta l + \sinh \beta l)^2 + (\sin \beta l - \sinh \beta l)^2 = 0$$

or

$$\sin \beta l \sinh \beta l = 0 \tag{11.47}$$

It can be observed that $\sinh \beta l$ is not equal to zero unless $\beta = 0$. The value of $\beta = 0$ need not be considered because it implies, according to Eq. (11.38), $\omega = 0$, which corresponds to the beam at rest. Thus, the frequency equation becomes

$$\sin \beta l = 0 \tag{11.48}$$

The roots of Eq. (11.48), $\beta_n l$, are given by

$$\beta_n l = n\pi, \qquad n = 1, 2, \ldots \tag{11.49}$$

and hence the natural frequencies of vibration become

$$\omega_n = (\beta_n l)^2 \left(\frac{EI}{\rho A l^4}\right)^{1/2} = n^2 \pi^2 \left(\frac{EI}{\rho A l^4}\right)^{1/2}, \qquad n = 1, 2, \ldots \tag{11.50}$$

Substituting Eq. (11.48) into Eq. (11.45), we find that $C_3 = C_4$. Hence, Eq. (11.37) gives the mode shape as

$$W_n(x) = C_n \sin \beta_n x = C_n \sin \frac{n\pi x}{l}, \qquad n = 1, 2, \ldots \tag{11.51}$$

The first four natural frequencies of vibration and the corresponding mode shapes are shown in Fig. 11.2. The normal modes of vibration are given by

$$w_n(x, t) = W_n(x)(A_n \cos \omega_n t + B_n \sin \omega_n t), \qquad n = 1, 2, \ldots \tag{11.52}$$

The total (free vibration) solution can be expressed as

$$w(x, t) = \sum_{n=1}^{\infty} w_n(x, t) = \sum_{n=1}^{\infty} \sin \frac{n\pi x}{l}(A_n \cos \omega_n t + B_n \sin \omega_n t) \tag{11.53}$$

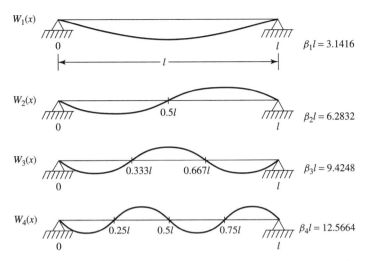

Figure 11.2 Natural frequencies and mode shapes of a beam simply supported at both ends. $\omega_n = (\beta_n l)^2 (EI/\rho A L^4)^{1/2}$, $\beta_n l = n\pi$.

If the initial conditions are given by

$$w(x, 0) = w_0(x) \tag{11.54}$$

$$\frac{\partial w}{\partial t}(x, 0) = \dot{w}_0(x) \tag{11.55}$$

Eqs. (11.53)–(11.55) lead to

$$\sum_{n=1}^{\infty} A_n \sin \frac{n\pi x}{l} = w_0(x) \tag{11.56}$$

$$\sum_{n=1}^{\infty} \omega_n B_n \sin \frac{n\pi x}{l} = \dot{w}_0(x) \tag{11.57}$$

Multiplying Eqs. (11.56) and (11.57) by $\sin(m\pi x/l)$ and integrating from 0 to l yields

$$A_n = \frac{2}{l} \int_0^l w_0(x) \sin \frac{n\pi x}{l} \, dx \tag{11.58}$$

$$B_n = \frac{2}{\omega_n l} \int_0^l \dot{w}_0(x) \sin \frac{n\pi x}{l} \, dx \tag{11.59}$$

11.5.2 Beam Fixed at Both Ends

At a fixed end, the transverse displacement and the slope of the displacement are zero. Hence, the boundary conditions are given by

$$W(0) = 0 \tag{11.60}$$

$$\frac{dW}{dx}(0) = 0 \tag{11.61}$$

$$W(l) = 0 \tag{11.62}$$

$$\frac{dW}{dx}(l) = 0 \tag{11.63}$$

When Eq. (11.37) is used, the boundary conditions (11.60) and (11.61) lead to

$$C_1 = C_3 = 0 \tag{11.64}$$

and the boundary conditions (11.62) and (11.63) yield

$$C_2(\cos \beta l - \cosh \beta l) + C_4(\sin \beta l - \sinh \beta l) = 0 \tag{11.65}$$

$$-C_2(\sin \beta l + \sinh \beta l) + C_4(\cos \beta l - \cosh \beta l) = 0 \tag{11.66}$$

Equations (11.65) and (11.66) denote a system of two homogeneous algebraic equations with C_2 and C_4 as unknowns. For a nontrivial solution of C_2 and C_4, we set the determinant of the coefficients of C_2 and C_4 in Eqs. (11.65) and (11.66) to zero to obtain

$$\begin{vmatrix} \cos \beta l - \cosh \beta l & \sin \beta l - \sinh \beta l \\ -(\sin \beta l + \sinh \beta l) & \cos \beta l - \cosh \beta l \end{vmatrix} = 0$$

or

$$(\cos \beta l - \cosh \beta l)^2 + (\sin^2 \beta l - \sinh^2 \beta l) = 0 \tag{11.67}$$

Equation (11.67) can be simplified to obtain the frequency equation as

$$\cos \beta l \cosh \beta l - 1 = 0 \tag{11.68}$$

Equation (11.65) gives

$$C_4 = -\frac{\cos \beta l - \cosh \beta l}{\sin \beta l - \sinh \beta l} C_2 \tag{11.69}$$

If $\beta_n l$ denotes the nth root of the transcendental equation (11.68), the corresponding mode shape can be obtained by substituting Eqs. (11.64) and (11.69) in Eq. (11.37) as

$$W_n(x) = C_n \left[(\cos \beta_n x - \cosh \beta_n x) - \frac{\cos \beta_n l - \cosh \beta_n l}{\sin \beta_n l - \sinh \beta_n l} (\sin \beta_n x - \sinh \beta_n x) \right] \tag{11.70}$$

The first four natural frequencies and the corresponding mode shapes are shown in Fig. 11.3. The nth normal mode of vibration can be expressed as

$$w_n(x, t) = W_n(x)(A_n \cos \omega_n t + B_n \sin \omega_n t) \tag{11.71}$$

and the free vibration solution as

$$w(x, t) = \sum_{n=1}^{\infty} w_n(x, t)$$

$$= \sum_{n=1}^{\infty} \left[(\cos \beta_n x - \cosh \beta_n x) - \frac{\cos \beta_n l - \cosh \beta_n l}{\sin \beta_n l - \sinh \beta_n l} (\sin \beta_n x - \sinh \beta_n x) \right]$$

$$\cdot (A_n \cos \omega_n t + B_n \sin \omega_n t) \tag{11.72}$$

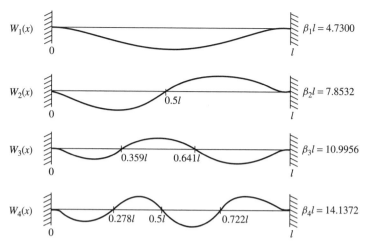

Figure 11.3 Natural frequencies and mode shapes of a beam fixed at both ends. $\omega_n = (\beta_n l)^2 (EI/\rho AL^4)^{1/2}$, $\beta_n l \simeq (2n+1)\pi/2$.

where the constants A_n and B_n in Eq. (11.72) can be determined from the known initial conditions as in the case of a beam with simply supported ends.

11.5.3   Beam Free at Both Ends

At a free end, the bending moment and shear force are zero. Hence, the boundary conditions of the beam can be stated as

$$EI\frac{d^2W(0)}{dx^2} = 0 \quad \text{or} \quad \frac{d^2W(0)}{dx^2} = 0 \tag{11.73}$$

$$EI\frac{d^3W(0)}{dx^3} = 0 \quad \text{or} \quad \frac{d^3W(0)}{dx^3} = 0 \tag{11.74}$$

$$EI\frac{d^2W(l)}{dx^2} = 0 \quad \text{or} \quad \frac{d^2W(l)}{dx^2} = 0 \tag{11.75}$$

$$EI\frac{d^3W(l)}{dx^3} = 0 \quad \text{or} \quad \frac{d^3W(l)}{dx^3} = 0 \tag{11.76}$$

By differentiating Eq. (11.37), we obtain

$$\frac{d^2W}{dx^2}(x) = \beta^2[C_1(-\cos\beta x + \cosh\beta x) + C_2(-\cos\beta x - \cosh\beta x)$$
$$+ C_3(-\sin\beta x + \sinh\beta x) + C_4(-\sin\beta x - \sinh\beta x)] \tag{11.77}$$

$$\frac{d^3W(x)}{dx^3} = \beta^3[C_1(\sin\beta x + \sinh\beta x) + C_2(\sin\beta x - \sinh\beta x)$$
$$+ C_3(-\cos\beta x + \cosh\beta x) + C_4(-\cos\beta x - \cosh\beta x)] \tag{11.78}$$

Equations (11.73) and (11.74) require that

$$C_2 = C_4 = 0 \qquad (11.79)$$

in Eq. (11.37), and Eqs. (11.75) and (11.76) lead to

$$C_1(-\cos \beta l + \cosh \beta l) + C_3(-\sin \beta l + \sinh \beta l) = 0 \qquad (11.80)$$

$$C_1(\sin \beta l + \sinh \beta l) + C_3(-\cos \beta l + \cosh \beta l) = 0 \qquad (11.81)$$

For a nontrivial solution of the constants C_1 and C_3 in Eqs. (11.80) and (11.81), the determinant formed by their coefficients is set equal to zero:

$$\begin{vmatrix} -\cos \beta l + \cosh \beta l & -\sin \beta l + \sinh \beta l \\ \sin \beta l + \sinh \beta l & -\cos \beta l + \cosh \beta l \end{vmatrix} = 0 \qquad (11.82)$$

Equation (11.82) can be simplified to obtain the frequency equation as

$$\cos \beta l \cosh \beta l - 1 = 0 \qquad (11.83)$$

Note that Eq. (11.83) is the same as Eq. (11.68) obtained for a beam fixed at both ends. The main difference is that a value of $\beta_0 l = 0$ leads to a rigid-body mode in the case of a free–free beam. By proceeding as in the preceding case, the nth mode shape of the beam can be expressed as

$$W_n(x) = (\cos \beta_n x + \cosh \beta_n x) - \frac{\cos \beta_n l - \cosh \beta_n l}{\sin \beta_n l - \sinh \beta_n l}(\sin \beta_n x + \sinh \beta_n x) \qquad (11.84)$$

The first five natural frequencies given by Eq. (11.83) and the corresponding mode shapes given by Eqs. (11.80) and (11.81) are shown in Fig. 11.4. The nth normal mode and the free vibration solution are given by

$$w_n(x, t) = W_n(x)(A_n \cos \omega_n t + B_n \sin \omega_n t) \qquad (11.85)$$

and

$$w(x, t) = \sum_{n=1}^{\infty} w_n(x, t)$$

$$= \sum_{n=1}^{\infty} \left[(\cos \beta_n x + \cosh \beta_n x) - \frac{\cos \beta_n l - \cosh \beta_n l}{\sin \beta_n l - \sinh \beta_n l}(\sin \beta_n x + \sinh \beta_n x) \right]$$

$$\cdot (A_n \cos \omega_n t + B_n \sin \omega_n t) \qquad (11.86)$$

11.5.4   Beam with One End Fixed and the Other Simply Supported

At a fixed end, the transverse deflection and slope of deflection are zero, and at a simply supported end, the transverse deflection and bending moment are zero. If the

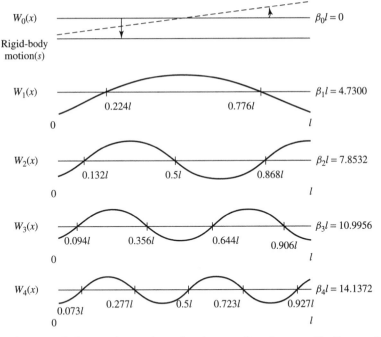

Figure 11.4 Natural frequencies and mode shapes of a beam with free ends. $\omega_n = (\beta_n l)^2 (EI/\rho AL^4)^{1/2}$, $\beta_n l \simeq (2n+1)\pi/2$.

beam is fixed at $x = 0$ and simply supported at $x = l$, the boundary conditions can be stated as

$$W(0) = 0 \tag{11.87}$$

$$\frac{dW}{dx}(0) = 0 \tag{11.88}$$

$$W(l) = 0 \tag{11.89}$$

$$EI\frac{d^2W}{dx^2}(l) = 0 \quad \text{or} \quad \frac{d^2W}{dx^2}(l) = 0 \tag{11.90}$$

Condition (11.87) leads to

$$C_1 + C_3 = 0 \tag{11.91}$$

in Eq.(11.36), and Eqs. (11.88) and (11.36) give

$$\frac{dW}{dx}\bigg|_{x=0} = \beta(-C_1 \sin \beta x + C_2 \cos \beta x + C_3 \sinh \beta x + C_4 \cosh \beta x)_{x=0} = 0$$

or

$$\beta(C_2 + C_4) = 0 \tag{11.92}$$

Thus, the solution, Eq. (11.36), becomes

$$W(x) = C_1(\cos \beta x - \cosh \beta x) + C_2(\sin \beta x - \sinh \beta x) \tag{11.93}$$

Applying conditions (11.89) and (11.90) to Eq. (11.93) yields

$$C_1(\cos \beta l - \cosh \beta l) + C_2(\sin \beta l - \sinh \beta l) = 0 \tag{11.94}$$

$$-C_1(\cos \beta l + \cosh \beta l) - C_2(\sin \beta l + \sinh \beta l) = 0 \tag{11.95}$$

For a nontrivial solution of C_1 and C_2 in Eqs. (11.94) and (11.95), the determinant of their coefficients must be zero:

$$\begin{vmatrix} \cos \beta l - \cosh \beta l & \sin \beta l - \sinh \beta l \\ -(\cos \beta l + \cosh \beta l) & -(\sin \beta l + \sinh \beta l) \end{vmatrix} = 0 \tag{11.96}$$

Expanding the determinant gives the frequency equation

$$\cos \beta l \sinh \beta l - \sin \beta l \cosh \beta l = 0$$

or

$$\tan \beta l = \tanh \beta l \tag{11.97}$$

The roots of this equation, $\beta_n l$, give the natural frequencies of vibration:

$$\omega_n = (\beta_n l)^2 \left(\frac{EI}{\rho A l^4} \right)^{1/2}, \qquad n = 1, 2, \ldots \tag{11.98}$$

If the value of C_2 corresponding to β_n is denoted as C_{2n}, it can be expressed in terms of C_{1n} from Eq. (11.94):

$$C_{2n} = -C_{1n} \frac{\cos \beta_n l - \cosh \beta_n l}{\sin \beta_n l - \sinh \beta_n l} \tag{11.99}$$

Hence, Eq. (11.93) can be written as

$$W_n(x) = C_{1n} \left[(\cos \beta_n x - \cosh \beta_n x) - \frac{\cos \beta_n l - \cosh \beta_n l}{\sin \beta_n l - \sinh \beta_n l} (\sin \beta_n x - \sinh \beta_n x) \right] \tag{11.100}$$

The first four natural frequencies and the corresponding mode shapes given by Eqs. (11.98) and (11.100) are shown in Fig. 11.5.

11.5.5   Beam Fixed at One End and Free at the Other

If the beam is fixed at $x = 0$ and free at $x = l$, the transverse deflection and its slope must be zero at $x = 0$ and the bending moment and shear force must be zero at $x = l$.

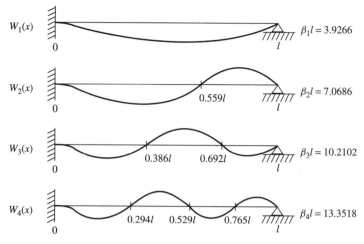

Figure 11.5   Natural frequencies and mode shapes of a fixed simply supported beam. $\omega_n = (\beta_n l)^2 (EI/\rho A L^4)^{1/2}$, $\beta_n l \simeq (4n+1)\pi/4$.

Thus, the boundary conditions become

$$W(0) = 0 \tag{11.101}$$

$$\frac{dW}{dx}(0) = 0 \tag{11.102}$$

$$EI\frac{d^2W}{dx^2}(l) = 0 \quad \text{or} \quad \frac{d^2W}{dx^2}(l) = 0 \tag{11.103}$$

$$EI\frac{d^3W}{dx^3}(l) = 0 \quad \text{or} \quad \frac{d^3W}{dx^3}(l) = 0 \tag{11.104}$$

When used in the solution of Eq. (11.37), Eqs. (11.101) and (11.102) yield

$$C_1 = C_3 = 0 \tag{11.105}$$

When used with Eq. (11.37) Eqs. (11.103) and (11.104) result in

$$C_2(\cos \beta l + \cosh \beta l) + C_4(\sin \beta l + \sinh \beta l) = 0 \tag{11.106}$$

$$C_2(-\sin \beta l + \sinh \beta l) + C_4(\cos \beta l + \cosh \beta l) = 0 \tag{11.107}$$

Equations (11.106) and (11.107) lead to the frequency equation

$$\cos \beta l \cosh \beta l + 1 = 0 \tag{11.108}$$

Using Eq. (11.106), we obtain

$$C_4 = -\frac{\cos \beta l + \cosh \beta l}{\sin \beta l + \sinh \beta l} C_2 \tag{11.109}$$

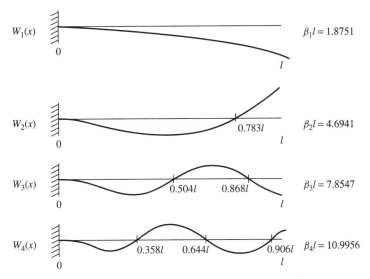

Figure 11.6 Natural frequencies and mode shapes of a fixed–free beam. $\omega_n = (\beta_n l)^2 (EI/\rho A L^4)^{1/2}$, $\beta_n l \simeq (2n-1)\pi/2$.

and hence the nth mode shape can be expressed as

$$W_n(x) = (\cos \beta_n x - \cosh \beta_n x) - \frac{\cos \beta_n l + \cosh \beta_n l}{\sin \beta_n l + \sinh \beta_n l} (\sin \beta_n x - \sinh \beta_n x) \quad (11.110)$$

The first four natural frequencies and the corresponding mode shapes given by Eqs. (11.108) and (11.110) are shown in Fig. 11.6.

Example 11.1   Determine the natural frequencies and mode shapes of transverse vibration of a uniform beam fixed at one end and a mass M attached at the other end.

SOLUTION   If the beam is fixed at $x = 0$, the transverse deflection and its slope are zero at $x = 0$. At the other end, $x = l$, the bending moment is zero and the shear force is equal to the inertia force due to the attached mass M. Thus, the boundary conditions can be stated as

$$W(0) = 0 \tag{E11.1.1}$$

$$\frac{dW}{dx}(0) = 0 \tag{E11.1.2}$$

$$EI\frac{d^2W}{dx^2}(l) = 0 \quad \text{or} \quad \frac{d^2W}{dx^2}(l) = 0 \tag{E11.1.3}$$

$$EI\frac{\partial^3 w(l,t)}{\partial x^3} = M\frac{\partial^2 w(l,t)}{\partial t^2}$$

or

$$EI\frac{d^3W}{dx^3}(l) = -M\omega^2 W(l) \tag{E11.1.4}$$

The boundary conditions (E11.1.1) and (E11.1.2) require that

$$C_1 = C_3 = 0 \tag{E11.1.5}$$

in Eq. (11.37). Thus, the solution becomes

$$W(x) = C_2(\cos \beta x - \cosh \beta x) + C_4(\sin \beta x - \sinh \beta x) \tag{E11.1.6}$$

When used in Eq. (E11.1.6), the conditions (E11.1.3) and (E11.1.4) lead to

$$C_2(\cos \beta l + \cosh \beta l) + C_4(\sin \beta l + \sinh \beta l) = 0 \tag{E11.1.7}$$

$$C_2[(-EI\beta^3)(\sin \beta l - \sinh \beta l) - M\omega^2(\cos \beta l - \cosh \beta l)]$$
$$+ C_4[(EI\beta^3)(\cos \beta l + \cosh \beta l) - M\omega^2(\sin \beta l - \sinh \beta l)] = 0 \tag{E11.1.8}$$

For a nontrivial solution of the constants C_2 and C_4, the determinant formed by their coefficients is set equal to zero:

$$\begin{vmatrix} \cos \beta l + \cosh \beta l & \sin \beta l + \sinh \beta l \\ [-EI\beta^3(\sin \beta l - \sinh \beta l) & [EI\beta^3(\cos \beta l + \cosh \beta l) \\ -M\omega^2(\cos \beta l - \cosh \beta l)] & -M\omega^2(\sin \beta l - \sinh \beta l)] \end{vmatrix} = 0 \tag{E11.1.9}$$

The simplification of Eq. (E11.1.9) leads to the frequency equation

$$1 + \frac{1}{\cos \beta l \cosh \beta l} - R\beta l(\tan \beta l - \tanh \beta l) = 0 \tag{E11.1.10}$$

$$R = \frac{M}{\rho A l} \tag{E11.1.11}$$

denotes the ratio of the attached mass M to the mass of the beam. Equation (E11.1.7) gives

$$C_4 = -\frac{\cos \beta l + \cosh \beta l}{\sin \beta l + \sinh \beta l} C_2 \tag{E11.1.12}$$

and hence the nth mode shape can be expressed as

$$W_n(x) = C_{2n}\left[(\cos \beta_n x - \cosh \beta_n x) \right.$$
$$\left. - \frac{\cos \beta_n l + \cosh \beta_n l}{\sin \beta_n l + \sinh \beta_n l}(\sin \beta_n x - \sinh \beta_n x) \right] \tag{E11.1.13}$$

where C_{2n} is a constant. The first two natural frequencies given by Eq. (E11.1.10) for different values of the mass ratio (R) are given in Table 11.1.2.

Table 11.1.2 Natural Frequencies of a Beam with One End Fixed and a Mass Attached at the Other

R	n	$\beta_n l$	R	n	$\beta_n l$
0.01	1	1.852	10	1	0.736
	2	4.650		2	3.938
0.1	1	1.723	100	1	0.416
	2	4.399		2	3.928
1	1	1.248	∞	1	0
	2	4.031		2	3.927

Example 11.2 The ends of a beam carry masses and are supported by linear springs and linear viscous dampers as shown in Fig. 11.7(a). State the boundary conditions of the beam using an equilibrium approach.

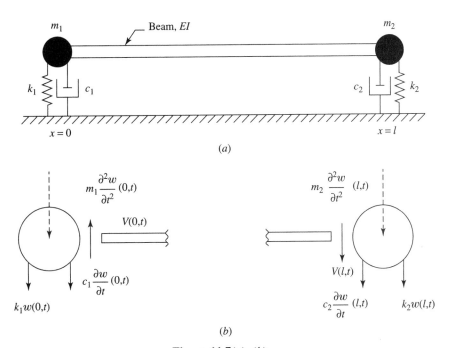

Figure 11.7(a), (b)

SOLUTION   If the transverse displacement, velocity, and acceleration of the beam at $x = 0$ are assumed to be positive with values $w(0, t)$, $\partial w(0, t)/\partial t$, and $\partial^2 w(0, t)/\partial t^2$, respectively, the spring force $k_1 w(0, t)$, damping force $c_1[\partial w(0, t)/\partial t]$, and the inertia force $m_1[\partial^2 w(0, t)/\partial t^2]$ act downward, as shown in Fig. 11.7(b). The positive shear force (V) at $x = 0$ is equal to the negative of the forces of spring, damper, and inertia

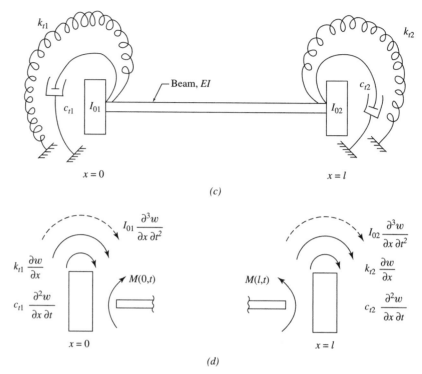

Figure 11.7(c), (d)

at $x = 0$. This boundary condition can be expressed as

$$V(x, t) = \frac{\partial}{\partial x}\left[EI(x)\frac{\partial^2 w(x, t)}{\partial x^2}\right]$$
$$= -\left[k_1 w(x, t) + c_1\frac{\partial w}{\partial t}(x, t) + m_1\frac{\partial^2 w(x, t)}{\partial t^2}\right] \quad \text{at} \quad x = 0 \quad \text{(E11.2.1)}$$

In addition, the bending moment must be zero at $x = 0$:

$$EI(x)\frac{\partial^2 w(x, t)}{\partial x^2} = 0 \quad \text{at} \quad x = 0 \quad \text{(E11.2.2)}$$

In a similar manner, the shear force boundary condition at $x = l$ can be expressed as

$$V(x, t) = \frac{\partial}{\partial x}\left[EI(x)\frac{\partial^2 w(x, t)}{\partial x^2}\right]$$
$$= k_2 w(x, t) + c_2\frac{\partial w(x, t)}{\partial t} + m_2\frac{\partial^2 w(x, t)}{\partial t^2} \quad \text{at} \quad x = l \quad \text{(E11.2.3)}$$

In addition, the bending moment must be zero at $x = l$:

$$EI(x)\frac{\partial^2 w(x, t)}{\partial x^2} = 0 \quad \text{at} \quad x = l \quad \text{(E11.2.4)}$$

Note: If the ends of the beam carry mass moments of inertia and are supported by torsional springs and torsional dampers as shown in Fig. 11.7(c), the reaction moments at the ends are shown in Fig. 11.7(d). Thus, the bending moment and shear force boundary conditions at the ends can be expressed as

$$M(x, t) = EI(x)\frac{\partial^2 w(x, t)}{\partial x^2} = -k_{t1}\frac{\partial w(x, t)}{\partial x}$$

$$-c_{t1}\frac{\partial^2 w(x, t)}{\partial x \partial t} - I_{01}\frac{\partial^3 w(x, t)}{\partial x \partial t^2} \qquad \text{at} \quad x = 0 \qquad \text{(E11.2.5)}$$

$$V(x, t) = \frac{\partial}{\partial x}\left[EI(x)\frac{\partial^2 w(x, t)}{\partial x^2}\right] = 0 \qquad \text{at} \quad x = 0 \qquad \text{(E11.2.6)}$$

$$M(x, t) = EI(x)\frac{\partial^2 w(x, t)}{\partial x^2} = k_{t2}\frac{\partial w(x, t)}{\partial x}$$

$$+c_{t2}\frac{\partial^2 w(x, t)}{\partial x \partial t} + I_{02}\frac{\partial^3 w(x, t)}{\partial x \partial t^2} \qquad \text{at} \quad x = l \qquad \text{(E11.2.7)}$$

$$V(x, t) = \frac{\partial}{\partial x}\left[EI(x)\frac{\partial^2 w(x, t)}{\partial x^2}\right] = 0 \qquad \text{at} \quad x = l \qquad \text{(E11.2.8)}$$

11.6 ORTHOGONALITY OF NORMAL MODES

The eigenvalue problem corresponding to a nonuniform beam can be obtained by assuming a harmonic solution with frequency ω in Eq. (11.23) as

$$\frac{d^2}{dx^2}\left[EI(x)\frac{d^2 W(x)}{dx^2}\right] = \omega^2 \rho A(x) W(x) \qquad (11.111)$$

To derive the orthogonality relations for beams, consider two eigenvalues ω_i^2 and ω_j^2 and the corresponding eigen or normal functions $W_i(x)$ and $W_j(x)$, respectively, so that

$$\frac{d^2}{dx^2}\left[EI(x)\frac{d^2 W_i(x)}{dx^2}\right] = \omega_i^2 \rho A(x) W_i(x) \qquad (11.112)$$

and

$$\frac{d^2}{dx^2}\left[EI(x)\frac{d^2 W_j(x)}{dx^2}\right] = \omega_j^2 \rho A(x) W_j(x) \qquad (11.113)$$

Multiply Eq. (11.112) by $W_j(x)$ and integrate over the length of the beam to obtain

$$\int_0^l W_j(x)\frac{d^2}{dx^2}\left[EI(x)\frac{d^2 W_i(x)}{dx^2}\right] dx = \omega_i^2 \int_0^l \rho A(x) W_j(x) W_i(x)\, dx \qquad (11.114)$$

Integrating the left-hand side of Eq. (11.114) by parts twice and using any combination of the boundary conditions among fixed, pinned, and free ends of the beam [given by

Eqs. (11.18)–(11.21)], we obtain

$$\int_0^l W_j(x)\frac{d^2}{dx^2}\left[EI(x)\frac{d^2 W_i(x)}{dx^2}\right]dx$$

$$= W_j(x)\frac{d}{dx}\left[EI(x)\frac{d^2 W_i(x)}{dx^2}\right]\Bigg|_0^l - \frac{dW_j(x)}{dx}EI(x)\frac{d^2 W_i(x)}{dx^2}\Bigg|_0^l$$

$$+ \int_0^l EI(x)\frac{d^2 W_j(x)}{dx^2}\frac{d^2 W_i(x)}{dx^2}dx$$

$$= \int_0^l EI(x)\frac{d^2 W_j(x)}{dx^2}\frac{d^2 W_i(x)}{dx^2}dx \tag{11.115}$$

Thus, Eq. (11.114) can be written as

$$\int_0^l EI(x)\frac{d^2 W_j(x)}{dx^2}\frac{d^2 W_i(x)}{dx^2}dx = \omega_i^2\int_0^l \rho A(x)W_j(x)W_i(x)dx \tag{11.116}$$

Similarly, by multiplying Eq. (11.113) by $W_i(x)$ and integrating over the length of the beam, we can obtain

$$\int_0^l EI(x)\frac{d^2 W_i(x)}{dx^2}\frac{d^2 W_j(x)}{dx^2}dx = \omega_j^2\int_0^l \rho A(x)W_i(x)W_j(x)dx \tag{11.117}$$

Noting that the order of the subscripts i and j under the integrals is immaterial and subtracting Eq. (11.117) from Eq. (11.116) yields

$$(\omega_i^2 - \omega_j^2)\int_0^l \rho A(x)W_i(x)W_j(x)dx = 0 \tag{11.118}$$

Since the eigenvalues are distinct, Eq. (11.118) gives

$$\int_0^l \rho A(x)W_i(x)W_j(x)dx = 0, \qquad i, j = 1, 2, \ldots, \quad \omega_i^2 \neq \omega_j^2 \tag{11.119}$$

In view of Eq. (11.119), Eq. (11.116) or (11.117) gives

$$\int_0^l EI(x)\frac{d^2 W_i(x)}{dx^2}\frac{d^2 W_j(x)}{dx^2}dx = 0, \qquad i, j = 1, 2, \ldots \quad \omega_i^2 \neq \omega_j^2 \tag{11.120}$$

Equation (11.119) is called the *orthogonality relation* for normal functions. Equation (11.120) represents another form of the orthogonality condition for the normal modes. In fact, by normalizing the ith normal mode as

$$\int_0^l \rho A(x)W_i^2(x)dx = 1, \qquad i = 1, 2, \ldots \tag{11.121}$$

Eqs. (11.119) and (11.121) can be expressed in the following form:

$$\int_0^l \rho A(x)W_i(x)W_j(x)dx = \delta_{ij} \tag{11.122}$$

where δ_{ij} is the Kronecker delta:

$$\delta_{ij} = \begin{cases} 0, & i \neq j \\ 1, & i = j \end{cases} \tag{11.123}$$

Using Eq. (11.122) in Eq. (11.114), another form of orthogonality relation can be derived as

$$\int_0^l W_j(x) \frac{d^2}{dx^2} \left[EI(x) \frac{d^2 W_i(x)}{dx^2} \right] dx = \omega_i^2 \delta_{ij} \tag{11.124}$$

According to the expansion theorem, any function $W(x)$ that satisfies the boundary conditions of the beam denotes a possible transverse displacement of the beam and can be expressed as a sum of eigenfunctions as

$$W(x) = \sum_{i=1}^{\infty} c_i W_i(x) \tag{11.125}$$

where the constants c_i are defined by

$$c_i = \int_0^l \rho A(x) W_i(x) W(x) \, dx, \qquad i = 1, 2, \ldots \tag{11.126}$$

and

$$c_i \omega_i^2 = \int_0^l W_i(x) \frac{d^2}{dx^2} \left[EI(x) \frac{d^2 W(x)}{dx^2} \right] dx, \qquad i = 1, 2, \ldots \tag{11.127}$$

Note that the derivative

$$\frac{d^2}{dx^2} \left[EI(x) \frac{d^2 W(x)}{dx^2} \right]$$

is assumed to be continuous in Eq. (11.127).

11.7   FREE VIBRATION RESPONSE DUE TO INITIAL CONDITIONS

The free vibration response of a beam can be expressed as a linear combination of all the natural or characteristic motions of the beam. The natural or characteristic motions consist of the natural modes multiplied by time-dependent harmonic functions with frequencies equal to the natural frequencies of the beam. To show this, consider the response in the following form:

$$w(x, t) = \sum_{i=1}^{\infty} W_i(x) \eta_i(t) \tag{11.128}$$

where $W_i(x)$ is the ith natural mode and $\eta_i(t)$ is a time-dependent function to be determined. By substituting Eq. (11.128) in the equation of motion of free vibration of the beam, Eq. (11.23), we obtain

$$\sum_{i=1}^{\infty} \frac{d^2}{dx^2} \left[EI(x) \frac{d^2 W_i(x)}{dx^2} \right] \eta_i(t) + \sum_{i=1}^{\infty} \rho A(x) W_i(x) \frac{d^2 \eta_i(t)}{dt^2} = 0 \tag{11.129}$$

Multiplying Eq. (11.129) by $W_j(x)$ and integrating over the length of the beam yields

$$\sum_{i=1}^{\infty} \left\{ \int_0^l W_j(x) \frac{d^2}{dx^2} \left[EI(x) \frac{d^2 W_i(x)}{dx^2} \right] dx \right\} \eta_i(t)$$

$$+ \sum_{i=1}^{\infty} \left[\int_0^l \rho A(x) W_j(x) W_i(x) \, dx \right] \frac{d^2 \eta_i(t)}{dt^2} = 0 \qquad (11.130)$$

In view of the orthogonality conditions, Eqs. (11.122) and (11.124), Eq. (11.130) gives the following equations, which are known as *modal equations*:

$$\frac{d^2 \eta_i(t)}{dt^2} + \omega_i^2 \eta_i(t) = 0, \qquad i = 1, 2, \dots \qquad (11.131)$$

where $\eta_i(t)$ is called the ith modal displacement (coordinate) and ω_i is the ith natural frequency of the beam. Each equation in (11.131) is similar to the equation of motion of a single-degree-of-freedom system whose solution can be expressed as

$$\eta_i(t) = A_i \cos \omega_i t + B_i \sin \omega_i t, \qquad i = 1, 2, \dots \qquad (11.132)$$

where A_i and B_i are constants that can be determined from the initial conditions. If

$$\eta_i(t = 0) = \eta_i(0) \quad \text{and} \quad \frac{d\eta_i}{dt}(t = 0) = \dot{\eta}_i(0) \qquad (11.133)$$

are the initial values of modal displacement and modal velocity, Eq. (11.132) can be expressed as

$$\eta_i(t) = \eta_i(0) \cos \omega_i t + \frac{\dot{\eta}_i(0)}{\omega_i} \sin \omega_i t, \qquad i = 1, 2, \dots \qquad (11.134)$$

If the initial displacement and velocity distributions of the beam are specified as

$$w(x, t = 0) = w_0(x), \qquad \frac{\partial w}{\partial t}(x, t = 0) = \dot{w}(x, 0) = \dot{w}_0(x) \qquad (11.135)$$

the initial values of modal displacement and modal velocity can be determined as follows. Using Eq. (11.135) in Eq. (11.128), we find

$$w(x, t = 0) = \sum_{i=1}^{\infty} W_i(x) \eta_i(0) = w_0(x) \qquad (11.136)$$

$$\frac{\partial w}{\partial t}(x, t = 0) = \sum_{i=1}^{\infty} W_i(x) \dot{\eta}_i(0) = \dot{w}_0(x) \qquad (11.137)$$

By multiplying Eq. (11.136) by $\rho A(x) W_j(x)$, integrating over the length of the beam, and using the orthogonality condition of Eq. (11.122), we obtain

$$\eta_i(0) = \int_0^l \rho A(x) W_i(x) w_0(x) \, dx, \qquad i = 1, 2, \dots \qquad (11.138)$$

A similar procedure with Eq. (11.137) leads to

$$\dot{\eta}_i(0) = \int_0^l \rho A(x) W_i(x) \dot{w}_0(x) \, dx, \qquad i = 1, 2, \ldots \tag{11.139}$$

Once $\eta_i(0)$ and $\dot{\eta}_i(0)$ are known, the response of the beam under the specified initial conditions can be computed using Eqs. (11.128), (11.134), (11.138), and (11.139). The procedure is illustrated in the following example.

Example 11.3   A uniformly distributed load of magnitude f_0 per unit length acts on the entire length of a uniform simply supported beam. Find the vibrations that ensue when the load is suddenly removed.

SOLUTION   The initial deflection of the beam under the distributed load of intensity f_0 is given by the static deflection curve [2]

$$w_0(x) = \frac{f_0}{24EI}(x^4 - 2lx^3 + l^3 x) \tag{E11.3.1}$$

and the initial velocity of the beam is assumed to be zero:

$$\dot{w}_0(x) = 0 \tag{E11.3.2}$$

For a simply supported beam, the normalized normal modes can be found from Eq. (11.121):

$$\int_0^l \rho A(x) W_i^2(x) \, dx = 1, \qquad i = 1, 2, \ldots \tag{E11.3.3}$$

In the present case, the normal modes are given by Eq. (11.51):

$$W_i(x) = C_i \sin \frac{i \pi x}{l}, \qquad i = 1, 2, \ldots \tag{E11.3.4}$$

where C_i is a constant. Equations (E11.3.3) and (E11.3.4) lead to

$$\rho A C_i^2 \int_0^l \sin^2 \frac{i \pi x}{l} \, dx = 1 \tag{E11.3.5}$$

or

$$C_i = \sqrt{\frac{2}{\rho Al}}, \qquad i = 1, 2, \ldots \tag{E11.3.6}$$

Thus, the normalized normal modes are given by

$$W_i(x) = \sqrt{\frac{2}{\rho Al}} \sin \frac{i \pi x}{l}, \qquad i = 1, 2, \ldots \tag{E11.3.7}$$

The response of a beam subject to initial conditions is given by Eqs. (11.128) and (11.134):

$$w(x, t) = \sum_{i=1}^{\infty} W_i(x) \left[\eta_i(0) \cos \omega_i t + \frac{\dot{\eta}_i(0)}{\omega_i} \sin \omega_i t \right] \tag{E11.3.8}$$

where $\eta_i(0)$ and $\dot{\eta}_i(0)$ are given by Eqs. (11.138) and (11.139):

$$\eta_i(0) = \int_0^l \rho A(x) W_i(x) w_0(x) \, dx = \rho A \sqrt{\frac{2}{\rho A l}} \int_0^l \sin \frac{i\pi x}{l} \frac{f_0}{24EI} (x^4 - 2lx^3 + l^3 x) \, dx$$

$$= \frac{2\sqrt{2\rho A l} \, f_0 l^4}{EI\pi^5 i^5}, \qquad i = 1, 3, 5, \ldots \tag{E11.3.9}$$

$$\dot{\eta}_i(0) = \int_0^l \rho A(x) W_i(x) \dot{w}_0(x) \, dx = 0, \qquad i = 1, 2, \ldots \tag{E11.3.10}$$

Thus, the response of the beam can be expressed as [Eq. (E11.3.8)]

$$w(x, t) = \sum_{i=1,3,\ldots}^{\infty} \sqrt{\frac{2}{\rho A l}} \sin \frac{i\pi x}{l} \frac{2\sqrt{2\rho A l} \, f_0 l^4}{EI\pi^5 i^5} \cos \omega_i t$$

$$= \frac{4 f_0 l^4}{EI\pi^5} \sum_{i=1,3,\ldots}^{\infty} \frac{1}{i^5} \sin \frac{i\pi x}{l} \cos \omega_i t \tag{E11.3.11}$$

11.8   FORCED VIBRATION

The equation of motion of a beam under distributed transverse force is given by [see Eq. (11.12)]

$$\frac{\partial^2}{\partial x^2} \left[EI(x) \frac{\partial^2 w(x, t)}{\partial x^2} \right] + \rho A(x) \frac{\partial^2 w(x, t)}{\partial t^2} = f(x, t) \tag{11.140}$$

Using the normal mode approach (modal analysis), the solution of Eq. (11.140) is assumed to be a linear combination of the normal modes of the beam as

$$w(x, t) = \sum_{i=1}^{\infty} W_i(x) \eta_i(t) \tag{11.141}$$

where $W_i(x)$ are the normal modes found by solving the equation (using the four boundary conditions of the beam)

$$\frac{d^2}{dx^2} \left[EI(x) \frac{d^2 W_i(x)}{dx^2} \right] - \rho A(x) \omega_i^2 W_i(x) = 0 \tag{11.142}$$

and $\eta_i(t)$ are the generalized coordinates or modal participation coefficients. Using Eq. (11.141), Eq. (11.140) can be expressed as

$$\sum_{i=1}^{\infty} \frac{d^2}{dx^2} \left[EI(x) \frac{d^2 W_i(x)}{dx^2} \right] \eta_i(t) + \rho A(x) \sum_{i=1}^{\infty} W_i(x) \frac{d^2 \eta_i(t)}{dt^2} = f(x, t) \tag{11.143}$$

Using Eq. (11.142), Eq. (11.143) can be rewritten as

$$\rho A(x) \sum_{i=1}^{\infty} \omega_i^2 W_i(x) \eta_i(t) + \rho A(x) \sum_{i=1}^{\infty} W_i(x) \frac{d^2 \eta_i(t)}{dt^2} = f(x, t) \tag{11.144}$$

Multiplying Eq. (11.144) by $W_j(x)$ and integrating from 0 to l results in

$$\sum_{i=1}^{\infty} \eta_i(t) \int_0^l \rho A(x)\omega_i^2 W_j(x)W_i(x)\, dx$$

$$+ \sum_{i=1}^{\infty} \frac{d^2\eta_i(t)}{dt^2} \int_0^l \rho A(x)W_j(x)W_i(x)\, dx = \int_0^l W_j(x)f(x,t)\, dx \quad (11.145)$$

In view of the orthogonality condition, Eq. (11.122), all terms in each of the summations on the left side of Eq. (11.145) vanish except for the one term for which $i = j$, leaving

$$\frac{d^2\eta_i(t)}{dt^2} + \omega_i^2\eta_i(t) = Q_i(t), \qquad i = 1, 2, \ldots \quad (11.146)$$

where $Q_i(t)$ is the generalized force corresponding to the ith mode given by

$$Q_i(t) = \int_0^l W_i(x)f(x,t)\, dx, \qquad i = 1, 2, \ldots \quad (11.147)$$

The complete solution of Eq. (11.146) can be expressed as [see Eq.(2.109)]

$$\eta_i(t) = A_i \cos \omega_i t + B_i \sin \omega_i t + \frac{1}{\omega_i} \int_0^t Q_i(\tau) \sin \omega_i(t - \tau)\, d\tau \quad (11.148)$$

Thus, the solution of Eq. (11.140) is given by Eqs.(11.141) and (11.148):

$$w(x,t) = \sum_{i=1}^{\infty} \left[A_i \cos \omega_i t + B_i \sin \omega_i t + \frac{1}{\omega_i} \int_0^t Q_i(\tau) \sin \omega_i(t - \tau)\, d\tau \right] W_i(x)$$

$$(11.149)$$

Note that the first two terms inside the brackets denote the free vibration, and the third term indicates the forced vibration of the beam. The constants A_i and B_i in Eq.(11.149) can be evaluated using the initial conditions of the beam.

Example 11.4   Find the response of a uniform simply supported beam subjected to a step-function force F_0 at $x = \xi$, as shown in Fig. 11.8. Assume the initial conditions of the beam to be zero.

SOLUTION   The natural frequencies and the normal modes of vibration of a uniform simply supported beam are given by

$$\omega_i = \frac{i^2\pi^2}{l^2}\sqrt{\frac{EI}{\rho A}} \quad (E11.4.1)$$

$$W_i(x) = C_i \sin\frac{i\pi x}{l} \quad (E11.4.2)$$

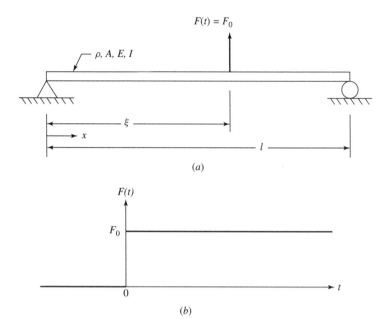

Figure 11.8

where C_i is a constant. If the normal modes are normalized according to Eq. (11.122), we have

$$\int_0^l \rho A(x) W_i^2(x)\, dx = \rho A C_i^2 \int_0^l \sin^2 \frac{i\pi x}{l}\, dx = 1$$

or

$$C_i = \sqrt{\frac{2}{\rho A l}} \qquad (\text{E11.4.3})$$

Thus, the normalized mode shapes are given by

$$W_i(x) = \sqrt{\frac{2}{\rho A l}} \sin \frac{i\pi x}{l} \qquad (\text{E11.4.4})$$

The force acting on the beam can be expressed as

$$f(x, t) = F_0 \delta(x - \xi) \qquad (\text{E11.4.5})$$

and the generalized force corresponding to the ith mode can be determined using Eq.(11.147) as

$$Q_i(t) = \int_0^l W_i(x) f(x, t)\, dx$$

$$= \sqrt{\frac{2}{\rho A l}} \int_0^l \sin \frac{i\pi x}{l} F_0 \delta(x - \xi)\, dx = \sqrt{\frac{2}{\rho A l}} F_0 \sin \frac{i\pi \xi}{l} \qquad (\text{E11.4.6})$$

The generalized coordinate in the ith mode is given by Eq.(11.148):

$$\eta_i(t) = A_i \cos \omega_i t + B_i \sin \omega_i t + \sqrt{\frac{2}{\rho A l}} F_0 \sin \frac{i\pi\xi}{l} \frac{1}{\omega_i} \int_0^t \sin \omega_i (t - \tau) \, d\tau$$

$$= A_i \cos \omega_i t + B_i \sin \omega_i t + F_0 \sqrt{\frac{2}{\rho A l}} \sin \frac{i\pi\xi}{l} \frac{1}{\omega_i^2} (1 - \cos \omega_i t) \quad \text{(E11.4.7)}$$

where the constants A_i and B_i can be determined from the initial conditions of the beam. In the present case, the initial conditions are zero and hence $\eta_i(t)$ can be expressed as

$$\eta_i(t) = F_0 \sqrt{\frac{2}{\rho A l}} \frac{l^4}{i^4 \pi^4} \frac{\rho A}{EI} \sin \frac{i\pi\xi}{l} (1 - \cos \omega_i t) \quad \text{(E11.4.8)}$$

Thus, the response of the beam is given by [see Eq.(11.149)]

$$w(x, t) = \frac{2 F_0 l^3}{\pi^4 EI} \sum_{i=1}^{\infty} \frac{1}{i^4} \sin \frac{i\pi x}{l} \sin \frac{i\pi\xi}{l} (1 - \cos \omega_i t) \quad \text{(E11.4.9)}$$

Example 11.5   A uniform beam is subjected to a concentrated harmonic force $F_0 \sin \Omega t$ at $x = \xi$ (Fig. 11.9).

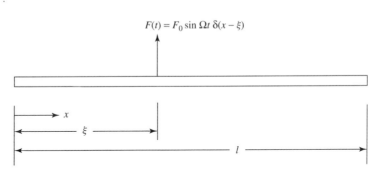

$$F(t) = F_0 \sin \Omega t \, \delta(x - \xi)$$

Figure 11.9

(a) Find an expression for the response of the beam valid for all support conditions.
(b) Find an expression for the response of a simply supported beam when $\xi = l/2$. Assume zero initial conditions.

SOLUTION   (a) The generalized force corresponding to the ith mode is given by Eq.(11.147):

$$Q_i(t) = F_0 \int_0^l W_i(x) \sin \Omega t \delta(x - \xi) \, dx = F_0 W_i(\xi) \sin \Omega t \quad \text{(E11.5.1)}$$

For zero initial conditions, the generalized coordinate in the ith mode becomes [see Eq.(11.148)]

$$\eta_i(t) = \frac{1}{\omega_i} \int_0^t Q_i(\tau) \sin \omega_i (t - \tau) \, d\tau = \frac{F_0}{\omega_i} \int_0^t W_i(\xi) \sin \Omega \tau \sin \omega_i (t - \tau) \, d\tau$$

$$= \frac{F_0}{\omega_i^2} W_i(\xi) \frac{1}{1 - \Omega^2/\omega_i^2} \left(\sin \Omega t - \frac{\Omega}{\omega_i} \sin \omega_i t \right) \tag{E11.5.2}$$

The response of the beam can be expressed as [Eq.(11.141)]

$$w(x, t) = F_0 \sum_{i=1}^{\infty} \frac{W_i(x) W_i(\xi)}{\omega_i^2 - \Omega^2} \left(\sin \Omega t - \frac{\Omega}{\omega_i} \sin \omega_i t \right) \tag{E11.5.3}$$

where $W_i(x)$ and $W_i(\xi)$ correspond to the normalized normal modes.

(b) For a simply supported beam, the normalized normal modes are given by [see Eq.(E11.4.4)]

$$W_i(x) = \sqrt{\frac{2}{\rho A l}} \sin \frac{i \pi x}{l} \tag{E11.5.4}$$

and hence

$$W_i\left(\xi = \frac{l}{2}\right) = \sqrt{\frac{2}{\rho A l}} \sin \frac{i \pi}{2} = \begin{cases} 0, & i = 2, 4, 6, \ldots \\ \sqrt{\frac{2}{\rho A l}}, & i = 1, 5, 9, \ldots \\ -\sqrt{\frac{2}{\rho A l}}, & i = 3, 7, 11, \ldots \end{cases} \tag{E11.5.5}$$

Thus, the response of the beam, Eq. (E11.5.3), becomes

$$w(x, t) = \frac{2F_0}{\rho A l} \left[\sum_{i=1,5,9,\ldots}^{\infty} \frac{\sin(i \pi x/l)}{\omega_i^2 - \Omega^2} \left(\sin \Omega t - \frac{\Omega}{\omega_i} \sin \omega_i t \right) \right.$$

$$\left. - \sum_{i=3,7,11,\ldots}^{\infty} \frac{\sin(i \pi x/l)}{\omega_i^2 - \Omega^2} \left(\sin \Omega t - \frac{\Omega}{\omega_i} \sin \omega_i t \right) \right] \tag{E11.5.6}$$

where

$$\omega_i = \frac{i^2 \pi^2}{l^2} \sqrt{\frac{EI}{\rho A}} \tag{E11.5.7}$$

Example 11.6   Find the dynamic response of a uniform beam simply supported at both ends and subjected to a harmonically varying load:

$$f(x, t) = f_0 \sin \frac{n \pi x}{l} \sin \omega t \tag{E11.6.1}$$

where f_0 is a constant, n is an integer, l is the length of the beam, and ω is the frequency of variation of the load. Assume the initial displacement and initial velocity of the beam to be zero.

SOLUTION   The equation of motion of the beam is given by [see Eq.(11.12)]

$$EI\frac{\partial^4 w(x,t)}{\partial x^4} + \rho A\frac{\partial^2 w(x,t)}{\partial t^2} = f_0 \sin\frac{n\pi x}{l}\sin\omega t \qquad \text{(E11.6.2)}$$

Although the solution of Eq. (E11.6.2) can be obtained using Eq. (11.141), a simpler approach can be used because of the nature of the load, Eq.(E11.6.1). The homogeneous (or free vibration) solution of Eq.(E11.6.2) can be expressed as

$$w(x,t) = \sum_{i=1}^{\infty} \sin\frac{i\pi x}{l}(C_i \cos\omega_i t + D_i \sin\omega_i t) \qquad \text{(E11.6.3)}$$

where ω_i is the natural frequency of the beam, given by [see Eq.(11.50)]

$$\omega_i = i^2\pi^2\sqrt{\frac{EI}{\rho Al^4}} \qquad \text{(E11.6.4)}$$

The particular integral of Eq.(E11.6.2) can be expressed as

$$w(x,t) = a_n \sin\frac{n\pi x}{l}\sin\omega t \qquad \text{(E11.6.5)}$$

where the expression for a_n can be found by substituting Eq.(E11.6.5) into Eq.(E11.6.2) as

$$a_n = \frac{f_0 l^4}{EI(n\pi)^4\left[1 - (\omega/\omega_n)^2\right]} \qquad \text{(E11.6.6)}$$

The total solution of Eq. (E11.6.2) is given by sum of its homogeneous solution and the particular integral:

$$w(x,t) = \sum_{i=1}^{\infty} \sin\frac{i\pi x}{l}(C_i \cos\omega_i t + D_i \sin\omega_i t) + a_n \sin\frac{n\pi x}{l}\sin\omega t \qquad \text{(E11.6.7)}$$

The initial conditions of the beam are given by

$$w(x,0) = 0 \qquad \text{(E11.6.8)}$$

$$\frac{\partial w}{\partial t}(x,0) = 0 \qquad \text{(E11.6.9)}$$

Substituting Eq. (E11.6.7) into Eqs. (E11.6.8) and (E11.6.9), we obtain

$$C_i = 0 \qquad \text{for all } i \qquad \text{(E11.6.10)}$$

$$D_i = \begin{cases} 0 & \text{for } i \neq n \\ -a_n\dfrac{\omega}{\omega_n} & \text{for } i = n \end{cases} \qquad \text{(E11.6.11)}$$

Thus, the solution of the beam becomes

$$w(x, t) = \frac{f_0 l^4}{EI (n\pi)^4 \left[1 - (\omega/\omega_n)^2\right]} \sin \frac{n\pi x}{l} \left(\sin \omega t - \frac{\omega}{\omega_n} \sin \omega_n t \right) \quad \text{(E11.6.12)}$$

11.9 RESPONSE OF BEAMS UNDER MOVING LOADS

Consider a uniform beam subjected to a concentrated load P that moves at a constant speed v_0 along the beam as shown in Fig. 11.10. The boundary conditions of the simply supported beam are given by

$$w(0, t) = 0 \quad \text{(11.150)}$$

$$\frac{\partial^2 w}{\partial x^2}(0, t) = 0 \quad \text{(11.151)}$$

$$w(l, t) = 0 \quad \text{(11.152)}$$

$$\frac{\partial^2 w}{\partial x^2}(l, t) = 0 \quad \text{(11.153)}$$

The beam is assumed to be at rest initially, so that the initial conditions can be written as

$$w(x, 0) = 0 \quad \text{(11.154)}$$

$$\frac{\partial w}{\partial t}(x, 0) = 0 \quad \text{(11.155)}$$

Instead of representing the concentrated force using a Dirac delta function, it will be represented using a Fourier series. For this, the concentrated load P acting at $x = d$ is assumed to be distributed uniformly over an elemental length $2\Delta x$ centered at $x = d$ as shown in Fig. 11.11. Now the distributed force, $f(x)$, can be defined as

$$f(x) = \begin{cases} 0 & \text{for } 0 < x < d - \Delta x \\ \frac{P}{2\Delta x} & \text{for } d - \Delta x \leq x \leq d + \Delta x \\ 0 & \text{for } d + \Delta x < x < l \end{cases} \quad \text{(11.156)}$$

From Fourier series analysis, it is known that if a function $f(x)$ is defined only over a finite interval (e.g., from x_0 to $x_0 + L$), the definition of the function $f(x)$ can be

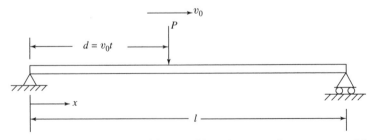

Figure 11.10   Simply supported beam subjected to a moving concentrated load.

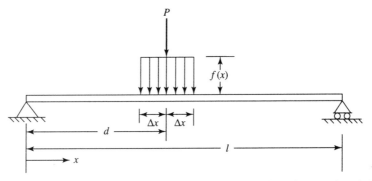

Figure 11.11   Concentrated load assumed to be uniformly distributed over a length $2\Delta x$.

extended for all values of x and can be considered to be periodic with period L. The Fourier series expansion of the extended periodic function converges to the function $f(x)$ in the original interval from x_0 to $x_0 + L$. As a specific case, if the function $f(x)$ is defined over the interval 0 to l, its Fourier series expansion in terms of only sine terms is given by

$$f(x) = \sum_{n=1}^{\infty} f_n \sin \frac{n\pi x}{l} \tag{11.157}$$

where the coefficients f_n are given by

$$f_n = \frac{2}{l} \int_0^l f(x) \sin \frac{n\pi x}{l} dx \tag{11.158}$$

In the present case, the Fourier coefficients f_n can be computed, using Eq. (11.156) for $f(x)$, as

$$f_n = \frac{2}{l} \left[\int_0^{d-\Delta x} (0) \left(\sin \frac{n\pi x}{l} \right) dx + \int_{d-\Delta x}^{d+\Delta x} \frac{P}{2\Delta x} \left(\sin \frac{n\pi x}{l} \right) dx \right.$$

$$\left. + \int_{d+\Delta x}^l (0) \left(\sin \frac{n\pi x}{l} \right) dx \right]$$

$$= \frac{P}{l\Delta x} \int_{d-\Delta x}^{d+\Delta x} \sin \frac{n\pi x}{l} dx = \frac{2P}{l} \sin \frac{n\pi d}{l} \frac{\sin(n\pi \Delta x/l)}{n\pi \Delta x/l} \tag{11.159}$$

Since P is actually a concentrated load acting at $x = d$, we let $\Delta x \to 0$ in Eq.(11.159) with

$$\lim_{\Delta x \to 0} \frac{\sin(n\pi \Delta x/l)}{(n\pi \Delta x/l)} = 1 \tag{11.160}$$

to obtain the coefficients

$$f_n = \frac{2P}{l} \sin \frac{n\pi d}{l} \tag{11.161}$$

Thus, the Fourier sine series expansion of the concentrated load acting at $x = d$ can be expressed as

$$f(x) = \frac{2P}{l} \sum_{n=1}^{\infty} \sin \frac{n\pi d}{l} \sin \frac{n\pi x}{l} \qquad (11.162)$$

Using $d = v_0 t$ in Eq. (11.162), the load distribution can be represented in terms of x and t as

$$f(x, t) = \frac{2P}{l} \left(\sin \frac{\pi x}{l} \sin \frac{\pi v_0 t}{l} + \sin \frac{2\pi x}{l} \sin \frac{2\pi v_0 t}{l} + \sin \frac{3\pi x}{l} \sin \frac{3\pi v_0 t}{l} + \cdots \right) \qquad (11.163)$$

The response of the beam under the nth component of the load represented by Eq. (11.163) can be obtained using Eq. (E11.6.12) as

$$w(x, t) = \frac{2Pl^3}{EI(n\pi)^4 \left[1 - (2\pi v_0 / l\omega_n)^2 \right]} \sin \frac{n\pi x}{l} \left(\sin \frac{2\pi v_0}{l} t - \frac{2\pi v_0}{\omega_n l} \sin \omega_n t \right) \qquad (11.164)$$

The total response of the beam considering all components (or harmonics) of the load, given by Eq.(11.163), can be expressed as

$$w(x, t) = \frac{2Pl^3}{EI\pi^4} \sum_{n=1}^{\infty} \frac{1}{n^4} \frac{1}{1 - (2\pi v_0 / l\omega_n)^2} \sin \frac{n\pi x}{l} \left(\sin \frac{2\pi v_0}{l} t - \frac{2\pi v_0}{\omega_n l} \sin \omega_n t \right) \qquad (11.165)$$

11.10 TRANSVERSE VIBRATION OF BEAMS SUBJECTED TO AXIAL FORCE

The problem of transverse vibration of beams subjected to axial force finds application in the study of vibration of cables, guy wires, and turbine blades. Although the vibration of a cable can be studied by modeling it as a taut string, many cables fail due to fatigue caused by alternating flexure induced by vortex shedding in a light wind. In turbines, blade failures are associated with combined transverse loads due to fluids flowing at high velocities and axial loads due to centrifugal action.

11.10.1 Derivation of Equations

Consider a beam undergoing transverse vibration under axial tensile force as shown in Fig. 11.12(a). The forces acting on an element of the beam of length dx are shown in Fig. 11.12(b). The change in the length of the beam element is given by

$$ds - dx = \left\{ (dx)^2 + \left[\frac{\partial w(x, t)}{\partial x} dx \right]^2 \right\}^{1/2} - dx \qquad (11.166)$$

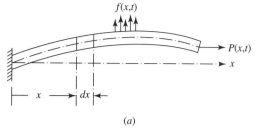

(a)

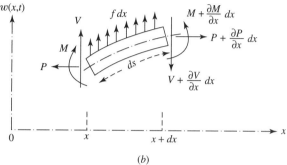

(b)

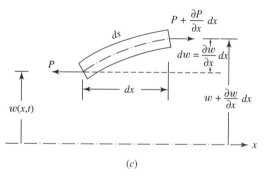

(c)

Figure 11.12

For small amplitudes of vibration, Eq. (11.166) can be approximated as [Fig. 11.12(c)]

$$ds - dx \approx \sqrt{dx^2 + dw^2} - dx \approx \frac{1}{2}\left(\frac{\partial w}{\partial x}\right)^2 dx \tag{11.167}$$

The small displacement values $w(x, t)$ are assumed to cause no changes in the axial force $P(x, t)$ and the transverse distributed force $f(x, t)$. The work done by the axial force against the change in the length of the element of the beam can be expressed as

$$W_P = -\frac{1}{2}\int_0^l P(x, t)\left[\frac{\partial w(x, t)}{\partial x}\right]^2 dx \tag{11.168}$$

The work done by the transverse load $f(x, t)$ is given by

$$W_f = \int_0^l f(x, t) w(x, t) \, dx \tag{11.169}$$

Thus, the total work done (W) is given by

$$W = W_P + W_f = -\frac{1}{2} \int_0^l P(x, t) \left[\frac{\partial w(x, t)}{\partial x} \right]^2 dx + \int_0^l f(x, t) w(x, t) \, dx \tag{11.170}$$

The strain and kinetic energies of the beam are given by

$$\pi = \frac{1}{2} \int_0^l EI(x) \left[\frac{\partial^2 w(x, t)}{\partial x^2} \right]^2 dx \tag{11.171}$$

$$T = \frac{1}{2} \int_0^l \rho A(x) \left[\frac{\partial w(x, t)}{\partial t} \right]^2 dx \tag{11.172}$$

The extended Hamilton's principle can be expressed as

$$\delta \int_{t_1}^{t_2} (T - \pi + W) \, dt = 0 \tag{11.173}$$

The various terms of Eq. (11.173) can be evaluated as follows:

$$
\begin{aligned}
\delta \int_{t_1}^{t_2} T \, dt &= \int_0^l \int_{t_1}^{t_2} \rho A \frac{\partial w}{\partial t} \frac{\partial}{\partial t} (\delta w) \, dt \, dx \\
&= \int_0^l \left[\rho A \frac{\partial w}{\partial t} \delta w \Big|_{t_1}^{t_2} - \int_{t_1}^{t_2} \frac{\partial}{\partial t} \left(\rho A \frac{\partial w}{\partial t} \right) \delta w \, dt \right] dx \\
&= -\int_{t_1}^{t_2} \int_0^l \rho A \frac{\partial^2 w}{\partial t^2} \delta w \, dx \, dt
\end{aligned} \tag{11.174}
$$

since $\delta w(x, t) = 0$ at $t = t_1$ and $t = t_2$.

$$
\begin{aligned}
\delta \int_{t_1}^{t_2} \pi \, dt &= \int_{t_1}^{t_2} \int_0^l EI \frac{\partial^2 w}{\partial x^2} \delta \left(\frac{\partial^2 w}{\partial x^2} \right) dx \, dt \\
&= \int_{t_1}^{t_2} \int_0^l EI \frac{\partial^2 w}{\partial x^2} \frac{\partial^2 (\delta w)}{\partial x^2} \, dx \, dt \\
&= \int_{t_1}^{t_2} \left[EI \frac{\partial^2 w}{\partial x^2} \frac{\partial (\delta w)}{\partial x} \Big|_0^l - \frac{\partial}{\partial x} \left(EI \frac{\partial^2 w}{\partial x^2} \right) \delta w \Big|_0^l \right. \\
&\quad \left. + \int_0^l \frac{\partial^2}{\partial x^2} \left(EI \frac{\partial^2 w}{\partial x^2} \right) \delta w \, dx \right] dt
\end{aligned} \tag{11.175}
$$

$$\delta \int_{t_1}^{t_2} W \, dt = \int_{t_1}^{t_2} \left[-P \frac{\partial w}{\partial x} \delta \left(\frac{\partial w}{\partial x} \right) dx + \int_0^l f \delta w \, dx \right] dt$$

$$= \int_{t_1}^{t_2} \left[-\int_0^l P \frac{\partial w}{\partial x} \frac{\partial}{\partial x} (\delta w) \, dx + \int_0^l f \delta w \, dx \right] dt$$

$$= \int_{t_1}^{t_2} \left[-P \frac{\partial w}{\partial x} \delta w \Big|_0^l + \int_0^l \frac{\partial}{\partial x} \left(P \frac{\partial w}{\partial x} \right) \delta w \, dx + \int_0^l f \delta w \, dx \right] dt$$

$$(11.176)$$

Substitution of Eqs. (11.174)–(11.176) into Eq. (11.173) leads to

$$-\int_{t_1}^{t_2} \int_0^l \left[\rho A \frac{\partial^2 w}{\partial t^2} + \frac{\partial^2}{\partial x^2} \left(EI \frac{\partial^2 w}{\partial x^2} \right) - \frac{\partial}{\partial x} \left(P \frac{\partial w}{\partial x} \right) - f \right] \delta w \, dt$$

$$-\int_{t_1}^{t_2} EI \frac{\partial^2 w}{\partial x^2} \delta \left(\frac{\partial w}{\partial x} \right) \Big|_0^l dt + \int_{t_1}^{t_2} \left[\frac{\partial}{\partial x} \left(EI \frac{\partial^2 w}{\partial x^2} \right) - P \frac{\partial w}{\partial x} \right] \delta w \Big|_0^l dt = 0 \quad (11.177)$$

Since δw is assumed to be an arbitrary (nonzero) variation in $0 < x < l$, the expression under the double integral in Eq. (11.177) is set equal to zero to obtain the differential equation of motion:

$$\frac{\partial^2}{\partial x^2} \left(EI \frac{\partial^2 w}{\partial x^2} \right) - \frac{\partial}{\partial x} \left(P \frac{\partial w}{\partial x} \right) + \rho A \frac{\partial^2 w}{\partial t^2} = f(x, t) \quad (11.178)$$

Setting the individual terms with single integrals in Eq. (11.177) equal to zero, we obtain the boundary conditions as

$$EI \frac{\partial^2 w}{\partial x^2} \delta \left(\frac{\partial w}{\partial x} \right) \Big|_0^l = 0 \quad (11.179)$$

$$\left[\frac{\partial}{\partial x} \left(EI \frac{\partial^2 w}{\partial x^2} \right) - P \frac{\partial w}{\partial x} \right] \delta w \Big|_0^l = 0 \quad (11.180)$$

Equation (11.179) indicates that the bending moment, $EI(\partial^2 w/\partial x^2)$, or the slope, $\partial w/\partial x$, must be zero at $x = 0$ as well as at $x = l$, while Eq. (11.180) denotes that either the total vertical force,

$$\frac{\partial}{\partial x} \left(EI \frac{\partial^2 w}{\partial x^2} \right) - P \frac{\partial w}{\partial x}$$

or the deflection, w, must be zero at $x = 0$ and also at $x = l$.

11.10.2 Free Vibration of a Uniform Beam

For a uniform beam with no transverse force, Eq. (11.178) becomes

$$EI \frac{\partial^4 w}{\partial x^4} + \rho A \frac{\partial^2 w}{\partial t^2} - P \frac{\partial^2 w}{\partial x^2} = 0 \quad (11.181)$$

The method of separation of variables is used to find the solution of Eq. (11.181):

$$w(x, t) = W(x)(A \cos \omega t + B \sin \omega t) \tag{11.182}$$

By substituting Eq. (11.182) into Eq. (11.181), we obtain

$$EI \frac{d^4 W}{dx^4} - P \frac{d^2 W}{dx^2} - \rho A \omega^2 W = 0 \tag{11.183}$$

By assuming the solution of $W(x)$ in the form

$$W(x) = C e^{sx} \tag{11.184}$$

where C is a constant, Eq. (11.183) gives the auxiliary equation

$$s^4 - \frac{P}{EI} s^2 - \frac{\rho A \omega^2}{EI} = 0 \tag{11.185}$$

The roots of Eq. (11.185) are given by

$$s_1^2, s_2^2 = \frac{P}{2EI} \pm \left(\frac{P^2}{4E^2 I^2} + \frac{\rho A \omega^2}{EI} \right)^{1/2} \tag{11.186}$$

Thus, the solution of Eq. (11.183) can be expressed as

$$W(x) = C_1 \cosh s_1 x + C_2 \sinh s_1 x + C_3 \cos s_2 x + C_4 \sin s_2 x \tag{11.187}$$

where the constants C_1 to C_4 are to be determined from the boundary conditions of the beam.

Example 11.7 Find the natural frequencies of a uniform simply supported beam subjected to an axial force P.

SOLUTION The boundary conditions of the beam are given by

$$W(0) = 0 \tag{E11.7.1}$$

$$\frac{d^2 W}{dx^2}(0) = 0 \tag{E11.7.2}$$

$$W(l) = 0 \tag{E11.7.3}$$

$$\frac{d^2 W}{dx^2}(l) = 0 \tag{E11.7.4}$$

When Eqs. (E11.7.1) and (E11.7.2) are used in the solution, Eq. (11.187), we obtain $C_1 = C_3 = 0$. This leads to

$$W(x) = C_2 \sinh s_1 x + C_4 \sin s_2 x \tag{E11.7.5}$$

Equations (E11.7.3)–(E11.7.5) yield

$$C_2 \sinh s_1 l + C_4 \sin s_2 l = 0 \tag{E11.7.6}$$

$$C_2 s_1^2 \sinh s_1 l - C_4 s_2^2 \sin s_2 l = 0 \tag{E11.7.7}$$

For a nontrivial solution of C_2 and C_4 in Eqs. (E11.7.6) and (E11.7.7), the determinant of their coefficient matrix is set equal to zero. This leads to the frequency equation

$$\sinh s_1 l \sin s_2 l = 0 \qquad \text{(E11.7.8)}$$

Noting that $s_1 l \geq 0$, the roots of Eq. (E11.7.8) are given by

$$s_2 l = n\pi, \qquad n = 0, 1, 2, \ldots \qquad \text{(E11.7.9)}$$

Equations (E11.7.9) and (11.186) yield the natural frequencies of the beam as

$$\omega_n = \frac{\pi^2}{l^2} \sqrt{\frac{EI}{\rho A}} \left(n^4 + \frac{n^2 P l^2}{\pi^2 EI} \right)^{1/2} \qquad \text{(E11.7.10)}$$

If the axial force is compressive (P is negative), Eq. (E11.7.10) can be rewritten as

$$\omega_n = \frac{\pi^2}{l^2} \left(\frac{EI}{\rho A} \right)^{1/2} \left(n^4 - n^2 \frac{P}{P_{\text{cri}}} \right)^{1/2} \qquad \text{(E11.7.11)}$$

where

$$P_{\text{cri}} = \frac{\pi^2 EI}{l^2} \qquad \text{(E11.7.12)}$$

is the smallest Euler buckling load of a simply supported beam under compressive load.

Notes:

1. If $P = 0$, Eq. (E11.7.11) reduces to Eq. (11.50), which gives the natural frequencies of vibration of a simply supported beam.
2. If $EI = 0$, Eq. (E11.7.10) reduces to Eq. (8.103), which gives the natural frequencies of vibration of a taut string fixed at both ends.
3. If $P \rightarrow P_{\text{cri}}$, the fundamental natural frequency of vibration approaches zero ($\omega_1 \rightarrow 0$).
4. If $P > 0$, the values of the natural frequencies increase due to the stiffening of the beam.

11.11   VIBRATION OF A ROTATING BEAM

Let a uniform beam rotate about an axis parallel to the z axis at a constant angular velocity Ω. The radius of the hub r is considered to be negligibly small (Fig. 11.13). The beam is assumed to be fixed at $x = 0$ and free at $x = l$. At any point x along the beam, the centrifugal force induces an axial force, $P(x)$, given by

$$P(x) = \int_x^l \rho A \Omega^2 \eta \, d\eta = \frac{1}{2} \rho A \Omega^2 l^2 \left(1 - \frac{x^2}{l^2} \right) \qquad \text{(11.188)}$$

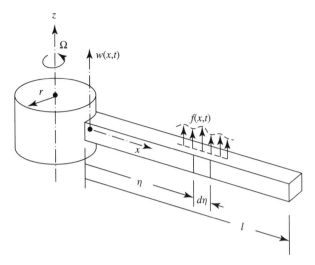

Figure 11.13

Note that the force induced due to the longitudinal elastic displacement of the beam is neglected in deriving Eq. (11.188). The equation of motion for the transverse vibration of the rotating beam can be obtained using Eq. (11.188) for P in Eq. (11.178):

$$EI\frac{\partial^4 w}{\partial x^4} + \rho A\frac{\partial^2 w}{\partial t^2} - \frac{1}{2}\rho A\Omega^2 l^2 \frac{\partial}{\partial x}\left[\left(1 - \frac{x^2}{l^2}\right)\frac{\partial w}{\partial x}\right] = f(x, t) \qquad (11.189)$$

Because of the coordinate system indicated in Fig. 11.13, the axial force, P, given by Eq. (11.188) will be zero at $x = l$, and hence the boundary conditions of the beam will be same as those of a nonrotating beam:

$$w = 0 \quad \text{and} \quad \frac{\partial w}{\partial x} = 0 \qquad \text{at } x = 0 \qquad (11.190)$$

and

$$EI\frac{\partial^2 w}{\partial x^2} = 0 \quad \text{and} \quad EI\frac{\partial^3 w}{\partial x^3} = 0 \quad \text{at} \quad x = l \qquad (11.191)$$

For the free vibration of the rotating beam, a harmonic solution of the form

$$w(x, t) = W(x)\cos(\omega t - \phi) \qquad (11.192)$$

is assumed. Using Eq. (11.192), Eqs. (11.189)–(11.191) can be expressed as

$$EI\frac{d^4 W}{dx^4} - \frac{1}{2}\rho A\Omega^2 l^2 \frac{d}{dx}\left[\left(1 - \frac{x^2}{l^2}\right)\frac{dW}{dx}\right] = \omega^2 \rho A W \qquad (11.193)$$

or

$$EI\frac{d^4W}{dx^4} - \frac{1}{2}\rho A\Omega^2(l^2 - x^2)\frac{d^2W}{dx^2} + \rho A\Omega^2 lx\frac{dW}{dx} - \rho A\omega^2 W = 0 \quad (11.194)$$

$$W(x) = \frac{dW(x)}{dx} = 0, \qquad x = 0 \tag{11.195}$$

$$\frac{d^2W(x)}{dx^2} = \frac{d^3W(x)}{dx^3} = 0, \qquad x = l \tag{11.196}$$

The exact solution of the problem defined by Eqs. (11.194)–(11.196) is difficult to find. However, approximate solutions can be found using the methods of Chapter 17.

11.12   NATURAL FREQUENCIES OF CONTINUOUS BEAMS ON MANY SUPPORTS

Consider a continuous beam supported at n points as shown in Fig. 11.14. For the vibration analysis of the beam, we consider the span (or section) between each pair of consecutive supports as a separate beam with its origin at the left support of the span. Hence the solution given by Eq. (11.36) or (11.37) is valid for each span of the beam. Thus, the characteristic function or normal mode of span i can be expressed, using Eq. (11.36), as

$$W_i(x) = A_i \cos \beta_i x + B_i \sin \beta_i x + C_i \cosh \beta_i x + D_i \sinh \beta_i x, \qquad i = 1, 2, \ldots, n-1 \tag{11.197}$$

where

$$\beta_i = \left(\frac{\rho_i A_i \omega^2}{E_i I_i}\right)^{1/4}, \qquad i = 1, 2, \ldots, n-1 \tag{11.198}$$

and $\rho_i, A_i, E_i,$ and I_i denote the values of $\rho, A, E,$ and I, respectively, for span i. The following conditions are used to evaluate the constants $A_i, B_i, C_i,$ and $D_i, i = 1, 2, \ldots, n-1$:

1. The deflection is zero at the origin of each span (except possibly the first span):

$$W_i(0) = 0 \tag{11.199}$$

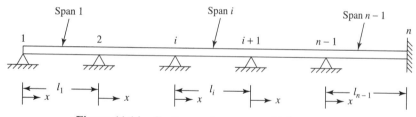

Figure 11.14   Continuous beam on multiple supports.

2. At the end of each span, the deflection is zero (except possibly when $i = n - 1$) since the deflection is zero at each intermediate support.

$$W_i(l_i) = 0 \tag{11.200}$$

3. Since the beam is continuous, the slope and bending moment just to the left and to the right of any intermediate support are the same. Thus,

$$\frac{d W_{i-1}(l_{i-1})}{dx} = \frac{d W_i(0)}{dx} \tag{11.201}$$

$$E_{i-1}I_{i-1}\frac{d^2 W_{i-1}(l_{i-1})}{dx^2} = E_i I_i \frac{d^2 W_i(0)}{dx^2} \tag{11.202}$$

4. At each of the end supports 1 and n, two boundary conditions can be written, depending on the nature of support (such as fixed, simply supported, or free condition). When Eqs. (11.199)–(11.202) are applied for each span of the beam ($i = 1, 2, \ldots, n - 1$), along with the boundary conditions at each end of the beam, we get a total of $4(n - 1)$ homogeneous algebraic equations in the unknown constants A_i, B_i, C_i, and D_i, $i = 1, 2, \ldots, n - 1$. Using the condition for the nontrivial solution of the constants, we can obtain the frequency equation of the system.

Example 11.8 Determine the natural frequencies and mode shapes of a beam resting on three simple supports as shown in Fig. 11.15(*a*). Assume the beam to be uniform with $l_1 = l_2 = l$.

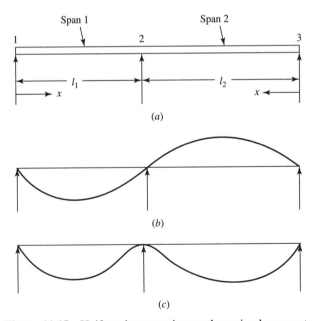

Figure 11.15 Uniform beam resting on three simple supports.

SOLUTION   The characteristic functions in the two spans of the beam can be expressed as

$$W_1(x) = A_1 \cos \beta_1 x + B_1 \sin \beta_1 x + C_1 \cosh \beta_1 x + D_1 \sinh \beta_1 x \tag{E11.8.1}$$

$$W_2(x) = A_2 \cos \beta_2 x + B_2 \sin \beta_2 x + C_2 \cosh \beta_2 x + D_2 \sinh \beta_2 x \tag{E11.8.2}$$

To simplify the computations, the x axis is taken from support 1 to the right for span 1 and from support 3 to the left for span 2. The simply supported end conditions at support 1 are given by

$$W_1(0) = 0 \tag{E11.8.3}$$

$$E_1 I_1 \frac{d^2 W_1(0)}{dx^2} = 0 \quad \text{or} \quad \frac{d^2 W_1(0)}{dx^2} = 0 \tag{E11.8.4}$$

Equations (E11.8.1), (E11.8.3), and (E11.8.4) give

$$A_1 + C_1 = 0$$

$$-A_1 + C_1 = 0$$

or

$$A_1 = C_1 = 0 \tag{E11.8.5}$$

Thus, Eq. (E11.8.1) reduces to

$$W_1(x) = B_1 \sin \beta_1 x + D_1 \sinh \beta_1 x \tag{E11.8.6}$$

Since the displacement is zero at support 2, $W_1(l_1) = 0$ and hence Eq. (E11.8.6) gives

$$B_1 \sin \beta_1 l_1 + D_1 \sinh \beta_1 l_1 = 0$$

or

$$D_1 = -B_1 \frac{\sin \beta_1 l_1}{\sinh \beta_1 l_1} \tag{E11.8.7}$$

Thus, Eq. (E11.8.6) can be written as

$$W_1(x) = B_1 \left(\sin \beta_1 x - \frac{\sin \beta_1 l_1}{\sinh \beta_1 l_1} \sinh \beta_1 x \right) \tag{E11.8.8}$$

Next, the simply supported end conditions at support 3 are given by

$$W_2(0) = 0 \tag{E11.8.9}$$

$$E_2 I_2 \frac{d^2 W_2(0)}{dx^2} = 0 \quad \text{or} \quad \frac{d^2 W_2(0)}{dx^2} = 0 \tag{E11.8.10}$$

Equations (E11.8.2), (E11.8.9), and (E11.8.10) yield

$$A_2 + C_2 = 0$$

$$-A_2 + C_2 = 0$$

or

$$A_2 = C_2 = 0 \qquad \text{(E11.8.11)}$$

Thus, Eq. (E11.8.2) reduces to

$$W_2(x) = B_2 \sin \beta_2 x + D_2 \sinh \beta_2 x \qquad \text{(E11.8.12)}$$

Using the condition that the displacement at support 2 is zero in Eq. (E11.8.12), we obtain

$$D_2 = -B_2 \frac{\sin \beta_2 l_2}{\sinh \beta_2 l_2} \qquad \text{(E11.8.13)}$$

and hence Eq. (E11.8.12) becomes

$$W_2(x) = B_2 \left(\sin \beta_2 x - \frac{\sin \beta_2 l_2}{\sinh \beta_2 l_2} \sinh \beta_2 x \right) \qquad \text{(E11.8.14)}$$

The slope is continuous at support 2. This yields

$$\frac{d W_1(l_1)}{dx} = -\frac{d W_2(l_2)}{dx} \qquad \text{(E11.8.15)}$$

which can be expressed, using Eqs. (E11.8.8) and (E11.8.14), as,

$$B_1 \beta_1 \left(\cos \beta_1 l_1 - \frac{\sin \beta_1 l_1}{\sinh \beta_1 l_1} \cosh \beta_1 l_1 \right) + B_2 \beta_2 \left(\cos \beta_2 l_2 - \frac{\sin \beta_2 l_2}{\sinh \beta_2 l_2} \cosh \beta_2 l_2 \right) = 0$$
$$\text{(E11.8.16)}$$

Finally, the bending moment is continuous at support 2. This leads to

$$E_1 I_1 \frac{d^2 W_1(l_1)}{dx^2} = E_2 I_2 \frac{d^2 W_2(l_2)}{dx^2} \qquad \text{(E11.8.17)}$$

which becomes, in view of Eqs. (E11.8.8) and (E11.8.14),

$$B_1 E_1 I_1 \beta_1^2 \sin \beta_1 l_1 - B_2 E_2 I_2 \beta_2^2 \sin \beta_2 l_2 = 0 \qquad \text{(E11.8.18)}$$

Equations (E11.8.16) and (E11.8.18) denote two simultaneous homogeneous algebraic equations in the unknown constants B_1 and B_2. For a nontrivial solution of these constants, the determinant of their coefficient matrix is set equal to zero. This gives the frequency equation as

$$\beta_1 \left(\cos \beta_1 l_1 - \frac{\sin \beta_1 l_1}{\sinh \beta_1 l_1} \cosh \beta_1 l_1 \right) E_2 I_2 \beta_2^2 \sin \beta_2 l_2$$

$$+ \beta_2 \left(\cos \beta_2 l_2 - \frac{\sin \beta_2 l_2}{\sinh \beta_2 l_2} \cosh \beta_2 l_2 \right) E_1 I_1 \beta_1^2 \sin \beta_1 l_1 = 0 \qquad \text{(E11.8.19)}$$

For a uniform beam with identical spans, $E_1 = E_2 = E$, $I_1 = I_2 = I$, $\beta_1 = \beta_2 = \beta$, and $l_1 = l_2 = l$, and Eq. (E11.8.19) reduces to

$$(\cos \beta l - \sin \beta l \coth \beta l) \sin \beta l = 0 \qquad \text{(E11.8.20)}$$

Equation (E11.8.20) will be satisfied when

$$\sin \beta l = 0 \qquad \text{(E11.8.21)}$$

or

$$\tan \beta l - \tanh \beta l = 0 \qquad \text{(E11.8.22)}$$

Case (i): When $\sin \beta l = 0$ This condition gives the natural frequencies as

$$\beta_n l = n\pi, \qquad n = 1, 2, \ldots$$

or

$$\omega_n = n^2 \pi^2 \sqrt{\frac{EI}{\rho A l^4}}, \qquad n = 1, 2, \ldots \qquad \text{(E11.8.23)}$$

These natural frequencies can be seen to be identical to those of a beam of length l simply supported at both ends. When Eq. (E11.8.21) is valid, Eq. (E11.8.16) gives

$$B_1 = -B_2 \qquad \text{(E11.8.24)}$$

and hence the mode shape becomes

$$W_{1n}(x) = C_{2n}\left(\sin \beta_n x - \frac{\sin \beta_n l}{\sinh \beta_n l} \sinh \beta_n x\right) \qquad \text{for span 1}$$

$$W_{2n}(x) = -C_{2n}\left(\sin \beta_n x - \frac{\sin \beta_n l}{\sinh \beta_n l} \sinh \beta_n x\right) \qquad \text{for span 2} \qquad \text{(E11.8.25)}$$

where the constants C_{2n} can be assumed to be 1, for simplicity. The mode shape given by Eq. (E11.8.25) denotes an antisymmetric mode with respect to the middle support 2, as shown in Fig. 11.15(b).

Case (ii): When $\tan \beta l - \tanh \beta l = 0$ This condition can be seen to correspond to a beam of length l fixed at one end and simply supported at the other end. The roots, $\beta_n l$, of Eq. (E11.8.22) are given in Fig. 11.4. When Eq. (E11.8.22) is valid, Eq. (E11.8.18) gives $B_1 = B_2$ and the mode shape becomes

$$W_{1n}(x) = W_{2n}(x) = C_{2n}\left(\sin \beta_n x - \frac{\sin \beta_n l}{\sinh \beta_n l} \sinh \beta_n x\right) \qquad \text{for spans 1 and 2}$$

$$\text{(E11.8.26)}$$

where the constant C_{2n} can be assumed to be 1, for simplicity. Notice that the mode shape given by Eq. (E11.8.26) indicates a symmetric mode with respect to the middle support 2, as shown in Fig. 11.15(c).

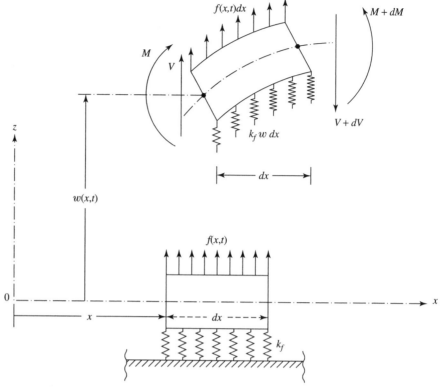

Figure 11.16 Free-body diagram of a beam on an elastic foundation.

11.13 BEAM ON AN ELASTIC FOUNDATION

Let a uniform beam rest on an elastic foundation, such as a rail track on soil, as shown in Fig. 11.16. The continuous elastic foundation is denoted by a large number of closely spaced translational springs. The load per unit length of the beam necessary to cause the foundation to deflect by a unit amount, called the *foundation modulus*, is assumed to be k_f. By considering a small element of the vibrating beam on an elastic foundation, the equation of motion of the beam can be expressed as

$$\frac{\partial^2}{\partial x^2}\left(EI\frac{\partial^2 w}{\partial x^2}\right) + \rho A\frac{\partial^2 w}{\partial t^2} + k_f w = f(x, t) \tag{11.203}$$

where $f(x, t)$ denotes the distributed load on the beam.

11.13.1 Free Vibration

For the free vibration of a uniform beam, Eq. (11.203) reduces to

$$EI\frac{\partial^4 w}{\partial x^4} + \rho A\frac{\partial^2 w}{\partial t^2} + k_f w = 0 \tag{11.204}$$

The free vibration solution of the beam is expressed as

$$w(x, t) = \sum_{i=1}^{\infty} W_i(x)(C_i \cos \omega_i t + D_i \sin \omega_i t) \tag{11.205}$$

where ω_i is the ith natural frequency and $W_i(x)$ is the corresponding natural mode shape of the beam. Substitution of the ith modal solution into Eq. (11.204) yields

$$\frac{d^4 W_i(x)}{dx^4} + \left(-\frac{\rho A \omega_i^2}{EI} + \frac{k_f}{EI} \right) W_i(x) = 0 \tag{11.206}$$

Defining

$$\alpha_i^4 = \left(-\frac{k_f}{EI} + \frac{\omega_i^2}{c^2} \right) \quad \text{and} \quad c = \sqrt{\frac{EI}{\rho A}} \tag{11.207}$$

Eq. (11.206) can be written as

$$\frac{d^4 W_i(x)}{dx^4} - \alpha_i^4 W_i(x) = 0 \tag{11.208}$$

The solution of Eq. (11.208) can be expressed as

$$W_i(x) = C_{1i} \cos \alpha_i x + C_{2i} \sin \alpha_i x + C_{3i} \cosh \alpha_i x + C_{4i} \sinh \alpha_i x \tag{11.209}$$

where the constants C_{1i}, C_{2i}, C_{3i}, and C_{4i} can be evaluated from the boundary conditions of the beam. Noting that Eq. (11.209) has the same form as that of a beam without a foundation [see Eq. (11.36)], the solutions obtained for beams with different end conditions in Section 11.5 are applicable to this case also if β_i is replaced by α_i. The natural frequencies of the beam on an elastic foundation are given by Eq. (11.207):

$$\omega_i = c\alpha_i^2 \sqrt{1 + \frac{k_f}{EI\alpha_i^4}} \tag{11.210}$$

Assuming the beam on elastic foundation to be simply supported at the ends, the normal modes can be expressed as

$$W_i(x) = C_i \sin \alpha_i x \tag{11.211}$$

where C_i is a constant. The natural frequencies of the beam can be found as

$$\alpha_i l = i\pi, \quad i = 1, 2, 3, \ldots$$

or

$$\omega_i = \frac{i^2 \pi^2 c}{l^2} \sqrt{1 + \frac{k_f l^4}{EI i^4 \pi^4}}, \quad i = 1, 2, 3, \ldots \tag{11.212}$$

The free-end forced response of a simply supported beam on an elastic foundation can be found from the corresponding results of a simply supported beam with no foundation by using Eq. (11.212) for ω_i instead of Eq.(11.50).

11.13.2 Forced Vibration

The forced transverse vibration of a uniform beam on elastic foundation is governed by the equation

$$\frac{\partial^2 w(x,t)}{\partial t^2} + \frac{k_f}{\rho A} w(x,t) + \frac{EI}{\rho A} \frac{\partial^4 w(x,t)}{\partial x^4} = \frac{f(x,t)}{\rho A} \tag{11.213}$$

When the normal mode method is used, the solution of Eq. (11.213) can be expressed as

$$w(x,t) = \sum_{n=1}^{\infty} W_n(x) \eta_n(t) \tag{11.214}$$

where $W_n(x)$ is the nth normal mode and η_n is the corresponding generalized coordinate of the beam. Noting that the normal mode $W_n(x)$ satisfies the relation [see Eq. (11.206)]

$$\frac{d^4 W_n(x)}{dx^4} = \left(\frac{\rho A \omega_n^2}{EI} - \frac{k_f}{EI} \right) W_n(x) \tag{11.215}$$

Eq. (11.213) can be reduced to

$$\frac{d^2 \eta_n(t)}{dt^2} + \omega_n^2 \eta_n(t) = Q_n(t), \qquad n = 1, 2, \ldots \tag{11.216}$$

where the natural frequency, ω_n, is given by Eq.(11.210) and the generalized force, $Q_n(t)$, by

$$Q_n(t) = \int_0^l W_n(x) f(x,t) \, dx \tag{11.217}$$

The solution of Eq. (11.216) can be expressed as

$$\eta_n(t) = \frac{1}{\omega_n} \int_0^t Q_n(\tau) \sin \omega_n(t - \tau) \, d\tau$$
$$+ \eta_n(0) \cos \omega_n t + \dot{\eta}_n(0) \frac{\sin \omega_n t}{\omega_n}, \qquad n = 1, 2, \ldots \tag{11.218}$$

where the initial values of the generalized displacement $\eta_n(0)$ and generalized velocity $\dot{\eta}_n(0)$ are determined from the initial conditions of $w(x,t)$ [see Eqs. (11.138) and (11.139)]:

$$\eta_n(0) = \int_0^l \rho A(x) W_n(x) w_0(x) \, dx \tag{11.219}$$

$$\dot{\eta}_n(0) = \int_0^l \rho A(x) W_n(x) \dot{w}_0(x) \, dx \tag{11.220}$$

where $w_0(x) = w(x,0)$ and $\dot{w}_0(x) = \dot{w}(x,0)$.

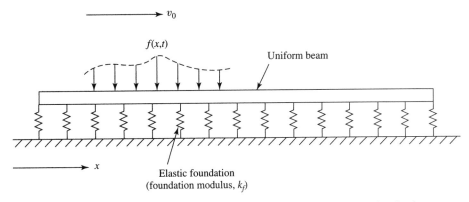

Figure 11.17   Beam on an elastic foundation subjected to a moving load.

11.13.3   Beam on an Elastic Foundation Subjected to a Moving Load

Let an infinitely long uniform beam on an elastic foundation be subjected to a distributed transverse load $f(x, t)$ traveling at a constant speed v_0 along the beam as shown in Fig. 11.17. The vibration response of a railroad track under the moving weight of the rail can be determined using the present analysis. Since the load moves at a constant speed along x, the distributed load can be denoted in terms of x and t as $f(x - v_0 t)$. The equation of motion for the transverse vibration of the beam is given by

$$EI\frac{\partial^4 w(x, t)}{\partial x^4} + \rho A\frac{\partial^2 w(x, t)}{\partial t^2} + k_f w(x, t) = f(x - v_0 t) \tag{11.221}$$

Using

$$z = x - v_0 t \tag{11.222}$$

Eq. (11.221) can be rewritten as

$$EI\frac{d^4 w(z)}{dz^4} + \rho A v_0^2\frac{d^2 w(z)}{dz^2} + k_f w(z) = f(z) \tag{11.223}$$

If $f(x, t)$ is a concentrated load F_0 moving along the beam with a constant velocity v_0, the equation of motion can be written as

$$EI\frac{d^4 w(z)}{dz^4} + \rho A v_0^2\frac{d^2 w(z)}{dz^2} + k_f w(z) = 0 \tag{11.224}$$

and the concentrated load F_0 is incorporated as a known shear force at $z = 0$ into the solution. The solution of Eq. (11.224) is assumed to be

$$w(z) = e^{\mu z} \tag{11.225}$$

Substitution of Eq. (11.225) into Eq. (11.224) yields the auxiliary equation

$$EI\mu^4 + \rho A v_0^2 \mu^2 + k_f = 0 \tag{11.226}$$

The roots of Eq. (11.226) can be expressed as

$$\mu_{1,2} = \pm i\sqrt{\alpha - \sqrt{\beta}} \tag{11.227}$$

$$\mu_{3,4} = \pm i\sqrt{\alpha + \sqrt{\beta}} \tag{11.228}$$

where

$$\alpha = \frac{\rho A v_0^2}{2EI} \tag{11.229}$$

$$\beta = \frac{\rho^2 A^2 v_0^4}{4E^2 I^2} - \frac{k_f}{EI} \tag{11.230}$$

Thus, the solution of Eq. (11.224) becomes

$$w(z) = C_1 e^{i\sqrt{\alpha - \sqrt{\beta}}z} + C_2 e^{-i\sqrt{\alpha - \sqrt{\beta}}z} + C_3 e^{i\sqrt{\alpha + \sqrt{\beta}}z} + C_4 e^{-i\sqrt{\alpha + \sqrt{\beta}}z} \tag{11.231}$$

where the constants C_1, C_2, C_3, and C_4 can be determined using the conditions

$$w = 0 \quad \text{at} \quad z = \infty \tag{11.232}$$

$$\frac{d^2 w}{dz^2} = 0 \quad \text{at} \quad z = \infty \tag{11.233}$$

$$\frac{dw}{dz} = 0 \quad \text{at} \quad z = 0 \tag{11.234}$$

$$EI\frac{d^3 w}{dz^3} = \frac{F_0}{2} \quad \text{at} \quad z = 0 \tag{11.235}$$

[Note that the concentrated load F_0 at $z = 0$ causes discontinuity of the shear force. By considering the shear forces immediately on the left- and right-hand sides of F_0, we obtain

$$-EI\frac{d^3 w}{dz^3}(z = 0^+) + EI\frac{d^3 w}{dz^3}(z = 0^-) = F_0 \tag{11.236}$$

Due to symmetry at $z = 0$, Eq. (11.236) yields Eq.(11.235) as $z \to 0$.]

To satisfy Eqs. (11.232) and (11.233), C_1 and C_3 must be zero in Eq. (11.231). This gives

$$w(z) = C_2 e^{-i\sqrt{\alpha - \sqrt{\beta}}z} + C_4 e^{-i\sqrt{\alpha + \sqrt{\beta}}z} \tag{11.237}$$

The use of conditions (11.234) and (11.235) in Eq. (11.237) yields

$$-iC_2\sqrt{\alpha - \sqrt{\beta}} - iC_4\sqrt{\alpha + \sqrt{\beta}} = 0 \tag{11.238}$$

$$iC_2\left(\sqrt{\alpha - \sqrt{\beta}}\right)^3 + iC_4\left(\sqrt{\alpha + \sqrt{\beta}}\right)^3 = \frac{F_0}{2} \tag{11.239}$$

The solution of Eqs. (11.238) and (11.239) is given by

$$C_2 = -\frac{F_0}{i \cdot 4EI\sqrt{\beta}\sqrt{\alpha - \sqrt{\beta}}} \tag{11.240}$$

$$C_4 = \frac{F_0}{i \cdot 4EI\sqrt{\beta}\sqrt{\alpha + \sqrt{\beta}}} \tag{11.241}$$

Thus, the solution of Eq. (11.224) can be expressed as

$$w(z) = -\frac{F_0}{i \cdot 4EI\sqrt{\beta}\sqrt{\alpha - \sqrt{\beta}}}e^{-i\sqrt{\alpha - \sqrt{\beta}}z} + \frac{F_0}{i \cdot 4EI\sqrt{\beta}\sqrt{\alpha + \sqrt{\beta}}}e^{-i\sqrt{\alpha + \sqrt{\beta}}z} \tag{11.242}$$

11.14 RAYLEIGH'S THEORY

The inertia due to the axial displacement of the beam is included in Rayleigh's theory. This effect is called *rotary* (or *rotatory*) *inertia*. The reason is that since the cross section remains plane during motion, the axial motion of points located in any cross section undergoes rotary motion about the y axis. Using $u = -z(\partial w/\partial x)$ from Eq. (11.1), the axial velocity is given by

$$\frac{\partial u}{\partial t} = -z\frac{\partial^2 w}{\partial t\,\partial x} \tag{11.243}$$

and hence the kinetic energy associated with the axial motion is given by

$$T_a = \frac{1}{2}\int_0^l \iint_A \rho\left(\frac{\partial u}{\partial t}\right)^2 dA\,dx = \frac{1}{2}\int_0^l \left(\iint_A z^2\,dA\right)\rho\left(\frac{\partial^2 w}{\partial t\,\partial x}\right)^2 dx$$

$$= \frac{1}{2}\int_0^l \rho I\left(\frac{\partial^2 w}{\partial t\,\partial x}\right)^2 dx \tag{11.244}$$

The term associated with T_a in Hamilton's principle can be evaluated as

$$I_a = \delta\int_{t_1}^{t_2} T_a\,dt = \delta\int_{t_1}^{t_2}\frac{1}{2}\int_0^l \rho I\left(\frac{\partial^2 w}{\partial t\,\partial x}\right)^2 dx\,dt$$

$$= \int_{t_1}^{t_2}\int_0^l \rho I\frac{\partial^2 w}{\partial t\,\partial x}\,\delta\left(\frac{\partial^2 w}{\partial t\,\partial x}\right) dx\,dt \tag{11.245}$$

Using integration by parts with respect to time, Eq. (11.245) gives

$$I_a = -\int_{t_1}^{t_2}\int_0^l \rho I\frac{\partial^3 w}{\partial t^2\,\partial x}\,\delta\left(\frac{\partial w}{\partial x}\right) dx\,dt \tag{11.246}$$

Using integration by parts with respect to x, Eq. (11.246) yields

$$I_a = \int_{t_1}^{t_2}\left[-\rho I\frac{\partial^3 w}{\partial t^2\,\partial x}\,\delta w\Big|_0^l + \int_0^l \frac{\partial}{\partial x}\left(\rho I\frac{\partial^3 w}{\partial t^2\,\partial x}\right)\delta w\,dx\right] dt \tag{11.247}$$

When $-I_a$ is added to Eq. (11.11), the equation of motion, Eq. (11.12), and the boundary conditions, Eq.(11.13), will be modified as follows (by neglecting the springs and masses at the ends):

$$\frac{\partial^2}{\partial x^2}\left(EI\frac{\partial^2 w}{\partial x^2}\right) - \frac{\partial}{\partial x}\left(\rho I\frac{\partial^3 w}{\partial t^2 \partial x}\right) + \rho A\frac{\partial^2 w}{\partial t^2} = f(x,t) \tag{11.248}$$

$$EI\frac{\partial^2 w}{\partial x^2}\delta\left(\frac{\partial w}{\partial x}\right)\Big|_0^l - \left[\frac{\partial}{\partial x}\left(EI\frac{\partial^2 w}{\partial x^2}\right) - \rho I\frac{\partial^3 w}{\partial t^2 \partial x}\right]\delta w\Big|_0^l = 0 \tag{11.249}$$

For a uniform beam, the equation of motion and the boundary conditions can be expressed as

$$EI\frac{\partial^4 w}{\partial x^4} + \rho A\frac{\partial^2 w}{\partial t^2} - \rho I\frac{\partial^4 w}{\partial x^2 \partial t^2} - f = 0 \tag{11.250}$$

$$EI\frac{\partial^2 w}{\partial x^2}\delta\left(\frac{\partial w}{\partial x}\right)\Big|_0^l = 0 \tag{11.251}$$

$$\left(EI\frac{\partial^3 w}{\partial x^3} - \rho I\frac{\partial^3 w}{\partial x \partial t^2}\right)\delta w\Big|_0^l = 0 \tag{11.252}$$

For free vibration, $f(x,t) = 0$, and Eq. (11.250) becomes

$$EI\frac{\partial^4 w}{\partial x^4} - \rho I\frac{\partial^4 w}{\partial x^2 \partial t^2} + \rho A\frac{\partial^2 w}{\partial t^2} = 0 \tag{11.253}$$

For harmonic oscillations, the solution is assumed as

$$w(x,t) = W(x)\cos(\omega t - \phi) \tag{11.254}$$

Equations (11.253) and (11.254) lead to

$$EI\frac{d^4 W}{dx^4} + \rho I\omega^2\frac{d^2 W}{dx^2} - \rho A\omega^2 W = 0 \tag{11.255}$$

Using

$$W(x) = e^{sx} \tag{11.256}$$

the auxiliary equation corresponding to Eq. (11.255) can be derived as

$$EIs^4 + \rho I\omega^2 s^2 - \rho A\omega^2 = 0 \tag{11.257}$$

Denoting the roots of Eq. (11.257) as s_1, s_2, s_3, and s_4, the solution of Eq. (11.255) can be expressed as

$$W(x) = \sum_{i=1}^{4} C_i e^{s_i x} \tag{11.258}$$

where the constants C_1, C_2, C_3, and C_4 can be determined from the boundary conditions.

Example 11.9 Determine the natural frequencies of vibration of a simply supported Rayleigh beam.

SOLUTION   By introducing the parameters

$$\alpha^2 = \frac{EI}{\rho A} \tag{E11.9.1}$$

$$r^2 = \frac{I}{A} \tag{E11.9.2}$$

the equation of motion, Eq. (11.253), can be written for free vibration as

$$\alpha^2 \frac{\partial^4 w}{\partial x^4} + \frac{\partial^2 w}{\partial t^2} - r^2 \frac{\partial^4 w}{\partial x^2 \partial t^2} = 0 \tag{E11.9.3}$$

The solution of Eq. (E11.9.3) can be assumed as

$$w(x, t) = C \, \sin \frac{n\pi x}{l} \cos \omega_n t \tag{E11.9.4}$$

where C is a constant and ω_n denotes the nth natural frequency of vibration. Equation (E11.9.4) can be seen to satisfy the boundary conditions of Eqs. (11.251) and (11.252). By substituting Eq. (E11.9.4) into Eq. (E11.9.3), we obtain the frequency equation as

$$\alpha^2 \left(\frac{n\pi}{l}\right)^4 - \omega_n^2 \left(1 + \frac{n^2 \pi^2 r^2}{l^2}\right) = 0 \tag{E11.9.5}$$

Equation (E11.9.5) gives the natural frequencies of vibration as

$$\omega_n^2 = \frac{\alpha^2 (n\pi/l)^4}{1 + (n^2 \pi^2/l^2) r^2}, \qquad n = 1, 2, \dots \tag{E11.9.6}$$

11.15   TIMOSHENKO'S THEORY

11.15.1   Equations of Motion

The effect of shear deformation, in addition to the effect of rotary inertia, is considered in this theory. To include the effect of shear deformation, first consider a beam undergoing only shear deformation as indicated in Fig. 11.18. Here a vertical section, such as PQ, before deformation remains vertical ($P'Q'$) after deformation but moves by a distance w in the z direction. Thus, the components of displacement of a point in the beam are given by

$$u = 0, \qquad v = 0, \qquad w = w(x, t) \tag{11.259}$$

The components of strain can be found as

$$\varepsilon_{xx} = \frac{\partial u}{\partial x} = 0, \qquad \varepsilon_{yy} = \frac{\partial v}{\partial y} = 0, \qquad \varepsilon_{zz} = \frac{\partial w}{\partial z} = 0 \qquad \varepsilon_{xy} = \frac{\partial u}{\partial y} + \frac{\partial v}{\partial x} = 0,$$

$$\varepsilon_{yz} = \frac{\partial v}{\partial z} + \frac{\partial w}{\partial y} = 0, \qquad \varepsilon_{zx} = \frac{\partial u}{\partial z} + \frac{\partial w}{\partial x} = \frac{\partial w}{\partial x} \tag{11.260}$$

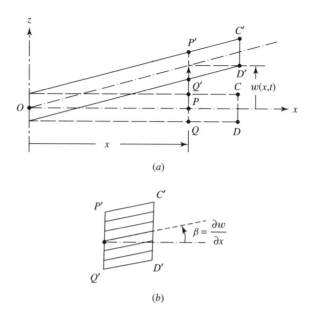

Figure 11.18   Beam in shear deformation.

The shear strain ε_{zx} is the same as the rotation $\beta(x, t) = \partial w / \partial x(x, t)$ experienced by any fiber located parallel to the centerline of the beam, as shown in Fig. 11.18(b). The components of stress corresponding to the strains indicated in Eq. (11.260) are given by

$$\sigma_{xx} = \sigma_{yy} = \sigma_{zz} = \sigma_{xy} = \sigma_{yz} = 0, \qquad \sigma_{zx} = G\varepsilon_{zx} = G\frac{\partial w}{\partial x} \tag{11.261}$$

Equation (11.261) states that the shear stress σ_{zx} is the same (uniform) at every point in the cross section of the beam. Since this is not true in reality, Timoshenko used a constant k, known as the *shear correction factor*, in the expression for σ_{zx} as

$$\sigma_{zx} = kG\frac{\partial w}{\partial x} \tag{11.262}$$

The total transverse displacement of the centerline of the beam is given by (see Fig. 11.19):

$$w = w_s + w_b \tag{11.263}$$

and hence the total slope of the deflected centerline of the beam is given by

$$\frac{\partial w}{\partial x} = \frac{\partial w_s}{\partial x} + \frac{\partial w_b}{\partial x} \tag{11.264}$$

Since the cross section of the beam undergoes rotation due only to bending, the rotation of the cross section can be expressed as

$$\phi = \frac{\partial w_b}{\partial x} = \frac{\partial w}{\partial x} - \frac{\partial w_s}{\partial x} = \frac{\partial w}{\partial x} - \beta \tag{11.265}$$

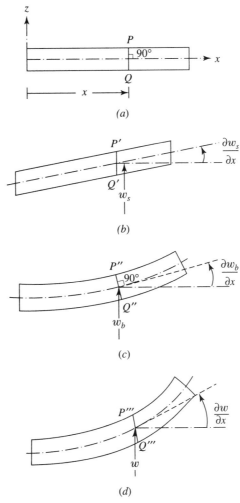

Figure 11.19   Bending and shear deformations: (*a*) element with no deformation; (*b*) Element with only shear deformation; (*c*) element with only bending deformation; (*d*) element with total deformation.

where $\beta = \partial w_s / \partial x$ is the shear deformation or shear angle. An element of fiber located at a distance z from the centerline undergoes axial displacement due only to the rotation of the cross section (shear deformation does not cause any axial displacement), and hence the components of displacement can be expressed as

$$u = -z\left(\frac{\partial w}{\partial x} - \beta\right) \equiv -z\phi(x, t)$$

$$v = 0 \tag{11.266}$$

$$w = w(x, t)$$

where the strains corresponding to the displacement field given by Eq. (11.266) are

$$\varepsilon_{xx} = \frac{\partial u}{\partial x} = -z\frac{\partial \phi}{\partial x}$$

$$\varepsilon_{yy} = \frac{\partial v}{\partial y} = 0$$

$$\varepsilon_{zz} = \frac{\partial w}{\partial z} = 0$$

$$\varepsilon_{xy} = \frac{\partial u}{\partial y} + \frac{\partial v}{\partial x} = 0 \qquad (11.267)$$

$$\varepsilon_{yz} = \frac{\partial w}{\partial y} + \frac{\partial v}{\partial z} = 0$$

$$\varepsilon_{zx} = \frac{\partial u}{\partial z} + \frac{\partial w}{\partial x} = -\phi + \frac{\partial w}{\partial x}$$

The components of stress corresponding to the strains of Eq. (11.267) are given by

$$\sigma_{xx} = -Ez\frac{\partial \phi}{\partial x}$$

$$\sigma_{zx} = kG\left(\frac{\partial w}{\partial x} - \phi\right) \qquad (11.268)$$

$$\sigma_{yy} = \sigma_{zz} = \sigma_{xy} = \sigma_{yz} = 0$$

The strain energy of the beam can be determined as

$$\pi = \frac{1}{2}\iiint_V (\sigma_{xx}\varepsilon_{xx} + \sigma_{yy}\varepsilon_{yy} + \sigma_{zz}\varepsilon_{zz} + \sigma_{xy}\varepsilon_{xy} + \sigma_{yz}\varepsilon_{yz} + \sigma_{zx}\varepsilon_{zx})\, dV$$

$$= \frac{1}{2}\int_0^l \iint_A \left[Ez^2\left(\frac{\partial \phi}{\partial x}\right)^2 + kG\left(\frac{\partial w}{\partial x} - \phi\right)^2\right] dA\, dx$$

$$= \frac{1}{2}\int_0^l \left[EI\left(\frac{\partial \phi}{\partial x}\right)^2 + kAG\left(\frac{\partial w}{\partial x} - \phi\right)^2\right] dx \qquad (11.269)$$

The kinetic energy of the beam, including rotary inertia, is given by

$$T = \frac{1}{2}\int_0^l \left[\rho A\left(\frac{\partial w}{\partial t}\right)^2 + \rho I\left(\frac{\partial \phi}{\partial t}\right)^2\right] dx \qquad (11.270)$$

The work done by the external distributed load $f(x, t)$ is given by

$$W = \int_0^l fw\, dx \qquad (11.271)$$

Application of extended Hamilton's principle gives

$$\delta \int_{t_1}^{t_2} (\pi - T - W)\,dt = 0$$

or

$$\int_{t_1}^{t_2} \left\{ \int_0^l \left[EI\frac{\partial \phi}{\partial x} \delta\left(\frac{\partial \phi}{\partial x}\right) + kAG\left(\frac{\partial w}{\partial x} - \phi\right)\delta\left(\frac{\partial w}{\partial x}\right) - kAG\left(\frac{\partial w}{\partial x} - \phi\right)\delta\phi \right] dx \right.$$
$$\left. - \int_0^l \left[\rho A\frac{\partial w}{\partial t} \delta\left(\frac{\partial w}{\partial t}\right) + \rho I\frac{\partial \phi}{\partial t}\delta\left(\frac{\partial \phi}{\partial t}\right) \right] dx - \int_0^l f\,\delta w\,dx \right\} dt = 0$$

(11.272)

The integrals in Eq. (11.272) can be evaluated using integration by parts (with respect to x or t) as follows.

$$\int_{t_1}^{t_2}\int_0^l EI\frac{\partial \phi}{\partial x}\delta\left(\frac{\partial \phi}{\partial x}\right)dx\,dt = \int_{t_1}^{t_2}\left[EI\frac{\partial \phi}{\partial x}\delta\phi\bigg|_0^l - \int_0^l \frac{\partial}{\partial x}\left(EI\frac{\partial \phi}{\partial x}\right)\delta\phi\,dx \right]dt$$

(11.273)

$$\int_{t_1}^{t_2}\int_0^l kAG\left(\frac{\partial w}{\partial x} - \phi\right)\delta\left(\frac{\partial w}{\partial x}\right)dx\,dt$$
$$= \int_{t_1}^{t_2}\left[kAG\left(\frac{\partial w}{\partial x} - \phi\right)\delta w\bigg|_0^l - \int_0^l kAG\frac{\partial}{\partial x}\left(\frac{\partial w}{\partial x} - \phi\right)\delta w\,dx \right]dt$$

(11.274)

$$-\int_{t_1}^{t_2}\int_0^l \rho A\frac{\partial w}{\partial t}\delta\left(\frac{\partial w}{\partial t}\right)dx\,dt = -\int_{t_1}^{t_2}\int_0^l \rho A\frac{\partial^2 w}{\partial t^2}\delta w\,dx\,dt \quad (11.275)$$

$$-\int_{t_1}^{t_2}\int_0^l \rho I\frac{\partial \phi}{\partial t}\delta\left(\frac{\partial \phi}{\partial t}\right)dx\,dt = \int_{t_1}^{t_2}\int_0^l \rho I\frac{\partial^2 \phi}{\partial t^2}\delta\phi\,dx\,dt \quad (11.276)$$

Substitution of Eqs. (11.273)–(11.276) into Eq. (11.269) leads to

$$\int_{t_1}^{t_2}\left\{ kAG\left(\frac{\partial w}{\partial x} - \phi\right)\delta w\bigg|_0^l + EI\frac{\partial \phi}{\partial x}\delta\phi\bigg|_0^l \right.$$
$$+ \int_0^l \left[-\frac{\partial}{\partial x}\left\langle kAG\left(\frac{\partial w}{\partial x} - \phi\right)\right\rangle + \rho A\frac{\partial^2 w}{\partial t^2} - f \right]\delta w\,dx$$
$$\left. + \int_0^l \left[-\frac{\partial}{\partial x}\left(EI\frac{\partial \phi}{\partial x}\right) - kAG\left(\frac{\partial w}{\partial x} - \phi\right) + \rho I\frac{\partial^2 \phi}{\partial t^2} \right]\delta\phi\,dx \right\}dt = 0 \quad (11.277)$$

Equation (11.277) gives the differential equations of motion for w and ϕ as

$$-\frac{\partial}{\partial x}\left[kAG\frac{\partial}{\partial x}\left(\frac{\partial w}{\partial x} - \phi\right)\right] + \rho A\frac{\partial^2 w}{\partial t^2} = f(x,t) \quad (11.278)$$

$$-\frac{\partial}{\partial x}\left(EI\frac{\partial \phi}{\partial x}\right) - kAG\left(\frac{\partial w}{\partial x} - \phi\right) + \rho I\frac{\partial^2 \phi}{\partial t^2} = 0 \quad (11.279)$$

and the boundary conditions as

$$kAG \left(\frac{\partial w}{\partial x} - \phi \right) \delta w \bigg|_0^l = 0 \tag{11.280}$$

$$EI \frac{\partial \phi}{\partial x} \delta \phi \bigg|_0^l = 0 \tag{11.281}$$

11.15.2 Equations for a Uniform Beam

Equations (11.278) and (11.279) can be combined into a single equation of motion for a uniform beam. For a uniform beam, Eqs. (11.278) and (11.279) reduce to

$$-kAG \frac{\partial^2 w}{\partial x^2} + kAG \frac{\partial \phi}{\partial x} + \rho A \frac{\partial^2 w}{\partial t^2} = f \tag{11.282}$$

or

$$\frac{\partial \phi}{\partial x} = \frac{\partial^2 w}{\partial x^2} - \frac{\rho}{kG} \frac{\partial^2 w}{\partial t^2} + \frac{f}{kAG} \tag{11.283}$$

and

$$-EI \frac{\partial^2 \phi}{\partial x^2} - kAG \frac{\partial w}{\partial x} + kAG\phi + \rho I \frac{\partial^2 \phi}{\partial t^2} = 0 \tag{11.284}$$

which upon differentiation with respect to x becomes

$$-EI \frac{\partial^2}{\partial x^2} \left(\frac{\partial \phi}{\partial x} \right) - kAG \frac{\partial^2 w}{\partial x^2} + kAG \frac{\partial \phi}{\partial x} + \rho I \frac{\partial^2}{\partial t^2} \left(\frac{\partial \phi}{\partial x} \right) = 0 \tag{11.285}$$

Substitution of Eq. (11.283) into Eq. (11.285) yields the desired equation:

$$EI \frac{\partial^4 w}{\partial x^4} + \rho A \frac{\partial^2 w}{\partial t^2} - \rho I \left(1 + \frac{E}{kG} \right) \frac{\partial^4 w}{\partial x^2 \partial t^2} + \frac{\rho^2 I}{kG} \frac{\partial^4 w}{\partial t^4}$$

$$+ \frac{EI}{kAG} \frac{\partial^2 f}{\partial x^2} - \frac{\rho I}{kAG} \frac{\partial^2 f}{\partial t^2} - f = 0 \tag{11.286}$$

For free vibration, Eq. (11.286) reduces to

$$EI \frac{\partial^4 w}{\partial x^4} + \rho A \frac{\partial^2 w}{\partial t^2} - \rho I \left(1 + \frac{E}{kG} \right) \frac{\partial^4 w}{\partial x^2 \partial t^2} + \frac{\rho^2 I}{kG} \frac{\partial^4 w}{\partial t^4} = 0 \tag{11.287}$$

The terms in Eq. (11.287) can be identified as follows. The first two terms are the same as those of the Euler–Bernoulli theory. The third term, $-\rho I \, (\partial^4 w/\partial x^2 \partial t^2)$, denotes the effect of rotary inertia. In fact, the first three terms are the same as those of the Rayleigh theory. Finally, the last two terms, involving kG in the denominators, represent the influence of shear deformation. Equations (11.280) and (11.281) will be satisfied by the following common support conditions. At a clamped or fixed end:

$$\phi = 0, \qquad w = 0 \tag{11.288}$$

At a pinned or simply supported end:

$$EI\frac{\partial\phi}{\partial x} = 0, \qquad w = 0 \tag{11.289}$$

At a free end:

$$EI\frac{\partial\phi}{\partial x} = 0, \qquad kAG\left(\frac{\partial w}{\partial x} - \phi\right) = 0 \tag{11.290}$$

11.15.3 Natural Frequencies of Vibration

The natural frequencies of vibration of uniform Timoshenko beams can be found by assuming a harmonic time variation of solution and solving Eq. (11.287) while satisfying the specific boundary conditions of the beam. In some cases, it is more convenient to solve the simultaneous differential equations in ϕ and w [Eqs. (11.278) and (11.279)] while satisfying the particular boundary conditions of the beam. Both these approaches are demonstrated in the following applications.

Simply Supported Beam By dividing Eq. (11.287) by ρA and defining

$$\alpha^2 = \frac{EI}{\rho A} \tag{11.291}$$

$$r^2 = \frac{I}{A} \tag{11.292}$$

Eq. (11.287) can be rewritten as

$$\alpha^2\frac{\partial^4 w}{\partial x^4} + \frac{\partial^2 w}{\partial t^2} - r^2\left(1 + \frac{E}{kG}\right)\frac{\partial^4 w}{\partial x^2\partial t^2} + \frac{\rho r^2}{kG}\frac{\partial^4 w}{\partial t^4} = 0 \tag{11.293}$$

The boundary conditions can be expressed as

$$w(x, t) = 0, \qquad x = 0, l \tag{11.294}$$

$$\frac{\partial\phi}{\partial x}(x, t) = 0, \qquad x = 0, l \tag{11.295}$$

Equation (11.295) can be expressed, using Eq. (11.283) with $f = 0$, as

$$\frac{\partial\phi}{\partial x}(x, t) = \frac{\partial^2 w(x, t)}{\partial x^2} - \frac{\rho}{kG}\frac{\partial^2 w(x, t)}{\partial t^2} = 0, \qquad x = 0, l \tag{11.296}$$

When harmonic time variations are assumed for $\phi(x, t)$ and $w(x, t)$ with frequency ω_n, the boundary condition of Eq. (11.296) will reduce to

$$\frac{\partial\phi}{\partial x}(x, t) = \frac{\partial^2 w(x, t)}{\partial x^2} = 0, \qquad x = 0, l \tag{11.297}$$

in view of Eq. (11.294). Thus, the boundary conditions can be stated in terms of w as

$$w(0, t) = 0, \qquad w(l, t) = 0 \tag{11.298}$$

$$\frac{d^2 w}{dx^2}(0, t) = 0, \qquad \frac{d^2 w}{dx^2}(l, t) = 0 \tag{11.299}$$

The solution of Eq. (11.293), which also satisfies the boundary conditions of Eqs. (11.298) and (11.299), is assumed as

$$w(x, t) = C \, \sin \frac{n \pi x}{l} \, \cos \omega_n t \tag{11.300}$$

where C is a constant and ω_n is the nth natural frequency of vibration. Substitution of Eq. (11.300) into Eq. (11.293) gives the frequency equation

$$\omega_n^4 \frac{\rho r^2}{kG} - \omega_n^2 \left(1 + \frac{n^2 \pi^2 r^2}{l^2} + \frac{n^2 \pi^2 r^2}{l^2} \frac{E}{kG} \right) + \frac{\alpha^2 n^4 \pi^4}{l^4} = 0 \tag{11.301}$$

Equation (11.301) is a quadratic equation in ω_n^2 and gives two values of ω_n^2 for any value of n. The smaller value of ω_n^2 corresponds to the bending deformation mode, and the larger value corresponds to the shear deformation mode.

Fixed–Fixed Beam   The boundary conditions can be expressed as

$$w(0, t) = 0, \qquad \phi(0, t) = 0 \tag{11.302}$$

$$w(l, t) = 0, \qquad \phi(l, t) = 0 \tag{11.303}$$

Since the expression for ϕ in terms of w is not directly available, Eqs. (11.283) and (11.284) are solved, with $f = 0$, simultaneously. For this, the solution is assumed to be of the form

$$w(x, t) = W(x) \, \cos \omega_n t \tag{11.304}$$

$$\phi(x, t) = \Phi(x) \, \cos \omega_n t \tag{11.305}$$

Substitution of Eqs. (11.304) and (11.305) into Eqs. (11.283) and (11.284) gives (by setting $f = 0$)

$$-kAG \frac{d^2 W}{dx^2} + kAG \frac{d\Phi}{dx} - \rho A \omega_n^2 W = 0 \tag{11.306}$$

$$-EI \frac{d^2 \Phi}{dx^2} - kAG \frac{dW}{dx} + kAG\Phi - \rho I \omega_n^2 \Phi = 0 \tag{11.307}$$

The solutions of Eqs. (11.306) and (11.307) are assumed to be

$$W(x) = C_1 \, \exp\left(\frac{ax}{l} \right) \tag{11.308}$$

$$\Phi(x) = C_2 \, \exp\left(\frac{ax}{l} \right) \tag{11.309}$$

where a, C_1, and C_2 are constants. Substitution of Eqs. (11.308) and (11.309) into Eqs. (11.306) and (11.307) leads to

$$\left(-kAG \frac{a^2}{l^2} - \rho A \omega_n^2 \right) C_1 + \left(kAG \frac{a}{l} \right) C_2 = 0 \tag{11.310}$$

$$\left(-kAG \frac{a}{l} \right) C_1 + \left(-EI \frac{a^2}{l^2} - kAG \frac{a}{l} + kAG - \rho I \, \omega_n^2 \right) C_2 = 0 \tag{11.311}$$

For a nontrivial solution of the constants C_1 and C_2, the determinant of their coefficients in Eqs. (11.310) and (11.311) is set equal to zero. This yields the equation

$$a^4 + \left[\omega_n^2 l^2 \left(\frac{\rho}{E} + \frac{\rho}{kG}\right)\right] a^2 + \left[\omega_n^2 l^4 \left(\frac{\omega_n^2 \rho^2}{kGE} - \frac{\rho A}{EI}\right)\right] = 0 \tag{11.312}$$

The roots of Eq. (11.312) are given by

$$a = \mp \left\{-\frac{a_1}{2} \mp \left[\left(\frac{a_1}{2}\right)^2 - a_2\right]^{1/2}\right\}^{1/2} \tag{11.313}$$

where

$$a_1 = \omega_n^2 l^2 \rho \left(\frac{1}{E} + \frac{1}{kG}\right) \tag{11.314}$$

$$a_2 = \omega_n^2 l^4 \rho \left(\frac{\omega_n^2 \rho}{kGE} - \frac{A}{EI}\right) \tag{11.315}$$

The four values of a given by Eq. (11.313) can be used to express $W(x)$ and $\Phi(x)$ in the form of trigonometric and hyperbolic functions. When the boundary conditions are used with the functions $W(x)$ and $\Phi(x)$, the characteristic equation for finding the natural frequencies of vibration, ω_n, can be found.

Example 11.10   Find the first three natural frequencies of vibration of a rectangular steel beam 1 m long, 0.05 m wide, and 0.15 m deep with simply supported ends using Euler–Bernoulli theory, Rayleigh theory, and Timoshenko theory. Assume that $E = 207 \times 10^9$ Pa, $G = 79.3 \times 10^9$ Pa, $\rho = 76.5 \times 10^3$ N/m^3, and $k = \frac{5}{6}$.

SOLUTION   The natural frequencies of vibration are given according to Timoshenko theory [Eq. (11.301)]:

$$\omega_n^4 \frac{\rho r^2}{kG} - \omega_n^2 \left(1 + \frac{n^2 \pi^2 r^2}{l^2} + \frac{n^2 \pi^2 r^2}{l^2} \frac{E}{kG}\right) + \frac{\alpha^2 n^4 \pi^4}{l^4} = 0 \tag{E11.10.1}$$

According to Rayleigh theory [terms involving k are to be deleted in Eq. (E11.10.1)],

$$-\omega_n^2 \left(1 + \frac{n^2 \pi^2 r^2}{l^2}\right) + \frac{\alpha^2 n^4 \pi^4}{l^4} = 0 \tag{E11.10.2}$$

By Euler–Bernoulli theory [term involving r^2 is to be deleted in Eq. (E11.10.2)],

$$-\omega_n^2 + \frac{\alpha^2 n^4 \pi^4}{l^4} = 0 \tag{E11.10.3}$$

For the given beam, $A = (0.05)(0.15) = 0.0075$ m^2, $I = \frac{1}{12}(0.05)(0.15)^3 = 14.063 \times 10^{-6}$ m^4, and Eqs. (E11.10.1)–(E11.10.3) become

$$2.1706 \times 10^{-9} \omega_n^4 - (1 + 76.4754 \times 10^{-3} n^2)\omega_n^2 + 494.2300 \times 10^3 n^4 = 0 \quad \text{(E11.10.4)}$$

$$-(1 + 18.5062 \times 10^{-3} n^2)\omega_n^2 + 494.2300 \times 10^3 n^4 = 0 \quad \text{(E11.10.5)}$$

$$-\omega_n^2 + 494.2300 \times 10^3 n^4 = 0 \quad \text{(E11.10.6)}$$

Table 11.3 Computation for Example 11.10

| | Natural frequency (rad/s) | | | |
| | Euler–Bernoulli | Rayleigh | Timoshenko | |
n			Bending	Shear
1	703.0149	696.5987	677.8909	22,259.102
2	2,812.0598	2,713.4221	2,473.3691	24,402.975
3	6,327.1348	5,858.0654	4,948.0063	27,446.297

The natural frequencies computed from Eqs. (E11.10.4)–(E11.10.6) are given in Table 11.3.

11.16 COUPLED BENDING–TORSIONAL VIBRATION OF BEAMS

In the transverse vibration of beams considered so far, it is implied that the cross section of the beam has two axes of symmetry (y and z axes). If the cross section of the beam has two axes of symmetry, the centroid and the shear center (or center of flexure) coincide and the bending and torsional vibrations are uncoupled. In all the cases considered so far, the transverse vibration of the beam is assumed to be in a plane of symmetry (xz plane). On the other hand, if the cross section of a beam has only one axis of symmetry, the shear center lies on the axis of symmetry. When the load does not act through the shear center, the beam will undergo twisting in addition to bending. Note that, in general, the shear center need not lie on a principal axis; it may lie outside the cross section of the beam as shown in Fig. 11.20 for a beam with channel section. In Fig. 11.20, the line GG' represents the centroidal axis, the x axis (OO') denotes the shear center axis, the z axis indicates the axis of symmetry, and the y axis represents a direction parallel to the web. For the thin-walled channel section, the locations of the shear center and the centroid from the center of the web are given by [3]

$$c = \frac{a^2 b^2 t}{I_z} \tag{11.316}$$

$$d = \frac{b^2}{2(b+a)} \tag{11.317}$$

$$I_z \approx \frac{2}{3}a^3 t + 2\left(\frac{1}{12}bt^3 + bta^2\right) \approx \frac{2}{3}a^3 t + 2a^2 bt \tag{11.318}$$

where $2a$, b, and t are the height of web, width of flanges, and thickness of web and flanges as shown in Fig. 11.20. The distance between the centroid and the shear center of the channel section, e, is given by

$$e = c + d = \frac{a^2 b^2 t}{I_z} + \frac{b^2}{2(a+b)} \tag{11.319}$$

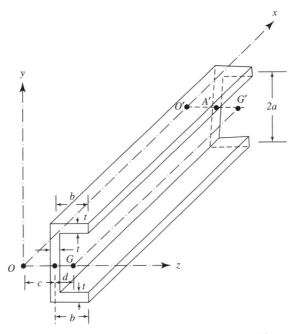

Figure 11.20 Beam with a channel section, GG' centroidal axis; OO' shear center axis.

11.16.1 Equations of Motion

Consider the beam with a channel section subjected to a distributed load $f(x)$ acting along the centroidal axis GG' as shown in Fig. 11.21. Since the load does not pass through the shear center axis OO', the beam will undergo both bending and torsional deflections. To study the resulting coupled bending–torsion motion of the beam, the load acting through the centroidal axis is replaced by the same load and a torque of magnitude fe, distributed along the shear center axis OO' (see Fig. 11.21). Then the equation governing the bending deflection of the beam in the xy plane can be written as

$$EI_z \frac{\partial^4 v}{\partial x^4} = f \qquad (11.320)$$

where v is the deflection of the beam in the y direction and EI_z is the bending rigidity of the cross section about the z axis. For the torsional deflection of the beam, the total torque acting on any cross section of the beam, $T(x)$, is written as the sum of the Saint-Venant torque (T_{sv}) and the torque arising from the restraint against warping (T_w) [3]:

$$T(x) = T_{sv}(x) + T_w(x) = GJ \frac{d\phi}{dx} - EJ_w \frac{d^3\phi}{dx^3} \qquad (11.321)$$

where ϕ is the angle of twist (rotation), GJ is the torsional rigidity under uniform torsion (in the absence of any warping restraint), EJ_w is the warping rigidity, and J_w

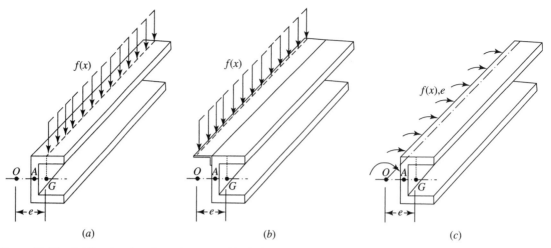

Figure 11.21 (a) Load acting through a centroid; (b) load acting through the shear center; (c) torque acting about the shear center.

is the sectional moment of inertia of the cross section with [3]

$$J = \frac{1}{3} \sum b_i t_i^3 = \frac{1}{3}(2bt^3 + 2at^3) = \frac{2}{3}t^3(a+b) \tag{11.322}$$

$$J_w = \frac{a^2 b^3 t}{3} \frac{4at + 3bt}{2at + 6bt} \tag{11.323}$$

By differentiating Eq. (11.321) with respect to x and using the relation $dT/dx = fe$, we obtain

$$GJ\frac{d^2\phi}{dx^2} - EJ_w\frac{d^4\phi}{dx^4} = fe \tag{11.324}$$

Note that the solution of Eqs. (11.320) and (11.324) gives the bending deflection, $v(x)$, and torsional deflection, $\phi(x)$, of the beam under a static load, $f(x)$. For the free vibration of the beam, the inertia forces acting in the y and ϕ directions are given by

$$-\rho A \frac{\partial^2}{\partial t^2}(v - e\phi) \tag{11.325}$$

and

$$-\rho I_G \frac{\partial^2 \phi}{\partial t^2} \tag{11.326}$$

where $v - e\phi$ denotes the net transverse deflection of the beam in the y direction, ρ is the density, A is the cross-sectional area, and I_G is the polar moment of inertia of the cross section about its centroidal axis. The equations of motion for the coupled bending–torsional vibration of the beam can be written, by using the term (11.325) in place of f and including the inertia force in Eq. (11.324), as

$$EI_z\frac{\partial^4 v}{\partial x^4} = -\rho A\frac{\partial^2}{\partial t^2}(v - e\phi) \tag{11.327}$$

$$GJ\frac{\partial^2 \phi}{\partial x^2} - EJ_w\frac{\partial^4 \phi}{\partial x^4} = -\rho Ae\frac{\partial^2}{\partial t^2}(v - e\phi) + \rho I_G\frac{\partial^2 \phi}{\partial t^2} \tag{11.328}$$

11.16.2   Natural Frequencies of Vibration

For free vibration of the beam, the solution is assumed to be in the form

$$v(x, t) = V(x)C_1 \cos(\omega t + \theta_1) \tag{11.329}$$

$$\phi(x, t) = \Phi(x)C_2 \cos(\omega t + \theta_2) \tag{11.330}$$

where $V(x)$ and $\Phi(x)$ are the normal modes, ω is the natural frequency of vibration, C_1 and C_2 are constants, and θ_1 and θ_2 are the phase angles. By substituting Eqs. (11.329) and (11.330) into Eqs. (11.327) and (11.328), we obtain

$$EI_z\frac{d^4 V}{dx^4} = \rho A\omega^2(X - e\Phi) \tag{11.331}$$

$$GJ\frac{d^2 \Phi}{dx^2} - EJ_w\frac{d^4 \Phi}{dx^4} = \rho Ae\omega^2(V - e\Phi) - \rho I_G\omega^2\Phi \tag{11.332}$$

The normal modes of the beam, $V(x)$ and $\Phi(x)$, can be found by satisfying not only Eqs. (11.331) and (11.332) but also the boundary conditions of the beam. The following example illustrates the procedure.

Example 11.11   Find the natural frequencies of coupled vibration of a beam simply supported at both ends. Assume the cross section of the beam to be a channel section, as shown in Fig. 11.20.

SOLUTION   The simply supported boundary conditions can be expressed as

$$V(x) = 0 \quad \text{at} \quad x = 0, \quad x = l \tag{E11.11.1}$$

$$\frac{d^2 V}{dx^2}(x) = 0 \quad \text{at} \quad x = 0, \quad x = l \tag{E11.11.2}$$

$$\Phi(x) = 0 \quad \text{at} \quad x = 0, \quad x = l \tag{E11.11.3}$$

$$\frac{d^2 \Phi}{dx^2}(x) = 0 \quad \text{at} \quad x = 0, \quad x = l \tag{E11.11.4}$$

The following functions can be seen to satisfy the boundary conditions of Eqs. (E11.11.1)–(E11.11.4):

$$V_j(x) = A_j \sin\frac{j\pi x}{l}, \quad j = 1, 2, 3, \ldots \tag{E11.11.5}$$

$$\Phi_j(x) = B_j \sin\frac{j\pi x}{l}, \quad j = 1, 2, 3, \ldots \tag{E11.11.6}$$

where A_j and B_j are constants. By substituting Eqs. (E11.11.5) and (E11.11.6) into Eqs. (11.331) and (11.332), we obtain the following equations for finding the jth natural frequency (ω_j):

$$EI_z \left(\frac{j\pi}{l} \right)^4 A_j = \rho A \omega_j^2 (A_j - eB_j) \tag{E11.11.7}$$

$$-GJ \left(\frac{j\pi}{l} \right)^2 B_j - EJ_w \left(\frac{j\pi}{l} \right)^4 B_j = \rho Ae\omega_j^2 (A_j - eB_j) - \rho I_G \omega_j^2 B_j \tag{E11.11.8}$$

Equations (E11.11.7) and (E11.11.8) can be rewritten as

$$(p^2 - \omega_j^2) A_j + (\omega_j^2 e) B_j = 0 \tag{E11.11.9}$$

$$(q^2 \omega_j^2) A_j + (r^2 - \omega_j^2) B_j = 0 \tag{E11.11.10}$$

where

$$p^2 = \frac{EI_z j^4 \pi^4}{\rho A l^4} \tag{E11.11.11}$$

$$q^2 = \frac{Ae}{I_G + Ae^2} \tag{E11.11.12}$$

$$r^2 = \frac{GJl^2 j^2 \pi^2 + EJ_w j^4 \pi^4}{\rho l^4 (I_G + Ae^2)} \tag{E11.11.13}$$

For a nontrivial solution of A_j and B_j, the determinant of their coefficient matrix must be equal to zero. This leads to

$$\begin{vmatrix} p^2 - \omega_j^2 & \omega_j^2 e \\ q^2 \omega_j^2 & r^2 - \omega_j^2 \end{vmatrix} = 0$$

or

$$\omega_j^4 (1 - q^2 e) - \omega_j^2 (p^2 + r^2) + p^2 r^2 = 0 \tag{E11.11.14}$$

The solution of Eq. (E11.11.14) gives

$$\omega_j^2 = \frac{p^2 + r^2 \mp [(p^2 - r^2)^2 + 4p^2 r^2 q^2 e]^{1/2}}{2(1 - eq^2)} \tag{E11.11.15}$$

Equation (E11.11.15) gives two values of ω_j^2, corresponding to two possible modes of the coupled bending–torsional vibration of the beam. The mode shapes corresponding to the two natural frequencies ω_j^2 can be determined by solving Eqs. (E11.11.9) and (E11.11.10). To find the physical significance of the two natural frequencies given by Eq. (E11.11.15), consider a beam with symmetric cross section with $e = 0$. For this case, Eq. (E11.11.15) gives

$$\omega_j^2 = r^2; \quad p^2 \tag{E11.11.16}$$

From Eqs. (E11.11.16), (E11.11.11), and (E11.11.13), we find that $\omega_j^2 = p^2$ corresponds to the flexural vibration mode and $\omega_j^2 = r^2$ corresponds to the torsional vibration mode. For a beam with a nonsymmetric cross section with $e \neq 0$, one of the natural frequencies given by Eq. (E11.11.15) will be smaller and the other will be larger than the values given by Eq. (E11.11.16).

11.17   TRANSFORM METHODS: FREE VIBRATION OF AN INFINITE BEAM

As indicated in Chapter 7, Laplace and Fourier transform methods can be used to solve free and forced vibration problems. The applicability of the methods to beam problems is illustrated in this section by considering the free vibration of an infinite beam. The equation of motion for the transverse vibration of a uniform beam is given by

$$c^2 \frac{\partial^4 w(x, t)}{\partial x^4} + \frac{\partial^2 w(x, t)}{\partial t^2} = 0 \tag{11.333}$$

where

$$c^2 = \frac{EI}{\rho} \tag{11.334}$$

Let the initial conditions of the beam be given by

$$w(x, t = 0) = w_0(x) \tag{11.335}$$

$$\frac{\partial w}{\partial t}(x, t = 0) = \dot{w}_0(x) \tag{11.336}$$

By taking the Laplace transform of Eq. (11.333), we obtain

$$c^2 \frac{d^4 W(x, s)}{dx^4} + s^2 W(x, s) = sw_0(x) + \dot{w}_0(x) \tag{11.337}$$

Next, we take Fourier transform of Eq. (11.337). For this, we multiply Eq. (11.337) by e^{ipx} and integrate the resulting equation with respect to x from $-\infty$ to ∞:

$$c^2 \int_{-\infty}^{\infty} \left[\frac{d^4 W(x, s)}{dx^4} + \frac{s^2}{c^2} W(x, s) \right] e^{ipx}\, dx = \int_{-\infty}^{\infty} [sw_0(x) + \dot{w}_0(x)] e^{ipx}\, dx \tag{11.338}$$

The first term of Eq. (11.338) can be integrated by parts to obtain

$$\int_{-\infty}^{\infty} \frac{d^4 W(x, s)}{dx^4} e^{ipx}\, dx = \left(\frac{d^3 W}{dx^3} - ip \frac{d^2 W}{dx^2} - p^2 \frac{dW}{dx} + ip^3 W \right) e^{ipx} \Bigg|_{-\infty}^{\infty}$$
$$+ p^4 \int_{-\infty}^{\infty} W e^{ipx}\, dx \tag{11.339}$$

Since $W(x, s)$, $dW(x, s)/dx$, $d^2W(x, s)/dx^2$ and $d^3W(x, s)/dx^3$ tend to zero as $|x| \to \infty$, Eq. (11.339) reduces to

$$\int_{-\infty}^{\infty} \frac{d^4 W}{dx^4} e^{ipx}\, dx = p^4 \int_{-\infty}^{\infty} W e^{ipx}\, dx \tag{11.340}$$

Defining the Fourier transforms of $W(x, s)$, $w_0(x)$, and $\dot{w}_0(x)$ as

$$\overline{W}(p, s) = \frac{1}{\sqrt{2\pi}} \int_{-\infty}^{\infty} W(x, s) e^{ipx}\, dx \tag{11.341}$$

$$\overline{W}_0(p) = \frac{1}{\sqrt{2\pi}} \int_{-\infty}^{\infty} w_0(x) e^{ipx}\, dx \tag{11.342}$$

$$\dot{\overline{W}}_0(p) = \frac{1}{\sqrt{2\pi}} \int_{-\infty}^{\infty} \dot{w}_0(x) e^{ipx}\, dx \tag{11.343}$$

Eq. (11.338) can be rewritten as

$$c^2 p^4 \overline{W}(p, s) + s^2 \overline{W}(p, s) = s \overline{W}_0(p) + \dot{\overline{W}}_0(p)$$

or

$$\overline{W}(p, s) = \frac{s \overline{W}_0(p) + \dot{\overline{W}}_0(p)}{c^2 p^4 + s^2} \tag{11.344}$$

The inverse Fourier transform of Eq. (11.344) yields

$$W(x, s) = \frac{1}{\sqrt{2\pi}} \int_{-\infty}^{\infty} \frac{s \overline{W}_0(p) + \dot{\overline{W}}_0(p)}{c^2 p^4 + s^2} e^{-ipx}\, dx \tag{11.345}$$

Finally, we take the inverse Laplace transform of $W(x, s)$. Noting that

$$L^{-1}[W(x, s)] = w(x, t) \tag{11.346}$$

$$L^{-1}\left[\frac{s}{c^2 p^4 + s^2} \right] = \cos p^2 ct \tag{11.347}$$

$$L^{-1}\left[\frac{1}{c^2 p^4 + s^2} \right] = \frac{1}{p^2 c} \sin p^2 ct \tag{11.348}$$

Eq. (11.345) gives

$$w(x, t) = \frac{1}{\sqrt{2\pi}} \int_{-\infty}^{\infty} \left[\overline{W}_0(p) \cos p^2 ct + \dot{\overline{W}}_0(p) \sin p^2 ct \right] e^{-ipx}\, dp \tag{11.349}$$

The convolution theorem for the Fourier transforms yields

$$\int_{-\infty}^{\infty} \overline{F}_1(p) \overline{F}_2(p) e^{-ipx}\, dp = \int_{-\infty}^{\infty} f_1(\eta) f_2(x - \eta)\, d\eta \tag{11.350}$$

Noting the validity of the relations

$$\frac{1}{\sqrt{2\pi}} \int_{-\infty}^{\infty} \cos p^2 ct e^{-ipx}\, dp = \frac{1}{2\sqrt{ct}} \left(\cos \frac{x^2}{4ct} + \sin \frac{x^2}{4ct} \right) \tag{11.351}$$

$$\frac{1}{\sqrt{2\pi}} \int_{-\infty}^{\infty} \sin p^2 ct e^{-ipx}\, dp = \frac{1}{2\sqrt{ct}} \left(\cos \frac{x^2}{4ct} - \sin \frac{x^2}{4ct} \right) \tag{11.352}$$

Eqs. (11.350)–(11.352) can be used to express Eq. (11.349) as

$$w(x,t) = \frac{1}{2\sqrt{2\pi ct}} \int_{-\infty}^{\infty} w_0(x-\eta) \left(\cos \frac{\eta^2}{4ct} + \sin \frac{\eta^2}{4ct} \right) d\eta$$

$$+ \frac{1}{2\sqrt{2\pi ct}} \int_{-\infty}^{\infty} \dot{w}_0(x-\eta) \left(\cos \frac{\eta^2}{4ct} - \sin \frac{\eta^2}{4ct} \right) d\eta \tag{11.353}$$

Introducing

$$\lambda^2 = \frac{\eta^2}{4ct} \tag{11.354}$$

Eq. (11.353) can be rewritten as

$$w(x,t) = \frac{1}{\sqrt{2\pi}} \int_{-\infty}^{\infty} [w_0(x - 2\lambda\sqrt{ct})(\cos \lambda^2 + \sin \lambda^2)$$

$$+ \dot{w}_0(x - 2\lambda\sqrt{ct})(\cos \lambda^2 - \sin \lambda^2)]\, d\lambda \tag{11.355}$$

11.18   RECENT CONTRIBUTIONS

Higher-Order Theories   The second-order theories account for the deformation of the beam due to shear by considering a second variable in deriving the governing differential equations. Some theories, such as Timoshenko theory, include a shear coefficient to account for the nonuniformity of shear deformation across the cross section and the accompanying cross-sectional warping. Other theories include a third variable to characterize the degree to which bending warping occurs. Ewing [6] presented a model of the latter type for application to constant-cross-section beams undergoing symmetric bending vibration (with no bending–torsion coupling).

Beams on an Elastic Foundation   A power series solution was presented for the free vibration of a simply supported Euler–Bernoulli beam resting on an elastic foundation having quadratic and cubic nonlinearities [7]. The problem was posed as a nonlinear eigenvalue problem by assuming the time dependence to be harmonic.

Multispan Beams   In Ref. [8], Wang determined the natural frequencies of continuous Timoshenko beams. A modal analysis procedure was proposed by Wang [9] to investigate the forced vibration of multispan Timoshenko beams. The effects of span number, rotary inertia, and shear deformation on the maximum moment, maximum deflection, and critical velocity of the beam are determined.

Self-Excited Vibration In tall buildings and pylons of suspension bridges, the flexibility and the transverse motion of the structures generates self-excited lift and drag aerodynamic forces, known as *galloping*, which cause nonlinear dynamic behavior. An approximate solution of this problem is presented by Nayfeh and Abdel-Rohman [10].

Damped Beams Wang et al. [11] studied the free vibration of a transmission-line conductor equipped with a number of Stockbridge dampers by modeling it as a tensioned beam acted on by concentrated frequency-dependent forces. An exact solution is obtained using integral transformation.

Beams with Tip Mass Zhou gave the exact analytical solution for the eigenfrequencies and mode shapes of a cantilever beam carrying a heavy tip mass with translational and rotational elastic supports in the context of antenna structures [12].

Beams under Axial Loads The vibrational behavior of initially imperfect simply supported beams subject to axial loading has been considered by Ilanko [13]. The natural frequencies of beams subjected to tensile axial loads are investigated by Bokaian [14].

Moving Loads and Bridge Structures The analysis of the vibrational behavior of structural elements traversed by moving forces or masses can be used to study the dynamics of a bridge traveled by a car or of rails traveled by a train. The problem was studied by Gutierrez and Laura [15].

Waves An analytical method for finding the vibrational response and the net transmitted power of bending wave fields in systems consisting of coupled finite beams has been studied by Hugin [16].

Beams with Discontinuities A method of finding the bending moments and shear forces as well as the free vibration characteristics of a Timoshenko beam having discontinuities was presented by Popplewell and Chang [17].

Beams with Variable Properties The eigenfunction method using shear theory was used by Gupta and Sharma [18] to analyze the forced motion of a rectangular beam whose thickness, density, and elastic properties along the length vary in any number of steps. A beam of two steps clamped at both edges and subjected to a constant or half-sine pulse load was considered.

Computation of Elastic Properties Larsson [19] discussed accuracy in the computation of Young's modulus and the longitudinal–transverse shear modulus from flexural vibration frequencies.

Coupled Vibrations Yaman [20] presented an exact analytical method, based on the wave propagation approach, for the forced vibrations of uniform open-section channels. Coupled wave numbers, various frequency response curves, and mode shapes were presented for undamped and structurally damped channels using the Euler–Bernoulli model. Bercin and Tanaka studied the coupled flexural–torsional vibrations of mono-symmetric beams [21]. The effects of warping stiffness, shear deformation and rotary inertia are taken into account in the formulations.

Frames   The vibration of frame structures according to the Timoshenko theory was studied by Wang and Kinsman [22]. A portal frame subjected to free and forced vibrations is used to illustrate the method.

REFERENCES

1. L. Fryba, *Vibration of Solids and Structures Under Moving Loads*, Noordhoff International Publishing, Groningen, The Netherlands, 1972.

2. J. E. Shigley and L. D. Mitchell, *Mechanical Engineering Design*, 4th ed., McGraw-Hill, New York, 1983.

3. W. B. Bickford, *Advanced Mechanics of Materials*, Addison-Wesley, Reading, MA, 1998.

4. S. S. Rao, Natural vibrations of systems of elastically connected Timoshenko beams, *Journal of the Acoustical Society of America*, Vol. 55, pp. 1232–1237, 1974.

5. C. F. Garland, Normal modes of vibration of beams having noncollinear elastic and mass axes, *Journal of Applied Mechanics*, Vol. 62, p. 97, 1940.

6. M. S. Ewing, Another second order beam vibration theory: explicit bending, warping flexibility and restraint, *Journal of Sound and Vibration*, Vol. 137, No. 1, pp. 43–51, 1990.

7. M. I. Qaisi, Normal modes of a continuous system with quadratic and cubic nonlinearities, *Journal of Sound and Vibration*, Vol. 265, No. 2, pp. 329–335, 2003.

8. T. M. Wang, Natural frequencies of continuous Timoshenko beams, *Journal of Sound and Vibration*, Vol. 13, No. 3, pp. 409–414, 1970.

9. R.-T. Wang, Vibration of multi-span Timoshenko beams to a moving force, *Journal of Sound and Vibration*, Vol. 207, No. 5, pp. 731–742, 1997.

10. A. H. Nayfeh and M. Abdel-Rohman, Analysis of galloping responses in cantilever beams, *Journal of Sound and Vibration*, Vol. 144, No.1, pp. 87–93, 1991.

11. H. Q. Wang, J. C. Miao, J. H. Luo, F. Huang, and L. G. Wang, The free vibration of long-span transmission line conductors with dampers, *Journal of Sound and Vibration*, Vol. 208, No. 4, pp. 501–516, 1997.

12. D. Zhou, The vibrations of a cantilever beam carrying a heavy tip mass with elastic supports, *Journal of Sound and Vibration*, Vol. 206, No. 2, pp. 275–279, 1997.

13. S. Ilanko, The vibration behaviour of initially imperfect simply supported beams subject to axial loading, *Journal of Sound and Vibration*, Vol. 142, No.2, pp. 355–359, 1990.

14. A. Bokaian, Natural frequencies of beams under tensile axial loads, *Journal of Sound and Vibration*, Vol. 142, No. 3, pp. 481–498, 1990.

15. R. H. Gutierrez and P. A. A. Laura, Vibrations of a beam of non-uniform cross-section traversed by a time varying concentrated force, *Journal of Sound and Vibration*, Vol. 207, No. 3, pp. 419–425, 1997.

16. C. T. Hugin, A physical description of the response of coupled beams, *Journal of Sound and Vibration*, Vol. 203, No. 4, pp. 563–580, 1997.

17. N. Popplewell and D. Chang, Free vibrations of a stepped, spinning Timoshenko beam, *Journal of Sound and Vibration*, Vol. 203, No. 4, pp. 717–722, 1997.

18. A. P. Gupta and N. Sharma, Effect of transverse shear and rotatory inertia on the forced motion of a stepped rectangular beam, *Journal of Sound and Vibration*, Vol. 209, No. 5, pp. 811–820, 1998.

19. P.–O. Larsson, Determination of Young's and shear moduli from flexural vibrations of beams, *Journal of Sound and Vibration*, Vol. 146, No. 1, pp. 111–123, 1991.

20. Y. Yaman, Vibrations of open-section channels: a coupled flexural and torsional wave analysis, *Journal of Sound and Vibration*, Vol. 204, No. 1, pp. 131–158, 1997.

21. A. N. Bercin and M. Tanaka, Coupled flexural–torsional vibrations of Timoshenko beams, *Journal of Sound and Vibration*, Vol. 207, No. 1, pp. 47–59, 1997.

22. M. Wang and T. A. Kinsman, Vibrations of frame structures according to the Timoshenko theory, *Journal of Sound and Vibration*, Vol. 14, No. 2, pp. 215–227, 1971.

23. S. S. Rao, *Mechanical Vibrations*, 4th ed., Prentice Hall, Upper Saddle River, NJ, 2004.

24. W. T. Thomson, *Laplace Transformation*, Second Edition, Prentice-Hall, Englewood Cliffs, NJ, 1960.

PROBLEMS

11.1 Derive the equation of motion and the boundary conditions of a Timoshenko beam resting on an elastic foundation using Newton's second law of motion. Assume that the beam is supported on a linear spring, of stiffness K_0, at $x = 0$ and a rotational spring, of stiffness K_{t0}, at $x = l$.

11.2 Derive the equation of motion and the boundary conditions of a Rayleigh beam resting on an elastic foundation using a variational approach. Assume that the beam is supported on a linear spring of stiffness K at $x = 0$ and carries a mass M at $x = l$.

11.3 Find the natural frequencies and normal modes of a uniform beam simply supported at $x = 0$ and free at $x = l$.

11.4 Derive the orthogonality relationships of the transverse vibration of a uniform beam that is fixed at $x = 0$ and carries a mass m_0 and mass moment of inertia I_0 at $x = l$, and is supported on a linear spring of stiffness K and torsional spring of stiffness K_t at $x = l$ as shown in Fig. 11.22.

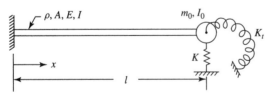

Figure 11.22

11.5 Determine the frequency equation for the two-span beam shown in Fig. 11.23. Assume that the two spans have identical values of mass density ρ, cross-sectional area A, and flexural rigidity EI.

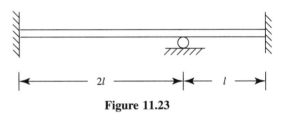

Figure 11.23

11.6 Compute the first four natural frequencies of transverse vibration of a uniform beam for the following boundary conditions: **(a)** fixed–fixed; **(b)** pinned–pinned; **(c)** fixed–free; **(d)** free–free; **(e)** fixed–pinned. Data: $E = 30 \times 10^6$ lb/in^2, $A = 2$ in^2, $I = \frac{1}{3}$ in^4, $\rho = 732.4 \times 10^{-6}$ lb-sec^2/in^4, and $l = 20$ in.

11.7 A uniform beam simply supported at both ends is deflected initially by a concentrated force F applied at the middle. Determine the free vibration of the beam when the force F is suddenly removed.

11.8 Determine the free vibration of a uniform beam simply supported at both ends that is subjected to an initial uniform transverse velocity v_0 at $0 < x < l$.

11.9 A uniform cantilever beam is deflected initially by a concentrated force F applied at the free end. Determine the free vibration of the beam when the force F is suddenly removed.

11.10 A uniform fixed–fixed beam is deflected initially by a concentrated force F applied at the middle. Determine the free vibration of the beam when the force F is suddenly removed.

11.11 A railway track can be modeled as an infinite beam resting on an elastic foundation with a soil stiffness of k per unit length. A railway car moving on the railway track can be modeled as a load F_0 moving at a constant velocity v_0. Derive the equation of motion of the beam.

11.12 Compare the natural frequencies of vibration of a uniform beam, simply supported at both ends, given by the Euler–Bernoulli, Rayleigh, and Timoshenko theories for the following data:

(a) $E = 30 \times 10^6 \text{psi}$, $\rho = 732.4 \times 10^{-6} \text{lb-sec}^2/\text{in}^4$,

$$A = 2\text{in}^2, I = \frac{1}{3}\text{in}^4, l = 20 \text{ in.,}$$

$$k = \frac{5}{6}, G = 11.5 \times 10^6 \text{psi}$$

(b) $E = 30 \times 10^6 \text{psi}$, $\rho = 732.4 \times 10^{-6} \text{lb-sec}^2/\text{in}^4$,

$$A = 8\text{in}^2, I = 10\frac{2}{3}\text{in}^4, l = 20 \text{ in.,}$$

$$k = \frac{5}{6}, G = 11.5 \times 10^6 \text{psi}$$

(c) $E = 10.3 \times 10^6 \text{psi}$, $\rho = 253.6 \times 10^{-6} \text{lb-sec}^2/\text{in}^4$,

$$A = 2\text{in}^2, I = \frac{1}{3}\text{in}^4, l = 20 \text{ in.,}$$

$$k = \frac{5}{6}, G = 3.8 \times 10^6 \text{psi}$$

(d) $E = 10.3 \times 10^6 \text{psi}$, $\rho = 253.6 \times 10^{-6} \text{lb-sec}^2/\text{in}^4$,

$$A = 8\text{in}^2, I = 10\frac{2}{3}\text{in}^4, l = 20 \text{ in.,}$$

$$k = \frac{5}{6}, G = 3.8 \times 10^6 \text{psi}$$

11.13 Plot the variations of ω_{Rn}/ω_{En} and ω_{Tn}/ω_{En} over the range $0 \le nr/l \le 1.0$ for $E/kG = 1.0, 2.5$ and 5.0, where ω_{En}, ω_{Rn}, and ω_{Tn} denote the nth natural frequency of vibration of a uniform beam given by the Euler–Bernoulli, Rayleigh, and Timoshenko theories, respectively.

11.14 Derive the equation of motion of a transversely vibrating beam subjected to an axial force P using Newton's second law of motion.

11.15 Find the first four natural frequencies of vibration of a uniform beam with a channel section as shown in Fig. 11.24 and simply supported at both ends.

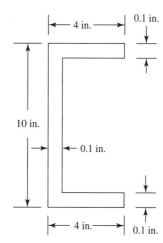

Figure 11.24

11.16 In the frequency equation of a simply supported Timoshenko beam, Eq. (11.301), the contribution of the first term, $\omega_n^4(\rho r^2/kG)$, can be shown to be negligibly small for $nr/l \ll 1$. By neglecting the first term, the approximate frequency of vibration can be found as

$$\omega_n^2 \approx \frac{\alpha^2 n^4 \pi^4/l^4}{1 + n^2\pi^2 r^2/l^2 + (n^2\pi^2 r^2/l^2)E/kG}$$

Find the first three natural frequencies of vibration of the beam considered in Example 11.10 using the approximate expression given above, and find the error involved compared to the values given by the exact equation (11.301).

11.17 Consider a beam subjected to a base motion such as one experienced during an earthquake shown in Fig. 11.25. If the elastic deflection of the beam, $w(x,t)$, is measured relative to the support motion, $w_g(t)$, the total displacement of the beam is given by

$$w_t(x,t) = w_g(t) + w(x,t)$$

(a) Derive the equation of motion of the beam, subjected to base motion, considering the inertia force associated with the total acceleration of the beam.

(b) Suggest a method of finding the response of a beam subjected to a specified base motion $w_g(t)$.

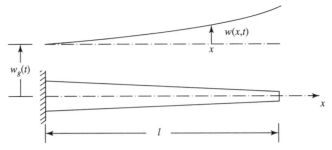

Figure 11.25 Beam subjected to support excitation.

11.18 A uniform simply supported beam is subjected to a step-function force F_0 at the midspan (at $x = l/2$). Assuming zero initial conditions, derive expressions for the following dynamic responses: (**a**) displacement distribution in the beam; (**b**) bending moment distribution in the beam; (**c**) bending stress distribution in the beam.

11.19 Determine the response of a simply supported beam subjected to a moving uniformly distributed load, as shown in Fig. 11.26.

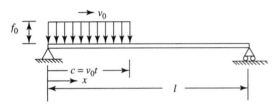

Figure 11.26 Simply supported beam subjected to a traveling distributed load.

11.20 Show that the constant a in Eq. (11.27) is positive.

11.21 Derive the expressions for the constants A_n and B_n in Eq. (11.72) in terms of the initial conditions of the beam [Eqs. (11.54) and (11.55)].

11.22 Using Hamilton's principle, derive the equation of motion, Eq. (11.203), and the boundary conditions of a beam resting on an elastic foundation.

11.23 Derive Eq. (11.216) from Eqs. (11.213) and (11.214) for the forced vibration of a beam on an elastic foundation.

11.24 Derive the characteristic equation for a fixed–fixed Timoshenko beam using the solution of Eqs. (11.308) and (11.309) and the boundary conditions of Eqs. (11.302) and (11.303).

11.25 Determine the natural frequencies of vibration of a fixed–fixed beam using Rayleigh's theory.

12

Vibration of Circular Rings and Curved Beams

12.1 INTRODUCTION

The problems of vibration of circular rings and curved beams (or rods) find application in several practical problems. The vibration of circular rings is encountered in an investigation of the frequencies and dynamic response of ring-stiffened cylinders such as those encountered in airplane fuselages, circular machine parts such as gears and pulleys, rotating machines, and stators of electrical machines. The vibration of a curved rod can be categorized into four types when the centerline of an undeformed rod is a plane curve and its plane is a principal plane of the rod at each point. In the first type, flexural vibrations take place in the plane of the ring without undergoing any extension of the centerline of the ring. In the second type, flexural vibrations, involving both displacement at right angles to the plane of the ring and twist, take place. In the third type, the curved rod or ring vibrates in modes similar to the torsional vibrations of a straight rod. In the fourth type, the ring possesses modes of vibration similar to the extensional vibration of a straight rod. It is assumed that the undeformed centerline of the ring has a radius R, the cross section of the ring is uniform, and the cross-sectional dimensions of the ring are small (for a thin ring) compared to the radius of the centerline of the ring. The vibration of curved beams is important in the study of the dynamic behavior of arches.

12.2 EQUATIONS OF MOTION OF A CIRCULAR RING

12.2.1 Three-Dimensional Vibrations of a Circular Thin Ring

Next, we derive the general equations of motion governing the three-dimensional vibrations of a thin rod which in the unstressed state forms a circular ring or a portion of such a ring [Fig. 12.1(a)]. The effects of rotary inertia and shear deformation are neglected. The following assumptions are made in the derivation:

1. The centerline of the ring in an undeformed state forms a full circle or an arc of a circle.
2. The cross section of the ring is constant around the circle.
3. No boundary constraints are introduced on the ring (i.e., the rim is assumed free).

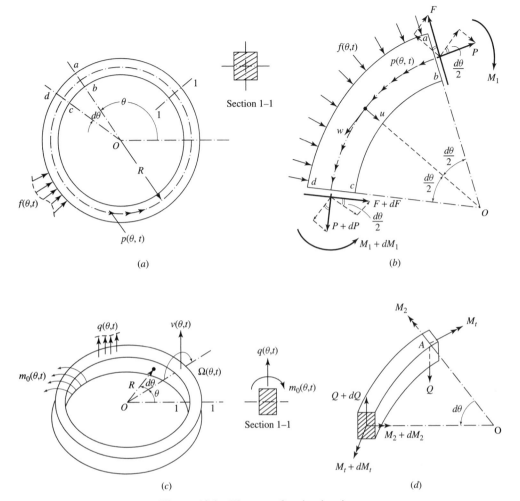

Section 1–1

Figure 12.1 Element of a circular ring.

Consider the free-body diagram of an element of a circular ring *abcd* shown in Fig. 12.1(*b*), where $M_1(\theta, t)$ is the in-plane bending moment, taken positive when it tends to reduce the radius of curvature of the beam, $F(\theta, t)$ is the shearing force, taken positive when it acts in a radially inward direction on a positive face, $P(\theta, t)$ is the tensile force, and $f(\theta, t)$ and $p(\theta, t)$ are the external radial and tangential forces, respectively, per unit length. The radial and tangential inertia forces acting on the element of the circular ring are given by

$$\rho A(\theta) R \, d\theta \, \frac{\partial^2 u(\theta, t)}{\partial t^2} \quad \text{and} \quad \rho A(\theta) R \, d\theta \, \frac{\partial^2 w(\theta, t)}{\partial t^2}$$

where ρ is the density, A is the cross-sectional area, and R is the radius of the centerline of the ring. The equations of motion in the radial and tangential directions can be expressed as (by neglecting small quantities of high order):

$$\frac{\partial F}{\partial \theta} + P + fR = \rho A R \frac{\partial^2 u}{\partial t^2} \tag{12.1}$$

$$\frac{\partial P}{\partial \theta} - F + pR = \rho A R \frac{\partial^2 w}{\partial t^2} \tag{12.2}$$

If $v(\theta, t)$ denotes the transverse displacement, $Q(\theta, t)$ the transverse shear force, and $q(\theta, t)$ the external distributed transverse force in a direction normal to the middle plane of the ring [Fig. 12.1(c) and (d)], the transverse inertia force acting on the element of the ring is

$$\rho A(\theta) R \, d\theta \frac{\partial^2 v(\theta, t)}{\partial t^2}$$

and the equation of motion in the transverse direction is given by

$$\frac{\partial Q}{\partial \theta} + qR = \rho A R \frac{\partial^2 v}{\partial t^2} \tag{12.3}$$

The moment equation of motion in the middle plane of the ring (about an axis normal to the middle plane of the ring) leads to

$$\frac{\partial M_1}{\partial \theta} + FR = 0 \tag{12.4}$$

Let $M_2(\theta, t)$ be the bending moment in the ring about the radial axis, $M_t(\theta, t)$ the torsional moment about the tangential axis, and $m_0(\theta, t)$ the distributed external torque acting on the ring. Then the dynamical moment equilibrium equation about the radial axis of the element of the ring is given by

$$\frac{\partial M_2}{\partial \theta} - QR + M_t = 0 \tag{12.5}$$

Similarly, the dynamical moment equilibrium equation about the tangential axis can be expressed as

$$\frac{\partial M_t}{\partial \theta} - M_2 + m_0 R = 0 \tag{12.6}$$

The six equations (12.1)–(12.6) are the equations of motion governing the three-dimensional vibrations of a ring in 10 unknowns: F, P, Q, M_1, M_2, M_t, u, v, w, and Ω [Ω and v appear in the expression of M_t, as shown in Eq. (12.18)]. To solve the equations, four more relations are required. These relations include the moment–displacement relations, along with the condition of inextensionality of the centerline of the ring.

12.2.2 Axial Force and Moments in Terms of Displacements

To express the forces and moments in the ring in terms of the deformation components, consider a typical element of the ring, ab, located at a distance x from the centroidal axis of the ring, as shown in Fig. 12.2(a). If the axial stress in the element is σ, the

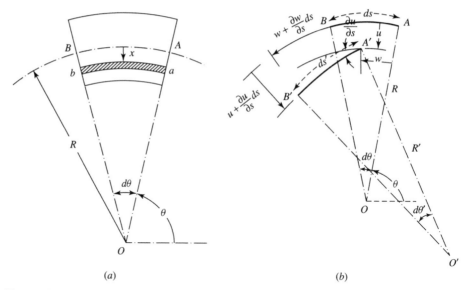

Figure 12.2   (*a*) Differential element of a curved beam; (*b*) Undeformed and deformed positions of the central axis of the curved beam.

axial force and the bending moment can be expressed as

$$P = \iint_A \sigma \, dA \tag{12.7}$$

$$M_1 = \iint_A \sigma x \, dA \tag{12.8}$$

where

$$\sigma = E\varepsilon \tag{12.9}$$

ε is the axial strain, E is Young's modulus, and A is the cross-sectional area of the ring. The strain in the element can be expressed as [1, 5, 6]

$$\varepsilon = \frac{1}{R}\left[-u + \frac{\partial w}{\partial \theta} - \frac{x}{R}\frac{\partial}{\partial \theta}\left(w + \frac{\partial u}{\partial \theta}\right)\right] \tag{12.10}$$

where u and w are, respectively, the radial and tangential displacements of a typical point A lying on the central axis of the ring as shown in Fig. 12.2(*b*). By substituting Eqs. (12.10) and (12.9) into Eqs. (12.7) and (12.8) and performing the integrations, we obtain

$$P = \frac{EA}{R}\left(-u + \frac{\partial w}{\partial \theta}\right) \tag{12.11}$$

$$M_1 = \frac{EI_1}{R^2}\frac{\partial}{\partial \theta}\left(w + \frac{\partial u}{\partial \theta}\right) \tag{12.12}$$

where I_1 denotes the moment of inertia of the cross section about an axis perpendicular to the middle plane of the ring and passing through the centroid:

$$I_1 = \iint_A x^2 \, dA \tag{12.13}$$

Note that the x contribution of Eq. (12.10) does not enter in Eq. (12.11) since the axis is the centroidal axis. The condition of inextensionality of the centerline of the ring is given by

$$\varepsilon = 0 \tag{12.14}$$

By neglecting products of small quantities in Eq. (12.10), Eq. (12.14) leads to

$$\frac{\partial w}{\partial \theta} = u \tag{12.15}$$

Equation (12.12) denotes the differential equation for the deflection curve of a thin curved bar with a circular centerline. If the inextensionality of the centerline of the ring, Eq. (12.15), is used, Eq. (12.12) takes the form

$$M_1 = \frac{E I_1}{R^2} \left(u + \frac{\partial^2 u}{\partial \theta^2} \right) \tag{12.16}$$

It can be seen that for infinitely large radius, Eq. (12.16) coincides with the equation for the bending moment of a straight beam.

Proceeding in a similar manner, two more relations for bending in a normal plane and for twisting of the ring can be derived as [5, 7]

$$M_2 = \frac{E I_2}{R^2} \left(R\Omega - \frac{d^2 v}{d\theta^2} \right) \tag{12.17}$$

$$M_t = \frac{G J}{R^2} \left(\frac{dv}{d\theta} + R\frac{d\Omega}{d\theta} \right) \tag{12.18}$$

where $E I_2$ is the flexural rigidity of the ring in a normal plane, GJ is the torsional rigidity of the ring (GJ is denoted as C for noncircular sections), and Ω is the angular displacement of the cross section of the ring due to torsion.

12.2.3   Summary of Equations and Classification of Vibrations

The 10 governing equations in 10 unknowns for the three-dimensional vibrations of a circular ring are given by Eqs. (12.1)–(12.6) and (12.15)–(12.18). It can be seen that these equations fall into two sets: one consisting of Eqs. (12.1), (12.2), (12.4), (12.15), and (12.16), where the variables v and Ω do not appear, and the other consisting of Eqs. (12.3), (12.5), (12.6), (12.17) and (12.18), where the variables u and w do not appear. In the first set of equations, the motion can be specified by the displacement u or w, and hence they represent flexural vibrations of the ring in its plane. In the second set of equations, the motion can be specified by v or Ω, and hence they represent flexural vibrations at right angles to the plane of the ring, also called coupled twist–bending vibrations involving both displacement at right angles to the plane of the ring and twist.

In general, it can be shown [5] that the vibrations of a curved rod fall into two such classes whenever the centerline of the undeformed rod is a plane curve and its plane is a principal plane of the rod at each point. In case the centerline is a curve of double curvature such as a helical rod, it is not possible to separate the modes of vibration into two classes, and the problem becomes extremely difficult. In addition to the two types of vibration stated, a curved rod also possesses modes of vibration analogous to the torsional and extensional vibrations of a straight rod. When $u = w = 0$ and v is assumed to be small compared to $R\Omega$, Eq. (12.6) becomes the primary equation governing the torsional vibrations of a ring. On the other hand, when $v = \Omega = 0$ and the inextensionality condition of Eq. (12.15) is not used, the ring undergoes extensional vibrations.

12.3 IN-PLANE FLEXURAL VIBRATIONS OF RINGS

12.3.1 Classical Equations of Motion

As indicated in Section 12.2.3, the equations of motion for the in-plane flexural vibrations of a thin ring are given by Eqs. (12.1), (12.2), (12.4), (12.15) and (12.16). These equations can be combined into a single equation as

$$\frac{\partial^6 w}{\partial \theta^6} + 2\frac{\partial^4 w}{\partial \theta^4} + \frac{\partial^2 w}{\partial \theta^2} - \frac{R^4}{EI_1}\left(\frac{\partial f}{\partial \theta} - p\right) + \frac{\rho AR^4}{EI_1}\frac{\partial^2}{\partial t^2}\left(\frac{\partial^2 w}{\partial \theta^2} - w\right) = 0 \quad (12.19)$$

Equation (12.19) represents the equation of motion for the in-plane vibrations of a thin ring in terms of the radial deflection w.

Natural Frequencies of Vibration   For free vibration, the external forces f and p are assumed to be zero. Then, by assuming the solution to be harmonic, as

$$w(\theta, t) = W(\theta)e^{i\omega t} \quad (12.20)$$

where ω is the frequency of vibration, Eq. (12.19) becomes

$$\frac{d^6 W}{d\theta^6} + 2\frac{d^4 W}{d\theta^4} + \frac{d^2 W}{d\theta^2} - \frac{\rho AR^4 \omega^2}{EI_1}\left(\frac{d^2 W}{d\theta^2} - W\right) = 0 \quad (12.21)$$

The solution of Eq. (12.21) is assumed as

$$W(\theta) = C_1 \sin(n\theta + \phi) \quad (12.22)$$

where C_1 and ϕ are constants. When Eq. (12.22) is substituted, Eq. (12.21) gives the natural frequencies of vibration as

$$\omega_n^2 = \frac{EI_1}{\rho AR^4}\frac{n^6 - 2n^4 + n^2}{n^2 + 1}, \quad n = 2, 3, \dots \quad (12.23)$$

Note that for a complete ring, n in Eq. (12.22) must be an integer, for this gives n complete waves of deflection W in the circumference of the ring and W must be a function whose values recur as θ increases by 2π. Also, n cannot take the value of 1 since it represents pure rigid-body oscillation without any alteration of shape, as shown in Fig. 12.3. Thus, n can only take the values 2, 3, ... The normal modes of the ring

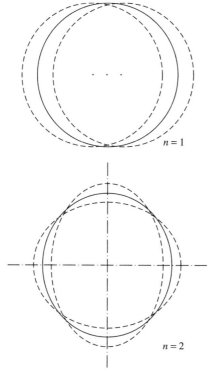

$n = 1$

$n = 2$

Figure 12.3   Mode shapes of a ring.

can be expressed as

$$w_n(\theta, t) = C_1 \sin(n\theta + \phi)e^{i\omega_n t} \tag{12.24}$$

where C_1 and ϕ can be determined from the initial conditions of the ring.

12.3.2   Equations of Motion That Include Effects of Rotary Inertia and Shear Deformation

A procedure similar to the one used in the ease of a straight beam by Timoshenko can be used to include the effects of shear deformation and rotatory inertia in the equations of motion of a ring. The slope of the deflection curve (ψ) depends not only on the rotation of cross sections of the ring, but also on the shear. If ϕ denotes the slope of the deflection curve when the shearing force is neglected and β denotes the angular deformation due to shear at the neutral axis in the same cross section, then the angle between the deformed and undeformed centerlines can be expressed as (Fig. 12.4)

$$\psi = \phi + \beta \tag{12.25}$$

But ψ can be expressed in terms of the displacement components u and w once the geometry of deformation is known. From Fig. 12.4, the total slope of the deflection curve can be expressed as

$$\psi = \frac{1}{R}\frac{\partial u}{\partial \theta} + \frac{w}{R} \tag{12.26}$$

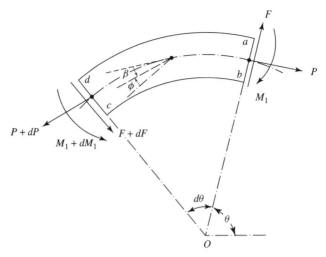

Figure 12.4 Composition of the slope of a deflection curve.

and hence Eq. (12.25) yields

$$\beta = \frac{1}{R}\left(\frac{\partial u}{\partial \theta} + w - R\phi\right) \tag{12.27}$$

As for the shear force, the exact manner of distribution of the shear stress over the cross section is not known. Hence, to account for the variation of β through the cross section, a numerical factor k that depends on the shape of the cross section is introduced and the shear force is expressed as

$$F = k\beta AG \tag{12.28}$$

where G is the rigidity modulus of the material of the ring. Since the exact determination of the factor k for rings involves consideration of the theory of elasticity, the value of k of a straight beam can be used in Eq. (12.28) as an approximation. Equations (12.27) and (12.28) can be combined to obtain

$$F = \frac{kAG}{R}\left(\frac{\partial u}{\partial \theta} + w - R\phi\right) \tag{12.29}$$

The bending moment can be expressed in terms of the displacement components as [Eq. (12.16) is modified in the presence of shear deformation]

$$M_1 = \frac{EI_1}{R}\frac{\partial \phi}{\partial \theta} \tag{12.30}$$

The differential equation of rotation of an element $abcd$, shown in Fig. 12.4, can be found as

$$\frac{\partial M_1}{\partial \theta} + RF = \rho I_1 R\frac{\partial^2 \phi}{\partial t^2} \tag{12.31}$$

The differential equations for the translatory motion of the element in the radial and tangential directions are given by

$$\frac{\partial F}{\partial \theta} + P + fR = \rho AR \frac{\partial^2 u}{\partial t^2} \tag{12.32}$$

$$\frac{\partial P}{\partial \theta} - F + pR = \rho AR \frac{\partial^2 w}{\partial t^2} \tag{12.33}$$

By eliminating P from Eqs. (12.32) and (12.33), we obtain

$$\frac{\partial^2 F}{\partial \theta^2} + F = \rho AR \left(\frac{\partial^3 u}{\partial \theta \partial t^2} - \frac{\partial^2 w}{\partial t^2} \right) + R \left(p - \frac{\partial f}{\partial \theta} \right) \tag{12.34}$$

The condition for inextensionality of the centerline is given by Eq. (12.15). Thus, the set of equations (12.29), (12.30), (12.31), (12.34), and (12.15) govern the in-plane flexural vibrations of a ring when the effects of rotary inertia and shear deformation are included. From these equations, a single equation of motion can be derived as

$$\frac{\partial^6 w}{\partial \theta^6} + 2 \frac{\partial^4 w}{\partial \theta^4} + \frac{\partial^2 w}{\partial \theta^2}$$

$$= \left(\frac{\rho R^2}{E} + \frac{\rho R^2}{kG} \right) \frac{\partial^6 w}{\partial \theta^4 \partial t^2} + \left(2 \frac{\rho R^2}{E} - \frac{\rho AR^4}{EI_1} - \frac{\rho R^2}{kG} \right) \frac{\partial^4 w}{\partial \theta^2 \partial t^2}$$

$$+ \left(\frac{\rho R^2}{E} + \frac{\rho AR^4}{EI_1} \right) \frac{\partial^2 w}{\partial t^2} + \left(\frac{\rho^2 R^4}{kEG} \right) \frac{\partial^4 w}{\partial t^4} - \left(\frac{\rho^2 R^4}{kEG} \right) \frac{\partial^6 w}{\partial \theta^2 \partial t^4}$$

$$- \frac{\rho R^4}{kEAG} \frac{\partial^2}{\partial t^2} \left(p - \frac{\partial f}{\partial \theta} \right) - \frac{R^4}{EI_1} \left(p - \frac{\partial f}{\partial \theta} \right) + \frac{R^2}{kAG} \left(p - \frac{\partial f}{\partial \theta} \right) \tag{12.35}$$

Equation (12.35) denotes the governing differential equation of motion of a ring that takes into account the effect of rotary inertia and shear deformation. If the effect of shear deformation is neglected, the terms involving k are to be neglected in Eq. (12.35). Similarly, if the effects of rotary inertia and shear deformation are neglected, Eq. (12.35) can be seen to reduce to Eq. (12.19). Finally, $R \to \infty$ and Eq. (12.35) reduces to the equation of motion of a Timoshenko beam.

Natural Frequencies of Vibration   For free vibration, the external forces f and p are assumed to be zero. By assuming a harmonic solution as in Eq. (12.24), the frequency equation can be derived as

$$K_2^2(-n^2 S_2^2 S_1 - S_2^2 S_1) + K_2(n^4 S_2 + n^4 S_2 S_1 - 2S_2 n^2 + n^2 S_2 S_1 + n^2 + S_2 + 1)$$

$$+ (-n^6 + 2n^4 - n^2) = 0 \tag{12.36}$$

where

$$K_2 = \frac{\rho AR^4}{EI_1} \omega^2, \qquad S_1 = \frac{E}{kG}, \qquad S_2 = \frac{I_1}{AR^2}$$

Equation (12.36) is a quadratic in K_2 and hence two frequency values are associated with each mode of vibration (i.e., for each value of n). The smaller of the two

ω values corresponds to the flexural mode, and the higher value corresponds to the thickness-shear mode. Similar behavior is exhibited by a Timoshenko beam as well. In Eq. (12.36), n must be an integer with a value greater than 1. The values of the natural frequencies given by Eq. (12.36) for various values of S_1 and S_2 are given in Tables 12.1. Note that if the terms involving S_2 are neglected in Eq. (12.36), we obtain a frequency equation that neglects the effect of shear deformation (but considers the effect of rotary inertia). Similarly, if the terms involving S_1 and S_2 are neglected, Eq. (12.36) reduces to Eq. (12.23), which neglects both the effects of rotary inertia and shear deformation.

12.4   FLEXURAL VIBRATIONS AT RIGHT ANGLES TO THE PLANE OF A RING

12.4.1   Classical Equations of Motion

As stated in Section 12.2.3, the equations of motion for the coupled twist–bending vibrations of a thin ring are given by Eqs. (12.3), (12.5), (12.6), (12.17), and (12.18). All these equations can be combined to obtain a single equation as

$$
\frac{\partial^6 v}{\partial \theta^6} + 2\frac{\partial^4 v}{\partial \theta^4} + \frac{\partial^2 v}{\partial \theta^2} + \frac{\rho A R^4}{E I_2}\frac{\partial^4 v}{\partial \theta^2 \partial t^2} - \frac{\rho A R^4}{C}\frac{\partial^2 v}{\partial t^2}
$$
$$
- \frac{R^4}{E I_2}\frac{\partial^2 q}{\partial \theta^2} + \frac{R^4}{C}q + R^3\left(\frac{1}{C} + \frac{1}{E I_2}\right)\frac{\partial^2 m_0}{\partial \theta^2} = 0 \qquad (12.37)
$$

where $C = GJ$. Thus, Eq. (12.37) denotes the classical equation for the flexural vibrations involving transverse displacement and twist of a thin ring. The twist Ω is related to the transverse deflection v by

$$
\frac{\partial^2 \Omega}{\partial \theta^2} = \frac{1}{E I_2 + C}\left(\frac{E I_2}{R}\frac{\partial^4 v}{\partial \theta^4} - \frac{C}{R}\frac{\partial^2 v}{\partial \theta^2} + \rho A R^3\frac{\partial^2 v}{\partial t^2} - R^3 q\right) \qquad (12.38)
$$

Table 12.1   In-Plane Flexural Vibrations of a Ring, Natural Frequencies, and Values of $\omega_n\sqrt{\rho A R^4/E I_1}$; Effects of Rotary Inertia and Shear Deformation Included

E/kG	I_1/AR^2	n	Flexural mode	Thickness-shear mode
1.0	0.02	2	2.543	52.759
		3	6.682	56.792
	0.10	2	2.167	12.382
		3	5.017	15.127
2.0	0.02	2	2.459	38.587
		3	6.289	42.667
3.0	0.02	2	2.380	32.400
		3	5.950	36.800
	0.25	2	1.321	4.695
		3	2.682	6.542

Natural Frequencies of Vibration   For free vibration, $q = m_0 = 0$, and Eq. (12.37) gives the natural frequencies of vibration, by assuming a harmonic solution, as

$$\omega_n^2 = \frac{EI_2}{\rho A R^4} \frac{n^6 - 2n^4 + n^2}{n^2 + EI_2/C}, \qquad n = 2, 3, \ldots \tag{12.39}$$

Note that the value of $n = 1$ corresponds to rigid-body motion and the normal modes of the ring are given by

$$v_n(\theta, t) = C_1 \sin(n\theta + \phi)e^{i\omega_n t}, \qquad n = 2, 3, \ldots \tag{12.40}$$

12.4.2   Equations of Motion That Include Effects of Rotary Inertia and Shear Deformation

Using a procedure similar to that of a Timosherko beam, the slope of the transverse deflection curve is expressed as

$$\frac{1}{R} \frac{\partial v}{\partial \theta} = \alpha + \beta \tag{12.41}$$

where α is the slope of the deflection curve when the shearing force is neglected and β is the angle of shear at the neutral axis in the same cross section. The transverse shearing force F is given by

$$Q = k\beta AG \tag{12.42}$$

where k is a numerical factor taken to account for the variation of β through the cross section and is a constant for any given cross section. Using Eq. (12.41), Q can be expressed as

$$Q = kAG \left(\frac{1}{R} \frac{\partial v}{\partial \theta} - \alpha \right) \tag{12.43}$$

The moment–displacement relations, with a consideration of the shear deformation effect, can be expressed as

$$M_2 = \frac{EI_2}{R} \left(\Omega - \frac{\partial \alpha}{\partial \theta} \right) \tag{12.44}$$

$$M_t = \frac{C}{R} \left(\alpha + \frac{\partial \Omega}{\partial \theta} \right) \tag{12.45}$$

The differential equation for the translatory motion of an element of the ring in the transverse direction is given by

$$\frac{\partial Q}{\partial \theta} + Rq = \rho A R \frac{\partial^2 v}{\partial t^2} \tag{12.46}$$

The equations of motion for the rotation of an element of the ring about the radial and tangential axes can be expressed as

$$\frac{\partial M_2}{\partial \theta} + M_t - QR = -\rho I_2 R \frac{\partial^2 \alpha}{\partial t^2} \tag{12.47}$$

$$\frac{\partial M_t}{\partial \theta} - M_2 + m_0 R = \rho J R \frac{\partial^2 \Omega}{\partial t^2} \tag{12.48}$$

In view of Eqs. (12.43)–(12.45), Eqs. (12.46)–(12.48) can be rewritten as

$$
\frac{\partial}{\partial\theta}\left[kAG\left(\frac{1}{R}\frac{\partial v}{\partial\theta}-\alpha\right)\right]+qR-\rho AR\frac{\partial^2 v}{\partial t^2}=0
$$

(12.49)

$$
\frac{\partial}{\partial\theta}\left[\frac{EI_2}{R}\left(\Omega-\frac{\partial\alpha}{\partial\theta}\right)\right]-kARG\left(\frac{1}{R}\frac{\partial v}{\partial\theta}-\alpha\right)+\frac{C}{R}\left(\alpha+\frac{\partial\Omega}{\partial\theta}\right)+\rho I_2 R\frac{\partial^2\alpha}{\partial t^2}=0
$$

(12.50)

$$
\frac{\partial}{\partial\theta}\left[\frac{C}{R}\left(\alpha+\frac{\partial\Omega}{\partial\theta}\right)\right]-\frac{EI_2}{R}\left(\Omega-\frac{\partial\alpha}{\partial\theta}\right)+m_0 R-\rho JR\frac{\partial^2\Omega}{\partial t^2}=0
$$

(12.51)

Equations (12.49)–(12.51) thus represent the equations of motion of a ring for the coupled twist–bending vibrations of a ring, including the effects of rotary inertia and shear deformation. These three equations can be combined to obtain a single equation in terms of the displacement variable v as [7, 9]

$$
\frac{\partial^6 v}{\partial\theta^6}+2\frac{\partial^4 v}{\partial\theta^4}+\frac{\partial^2 v}{\partial\theta^2}-\frac{\rho R^2}{G}\left(\frac{1}{k}+\frac{AR^2 G}{C}\right)\frac{\partial^2 v}{\partial t^2}
$$

$$
-\frac{\rho^2 R^4}{G^2}\left(\frac{I_2 G}{kC}+\frac{GJ}{kEI_2}+\frac{AR^2 G^2 J}{CEI_2}\right)\frac{\partial^4 v}{\partial t^4}
$$

$$
-\frac{\rho^3 R^6 J}{GEkC}\frac{\partial^6 v}{\partial t^6}+\frac{\rho R^2}{G}\left(\frac{AR^2 G}{EI_2}-\frac{2}{k}+\frac{I_2 G}{kC}+\frac{GJ}{EI_2}\right)\frac{\partial^4 v}{\partial\theta^2\partial t^2}
$$

$$
-\frac{\rho R^2}{G}\left(\frac{GJ}{C}+\frac{1}{k}+\frac{G}{E}\right)\frac{\partial^6 v}{\partial\theta^4\partial t^2}
$$

$$
+\frac{\rho^2 R^4}{G^2}\left(\frac{GJ}{kC}+\frac{G^2 J}{EC}+\frac{G}{kE}\right)\frac{\partial^6 v}{\partial\theta^2\partial t^4}+\frac{R^2}{kAG}\frac{\partial^4 q}{\partial\theta^4}+\frac{R^2}{kAG}\left(2-\frac{kAGR^2}{EI_2}\right)\frac{\partial^2 q}{\partial\theta^2}
$$

$$
-\frac{\rho R^4}{kAG}\left(\frac{1}{E}+\frac{J}{C}\right)\frac{\partial^4 q}{\partial\theta^2\partial t^2}+R^2\left(\frac{1}{kAG}+\frac{R^2}{C}\right)q
$$

$$
+\frac{\rho R^4}{kAG}\left(\frac{I_2}{C}+\frac{J}{EI_2}+\frac{kAR^2 GJ}{EI_2 C}\right)\frac{\partial^2 q}{\partial t^2}+\frac{\rho^2 R^6 J}{kAGCE}\frac{\partial^4 q}{\partial t^4}
$$

$$
+R^3\left(\frac{1}{EI_2}+\frac{1}{C}\right)\frac{\partial^2 m_0}{\partial\theta^2}=0
$$

(12.52)

In this case, the twist Ω is related to v as

$$
\frac{\partial^2\Omega}{\partial\theta}=\frac{1}{EI_2+C}\left[\frac{EI_2}{R}\frac{\partial^4 v}{\partial\theta^4}-\frac{C}{R}\frac{\partial^2 v}{\partial\theta^2}+\rho R\left(\frac{C}{kG}+AR^2\right)\frac{\partial^2 v}{\partial t^2}+\frac{\rho^2 I_2 R^3}{kG}\frac{\partial^4 v}{\partial t^4}\right.
$$

$$
\left.-\rho I_2 R\left(1+\frac{E}{kG}\right)\frac{\partial^4 v}{\partial\theta^2\partial t^2}-R\left(R^2+\frac{C}{kAG}\right)q+\frac{EI_2 R}{kAG}\frac{\partial^2 q}{\partial\theta^2}-\frac{\rho I_2 R^3}{kAG}\frac{\partial^2 q}{\partial t^2}\right]
$$

(12.53)

Various special cases can be derived from Eqs. (12.52) and (12.53) as follows:

1. When the effect of shear deformation is considered without the effect of rotary inertia:

$$\frac{\partial^6 v}{\partial \theta^6} + 2\frac{\partial^4 v}{\partial \theta^4} + \frac{\partial^2 v}{\partial \theta^2} - \left(\frac{\rho R^2}{kG} + \frac{\rho A R^4}{C}\right)\frac{\partial^2 v}{\partial t^2} + \left(\frac{\rho A R^4}{EI_2} - 2\frac{\rho R^2}{kG}\right)\frac{\partial^4 v}{\partial \theta^2 \partial t^2}$$

$$- \frac{\rho R^2}{kG}\frac{\partial^6 v}{\partial \theta^4 \partial t^2} + \frac{R^2}{kAG}\frac{\partial^4 q}{\partial \theta^4} + \left(\frac{2R^2}{kAG} - \frac{R^4}{EI_2}\right)\frac{\partial^2 q}{\partial \theta^2} + \left(\frac{R^2}{kAG} + \frac{R^4}{C}\right)q$$

$$+ \left(\frac{R^3}{EI_2} + \frac{R^3}{C}\right)\frac{\partial^2 m_0}{\partial \theta^2} = 0 \tag{12.54}$$

$$\frac{\partial^2 \Omega}{\partial \theta^2} = \frac{1}{EI_2 + C}\left[\frac{EI_2}{R}\frac{\partial^4 v}{\partial \theta^4} - \frac{C}{R}\frac{\partial^2 v}{\partial \theta^2} + \left(\frac{\rho R C}{kG} + \rho A R^3\right)\frac{\partial^2 v}{\partial t^2}\right.$$

$$\left.- \frac{EI_2 \rho R}{kG}\frac{\partial^4 v}{\partial \theta^2 \partial t^2} + \frac{EI_2 R}{kAG}\frac{\partial^2 q}{\partial \theta^2} - \left(\frac{RC}{kAG} + R^3\right)q\right] \tag{12.55}$$

2. When the effect of rotary inertia is considered without the effect of shear deformation:

$$\frac{\partial^6 v}{\partial \theta^6} + 2\frac{\partial^4 v}{\partial \theta^4} + \frac{\partial^2 v}{\partial \theta^2} - \frac{\rho A R^4}{C}\frac{\partial^2 v}{\partial t^2} - \frac{\rho^2 A R^6 J}{EI_2 C}\frac{\partial^4 v}{\partial t^4}$$

$$+ \frac{\rho R^2}{G}\left(\frac{A R^2 G}{EI_2} + \frac{I_2 G}{C} + \frac{GJ}{EI_2}\right)\frac{\partial^4 v}{\partial \theta^2 \partial t^2} - \frac{\rho R^2}{G}\left(\frac{GJ}{C} + \frac{G}{E}\right)\frac{\partial^6 v}{\partial \theta^4 \partial t^2}$$

$$+ \frac{\rho^2 R^4 J}{EC}\frac{\partial^6 v}{\partial \theta^2 \partial t^4} - \frac{R^4}{EI_2}\frac{\partial^2 q}{\partial \theta^2} + \frac{R^4}{C}q + \frac{\rho R^6 J}{EI_2 C}\frac{\partial^2 q}{\partial t^2}$$

$$+ R^3\left(\frac{1}{EI_2} + \frac{1}{C}\right)\frac{\partial^2 m_0}{\partial \theta^2} = 0 \tag{12.56}$$

$$\frac{\partial^2 \Omega}{\partial \theta^2} = \frac{1}{EI_2 + C}\left(\frac{EI_2}{R}\frac{\partial^4 v}{\partial \theta^4} - \frac{C}{R}\frac{\partial^2 v}{\partial \theta^2} + \rho A R^3 \frac{\partial^2 v}{\partial t^2} - \rho I_2 R \frac{\partial^4 v}{\partial \theta^2 \partial t^2} - R^3 q\right) \tag{12.57}$$

3. When the effects of both rotary inertia and shear deformation are neglected:

$$\frac{\partial^6 v}{\partial \theta^6} + 2\frac{\partial^4 v}{\partial \theta^4} + \frac{\partial^2 v}{\partial \theta^2} + \frac{\rho A R^4}{EI_2}\frac{\partial^4 v}{\partial \theta^2 \partial t^2} - \frac{\rho A R^4}{C}\frac{\partial^2 v}{\partial t^2} - \frac{R^4}{EI_2}\frac{\partial^2 q}{\partial \theta^2}$$

$$+ \frac{R^4}{C}q + R^3\left(\frac{1}{EI_2} + \frac{1}{C}\right)\frac{\partial^2 m_0}{\partial \theta^2} = 0 \tag{12.58}$$

$$\frac{\partial^2 \Omega}{\partial \theta^2} = \frac{1}{EI_2 + C}\left(\frac{EI_2}{R}\frac{\partial^4 v}{\partial \theta^4} - \frac{C}{R}\frac{\partial^2 v}{\partial \theta^2} + \rho A R^3 \frac{\partial^2 v}{\partial t^2} - R^3 q\right) \tag{12.59}$$

Further, if R is made equal to infinity and the terms involving the torsional motion are neglected, Eq. (12.52) will reduce to the Timoshenko beam equation.

Natural Frequencies of Vibration   Setting $q = m_0 = 0$ for free vibration with the harmonic solution and introducing

$$T = \frac{\rho A R^4 \omega^2}{E I_2}, \quad S_1 = \frac{E}{G}, \quad S_2 = \frac{I_2}{A R^2}, \quad S_3 = \frac{I_2}{J}, \quad S_4 = \frac{1}{k}, \quad S_5 = \frac{E I_2}{C}$$

$$(12.60)$$

the frequency equation can be derived as

$$T^3 \frac{S_1 S_2^3 S_4 S_5}{S_3} - T^2 \left[n^2 S_2^2 \left(S_1 S_4 + \frac{S_5}{S_3} + \frac{S_1 S_4 S_5}{S_3} \right) + \frac{S_1 S_2^2 S_4}{S_3} + S_1 S_2^2 S_4 S_5 + \frac{S_2 S_5}{S_3} \right]$$

$$+ T \left[S_1 S_2 S_4 + S_5 + n^2 \left(1 + \frac{S_2}{S_3} - 2 S_1 S_2 S_4 + S_2 S_5 \right) + n^4 S_2 \left(1 + S_1 S_4 + \frac{S_5}{S_3} \right) \right]$$

$$+ (2n^4 - n^6 - n^2) = 0 \tag{12.61}$$

This equation is a cubic in T and gives three frequency values for each mode number n with the lowest one corresponding to the flexural mode. The two higher frequencies correspond to the torsional and transverse thickness–shear modes. For a complete ring, n must be an integer with values greater than 1. Corresponding to any ω_n given by Eq. (12.61), the normal mode can be expressed as

$$v_n(\theta, t) = C_1 \sin(n\theta + \phi) e^{i \omega_n t} \tag{12.62}$$

In this case, the solution of Eq. (12.53) can be expressed as (by considering the periodic nature of Ω):

$$\Omega_n(\theta, t) = -\frac{1}{(E I_2 + C) n^2} \left[n^4 \frac{E I_2}{R} + n^2 \frac{C}{R} - \rho R \omega_n^2 \left(\frac{C}{kG} + A R^2 \right) \right.$$

$$\left. - n^2 \rho I_2 R \omega_n^2 \left(1 + \frac{E}{kG} \right) + \frac{\rho^2 R^3 I_2}{kG} \omega_n^4 \right] C_1 \sin(n\theta + \phi) e^{i \omega_n t}$$

$$(12.63)$$

Note that if the effect of rotatory inertia only is under consideration, terms involving k in Eqs. (12.61) and (12.63) are to be omitted. The in-plane and normal-to-plane natural frequencies of vibration of a ring with circular cross section are compared in Table 12.2.

12.5   TORSIONAL VIBRATIONS

For the torsional vibration of a circular ring, the in-plane displacement components u and w are assumed to be zero. In addition, the transverse displacement v, perpendicular to the plane of the ring, is assumed to be small compared to $R\Omega$, where R is the radius of the undeformed centerline of the ring and Ω is the angular deformation of the cross section of the ring (Fig. 12.5). In this case, ignoring the terms involving α, Eq. (12.51) gives

$$\frac{C}{R^2} \frac{\partial^2 (R\Omega)}{\partial \theta^2} - \frac{E I_2}{R^2} (R\Omega) = \rho J \frac{\partial^2 (R\Omega)}{\partial t^2} \tag{12.64}$$

Table 12.2   Comparison of In-Plane and Normal-to-Plane Frequencies of a Ring[a]

			Value of $\omega_n\sqrt{mAR^4/EI}$ according to:					
	Classical theory		Rotary inertia considered		Shear deformation considered		Both shear and rotary inertia considered	
Mode shape, n	In-plane	Normal-to-plane	In-plane	Normal-to-plane	In-plane	Normal-to-plane	In-plane	Normal-to-plane
2	2.683	2.606	2.544	2.011	2.011	2.259	1.975	1.898
3	7.589	7.478	6.414	4.860	4.572	5.028	4.446	4.339
4	14.552	14.425	10.765	7.875	7.169	7.671	6.974	6.886
5	23.534	23.399	15.240	10.796	9.709	10.192	9.478	9.431
6	34.524	34.385	19.706	13.590	12.191	12.636	11.948	11.946

[a]Ring with circular cross section, $r/R = 0.5$, $k = 0.833$, $E/G = 2.6$.

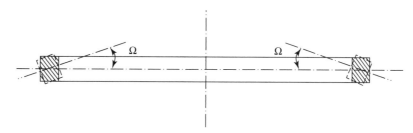

Figure 12.5   Torsional vibration of a ring

For a complete circular ring, the free vibrations involving n wavelengths in the circumference and frequency ω are assumed to be of the form

$$\Omega(\theta, t) = C_1 \sin(n\theta + \phi)e^{i\omega t} \qquad (12.65)$$

Substituting Eq. (12.65) into (12.64), we obtain

$$\omega_n^2 = \frac{Cn^2 + EI_2}{\rho JR^2} \qquad (12.66)$$

12.6   EXTENSIONAL VIBRATIONS

For the extensional vibration of a circular ring, we assume v and Ω to be zero. In this case, the centerline of the ring extends by $(1/R)[(\partial w/\partial \theta) - u]$ and tension developed is given by [1, 5]

$$P = \frac{EA}{R}\left(\frac{\partial w}{\partial \theta} - u\right) \qquad (12.67)$$

The equations of motion governing u and w are given by Eqs. (12.1) and (12.2), which can be rewritten, for free vibration, as

$$\frac{\partial F}{\partial \theta} + P = \rho A R \frac{\partial^2 u}{\partial t^2} \tag{12.68}$$

$$\frac{\partial P}{\partial \theta} - F = \rho A R \frac{\partial^2 w}{\partial t^2} \tag{12.69}$$

Neglecting F, Eqs. (12.68) and (12.69) can be written as

$$P = \frac{EA}{R}\left(\frac{\partial w}{\partial \theta} - u\right) = \rho A R \frac{\partial^2 u}{\partial t^2} \tag{12.70}$$

$$\frac{\partial P}{\partial \theta} = \frac{EA}{R}\left(\frac{\partial^2 w}{\partial \theta^2} - \frac{\partial u}{\partial \theta}\right) = \rho A R \frac{\partial^2 w}{\partial t^2} \tag{12.71}$$

During free vibration, the displacements u and w can be assumed to be of the form

$$u(\theta, t) = (C_1 \sin n\theta + C_2 \cos n\theta)e^{i\omega t} \tag{12.72}$$

$$w(\theta, t) = n(C_1 \cos n\theta - C_2 \sin n\theta)e^{i\omega t} \tag{12.73}$$

Using Eqs. (12.72) and (12.73), Eqs. (12.70) and (12.71) yield the frequency of extensional vibrations of the ring as

$$\omega_n^2 = \frac{E}{\rho R^2}(1 + n^2) \tag{12.74}$$

12.7   VIBRATION OF A CURVED BEAM WITH VARIABLE CURVATURE

12.7.1   Thin Curved Beam

Consider a thin uniform curved beam with variable curvature as shown in Fig. 12.6. The curved beam is assumed to have a span l and height h and its center line (middle surface) is defined by the equation $y = y(x)$, where the x axis is defined by the line joining the two endpoints (supports) of the curved beam. The curvature of its centerline is defined by $\rho(\theta)$ or $\rho(x)$, where the angle θ is indicated in Fig. 12.6(a). The radial and tangential displacements of the centerline are denoted u and w, respectively, and the rotation of the cross section of the beam is denoted ϕ, with the positive directions of the displacements as indicated in Fig. 12.6(a).

Equations of Motion   The dynamic equilibrium approach will be used for derivation of the equations of motion by including the effect of rotary inertia [21, 22]. For this we consider the free-body diagram of an element of the curved beam shown in Fig. 12.6(b). By denoting the inertia forces per unit length in the radial and tangential directions, respectively, as F_i and P_i and the inertia moment (rotary inertia) per unit length as M_i,

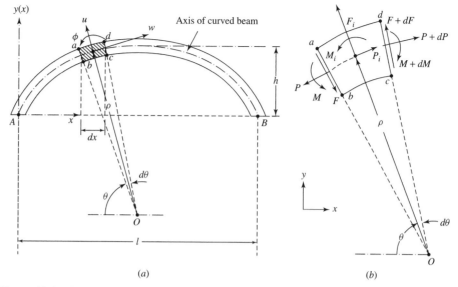

Figure 12.6   Curved beam analysis: (a) geometry of the curved beam; (b) free-body diagram of an element of the curved beam.

the equilibrium equations can be expressed as follows [22, 23]. Equilibrium of forces in the radial direction:

$$\frac{\partial F}{\partial \theta} - P + \rho F_i = 0 \tag{12.75}$$

Equilibrium of forces in the tangential direction:

$$\frac{\partial P}{\partial \theta} + F + \rho P_i = 0 \tag{12.76}$$

Equilibrium of moments in the xy plane:

$$\frac{\partial M}{\partial \theta} - \rho F - \rho M_i = 0 \tag{12.77}$$

The tangential force (P) and the moment (M) can be expressed in terms of the displacement components u and w as [5]

$$P = \frac{EA}{\rho}\left[\frac{dw}{d\theta} + w + \frac{I}{A\rho^2}\left(\frac{d^2u}{d\theta^2} + u\right)\right] \tag{12.78}$$

$$M = -\frac{EI}{\rho^2}\left(\frac{d^2u}{d\theta^2} + u\right) \tag{12.79}$$

The rotation of the cross section (ϕ) is related to the displacement components u and w as [13, 21]

$$\phi = \frac{1}{\rho}\left(\frac{du}{d\theta} - w\right) \tag{12.80}$$

During free vibration, u and w can be expressed as

$$u(\theta, t) = U(\theta) \cos \omega t \tag{12.81}$$

$$w(\theta, t) = W(\theta) \cos \omega t \tag{12.82}$$

where ω is the frequency of vibration and $U(\theta)$ and $W(\theta)$ denote the time-independent variations of the amplitudes of $u(\theta, t)$ and $w(\theta, t)$, respectively. Thus, the inertia forces and the inertia moment during free vibration of the curved beam are given by

$$F_i = m\omega^2 U \tag{12.83}$$

$$P_i = m\omega^2 W \tag{12.84}$$

$$M_i = \frac{m\omega^2 I}{A\rho} \left(\frac{dU}{d\theta} - W \right) \tag{12.85}$$

where m is the mass of the beam per unit length.

Equations (12.78), (12.79), (12.77), (12.83)–(12.85), (12.81), and (12.82) can be used in Eqs. (12.75) and (12.76) to obtain

$$- EI \left\{ \frac{1}{\rho^3} \left(\frac{d^4U}{d\theta^4} + \frac{d^2U}{d\theta^2} \right) - \frac{5}{\rho^4} \frac{d\rho}{d\theta} \left(\frac{d^3U}{d\theta^3} + \frac{dU}{d\theta} \right) \right.$$
$$\left. + 2 \left[\frac{4}{\rho^5} \left(\frac{d\rho}{d\theta} \right)^2 - \frac{1}{\rho^4} \frac{d^2\rho}{d\theta^2} \right] \left(\frac{d^2U}{d\theta^2} - \frac{dW}{d\theta} \right) \right\}$$
$$- \frac{cm\omega^2 I}{A} \left[\frac{1}{\rho} \left(\frac{d^2U}{d\theta^2} - \frac{dW}{d\theta} \right) - \frac{1}{\rho^2} \frac{d\rho}{d\theta} \left(\frac{dU}{d\theta} - W \right) \right]$$
$$- \frac{EA}{\rho} \left[\frac{dW}{d\theta} + W + \frac{I}{A\rho^2} \left(\frac{d^2U}{d\theta^2} + U \right) \right] + \rho m\omega^2 U = 0 \tag{12.86}$$

$$EA \left[\frac{1}{\rho} \left(\frac{d^2W}{d\theta^2} + \frac{dU}{d\theta} \right) + \frac{I}{A\rho^3} \left(\frac{d^3U}{d\theta^3} + \frac{dU}{d\theta} \right) - \frac{1}{\rho^2} \frac{d\rho}{d\theta} \left(\frac{dW}{d\theta} + U \right) \right.$$
$$\left. - \frac{3I}{A\rho^4} \frac{d\rho}{d\theta} \left(\frac{d^2U}{d\theta^2} + U \right) \right] - EI \left[\frac{1}{\rho^3} \left(\frac{d^3U}{d\theta^3} + \frac{dU}{d\theta} \right) \right.$$
$$\left. - \frac{2}{\rho^4} \frac{d\rho}{d\theta} \left(\frac{d^2U}{d\theta^2} + U \right) \right] - \frac{cm\omega^2 I}{A\rho} \left(\frac{dU}{d\theta} - W \right) + \rho m\omega^2 W = 0 \tag{12.87}$$

where c is a constant set equal to 0 or 1 if rotary inertia effect (M_i) is excluded or included, respectively. It can be seen that Eqs. (12.86) and (12.87) denote two coupled differential equations in the displacement variables $U(\theta)$ and $W(\theta)$. Introducing the nondimensional parameters

$$\xi = \frac{x}{l}, \qquad \eta = \frac{y}{l}, \qquad \underset{\sim}{U} = \frac{U}{l}, \qquad \underset{\sim}{W} = \frac{W}{l} \tag{12.88}$$

$$\zeta = \frac{\rho}{l}, \qquad \underset{\sim}{h} = \frac{h}{l}, \qquad \underset{\sim}{r} = \frac{1}{l}\sqrt{\frac{A}{I}}, \qquad \Omega_i = \omega_i \sqrt{\frac{ml^4}{EI}} \tag{12.89}$$

where ω_i denotes the ith frequency; $i = 1, 2, \ldots$, Eqs. (12.86) and (12.87) can be expressed as [21, 22]

$$\frac{d^4 U}{d\theta^4} = p_1 \frac{d^3 U}{d\theta^3} + \left(p_2 + \frac{c p_3}{r^4} \Omega_i^2 \right) \frac{d^2 U}{d\theta^2} + \left(p_1 - \frac{c p_4}{r^4} \Omega_i^2 \right) \frac{d U}{d\theta}$$

$$+ \left(p_5 + \frac{p_6}{r^4} \Omega_i^2 \right) U + \left(1 - \frac{c}{r^4} \Omega_i^2 \right) p_3 \frac{d W}{d\theta} + \frac{c p_4}{r^4} \Omega_i^2 W \qquad (12.90)$$

$$\frac{d^2 W}{d\theta^2} = p_7 \frac{d^2 U}{d\theta^2} + \left(\frac{c}{r^4} \Omega_i^2 - 1 \right) \frac{d U}{d\theta} + p_8 U + p_9 \frac{d W}{d\theta}$$

$$+ \frac{p_3 - c}{r^4} \Omega_i^2 W \qquad (12.91)$$

where

$$p_1 = \frac{5}{\zeta} \frac{d\zeta}{d\theta} \qquad (12.92)$$

$$p_2 = \frac{2}{\zeta} \frac{d^2 \zeta}{d\theta^2} - \frac{8}{\zeta^2} \left(\frac{d\zeta}{d\theta} \right)^2 - 2 \qquad (12.93)$$

$$p_3 = -r^2 \zeta^2 \qquad (12.94)$$

$$p_4 = -r^2 \zeta \frac{d\zeta}{d\theta} \qquad (12.95)$$

$$p_5 = \frac{2}{\zeta} \frac{d^2 \zeta}{d\theta^2} - \frac{8}{\zeta^2} \left(\frac{d\zeta}{d\theta} \right)^2 - r^2 \zeta^2 - 1 \qquad (12.96)$$

$$p_6 = r^4 \zeta^4 \qquad (12.97)$$

$$p_7 = \frac{1}{r^2 \zeta^3} \frac{d\zeta}{d\theta} \qquad (12.98)$$

$$p_8 = \frac{1}{\zeta} \frac{d\zeta}{d\theta} \left(1 + \frac{1}{r^2 \zeta^2} \right) \qquad (12.99)$$

$$p_9 = \frac{1}{\zeta} \frac{d\zeta}{d\theta} \qquad (12.100)$$

The coefficients p_1 to p_9 can be computed by starting from the equation of the curved beam, $y = y(x)$. Using the nondimensional parameters defined in Eqs. (12.88) and (12.89) for ξ, η, ζ and r, we can express the equation of the curved beam in nondimensional form as $\eta = \eta(\xi)$. Using the relation

$$\theta = \frac{\pi}{2} - \tan^{-1} \frac{d\eta}{d\xi} \qquad (12.101)$$

we obtain

$$\frac{1}{\zeta} = \frac{d^2 \eta}{d\xi^2} \left[1 + \left(\frac{d\eta}{d\xi} \right)^2 \right]^{-3/2} \qquad (12.102)$$

Equations (12.101) and (12.102) can be used to compute the first and second derivatives of ζ as

$$\frac{d\zeta}{d\theta} = \frac{d\zeta}{d\xi}\frac{d\xi}{d\theta} \tag{12.103}$$

$$\frac{d^2\zeta}{d\theta^2} = \frac{d}{d\xi}\left(\frac{d\zeta}{d\theta}\right)\frac{d\xi}{d\theta} \tag{12.104}$$

The coefficients p_1 to p_9, defined in Eqs. (12.92)–(12.100), can be computed using Eqs. (12.101)–(12.104).

Numerical Solution   The boundary conditions of the curved beam or arch can be stated as follows. For a clamped or fixed end:

$$w = 0 \quad \text{or} \quad \underset{\sim}{W} = 0 \tag{12.105}$$

$$u = 0 \quad \text{or} \quad \underset{\sim}{U} = 0 \tag{12.106}$$

$$\frac{\partial u}{\partial \theta} = 0 \quad \text{or} \quad \frac{d\underset{\sim}{U}}{d\theta} = 0 \tag{12.107}$$

For a pinned or hinged end:

$$w = 0 \quad \text{or} \quad \underset{\sim}{W} = 0 \tag{12.108}$$

$$u = 0 \quad \text{or} \quad \underset{\sim}{U} = 0 \tag{12.109}$$

$$\frac{\partial^2 u}{\partial \theta^2} = 0 \quad \text{or} \quad \frac{d^2\underset{\sim}{U}}{d\theta^2} = 0 \tag{12.110}$$

The equations of motion, Eqs. (12.90) and (12.91), can be solved numerically to find the frequency parameter Ω_i and the mode shape defined by $U_i(\xi)$ and $W_i(\xi)$. Numerical results are obtained for three types of curved beams: parabolic, sinusoidal, and elliptic-shaped beams [22].

For a parabolic-shaped curved beam, the equation of the beam is given by

$$y(x) = -\frac{4hx}{l^2}(x - l), \qquad 0 \le x \le l \tag{12.111}$$

or

$$\eta(\xi) = -4\xi(\xi - 1)\underset{\sim}{h}, \qquad 0 \le \xi \le 1 \tag{12.112}$$

For a sinusoidal-shaped curved beam, shown in Fig. 12.7, the equation of the curved beam is given by

$$\overline{y} = \sin\frac{\pi\overline{x}}{L} \tag{12.113}$$

$$\overline{x} = \alpha l + x, \qquad \overline{y} = H - h + y \tag{12.114}$$

or

$$\eta = \underset{\sim}{h} - d + d\sin(e\xi + e\alpha), \qquad 0 \le \xi \le 1 \tag{12.115}$$

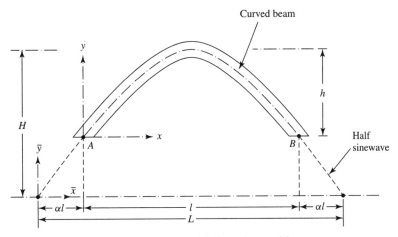

Figure 12.7   Sinusoidal-shaped curved beam.

where

$$d = \frac{\underset{\sim}{h}}{1 - \sin \alpha e} \tag{12.116}$$

$$e = \frac{\pi}{1 + 2\alpha} \tag{12.117}$$

For the elliptic-shaped curved beam shown in Fig. 12.8, the equation of the curved beam is given by

$$\eta = \underset{\sim}{h} - f + \frac{f}{g}\left[g^2 - \left(\xi - \frac{1}{2}\right)^2\right]^{1/2}, \qquad 0 \le \xi \le 1 \tag{12.118}$$

where

$$f = \frac{g\underset{\sim}{h}}{g - (\alpha + \alpha^2)^{1/2}} \tag{12.119}$$

$$g = \tfrac{1}{2}(1 + 2\alpha) \tag{12.120}$$

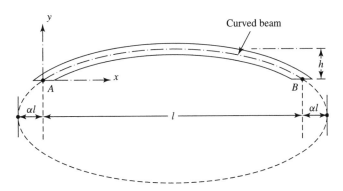

Figure 12.8   Elliptic-shaped curved beam.

Table 12.3   Natural Frequency Parameters of Curved Beams

Geometry and boundary conditions of the curved beam	Frequency parameter, Ω_i					
	$i = 1$		$i = 2$		$i = 3$	
	Without rotary inertia, $c = 0$	With rotary inertia, $c = 1$	Without rotary inertia, $c = 0$	With rotary inertia, $c = 1$	Without rotary inertia, $c = 0$	With rotary inertia, $c = 1$
Parabolic shape, pinned–pinned: $h = 0.1, \underset{\sim}{r} = 10$	11.47	10.94	29.49	28.72	38.69	33.73
Sinusoidal shape, pinned–fixed: $\alpha = 0.5, h = 0.3, \underset{\sim}{r} = 10$	16.83	16.37	23.32	22.88	35.61	31.78
Elliptic shape, fixed–fixed: $\alpha = 0.5, \underset{\sim}{h} = 0.5, \underset{\sim}{r} = 10$	16.92	16.69	17.24	16.98	28.49	26.20

Source: Ref. 22.

Frequency parameters corresponding to the three types of curved beams with different boundary conditions are given in Table 12.3.

12.7.2   Curved Beam Analysis, Including the Effect of Shear Deformation

Equations of Motion   The dynamic equilibrium approach will be used to derive the equations of motion by considering the free-body diagram of an element of the curved beam as shown in Fig. 12.6(*b*). When the effects of rotary inertia and shear deformation are considered, the equilibrium equations can be obtained, similar to those of Section 12.7.1, as follows [21]. Equilibrium of forces in the radial direction:

$$\frac{\partial F}{\partial \theta} - P - \rho F_i = 0 \tag{12.121}$$

Equilibrium of forces in the tangential direction:

$$\frac{\partial P}{\partial \theta} + F - \rho P_i = 0 \tag{12.122}$$

Equilibrium of moments in the *xy* plane:

$$\frac{1}{\rho}\frac{\partial M}{\partial \theta} - F + M_i = 0 \tag{12.123}$$

Rotation of the tangent to the centroidal axis is given by [21]

$$\gamma = \frac{1}{\rho}\left(\frac{\partial u}{\partial \theta} - w\right) \tag{12.124}$$

When shear deformation is considered, the rotation angle γ can be expressed as

$$\gamma = \psi + \beta \tag{12.125}$$

where ψ is the rotation angle with no shear deformation and β is the angular deformation due to shear. Equations (12.124) and (12.125) yield

$$\beta = \frac{1}{\rho}\left(\frac{\partial u}{\partial \theta} - w - \rho\psi\right) \tag{12.126}$$

When the effects of rotary inertia, shear deformation, and axial deformation are considered, the bending moment (M), normal force (P), and shear force (F) in the curved beam are given by [19, 20]

$$M = -\frac{EI}{\rho}\frac{d\psi}{d\theta} \tag{12.127}$$

$$P = \frac{EA}{\rho}\left(\frac{dw}{d\theta} + u\right) + \frac{EI}{\rho^2}\frac{d\psi}{d\theta} \tag{12.128}$$

$$F = kAG\beta = \frac{kAG}{\rho}\left(\frac{du}{d\theta} - w - \rho\psi\right) \tag{12.129}$$

where k is the shear factor. To find the natural frequencies of vibration of the curved beam, all the displacement components are assumed to be harmonic with frequency ω, so that

$$w(\theta, t) = W(\theta)\cos\omega t \tag{12.130}$$

$$u(\theta, t) = U(\theta)\cos\omega t \tag{12.131}$$

$$\psi(\theta, t) = \Psi(\theta)\cos\omega t \tag{12.132}$$

Thus, the inertia forces are given by

$$P_i(\theta) = -\overline{m}A\omega^2 W(\theta) \tag{12.133}$$

$$F_i(\theta) = -\overline{m}A\omega^2 U(\theta) \tag{12.134}$$

$$M_i(\theta) = -\overline{m}I\omega^2 \Psi(\theta) \tag{12.135}$$

where $\overline{m}$ is the mass density of the curved beam. Using a procedure similar to that of Section 12.7.1, the equations of motion can be expressed in nondimensional form as [21]:

$$\frac{d^2 U}{d\theta^2} = \frac{1}{\zeta}\frac{d\zeta}{d\theta}\frac{dU}{d\theta} + \frac{1}{\mu}\left(1 - \frac{\zeta^2}{r^2}\Psi_i^2\right)U + \left(1 + \frac{1}{\mu}\right)\frac{dW}{d\theta}$$
$$- \frac{1}{\zeta}\frac{d\zeta}{d\theta}W + \left(\zeta + \frac{1}{\mu\zeta r^2}\right)\frac{d\Psi}{d\theta} \tag{12.136}$$

$$\frac{d^2W}{d\theta^2} = \frac{1}{\zeta}\frac{d\zeta}{d\theta}\frac{dW}{d\theta} + \left(\mu - \frac{\zeta^2}{r^2}\Omega_i^2\right)W - (1+\mu)\frac{dU}{d\theta}$$

$$+ \frac{1}{\zeta}\frac{d\zeta}{d\theta}U - \frac{1}{\zeta r^2}\frac{d^2\Psi}{d\theta^2} + \frac{2}{\zeta^2 r^2}\frac{d\zeta}{d\theta}\frac{d\Psi}{d\theta} + \mu\zeta\Psi \tag{12.137}$$

$$\frac{d^2\Psi}{d\theta^2} = \frac{1}{\zeta}\frac{d\zeta}{d\theta}\frac{d\Psi}{d\theta} + \left(\mu r^2 - \frac{\Omega_i^2}{r^2}\right)\zeta^2\Psi - \zeta\mu r^2\frac{dU}{d\theta} + \zeta\mu r^2 W \tag{12.138}$$

where

$$\mu = \frac{kG}{E} \tag{12.139}$$

$$\Omega_i = \omega_i r l\sqrt{\frac{m}{E}} \tag{12.140}$$

and the other symbols are defined by Eqs. (12.88) and (12.89).

Numerical Solution   The boundary conditions of the curved beam are as follows. For a clamped or fixed end:

$$u = 0 \quad \text{or} \quad U = 0 \tag{12.141}$$

$$w = 0 \quad \text{or} \quad W = 0 \tag{12.142}$$

$$\psi = 0 \quad \text{or} \quad \Psi = 0 \tag{12.143}$$

For a pinned or hinged end:

$$u = 0 \quad \text{or} \quad U = 0 \tag{12.144}$$

$$w = 0 \quad \text{or} \quad W = 0 \tag{12.145}$$

$$\frac{\partial\psi}{\partial\theta} = 0 \quad \text{or} \quad \frac{d\Psi}{d\theta} = 0 \tag{12.146}$$

The frequency parameters corresponding to three types of curved beams with different boundary conditions are given in Table 12.4.

12.8   RECENT CONTRIBUTIONS

Curved Beams and Rings   An analytical procedure was proposed by Stavridis and Michaltsos [12] for evaluation of the eigenfrequencies of a thin-walled beam curved in plan in response to transverse bending and torsion with various boundary conditions. Wasserman [13] derived an exact formula for the lowest natural frequencies and critical loads of elastic circular arches with flexibly supported ends for symmetric vibration in a direction perpendicular to the initial curvature of the arch. The values of frequencies and critical loads were shown to be dependent on the opening angle of the arch, on the stiffness of the flexibly supported ends, and on the ratio of the flexural rigidity to the torsional rigidity of the cross section. Bickford and Maganty [14] obtained the expressions for out-of-plane modal frequencies of a thick ring, which accounts for the variations in curvature across the cross section.

Table 12.4   Natural Frequencies of Curved Beams, Including the Effect of Shear Deformation[a]

Geometry and boundary conditions of the curved beam	Frequency parameter, Ω_i		
	$i = 1$	$i = 2$	$i = 3$
Parabolic shape, Pinned-pinned: $h = 0.3, r = 75$	21.83	56.00	102.3
Sinusoidal shape, fixed–fixed: $\alpha = 0.5, h = 0.1, r = 100$	56.30	66.14	114.3
Elliptic shape, fixed–pinned: $\alpha = 0.5, h = 0.2, r = 50$	35.25	57.11	83.00

Source: Ref. 21.
[a]For shear coefficient, $\mu = 0.3$.

Vibration of Multispan Curved Beams   Culver and Oestel [15] presented a method of analysis for determining the natural frequencies of multispan horizontally curved girders used in bridge structures. The method was illustrated by deriving the frequency equation of a two-span curved girder. Numerical results and comparison with existing solutions were also given.

Vibration of Helical Springs   The longitudinal and torsional vibrations of helical springs of finite length with small pitch were analyzed by Kagawa [16] on the basis of Love's formulation for a naturally curved thin rod of small deformation. The driving-point impedance at one end of the spring while the other end is free or supported was discussed.

Vibration of Gears   Ring gear structural modes of planetary gears used in modern automotive, aerospace, marine, and other industrial drivetrain systems often contribute significantly to the severity of the gear whine problem caused by transmission error excitation. The dynamics and modes of ring gears have been studied utilizing the analytical and computational solutions of smooth rings having nearly the same nominal dimensions but without the explicit presence of the spline and tooth geometries by Tanna and Lim [17].

Vibration of Frames of Electrical Machines   The frames of electrical machines such as motors and generators can be modeled as circular arcs with partly built-in ends. The actual frequencies of vibration of these frames are expected to lie within the limits given by those of an arc with hinged ends and of an arc with fixed ends. The vibration of the frames of electrical machines were studied by Erdelyi and Horvay [18].

Vibration of Arches and Frames   General relations between the forces and moments developed in arches and frames as well as the dynamic equilibrium equations have been presented by Borg and Gennaro [19] and Henrych [20]. The natural frequencies of noncircular arches, considering the effects of rotary inertia and shear deformation, have been investigated by Oh et al. [21].

REFERENCES

1. L. L. Philipson, On the role of extension in the flexural vibrations of rings, *Journal of Applied Mechanics*, Vol. 23 (*Transactions of ASME*, Vol. 78), p. 364, 1956.

2. B. S. Seidel and E. A. Erdelyi, On the vibration of a thick ring in its own plane, *Journal of Engineering for Industry* (*Transactions of ASME*, Ser. B), Vol. 86, p. 240, 1964.

3. S. Timoshenko, On the correction for shear of the differential equation for transverse vibration of prismatic bars, *Philosophical Magazine,* Ser. 6, Vol. 41, p. 744, 1921.

4. J. P. Den Hartog, Vibrations of frames of electrical machines, *Journal of Applied Mechanics, Transactions of ASME*, Vol. 50, APM-50-11, 1928.

5. A. E. H. Love, *A Treatise on the Mathematical Theory of Elasticity*, 4th ed., Dover, New York, 1944.

6. S. S. Rao and V. Sundararajan, Inplane flexural vibrations of circular rings, *Journal of Applied Mechanics*, Vol. 36, No. 3, pp. 620–625, 1969.

7. S. S. Rao, Effects of transverse shear and rotatory inertia on the coupled twist-bending vibrations of circular rings, *Journal of Sound and Vibration,* Vol. 16, No. 4, pp. 551–566, 1971.

8. S. S. Rao, On the natural vibrations of systems of elastically connected concentric thick rings, *Journal of Sound and Vibration.* Vol. 32, No. 3, pp. 467–479, 1974.

9. S. S. Rao, Three dimensional vibration of a ring on an elastic foundation, *Aeronautical Journal of the Royal Aeronautical Society, London*, Vol. 75, pp. 417–419, 1971.

10. R. E. Peterson, An experimental investigation of ring vibration in connection with the study of gear-noise, *Transactions of the American Society of Mechanical Engineers*, Vol. 52, APM-52-1, 1930.

11. E. R. Kaiser, Acoustical vibrations of rings, *Journal of the Acoustical Society of America*, Vol. 25, p. 617, 1953.

12. L. T. Stavridis and G. T. Michaltsos, Eigenfrequency analysis of thin-walled girders curved in plan, *Journal of Sound and Vibration*, Vol. 227, No. 2, pp. 383–396, 1999.

13. Y. Wasserman, Spatial symmetrical vibrations and stability of circular arches with flexibly supported ends, *Journal of Sound and Vibration*, Vol. 59, No. 2, pp. 181–194, 1978.

14. W. B. Bickford and S. P. Maganty, On the out-of-plane vibrations of thick rings, *Journal of Sound and Vibration*, Vol. 108, No. 3, pp. 503–507, 1986.

15. C. G. Culver and D. J. Oestel, Natural frequencies of multispan curved beams, *Journal of Sound and Vibration*, Vol. 10, No. 3, pp. 380–389, 1969.

16. Y. Kagawa, On the dynamical properties of helical springs of finite length with small pitch, *Journal of Sound and Vibration*, Vol. 8, No. 1, pp. 1–15, 1968.

17. R. P. Tanna and T. C. Lim, Modal frequency deviations in estimating ring gear modes using smooth ring solutions, *Journal of Sound and Vibration*, Vol. 269, No. 3–5, pp. 1099–1110, 2004.

18. E. Erdelyi and G. Horvay, Vibration modes of stators of induction motors, *Journal of Applied Mechanics*, Vol. 24 (*Transactions of ASME*, Vol. 79), p. 39, 1957.

19. S. F. Borg and J. J. Gennaro, *Modern Structural Analysis*, Van Nostrand Reinhold, New York, 1969.

20. J. Henrych, *The Dynamics of Arches and Frames*, Elsevier, Amsterdam, 1981.

21. S. J. Oh, B. K. Lee, and I. W. Lee, Natural frequencies of non-circular arches with rotatory inertia and shear deformation, *Journal of Sound and Vibration*, Vol. 219, No. 1, pp. 23–33, 1999.

22. B. K. Lee and J. F. Wilson, Free vibrations of arches with variable curvature, *Journal of Sound and Vibration*, Vol. 136, No. 1, pp. 75–89, 1989.

23. K. F. Graff, *Wave Motion in Elastic Solids*, Ohio State University Press, Columbus, OH, 1975.

PROBLEMS

12.1 Derive Eqs. (12.17) and (12.18).

12.2 Derive Eq. (12.24) from Eqs. (12.19)–(12.23).

12.3 Derive Eq. (12.26) from Eqs. (12.19), (12.20), (12.22), (12.23), and (12.25).

12.4 Derive Eq. (12.10) by considering the deformations shown in Fig. 12.2(*b*).

12.5 Find the first five natural frequencies of inplane flexural vibrations of a circular ring with $I_1/AR^2 = 0.25$ and $E/kG = 2.0$ (**a**) according to classical theory; (**b**) by considering the effect of rotary inertia only; (**c**) by considering the effects of both rotary inertia and shear deformation.

12.6 Derive Eq. (12.50) from Eqs. (12.45)–(12.49).

12.7 Derive Eq. (12.61) from Eqs. (12.45), (12.48), (12.49), (12.59), and (12.60).

12.8 Derive Eq. (12.78) from Eqs. (12.69)–(12.71) and (12.72)–(12.74).

12.9 Find the first five natural frequencies of coupled twist–bending vibrations of a circular ring with $r/R = 0.25$, $k = 0.8333$, and $E/G = 3.0$ (**a**) according to classical theory; (**b**) by considering the effect of rotary inertia only; (**c**) by considering the effects of both rotary inertia and shear deformation.

12.10 Find the first five natural frequencies of pure torsional vibrations of a circular ring with a circular cross section for the following data: radius of the cross section $= 1$ cm, radius of the centerline of the ring $= 15$ cm, Young's modulus $= 207$ GPa, shear modulus $= 79.3$ GPa, and unit weight $= 76.5$ kN/m^3.

12.11 Find the first five natural frequencies of extensional vibrations of a circular ring with circular cross section for the following data: radius of cross section $= 1$ cm, radius of the centerline of the ring $= 15$ cm, Young's modulus $= 207$ GPa, and unit weight $= 76.5$ kN/m^3.

12.12 Derive Eqs. (12.131) and (12.132).

12.13 Derive Eqs. (12.136)–(12.138).

12.14 Derive the following differential equations for determining the radial and tangential components of displacement (u and v) of an arch:

$$\frac{u}{\rho} + \frac{dv}{ds} = \frac{N}{EA} - \frac{M}{\rho EA} + \frac{Me}{EI}$$

$$\frac{d^2u}{ds^2} + \frac{u}{\rho^2} = -\frac{M}{EI}$$

where N is the tangential force, M is the bending moment, E is Young's modulus, A is the cross-sectional area, I is the area moment of inertia of the cross section, ρ is the radius of curvature of the centroidal axis, e is the distance between the centroidal axis and the neutral axis of the cross section, and s is the tangential coordinate.

13

Vibration of Membranes

13.1 INTRODUCTION

A membrane is a perfectly flexible thin plate or lamina that is subjected to tension. It has negligible resistance to shear or bending forces, and the restoring forces arise exclusively from the in-plane stretching or tensile forces. The drumhead and diaphragms of condenser microphones are examples of membranes.

13.2 EQUATION OF MOTION

13.2.1 Equilibrium Approach

Consider a homogeneous and perfectly flexible membrane bounded by a plane curve C in the xy plane in the undeformed state. It is subjected to a pressure loading of intensity $f(x, y, t)$ per unit area in the transverse or z direction and tension of magnitude P per unit length along the edge as in the case of a drumhead. Each point of the membrane is assumed to move only in the z direction, and the displacement, $w(x, y, t)$, is assumed to be very small compared to the dimensions of the membrane. Consider an elemental area of the membrane, $dx\, dy$, with tensile forces of magnitude $P\, dx$ and $P\, dy$ acting on the sides parallel to the x and y axes, respectively, as shown in Fig. 13.1. After deformation, the net forces acting on the element of the membrane along the z direction due to the forces $P\, dx$ and $P\, dy$ will be [see Fig. 13.1(d)]

$$\left(P \frac{\partial^2 w}{\partial y^2}\, dx\, dy \right) \quad \text{and} \quad \left(P \frac{\partial^2 w}{\partial x^2}\, dx\, dy \right)$$

The pressure force acting on the element of the membrane in the z direction is $f(x, y)\, dx\, dy$. The inertia force on the element is given by

$$\rho(x, y) \frac{\partial^2 w}{\partial t^2}\, dx\, dy$$

where $\rho(x, y)$ is the mass per unit area. The application of Newton's second law of motion yields the equation of motion for the forced transverse vibration of the membrane as

$$P \left(\frac{\partial^2 w}{\partial x^2} + \frac{\partial^2 w}{\partial y^2} \right) + f = \rho \frac{\partial^2 w}{\partial t^2} \tag{13.1}$$

When the external force $f(x, y) = 0$, the free vibration equation can be expressed as

$$c^2 \left(\frac{\partial^2 w}{\partial x^2} + \frac{\partial^2 w}{\partial y^2} \right) = \frac{\partial^2 w}{\partial t^2} \tag{13.2}$$

where

$$c = \left(\frac{P}{\rho} \right)^{1/2} \tag{13.3}$$

Equation (13.2) is also known as the *two-dimensional wave equation*, with c denoting the wave velocity.

Initial and Boundary Conditions   Since the equation of motion, Eq. (13.1) or (13.2), involves second-order partial derivatives with respect to each of t, x, and y, we need to specify two initial conditions and four boundary conditions to find a unique solution

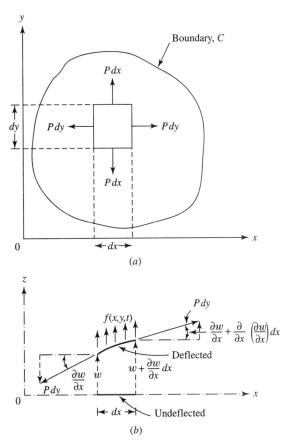

Figure 13.1   (*a*) Undeformed membrane in the xy plane; (*b*) deformed membrane as seen in the xz plane; (*c*) deformed membrane as seen in the yz plane; (*d*) forces acting on an element of the membrane.

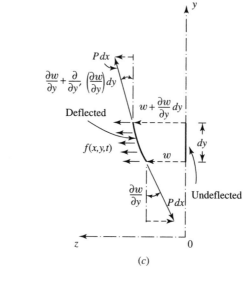

(c)

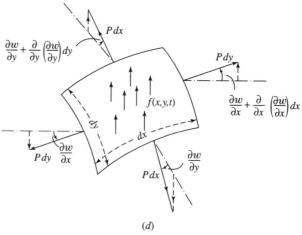

(d)

Figure 13.1   (*continued*)

of the problem. Usually, the displacement and velocity of the membrane at $t = 0$ are specified as $w_0(x)$ and $\dot{w}_0(x)$, respectively. Thus, the initial conditions are given by

$$w(x, y, 0) = w_0(x, y) \tag{13.4}$$

$$\frac{\partial w}{\partial t}(x, y, 0) = \dot{w}_0(x, y) \tag{13.5}$$

The boundary conditions of the membrane can be stated as follows:

1. If the membrane is fixed at any point (x_1, y_1) on the boundary, the deflection must be zero, and hence

$$w(x_1, y_1, t) = 0, \qquad t \geq 0 \tag{13.6}$$

2. If the membrane is free to deflect transversely (in the z direction) at any point (x_2, y_2) of the boundary, there cannot be any force at the point in the z direction. Thus,

$$P \frac{\partial w}{\partial n}(x_2, y_2, t) = 0, \qquad t \geq 0 \tag{13.7}$$

where $\partial w / \partial n$ indicates the derivative of w with respect to a direction n normal to the boundary at the point (x_2, y_2).

13.2.2   Variational Approach

To derive the equation of motion of a membrane using the extended Hamilton's principle, the expressions for the strain and kinetic energies as well as the work done by external forces are needed. The strain and kinetic energies of a membrane can be expressed as

$$\pi = \frac{1}{2} \iint_A P \left[\left(\frac{\partial w}{\partial x} \right)^2 + \left(\frac{\partial w}{\partial y} \right)^2 \right] dA \tag{13.8}$$

$$T = \frac{1}{2} \iint_A \rho \left(\frac{\partial w}{\partial t} \right)^2 dA \tag{13.9}$$

The work done by the distributed pressure loading $f(x, y, t)$ is given by

$$W = \iint_A f w \, dA \tag{13.10}$$

The application of Hamilton's principle gives

$$\delta \int_{t_1}^{t_2} (\pi - T - W) \, dt = 0 \tag{13.11}$$

or

$$\delta \int_{t_1}^{t_2} \left\{ \frac{1}{2} \iint_A P \left[\left(\frac{\partial w}{\partial x} \right)^2 + \left(\frac{\partial w}{\partial y} \right)^2 \right] dA - \frac{1}{2} \iint_A \rho \left(\frac{\partial w}{\partial t} \right)^2 dA - \iint_A f w \, dA \right\} dt = 0 \tag{13.12}$$

The variations in Eq. (13.12) can be evaluated using integration by parts as follows:

$$I_1 = \delta \int_{t_1}^{t_2} \frac{P}{2} \iint_A \left(\frac{\partial w}{\partial x} \right)^2 dA \, dt = P \int_{t_1}^{t_2} \iint_A \frac{\partial w}{\partial x} \frac{\partial}{\partial x} (\delta w) \, dA \, dt$$

$$= P \int_{t_1}^{t_2} \left[\oint_C \frac{\partial w}{\partial x} \delta w l_x \, dC - \iint_A \frac{\partial}{\partial x} \left(\frac{\partial w}{\partial x} \right) \delta w \, dA \right] dt \tag{13.13}$$

$$I_2 = \delta \int_{t_1}^{t_2} \frac{P}{2} \iint_A \left(\frac{\partial w}{\partial y}\right)^2 dA\, dt = P \int_{t_1}^{t_2} \iint_A \frac{\partial w}{\partial y}\frac{\partial}{\partial y}(\delta w)\, dA\, dt$$

$$= P \int_{t_1}^{t_2} \left[\oint_C \frac{\partial w}{\partial y}\delta w l_y\, dC - \iint_A \frac{\partial}{\partial y}\left(\frac{\partial w}{\partial y}\right)\delta w\, dA \right] dt \tag{13.14}$$

$$I_3 = \delta \int_{t_1}^{t_2} \frac{\rho}{2} \iint_A \left(\frac{\partial w}{\partial t}\right)^2 dA\, dt = \frac{\rho}{2} \iint_A \delta \int_{t_1}^{t_2} \left(\frac{\partial w}{\partial t}\right)^2 dA\, dt \tag{13.15}$$

By using integration by parts with respect to time, the integral I_3 can be written as

$$I_3 = \frac{\rho}{2} \iint_A \delta \int_{t_1}^{t_2} \left(\frac{\partial w}{\partial t}\right)^2 dt\, dA = \rho \iint_A \left[\frac{\partial w}{\partial t}\delta w \Big|_{t_1}^{t_2} - \int_{t_1}^{t_2} \frac{\partial}{\partial t}\left(\frac{\partial w}{\partial t}\right)\delta w\, dt \right] dA \tag{13.16}$$

Since δw vanishes at t_1 and t_2, Eq. (13.16) reduces to

$$I_3 = -\rho \int_{t_1}^{t_2} \iint_A \frac{\partial^2 w}{\partial t^2}\delta w\, dA\, dt \tag{13.17}$$

$$I_4 = \delta \int_{t_1}^{t_2} \iint_A fw\, dA\, dt = \int_{t_1}^{t_2} \iint_A f\delta w\, dA\, dt \tag{13.18}$$

Using Eqs. (13.13), (13.14), (13.17), and (13.18), Eq. (13.12) can be expressed as

$$\int_{t_1}^{t_2} \left[-\iint_A \left(P\frac{\partial^2 w}{\partial x^2} + P\frac{\partial^2 w}{\partial y^2} + f - \rho\frac{\partial^2 w}{\partial t^2} \right)\delta w\, dA \right] dt$$

$$+ \int_{t_1}^{t_2} \left[\oint_C P\left(\frac{\partial w}{\partial x}l_x + \frac{\partial w}{\partial y}l_y \right)\delta w\, dC \right] dt = 0 \tag{13.19}$$

By setting each of the expressions under the brackets in Eq. (13.19) equal to zero, we obtain the differential equation of motion for the transverse vibration of the membrane as

$$P\left(\frac{\partial^2 w}{\partial x^2} + \frac{\partial^2 w}{\partial y^2} \right) + f = \rho\frac{\partial^2 w}{\partial t^2} \tag{13.20}$$

and the boundary condition as

$$\oint_C P\left(\frac{\partial w}{\partial x}l_x + \frac{\partial w}{\partial y}l_y \right)\delta w\, dC = 0 \tag{13.21}$$

Note that Eq. (13.21) will be satisfied for any combination of boundary conditions for a rectangular membrane. For a fixed edge:

$$w = 0 \quad \text{and hence} \quad \delta w = 0 \tag{13.22}$$

For a free edge with $x = 0$ or $x = a$, $l_y = 0$ and $l_x = 1$:

$$P \frac{\partial w}{\partial x} = 0 \tag{13.23}$$

With $y = 0$ or $y = b$, $l_x = 0$ and $l_y = 1$:

$$P \frac{\partial w}{\partial y} = 0 \tag{13.24}$$

For arbitrary geometries of the membrane, Eq. (13.21) can be expressed as

$$\oint_C P \frac{\partial w}{\partial n} \delta w \, dC = 0 \tag{13.25}$$

which will be satisfied when either the edge is fixed with

$$w = 0 \quad \text{and hence} \quad \delta w = 0 \tag{13.26}$$

or the edge is free with

$$P \frac{\partial w}{\partial n} = 0 \tag{13.27}$$

13.3   WAVE SOLUTION

The functions

$$w_1(x, y, t) = f(x - ct) \tag{13.28}$$

$$w_2(x, y, t) = f(x + ct) \tag{13.29}$$

$$w_3(x, y, t) = f(x \cos \theta + y \sin \theta - ct) \tag{13.30}$$

can be verified to be the solutions of the two-dimensional wave equation, Eq. (13.2). For example, consider the function w_3 given by Eq. (13.30). The partial derivatives of w_3 with respect to x, y, and t are given by

$$\frac{\partial w_3}{\partial x} = f' \cos \theta \tag{13.31}$$

$$\frac{\partial^2 w_3}{\partial x^2} = f'' \cos^2 \theta \tag{13.32}$$

$$\frac{\partial w_3}{\partial y} = f' \sin \theta \tag{13.33}$$

$$\frac{\partial^2 w_3}{\partial y^2} = f'' \sin^2 \theta \tag{13.34}$$

$$\frac{\partial w_3}{\partial t} = -cf' \tag{13.35}$$

$$\frac{\partial^2 w_3}{\partial t^2} = c^2 f'' \tag{13.36}$$

where a prime denotes a derivative with respect to the argument of the function. When w_3 is substituted for w using the relations (13.32), (13.34), and (13.36), Eq. (13.2) can be seen to be satisfied. The solutions given by Eqs. (13.28) and (13.29) are the same as those of a string. Equation (13.28) denotes a wave moving in the positive x direction at velocity c with its crests parallel to the y axis. The shape of the wave is independent of y and the membrane behaves as if it were made up of an infinite number of strips, all parallel to the x axis. Similarly, Eq. (13.29) denotes a wave moving in the negative x direction with crests parallel to the x axis and the shape independent of x. Equation (13.30) denotes a parallel wave moving in a direction at an angle θ to the x axis with a velocity c.

13.4   FREE VIBRATION OF RECTANGULAR MEMBRANES

The free vibration of a rectangular membrane of sides a and b (Fig. 13.2) can be determined using the method of separation of variables. Thus, the displacement $w(x, y, t)$ is expressed as a product of three functions as

$$w(x, y, t) = W(x, y)T(t) \equiv X(x)Y(y)T(t) \tag{13.37}$$

where W is a function of x and y, and X, Y, and T are functions of x, y, and t, respectively. Substituting Eq. (13.37) into the free vibration equation, Eq.(13.2), and dividing the resulting expression through by $X(x)Y(y)T(t)$, we obtain

$$c^2 \left[\frac{1}{X(x)} \frac{d^2 X(x)}{dx^2} + \frac{1}{Y(y)} \frac{d^2 Y(y)}{dy^2} \right] = \frac{1}{T(t)} \frac{d^2 T(t)}{dt^2} \tag{13.38}$$

Since the left-hand side of Eq. (13.38) is a function of x and y only, and the right-hand side is a function of t only, each side must be equal to a constant, say, k[1]:

$$c^2 \left[\frac{1}{X(x)} \frac{d^2 X(x)}{dx^2} + \frac{1}{Y(y)} \frac{d^2 Y(y)}{dy^2} \right] = \frac{1}{T(t)} \frac{d^2 T(t)}{dt^2} = k = -\omega^2 \tag{13.39}$$

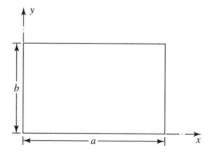

Figure 13.2   Rectangular membrane.

---

[1] The constant k can be shown to be a negative quantity by proceeding as in the case of free vibration of strings (Problem 13.3). Thus, we can write $k = -\omega^2$, where ω is another constant.

Equation (13.39) can be rewritten as two separate equations:

$$\frac{1}{X(x)} \frac{d^2 X(x)}{dx^2} + \frac{\omega^2}{c^2} = -\frac{1}{Y(y)} \frac{d^2 Y(y)}{dy^2} \tag{13.40}$$

$$\frac{d^2 T(t)}{dt^2} + \omega^2 T(t) = 0 \tag{13.41}$$

It can be noted, again, that the left-hand side of Eq. (13.40) is a function of x only and the right-hand side is a function of y only. Hence, Eq. (13.40) can be rewritten as two separate equations:

$$\frac{d^2 X(x)}{dx^2} + \alpha^2 X(x) = 0 \tag{13.42}$$

$$\frac{d^2 Y(y)}{dy^2} + \beta^2 Y(y) = 0 \tag{13.43}$$

where α^2 and β^2 are new constants related to ω^2 as

$$\beta^2 = \frac{\omega^2}{c^2} - \alpha^2 \tag{13.44}$$

Thus, the problem of solving a partial differential equation involving three variables, Eq. (13.2), has been reduced to the problem of solving three second-order ordinary differential equations, Eqs. (13.41)–(13.43). The solutions of Eqs. (13.41)–(13.43) can be expressed as[2]

$$T(t) = A \cos \omega t + B \sin \omega t \tag{13.45}$$

$$X(x) = C_1 \cos \alpha x + C_2 \sin \alpha x \tag{13.46}$$

$$Y(y) = C_3 \cos \beta y + C_4 \sin \beta y \tag{13.47}$$

where the constants A and B can be determined from the initial conditions and the constants C_1 to C_4 can be found from the boundary conditions of the membrane.

---

[2]The solution given by Eqs. (13.45)–(13.47) can also be obtained by proceeding as follows. The equation governing the free vibration of a rectangular membrane can be expressed, setting $f = 0$ in Eq. (13.1), as

$$\frac{\partial^2 w}{\partial x^2} + \frac{\partial^2 w}{\partial y^2} = \frac{1}{c^2} \frac{\partial^2 w}{\partial t^2} \tag{a}$$

By assuming a harmonic solution at frequency ω as

$$w(x, y, t) = W(x, y) e^{i\omega t} \tag{b}$$

Eq. (a) can be expressed as

$$\frac{\partial^2 W(x, y)}{\partial x^2} + \frac{\partial^2 W(x, y)}{\partial y^2} + \frac{\omega^2}{c^2} W(x, y) = 0 \tag{c}$$

By assuming the solution of $W(x, y)$ in the form

$$W(x, y) = X(x) Y(y) \tag{d}$$

and proceeding as indicated earlier, the solution shown in Eqs. (13.45)–(13.47) can be obtained.

13.4.1   Membrane with Clamped Boundaries

If a rectangular membrane is clamped or fixed on all the edges, the boundary conditions can be stated as

$$w(0, y, t) = 0, \qquad 0 \leq y \leq b, \quad t \geq 0 \tag{13.48}$$

$$w(a, y, t) = 0, \qquad 0 \leq y \leq b, \quad t \geq 0 \tag{13.49}$$

$$w(x, 0, t) = 0, \qquad 0 \leq x \leq a, \quad t \geq 0 \tag{13.50}$$

$$w(x, b, t) = 0, \qquad 0 \leq x \leq a, \quad t \geq 0 \tag{13.51}$$

In view of Eq. (13.37), the boundary conditions of Eqs. (13.48)–(13.51) can be restated as

$$X(0) = 0 \tag{13.52}$$

$$X(a) = 0 \tag{13.53}$$

$$Y(0) = 0 \tag{13.54}$$

$$Y(b) = 0 \tag{13.55}$$

The conditions $X(0) = 0$ and $Y(0) = 0$ [Eqs. (13.52) and (13.54)] require that $C_1 = 0$ in Eq. (13.46) and $C_3 = 0$ in Eq. (13.47). Thus, the functions $X(x)$ and $Y(y)$ become

$$X(x) = C_2 \sin \alpha x \tag{13.56}$$

$$Y(y) = C_4 \sin \beta y \tag{13.57}$$

For nontrivial solutions of $X(x)$ and $Y(y)$, the conditions $X(a) = 0$ and $Y(b) = 0$ [Eqs. (13.53) and (13.55)] require that

$$\sin \alpha a = 0 \tag{13.58}$$

$$\sin \beta b = 0 \tag{13.59}$$

Equations (13.58) and (13.59) together define the eigenvalues of the membrane through Eq. (13.44). The roots of Eqs. (13.58) and (13.59) are given by

$$\alpha_m a = m\pi, \qquad m = 1, 2, \ldots \tag{13.60}$$

$$\beta_n b = n\pi, \qquad n = 1, 2, \ldots \tag{13.61}$$

The natural frequencies of the membrane, ω_{mn}, can be determined using Eq. (13.44) as

$$\omega_{mn}^2 = c^2(\alpha_m^2 + \beta_n^2)$$

or

$$\omega_{mn} = \pi c \left[\left(\frac{m}{a} \right)^2 + \left(\frac{n}{b} \right)^2 \right]^{1/2}, \qquad m = 1, 2, \ldots, \quad n = 1, 2, \ldots \tag{13.62}$$

The following observations can be made from Eq. (13.62):

1. For any given mode of vibration, the natural frequency will decrease if either side of the rectangle is increased.

2. The fundamental natural frequency, ω_{11}, is most influenced by changes in the shorter side of the rectangle.

3. For an elongated rectangular membrane (with $b \gg a$), the fundamental natural frequency, ω_{11}, is negligibly influenced by variations in the longer side.

The eigenfunction or mode shape, $W_{mn}(x, y)$, of the membrane corresponding to the natural frequency ω_{mn} is given by

$$W_{mn}(x, y) = X_m(x)Y_n(y) = C_{2m} \sin \frac{m\pi x}{a} C_{4n} \sin \frac{n\pi y}{b}$$

$$= C_{mn} \sin \frac{m\pi x}{a} \sin \frac{n\pi y}{b}, \qquad m, n = 1, 2, \ldots \qquad (13.63)$$

where $C_{mn} = C_{2m}C_{4n}$ is a constant. Thus, the natural mode of vibration corresponding to ω_{mn} can be expressed as

$$w_{mn}(x, y, t) = \sin \frac{m\pi x}{a} \sin \frac{n\pi y}{b}(A_{mn} \cos \omega_{mn}t + B_{mn} \sin \omega_{mn}t) \qquad (13.64)$$

where $A_{mn} = C_{mn}A$ and $B_{mn} = C_{mn}B$ are new constants. The general solution of Eq. (13.64) is given by the sum of all the natural modes as

$$w_{mn}(x, y, t) = \sum_{m=1}^{\infty}\sum_{n=1}^{\infty} \sin \frac{m\pi x}{a} \sin \frac{n\pi y}{b}(A_{mn} \cos \omega_{mn}t + B_{mn} \sin \omega_{mn}t) \qquad (13.65)$$

The constants A_{mn} and B_{mn} in Eq. (13.65) can be determined using the initial conditions stated in Eqs. (13.4) and (13.5). Substituting Eq. (13.65) into Eqs. (13.4) and (13.5), we obtain

$$\sum_{m=1}^{\infty}\sum_{n=1}^{\infty} A_{mn} \sin \frac{m\pi x}{a} \sin \frac{n\pi y}{b} = w_0(x, y) \qquad (13.66)$$

$$\sum_{m=1}^{\infty}\sum_{n=1}^{\infty} B_{mn}\omega_{mn} \sin \frac{m\pi x}{a} \sin \frac{n\pi y}{b} = \dot{w}_0(x, y) \qquad (13.67)$$

Equations (13.66) and (13.67) denote the double Fourier sine series expansions of the functions $w_0(x, y)$ and $\dot{w}_0(x, y)$, respectively. Multiplying Eqs. (13.66) and (13.67) by $\sin(m\pi x/a) \sin(n\pi y/b)$ and integrating over the area of the membrane leads to the relations

$$A_{mn} = \frac{4}{ab} \int_0^a \int_0^b w_0(x, y) \sin \frac{m\pi x}{a} \sin \frac{n\pi y}{b} \, dx \, dy \qquad (13.68)$$

$$B_{mn} = \frac{4}{ab\omega_{mn}} \int_0^a \int_0^b \dot{w}_0(x, y) \sin \frac{m\pi x}{a} \sin \frac{n\pi y}{b} \, dx \, dy \qquad (13.69)$$

Example 13.1   Find the free vibration response of a rectangular membrane when it is struck such that the middle point experiences a velocity V_0 at $t = 0$.

SOLUTION Assuming the initial conditions as

$$w_0(x, y) = 0 \tag{E13.1.1}$$

$$\dot{w}_0(x, y) = V_0 \delta\left(x - \frac{a}{2}\right) \delta\left(y - \frac{b}{2}\right) \tag{E13.1.2}$$

the constants A_{mn} and B_{mn} given by Eqs. (13.68) and (13.69) can be evaluated as

$$A_{mn} = 0 \tag{E13.1.3}$$

$$B_{mn} = \frac{4}{ab\omega_{mn}} \int_0^a \int_0^b V_0 \delta\left(x - \frac{a}{2}\right) \delta\left(y - \frac{b}{2}\right) \sin\frac{m\pi x}{a} \sin\frac{n\pi y}{b} \, dx \, dy$$

$$= \frac{4V_0}{ab\omega_{mn}} \sin\frac{m\pi}{2} \sin\frac{n\pi}{2} \tag{E13.1.4}$$

Thus, the free vibration response of the membrane is given by Eq. (13.65):

$$w(x, y, t) = \frac{4V_0}{ab} \sum_{m=1}^{\infty} \sum_{n=1}^{\infty} \frac{1}{\omega_{mn}} \sin\frac{m\pi x}{a} \sin\frac{n\pi y}{b} \sin\frac{m\pi}{2} \sin\frac{n\pi}{2} \sin\omega_{mn}t \tag{E13.1.5}$$

The displacements of the membrane given by Eq. (E13.1.5) using values of each of m and n up to 10 at different instants of time are shown in Fig. 13.3.

13.4.2 Mode Shapes

Equation (13.64) describes a possible displacement variation of a membrane clamped at the boundary. Each point of the membrane moves harmonically with circular frequency ω_{mn} and amplitude given by the eigenfunction W_{mn} of Eq. (13.63). The following observations can be made regarding the characteristics of mode shapes [8].

1. The fundamental or lowest mode shape of the membrane corresponds to $m = n = 1$. In this modal pattern, the deflected surface of the membrane will consist of one half of a sine wave in each of the x and y directions. The higher values of m and n correspond to mode shapes with m and n half sine waves along the x and y directions, respectively. Thus, for values of m and n larger than 1, the deflection (mode) shapes will consist of lines within the membrane along which the deflection is zero. The lines along which the deflection is zero during vibration are called *nodal lines*. For specificity, the nodal lines corresponding to $m, n = 1, 2$ are shown in Fig. 13.4. For example, for $m = 2$ and $n = 1$, the nodal line will be parallel to the y axis at $x = a/2$, as shown in Fig. 13.5(a). Note that a specific natural frequency is associated with each combination of m and n values.

2. Equation (13.62) indicates that some of the higher natural frequencies (ω_{mn}) are integral multiples of the fundamental natural frequency ($\omega_{pp} = p\omega_{11}$), where p is an integer, whereas some higher frequencies are not integral multiples of ω_{11}. For example, $\omega_{12}, \omega_{21}, \omega_{13}$, and ω_{31} are not integral multiples of ω_{11}.

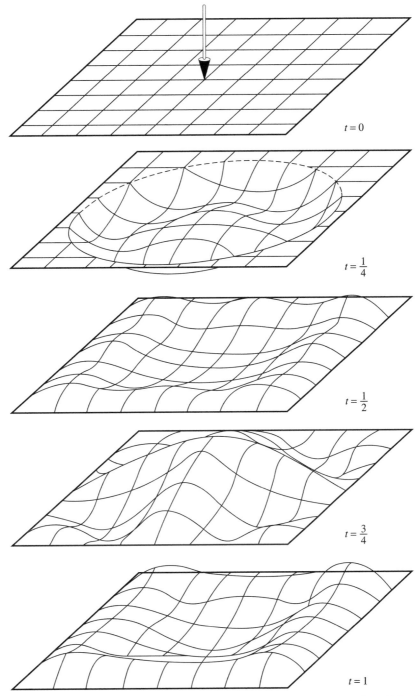

Figure 13.3 Deflection of a membrane at different times, initial velocity at the middle. Times given, t, are in terms of fractions of the fundamental natural period of vibration. (Source: Ref. [10].)

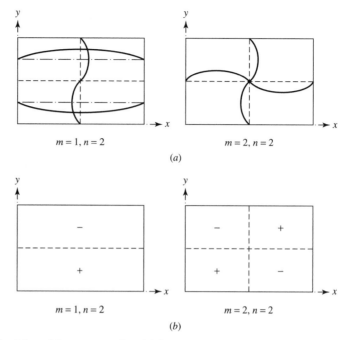

Figure 13.4   (a) modal patterns and nodal lines; (b) schematic illustration of mode shapes (+, positive deflection; −, negative deflection).

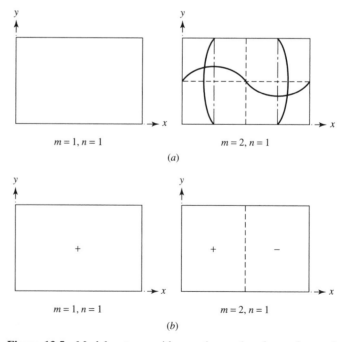

Figure 13.5   Modal patterns with $m = 1, n = 1$ and $m = 2, n = 1$.

3. It can be seen that when a^2 and b^2 are incommensurable, no two pairs of values of m and n can result in the same natural frequency. However, when a^2 and b^2 are commensurable, two or more values of ω_{mn} may have the same magnitude. If the ratio of sides $K = a/b$ is a rational number, the eigenvalues ω_{mn} and ω_{ij} will have the same magnitude if

$$m^2 + K^2 n^2 = i^2 + K^2 j^2 \tag{13.70}$$

For example, $\omega_{35} = \omega_{54}$, $\omega_{53} = \omega_{46}$, etc. when $K = \frac{4}{3}$, and $\omega_{13,4} = \omega_{12,5}$, etc. when $K = \frac{5}{3}$.

4. If the membrane is square, $a = b$ and Eq. (13.70) reduces to

$$m^2 + n^2 = i^2 + j^2 \tag{13.71}$$

and the magnitudes of ω_{mn} and ω_{nm} will be the same. This means that two different eigenfunctions $W_{mn}(x, y)$ and $W_{nm}(x, y)$ correspond to the same frequency $\omega_{mn}(= \omega_{nm})$; thus, there will be fewer frequencies than modes. Such cases are called *degenerate cases*. If the natural frequencies are repeated with $\omega_{mn} = \omega_{nm}$, any linear combination of the corresponding natural modes W_{mn} and W_{nm} can also be shown to be a natural mode of the membrane. Thus, for these cases a large variety of nodal patterns occur.

5. To find the modal patterns and nodal lines of a square membrane corresponding to repeated frequencies, consider, as an example, the case of ω_{mn} with $m = 1$ and $n = 2$. For this case, $\omega_{12} = \omega_{21} = \sqrt{5}\pi c/a$ and the corresponding distinct mode shapes can be expressed as (with $a = b$)

$$w_{12}(x, y, t) = \sin\frac{\pi x}{a}\sin\frac{2\pi y}{b}\left(A_{12}\cos\frac{\sqrt{5}\pi ct}{a} + B_{12}\sin\frac{\sqrt{5}\pi ct}{a}\right) \tag{13.72}$$

$$w_{21}(x, y, t) = \sin\frac{2\pi x}{a}\sin\frac{\pi y}{b}\left(A_{21}\cos\frac{\sqrt{5}\pi ct}{a} + B_{21}\sin\frac{\sqrt{5}\pi ct}{a}\right) \tag{13.73}$$

Since the frequencies are the same, it will be of interest to consider a linear combination of the maximum deflection patterns given by Eqs. (13.72) and (13.73) as

$$w = A\sin\frac{\pi x}{a}\sin\frac{2\pi y}{b} + B\sin\frac{2\pi x}{a}\sin\frac{\pi y}{b} \tag{13.74}$$

where A and B are constants. The deflection shapes given by Eq. (13.74) for specific combinations of values of A and B are shown in Fig. 13.6. Figure 13.6(a) to 13.6(d) correspond to values of $B = 0$, $A = 0$, $A = B$, and $A = -B$, respectively. When $B = 0$, the deflection shape given by Eq. (13.74) consists of one-half sine wave along the x direction and two half sine waves along the y direction with a nodal line at $y = b/2$. Similarly, when $A = 0$, the

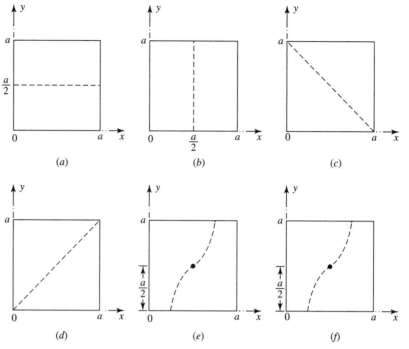

Figure 13.6   Deflection shapes given by Eq. (13.74): (*a*) $B = 0$; (*b*) $A = 0$; (*c*) $A = B$; (*d*) $A = -B$; (*e*) $A = B/2$; (*f*) $A = 2B$.

nodal line will be at $x = a/2$. When $A = B$, Eq. (13.74) becomes

$$
\begin{aligned}
w &= A\left(\sin\frac{\pi x}{a}\sin\frac{2\pi y}{b} + \sin\frac{2\pi x}{a}\sin\frac{\pi y}{b}\right) \\
&= 2A\sin\frac{\pi x}{a}\sin\frac{\pi y}{b}\left(\cos\frac{\pi x}{a} + \cos\frac{\pi y}{b}\right)
\end{aligned}
\tag{13.75}
$$

It can be seen that $w = 0$ in Eq. (13.75) when

$$
\sin\frac{\pi x}{a} = 0 \quad\text{or}\quad \sin\frac{\pi y}{b} = 0
\tag{13.76}
$$

or

$$
\cos\frac{\pi x}{a} + \cos\frac{\pi y}{b} = 0
\tag{13.77}
$$

The cases in Eq. (13.76) correspond to $w = 0$ along the edges of the membrane, while the case in Eq. (13.77) gives $w = 0$ at which

$$
\frac{\pi x}{a} = \pi - \frac{\pi y}{a} \quad\text{or}\quad x + y = a
\tag{13.78}
$$

Equation (13.78) indicates that the nodal line is a diagonal of the square as shown in Fig. 13.6(*c*). Similarly, the case $A = -B$ gives the nodal line along the other diagonal of the square as indicated in Fig. 13.6(d). For arbitrary values of A and B, Eq. (13.74) can be written as

$$w = A \left(\sin \frac{\pi x}{a} \sin \frac{2\pi y}{b} + R \sin \frac{2\pi x}{a} \sin \frac{\pi y}{b} \right) \tag{13.79}$$

where $R = B/A$ is a constant. Different nodal lines can be obtained based on the value of R. For example, the nodal line [along which $w = 0$ in Eq. (13.79)] corresponding to $K = 2$ is shown in Fig. 13.6(e) and (f).
The following observations can be made from the discussion above:

(a) A large variety of nodal patterns can exist for any repeated frequency in a square or rectangular membrane. Thus, it is not possible to associate a mode shape uniquely with a frequency in a membrane problem.
(b) The nodal lines need not be straight lines. It can be shown that all the nodal lines of a square membrane pass through the center, $x = y = a/2$, which is called a *pole*.

6. For a square membrane, the modal pattern corresponding to $m = n = 1$ consists of one-half of a sine wave along each of the x and y directions. For $m = n = 2$, no other pair of integers i and j give the same natural frequency, ω_{22}. In this case the maximum modal deflection can be expressed as

$$w = \sin \frac{2\pi x}{a} \sin \frac{2\pi y}{b} = 4 \sin \frac{\pi x}{a} \cos \frac{\pi x}{a} \sin \frac{\pi y}{b} \cos \frac{\pi y}{b} \tag{13.80}$$

The nodal lines corresponding to this mode are determined by the equation

$$\sin \frac{\pi x}{a} \cos \frac{\pi x}{a} \sin \frac{\pi y}{b} \cos \frac{\pi y}{b} = 0 \tag{13.81}$$

Equation (13.81) gives the nodal lines as (in addition to the edges)

$$x = \frac{a}{2}, \qquad y = \frac{a}{2} \tag{13.82}$$

which are shown in Fig. 13.7.

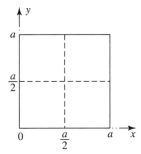

Figure 13.7   Nodal lines corresponding to ω_{22} of a square membrane.

7. Next, consider the case of $m = 3$ and $n = 1$ for a square membrane. In this case, $\omega_{31} = \omega_{13} = \sqrt{10}(\pi c/a)$ and the corresponding distinct mode shapes can be expressed as

$$w_{31}(x, y, t) = \sin \frac{3\pi x}{a} \sin \frac{\pi y}{b} \left(A_{31} \cos \frac{\sqrt{10}\pi ct}{a} + B_{31} \sin \frac{\sqrt{10}\pi ct}{a} \right)$$

$$(13.83)$$

$$w_{13}(x, y, t) = \sin \frac{\pi x}{a} \sin \frac{3\pi y}{b} \left(A_{13} \cos \frac{\sqrt{10}\pi ct}{a} + B_{13} \sin \frac{\sqrt{10}\pi ct}{a} \right)$$

$$(13.84)$$

Since the frequencies are the same, a linear combination of the maximum deflection patterns given by Eqs. (13.83) and (13.84) can be represented as

$$w = A \sin \frac{3\pi x}{a} \sin \frac{\pi y}{a} + B \sin \frac{\pi x}{a} \sin \frac{3\pi y}{a} \qquad (13.85)$$

where A and B are constants. The nodal lines corresponding to Eq. (13.85) are defined by $w = 0$, which can be rewritten as

$$\sin \frac{\pi x}{a} \sin \frac{\pi y}{a} \left[A \left(4\cos^2 \frac{\pi x}{a} - 1 \right) + B \left(4\cos^2 \frac{\pi y}{a} - 1 \right) \right] = 0 \qquad (13.86)$$

Neglecting the factor $\sin(\pi x/a) \sin(\pi y/a)$, which corresponds to nodal lines along the edges, Eq. (13.86) can be expressed as

$$A \left(4\cos^2 \frac{\pi x}{a} - 1 \right) + B \left(4\cos^2 \frac{\pi y}{a} - 1 \right) = 0 \qquad (13.87)$$

It can be seen from Eq. (13.87) that:

(a) When $A = 0$, $y = a/3$ and $2a/3$ denote the nodal lines.
(b) When $B = 0$, $x = a/3$ and $2a/3$ denote the nodal lines.
(c) When $A = -B$, Eq. (13.87) reduces to

$$\cos \frac{\pi x}{a} = \pm \cos \frac{\pi y}{a} \qquad (13.88)$$

or

$$x = y \quad \text{and} \quad x = a - y \qquad (13.89)$$

denote the nodal lines.
(d) When $A = B$, Eq. (13.87) reduces to

$$\cos^2 \frac{\pi x}{a} + \cos^2 \frac{\pi y}{a} = \frac{1}{2} \quad \text{or} \quad \cos \frac{2\pi x}{a} + \cos \frac{2\pi y}{a} + 1 = 0 \qquad (13.90)$$

which represents a circle. The nodal lines in each of these cases are shown in Fig. 13.8.

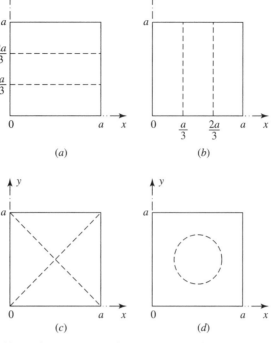

Figure 13.8   Nodal lines of a square membrane corresponding to $\omega_{31} = \omega_{13}$: (a) $A = 0$; (b) $B = 0$; (c) $A = -B$; (d) $A = B$.

8. Whenever, in a vibrating system, including a membrane, certain parts or points remain at rest, they can be assumed to be absolutely fixed and the result may be applicable to another system. For example, at a particular natural frequency ω, if the modal pattern of a square membrane consists of a diagonal line as a nodal line, the solution will also be applicable for a membrane whose boundary is an isosceles right triangle. In addition, it can be observed that each possible mode of vibration of the isosceles triangle corresponds to some natural mode of the square. Accordingly, the fundamental natural frequency of vibration of an isosceles right triangle will be equal to the natural frequency of a square with $m = 1$ and $n = 2$:

$$\omega = \frac{\sqrt{5}\pi c}{a} \tag{13.91}$$

The second natural frequency of the isosceles right triangle will be equal to the natural frequency of a square plate with $m = 3$ and $n = 1$:

$$\omega = \frac{\sqrt{10}\pi c}{a} \tag{13.92}$$

The mode shapes corresponding to the natural frequencies of Eqs. (13.91) and (13.92) are shown in Fig. 13.9.

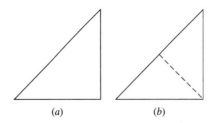

Figure 13.9   (a) $\omega = \sqrt{5}\pi c/a$; (b) $\omega = \sqrt{10}\pi c/a$.

13.5   FORCED VIBRATION OF RECTANGULAR MEMBRANES

13.5.1   Modal Analysis Approach

The equation of motion governing the forced vibration of a rectangular membrane is given by Eq. (13.1):

$$-P\left[\frac{\partial^2 w(x, y, t)}{\partial x^2} + \frac{\partial^2 w(x, y, t)}{\partial y^2}\right] + \rho\frac{\partial^2 w(x, y, t)}{\partial t^2} = f(x, y, t) \qquad (13.93)$$

We can find the solution of Eq. (13.93) using the modal analysis procedure. Accordingly, we assume the solution of Eq. (13.93) as

$$w(x, y, t) = \sum_{m=1}^{\infty}\sum_{n=1}^{\infty} W_{mn}(x, y)\eta_{mn}(t) \qquad (13.94)$$

where $W_{mn}(x, y)$ are the natural modes of vibration and $\eta_{mn}(t)$ are the corresponding generalized coordinates. For specificity we consider a membrane with clamped edges. For this, the eigenfunctions are given by Eq. (13.63):

$$W_{mn}(x, y) = C_{mn} \sin\frac{m\pi x}{a} \sin\frac{n\pi y}{b}, \qquad m, n = 1, 2, \ldots \qquad (13.95)$$

The eigenfunctions (or normal modes) can be normalized as

$$\int_0^a \int_0^b \rho W_{mn}^2(x, y)\, dx\, dy = 1 \qquad (13.96)$$

or

$$\int_0^a \int_0^b \rho C_{mn}^2 \sin^2\frac{m\pi x}{a} \sin^2\frac{n\pi y}{b}\, dx\, dy = 1 \qquad (13.97)$$

The simplification of Eq. (13.97) yields

$$C_{mn} = \frac{2}{\sqrt{\rho ab}} \qquad (13.98)$$

Thus, the normal modes take the form

$$W_{mn}(x, y) = \frac{2}{\sqrt{\rho ab}} \sin\frac{m\pi x}{a} \sin\frac{n\pi y}{b}, \qquad m, n = 1, 2, \ldots \qquad (13.99)$$

Substituting Eq. (13.94) into Eq. (13.93), multiplying the resulting equation throughout by $W_{mn}(x)$, and integrating over the area of the membrane leads to the equation

$$\ddot{\eta}_{mn}(t) + \omega_{mn}^2 \eta_{mn}(t) = N_{mn}(t), \qquad m, n = 1, 2, \ldots \tag{13.100}$$

where

$$\omega_{mn} = \pi \sqrt{\frac{P}{\rho}\left[\left(\frac{m}{a}\right)^2 + \left(\frac{n}{b}\right)^2\right]} \tag{13.101}$$

$$N_{mn}(t) = \int_0^a \int_0^b W_{mn}(x, y) f(x, y, t)\, dx\, dy$$

$$= \frac{2}{\sqrt{\rho ab}} \int_0^a \int_0^b f(x, y, t) \sin\frac{m\pi x}{a} \sin\frac{n\pi y}{b}\, dx\, dy \tag{13.102}$$

The solution of Eq. (13.100) can be written as [see Eq. (2.109)]

$$\eta_{mn}(t) = \frac{1}{\omega_{mn}} \int_0^t N_{mn}(\tau) \sin\omega_{mn}(t - \tau)\, d\tau + \eta_{mn}(0) \cos\omega_{mn}t + \frac{\dot{\eta}_{mn}(0)}{\omega_{mn}} \sin\omega_{mn}t \tag{13.103}$$

The solution of Eq. (13.93) becomes, in view of Eqs. (13.94) and (13.103),

$$w(x, y, t) = \left\{ \frac{2}{\sqrt{\rho ab}} \sum_{m=1}^{\infty}\sum_{n=1}^{\infty} \sin\frac{m\pi x}{a}\sin\frac{n\pi y}{b}\left[\eta_{mn}(0)\cos\omega_{mn}t + \frac{\dot{\eta}_{mn}(0)}{\omega_{mn}}\sin\omega_{mn}t\right]\right\}$$

$$+ \left[\frac{2}{\sqrt{\rho ab}} \sum_{m=1}^{\infty}\sum_{n=1}^{\infty} \frac{1}{\omega_{mn}} \sin\frac{m\pi x}{a}\sin\frac{n\pi y}{b}\right.$$

$$\left. \times \int_0^t N_{mn}(\tau)\sin\omega_{mn}(t - \tau)\, d\tau\right] \tag{13.104}$$

It can be seen that the quantity inside the braces represents the free vibration response (due to the initial conditions) and the quantity in the second set of brackets denotes the forced vibrations of the membrane.

Example 13.2   Find the forced vibration response of a rectangular membrane of sides a and b subjected to a harmonic force $F_0 \sin\Omega t$ at the center of the membrane. Assume all edges of the membrane to be fixed and the initial conditions to be zero.

SOLUTION   The applied force can be described as (see Fig. 13.10)

$$f(x, y, t) = F_0 \sin\Omega t\, \delta\left(x - \frac{a}{2}, y - \frac{b}{2}\right) \tag{E13.2.1}$$

where $F_0 \sin\Omega t$ denotes the time-dependent amplitude of the concentrated force and $\delta(x - a/2, y - b/2)$ is a two-dimensional spatial Dirac delta function defined as

$$\left.\begin{array}{l}\delta\left(x - \dfrac{a}{2}, y - \dfrac{b}{2}\right) = 0, \qquad x \neq \dfrac{a}{2}\ \text{and/or}\ y \neq \dfrac{b}{2} \\[2ex] \displaystyle\int_0^a \int_0^b \delta\left(x - \dfrac{a}{2}, y - \dfrac{b}{2}\right) dx\, dy = 1\end{array}\right\} \tag{E13.2.2}$$

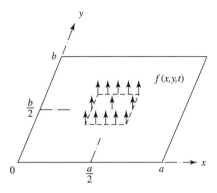

Figure 13.10 Harmonic force at center.

The normalized eigenfunctions or normal modes of the membrane, with all edges fixed, are given by [see Eq. (13.99)]

$$W_{mn}(x, y) = \frac{2}{\sqrt{\rho ab}} \sin \frac{m\pi x}{a} \sin \frac{n\pi y}{b}, \qquad m, n = 1, 2, \ldots \qquad \text{(E13.2.3)}$$

and the corresponding natural frequencies by [see Eq. (13.101)]

$$\omega_{mn} = \pi \sqrt{\frac{P}{\rho} \left[\left(\frac{m}{a}\right)^2 + \left(\frac{n}{b}\right)^2 \right]}, \qquad m, n = 1, 2, \ldots \qquad \text{(E13.2.4)}$$

The response or deflection of the membrane can be expressed, using Eq. (13.94), as

$$w(x, y, t) = \sum_{m=1}^{\infty} \sum_{n=1}^{\infty} W_{mn}(x, y) \eta_{mn}(t) \qquad \text{(E13.2.5)}$$

where the generalized coordinates $\eta_{mn}(t)$ can be determined using Eq. (13.103). For zero initial conditions, Eq. (13.103) gives

$$\eta_{mn}(t) = \frac{1}{\omega_{mn}} \int_0^t N_{mn}(\tau) \sin \omega_{mn}(t - \tau) \, d\tau \qquad \text{(E13.2.6)}$$

The generalized force $N_{mn}(t)$ can be found as [see Eq. (13.102)]

$$N_{mn}(t) = \frac{2}{\sqrt{\rho ab}} \int_0^a \int_0^b f(x, y, t) \sin \frac{m\pi x}{a} \sin \frac{n\pi y}{b} \, dx \, dy \qquad \text{(E13.2.7)}$$

Using Eqs. (E13.2.1) and (E13.2.2) in Eq. (E13.2.7), we obtain

$$N_{mn}(t) = \frac{2F_0}{\sqrt{\rho ab}} \sin \Omega t \int_0^a \int_0^b \sin \frac{m\pi x}{a} \sin \frac{n\pi y}{b} \delta \left(x - \frac{a}{2}, y - \frac{b}{2} \right) dx \, dy$$

$$= \frac{2F_0}{\sqrt{\rho ab}} \sin \Omega t \sin \frac{m\pi}{2} \sin \frac{n\pi}{2} \qquad \text{(E13.2.8)}$$

Equations (E13.2.6) and (E13.2.8) yield

$$\eta_{mn}(t) = \frac{2F_0}{\omega_{mn} \sqrt{\rho ab}} \sin \frac{m\pi}{2} \sin \frac{n\pi}{2} \int_0^t \sin \Omega \tau \sin \omega_{mn}(t - \tau) \, d\tau \qquad \text{(E13.2.9)}$$

The integral in Eq. (E13.2.9) can be evaluated using a trigonometric identity for a product as

$$\int_0^t \sin \Omega \tau \sin \omega_{mn}(t-\tau)\, d\tau$$

$$= \frac{1}{2} \int_0^t \{\cos[(\Omega + \omega_{mn})\tau - \omega_{mn}t] - \cos[(\Omega - \omega_{mn})\tau + \omega_{mn}t]\}\, d\tau$$

$$= \frac{1}{2}\left[\frac{\sin[(\Omega + \omega_{mn})\tau - \omega_{mn}t]}{\Omega + \omega_{mn}} - \frac{\sin[(\Omega - \omega_{mn})\tau + \omega_{mn}t]}{\Omega - \omega_{mn}} \right]_0^t$$

$$= \frac{1}{2}\left(\frac{\sin \Omega t}{\Omega + \omega_{mn}} - \frac{\sin \Omega t}{\Omega - \omega_{mn}} + \frac{\sin \omega_{mn}t}{\Omega + \omega_{mn}} + \frac{\sin \omega_{mn}t}{\Omega - \omega_{mn}} \right)$$

$$= \frac{1}{\omega_{mn}[1 - (\Omega/\omega_{mn})^2]}\left(\sin \Omega t - \frac{\Omega}{\omega_{mn}} \sin \omega_{mn}t \right) \qquad \text{(E13.2.10)}$$

Equations (E13.2.9) and (E13.2.10) give

$$\eta_{mn}(t) = \frac{2F_0}{\sqrt{\rho a b}\, \omega_{mn}^2 (1 - \Omega^2/\omega_{mn}^2)} \sin \frac{m\pi}{2} \sin \frac{n\pi}{2} \left(\sin \Omega t - \frac{\Omega}{\omega_{mn}} \sin \omega_{mn}t \right) \qquad \text{(E13.2.11)}$$

Thus, the steady-state response of the membrane can be expressed as [see Eq. (E13.2.5)]

$$w(x, y, t) = \frac{4F_0}{\rho a b} \sum_{m=1}^{\infty} \sum_{n=1}^{\infty} \frac{\sin(m\pi/2)\sin(n\pi/2)\sin(m\pi x/a)\sin(n\pi y/b)}{\omega_{mn}^2 (1 - \Omega^2/\omega_{mn}^2)}$$

$$\times \left(\sin \Omega t - \frac{\Omega}{\omega_{mn}} \sin \omega_{mn}t \right) \qquad \text{(E13.2.12)}$$

with ω_{mn} given by Eq. (E13.2.4).

13.5.2   Fourier Transform Approach

The governing equation can be expressed as [see Eqs. (13.1)–(13.3)]

$$\frac{1}{c^2}\frac{\partial^2 w}{\partial t^2} = \frac{\partial^2 w}{\partial x^2} + \frac{\partial^2 w}{\partial y^2} + \frac{f(x, y, t)}{P}, \qquad 0 \le x \le a, \qquad 0 \le y \le b \qquad (13.105)$$

where $w(x, y, t)$ is the transverse displacement. Let the membrane be fixed along all the edges, $x = 0, x = a, y = 0$, and $y = b$. Multiplying Eq. (13.105) by $\sin(m\pi x/a)\sin(n\pi y/b)$ and integrating the resulting equation over the area of the membrane yields

$$\frac{1}{c^2}\frac{d^2 W(m, n, t)}{dt^2} + \pi^2\left(\frac{m^2}{a^2} + \frac{n^2}{b^2} \right) W(m, n, t) = \frac{1}{P}F(m, n, t) \qquad (13.106)$$

where $W(m, n, t)$ and $F(m, n, t)$ denote the double finite Fourier sine transforms of $w(x, y, t)$ and $f(x, y, t)$, respectively:

$$W(m, n, t) = \int_0^a \int_0^b w(x, y, t) \sin \frac{m\pi x}{a} \sin \frac{n\pi y}{b} \, dx \, dy \qquad (13.107)$$

$$F(m, n, t) = \int_0^a \int_0^b f(x, y, t) \sin \frac{m\pi x}{a} \sin \frac{n\pi y}{b} \, dx \, dy \qquad (13.108)$$

The solution of Eq. (13.106) can be expressed as

$$W(m, n, t) = W_0(m, n) \cos \pi c\alpha_{mn} t + \frac{\dot{W}_0(m, n)}{\pi c\alpha_{mn}} \sin \pi c\alpha_{mn} t$$

$$+ \frac{c}{\pi^2 P\alpha_{mn}} \int_0^t F(m, n, \tau) \sin \pi c\alpha_{mn}(t - \tau) \, d\tau \qquad (13.109)$$

where $W_0(m, n)$ and $\dot{W}_0(m, n)$ are the double finite Fourier sine transforms of the initial values of w and $\partial w/\partial t$:

$$w(x, y, t = 0) = w_0(x, y) \qquad (13.110)$$

$$\frac{\partial w}{\partial t}(x, y, t = 0) = \dot{w}_0(x, y) \qquad (13.111)$$

$$W_0(m, n) = \int_0^a \int_0^b w_0(x, y) \sin \frac{m\pi x}{a} \sin \frac{n\pi y}{b} \, dx \, dy \qquad (13.112)$$

$$\dot{W}_0(m, n) = \int_0^a \int_0^b \dot{w}_0(x, y) \sin \frac{m\pi x}{a} \sin \frac{n\pi y}{b} \, dx \, dy \qquad (13.113)$$

where

$$\alpha_{mn} = \left(\frac{m^2}{a^2} + \frac{n^2}{b^2} \right)^{1/2} \qquad (13.114)$$

The displacement of the membrane can be found by taking the double inverse finite sine transform of Eq. (13.109). The procedure is illustrated for a simple case in the following example.

Example 13.3   Find the response of a rectangular membrane subjected to an impulse $\hat{G}$ applied at $(x = \xi, y = \eta)$ in the transverse direction. Assume the initial displacement of the membrane to be zero.

SOLUTION   The initial conditions of the membrane can be expressed as

$$w_0(x, y) = 0 \qquad (E13.3.1)$$

$$\dot{w}_0(x, y) = \frac{\hat{G}}{\rho} \delta(x - \xi)\delta(y - \eta) \qquad (E13.3.2)$$

where ρ is the mass per unit area of the membrane. The free vibration response of the membrane can be obtained from Eq. (13.109) by setting $F = 0$:

$$W(m, n, t) = W_0(m, n) \cos \pi c \alpha_{mn} t + \frac{\dot{W}_0(m, n)}{\pi c \alpha_{mn}} \sin \pi c \alpha_{mn} t$$

$$= \cos \pi c \alpha_{mn} t \int_0^a \int_0^b w_0(x, y, t) \sin \frac{m \pi x}{a} \sin \frac{n \pi y}{b} \, dx \, dy$$

$$+ \frac{\sin \pi c \alpha_{mn} t}{\pi c \alpha_{mn}} \int_0^a \int_0^b \dot{w}_0(x, y) \sin \frac{m \pi x}{a} \sin \frac{n \pi y}{b} \, dx \, dy$$

$$\text{(E13.3.3)}$$

Taking the inverse transform of Eq. (E13.3.3), we obtain[3]

$$w(x, y, t) = \frac{4}{ab} \sum_{m=1}^{\infty} \sum_{n=1}^{\infty} \sin \frac{m \pi x}{a} \sin \frac{n \pi y}{b} \cos \pi c \alpha_{mn} t$$

$$\times \left[\int_{x'=0}^a \int_{y'=0}^b w_0(x', y') \sin \frac{m \pi x'}{a} \sin \frac{n \pi y'}{b} \, dx' \, dy' \right]$$

$$+ \frac{4}{\pi c} \sum_{m=1}^{\infty} \sum_{n=1}^{\infty} \sin \frac{m \pi x}{a} \sin \frac{n \pi y}{b} \frac{\sin \pi c \alpha_{mn} t}{(m^2 b^2 + n^2 a^2)^{1/2}}$$

$$\times \int_{x'=0}^a \int_{y'=0}^b \dot{w}_0(x', y') \sin \frac{m \pi x'}{a} \sin \frac{n \pi y'}{b} \, dx' \, dy' \quad \text{(E13.3.4)}$$

By substituting the initial conditions, Eqs. (E13.3.1) and (E13.3.2), and noting that

$$\int_0^a \int_0^b \dot{w}_0(x, y) \sin \frac{m \pi x}{a} \sin \frac{n \pi y}{b} \, dx \, dy = \frac{\hat{G}}{\rho} \sin \frac{m \pi \xi}{a} \sin \frac{n \pi \eta}{b} \quad \text{(E13.3.5)}$$

we find the response of the membrane as

$$w(x, y, t)$$

$$= \frac{4 \hat{G}}{\pi \rho c} \sum_{m=1}^{\infty} \sum_{n=1}^{\infty} \frac{1}{(m^2 b^2 + n^2 a^2)^{1/2}} \sin \left[\pi c \left(\frac{m^2}{a^2} + \frac{n^2}{b^2} \right)^{1/2} t \right]$$

$$\times \sin \frac{m \pi \xi}{a} \sin \frac{n \pi \eta}{b} \sin \frac{m \pi x}{a} \sin \frac{n \pi y}{b} \quad \text{(E13.3.6)}$$

---

[3]If $p(x, y)$ denotes a function of the variables x and y that satisfies the Dirichlet's condition over the region $0 \le x \le a, 0 \le y \le b$, its finite double Fourier sine transform, $P(m, n)$, is defined by

$$F(m, n) = \int_{x=0}^a \int_{y=0}^b p(x, y) \sin \frac{m \pi x}{a} \sin \frac{n \pi y}{b} \, dx \, dy \quad \text{(a)}$$

The inverse transform of $F(m, n)$, given by Eq. (a), can be expressed as

$$p(x, y) = \frac{4}{ab} \sum_{m=1}^{\infty} \sum_{n=1}^{\infty} P(m, n) \sin \frac{m \pi x}{a} \sin \frac{n \pi y}{b} \, dx \, dy \quad \text{(b)}$$

13.6   FREE VIBRATION OF CIRCULAR MEMBRANES

13.6.1   Equation of Motion

Noting that the Laplacian operator in rectangular coordinates is defined by

$$\nabla^2 = \frac{\partial^2}{\partial x^2} + \frac{\partial^2}{\partial y^2} \tag{13.115}$$

the equation of motion of a rectangular membrane, Eq. (13.1), can be expressed as

$$P\nabla^2 w(x, y, t) + f(x, y, t) = \rho \frac{\partial^2 w(x, y, t)}{\partial t^2} \tag{13.116}$$

For a circular membrane, the governing equation of motion can be derived using an equilibrium approach by considering a differential element in the polar coordinates r and θ (see Problem 13.1). Alternatively, a coordinate transformation using the relations

$$x = r\cos\theta, \qquad y = r\sin\theta \tag{13.117}$$

can be used to express the Laplacian operator in polar coordinates as (see Problem 13.2)

$$\nabla = \frac{1}{r}\frac{\partial}{\partial r}\left(r\frac{\partial}{\partial r}\right) + \frac{1}{r^2}\frac{\partial^2}{\partial\theta^2} = \frac{\partial^2}{\partial r^2} + \frac{1}{r}\frac{\partial}{\partial r} + \frac{1}{r^2}\frac{\partial^2}{\partial\theta^2} \tag{13.118}$$

Thus, the equation of motion for the forced vibration of a circular membrane can be expressed as

$$P\left[\frac{\partial^2 w(r, \theta, t)}{\partial r^2} + \frac{1}{r}\frac{\partial w(r, \theta, t)}{\partial r} + \frac{1}{r^2}\frac{\partial^2 w(r, \theta, t)}{\partial\theta^2}\right] + f(r, \theta, t) = \rho\frac{\partial^2 w(r, \theta, t)}{\partial t^2} \tag{13.119}$$

For free vibration, Eq. (13.119) reduces to

$$c^2\left[\frac{\partial^2 w(r, \theta, t)}{\partial r^2} + \frac{1}{r}\frac{\partial w(r, \theta, t)}{\partial r} + \frac{1}{r^2}\frac{\partial^2 w(r, \theta, t)}{\partial\theta^2}\right] = \frac{\partial^2 w(r, \theta, t)}{\partial t^2} \tag{13.120}$$

where c is given by Eq. (13.3). As the displacement, w, is now a function of r, θ, and t, we use the method of separation of variables and express the solution as

$$w(r, \theta, t) = R(r)\Theta(\theta)T(t) \tag{13.121}$$

where R, Θ, and T are functions of only r, θ, and t, respectively. By substituting Eq. (13.121) into Eq. (13.120), we obtain

$$R\Theta\frac{d^2 T}{dt^2} = c^2\left(\frac{d^2 R}{dr^2}\Theta T + \frac{1}{r}\frac{dR}{dr}\Theta T + \frac{1}{r^2}R\frac{d^2\Theta}{d\theta^2}T\right) \tag{13.122}$$

which, upon division by $R\Theta T$, becomes

$$\frac{1}{c^2}\frac{1}{T}\frac{d^2 T}{dt^2} = \frac{1}{R}\frac{d^2 R}{dr^2} + \frac{1}{rR}\frac{dR}{dr} + \frac{1}{r^2\Theta}\frac{d^2\Theta}{d\theta^2} \tag{13.123}$$

Noting that each side of Eq. (13.123) must be a constant with a negative value (see Problem 13.3), the constant is taken as $-\omega^2$, where ω is any number, and Eq. (13.123) is rewritten as

$$\frac{d^2T}{dt^2} + c^2\omega^2 T = 0 \tag{13.124}$$

$$\frac{1}{R}\left(r^2\frac{d^2R}{dr^2} + r\frac{dR}{dr}\right) + \omega^2 r^2 = -\frac{1}{\Theta}\frac{d^2\Theta}{d\theta^2} \tag{13.125}$$

Again, we note that each side of Eq. (13.125) must be a constant. Using α^2 as the constant, Eq. (13.125) is rewritten as

$$\frac{d^2R(r)}{dr^2} + \frac{1}{r}\frac{dR(r)}{dr} + \left(\omega^2 - \frac{\alpha^2}{r^2}\right)R(r) = 0 \tag{13.126}$$

$$\frac{d^2\Theta}{d\theta^2} + \alpha^2\Theta = 0 \tag{13.127}$$

Since the constant α^2 must yield the displacement w as a periodic function of θ with a period 2π [i.e., $w(r, \theta, t) = w(r, \theta + 2\pi, t)$] α must be an integer:

$$\alpha = m, \qquad m = 0, 1, 2, \ldots \tag{13.128}$$

The solutions of Eqs. (13.124) and (13.127) can be expressed as

$$T(t) = A_1\cos\omega t + A_2\sin\omega t \tag{13.129}$$

$$\Theta(\theta) = C_{1m}\cos m\theta + C_{2m}\sin m\theta, \qquad m = 0, 1, 2, \ldots \tag{13.130}$$

Equation (13.126) can be rewritten as

$$r^2\frac{d^2R}{dr^2} + r\frac{dR}{dr} + (r^2\omega^2 - m^2)R = 0 \tag{13.131}$$

which can be identified as Bessel's equation of order m with the parameter ω. The solution of Eq. (13.131) is given by [9]:

$$R(r) = B_1 J_m(\omega r) + B_2 Y_m(\omega r) \tag{13.132}$$

where B_1 and B_2 are constants to be determined from the boundary conditions, and J_m and Y_m are Bessel functions of first and second kind, respectively. The Bessel functions are in the form of infinite series and are studied extensively and tabulated in the literature [2],[3] because of their importance in the study of problems involving circular geometry. For a circular membrane, $w(r, \theta, t)$ must be finite (bounded) everywhere. However, $Y_m(\omega r)$ approaches infinity at the origin ($r = 0$). Hence, the constant B_2 must be zero in Eq. (13.132). Thus, Eq. (13.132) reduces to

$$R(r) = B_1 J_m(\omega r) \tag{13.133}$$

and the complete solution can be expressed as

$$w(r, \theta, t) = W_m(r, \theta)(A_1\cos\omega t + A_2\sin\omega t) \tag{13.134}$$

where

$$W_m(r, \theta) = J_m(\omega r)(C_{1m} \cos m\theta + C_{2m} \sin m\theta), \qquad m = 0, 1, 2, \ldots \qquad (13.135)$$

with C_{1m} and C_{2m} denoting some new constants.

13.6.2 Membrane with a Clamped Boundary

If the membrane is clamped or fixed at the boundary $r = a$, the boundary conditions can be stated as

$$W_m(a, \theta) = 0, \qquad m = 0, 1, 2, \ldots \qquad (13.136)$$

Using Eq. (13.136), Eq. (13.135) can be expressed as

$$W_m(a, \theta) = C_{1m} J_m(\omega a) \cos m\theta + C_{2m} J_m(\omega a) \sin m\theta = 0, \qquad m = 0, 1, 2, \ldots \qquad (13.137)$$

The Bessel function of the first kind, $J_m(\omega r)$, is given by [2],[3]

$$J_m(\omega r) = \sum_{i=0}^{\infty} \frac{(-1)^i}{i! \, \Gamma(m + i + 1)} \left(\frac{\omega r}{2}\right)^{m+2i} \qquad (13.138)$$

Equation (13.137) has to be satisfied for all values of θ. It can be satisfied only if

$$J_m(\omega a) = 0, \qquad m = 0, 1, 2, \ldots \qquad (13.139)$$

This is the frequency equation, which has an infinite number of discrete solutions, ω_{mn}, for each value of m. Although $\omega a = 0$ is a root of Eq. (13.139) for $m \geq 1$, this leads to the trivial solutions $w = 0$, and hence we need to consider roots with $\omega a \geq 0$. Some of the roots of Eq. (13.139) are given below [2],[3].
For $m = 0$, $J_0(\omega a) = 0$:

$$\gamma = \omega a = 2.405, 5.520, 8.654, 11.792, 14.931, 18.071,$$

$$21.212, 24.353, \ldots$$

For $m = 1$, $J_1(\omega a) = 0$:

$$\gamma = \omega a = 3.832, 7.016, 10.173, 13.323, 16.470,$$

$$19.616, 22.760, 25.903, \ldots$$

For $m = 2$, $J_2(\omega a) = 0$:

$$\gamma = \omega a = 5.135, 8.417, 11.620, 14.796, 17.960,$$

$$21.117, 24.270, 27.421, \ldots$$

For $m = 3$, $J_3(\omega a) = 0$:

$$\gamma = \omega a = 6.379, 9.760, 13.017, 16.224, 19.410,$$

$$22.583, 25.749, 28.909, \ldots$$

For $m = 4$, $J_4(\omega a) = 0$:

$$\gamma = \omega a = 7.586, 11.064, 14.373, 17.616, 20.827,$$

$$24.018, 27.200, 30.371, \ldots$$

For $m = 5$, $J_5(\omega a) = 0$:

$$\gamma = \omega a = 8.780, 12.339, 15.700, 18.982, 22.220,$$

$$25.431, 28.628, 31.813, \ldots$$

It can be seen from Eqs. (13.134) and (13.135) that the general solution of $w(r, \theta, t)$ becomes complicated in view of the various combinations of J_m, $\sin m\theta$, $\cos m\theta$, $\sin \omega_{mn} t$, and $\cos \omega_{mn} t$ involved for each value of $m = 0, 1, 2, \ldots$ Hence, the solution is usually expressed in terms of two characteristic functions $W_{mn}^{(1)}(r, \theta)$ and $W_{mn}^{(2)}(r, \theta)$ as indicated below. If γ_{mn} denotes the nth solution or root of $J_m(\gamma) = 0$, the natural frequencies can be expressed as

$$\omega_{mn} = \frac{\gamma_{mn}}{a} \tag{13.140}$$

Two characteristic functions $W_{mn}^{(1)}(r, \theta)$ and $W_{mn}^{(2)}(r, \theta)$ can be defined for any ω_{mn} as

$$W_{mn}^{(1)}(r, \theta) = C_{1mn} J_m(\omega_{mn} r) \cos m\theta$$

$$W_{mn}^{(2)}(r, \theta) = C_{2mn} J_m(\omega_{mn} r) \sin m\theta \tag{13.141}$$

It can be seen that for any given values of m and $n(m \neq 0)$ the two characteristic functions will have the same shape; they differ from one another only by an angular rotation of $90°$. Thus, the two natural modes of vibration corresponding to ω_{mn} are given by

$$w_{mn}^{(1)}(r, \theta, t) = J_m(\omega_{mn} r) \cos m\theta [A_{mn}^{(1)} \cos \omega_{mn} t + A_{mn}^{(2)} \sin \omega_{mn} t]$$

$$w_{mn}^{(2)}(r, \theta, t) = J_m(\omega_{mn} r) \sin m\theta [A_{mn}^{(3)} \cos \omega_{mn} t + A_{mn}^{(4)} \sin \omega_{mn} t] \tag{13.142}$$

The general solution of Eq. (13.120) can be expressed as

$$w(r, \theta, t) = \sum_{m=0}^{\infty} \sum_{n=0}^{\infty} [w_{mn}^{(1)}(r, \theta, t) + w_{mn}^{(2)}(r, \theta, t)] \tag{13.143}$$

where the constants $A_{mn}^{(1)}, \ldots, A_{mn}^{(4)}$ can be determined from the initial conditions.

13.6.3   Mode Shapes

It can be seen from Eq. (13.141) that the characteristic functions or normal modes will have the same shape for any given values of m and n, and differ from one another only by an angular rotation of $90°$. In the mode shapes given by Eq. (13.141), the value of m determines the number of nodal diameters. The value of n, indicating the order of the root or the zero of the Bessel function, denotes the number of nodal circles in the mode shapes. The nodal diameters and nodal circles corresponding to $m, n = 0, 1$, and

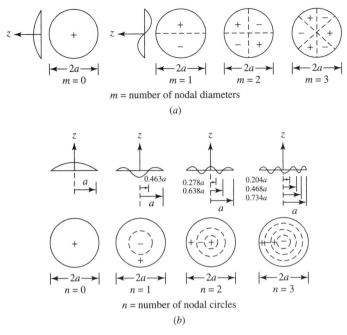

m = number of nodal diameters

(a)

n = number of nodal circles

(b)

Figure 13.11 Mode shapes of a clamped circular membrane.

2 are shown in Fig. 13.11. When $m = 0$, there will be no diametrical nodal lines but there will be n circular nodal lines, including the boundary of the membrane. When $m = 1$, there will be one diametrical node and n circular nodes, including the boundary. In general, the mode shape W_{mn} has m equally spaced diametrical nodes and n circular nodes (including the boundary) of radius $r_i = (\omega_{mi}/\omega_{mn})a$, $i = 1, 2, \ldots, n$. The mode shapes corresponding to a few combinations of m and n are shown in Fig. 13.11.

13.7 FORCED VIBRATION OF CIRCULAR MEMBRANES

The equation of motion governing the forced vibration of a circular membrane is given by Eq. (13.119):

$$P\left(\frac{\partial^2 w}{\partial r^2} + \frac{1}{r}\frac{\partial w}{\partial r} + \frac{1}{r^2}\frac{\partial^2 w}{\partial \theta^2}\right) + f = \rho\frac{\partial^2 w}{\partial t^2} \qquad (13.144)$$

Using modal analysis, the solution of Eq. (13.144) is assumed in the form

$$w(r, \theta, t) = \sum_{m=0}^{\infty} \sum_{n=0}^{\infty} W_{mn}(r, \theta)\eta_{mn}(t) \qquad (13.145)$$

where $W_{mn}(r, \theta)$ are the natural modes of vibration and $\eta_{mn}(t)$ are the corresponding generalized coordinates. For specificity we consider a circular membrane clamped at the edge, $r = a$. For this, the eigenfunctions are given by Eq. (13.141), since two eigenfunctions are used for any ω_{mn}, the modes will be degenerate except when $m = 0$, and we rewrite Eq. (13.145) as

$$w(r, \theta, t) = \sum_{n=1}^{\infty} W_{0n}^{(1)}(r, \theta)\eta_{0n}^{(1)}(t) + \sum_{m=1}^{\infty}\sum_{n=1}^{\infty} W_{mn}^{(1)}(r, \theta)\eta_{mn}^{(1)}(t)$$

$$+ \sum_{m=1}^{\infty}\sum_{n=1}^{\infty} W_{mn}^{(2)}(r, \theta)\eta_{mn}^{(2)}(t) \tag{13.146}$$

where the normal modes $W_{0n}^{(1)}$, $W_{mn}^{(1)}$, and $W_{mn}^{(2)}$ are given by Eq. (13.141). The normal modes can be normalized as

$$\iint_A \rho[W_{mn}^{(1)}(r, \theta)]^2 \, dA = \int_0^{2\pi}\int_0^a \rho C_{1mn}^2 J_m^2(\omega_{mn}r)\cos^2 m\theta \, r \, dr \, d\theta$$

$$= \frac{\pi}{2}\rho C_{1mn}^2 a^2 J_{m+1}^2(\omega_{mn}a) = 1 \tag{13.147}$$

where A is the area of the circular membrane. Equation (13.147) yields

$$C_{1mn}^2 = \frac{2}{\pi\rho a^2 J_{m+1}^2(\omega_{mn}a)} \tag{13.148}$$

$$\iint_A \rho[W_{mn}^{(2)}(r, \theta)]^2 \, dA = \int_0^{2\pi}\int_0^a \rho C_{2mn}^2 J_m^2(\omega_{mn}r)\sin^2 m\theta \, r \, dr \, d\theta$$

$$= \frac{\pi}{2}\rho C_{2mn}^2 a^2 J_{m+1}^2(\omega_{mn}a) = 1 \tag{13.149}$$

or

$$C_{2mn}^2 = \frac{2}{\pi\rho a^2 J_{m+1}^2(\omega_{mn}a)} \tag{13.150}$$

Thus, the normalized normal modes can be expressed as

$$\left.\begin{matrix} W_{mn}^{(1)} \\ W_{mn}^{(2)} \end{matrix}\right\} = \frac{\sqrt{2}}{\sqrt{\pi\rho}\, a J_{m+1}(\omega_{mn}a)} J_m(\omega_{mn}r) \left.\begin{matrix} \cos m\theta \\ \sin n\theta \end{matrix}\right\},$$

$$m = 0, 1, 2, \ldots, \quad n = 1, 2, 3, \ldots \tag{13.151}$$

When Eq. (13.145) is used, Eq. (13.144) yields the equations

$$\ddot{\eta}_{mn}(t) + \omega_{mn}^2\eta_{mn}(t) = N_{mn}(t) \tag{13.152}$$

where $N_{mn}(t)$ denotes the generalized force given by

$$N_{mn}(t) = \int_0^{2\pi}\int_0^a W_{mn}(r, \theta)f(r, \theta, t)r \, dr \, d\theta \tag{13.153}$$

Neglecting the contribution due to initial conditions, the generalized coordinates can be expressed as [see Eq. (2.109)]

$$\eta_{0n}(t) = \frac{1}{\omega_{0n}} \int_0^t N_{0n}^{(1)}(\tau) \sin \omega_{0n}(t - \tau)\, d\tau \tag{13.154}$$

$$\eta_{mn}^{(1)}(t) = \frac{1}{\omega_{mn}} \int_0^t N_{mn}^{(1)}(\tau) \sin \omega_{mn}(t - \tau)\, d\tau \tag{13.155}$$

$$\eta_{mn}^{(2)}(t) = \frac{1}{\omega_{mn}} \int_0^t N_{mn}^{(2)}(\tau) \sin \omega_{mn}(t - \tau)\, d\tau \tag{13.156}$$

where the natural frequencies ω_{mn} are given by Eq. (13.140), and the generalized forces by Eq. (13.153).

Example 13.4   Find the steady-state response of a circular membrane of radius a subjected to a harmonically varying uniform pressure all over the surface area, as shown in Fig. 13.12. Assume that the membrane is fixed at the boundary, $r = a$.

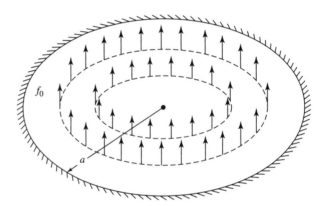

Figure 13.12   Circular membrane subjected to uniform pressure.

SOLUTION   The load acting on the membrane can be described as

$$f(r, \theta, t) = f_0 \cos \Omega t, \qquad 0 \le r \le a, \quad 0 \le \theta \le 2\pi \tag{E13.4.1}$$

The transverse displacement of the membrane can be expressed in terms of the normal modes, $W_{mn}(r, \theta)$, as

$$w(r, \theta, t) = \sum_{m=0}^{\infty} \sum_{n=1}^{\infty} W_{mn}(r, \theta) \eta_{mn}(t) \tag{E13.4.2}$$

where the $W_{mn}(r, \theta)$ for a circular membrane clamped at $r = a$ are given by Eq. (13.151). Equation (E10.4.2) can be rewritten in view of Eq. (13.141) as

$$w(r, \theta, t) = \sum_{n=1}^{\infty} W_{0n}^{(1)}(r, \theta) \eta_{0n}^{(1)}(t) + \sum_{m=1}^{\infty} \sum_{n=1}^{\infty} W_{mn}^{(1)}(r, \theta) \eta_{mn}^{(1)}(t)$$

$$+ \sum_{m=1}^{\infty} \sum_{n=1}^{\infty} W_{mn}^{(2)}(r, \theta) \eta_{mn}^{(2)}(t) \tag{E13.4.3}$$

where the generalized coordinates $\eta_{0n}^{(1)}(t)$, $\eta_{mn}^{(1)}(t)$, and $\eta_{mn}^{(2)}(t)$ are given by Eqs. (13.154)–(13.156). Using, from Eq. (13.151),

$$W_{0n}^{(1)} = \frac{\sqrt{2}}{\sqrt{\pi\rho} a J_1(\omega_{0n} a)} J_0(\omega_{0n} r)) \tag{E13.4.4}$$

$$W_{mn}^{(1)} = \frac{\sqrt{2}}{\sqrt{\pi\rho} a J_{m+1}(\omega_{mn} a)} J_m(\omega_{mn} r) \cos m\theta \tag{E13.4.5}$$

$$W_{mn}^{(2)} = \frac{\sqrt{2}}{\sqrt{\pi\rho} a J_{m+1}(\omega_{mn} a)} J_m(\omega_{mn} r) \sin n\theta \tag{E13.4.6}$$

the generalized forces can be evaluated as

$$N_{0n}^{(1)}(t) = \int_0^{2\pi} \int_0^a W_{0n}^{(1)}(r, \theta) f(r, \theta, t) r \, dr \, d\theta = \int_0^{2\pi} \int_0^a W_{0n}^{(1)}(r, \theta) f_0 \cos \Omega t \, r \, dr \, d\theta$$

$$= \int_0^{2\pi} \int_0^a \frac{\sqrt{2} f_0}{\sqrt{\pi\rho} a J_1(\omega_{0n} a)} J_0(\omega_{0n} r) \cos \Omega t \, r \, dr \, d\theta$$

$$= \frac{2\pi f_0 \cos \Omega t}{\sqrt{\pi\rho} J_1(\omega_{0n} a)} \frac{1}{\omega_{0n}} J_0(\omega_{0n} a) \tag{E13.4.7}$$

$$N_{mn}^{(1)}(t) = \int_0^{2\pi} \int_0^a W_{mn}^{(1)}(r, \theta) f(r, \theta, t) r \, dr \, d\theta$$

$$= \int_0^{2\pi} \int_0^a \frac{\sqrt{2} f_0}{\sqrt{\pi\rho} a J_{m+1}(\omega_{mn} a)} J_m(\omega_{mn} r) \cos m\theta \cos \Omega t \, r \, dr \, d\theta = 0 \tag{E13.4.8}$$

$$N_{mn}^{(2)}(t) = \int_0^{2\pi} \int_0^a W_{mn}^{(2)}(r, \theta) f(r, \theta, t) r \, dr \, d\theta$$

$$= \int_0^{2\pi} \int_0^a \frac{\sqrt{2} f_0}{\sqrt{\pi\rho} a J_{m+1}(\omega_{mn} a)} J_m(\omega_{mn} r) \sin n\theta \cos \Omega t \, r \, dr \, d\theta = 0 \tag{E13.4.9}$$

In view of Eqs. (E13.4.7)–(E13.4.9), Eqs. (13.154)–(13.156) become

$$\eta_{0n}^{(1)}(t) = \frac{2\pi f_0}{\sqrt{\pi\rho} \omega_{0n}^2} \frac{J_0(\omega_{0n} a)}{J_1(\omega_{0n} a)} \int_0^t \cos \Omega \tau \sin \omega_{0n}(t - \tau) \, d\tau$$

$$= \frac{2\pi f_0 J_0(\omega_{0n} a)(\cos \omega_{0n} t - \cos \Omega t)}{\sqrt{\pi\rho} \omega_{0n}(\Omega^2 - \omega_{0n}^2) J_1(\omega_{0n} a)} \tag{E13.4.10}$$

$$\eta_{mn}^{(1)}(t) = 0 \tag{E13.4.11}$$

$$\eta_{mn}^{(2)}(t) = 0 \tag{E13.4.12}$$

Thus, the response of the membrane can be expressed, using Eqs. (13.145), (13.151), and (E13.4.10)–(E13.4.12), as

$$w(r, \theta, t) = \frac{2\sqrt{2} f_0}{\rho a} \sum_{n=1}^{\infty} \frac{J_0(\omega_{0n} a) J_0(\omega_{0n} r)}{J_1^2(\omega_{0n} a) \omega_{0n} (\Omega^2 - \omega_{0n}^2)} (\cos \omega_{0n} t - \cos \Omega t) \tag{E13.4.13}$$

13.8 MEMBRANES WITH IRREGULAR SHAPES

The known natural frequencies of vibration of rectangular and circular membranes can be used to estimate the natural frequencies of membranes having irregular boundaries. For example, the natural frequencies of a regular polygon are expected to lie in between those of the inscribed and circumscribed circles. Rayleigh presented an analysis to find the effect of a departure from the exact circular shape on the natural frequencies of vibration of uniform membranes. The results of the analysis indicate that for membranes of fairly regular shape, the fundamental or lowest natural frequency of vibration can be approximated as

$$f = \alpha \sqrt{\frac{T}{\rho A}} \tag{13.157}$$

where T is the tension, ρ is the density per unit area, A is the surface area, and α is a factor whose values are given in Table 13.1 for several irregular shapes. The factors given in the table 13.1 indicate, for instance, that for the same values of tension, density, and surface area, the fundamental natural frequency of vibration of a square membrane is larger than that of a circular membrane by the factor $4.443/4.261 = 1.043$.

Table 13.1 Values of the Factor α in Eq. (13.157)

Shape of the membrane	α
Square	4.443
Rectangle with $b/a = 2$	4.967
Rectangle with $b/a = 3$	5.736
Rectangle with $b/a = 3/2$	4.624
Circle	4.261
Semicircle	4.803
Quarter circle	4.551
$60°$ sector of a circle	4.616
Equilateral triangle	4.774
Isosceles right triangle	4.967

Source:
Ref. [8]

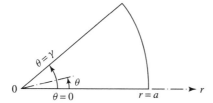

Figure 13.13   Circular sector membrane.

13.9   PARTIAL CIRCULAR MEMBRANES

Consider a membrane in the form of a circular sector of radius a as shown in Fig. 13.13. Let the membrane be fixed on all three edges. When the zero-displacement conditions along the edges $\theta = 0$ and $\theta = \gamma$ are used in the general solution of Eq. (13.134), we find that the solution becomes

$$w(r, \theta) = C J_{n\pi/\gamma}(\lambda r) \sin \frac{n\pi\theta}{\beta} \cos(c\lambda t + \delta) \qquad (13.158)$$

where C is a constant and n is an integer. To satisfy the boundary condition along the edge $r = a$, Eq. (13.158) is set equal to zero at $r = a$. This leads to the frequency equation

$$J_{n\pi/\gamma}(\lambda a) = 0 \qquad (13.159)$$

If $\lambda_i a$ denotes the ith root of Eq. (13.159), the corresponding natural frequency of vibration can be computed as

$$\omega_i = \frac{\lambda_i a}{a} = \lambda_i \qquad (13.160)$$

For example, for a semicircular membrane, $\gamma = \pi$ and for $n = 1$, Eq. (13.159) becomes

$$J_1(\lambda a) = 0 \qquad (13.161)$$

whose roots are given by $\lambda_1 a = 3.832$, $\lambda_2 a = 7.016$, $\lambda_3 a = 10.173$, and $\lambda_4 a = 13.324$, …. Thus, the natural frequencies of vibration of the semicircular membrane will be $\omega_1 = 3.832/a$, $\omega_2 = 7.016/a$, $\omega_3 = 10.173/a$, ….

13.10   RECENT CONTRIBUTIONS

Spence and Horgan [11] derived bounds on the natural frequencies of composite circular membranes using an integral equation method. The membrane was assumed to have a stepped radial density. Although such problems, involving discontinuous coefficients in the differential equation, can be treated using the classical variational methods, it was shown that an eigenvalue estimation technique based on an integral formulation is more efficient. For a comparable amount of effort, the integral equation method is expected to provide more accurate bounds on the natural frequencies.

The transient response of hanging curtains clamped at three edges was considered by Yamada et al. [12]. A hanging curtain was replaced by an equivalent membrane for deriving the equation of motion. The free vibration of the membrane was analyzed theoretically and its transient response when subjected to a rectangularly varying point force was also studied by using Galerkin's method. The forced vibration response of a uniform membrane of arbitrary shape under an arbitrary distribution of time-dependent excitation with arbitrary initial conditions and time-dependent boundary conditions was found by Olcer [13]. Leissa and Ghamat-Rezaei [14] presented the vibration frequencies and mode shapes of rectangular membranes subjected to shear stresses and/or nonuniform tensile stresses. The solution is found using the Ritz method, with the transverse displacement in the form of a double series of trigonometric functions.

The scalar wave equation of an annular membrane in which the motion is symmetrical about the origin was solved for arbitrary initial and boundary conditions by Sharp [15]. The solution was obtained using a finite Hankel transform. A simple example was given and its solution was compared with one given by the method of separation of variables. The vibration of a loaded kettledrum was considered by De [16]. In this work, the author discussed the effect of the applied mass load at a point on the frequency of a vibrating kettledrum. In a method of obtaining approximations of the natural frequencies of membranes was developed by Torvik and Eastep [17], an approximate expression for the radius of the bounding curve is first written as a truncated Fourier series. The deflection, expressed as a superposition of the modes of the circular membrane, is forced to vanish approximately on the approximated boundary. This leads to a system of linear homogeneous equations in terms of the amplitudes of the modes of the circular membrane. By equating the determinant of coefficients to zero, the approximate frequencies are found. Some exact solutions of the vibration of nonhomogeneous membranes were presented by Wang [18], including the exact solutions of a rectangular membrane with a linear density variation and a nonhomogeneous annular membrane with inverse square density distribution.

REFERENCES

1. K. F. Graff, *Wave Motion in Elastic Solids,* Ohio State University Press, Columbus, OH, 1975.

2. N. W. McLachlan, *Bessel Functions for Engineers,* Oxford University Press, New York, 1934.

3. G. N. Watson, *Theory of Bessel Functions,* Cambridge University Press, London, 1922.

4. L. Meirovitch, *Analytical Methods in Vibrations*, Macmillan, New York, 1967.

5. S. Timoshenko, D. H. Young, and W. Weaver, Jr., *Vibration Problems in Engineering,* 4th ed., Wiley, New York, 1974.

6. E. Volterra and E. C. Zachmanoglou, *Dynamics of Vibrations,* Charles E. Merrill, Columbus, OH, 1965.

7. S. K. Clark, *Dynamics of Continuous Elements*, Prentice-Hall, Englewood Cliffs, NJ, 1972.

8. Lord Rayleigh, *The Theory of Sound,* 2nd ed., Vol. 1, Dover, New York, 1945.

9. A. Jeffrey, *Advanced Engineering Mathematics*, Harcourt/Academic Press, San Diego, CA, 2002.

10. P. M. Morse, *Vibration and Sound*, McGraw-Hill, New York, 1936.

11. J. P. Spence and C. O. Horgan, Bounds on natural frequencies of composite circular membranes: integral equation methods, *Journal of Sound and Vibration*, Vol. 87, No. 1, pp. 71–81, 1983.

12. G. Yamada, Y. Kobayashi, and H. Hamaya, Transient response of a hanging curtain, *Journal of Sound and Vibration*, Vol. 130, No. 2, pp. 223–235, 1989.

13. N. Y. Olcer, General solution to the equation of the vibrating membrane, *Journal of Sound and Vibration*, Vol. 6, No. 3, pp. 365–374, 1967.

14. A. W. Leissa and A. Ghamat-Rezaei, Vibrations of rectangular membranes subjected to shear and nonuniform tensile stresses, *Journal of the Acoustical Society of America*, Vol. 88, No. 1, pp. 231–238, 1990.

15. G. R. Sharp, Finite transform solution of the symmetrically vibrating annular membrane, *Journal of Sound and Vibration*, Vol. 5, No. 1, pp. 1–8, 1967.

16. S. De, Vibrations of a loaded kettledrum, *Journal of Sound and Vibration*, Vol. 20, No. 1, pp. 79–92, 1972.

17. P. J. Torvik and F. E. Eastep, A method for improving the estimation of membrane frequencies, *Journal of Sound and Vibration*, Vol. 21, No. 3, pp. 285–294, 1972.

18. C. Y. Wang, Some exact solutions of the vibration of non-homogeneous membranes, *Journal of Sound and Vibration*, Vol. 210, No. 4, pp. 555–558, 1998.

PROBLEMS

13.1 Starting from the free-body diagram of an element of a membrane in polar coordinates, derive the equation of motion of a vibrating membrane in polar coordinates using the equilibrium approach.

13.2 Derive the expression for the Laplacian operator in polar coordinates starting from the relation

$$\nabla^2 = \frac{\partial^2}{\partial x^2} + \frac{\partial^2}{\partial y^2}$$

and using the coordinate transformation relations $x = r\cos\theta$ and $y = r\sin\theta$.

13.3 Show that each side of Eqs. (13.38) and (13.123) is equal to a negative constant.

13.4 Consider a rectangular membrane with the ratio of sides a and b equal to $K = a/b = \frac{4}{3}$. Find the distinct natural frequencies ω_{mn} and ω_{ij} that will have the same magnitude.

13.5 Find the forced vibration response of a rectangular membrane of sides a and b subjected to a suddenly applied uniformly distributed pressure f_0 per unit area. Assume zero initial conditions and the membrane to be fixed around all the edges.

13.6 Find the steady-state response of a rectangular membrane of sides a and b subjected to a harmonic force

$F_0 \cos \Omega t$ at the point $(x = 2a/3, y = b/3)$. Assume the membrane to be clamped on all four edges.

13.7 Derive the equation of motion of a membrane in polar coordinates using a variational approach.

13.8 A thin sheet of steel of thickness 0.01 mm is stretched over a rectangular metal framework of size 25 mm × 50 mm under a tension of 2 kN per unit length of periphery. Determine the first three natural frequencies of vibration and the corresponding mode shapes of the sheet. Assume the density of steel sheet to be 76.5 kN/m³.

13.9 Solve Problem 13.8 by assuming the sheet to be aluminum instead of steel, with a density of 26.5 kN/m³.

13.10 A thin sheet of steel of thickness 0.01 mm is stretched over a circular metal framework of diameter 50 mm under a tension of 2 kN per unit length of periphery. Determine the first three natural frequencies of vibration and the corresponding mode shapes of symmetric vibration of the sheet. Assume the density of the steel sheet to be 76.5 kN/m³.

13.11 Solve Problem 13.10 by assuming the sheet to be aluminum instead of steel, with a density of 26.5 kN/m³.

13.12 Derive the frequency equation of an annular membrane of inner radius r_1 and outer radius r_2 fixed at both edges.

13.13 Find the free vibration response of a rectangular membrane of sides a and b subjected to the following initial conditions:

$$w(x, y, 0) = w_0 \sin \frac{\pi x}{a} \sin \frac{\pi y}{b}, \qquad \frac{\partial w}{\partial t}, (x, y, 0) = 0$$

Assume that the membrane is fixed on all the sides.

13.14 Find the free vibration response of a rectangular membrane of sides a and b subjected to the following initial conditions:

$$w(x, y, 0) = 0, \qquad \frac{\partial w}{\partial t}(x, y, 0) = \dot{w}_0$$

Assume that the membrane is fixed on all sides.

13.15 Find the steady-state response of a rectangular membrane fixed on all sides subjected to the force $f(x, y, t) = f_0 \sin \Omega t$.

13.16 Find the free vibration response of a circular membrane of radius a subjected to the following conditions:

$$w(r, \theta, 0) = w_0 r, \qquad \frac{\partial w}{\partial t}(r, \theta, 0) = 0$$

Assume that the membrane is fixed at the outer edge, $r = a$.

13.17 Find the response of a rectangular membrane of sides a and b fixed on all four sides when subjected to an impulse $\hat{G}$ at $(x = x_0, y = y_0)$.

13.18 Find the steady-state response of a circular membrane of radius a when subjected to the force $F_0 \sin \Omega t$ at $r = 0$.

13.19 Derive the frequency equation of an annular membrane of inner radius b and outer radius a assuming a clamped inner edge and free outer edge.

13.20 Derive the frequency equation of an annular membrane of inner radius b and outer radius a assuming that it is free at both the inner and outer edges.

14

Transverse Vibration of Plates

14.1 INTRODUCTION

A plate is a solid body bounded by two surfaces. The distance between the two surfaces defines the thickness of the plate, which is assumed to be small compared to the lateral dimensions, such as the length and width in the case of a rectangular plate and the diameter in the case of a circular plate. A plate is usually considered to be thin when the ratio of its thickness to the smaller lateral dimension (such as width in the case of a rectangular plate and diameter in the case of a circular plate) is less than $\frac{1}{20}$. The vibration of plates is important in the study of practical systems such as bridge decks, hydraulic structures, pressure vessel covers, pavements of highways and airport runways, ship decks, airplanes, missiles, and machine parts. The theory of elastic plates is an approximation of the three-dimensional elasticity theory to two dimensions, which permits a description of the deformation of every point in the plate in terms of only the deformation of the midplane of the plate. The equations of motion of plates are derived using the thin plate theory as well as Mindlin theory, which considers the effects of rotary inertia and shear deformation. Free and forced vibration of rectangular and circular plates are considered. The vibration of plates with variable thickness, of plates on elastic foundation, and of plates subjected to in-plane loads is also outlined.

14.2 EQUATION OF MOTION: CLASSICAL PLATE THEORY

14.2.1 Equilibrium Approach

The small deflection theory of thin plates, called *classical plate theory* or *Kirchhoff theory*, is based on assumptions similar to those used in thin beam or Euler–Bernoulli beam theory. The following assumptions are made in thin or classical plate theory:

1. The thickness of the plate (h) is small compared to its lateral dimensions.
2. The middle plane of the plate does not undergo in-plane deformation. Thus, the midplane remains as the neutral plane after deformation or bending.
3. The displacement components of the midsurface of the plate are small compared to the thickness of the plate.
4. The influence of transverse shear deformation is neglected. This implies that plane sections normal to the midsurface before deformation remain normal to the midsurface even after deformation or bending. This assumption implies that the transverse shear strains, ε_{xz} and ε_{yz}, are negligible, where z denotes the thickness direction.
5. The transverse normal strain ε_{zz} under transverse loading can be neglected. The transverse normal stress σ_{zz} is small and hence can be neglected compared to the other components of stress.

The equation of motion for the transverse vibration of a thin plate has been derived using an equilibrium approach in Section 3.6. Some of the important relations and equations are summarized below for a rectangular plate (see Fig. 3.3).

Moment resultant–transverse displacement relations:

$$M_x = -D\left(\frac{\partial^2 w}{\partial x^2} + \nu\frac{\partial^2 w}{\partial y^2}\right) \tag{14.1}$$

$$M_y = -D\left(\frac{\partial^2 w}{\partial y^2} + \nu\frac{\partial^2 w}{\partial x^2}\right) \tag{14.2}$$

$$M_{xy} = M_{yx} = -(1-\nu)D\frac{\partial^2 w}{\partial x\,\partial y} \tag{14.3}$$

in which D represents the flexural rigidity of the plate:

$$D = \frac{Eh^3}{12(1-\nu^2)} \tag{14.4}$$

where h is the thickness, E is Young's modulus, and ν is Poisson's ratio of the plate.

Shear force resultants:

$$Q_x = \frac{\partial M_x}{\partial x} + \frac{\partial M_{xy}}{\partial y} = -D\frac{\partial}{\partial x}\left(\frac{\partial^2 w}{\partial x^2} + \frac{\partial^2 w}{\partial y^2}\right) \tag{14.5}$$

$$Q_y = \frac{\partial M_y}{\partial y} + \frac{\partial M_{xy}}{\partial x} = -D\frac{\partial}{\partial y}\left(\frac{\partial^2 w}{\partial x^2} + \frac{\partial^2 w}{\partial y^2}\right) \tag{14.6}$$

Equation of motion (force equilibrium in the z direction):

$$\frac{\partial Q_x}{\partial x} + \frac{\partial Q_y}{\partial y} + f(x,y,t) = \rho h\frac{\partial^2 w}{\partial t^2} \tag{14.7}$$

or

$$D\left(\frac{\partial^4 w}{\partial x^4} + 2\frac{\partial^4 w}{\partial x^2\,\partial y^2} + \frac{\partial^4 w}{\partial y^4}\right) + \rho h\frac{\partial^2 w}{\partial t^2} = f(x,y,t) \tag{14.8}$$

where ρ is the density of the plate and f is the distributed transverse load acting on the plate per unit area.

14.2.2 Variational Approach

Because of assumptions 4 and 5 in Section 14.2.1, the state of stress in a thin plate can be assumed to be plane stress. Thus, the nonzero stresses induced in a thin plate are given by σ_{xx}, σ_{yy}, and σ_{xy}. The strain energy density (π_0) of the plate can be expressed as

$$\pi_0 = \tfrac{1}{2}(\sigma_{xx}\varepsilon_{xx} + \sigma_{yy}\varepsilon_{yy} + \sigma_{xy}\varepsilon_{xy}) \tag{14.9}$$

The strain components can be expressed in terms of the transverse displacement of the middle surface of the plate, $w(x,y)$, as follows:

$$\varepsilon_{xx} = \frac{\partial u}{\partial x} = -z\frac{\partial^2 w}{\partial x^2} \tag{14.10}$$

$$\varepsilon_{yy} = \frac{\partial v}{\partial y} = -z\frac{\partial^2 w}{\partial y^2} \tag{14.11}$$

$$\varepsilon_{xy} = \frac{\partial u}{\partial y} + \frac{\partial v}{\partial x} = -2z\frac{\partial^2 w}{\partial x\,\partial y} \tag{14.12}$$

$$\varepsilon_{zz} = \frac{\partial w}{\partial z} \approx 0, \qquad \varepsilon_{xz} = \frac{\partial u}{\partial z} + \frac{\partial w}{\partial x} = 0, \qquad \varepsilon_{yz} = \frac{\partial v}{\partial z} + \frac{\partial w}{\partial y} = 0 \tag{14.13}$$

The stress–strain relations permit stresses to be expressed in terms of the transverse displacement, $w(x,y)$, a

$$\sigma_{xx} = \frac{E}{1 - v^2}(\varepsilon_{xx} + v\varepsilon_{yy}) = -\frac{Ez}{1 - v^2}\left(\frac{\partial^2 w}{\partial x^2} + v\frac{\partial^2 w}{\partial y^2}\right) \tag{14.14}$$

$$\sigma_{yy} = \frac{E}{1 - v^2}(\varepsilon_{yy} + v\varepsilon_{xx}) = -\frac{Ez}{1 - v^2}\left(\frac{\partial^2 w}{\partial y^2} + v\frac{\partial^2 w}{\partial x^2}\right) \tag{14.15}$$

$$\sigma_{xy} = G\varepsilon_{xy} = \frac{E}{2(1 + v)}\varepsilon_{xy} = -2Gz\frac{\partial^2 w}{\partial x\,\partial y} = -\frac{Ez}{1 + v}\frac{\partial^2 w}{\partial x\,\partial y} \tag{14.16}$$

By substituting Eqs. (14.10)–(14.12) and Eqs. (14.14)–(14.16) into Eq. (14.9), the strain energy density can be written in terms of w as

$$\pi_0 = \frac{Ez^2}{2(1 - v^2)}\left[\left(\frac{\partial^2 w}{\partial x^2}\right)^2 + \left(\frac{\partial^2 w}{\partial y^2}\right)^2 + 2v\frac{\partial^2 w}{\partial x^2}\frac{\partial^2 w}{\partial y^2} + 2(1 - v)\left(\frac{\partial^2 w}{\partial x\,\partial y}\right)^2\right] \tag{14.17}$$

Integrating Eq. (14.17) over the volume of the plate (V), the strain energy of bending can be obtained as

$$\pi = \iiint_V \pi_0\,dV = \iint_A dA \int_{z=-h/2}^{h/2} \pi_0\,dz = \frac{E}{2(1 - v^2)}\iint_A\left[\left(\frac{\partial^2 w}{\partial x^2}\right)^2 + \left(\frac{\partial^2 w}{\partial y^2}\right)^2\right.$$

$$\left. + 2v\frac{\partial^2 w}{\partial x^2}\frac{\partial^2 w}{\partial y^2} + 2(1 - v)\left(\frac{\partial^2 w}{\partial x\,\partial y}\right)^2\right]dA \int_{z=-h/2}^{h/2} z^2\,dz \tag{14.18}$$

where $dV = dA\,dz$ denotes the volume of an infinitesimal element of the plate. Noting that

$$\frac{E}{1 - v^2}\int_{z=-h/2}^{h/2} z^2\,dz = \frac{Eh^3}{12(1 - v^2)} \tag{14.19}$$

is the flexural rigidity of the plate (D), Eq. (14.18) can be rewritten as

$$\pi = \frac{D}{2}\iint_A\left\{\left(\frac{\partial^2 w}{\partial x^2} + \frac{\partial^2 w}{\partial y^2}\right)^2 - 2(1 - v)\left[\frac{\partial^2 w}{\partial x^2}\frac{\partial^2 w}{\partial y^2} - \left(\frac{\partial^2 w}{\partial x\,\partial y}\right)^2\right]\right\}dx\,dy \tag{14.20}$$

Considering only the transverse motion and neglecting the effect of rotary inertia, the kinetic energy of the plate (T) can be expressed as

$$T = \frac{\rho h}{2} \iint_A \left(\frac{\partial w}{\partial t}\right)^2 dx\, dy \tag{14.21}$$

If there is a distributed transverse load, $f(x, y, t)$, acting on the plate, the work done by the external load (W) is given by

$$W = \iint_A fw\, dx\, dy \tag{14.22}$$

The generalized Hamilton's principle can be used to derive the equations of motion:

$$\delta \int_{t_1}^{t_2} L\, dt = \delta \int_{t_1}^{t_2} (\pi - W - T)\, dt = 0 \tag{14.23}$$

Substituting Eqs. (14.20)–(14.22) into Eq. (14.23), Hamilton's principle can be written as

$$\delta \int_{t_1}^{t_2} \left(\frac{D}{2} \iint_A \left\{ (\nabla^2 w)^2 - 2(1-\nu) \left[\frac{\partial^2 w}{\partial x^2} \frac{\partial^2 w}{\partial y^2} \right. \right. \right.$$
$$\left. \left. \left. - \left(\frac{\partial^2 w}{\partial x \partial y} \right)^2 \right] \right\} dx\, dy - \frac{\rho h}{2} \iint_A \left(\frac{\partial w}{\partial t} \right)^2 dx\, dy - \iint_A fw\, dx\, dy \right) dt = 0$$
$$\tag{14.24}$$

where ∇^2 denotes the harmonic operator with

$$\nabla^2 w = \frac{\partial^2 w}{\partial x^2} + \frac{\partial^2 w}{\partial y^2} \tag{14.25}$$

Performing the variation of the first integral term in Eq. (14.24), we have

$$I_1 = \delta \int_{t_1}^{t_2} \frac{D}{2} \iint_A (\nabla^2 w)^2\, dx\, dy\, dt = D \int_{t_1}^{t_2} \iint_A \nabla^2 w \nabla^2 \delta w\, dx\, dy\, dt \tag{14.26}$$

By using the two-dimensional Green's theorem [40], Eq. (14.26) can be written as

$$I_1 = D \int_{t_1}^{t_2} \left\{ \iint_A \nabla^4 w \delta w\, dx\, dy + \int_C \left[\nabla^2 w \frac{\partial(\delta w)}{\partial n} - \delta w \frac{\partial(\nabla^2 w)}{\partial n} \right] dC \right\} dt \tag{14.27}$$

where C denotes the boundary of the plate, n indicates the outward drawn normal to the boundary, and ∇^4 represents the biharmonic operator, so that

$$\nabla^4 w = \nabla^2 (\nabla^2 w) = \frac{\partial^4 w}{\partial x^4} + 2\frac{\partial^4 w}{\partial x^2 \partial y^2} + \frac{\partial^4 w}{\partial y^4} \tag{14.28}$$

Note that the integration on the boundary, $\int_C$, extends all around the boundary of the plate. Similarly, we can express the variation of the second integral term in

Eq. (14.24) as

$$I_2 = \delta \int_{t_1}^{t_2} -D(1-v) \iint_A \left[\frac{\partial^2 w}{\partial x^2} \frac{\partial^2 w}{\partial y^2} - \left(\frac{\partial^2 w}{\partial x \, \partial y} \right)^2 \right] dx \, dy \, dt$$

$$= -D(1-v) \int_{t_1}^{t_2} \iint_A \left[\frac{\partial^2 w}{\partial x^2} \delta \left(\frac{\partial^2 w}{\partial y^2} \right) + \frac{\partial^2 w}{\partial y^2} \delta \left(\frac{\partial^2 w}{\partial x^2} \right) \right.$$

$$\left. -2 \frac{\partial^2 w}{\partial x \, \partial y} \delta \left(\frac{\partial^2 w}{\partial x \, \partial y} \right) \right] dx \, dy \, dt$$

$$= -D(1-v) \int_{t_1}^{t_2} \iint_A \left[\frac{\partial^2 w}{\partial x^2} \frac{\partial^2 (\delta w)}{\partial y^2} + \frac{\partial^2 w}{\partial y^2} \frac{\partial^2 (\delta w)}{\partial x^2} - 2 \frac{\partial^2 w}{\partial x \, \partial y} \frac{\partial^2 (\delta w)}{\partial x \, \partial y} \right] dx \, dy \, dt$$

$$(14.29)$$

Noting that the quantity under the sign of the area integral in Eq. (14.29) can be written as

$$\frac{\partial}{\partial x} \left[\frac{\partial^2 w}{\partial y^2} \frac{\partial (\delta w)}{\partial x} - \frac{\partial^2 w}{\partial x \, \partial y} \frac{\partial (\delta w)}{\partial y} \right] + \frac{\partial}{\partial y} \left[\frac{\partial^2 w}{\partial x^2} \frac{\partial (\delta w)}{\partial y} - \frac{\partial^2 w}{\partial x \, \partial y} \frac{\partial (\delta w)}{\partial x} \right]$$

$$= \frac{\partial h_1}{\partial x} + \frac{\partial h_2}{\partial y} \tag{14.30}$$

where

$$h_1 = \frac{\partial^2 w}{\partial y^2} \frac{\partial (\delta w)}{\partial x} - \frac{\partial^2 w}{\partial x \, \partial y} \frac{\partial (\delta w)}{\partial y} \tag{14.31}$$

$$h_2 = \frac{\partial^2 w}{\partial x^2} \frac{\partial (\delta w)}{\partial y} - \frac{\partial^2 w}{\partial x \, \partial y} \frac{\partial (\delta w)}{\partial x} \tag{14.32}$$

Equation (14.29) can be expressed as

$$I_2 = -D(1-v) \int_{t_1}^{t_2} \iint_A \left(\frac{\partial h_1}{\partial x} + \frac{\partial h_2}{\partial y} \right) dx \, dy \, dt \tag{14.33}$$

Using the relations

$$\iint_A \frac{\partial f}{\partial x} dx \, dy = \int_C f \cos \theta \, dC \tag{14.34}$$

$$\iint_A \frac{\partial f}{\partial y} dx \, dy = \int_C f \sin \theta \, dC \tag{14.35}$$

Eq. (14.33) can be rewritten as

$$I_2 = -D(1-v) \int_{t_1}^{t_2} \int_C (h_1 \cos \theta + h_2 \sin \theta) \, dC \tag{14.36}$$

where θ is the angle between the outward drawn normal to the boundary (n) and the x axis as shown in Fig. 14.1. The quantities $\partial (\delta w)/\partial x$ and $\partial (\delta w)/\partial y$ appearing in h_1

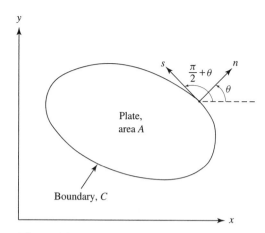

Figure 14.1   Normal to the boundary of plate.

and h_2 [Eqs. (14.31) and (14.32)] can be transformed into the new coordinate system (n, s) as follows:

$$\frac{\partial(\delta w)}{\partial x} = \frac{\partial(\delta w)}{\partial n}\frac{\partial n}{\partial x} + \frac{\partial(\delta w)}{\partial s}\frac{\partial s}{\partial x} = \frac{\partial(\delta w)}{\partial n}\cos\theta - \frac{\partial(\delta w)}{\partial s}\sin\theta \qquad (14.37)$$

$$\frac{\partial(\delta w)}{\partial y} = \frac{\partial(\delta w)}{\partial n}\frac{\partial n}{\partial y} + \frac{\partial(\delta w)}{\partial s}\frac{\partial s}{\partial y} = \frac{\partial(\delta w)}{\partial n}\sin\theta + \frac{\partial(\delta w)}{\partial s}\cos\theta \qquad (14.38)$$

where s is the tangential direction to the boundary. In view of Eqs. (14.37) and (14.38), Eqs. (14.31) and (14.32) can be expressed as

$$h_1 = \frac{\partial^2 w}{\partial y^2}\left[\frac{\partial(\delta w)}{\partial n}\cos\theta - \frac{\partial(\delta w)}{\partial s}\sin\theta\right] - \frac{\partial^2 w}{\partial x\,\partial y}\left[\frac{\partial(\delta w)}{\partial n}\sin\theta + \frac{\partial(\delta w)}{\partial s}\cos\theta\right]$$
$$(14.39)$$

$$h_2 = \frac{\partial^2 w}{\partial x^2}\left[\frac{\partial(\delta w)}{\partial n}\sin\theta + \frac{\partial(\delta w)}{\partial s}\cos\theta\right] - \frac{\partial^2 w}{\partial x\,\partial y}\left[\frac{\partial(\delta w)}{\partial n}\cos\theta - \frac{\partial(\delta w)}{\partial s}\sin\theta\right]$$
$$(14.40)$$

When Eqs. (14.39) and (14.40) are substituted, Eq. (14.36) becomes

$$I_2 = -D(1-v)\int_{t_1}^{t_2}\left\{\int_C \frac{\partial\delta w}{\partial n}\left(\frac{\partial^2 w}{\partial y^2}\cos^2\theta + \frac{\partial^2 w}{\partial x^2}\sin^2\theta - 2\frac{\partial^2 w}{\partial x\,\partial y}\sin\theta\cos\theta\right)dC\right.$$
$$\left. + \int_C \frac{\partial\delta w}{\partial s}\left[\left(\frac{\partial^2 w}{\partial x^2} - \frac{\partial^2 w}{\partial y^2}\right)\cos\theta\sin\theta + \frac{\partial^2 w}{\partial x\,\partial y}(\sin^2\theta - \cos^2\theta)\right]dC\right\}dt$$
$$(14.41)$$

The second integral involving integration with respect to C can be integrated by parts using the relation

$$\int_C \frac{\partial\delta w}{\partial s}g(x,y)\,dC = g(x,y)\delta w|_C - \int_C \delta w\frac{\partial g}{\partial s}\,dC \qquad (14.42)$$

where in the present case, $g(x, y)$ is given by

$$g(x, y) = \left(\frac{\partial^2 w}{\partial x^2} - \frac{\partial^2 w}{\partial y^2}\right) \cos\theta \sin\theta + \frac{\partial^2 w}{\partial x\, \partial y}(\sin^2\theta - \cos^2\theta) \tag{14.43}$$

Since the variation of displacement, δw, on the boundary is zero, Eq. (14.42) reduces to

$$\int_C \frac{\partial \delta w}{\partial s} g(x,y)\, dC = -\int_C \delta w \frac{\partial g}{\partial s}\, dC \tag{14.44}$$

Inserting Eq. (14.43) in (14.44) and the result in Eq. (14.41), we obtain

$$I_2 = -D(1-v) \int_{t_1}^{t_2} \left\{ \int_C \frac{\partial(\delta w)}{\partial n} \right.$$

$$\times \left(\frac{\partial^2 w}{\partial y^2} \cos^2\theta + \frac{\partial^2 w}{\partial x^2} \sin^2\theta - 2\frac{\partial^2 w}{\partial x\, \partial y} \sin\theta \cos\theta \right) dC$$

$$\left. + \int_C \delta w \frac{\partial}{\partial s}\left[\left(\frac{\partial^2 w}{\partial x^2} - \frac{\partial^2 w}{\partial y^2}\right)\cos\theta \sin\theta + \frac{\partial^2 w}{\partial x\, \partial y}(\sin^2\theta - \cos^2\theta)\right] dC \right\} dt \tag{14.45}$$

The variation of the third integral term in Eq. (14.24) can be expressed as

$$I_3 = -\delta \int_{t_1}^{t_2} \frac{\rho h}{2} \iint_A \left(\frac{\partial w}{\partial t}\right)^2 dx\, dy\, dt = -\frac{\rho h}{2} \iint_A \delta \int_{t_1}^{t_2} \left(\frac{\partial w}{\partial t}\right)^2 dx\, dy\, dt \tag{14.46}$$

By using integration by parts with respect to time, the integral I_3 can be written as

$$I_3 = -\rho h \iint_A \int_{t_1}^{t_2} \frac{\partial w}{\partial t} \frac{\partial(\delta w)}{\partial t}\, dx\, dy\, dt$$

$$= -\rho h \iint_A \left[\frac{\partial w}{\partial t} \delta w \Big|_{t_1}^{t_2} - \int_{t_1}^{t_2} \frac{\partial}{\partial t}\left(\frac{\partial w}{\partial t}\right) \delta w\, dt\right] dx\, dy \tag{14.47}$$

Since the variation of the displacement (δw) is zero at t_1 and t_2, Eq. (14.47) reduces to

$$I_3 = \rho h \iint_A \int_{t_1}^{t_2} \ddot{w}\, \delta w\, dt\, dx\, dy \tag{14.48}$$

where $\ddot{w} = \partial^2 w/\partial t^2$. The variation of the last integral term in Eq. (14.24) yields

$$I_4 = -\delta \int_{t_1}^{t_2} \iint_A f w\, dx\, dy\, dt = -\int_{t_1}^{t_2} \iint_A f\, \delta w\, dx\, dy\, dt \tag{14.49}$$

Using Eqs. (14.27), (14.45), (14.48), and (14.49), Hamilton's principle of Eq. (14.24) can be expressed as

$$
\int_{t_1}^{t_2} \left(\iint_A (D\nabla^4 w + \rho h \ddot{w} - f)\delta w \, dx \, dy \right.
$$

$$
+ D \int_C \left[\nabla^2 w - (1-v) \left(\frac{\partial^2 w}{\partial x^2} \sin^2 \theta - 2\frac{\partial^2 w}{\partial x \partial y} \sin \theta \cos \theta + \frac{\partial^2 w}{\partial y^2} \cos^2 \theta \right) \right]
$$

$$
\frac{\partial \delta w}{\partial n} \, dC
$$

$$
- D \int_C \left\{ \frac{\partial \nabla^2 w}{\partial n} - (1-v)\frac{\partial}{\partial s} \left[\left(\frac{\partial^2 w}{\partial x^2} - \frac{\partial^2 w}{\partial y^2} \right) \cos \theta \sin \theta \right. \right.
$$

$$
\left. \left. + \frac{\partial^2 w}{\partial x \partial y}(\sin^2 \theta - \cos^2 \theta) \right] \right\} \delta w \, dC \right) \, dt = 0
\tag{14.50}
$$

To satisfy Eq. (14.50), each of the three terms within the outside parentheses must be zero. Furthermore, since δw is arbitrary, the expression inside the parentheses under the area integral must be zero. This leads to the relations

$$
D\nabla^4 w + \rho h \ddot{w} - f = 0 \qquad \text{in } A
\tag{14.51}
$$

$$
\left\{ \nabla^2 w - (1-v) \left(\frac{\partial^2 w}{\partial x^2} \sin^2 \theta - 2\frac{\partial^2 w}{\partial x \partial y} \sin \theta \cos \theta + \frac{\partial^2 w}{\partial y^2} \cos^2 \theta \right) \right\}
$$

$$
\frac{\partial \delta w}{\partial n} = 0 \qquad \text{on } C
\tag{14.52}
$$

$$
\left\{ \frac{\partial \nabla^2 w}{\partial n} - (1-v)\frac{\partial}{\partial s} \left[\left(\frac{\partial^2 w}{\partial x^2} - \frac{\partial^2 w}{\partial y^2} \right) \cos \theta \sin \theta + \frac{\partial^2 w}{\partial x \partial y}(\sin^2 \theta - \cos^2 \theta) \right] \right\}
$$

$$
\delta w = 0 \qquad \text{on } C
\tag{14.53}
$$

It can be seen that Eq. (14.51) is the equation of motion for the transverse vibration of a plate and Eqs. (14.52) and (14.53) are the boundary conditions. Note that for a clamped or fixed edge, the deflection and the slope of deflection normal to the edge must be zero (Fig. 14.2):

$$
w = 0, \qquad \frac{\partial w}{\partial n} = 0
$$

Thus, $\delta w = 0$, $\partial \delta w / \partial n = 0$ in Eqs. (14.52) and (14.53). For a simply supported edge, the deflection is zero and the slope of deflection normal to the edge is not zero (Fig. 14.2):

$$
w = 0, \qquad \frac{\partial w}{\partial n} \neq 0
$$

Thus, $\delta w = 0$ in Eq. (14.53) and $\partial \delta w / \partial n$ is arbitrary in Eq. (14.52). Hence, the expression in braces, which will later be shown to be equal to the bending moment on the edge, must be zero. For a free edge, there is no restriction on the values of w, and $\partial w / \partial n$ and hence δw and $\partial \delta w / \partial n$ are arbitrary. Hence, the expressions inside braces

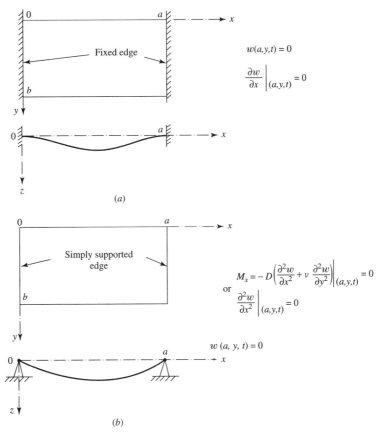

Figure 14.2   Boundary conditions: (a) fixed edge; (b) simply supported edge; (c) free edge; (d) edge supported on a linear elastic spring; (e) edge supported on a torsional elastic spring.

in Eqs. (14.53) and (14.52), which will later be shown to be equal to the effective shear force and bending moment, respectively, on the edge, must be zero.

14.3   BOUNDARY CONDITIONS

The equation of motion governing the transverse vibration of a plate is a fourth-order partial differential equation. As such, the solution of the equation requires two boundary conditions on each edge for a rectangular plate. If the edges of the rectangular plate are parallel to the x and y axes, the following boundary conditions are valid.

1. *Clamped, fixed, or built-in edge.* If the edge $x = a$ is clamped, the deflection and slope (normal to the edge) must be zero [Fig. 14.2(a)]:

$$w|_{x=a} = 0 \tag{14.54}$$

$$\frac{\partial w}{\partial x}\bigg|_{x=a} = 0 \tag{14.55}$$

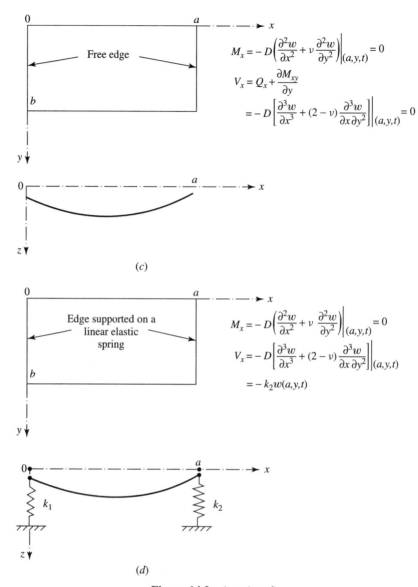

Figure 14.2   (*continued*)

2. *Simply supported edge.* If the edge $x = a$ is simply supported, the deflection and bending moment must be zero [Fig. 14.2(*b*)]:

$$w|_{x=a} = 0 \tag{14.56}$$

$$M_x = -D\left(\frac{\partial^2 w}{\partial x^2} + v\frac{\partial^2 w}{\partial y^2}\right)\bigg|_{x=a} = 0 \tag{14.57}$$

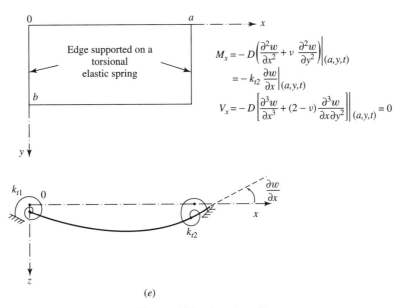

Edge supported on a
torsional
elastic spring

$$M_x = -D\left(\frac{\partial^2 w}{\partial x^2} + v\,\frac{\partial^2 w}{\partial y^2}\right)\bigg|_{(a,y,t)}$$

$$= -k_{t2}\,\frac{\partial w}{\partial x}\bigg|_{(a,y,t)}$$

$$V_x = -D\left[\frac{\partial^3 w}{\partial x^3} + (2-v)\frac{\partial^3 w}{\partial x\,\partial y^2}\right]\bigg|_{(a,y,t)} = 0$$

(e)

Figure 14.2   (*continued*)

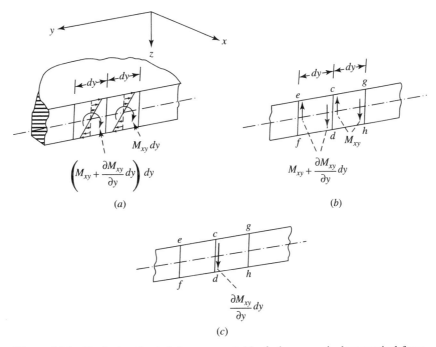

(a)

(b)

(c)

Figure 14.3   Replacing the twisting moment $M_{xy}dy$ by an equivalent vertical force.

Since $w = 0$ along the edge $x = a$, all the derivatives of w with respect to y are also zero. Thus,

$$\frac{\partial w}{\partial y}\bigg|_{x=a} = 0 \tag{14.58}$$

$$\frac{\partial^2 w}{\partial y^2}\bigg|_{x=a} = 0 \tag{14.59}$$

In view of Eq. (14.59), Eq. (14.57) can be rewritten as

$$\frac{\partial^2 w}{\partial x^2}\bigg|_{x=a} = 0 \tag{14.60}$$

3. *Free edge.* If the edge $x = a$ is free, there will be no stresses on the edge [Fig. 14.2(c)]. Hence, it appears that all the force and moment resultants on the edge are zero; that is,

$$M_x|_{x=a} = 0 \tag{14.61}$$

$$Q_x|_{x=a} = 0 \tag{14.62}$$

$$M_{xy}\big|_{x=a} = 0 \tag{14.63}$$

Equations (14.61)–(14.63) represent three boundary conditions, whereas the equation of motion requires only two. Although Poisson formulated Eqs. (14.61)–(14.63), Kirchhoff showed that the conditions on the shear force Q_x and the twisting moment M_{xy} given by Eqs. (14.62) and (14.63) are not independent and can be combined into only one boundary condition.

To combine the two conditions given by Eqs. (14.62) and (14.63) into one condition, consider two adjacent elements, each of length dy, along the edge $x = a$ as shown in Fig. 14.3(a). Because of the shear stresses τ_{xy} acting on the edge, a twisting moment $M_{xy}dy$ is developed on the element $cghd$ and a twisting moment $[M_{xy} + (\partial M_{xy}/\partial y)dy]dy$ is developed on the element $ecdf$. These moments can be replaced by vertical forces of magnitude M_{xy} and $M_{xy} + (\partial M_{xy}/\partial y)dy$, respectively, on the elements $cghd$ and $ecdf$, each with a moment arm dy as indicated in Fig. 14.3(b). Noting that such forces can be visualized for all elements of length dy along the entire edge $x = a$, we find that an unbalanced force of magnitude $(\partial M_{xy}/\partial y)dy$ exists at the boundary between two adjacent elements, such as line cd [Fig. 14.3(c)]. When this unbalanced force per unit length, $\partial M_{xy}/\partial y$, is added to the shear force resultant Q_x that is present on the edge $x = a$, we obtain the effective shear force V_x per unit length as

$$V_x = Q_x + \frac{\partial M_{xy}}{\partial y} \tag{14.64}$$

In a similar manner, the effective shear force V_y per unit length along the free edge $y = b$ can be expressed as

$$V_y = Q_y + \frac{\partial M_{yx}}{\partial x} \tag{14.65}$$

The effective shear force resultant, along with the bending moment M_x, is set equal to zero along the free edge $x = a$. Thus, the two boundary conditions, also known as *Kirchhoff boundary conditions*, valid for a free edge $x = a$ are given by [Fig. 14.2(c)]

$$M_x = -D\left(\frac{\partial^2 w}{\partial x^2} + v\frac{\partial^2 w}{\partial y^2}\right)\bigg|_{x=a} = 0 \tag{14.66}$$

$$V_x = Q_x + \frac{\partial M_{xy}}{\partial y} = -D\left[\frac{\partial^3 w}{\partial x^3} + (2-v)\frac{\partial^3 w}{\partial x\partial y^2}\right]\bigg|_{x=a} = 0 \tag{14.67}$$

4. *Edge resting on a linear elastic spring.* If the edge $x = a$, otherwise unloaded, is supported on a linear elastic spring that offers resistance to transverse displacement, the restoring force will be $k_2 w$, as shown in Fig. 14.2(d). The effective shear force V_x at the edge must be equal to the restoring force of the spring. Also, the bending moment on the edge must be zero. Thus, the boundary conditions can be stated as

$$M_x|_{x=a} = -D\left(\frac{\partial^2 w}{\partial x^2} + v\frac{\partial^2 w}{\partial y^2}\right)\bigg|_{x=a} = 0$$

$$V|_{x=a} = Q_x + \frac{\partial M_{xy}}{\partial y}\bigg|_{x=a} = -k_2 w|_{x=a} \tag{14.68}$$

or

$$-D\left[\frac{\partial^3 w}{\partial x^3} + (2-v)\frac{\partial^3 w}{\partial x\,\partial y^2}\right]\bigg|_{x=a} = -k_2 w|_{x=a} \tag{14.69}$$

5 . *Edge resting on an elastic torsional spring.* If the edge $x = a$, otherwise unloaded, is supported on a torsional spring that offers resistance to the rotation of the edge, the restoring moment will be $k_{t_2}(\partial w/\partial x)$, as shown in Fig. 14.2(e). The bending moment, M_x, at the edge must be equal to the resisting moment of the spring. Also, the effective shear force on the edge, V_x, must be zero. Thus, the boundary conditions can be expressed as

$$M_x|_{x=a} = -D\left(\frac{\partial^2 w}{\partial x^2} + v\frac{\partial^2 w}{\partial y^2}\right)\bigg|_{x=a} = k_{t_2}\frac{\partial w}{\partial x}\bigg|_{x=a} \tag{14.70}$$

$$V_x|_{x=a} = \left(Q_x + \frac{\partial M_{xy}}{\partial y}\right)\bigg|_{x=a} = -D\left[\frac{\partial^3 w}{\partial x^3} + (2-v)\frac{\partial^3 w}{\partial x\,\partial y^2}\right]\bigg|_{x=a} = 0 \tag{14.71}$$

6. *Boundary conditions on a skew edge.* Let an edge of the plate be skewed with its outward drawn normal (n) making an angle θ with the $+x$ axis and s indicating the tangential direction as shown in Fig. 14.4. The positive directions of the shear force Q_n, normal bending moment M_n, and twisting moment M_{ns} acting on the edge are also indicated in Fig. 14.4(a). Noting that the normal and twisting moments on the skew edge are defined as

$$M_n = \int_{-h/2}^{h/2} \sigma_n z\, dz \tag{14.72}$$

$$M_{ns} = \int_{-h/2}^{h/2} \tau_{ns} z\, dz \tag{14.73}$$

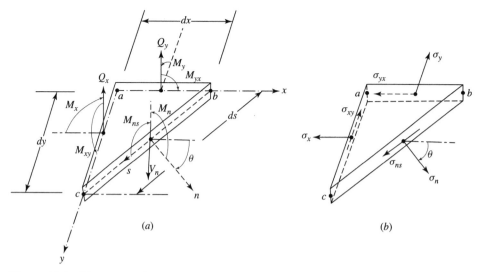

Figure 14.4 Skew edge of a plate: (*a*) shear force and moment resultants; (*b*) stresses acting on the edges.

where σ_n and τ_{ns} denote, respectively, the normal and shear stresses acting on the edge as shown in Fig. 14.4(*b*). The stresses σ_n and τ_{ns} acting on the skew edge (*bc*) can be expressed in terms of the stresses acting on the edges *ab* and *ac* using the stress transformation relations as [41]

$$\sigma_n = \sigma_x \cos^2 \theta + \sigma_y \sin^2 \theta + \tau_{xy} \sin 2\theta \tag{14.74}$$

$$\tau_{ns} = \tau_{xy} \cos 2\theta - \tfrac{1}{2}(\sigma_x - \sigma_y) \sin 2\theta \tag{14.75}$$

By substituting Eqs. (14.74) and (14.75) into Eqs. (14.72) and (14.73), respectively, and carrying out the indicated integrations over the thickness of the plate, we obtain

$$M_n = M_x \cos^2 \theta + M_y \sin^2 \theta + M_{xy} \sin 2\theta \tag{14.76}$$

$$M_{ns} = M_{xy} \cos 2\theta - \tfrac{1}{2}(M_x - M_y) \sin 2\theta \tag{14.77}$$

The vertical force equilibrium of the element of the plate shown in Fig. 14.4(*a*) yields

$$Q_n ds - Q_x dy - Q_y dx = 0$$

or

$$Q_n = Q_x \frac{dy}{ds} + Q_y \frac{dx}{ds} = Q_x \cos \theta + Q_y \sin \theta \tag{14.78}$$

The effective shear force resultant on the skew edge V_n can be found as

$$V_n = Q_n + \frac{\partial M_{ns}}{\partial s} \tag{14.79}$$

The different boundary conditions on the skew edge can be stated as follows:

(a) *Clamped, fixed, or built-in edge.* The deflection and slope (normal to the edge) must be zero:

$$w = 0 \tag{14.80}$$

$$\frac{\partial w}{\partial n} = 0 \tag{14.81}$$

(b) *Hinged or simply supported edge.* The deflection and the normal bending moment resultant on the edge must be zero:

$$w = 0 \tag{14.82}$$

$$M_n = 0 \tag{14.83}$$

(c) *Free edge.* The normal bending moment and effective shear force resultants must be zero on the edge:

$$M_n = 0 \tag{14.84}$$

$$V_n = M_n + \frac{\partial M_{ns}}{\partial s} = 0 \tag{14.85}$$

Note that the boundary conditions of Eqs. (14.83)–(14.85) can be expressed in terms of deflection, w, using Eqs. (14.76)–(14.79) and the known expressions of M_x, M_y, M_{xy}, Q_x, and Q_y in terms of w from Eqs. (14.1)–(14.6).

14.4 FREE VIBRATION OF RECTANGULAR PLATES

Let the boundaries of the rectangular plate be defined by the lines $x = 0, a$ and $y = 0, b$. To find the solution of the free vibration equation, Eq. (14.8) with $f = 0$, we assume the solution to be of the type

$$w(x,y,t) = W(x, y)T(t) \tag{14.86}$$

and obtain the following equations from Eq. (14.8):

$$\frac{1}{T(t)} \frac{d^2 T(t)}{dt^2} = -\omega^2 \tag{14.87}$$

$$-\frac{\beta_1^2}{W(x, y)} \nabla^4 W(x, y) = -\omega^2 \tag{14.88}$$

where ω^2 is a constant and

$$\beta_1^2 = \frac{D}{\rho h} \tag{14.89}$$

Equations (14.87) and (14.88) can be rewritten as

$$\frac{d^2 T(t)}{dt^2} + \omega^2 T(t) = 0 \tag{14.90}$$

$$\nabla^4 W(x, y) - \lambda^4 W(x, y) = 0 \tag{14.91}$$

where

$$\lambda^4 = \frac{\omega^2}{\beta_1^2} = \frac{\rho h \omega^2}{D} \tag{14.92}$$

The general solution of Eq. (14.90) is

$$T(t) = A \cos \omega t + B \sin \omega t \tag{14.93}$$

and Eq. (14.91) can be expressed as

$$(\nabla^4 - \lambda^4) W(x, y) = (\nabla^2 + \lambda^2)(\nabla^2 - \lambda^2) W(x, y) = 0 \tag{14.94}$$

By the theory of linear differential equations, the complete solution of Eq. (14.94) can be obtained by superposing the solutions of the following equations:

$$(\nabla^2 + \lambda^2) W_1(x, y) = \frac{\partial^2 W_1}{\partial x^2} + \frac{\partial^2 W_1}{\partial y^2} + \lambda^2 W_1(x, y) = 0 \tag{14.95}$$

$$(\nabla^2 - \lambda^2) W_2(x, y) = \frac{\partial^2 W_2}{\partial x^2} + \frac{\partial^2 W_2}{\partial y^2} - \lambda^2 W_2(x, y) = 0 \tag{14.96}$$

Equation (14.95) is similar to the equation obtained in the case of free vibration of a membrane [Eq. (c) of the footnote following Eq. (13.44)], whose solution is given by the product of Eqs. (13.46) and (13.47) as

$$W_1(x, y) = A_1 \sin \alpha x \sin \beta y + A_2 \sin \alpha x \cos \beta y$$
$$+ A_3 \cos \alpha x \sin \beta y + A_4 \cos \alpha x \cos \beta y \tag{14.97}$$

where $\lambda^2 = \alpha^2 + \beta^2$. The solution of Eq. (14.96) can be obtained as in the case of solution of Eq. (14.95) except that λ is to be replaced by $i\lambda$. Hence the solution of Eq. (14.96) will be composed of products of sinh and cosh terms. Thus, the general solution of Eq. (14.91) can be expressed as

$$W(x, y) = A_1 \sin \alpha x \sin \beta y + A_2 \sin \alpha x \cos \beta y$$
$$+ A_3 \cos \alpha x \sin \beta y + A_4 \cos \alpha x \cos \beta y$$
$$+ A_5 \sinh \theta x \sinh \phi y + A_6 \sinh \theta x \cosh \phi y$$
$$+ A_7 \cosh \theta x \sinh \phi y + A_8 \cosh \theta x \cosh \phi y \tag{14.98}$$

where

$$\lambda^2 = \alpha^2 + \beta^2 = \theta^2 + \phi^2 \tag{14.99}$$

14.4.1 Solution for a Simply Supported Plate

For a plate simply supported on all the sides, the boundary conditions to be satisfied are

$$
\left.\begin{array}{ll}
w(x,y,t) = M_x(x,y,t) = 0 & \text{for } x = 0 \text{ and } a \\
w(x,y,t) = M_y(x,y,t) = 0 & \text{for } y = 0 \text{ and } b
\end{array}\right\}, \quad t \geq 0 \tag{14.100}
$$

These boundary conditions can be expressed in terms of W, using Eq. (14.86), as

$$
\begin{aligned}
W(0,\,y) = 0, & \qquad \left.\left(\frac{d^2 W}{dx^2} + \nu\frac{d^2 W}{dy^2}\right)\right|_{(0,y)} = 0, \\
W(a,\,y) = 0, & \qquad \left.\left(\frac{d^2 W}{dx^2} + \nu\frac{d^2 W}{dy^2}\right)\right|_{(a,y)} = 0, \\
W(x,\,0) = 0, & \qquad \left.\left(\frac{d^2 W}{dy^2} + \nu\frac{d^2 W}{dx^2}\right)\right|_{(x,0)} = 0, \\
W(x,\,b) = 0, & \qquad \left.\left(\frac{d^2 W}{dy^2} + \nu\frac{d^2 W}{dx^2}\right)\right|_{(x,b)} = 0,
\end{aligned} \tag{14.101}
$$

As W is a constant along the edges $x = 0$ and $x = a$, $d^2 W/dy^2$ will be zero along these edges. Similarly, $d^2 W/dx^2$ will be zero along the edges $y = 0$ and $y = b$. Thus, Eqs. (14.101) will be simplified as

$$
\begin{aligned}
W(0,\,y) = \frac{d^2 W}{dx^2}(0,\,y) = W(a,\,y) = \frac{d^2 W}{dx^2}(a,\,y) = 0 \\
W(x,\,0) = \frac{d^2 W}{dy^2}(x,\,0) = W(x,\,b) = \frac{d^2 W}{dy^2}(x,\,b) = 0
\end{aligned} \tag{14.102}
$$

When these boundary conditions are used, we find that all the constants A_i, except A_1, in Eq. (14.98) are zero; in addition, we obtain two equations that α and β must satisfy:

$$
\begin{aligned}
\sin \alpha a = 0 \\
\sin \beta b = 0
\end{aligned} \tag{14.103}
$$

Equations (14.103) represent the frequency equations whose solution is given by

$$
\begin{aligned}
\alpha_m a = m\pi, & \qquad m = 1, 2, \ldots \\
\beta_n b = n\pi, & \qquad n = 1, 2, \ldots
\end{aligned} \tag{14.104}
$$

Thus, we obtain from Eqs. (14.104), (14.99), and (14.92) the natural frequencies of the plate as

$$
\omega_{mn} = \lambda_{mn}^2\left(\frac{D}{\rho h}\right)^{1/2} = \pi^2\left[\left(\frac{m}{a}\right)^2 + \left(\frac{n}{b}\right)^2\right]\left(\frac{D}{\rho h}\right)^{1/2}, \qquad m, n = 1, 2, \ldots \tag{14.105}
$$

The characteristic function $W_{mn}(x,y)$ corresponding to ω_{mn} can be expressed as

$$
W_{mn}(x,y) = A_{1mn}\sin\frac{m\pi x}{a}\sin\frac{n\pi y}{b}, \qquad m, n = 1, 2, \ldots \tag{14.106}
$$

and the natural mode as

$$
w_{mn}(x,y,t) = \sin\frac{m\pi x}{a}\sin\frac{n\pi y}{b}(A_{mn}\cos\omega_{mn}t + B_{mn}\sin\omega_{mn}t) \tag{14.107}
$$

The general solution of Eq. (14.8) with $f = 0$ is given by the sum of the natural modes:

$$w(x,y,t) = \sum_{m=1}^{\infty} \sum_{n=1}^{\infty} \sin \frac{m\pi x}{a} \sin \frac{n\pi y}{b} (A_{mn} \cos \omega_{mn} t + B_{mn} \sin \omega_{mn} t) \qquad (14.108)$$

Let the initial conditions of the plate be given by

$$w(x,y,0) = w_0(x,y)$$

$$\frac{\partial w}{\partial t}(x,y,0) = \dot{w}_0(x,y) \qquad (14.109)$$

By substituting Eq. (14.108) into Eqs. (14.109), we obtain

$$\sum_{m=1}^{\infty} \sum_{n=1}^{\infty} A_{mn} \sin \frac{m\pi x}{a} \sin \frac{n\pi y}{b} = w_0(x,y)$$

$$\sum_{m=1}^{\infty} \sum_{n=1}^{\infty} B_{mn} \omega_{mn} \sin \frac{m\pi x}{a} \sin \frac{n\pi y}{b} = \dot{w}_0(x,y) \qquad (14.110)$$

Multiplying each of the equations in Eq. (14.110) by $\sin(m\pi x/a) \sin(n\pi y/b)$ and integrating over the area of the plate leads to

$$A_{mn} = \frac{4}{ab} \int_0^a \int_0^b w_0(x,y) \sin \frac{m\pi x}{a} \sin \frac{n\pi y}{b} \, dx \, dy$$

$$B_{mn} = \frac{4}{ab\omega_{mn}} \int_0^a \int_0^b \dot{w}_0(x,y) \sin \frac{m\pi x}{a} \sin \frac{n\pi y}{b} \, dx \, dy \qquad (14.111)$$

The mode shape, $W_{mn}(x,y)$, given by Eq. (14.106) consists of m half sine waves in the x direction and n half sine waves in the y direction of the plate. The first few modes of vibration corresponding to the natural frequencies ω_{mn}, given by Eq. (14.105), are shown in Fig. 14.5.

14.4.2 Solution for Plates with Other Boundary Conditions

To solve Eq. (14.91) for a plate with arbitrary boundary conditions, we use the separation-of-variables technique as

$$W(x,y) = X(x)Y(y) \qquad (14.112)$$

Substitution of Eq. (14.112) into Eq. (14.91) leads to

$$X''''Y + 2X''Y'' + XY'''' - \lambda^4 XY = 0 \qquad (14.113)$$

where a prime indicates a derivative with respect to its argument. The functions $X(x)$ and $Y(y)$ can be separated in Eq. (14.113) provided that either

$$Y''(y) = -\beta^2 Y(y), \qquad Y''''(y) = -\beta^2 Y''(y)$$

or

$$X''(x) = -\alpha^2 X(x), \qquad X''''(x) = -\alpha^2 X''(x) \qquad (14.114)$$

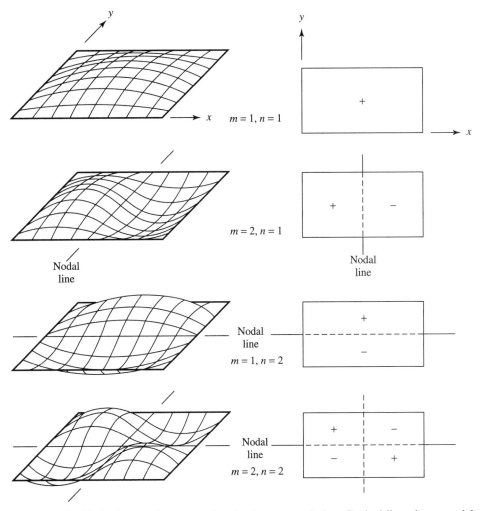

Figure 14.5   Mode shapes of a rectangular simply supported plate. Dashed lines denote nodal lines other than the edges.

or both are satisfied. Equations (14.114) can be satisfied only by the trigonometric functions

$$\begin{Bmatrix} \sin \alpha_m x \\ \cos \alpha_m x \end{Bmatrix} \quad \text{or} \quad \begin{Bmatrix} \sin \beta_n y \\ \cos \beta_n y \end{Bmatrix} \tag{14.115}$$

with

$$\alpha_m = \frac{m\pi}{a}, \quad m = 1, 2, \dots, \qquad \beta_n = \frac{n\pi}{b}, \quad n = 1, 2, \dots \tag{14.116}$$

We assume that the plate is simply supported along edges $x = 0$ and $x = a$. This implies that

$$X_m(x) = A \sin \alpha_m x, \qquad m = 1, 2, \dots \tag{14.117}$$

where A is a constant. Equation (14.117) satisfies the conditions

$$X_m(0) = X_m(a) = X_m''(0) = X_m''(a) = 0 \tag{14.118}$$

for any integer m, and hence the boundary conditions

$$w(0,y,t) = w(a,y,t) = \nabla^2 w(0,y,t) = \nabla^2 w(a,y,t) = 0 \tag{14.119}$$

Using the solution of Eq. (14.117), Eq. (14.113) becomes

$$Y''''(y) - 2\alpha_m^2 Y''(y) - (\lambda^4 - \alpha_m^4)Y(y) = 0 \tag{14.120}$$

It can be observed that there are six possible combinations of simple boundary conditions when the edges $x = 0$ and $x = a$ are simply supported. The various boundary conditions can be stated, using the abbreviations SS, F, and C for simply supported, free, and clamped edges, respectively, as SS–SS–SS–SS, SS–C–SS–C, SS–F–SS–F, SS–C–SS–SS, SS–F–SS–SS, and SS–F–SS–C.

Assuming that $\lambda^4 > \alpha_m^4$ in Eq. (14.120), its solution is taken in the form

$$Y(y) = e^{sy} \tag{14.121}$$

which yields the auxiliary equation:

$$s^4 - 2s^2\alpha_m^2 - (\lambda^4 - \alpha_m^4) = 0 \tag{14.122}$$

The roots of Eq. (14.122) are given by

$$s_{1,2} = \pm\sqrt{\lambda^2 + \alpha_m^2}, \qquad s_{3,4} = \pm i\sqrt{\lambda^2 - \alpha_m^2} \tag{14.123}$$

Thus, the general solution of Eq. (14.120) can be expressed as

$$Y(y) = C_1 \sin\delta_1 y + C_2 \cos\delta_1 y + C_3 \sinh\delta_2 y + C_4 \cosh\delta_2 y \tag{14.124}$$

where

$$\delta_1 = \sqrt{\lambda^2 - \alpha_m^2}, \qquad \delta_2 = \sqrt{\lambda^2 + \alpha_m^2} \tag{14.125}$$

Equation (14.124), when substituted into the boundary condition relations on the edges $y = 0$ and $y = b$, we obtain four homogeneous algebraic equations for the coefficients C_1, C_2, C_3, and C_4. By setting the determinant of the coefficient matrix equal to zero, we can derive the frequency equation. The procedure is illustrated below for simply supported and clamped boundary conditions.

1. *When edges $y = 0$ and $y = b$ are simply supported.* The boundary conditions can be stated as

$$W(x, 0) = 0 \tag{14.126}$$

$$W(x, b) = 0 \tag{14.127}$$

$$M_y(x, 0) = -D\left(\frac{\partial^2 W}{\partial y^2} + \nu\frac{\partial^2 W}{\partial x^2}\right)\bigg|_{(x,0)} = 0 \tag{14.128}$$

$$M_y(x, b) = -D\left(\frac{\partial^2 W}{\partial y^2} + \nu\frac{\partial^2 W}{\partial x^2}\right)\bigg|_{(x,b)} = 0 \tag{14.129}$$

Since $W = 0$ along the edges $y = 0$ and $y = b$, $\partial W/\partial x = \partial^2 W/\partial x^2 = 0$ will also be zero. Thus, the boundary conditions of Eqs. (14.126)–(14.129) can be restated as follows:

$$Y(0) = 0 \qquad\qquad (14.130)$$

$$Y(b) = 0 \qquad\qquad (14.131)$$

$$\frac{d^2Y(0)}{dy^2} = 0 \qquad\qquad (14.132)$$

$$\frac{d^2Y(b)}{dy^2} = 0 \qquad\qquad (14.133)$$

Since Eqs. (14.132) and (14.133) involve the second derivative of Y, we find from Eq. (14.124),

$$\frac{d^2Y(y)}{dy^2} = -\delta_1^2 C_1 \sin\delta_1 y - \delta_1^2 C_2 \cos\delta_1 y + \delta_2^2 C_3 \sinh\delta_2 y + \delta_2^2 C_4 \cosh\delta_2 y \quad (14.134)$$

Using Eqs. (14.124) and (14.134), the boundary conditions of Eqs. (14.130)–(14.133) can be expressed as

$$C_2 + C_4 = 0 \qquad (14.135)$$

$$C_1 \sin\delta_1 b + C_2 \cos\delta_1 b + C_3 \sinh\delta_2 b + C_4 \cosh\delta_2 b = 0 \qquad (14.136)$$

$$-\delta_1^2 C_2 + \delta_2^2 C_4 = 0 \qquad (14.137)$$

$$-C_1 \delta_1^2 \sin\delta_1 b - C_2 \delta_1^2 \cos\delta_1 b + C_3 \delta_2^2 \sinh\delta_2 b + C_4 \delta_2^2 \cosh\delta_2 b = 0 \qquad (14.138)$$

Equations (14.135) and (14.137) yield

$$C_2 = C_4 = 0 \qquad\qquad (14.139)$$

In view of Eq. (14.139), Eqs. (14.136) and (14.138) reduce to

$$C_1 \sin\delta_1 b + C_3 \sinh\delta_2 b = 0$$

or

$$C_3 \sinh\delta_2 b = -C_1 \sin\delta_1 b \qquad\qquad (14.140)$$

and

$$-C_1 \delta_1^2 \sin\delta_1 b + C_3 \delta_2^2 \sinh\delta_2 b = 0 \qquad\qquad (14.141)$$

Using Eq. (14.140), Eq. (14.141) can be written as

$$C_1(\delta_1^2 + \delta_2^2) \sin\delta_1 b = 0 \qquad\qquad (14.142)$$

For a nontrivial solution, we should have

$$\sin\delta_1 b = 0$$

or

$$\delta_1 = \frac{n\pi}{b}, \qquad n = 1, 2, \ldots . \qquad\qquad (14.143)$$

with the corresponding mode shapes as

$$Y_n(y) = C_1 \sin \delta_1 y = C_1 \sin \frac{n\pi y}{b} \tag{14.144}$$

Since $\delta_1 = \sqrt{\lambda^2 - \alpha_m^2}$, we have

$$\lambda_{mn}^2 = \alpha_m^2 + \beta_n^2, \qquad \beta_n = \frac{n\pi}{b} \tag{14.145}$$

This leads to the result

$$\omega_{mn} = \lambda_{mn}^2 \sqrt{\frac{D}{\rho h}} = (\alpha_m^2 + \beta_n^2) \sqrt{\frac{D}{\rho h}}$$

$$= \pi^2 \left[\left(\frac{m}{a}\right)^2 + \left(\frac{n}{b}\right)^2 \right] \sqrt{\frac{D}{\rho h}}, \qquad m, n = 1, 2, \dots \tag{14.146}$$

The mode shapes $W_{mn}(x,y) = X_m(x)Y_n(y)$, corresponding to the natural frequencies ω_{mn} of Eq. (14.146), are given by

$$W_{mn}(x,y) = C_{mn} \sin \alpha_m x \, \sin \beta_n y, \qquad m, n = 1, 2, \dots \tag{14.147}$$

where C_{mn} is a constant. This solution can be seen to be the same as the one given in Section 14.4.1.

2. *When edges $y = 0$ and $y = b$ are clamped.* The boundary conditions can be stated as

$$Y(0) = 0 \tag{14.148}$$

$$\frac{dY}{dy}(0) = 0 \tag{14.149}$$

$$Y(b) = 0 \tag{14.150}$$

$$\frac{dY}{dy}(b) = 0 \tag{14.151}$$

Using Eq. (14.124) and

$$\frac{dY(y)}{dy} = C_1\delta_1 \cos \delta_1 y - C_2\delta_1 \sin \delta_1 y + C_3\delta_2 \cosh \delta_2 y + C_4\delta_2 \sinh \delta_2 y \tag{14.152}$$

the boundary conditions of Eqs. (14.148)–(14.151) can be expressed as

$$C_2 + C_4 = 0 \tag{14.153}$$

$$C_1\delta_1 + C_3\delta_2 = 0 \tag{14.154}$$

$$C_1 \sin \delta_1 b + C_2 \cos \delta_1 b + C_3 \sinh \delta_2 b + C_4 \cosh \delta_2 b = 0 \tag{14.155}$$

$$C_1\delta_1 \cos \delta_1 b - C_2\delta_1 \sin \delta_1 b + C_3\delta_2 \cosh \delta_2 b + C_4\delta_2 \sinh \delta_2 b = 0 \tag{14.156}$$

Equations (14.153)–(14.156) can be written in matrix form as

$$
\begin{bmatrix}
0 & 1 & 0 & 1 \\
\delta_1 & 0 & \delta_2 & 0 \\
\sin \delta_1 b & \cos \delta_1 b & \sinh \delta_2 b & \cosh \delta_2 b \\
\delta_1 \cos \delta_1 b & -\delta_1 \sin \delta_1 b & \delta_2 \cosh \delta_2 b & \delta_2 \sinh \delta_2 b
\end{bmatrix}
\begin{Bmatrix}
C_1 \\ C_2 \\ C_3 \\ C_4
\end{Bmatrix}
=
\begin{Bmatrix}
0 \\ 0 \\ 0 \\ 0
\end{Bmatrix}
\tag{14.157}
$$

By setting the determinant of the matrix in Eq. (14.157) equal to zero, we obtain the frequency equation, after simplification, as

$$
2\delta_1 \delta_2 (\cos \delta_1 b \ \cosh \delta_2 b - 1) - \alpha_m^2 \sin \delta_1 b \ \sinh \delta_2 b = 0
\tag{14.158}
$$

For any specific value of m, there will be successive values of λ and hence ω that satisfy the frequency equation (14.158). The natural frequencies can be denoted as $\omega_{11}, \omega_{12}, \omega_{13}, \ldots, \omega_{21}, \omega_{22}, \omega_{23}, \ldots$, whose values depend on the material properties E, ν, and ρ and the geometry h and b/a of the plate. The mode shape corresponding to the nth root of Eq. (14.158) can be expressed as

$$
Y_n(y) = C_n[(\cosh \delta_2 b - \cos \delta_1 b) (\delta_1 \sinh \delta_2 y - \delta_2 \sin \delta_1 y)
$$
$$
- (\delta_1 \sinh \delta_2 b - \delta_2 \sin \delta_1 b) (\cosh \delta_2 y - \cos \delta_1 y)]
\tag{14.159}
$$

where C_n is an arbitrary constant. Thus, the complete mode shape $W_{mn} = X_m(x)Y_n(y)$, corresponding to the natural frequency ω_{mn}, becomes

$$
W_{mn}(x,y) = C_{mn} Y_n(y) \ \sin \alpha_m x
\tag{14.160}
$$

where $Y_n(y)$ is given by Eq. (14.159) and C_{mn} is a new (arbitrary) constant.

The frequency equations and the mode shapes for other cases (with other edge conditions at $y = 0$ and $y = b$) can be derived in a similar manner. The results for the six combinations of boundary conditions are summarized in Table 14.1.

14.5   FORCED VIBRATION OF RECTANGULAR PLATES

We consider in this section the response of simply supported rectangular plates subjected to external pressure $f(x,y,t)$ using a modal analysis procedure. Accordingly, the transverse displacement of the plate, $w(x,y,t)$, is represented as

$$
w(x,y,t) = \sum_{m=1}^{\infty} \sum_{n=1}^{\infty} W_{mn}(x,y)\eta_{mn}(t)
\tag{14.161}
$$

where the normal modes are given by [Eq. (14.106)]

$$
W_{mn}(x,y) = A_{1mn} \sin \frac{m\pi x}{a} \sin \frac{n\pi y}{b}, \qquad m, n = 1, 2, \ldots
\tag{14.162}
$$

The normal modes are normalized to satisfy the normalization condition

$$
\int_0^a \int_0^b \rho h W_{mn}^2 \, dx \, dy = 1
$$

Table 14.1 Frequency Equations and Mode Shapes of Rectangular Plates with Different Boundary Conditions[a]

Case	Boundary conditions	Frequency equation	y-mode shape, $Y_n(y)$ without a multiplication factor, where $W_{mn}(x,y) = C_{mn} X_m(x) Y_n(y)$, with $X_m(x) = \sin \alpha_m x$
1	SS-SS-SS-SS	$\sin \delta_1 b = 0$	$Y_n(y) = \sin \beta_n y$
2	SS-C-SS-C	$2\delta_1\delta_2(\cos \delta_1 b \cosh \delta_2 b - 1) - \alpha_m^2 \sin \delta_1 b \sinh \delta_2 b = 0$	$Y_n(y) = (\cos \delta_2 b - \cos \delta_1 b)(\delta_1 \sinh \delta_2 y - \delta_2 \sin \delta_1 y)$ $-(\delta_1 \sinh \delta_2 b - \delta_2 \sin \delta_1 b)(\cosh \delta_2 y - \cos \delta_1 y)$
3	SS-F-SS-F	$\sinh \delta_2 b \sin \delta_1 b \{\delta_2^2[\lambda^2 - \alpha_m^2(1-\nu)]^4$ $\quad -\delta_1^2[\lambda^2 + \alpha_m^2(1-\nu)]^4\}$ $\quad -2\delta_1\delta_2[\lambda^4 - \alpha_m^4(1-\nu)^2]^2 (\cosh \delta_2 b \cos \delta_1 b - 1) = 0$	$Y_n(y) = -(\cosh \delta_2 b - \cos \delta_1 b)[\lambda^4 - \alpha_m^4(1-\nu)^2]$ $\{\delta_1[\lambda^2 + \alpha_m^2(1-\nu)]\sin \delta_{1y}\} + \{\delta_1[\lambda^2 + \alpha_m^2(1-\nu)]^2 \sinh \delta_{2b}$ $+\delta_2[\lambda^2 - \alpha_m^2(1-\nu)]\sin \delta_{1y}\} + \{\delta_1[\lambda^2 - \alpha_m^2(1-\nu)]^2 \sin \delta_1 b\}[\lambda^2 - \alpha_m^2(1-\nu)]^2 \sinh \delta_{2b}$ $-\delta_2[\lambda^2 - \alpha_m^2(1-\nu)]^2 \sin \delta_1 b]^2 \cosh \delta_{2y}$ $+[\lambda^2 + \alpha_m^2(1-\nu)] \cos \delta_{1y}\}$
4	SS-C-SS-SS	$\delta_2 \cosh \delta_2 b \sin \delta_1 b - \delta_1 \sinh \delta_2 b \cos \delta_1 b = 0$	$Y_n(y) = \sin \delta_1 b \sinh \delta_{2y} - \sinh \delta_2 b \sin \delta_{1y}$
5	SS-F-SS-SS	$\delta_2[\lambda^2 - \alpha_m^2(1-\nu)]^2 \cosh \delta_2 b \sin \delta_1 b$ $\quad -\delta_1[\lambda^2 + \alpha_m^2(1-\nu)]^2 \sinh \delta_2 b \cos \delta_1 b = 0$	$Y_n(y) = [\lambda^2 - \alpha_m^2(1-\nu)] \sin \delta_1 b \sinh \delta_{2y}$ $+[\lambda^2 + \alpha_m^2(1-\nu)] \sinh \delta_2 b \sin \delta_{1y}$
6	SS-F-SS-C	$\delta_1\delta_2[\lambda^4 - \alpha_m^4(1-\nu)^2] + \delta_1\delta_2[\lambda^4 + \alpha_m^4(1-\nu)^2]$ $\quad \cdot \cosh \delta_2 b \cos \delta_1 b + \alpha_m^2[\lambda^4(1-2\nu) - \alpha_m^4(1-\nu)^2]$ $\quad \cdot \sinh \delta_2 b \sin \delta_1 b = 0$	$Y_n(y) = \{[\lambda^2 + \alpha_m^2(1-\nu)] \cosh \delta_2 b + [\lambda^2 - \alpha_m^2(1-\nu)] \cos \delta_2 b\}$ $\cdot (\delta_2 \sin \delta_1 y - \delta_1 \sinh \delta_2 y) + \{\delta_1[\lambda^2 + \alpha_m^2(1-\nu)] \sinh \delta_2 b$ $+\delta_2[\lambda^2 - \alpha_m^2(1-\nu)] \sin \delta_1 b\}(\cosh \delta_2 y - \cos \delta_1 y)$

Source: Refs. [1] and [2].

[a] Edges $x = 0$ and $x = a$ simply supported.

which yields $A_{1mn} = 2/\sqrt{\rho h a b}$. By using the normalized normal modes in Eq. (14.161) and substituting the result into the equation of motion, Eq. (14.8), we can derive the equation governing the generalized coordinates $\eta_{mn}(t)$ as

$$\ddot{\eta}_{mn}(t) + \omega_{mn}^2 \eta_{mn}(t) = N_{mn}(t), \qquad m, n = 1, 2, \ldots \tag{14.163}$$

where the generalized force $N_{mn}(t)$ is given by

$$N_{mn}(t) = \int_0^a \int_0^b W_{mn}(x,y) f(x,y,t)\, dx\, dy \tag{14.164}$$

and the natural frequencies by [Eq. (14.105)]

$$\omega_{mn} = \pi^2 \left(\frac{D}{\rho h}\right)^{1/2} \left[\left(\frac{m}{a}\right)^2 + \left(\frac{n}{b}\right)^2\right], \qquad m, n = 1, 2, \ldots \tag{14.165}$$

The solution of Eq. (14.163) can be expressed as [see Eq. (2.109)]

$$\eta_{mn}(t) = \eta_{mn}(0) \cos \omega_{mn} t + \frac{\dot{\eta}_{mn}(0)}{\omega_{mn}} \sin \omega_{mn} t + \frac{1}{\omega_{mn}} \int_0^t N_{mn}(\tau) \sin \omega_{mn}(t - \tau)\, d\tau \tag{14.166}$$

Thus, the final solution can be written as

$$
\begin{aligned}
w(x,y,t) = &\sum_{m=1}^{\infty}\sum_{n=1}^{\infty} \eta_{mn}(0) \sin \frac{m\pi x}{a} \sin \frac{n\pi y}{b} \cos\left[\pi^2 \sqrt{\frac{D}{\rho h}}\left(\frac{m^2}{a^2} + \frac{n^2}{b^2}\right) t\right] \\
&+ \sum_{m=1}^{\infty}\sum_{n=1}^{\infty} \frac{\dot{\eta}_{mn}(0)(\rho h)^{1/2}}{\pi^2 (D)^{1/2}} \frac{1}{m^2/a^2 + n^2/b^2} \sin \frac{m\pi x}{a} \sin \frac{n\pi y}{b} \\
&\qquad \sin\left[\pi^2 \sqrt{\frac{D}{\rho h}}\left(\frac{m^2}{a^2} + \frac{n^2}{b^2}\right) t\right] \\
&+ \sum_{m=1}^{\infty}\sum_{n=1}^{\infty} \frac{(\rho h)^{1/2}}{\pi^2 D^{1/2}} \frac{1}{m^2/a^2 + n^2/b^2} \sin \frac{m\pi x}{a} \sin \frac{n\pi y}{b} \int_0^t N_{mn}(\tau) \\
&\qquad \sin\left[\pi^2 \sqrt{\frac{D}{\rho h}}\left(\frac{m^2}{a^2} + \frac{n^2}{b^2}\right)(t - \tau)\right] d\tau
\end{aligned} \tag{14.167}
$$

Example 14.1   Find the response of a simply supported uniform plate subjected to a concentrated force $F(t)$ at the point $x = x_0$, $y = y_0$. Assume the initial conditions to be zero.

SOLUTION   Since the initial conditions are zero, the response is given by the steady-state solution:

$$w(x,y,t) = \sum_{m=1}^{\infty}\sum_{n=1}^{\infty} W_{mn}(x,y)\eta_{mn}(t) \tag{E14.1.1}$$

where

$$\eta_{mn}(t) = \frac{1}{\omega_{mn}} \int_0^t N_{mn}(\tau) \sin \omega_{mn}(t - \tau) \, d\tau \qquad \text{(E14.1.2)}$$

$$N_{mn}(\tau) = \int_0^a \int_0^b W_{mn}(x,y) f(x,y,t) \, dx \, dy \qquad \text{(E14.1.3)}$$

The concentrated force $F(t)$ can be expressed as

$$f(x,y,t) = F(t)\delta(x - x_0, y - y_0) \qquad \text{(E14.1.4)}$$

where $\delta(x - x_0, y - y_0)$ is a two-dimensional spatial Dirac delta function defined by

$$\delta(x - x_0, y - y_0) = 0 \qquad \text{for } x \neq x_0 \text{ and/or } y \neq y_0$$

$$\int_0^a \int_0^b \delta(x - x_0, y - y_0) \, dx \, dy = 1 \qquad \text{(E14.1.5)}$$

The natural frequencies ω_{mn} and the normal modes $W_{mn}(x,y)$ are given by Eqs. (14.165) and (14.162), respectively. By substituting Eqs. (14.162) and (E14.1.5), into Eq. (E14.1.3), we obtain

$$N_{mn}(\tau) = \frac{2}{\sqrt{\rho hab}} F(\tau) \int_0^a \int_0^b \sin \frac{m\pi x}{a} \sin \frac{n\pi y}{b} \delta(x - x_0, y - y_0) \, dx \, dy$$

$$= \frac{2F(\tau)}{\sqrt{\rho hab}} \sin \frac{m\pi x_0}{a} \sin \frac{n\pi y_0}{b} \qquad \text{(E14.1.6)}$$

With the help of Eq. (E14.1.6), Eq. (E14.1.2) can be written as

$$\eta_{mn}(t) = \frac{1}{\omega_{mn}} \int_0^t \frac{2F(\tau)}{\sqrt{\rho hab}} \sin \frac{m\pi x_0}{a} \sin \frac{n\pi y_0}{b} \sin \omega_{mn}(t - \tau) \, d\tau$$

$$= \frac{2}{\omega_{mn}\sqrt{\rho hab}} \sin \frac{m\pi x_0}{a} \sin \frac{n\pi y_0}{b} \int_0^t F(\tau) \sin \omega_{mn}(t - \tau) \, d\tau \qquad \text{(E14.1.7)}$$

If $F(t) = F_0 = \text{constant}$, Eq. (E14.1.7) becomes

$$\eta_{mn}(t) = \frac{2F_0}{\omega_{mn}^2 \sqrt{\rho hab}} \sin \frac{m\pi x_0}{a} \sin \frac{n\pi y_0}{b} (1 - \cos \omega_{mn}t) \qquad \text{(E14.1.8)}$$

If $F(t) = F_0 \sin \Omega t$, Eq. (E14.1.7) becomes

$$\eta_{mn}(t) = \frac{2F_0}{(\omega_{mn}^2 - \Omega^2)\sqrt{\rho hab}} \sin \frac{m\pi x_0}{a} \sin \frac{n\pi y_0}{b} (\omega_{mn} \sin \Omega t - \Omega \sin \omega_{mn}t)$$

$$\text{(E14.1.9)}$$

Once $\eta_{mn}(t)$ is known, the response can be found from Eq. (E14.1.1).

Example 14.2   A rectangular plate simply supported along all the edges is subjected to a harmonically varying pressure distribution given by

$$f(x,y,t) = f_0(x,y) \sin \Omega t \qquad \text{(E14.2.1)}$$

where Ω is the frequency of the applied force. Find the steady-state response of the plate.

SOLUTION   The equation of motion for the forced vibration of a rectangular plate can be expressed as

$$D\nabla^4 w + \rho h \frac{\partial^2 w}{\partial t^2} = f(x,y,t) \tag{E14.2.2}$$

where

$$f(x,y,t) = f_0(x,y) \sin \Omega t \tag{E14.2.3}$$

We assume the response of the plate, $w(x,y,t)$, also to be harmonic with

$$w(x,y,t) = W(x,y) \sin \Omega t \tag{E14.2.4}$$

where $W(x,y)$ indicates the harmonically varying displacement distribution. Using Eq. (E14.2.3), Eq. (E14.2.2) can be written as

$$\nabla^4 W - \lambda^4 W = \frac{f_0(x,y)}{D} \tag{E14.2.5}$$

where

$$\lambda^4 = \frac{\Omega^2 \rho h}{D} \tag{E14.2.6}$$

We express the pressure distribution $f_0(x,y)$ and the displacement distribution $W(x,y)$ in terms of the mode shapes or eigenfunctions of the plate $W_{mn}(x,y)$ as

$$W(x,y) = \sum_{m=1}^{\infty} \sum_{n=1}^{\infty} A_{mn} W_{mn}(x,y) \tag{E14.2.7}$$

$$f_0(x,y) = \sum_{m=1}^{\infty} \sum_{n=1}^{\infty} B_{mn} W_{mn}(x,y) \tag{E14.2.8}$$

where

$$A_{mn} = \int_0^a \int_0^b w(x,y) \, W_{mn}(x,y) \, dx \, dy \tag{E14.2.9}$$

$$B_{mn} = \int_0^a \int_0^b f_0(x,y) \, W_{mn}(x,y) \, dx \, dy \tag{E14.2.10}$$

The eigenvalue problem corresponding to Eq. (E14.2.2) can be expressed as

$$\nabla^4 W_{mn} - \lambda_{mn}^4 W_{mn} = 0, \qquad m, n = 1, \ 2, \ \ldots \tag{E14.2.11}$$

where

$$\lambda_{mn}^4 = \frac{\omega_{mn}^2 \rho h}{D} \tag{E14.2.12}$$

Let the eigenfunctions of the plate $W_{mn}(x,y)$ be normalized as

$$\int_0^a \int_0^b W_{mn}^2(x,y)\,dx\,dy = 1 \qquad (E14.2.13)$$

Multiply both sides of Eq. (E14.1.4) by $W_{mn}(x,y)$ and integrate over the area of the plate to obtain

$$\int_0^a \int_0^b [\nabla^4 W(x,y) - \lambda^4 W(x,y)] W_{mn}(x,y)\,dx\,dy$$

$$= \frac{1}{D} \int_0^a \int_0^b f_0(x,y)\,W_{mn}(x,y)\,dx\,dy \qquad (E14.2.14)$$

which upon integration by parts yields

$$A_{mn}(\lambda_{mn}^4 - \lambda^4) = \frac{B_{mn}}{D}$$

or

$$A_{mn} = \frac{B_{mn}}{D(\lambda_{mn}^4 - \lambda^4)} \qquad (E14.2.15)$$

By substituting Eqs. (E14.2.15) and (E14.2.10) into Eq. (E14.2.7), we obtain the displacement distribution of the plate as

$$W(x,y) = \sum_{m=1}^{\infty} \sum_{n=1}^{\infty} \frac{B_{mn} W_{mn}(x,y)}{D(\lambda_{mn}^4 - \lambda^4)}$$

$$= \frac{1}{\rho h} \sum_{m=1}^{\infty} \sum_{n=1}^{\infty} \frac{W_{mn}(x,y) \int_0^a \int_0^b f_0(x',y')\,W_{mn}(x',y')\,dx'\,dy'}{\omega_{mn}^2 - \Omega^2} \qquad (E14.2.16)$$

Note that Eq. (E14.2.16) is applicable to plates with arbitrary boundary conditions. In the case of a simply supported plate, the natural frequencies and normalized eigenfunctions are given by

$$\omega_{mn} = \pi^2 \sqrt{\frac{D}{\rho h}} \left[\left(\frac{m}{a}\right)^2 + \left(\frac{n}{b}\right)^2 \right], \qquad m,\,n = 1,\,2,\,\ldots \qquad (E14.2.17)$$

$$W_{mn}(x,y) = \frac{2}{\sqrt{\rho h a b}} \sin \frac{m\pi x}{a} \sin \frac{n\pi y}{b} \qquad (E14.2.18)$$

If $f_0(x,y) = f_0 = $ constant, the double integral in Eq. (E14.2.16) can be evaluated as

$$\int_0^a \int_0^b f_0(x',y')\,W_{mn}(x',y')\,dx'\,dy' = \frac{2f_0}{\sqrt{\rho h a b}} \int_0^a \int_0^b \sin \frac{m\pi x'}{a} \sin \frac{n\pi y'}{b}\,dx'\,dy'$$

$$= \frac{2f_0}{\sqrt{\rho h a b}} \frac{ab}{\pi^2 mn} (1 - \cos m\pi)\,(1 - \cos n\pi)$$

$$= \begin{cases} 0 & \text{if } m \text{ is even or } n \text{ is even} \\ \dfrac{8f_0\sqrt{ab}}{\pi^2 m\,n} & \text{if } m \text{ is odd and } n \text{ is odd} \end{cases}$$

$$(E14.2.19)$$

Thus, the response of the simply supported plate can be expressed, using Eqs. (E14.2.19), (E14.2.16) and (E14.2.4), as

$$w(x,y,t) = \frac{16 f_0}{\pi^2} \sum_{m=1,3,\dots}^{\infty} \sum_{n=1,3,\dots}^{\infty} \frac{\sin(m\pi x/a)\sin(n\pi y/b)\sin\Omega t}{mn\,\{\pi^4 D\,[(m/a)^2 + (n/b)^2]^2 - \rho h\Omega^2\}} \quad \text{(E14.2.20)}$$

14.6   CIRCULAR PLATES

14.6.1   Equation of Motion

Consider an infinitesimal element of the plate in polar coordinates as shown in Fig. 14.6. In this figure the radial moment M_r, tangential moment M_θ, twisting moments $M_{r\theta}$ and $M_{\theta r}$, and the transverse shear forces Q_r and Q_θ are shown on the positive and negative edges of the element. The equations of motion of the plate can be derived in polar coordinates by considering the dynamic equilibrium of the element shown in Fig. 14.6 as follows (see Problem 14.9): Moment equilibrium about the tangential (θ) direction:

$$\frac{\partial M_r}{\partial r} + \frac{1}{r}\frac{\partial M_{r\theta}}{\partial \theta} + \frac{M_r - M_\theta}{r} - Q_r = 0 \quad \text{(14.168)}$$

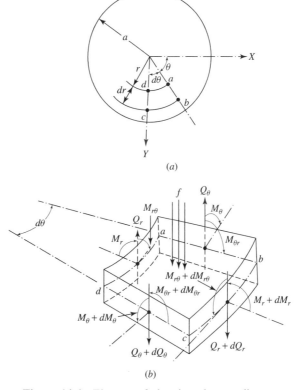

(a)

(b)

Figure 14.6   Element of plate in polar coordinates.

Moment equilibrium about the radial (R) direction:

$$\frac{\partial M_{r\theta}}{\partial r} + \frac{1}{r}\frac{\partial M_\theta}{\partial \theta} + \frac{2}{r}M_{r\theta} - Q_\theta = 0 \qquad (14.169)$$

Force equilibrium in the z direction:

$$\frac{\partial Q_r}{\partial r} + \frac{1}{r}\frac{\partial Q_\theta}{\partial \theta} + \frac{Q_r}{r} + f - \rho h\frac{\partial^2 w}{\partial t^2} = 0 \qquad (14.170)$$

Equations (14.168)–(14.170) can be combined to derive a single equation of motion in terms of the moment resultants M_r, M_θ, and $M_{r\theta}$. By substituting the moment resultants in terms of the transverse displacement w, the final equation of motion, shown in Eq. (14.183), can be obtained.

The coordinate transformation technique can also be used to derive the equation of motion in polar coordinates from the corresponding equation in Cartesian coordinates, as indicated below.

14.6.2   Transformation of Relations

The Cartesian and polar coordinates of a point P are related as (Fig. 14.7)

$$x = r\,\cos\theta, \qquad y = r\,\sin\theta \qquad (14.171)$$

$$r^2 = x^2 + y^2 \qquad (14.172)$$

$$\theta = \tan^{-1}\frac{y}{x} \qquad (14.173)$$

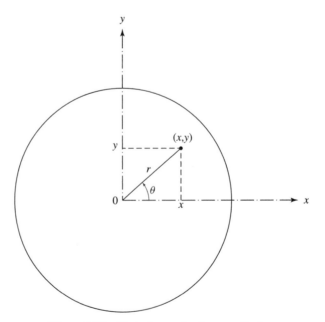

Figure 14.7   Cartesian and polar coordinates.

From Eqs. (14.172) and (14.171), we obtain

$$\frac{\partial r}{\partial x} = \frac{x}{r} = \cos\theta, \qquad \frac{\partial r}{\partial y} = \frac{y}{r} = \sin\theta \qquad (14.174)$$

Similarly, Eqs. (14.173) and (14.171) give

$$\frac{\partial\theta}{\partial r} = -\frac{y}{r^2} = -\frac{\sin\theta}{r}, \qquad \frac{\partial\theta}{\partial y} = \frac{x}{r^2} = \frac{\cos\theta}{r} \qquad (14.175)$$

Since the deflection of the plate w is a function of r and θ, the chain rule of differentiation yields

$$\frac{\partial w}{\partial x} = \frac{\partial w}{\partial r}\frac{\partial r}{\partial x} + \frac{\partial w}{\partial\theta}\frac{\partial\theta}{\partial x} = \frac{\partial w}{\partial r}\cos\theta - \frac{1}{r}\frac{\partial w}{\partial\theta}\sin\theta \qquad (14.176)$$

$$\frac{\partial w}{\partial y} = \frac{\partial w}{\partial r}\frac{\partial r}{\partial y} + \frac{\partial w}{\partial\theta}\frac{\partial\theta}{\partial y} = \frac{\partial w}{\partial r}\sin\theta + \frac{\partial w}{\partial\theta}\frac{\cos\theta}{r} \qquad (14.177)$$

For the expressions $\partial^2 w/\partial x^2$, $\partial^2 w/\partial x\,\partial y$ and $\partial^2 w/\partial y^2$, the operations $\partial/\partial x$ and $\partial/\partial y$ of Eqs. (14.176) and (14.177) are repeated to obtain

$$\frac{\partial^2 w}{\partial x^2} = \frac{\partial}{\partial x}\left(\frac{\partial w}{\partial x}\right) = \frac{\partial}{\partial r}\left(\frac{\partial w}{\partial x}\right)\cos\theta - \frac{1}{r}\frac{\partial}{\partial\theta}\left(\frac{\partial w}{\partial x}\right)\sin\theta$$

$$= \frac{\partial^2 w}{\partial r^2}\cos^2\theta - \frac{\partial^2 w}{\partial\theta\partial r}\frac{\sin 2\theta}{r} + \frac{\partial w}{\partial r}\frac{\sin^2\theta}{r} + \frac{\partial w}{\partial\theta}\frac{\sin 2\theta}{r^2} + \frac{\partial^2 w}{\partial\theta^2}\frac{\sin^2\theta}{r^2}$$
$$(14.178)$$

$$\frac{\partial^2 w}{\partial y^2} = \frac{\partial}{\partial y}\left(\frac{\partial w}{\partial y}\right) = \frac{\partial}{\partial r}\left(\frac{\partial w}{\partial y}\right)\sin\theta + \frac{\partial}{\partial\theta}\left(\frac{\partial w}{\partial y}\right)\frac{\cos\theta}{r}$$

$$= \frac{\partial^2 w}{\partial r^2}\sin^2\theta + \frac{\partial^2 w}{\partial r\partial\theta}\frac{\sin 2\theta}{r} + \frac{\partial w}{\partial r}\frac{\cos^2\theta}{r} - \frac{\partial w}{\partial\theta}\frac{\sin 2\theta}{r^2} + \frac{\partial^2 w}{\partial\theta^2}\frac{\cos^2\theta}{r^2}$$
$$(14.179)$$

$$\frac{\partial^2 w}{\partial x\,\partial y} = \frac{\partial}{\partial x}\left(\frac{\partial w}{\partial y}\right) = \frac{\partial}{\partial r}\left(\frac{\partial w}{\partial y}\right)\cos\theta - \frac{1}{r}\frac{\partial}{\partial\theta}\left(\frac{\partial w}{\partial y}\right)\sin\theta$$

$$= \frac{\partial^2 w}{\partial r^2}\frac{\sin 2\theta}{2} + \frac{\partial^2 w}{\partial r\partial\theta}\frac{\cos 2\theta}{r} - \frac{\partial w}{\partial\theta}\frac{\cos 2\theta}{r^2} - \frac{\partial w}{\partial r}\frac{\sin 2\theta}{2r} - \frac{\partial^2 w}{\partial\theta^2}\frac{\sin 2\theta}{2r^2}$$
$$(14.180)$$

By adding Eqs. (14.178) and (14.179), we obtain

$$\nabla^2 w = \frac{\partial^2 w}{\partial x^2} + \frac{\partial^2 w}{\partial y^2} = \frac{\partial^2 w}{\partial r^2} + \frac{1}{r}\frac{\partial w}{\partial r} + \frac{1}{r^2}\frac{\partial^2 w}{\partial\theta^2} \qquad (14.181)$$

By repeating the operation ∇^2 twice, we can express

$$\nabla^4 w = \nabla^2(\nabla^2 w) = \left(\frac{\partial^2}{\partial r^2} + \frac{1}{r}\frac{\partial}{\partial r} + \frac{1}{r^2}\frac{\partial^2}{\partial\theta^2}\right)\left(\frac{\partial^2 w}{\partial r^2} + \frac{1}{r}\frac{\partial w}{\partial r} + \frac{1}{r^2}\frac{\partial^2 w}{\partial\theta^2}\right)$$

$$= \frac{\partial^4 w}{\partial r^4} + \frac{2}{r}\frac{\partial^3 w}{\partial r^3} - \frac{1}{r^2}\frac{\partial^2 w}{\partial r^2} + \frac{1}{r^3}\frac{\partial w}{\partial r} + \frac{2}{r^2}\frac{\partial^4 w}{\partial r^2\partial\theta^2} - \frac{2}{r^3}\frac{\partial^3 w}{\partial\theta^2\partial r}$$

$$+ \frac{4}{r^4}\frac{\partial^2 w}{\partial\theta^2} + \frac{1}{r^4}\frac{\partial^4 w}{\partial\theta^4} \tag{14.182}$$

Using Eqs. (14.178), (14.179), and (14.180) in (14.8), the equation of motion for the forced transverse vibration of a circular plate can be expressed as

$$D\nabla^4 w + \rho h \frac{\partial^2 w}{\partial t^2} = f$$

or

$$D\left(\frac{\partial^4 w}{\partial r^4} + \frac{2}{r}\frac{\partial^3 w}{\partial r^3} - \frac{1}{r^2}\frac{\partial^2 w}{\partial r^2} + \frac{1}{r^3}\frac{\partial w}{\partial r} + \frac{2}{r^2}\frac{\partial^4 w}{\partial r^2\partial\theta^2}\right.$$

$$\left. - \frac{2}{r^3}\frac{\partial^3 w}{\partial r\partial\theta^2} + \frac{4}{r^4}\frac{\partial^2 w}{\partial\theta^2} + \frac{1}{r^4}\frac{\partial^4 w}{\partial\theta^4}\right) + \rho h \frac{\partial^2 w}{\partial t^2} = f(r, \theta, t) \tag{14.183}$$

14.6.3   Moment and Force Resultants

Using the transformation procedure, the moment resultant–transverse displacement relations can be obtained as (see Problem 14.17)

$$M_r = -D\left[\frac{\partial^2 w}{\partial r^2} + \nu\left(\frac{1}{r}\frac{\partial w}{\partial r} + \frac{1}{r^2}\frac{\partial^2 w}{\partial\theta^2}\right)\right] \tag{14.184}$$

$$M_\theta = -D\left(\frac{1}{r}\frac{\partial w}{\partial r} + \frac{1}{r^2}\frac{\partial^2 w}{\partial\theta^2} + \nu\frac{\partial^2 w}{\partial r^2}\right) \tag{14.185}$$

$$M_{r\theta} = M_{\theta r} = -(1-\nu)D\frac{\partial}{\partial r}\left(\frac{1}{r}\frac{\partial w}{\partial\theta}\right) \tag{14.186}$$

Similarly, the shear force resultants can be expressed as

$$Q_r = \frac{1}{r}\left[\frac{\partial}{\partial r}(rM_r) - M_\theta + \frac{\partial M_{r\theta}}{\partial\theta}\right] \tag{14.187a}$$

$$= -D\frac{\partial}{\partial r}\left(\frac{\partial^2 w}{\partial r^2} + \frac{1}{r}\frac{\partial w}{\partial r} + \frac{1}{r^2}\frac{\partial^2 w}{\partial\theta^2}\right) = -D\frac{\partial}{\partial r}(\nabla^2 w) \tag{14.187b}$$

$$Q_\theta = \frac{1}{r}\left[\frac{\partial}{\partial r}(rM_{r\theta}) + \frac{\partial M_\theta}{\partial\theta} + M_{r\theta}\right] \tag{14.188a}$$

$$= -D\frac{1}{r}\frac{\partial}{\partial\theta}\left(\frac{\partial^2 w}{\partial r^2} + \frac{1}{r}\frac{\partial w}{\partial r} + \frac{1}{r^2}\frac{\partial^2 w}{\partial\theta^2}\right) = -D\frac{1}{r}\frac{\partial}{\partial\theta}(\nabla^2 w) \tag{14.188b}$$

The effective transverse shear forces can be written as

$$V_r = Q_r + \frac{1}{r}\frac{\partial M_{r\theta}}{\partial \theta} = -D\left[\frac{\partial}{\partial r}(\nabla^2 w) + \frac{1-\nu}{r}\frac{\partial}{\partial \theta}\left(\frac{1}{r}\frac{\partial^2 w}{\partial r\,\partial \theta} - \frac{1}{r^2}\frac{\partial w}{\partial \theta}\right)\right] \quad (14.189)$$

$$V_\theta = Q_\theta + \frac{\partial M_{r\theta}}{\partial r} = -D\left[\frac{1}{r}\frac{\partial}{\partial \theta}(\nabla^2 w) + (1-\nu)\frac{\partial}{\partial r}\left(\frac{1}{r}\frac{\partial^2 w}{\partial r\,\partial \theta} - \frac{1}{r^2}\frac{\partial w}{\partial \theta}\right)\right]$$
$$(14.190)$$

Note that the Laplacian operator appearing in Eqs. (14.187)–(14.190) is given in polar coordinates by Eq. (14.181).

14.6.4 Boundary Conditions

1. *Clamped, fixed, or built-in edge.* The deflection and slope (normal to the boundary) must be zero:

$$w = 0 \qquad\qquad (14.191)$$

$$\frac{\partial w}{\partial r} = 0 \qquad\qquad (14.192)$$

where r denotes the radial (normal) direction to the boundary.

2. *Simply supported edge.* The deflection and bending moment resultant must be zero:

$$w = 0 \qquad\qquad (14.193)$$

$$M_r = -D\left[\frac{\partial^2 w}{\partial r^2} + \nu\left(\frac{1}{r}\frac{\partial w}{\partial r} + \frac{1}{r^2}\frac{\partial^2 w}{\partial \theta^2}\right)\right] = 0 \qquad (14.194)$$

3. *Free edge.* The bending moment resultant and the effective shear force resultant on the edge must be zero:

$$M_r = -D\left[\frac{\partial^2 w}{\partial r^2} + \nu\left(\frac{1}{r}\frac{\partial w}{\partial r} + \frac{1}{r^2}\frac{\partial^2 w}{\partial \theta^2}\right)\right] = 0 \qquad (14.195)$$

$$V_r = Q_r + \frac{1}{r}\frac{\partial M_{r\theta}}{\partial \theta} = 0 \qquad (14.196)$$

or

$$-D\left[\frac{\partial}{\partial r}(\nabla^2 w) + \frac{1-\nu}{r}\frac{\partial}{\partial \theta}\left(\frac{1}{r}\frac{\partial^2 w}{\partial r\,\partial \theta} - \frac{1}{r^2}\frac{\partial w}{\partial \theta}\right)\right] = 0 \qquad (14.197)$$

4. *Edge supported on elastic springs.* If the edge is supported on linear and torsional springs all around as shown in Fig. 14.8, the boundary conditions can be stated as follows:

$$M_r = -k_{t_0}\frac{\partial w}{\partial r}$$

or

$$-D\left[\frac{\partial^2 w}{\partial r^2} + \nu\left(\frac{1}{r}\frac{\partial w}{\partial r} + \frac{1}{r^2}\frac{\partial^2 w}{\partial \theta^2}\right)\right] = -k_{t_0}\frac{\partial w}{\partial r} \qquad (14.198)$$

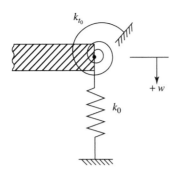

Figure 14.8   Edge supported on elastic springs.

$$V_r = -k_0 w$$

or

$$-D\left[\frac{\partial}{\partial r}(\nabla^2 w) + \frac{1-\nu}{r}\frac{\partial}{\partial \theta}\left(\frac{1}{r}\frac{\partial^2 w}{\partial r\,\partial \theta} - \frac{1}{r^2}\frac{\partial w}{\partial \theta}\right)\right] = -k_0 w \qquad (14.199)$$

14.7   FREE VIBRATION OF CIRCULAR PLATES

The equation of motion of a circular plate is given by Eq. (14.183):

$$D\nabla^4 w + \rho h\frac{\partial^2 w}{\partial t^2} = f \qquad (14.200)$$

where

$$\nabla^2 = \frac{\partial^2}{\partial r^2} + \frac{1}{r}\frac{\partial}{\partial r} + \frac{1}{r^2}\frac{\partial^2}{\partial \theta^2} \qquad (14.201)$$

For free vibrations of the plate, Eq. (14.200) gives, after separation of variables,

$$\frac{d^2 T(t)}{dt^2} + \omega^2 T(t) = 0 \qquad (14.202)$$

$$\nabla^4 W(r,\theta) - \lambda^4 W(r,\theta) = 0 \qquad (14.203)$$

where

$$\lambda^4 = \frac{\rho h\omega^2}{D} \qquad (14.204)$$

Using Eq. (14.201), Eq. (14.203) can be written as two separate equations:

$$\frac{\partial^2 W}{\partial r^2} + \frac{1}{r}\frac{\partial W}{\partial r} + \frac{1}{r^2}\frac{\partial^2 W}{\partial \theta^2} + \lambda^2 W = 0 \qquad (14.205)$$

$$\frac{\partial^2 W}{\partial r^2} + \frac{1}{r}\frac{\partial W}{\partial r} + \frac{1}{r^2}\frac{\partial^2 W}{\partial \theta^2} - \lambda^2 W = 0 \qquad (14.206)$$

By expressing $W(r, \theta) = R(r)\Theta(\theta)$, Eqs. (14.205) and (14.206) can be rewritten [after dividing each equation by $R(r)\Theta(\theta)/r^2$] as

$$\frac{r^2}{R(r)} \left[\frac{d^2 R(r)}{dr^2} + \frac{1}{r} \frac{d R(r)}{dr} \pm \lambda^2 \right] = -\frac{1}{\Theta(\theta)} \frac{d^2 \Theta}{d\theta^2} = \alpha^2$$

where α^2 is a constant. Thus,

$$\frac{d^2 \Theta}{d\theta^2} + \alpha^2 \Theta = 0 \tag{14.207}$$

$$\frac{d^2 R}{dr^2} + \frac{1}{r} \frac{d R}{dr} + \left(\pm \lambda^2 - \frac{\alpha^2}{r^2} \right) R = 0 \tag{14.208}$$

The solution of Eq. (14.207) is

$$\Theta(\theta) = A \cos \alpha\theta + B \sin \alpha\theta \tag{14.209}$$

Since $W(r, \theta)$ has to be a continuous function, $\Theta(\theta)$ must be a periodic function with a period of 2π so that $W(r, \theta) = W(r, \theta + 2\pi)$. Thus, α must be an integer:

$$\alpha = m, \qquad m = 0, 1, 2, \ldots \tag{14.210}$$

Equation (14.208) can be rewritten as two separate equations:

$$\frac{d^2 R}{dr^2} + \frac{1}{r} \frac{d R}{dr} + \left(\lambda^2 - \frac{\alpha^2}{r^2} \right) R = 0 \tag{14.211}$$

$$\frac{d^2 R}{dr^2} + \frac{1}{r} \frac{d R}{dr} - \left(\lambda^2 + \frac{\alpha^2}{r^2} \right) R = 0 \tag{14.212}$$

Equation (14.211) can be seen to be a Bessel differential equation [as in the case of a circular membrane, (Eq. (13.131))] of order $m (= \alpha)$ with argument λr whose solution is given by

$$R_1(r) = C_1 J_m(\lambda r) + C_2 Y_m(\lambda r) \tag{14.213}$$

where J_m and Y_m are Bessel functions of order m of the first and second kind, respectively. Equation (14.212) is a Bessel differential equation of order $m (= \alpha)$ with the imaginary argument $i\lambda r$ whose solution is given by

$$R_2(r) = C_3 I_m(\lambda r) + C_4 K_m(\lambda r) \tag{14.214}$$

where I_m and K_m are the hyperbolic or modified Bessel functions of order m of the first and second kind, respectively. The general solution of Eq. (14.203) can be expressed as

$$W(r, \theta) = [C_m^{(1)} J_m(\lambda r) + C_m^{(2)} Y_m(\lambda r) + C_m^{(3)} I_m(\lambda r)$$
$$+ C_m^{(4)} K_m(\lambda r)](A_m \cos m\theta + B_m \sin m\theta), \qquad m = 0, 1, 2, \ldots \tag{14.215}$$

where the constants $C_m^{(1)}, \ldots, C_m^{(4)}$, A_m, B_m, and λ depend on the boundary conditions of the plate. The boundary conditions of the plate are given in Section 14.6.4.

14.7.1   Solution for a Clamped Plate

For a clamped plate of radius a the boundary conditions, in terms of $W(r, \theta)$, are

$$W(a, \theta) = 0 \tag{14.216}$$

$$\frac{\partial W}{\partial r}(a, \theta) = 0 \tag{14.217}$$

In addition, the solution $W(r, \theta)$ at all points inside the plate must be finite. For this, the constants $C_m^{(2)}$ and $C_m^{(4)}$ must be zero as the Bessel functions of the second kind, $Y_m(\lambda r)$ and $K_m(\lambda r)$, become infinite at $r = 0$. Thus, Eq. (14.215) reduces to

$$W(r, \theta) = [C_m^{(1)} J_m(\lambda r) + C_m^{(3)} I_m(\lambda r)](A_m \cos m\theta + B_m \sin m\theta), \qquad m = 0, 1, 2, \ldots \tag{14.218}$$

The boundary condition of Eq. (14.216) gives

$$C_m^{(3)} = -\frac{J_m(\lambda a)}{I_m(\lambda a)} C_m^{(1)} \tag{14.219}$$

so that

$$W(r, \theta) = \left[J_m(\lambda r) - \frac{J_m(\lambda a)}{I_m(\lambda a)} I_m(\lambda r) \right] (A_m \cos m\theta + B_m \sin m\theta), \qquad m = 0, 1, 2, \ldots \tag{14.220}$$

where A_m and B_m are new constants. Finally, Eqs. (14.217) and (14.220) give the frequency equations:

$$\left[\frac{d}{dr} J_m(\lambda r) - \frac{J_m(\lambda a)}{I_m(\lambda a)} \frac{d}{dr} I_m(\lambda r) \right]_{r=a} = 0, \qquad m = 0, 1, 2, \ldots \tag{14.221}$$

From the known relations [13, 36]

$$\frac{d}{dr} J_m(\lambda r) = \lambda J_{m-1}(\lambda r) - \frac{m}{r} J_m(\lambda r) \tag{14.222}$$

$$\frac{d}{dr} I_m(\lambda r) = \lambda I_{m-1}(\lambda r) - \frac{m}{r} I_m(\lambda r) \tag{14.223}$$

the frequency equations can be expressed as

$$I_m(\lambda a) J_{m-1}(\lambda a) - J_m(\lambda a) I_{m-1}(\lambda a) = 0, \qquad m = 0, 1, 2, \ldots \tag{14.224}$$

For a given value of m, we have to solve Eq. (14.224) and find the roots (eigenvalues) λ_{mn} from which the natural frequencies can be computed, using Eq. (14.204), as

$$\omega_{mn} = \lambda_{mn}^2 \left(\frac{D}{\rho h} \right)^{1/2} \tag{14.225}$$

As in the case of membranes, we find that for each frequency ω_{mn}, there are two natural modes (except for $m = 0$ for which there is only one mode). Hence, all the natural modes (except for $m = 0$) are degenerate. The two mode shapes are given by

$$W_{mn}^{(1)}(r, \theta) = [J_m(\lambda r) I_m(\lambda a) - J_m(\lambda a) I_m(\lambda r)] \cos m\theta \tag{14.226}$$

$$W_{mn}^{(2)}(r, \theta) = [J_m(\lambda r) I_m(\lambda a) - J_m(\lambda a) I_m(\lambda r)] \sin m\theta \tag{14.227}$$

The two natural modes of vibration corresponding to ω_{mn} are given by

$$w_{mn}^{(1)}(r, \theta, t) = \cos m\theta [I_m(\lambda_{mn} a) J_m(\lambda_{mn} r) - J_m(\lambda_{mn} a) I_m(\lambda_{mn} r)]$$
$$\cdot (A_{mn}^{(1)} \cos \omega_{mn} t + A_{mn}^{(2)} \sin \omega_{mn} t) \tag{14.228}$$

$$w_{mn}^{(2)}(r, \theta, t) = \sin m\theta [I_m(\lambda_{mn} a) J_m(\lambda_{mn} r) - J_m(\lambda_{mn} a) I_m(\lambda_{mn} r)]$$
$$\cdot (A_{mn}^{(3)} \cos \omega_{mn} t + A_{mn}^{(4)} \sin \omega_{mn} t) \tag{14.229}$$

The general solution of Eq. (14.200) with $f = 0$ can be expressed as

$$w(r, \theta, t) = \sum_{m=0}^{\infty} \sum_{n=0}^{\infty} [w_{mn}^{(1)}(r, \theta, t) + w_{mn}^{(2)}(r, \theta, t)] \tag{14.230}$$

and the constants $A_{mn}^{(1)}, \ldots, A_{mn}^{(4)}$ can be determined from the initial conditions. Some of the first few roots of Eq. (14.224) are $\lambda_{01} a = 3.196$, $\lambda_{02} a = 6.306$, $\lambda_{03} a = 9.439$, $\lambda_{11} a = 4.611$, $\lambda_{12} a = 7.799$, $\lambda_{13} a = 10.958$, $\lambda_{21} a = 5.906$, $\lambda_{22} a = 9.197$, and $\lambda_{23} a = 12.402$. Note that the mode shape $W_{mn}(r, \theta)$ will have m nodal diameters and n nodal circles, including the boundary of the circular plate. The first few mode shapes of the clamped circular plate are shown in Fig. 14.9.

14.7.2   Solution for a Plate with a Free Edge

For a circular plate of radius a with a free edge, the boundary conditions are given by Eqs. (14.195) and (14.197):

$$\frac{\partial^2 W}{\partial r^2} + \frac{\nu}{r} \frac{\partial W}{\partial r} + \frac{\nu}{r^2} \frac{\partial^2 W}{\partial \theta^2} = 0 \quad \text{at} \quad r = a \tag{14.231}$$

$$\frac{\partial}{\partial r} \left(\frac{\partial^2 W}{\partial r^2} + \frac{1}{r} \frac{\partial W}{\partial r} + \frac{1}{r^2} \frac{\partial^2 W}{\partial \theta^2} \right) + \frac{1 - \nu}{r^2} \frac{\partial^2}{\partial \theta^2} \left(\frac{\partial W}{\partial r} - \frac{W}{r} \right) = 0 \quad \text{at} \quad r = a \tag{14.232}$$

By substituting Eq. (14.215) into Eqs. (14.231) and (14.232), the frequency equation can be derived as [1]

$$\frac{(\lambda a)^2 J_m(\lambda a) + (1 - \nu)[\lambda a J_m'(\lambda a) - m^2 J_m(\lambda a)]}{(\lambda a)^2 I_m(\lambda a) - (1 - \nu)[\lambda a I_m'(\lambda a) - m^2 I_m(\lambda a)]}$$
$$= \frac{(\lambda a)^2 I_m'(\lambda a) + (1 - \nu)m^2[\lambda a J_m'(\lambda a) - J_m(\lambda a)]}{(\lambda a)^2 I_m'(\lambda a) - (1 - \nu)m^2[\lambda a I_m'(\lambda a) - I_m(\lambda a)]} \tag{14.233}$$

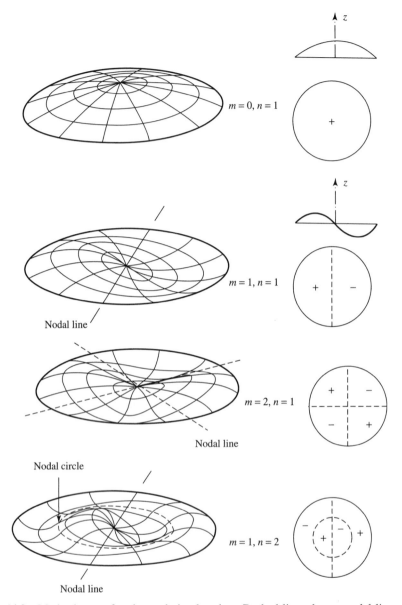

Figure 14.9 Mode shapes of a clamped circular plate. Dashed lines denote nodal lines within the plate.

where a prime denotes derivative with respect to the argument. When $\lambda a > m$, the frequency equation (14.233) can be approximated by the equation [16]

$$\frac{J_m(\lambda a)}{J'_m(\lambda a)} \approx \frac{[(\lambda a)^2 + 2(1 - \nu)m^2][I_m(\lambda a)/I'_m(\lambda a)] - 2\lambda a(1 - \nu)}{(\lambda a)^2 - 2(1 - \nu)m^2} \tag{14.234}$$

The first few roots of Eq. (14.234) are given in Table 14.2.

Table 14.2   Natural Frequencies of Vibration of a Free Circular Plate with $v = 0.33^a$

	Number of nodal diameters, m			
Number of nodal circles, n	0	1	2	3
0	—	—	5.253	12.23
1	9.084	20.52	35.25	52.91
2	38.55	59.86	83.9	111.3
3	87.80	119.0	154.0	192.1

Source: Data from Refs. [1] and [16].
^{a} Values of $(\lambda a)^2 = \omega a^2 \sqrt{\rho h / D}$.

14.8   FORCED VIBRATION OF CIRCULAR PLATES

We shall consider the axisymmetric vibrations of a circular plate in this section. The equation of motion governing the axisymmetric vibrations of a circular plate is given by

$$p^2 \left(\frac{\partial^2}{\partial r^2} + \frac{1}{r} \frac{\partial}{\partial r} \right)^2 w(r, t) + \ddot{w}(r, t) = \frac{1}{\rho h} f(r, t) \tag{14.235}$$

where

$$p^2 = \frac{D}{\rho h} \tag{14.236}$$

and a dot over w denotes a partial derivative with respect to time. The boundary conditions for a plate simply supported around the boundary $r = a$ are given by

$$w = 0 \quad \text{at} \quad r = a \tag{14.237}$$

$$\frac{\partial^2 w}{\partial r^2} + \frac{v}{r} \frac{\partial w}{\partial r} = 0 \quad \text{at} \quad r = a \tag{14.238}$$

To simplify the solution, the boundary condition of Eq. (14.238) is taken approximately as [14]

$$\frac{\partial^2 w}{\partial r^2} + \frac{1}{r} \frac{\partial w}{\partial r} = 0 \quad \text{at} \quad r = a \tag{14.239}$$

Equations (14.237) and (14.239) imply that the plate is supported at the boundary, $r = a$, such that the deflection and the curvature are zero. Thus, Eq. (14.239) will be satisfied to a greater extent for larger plates (with large values of a) than for smaller plates (with small values of a).

14.8.1   Harmonic Forcing Function

The forcing function, $f(r, t)$, is assumed to be harmonic, with frequency Ω, as

$$f(r, t) = F(r) e^{i \Omega t} \tag{14.240}$$

The solution of Eq. (14.235) is assumed to be of the form

$$w(x, t) = W(r)e^{i\Omega t} \tag{14.241}$$

Using Eqs. (14.240) and (14.241), the equation of motion, Eq. (14.235), can be expressed as

$$\left(\frac{d^2}{dr^2} + \frac{1}{r}\frac{d}{dr}\right)^2 W(r) - \lambda^4 W(r) = \frac{1}{D}F(r) \tag{14.242}$$

where

$$\lambda^4 = \frac{\Omega^2}{p^2} = \frac{\Omega^2 \rho h}{D} \tag{14.243}$$

Equation (14.242) can be solved conveniently by applying Hankel transforms. For this, we multiply both sides of Eq. (14.242) by $r J_0(\lambda r)$ and integrate with respect to r from 0 to a to obtain

$$\int_0^a \left(\frac{d^2}{dr^2} + \frac{1}{r}\frac{d}{dr}\right)^2 W(r) r J_0(\lambda r)\, dr - \lambda^4 \overline{W}(\lambda) = \frac{1}{D}\overline{F}(\lambda) \tag{14.244}$$

where $\overline{W}(\lambda)$ and $\overline{F}(\lambda)$ are called the finite Hankel transforms of $W(r)$ and $F(r)$, respectively, and are defined as

$$\overline{W}(\lambda) = \int_0^a r W(r) J_0(\lambda r)\, dr \tag{14.245}$$

$$\overline{F}(\lambda) = \int_0^a r F(r) J_0(\lambda r)\, dr \tag{14.246}$$

To simplify Eq. (14.244), first consider the integral

$$I = \int_0^a r \left(\frac{d^2 W}{dr^2} + \frac{1}{r}\frac{dW}{dr}\right) J_0(\lambda r)\, dr \tag{14.247}$$

Using integration by parts, this integral can be evaluated as

$$I = \left[r\frac{dW}{dr} J_0(\lambda r) - \lambda r W J_0'(\lambda r)\right]_0^a - \lambda^2 \int_0^a r W J_0(\lambda r)\, dr \tag{14.248}$$

The expression in brackets in Eq. (14.248) will always be zero at $r = 0$ and will be zero at $r = a$ if λ is chosen to satisfy the relation

$$J_0(\lambda a) = 0 \tag{14.249}$$

or

$$J_0(\lambda_i a) = 0, \qquad i = 1, 2, \ldots \tag{14.250}$$

where $\lambda_i a$ is the ith root of Eq. (14.249). Thus, Eq. (14.247) takes the form

$$\int_0^a r \left(\frac{d^2 W}{dr^2} + \frac{1}{r}\frac{dW}{dr}\right) J_0(\lambda r)\, dr = -\lambda_i^2 \overline{W}(\lambda_i) \tag{14.251}$$

When the result of Eq. (14.251) is applied twice, we obtain

$$\int_0^a r \left(\frac{d^2}{dr^2} + \frac{1}{r} \frac{d}{dr} \right)^2 W(r) r J_0(\lambda r) \, dr = \lambda_i^4 \overline{W}(\lambda_i) \tag{14.252}$$

In view of Eq. (14.252), Eq. (14.244) yields

$$\overline{W}(\lambda_i) = \frac{1}{D} \frac{\overline{F}(\lambda_i)}{\lambda_i^4 - \lambda^4} \tag{14.253}$$

By taking the inverse Hankel transforms, we obtain [13–15]

$$W(r) = \frac{2}{a^2} \sum_{i=1,2,\cdots} \overline{W}(\lambda_i) \frac{J_0(\lambda_i r)}{[J_1(\lambda_i a)]^2} = \frac{2}{a^2 D} \sum_{i=1,2,\ldots} \frac{J_0(\lambda_i r) \overline{F}(\lambda_i)}{[J_1(\lambda_1 a)]^2 (\lambda_i^4 - \lambda^4)} \tag{14.254}$$

Note that as the forcing frequency Ω approaches the ith natural frequency of vibration of the plate, $\lambda_i^2 \sqrt{D/\rho h}$, the deflection of the plate $W(r) \to \infty$, thereby causing resonance.

14.8.2   General Forcing Function

For a general forcing function $f(r, t)$, the Hankel transforms of $w(r, t)$ and $f(r, t)$ are defined as

$$\overline{W}(\lambda) = \int_0^a r w(r, t) J_0(\lambda r) \, dr \tag{14.255}$$

$$\overline{F}(\lambda) = \int_0^a r f(r, t) J_0(\lambda r) \, dr \tag{14.256}$$

Using a procedure similar to the one used in the case of a harmonic forcing function, we obtain from Eq. (14.235),

$$\frac{d^2 \overline{W}(\lambda_i)}{dt^2} + \lambda_i^4 \overline{W}(\lambda_i) = \frac{1}{D} \overline{F}(\lambda_i) \tag{14.257}$$

The solution of Eq. (14.257), after taking the inverse Hankel transforms, can be expressed as [14, 15]

$$w(r, t) = \frac{2}{a^2} \sum_{i=1,2,\ldots} \frac{J_0(\lambda_i r)}{[J_1(\lambda_i a)]^2} \int_0^a r J_0(\lambda_i r) \left[w_0(r) \cos \omega_i t + \frac{\dot{w}_0(r)}{\omega_i} \sin \omega_i t \right] dr$$
$$+ \frac{2}{a^2 \rho h} \sum_{i=1,2,\ldots} \frac{J_0(\lambda_i r)}{[J_1(\lambda_i a)]^2} \int_0^a r J_0(\lambda_i r) \, dr \int_0^t \frac{f(r, \tau)}{\omega_i} \sin \omega_i (t - \tau) \, d\tau \tag{14.258}$$

where

$$w_0(r) = w(r, t = 0) \tag{14.259}$$

$$\dot{w}_0(r) = \frac{\partial w}{\partial t}(r, t = 0) \tag{14.260}$$

denote the known initial conditions of the plate and

$$\omega_i = p\lambda_i^2 = \sqrt{\frac{D}{\rho h}}\lambda_i^2 \tag{14.261}$$

Note that the first term on the right-hand side of Eq. (14.258) denotes transient vibrations due to the initial conditions, and the second term represents the steady-state vibrations due to the forcing function specified.

To illustrate the use of Eq. (14.258) for free (transient) vibration response, consider a plate subjected to the initial displacement

$$w_0(r) = b\left(1 - \frac{r^2}{a^2}\right), \qquad 0 \le r \le a \tag{14.262}$$

where b indicates the displacement of the center of the plate (assumed to be small) and $\dot{w}_0(r) = f(r, t) = 0$. Using the initial condition of Eq. (14.262) and noting that

$$\int_0^a r w_0(r) J_0(\lambda_i r)\, dr = \frac{b}{a^2} \int_0^a (a^2 - r^2) r J_0(\lambda_i r)\, dr = \frac{4b J_1(\lambda_i a)}{a \lambda_i^3} \tag{14.263}$$

the free (transient) vibration response of the plate can be obtained from Eq. (14.258) as

$$w(r, t) = \frac{8b}{a^3} \sum_{i=1,2,\dots} \frac{J_0(\lambda_i r)}{J_1(\lambda_1 a)} \frac{\cos \omega_i t}{\lambda_i^3} \tag{14.264}$$

To illustrate the use of Eq. (14.258) for forced vibration response, consider the steady-state response of a plate subjected to a constant distributed force of magnitude d_0 acting on a circle of radius c_0 suddenly applied at $t = 0$. In this case,

$$f(r, t) = d_0 H(c_0 - r) H(t) \tag{14.265}$$

where $H(x)$ denotes the Heaviside unit function defined by

$$H(x) = \begin{cases} 0 & \text{for } x < 0 \\ 1 & \text{for } x \ge 0 \end{cases} \tag{14.266}$$

Using the relation

$$\int_0^a r J_0(\lambda_i r)\, dr \int_0^t \frac{f(r, \tau)}{\omega_i} \sin \omega_i(t - \tau)\, d\tau$$

$$= d_0 \int_0^a r J_0(\lambda_i r) H(c_0 - r)\, dr \int_0^t \frac{H(\tau)}{\omega_i} \sin \omega_i(t - \tau)\, d\tau = d_0 \frac{1 - \cos \omega_i t}{\omega_i^2}$$

$$\int_0^{c_0} r J_0(\lambda_i r)\, dr$$

$$= \frac{d_0 c_0 a J_1(\lambda_i c_0)}{\lambda_i \omega_i^2}(1 - \cos \omega_i t) \tag{14.267}$$

the forced response of the plate, given by Eq. (14.258), can be expressed as

$$w(r, t) = \frac{2d_0 c_0}{aD} \sum_{i=1,2,\dots} \frac{J_0(\lambda_i r) J_1(\lambda_i c_0)}{\lambda_i^5 [J_1(\lambda_i a)]^2} (1 - \cos p\lambda_i^2 t) \qquad (14.268)$$

14.9   EFFECTS OF ROTARY INERTIA AND SHEAR DEFORMATION

In the derivation of Eq. (14.8) we assumed that the thickness of the plate is small compared to its other dimensions, and the effects of rotary inertia and shear deformation are small. In the case of beams a method of accounting for the effects of rotary inertia and shear deformation, according to Timoshenko beam theory, was presented in Section 11.15. Mindlin extended the Timoshenko beam theory and presented a method of including the effects of rotary inertia and shear deformation in the dynamic analysis of plates [3, 4]. We discuss the essential features of Mindlin plate theory in this section.

14.9.1   Equilibrium Approach

Strain–Displacement Relations   We assume that the middle plane of the plate lies in the xy plane before deformation and its deflection is given by $w(x, y, t)$. A fiber AB oriented in the z direction takes the positions $A'B'$ and $A''B''$ due to bending and shear deformations, respectively, in the xz plane, as shown in Fig. 14.10. Thus, if ϕ_x

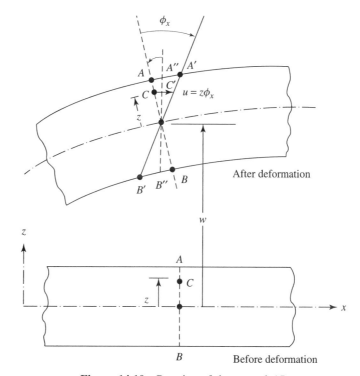

Figure 14.10   Rotation of the normal AB.

denotes the rotation in the xz plane of a line originally normal to the midplane before deformation, the displacement of a point C located at a distance z from the midplane in the direction of the x axis is $+z\phi_x$. It can be seen that point C will not have any x component of displacement due to shear deformation. Similarly, point C will have a component of displacement parallel to the y axis, due to bending of the plate in the yz plane. Its value is $+z\phi_y$, where ϕ_y denotes the rotation in the yz plane of a line originally normal to the midplane before deformation. Thus, the complete displacement state of any point (like C) in the plate is given by

$$
\begin{aligned}
u(x,y,t) &= +z\phi_x(x,y,t) \\
v(x,y,t) &= +z\phi_y(x,y,t) \\
w(x,y,t) &= w(x,y,t)
\end{aligned}
\tag{14.269}
$$

As in the case of a beam, the slope of the deflection surface in the xz and yz planes ($\partial w/\partial x$ and $\partial w/\partial y$) will be increased by the shear angles γ_x and γ_y, respectively, so that

$$
\gamma_x = \phi_x + \frac{\partial w}{\partial x}, \qquad \gamma_y = \phi_y + \frac{\partial w}{\partial y}
\tag{14.270}
$$

Note that the classical or Kirchhoff plate theory can be obtained by setting $\gamma_x = \gamma_y = 0$ or $\phi_x = -\partial w/\partial x$ and $\phi_y = -\partial w/\partial y$. The linear strain–displacement relations can be expressed as

$$
\begin{aligned}
\varepsilon_{xx} &= \frac{\partial u}{\partial x} = \frac{\partial}{\partial x}(+z\phi_x) = +z\frac{\partial \phi_x}{\partial x} \\
\varepsilon_{yy} &= \frac{\partial v}{\partial y} = \frac{\partial}{\partial y}(+z\phi_y) = +z\frac{\partial \phi_y}{\partial y} \\
\varepsilon_{xy} &= \frac{\partial u}{\partial y} + \frac{\partial v}{\partial x} = \frac{\partial}{\partial y}(+z\phi_x) + \frac{\partial}{\partial x}(+z\phi_y) = +z\left(\frac{\partial \phi_x}{\partial y} + \frac{\partial \phi_y}{\partial x}\right) \\
\varepsilon_{yz} &= \frac{\partial v}{\partial z} + \frac{\partial w}{\partial y} = \frac{\partial}{\partial z}(+z\phi_y) + \frac{\partial w}{\partial y} = +\phi_y + \frac{\partial w}{\partial y} \\
\varepsilon_{xz} &= \frac{\partial u}{\partial z} + \frac{\partial w}{\partial x} = \frac{\partial}{\partial z}(+z\phi_x) + \frac{\partial w}{\partial x} = +\phi_x + \frac{\partial w}{\partial x} \\
\varepsilon_{zz} &= 0
\end{aligned}
\tag{14.271}
$$

Stress Resultants   As in thin plates, the nonzero stress components are σ_{xx}, σ_{yy}, σ_{xy}, σ_{yz}, and σ_{xz}. The force and moment resultants per unit length, Q_x, Q_y, M_x, M_y, and M_{xy}, are defined as in Eq. (3.27). Using the stress–strain relations

$$
\begin{aligned}
\sigma_{xz} &= \frac{E}{1-\nu^2}(\varepsilon_{xx} + \nu\varepsilon_{yy}) \\
\sigma_{yy} &= \frac{E}{1-\nu^2}(\varepsilon_{yy} + \nu\varepsilon_{xx}) \\
\sigma_{xy} &= G\varepsilon_{xy} \\
\sigma_{yz} &= G\varepsilon_{yz} \\
\sigma_{xz} &= G\varepsilon_{xz}
\end{aligned}
\tag{14.272}
$$

and the strain–displacement relations of Eq. (14.271), the force and moment resultants can be written as

$$Q_x = \int_{-h/2}^{h/2} \sigma_{xz}\, dz = \int_{-h/2}^{h/2} G\varepsilon_{xz}\, dz = G \int_{-h/2}^{h/2} \left(+\phi_x + \frac{\partial w}{\partial x} \right) dz$$

$$= k^2 G h \left(+\phi_x + \frac{\partial w}{\partial x} \right)$$

$$Q_y = \int_{-h/2}^{h/2} \sigma_{yz}\, dz = \int_{-h/2}^{h/2} G\varepsilon_{yz}\, dz = G \int_{-h/2}^{h/2} \left(+\phi_y + \frac{\partial w}{\partial y} \right) dz$$

$$= k^2 G h \left(+\phi_y + \frac{\partial w}{\partial y} \right)$$

$$M_x = \int_{-h/2}^{h/2} \sigma_{xx} z\, dz = \int_{-h/2}^{h/2} \frac{E}{1-\nu^2} (\varepsilon_{xx} + \nu\varepsilon_{yy}) z\, dz$$

$$= \frac{E}{1-\nu^2} \int_{-h/2}^{h/2} z^2 \left(\frac{\partial\phi_x}{\partial x} + \nu\frac{\partial\phi_y}{\partial y} \right) dz = +D \left(\frac{\partial\phi_x}{\partial x} + \nu\frac{\partial\phi_y}{\partial y} \right)$$

$$M_y = \int_{-h/2}^{h/2} \sigma_{yy} z\, dz = \int_{-h/2}^{h/2} \frac{E}{1-\nu^2} (\varepsilon_{yy} + \nu\varepsilon_{xx}) z\, dz$$

$$= \frac{E}{1-\nu^2} \int_{-h/2}^{h/2} z^2 \left(\frac{\partial\phi_y}{\partial y} + \nu\frac{\partial\phi_x}{\partial x} \right) dz = +D \left(\frac{\partial\phi_y}{\partial y} + \nu\frac{\partial\phi_x}{\partial x} \right)$$

$$M_{xy} = \int_{-h/2}^{h/2} \sigma_{xy} z\, dz = \int_{-h/2}^{h/2} G\varepsilon_{xy} z\, dz = +G \int_{-h/2}^{h/2} z^2 \left(\frac{\partial\phi_x}{\partial y} + \frac{\partial\phi_y}{\partial x} \right) dz$$

$$= \frac{D(1-\nu)}{2} \left(\frac{\partial\phi_x}{\partial y} + \frac{\partial\phi_y}{\partial x} \right) = M_{yx}$$

(14.273)

where

$$G = \frac{E}{2(1+\nu)} \tag{14.274}$$

is the shear modulus of the plate. Note that the quantity k^2 in the expressions of Q_x and Q_y is similar to the Timoshenko shear coefficient k and is introduced to account for the fact that the shear stresses σ_{xz} and σ_{yz} are not constant over the thickness of the plate, $-h/2 \le z \le h/2$. The value of the constant k^2 was taken as $\frac{5}{6}$ by Reissner for static problems, while Mindlin chose the value of k^2 as $\pi^2/12$ so as to make the dynamic theory consistent with the known exact frequency for the fundamental "thickness shear" mode of vibration.

Equilibrium Equations   The equilibrium equations for the Mindlin plate can be derived with reference to Fig. 14.11 and by considering the effect of rotatory inertia as follows:

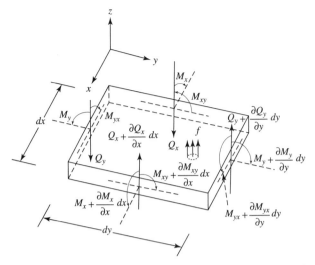

Figure 14.11   Moment and shear force resultants on an element of a plate.

1. Vertical force equilibrium in the z direction:

$$\left(Q_x + \frac{\partial Q_x}{\partial x}\,dx\right)dy + \left(Q_y + \frac{\partial Q_y}{\partial y}\,dy\right)dx + f\,dx\,dy - Q_x\,dy - Q_y\,dx$$

$$= \text{inertia force in the } z \text{ direction} = \rho h\,dx\,dy\,\frac{\partial^2 w}{\partial t^2}$$

or

$$\frac{\partial Q_x}{\partial x} + \frac{\partial Q_y}{\partial y} + f = \rho h \frac{\partial^2 w}{\partial t^2} \tag{14.275}$$

where ρ is the mass density of the plate and $f(x,y,t)$ is the intensity of the external distributed force.

2. Moment equilibrium about the x axis:

$$-\left(Q_y + \frac{\partial Q_y}{\partial y}\,dy\right)dx\,dy + \left(M_y + \frac{\partial M_y}{\partial y}\,dy\right)dx + \left(M_{xy} + \frac{\partial M_{xy}}{\partial x}\,dx\right)dy$$

$$- M_y\,dx - M_{xy}\,dy + f\,dx\,dy\,\frac{dy}{2}$$

$$= \text{inertia moment due to rotation } \phi_y = \rho I_x\,dx\,dy\frac{\partial^2 \phi_y}{\partial t^2} = \frac{\rho h^3}{12}\frac{\partial^2 \phi_y}{\partial t^2}\,dx\,dy$$

or

$$-Q_y + \frac{\partial M_y}{\partial y} + \frac{\partial M_{xy}}{\partial x} = \frac{\rho h^3}{12}\frac{\partial^2 \phi_y}{\partial t^2} \tag{14.276}$$

where the terms involving products of small quantities are neglected, and $(\rho I_x\,dx\,dy)$ is the mass moment of inertia of the element about the x axis, with $I_x = \frac{1}{12}(1)h^3 = h^3/12$ denoting the area moment of inertia per unit width of the plate about the x axis.

3. Moment equilibrium about the y axis:

$$-\left(Q_x + \frac{\partial Q_x}{\partial x}\,dx\right)dy\,dx + \left(M_x + \frac{\partial M_x}{\partial x}\,dx\right)dy + \left(M_{yx} + \frac{\partial M_{yx}}{\partial y}\,dy\right)dx$$

$$-\,M_x\,dy - M_{yx}\,dx + f\,dx\,dy\frac{dx}{2}$$

$$= \text{inertia moment due to rotation } \phi_x = \rho I_y\,dx\,dy\frac{\partial^2 \phi_x}{\partial t^2} = \frac{\rho h^3}{12}\frac{\partial^2 \phi_x}{\partial t^2}\,dx\,dy$$

or

$$-Q_x + \frac{\partial M_x}{\partial x} + \frac{\partial M_{xy}}{\partial y} = \frac{\rho h^3}{12}\frac{\partial^2 \phi_x}{\partial t^2} \tag{14.277}$$

by neglecting products of small quantities, and $(\rho I_y\,dx\,dy)$ is the mass moment of inertia of the element about the y axis, with $I_y = \frac{1}{12}(1)h^3 = h^3/12$ representing the area moment of inertia of the plate per unit width about the y axis.

Substituting Eqs. (14.273) into Eqs. (14.275) to (14.277) yields the final equations of motion in terms of the displacement unknowns w, ϕ_x, and ϕ_y as

$$k^2 Gh\left(\nabla^2 w + \frac{\partial \phi_x}{\partial x} + \frac{\partial \phi_y}{\partial y}\right) + f = \rho h\frac{\partial^2 w}{\partial t^2} \tag{14.278}$$

$$\frac{D}{2}\left[(1-v)\nabla^2\phi_x + (1+v)\frac{\partial}{\partial x}\left(\frac{\partial \phi_x}{\partial x} + \frac{\partial \phi_y}{\partial y}\right)\right] - k^2 Gh\left(\phi_x + \frac{\partial w}{\partial x}\right) = \frac{\rho h^3}{12}\frac{\partial^2 \phi_x}{\partial t^2} \tag{14.279}$$

$$\frac{D}{2}\left[(1-v)\nabla^2\phi_y + (1+v)\frac{\partial}{\partial y}\left(\frac{\partial \phi_x}{\partial x} + \frac{\partial \phi_y}{\partial y}\right)\right] - k^2 Gh\left(\phi_y + \frac{\partial w}{\partial y}\right) = \frac{\rho h^3}{12}\frac{\partial^2 \phi_y}{\partial t^2} \tag{14.280}$$

where $\nabla^2 = \partial^2/\partial x^2 + \partial^2/\partial y^2$ is the Laplacian operator. Equations (14.278) to (14.280) can be rewritten as

$$k^2 Gh(\nabla^2 w + \Phi) + f = \rho h\frac{\partial^2 w}{\partial t^2} \tag{14.281}$$

$$\frac{D}{2}\left[(1-v)\nabla^2\phi_x + (1+v)\frac{\partial \Phi}{\partial x}\right] - k^2 Gh\left(\phi_x + \frac{\partial w}{\partial x}\right) = \frac{\rho h^3}{12}\frac{\partial^2 \phi_x}{\partial t^2} \tag{14.282}$$

$$\frac{D}{2}\left[(1-v)\nabla^2\phi_y + (1+v)\frac{\partial \Phi}{\partial y}\right] - k^2 Gh\left(\phi_y + \frac{\partial w}{\partial y}\right) = \frac{\rho h^3}{12}\frac{\partial^2 \phi_y}{\partial t^2} \tag{14.283}$$

where

$$\Phi = \frac{\partial \phi_x}{\partial x} + \frac{\partial \phi_y}{\partial y} \tag{14.284}$$

We can eliminate ϕ_x and ϕ_y from Eqs. (14.282) and (14.283) by first differentiating them with respect to x and y, respectively, and then adding them to obtain

$$D\nabla^2\Phi - k^2 Gh\Phi - k^2 Gh\nabla^2 w = \frac{\rho h^3}{12}\frac{\partial^2 \Phi}{\partial t^2} \tag{14.285}$$

By eliminating Φ from Eqs. (14.281) and (14.285), we obtain a single equation in terms of w as

$$\left(\nabla^2 - \frac{\rho}{k^2 G}\frac{\partial^2}{\partial t^2}\right)\left(D\nabla^2 - \frac{\rho h^3}{12}\frac{\partial^2}{\partial t^2}\right)w + \rho h\frac{\partial^2 w}{\partial t^2}$$

$$= \left(1 - \frac{D}{k^2 Gh}\nabla^2 + \frac{\rho h^2}{12k^2 G}\frac{\partial^2}{\partial t^2}\right)f \tag{14.286}$$

If shear deformation only is considered, the right-hand-side terms in Eqs. (14.279) and (14.280) will be zero, and Eq. (14.286) reduces to

$$D\left(\nabla^2 - \frac{\rho}{k^2 G}\frac{\partial^2}{\partial t^2}\right)\nabla^2 w + \rho h\frac{\partial^2 w}{\partial t^2} = \left(1 - \frac{D}{k^2 Gh}\nabla^2\right)f \tag{14.287}$$

If rotary inertia only is to be considered, terms involving k^2 should be neglected in Eqs. (14.278) to (14.280), in which case Eq. (14.286) becomes

$$\left(D\nabla^2 - \frac{\rho h^3}{12}\frac{\partial^2}{\partial t^2}\right)\nabla^2 w + \rho h\frac{\partial^2 w}{\partial t^2} = f \tag{14.288}$$

If both the effects of rotary inertia and shear deformation are neglected, terms involving k^2 and $\rho h^3/12$ will be zero in Eqs. (14.278) to (14.280), and ϕ_x and ϕ_y will be replaced by $-\partial w/\partial x$ and $-\partial w/\partial y$, respectively, and Eq. (14.286) reduces to the classical thin plate equation:

$$D\nabla^4 w + \rho h\frac{\partial^2 w}{\partial t^2} = f \tag{14.289}$$

For a plate with constant thickness, the equation of motion, Eq. (14.286), can be expressed as

$$D\nabla^4 w - \frac{\rho D}{k^2 G}\frac{\partial^2}{\partial t^2}(\nabla^2 w) - \frac{\rho h^3}{12}\nabla^2\frac{\partial^2 w}{\partial t^2} + \frac{\rho^2 h^3}{12k^2 G}\frac{\partial^4 w}{\partial t^4} + \rho h\frac{\partial^2 w}{\partial t^2}$$

$$= f - \frac{D}{k^2 Gh}\nabla^2 f + \frac{\rho h^2}{12k^2 G}\frac{\partial^2 f}{\partial t^2} \tag{14.290}$$

For free vibration, Eq. (14.290) becomes

$$\nabla^4 w - \frac{\rho h}{k^2 Gh}\frac{\partial^2}{\partial t^2}(\nabla^2 w) - \frac{\rho h^3}{12D}\nabla^2\frac{\partial^2 w}{\partial t^2} + \frac{\rho h}{D}\frac{\rho h^3}{12k^2 Gh}\frac{\partial^4 w}{\partial t^4} + \frac{\rho h}{D}\frac{\partial^2 w}{\partial t^2} = 0 \tag{14.291}$$

For harmonic motion at frequency ω,

$$w(x,y,t) = W(x,y)e^{i\omega t} \tag{14.292}$$

and Eq. (14.291) becomes

$$\nabla^4 W + \left(\frac{\rho h}{k^2 Gh} + \frac{\rho h^3}{12D}\right)\omega^2\nabla^2 W + \frac{\rho h}{D}\left(\frac{\rho h^3\omega^2}{12k^2 Gh} - 1\right)\omega^2 W = 0 \tag{14.293}$$

14.9.2   Variational Approach

To derive the equations of motion and the associated boundary conditions using the variational approach, we first note that the displacement components are given by Eq. (14.269). The strain–displacement relations are given by Eq. (14.271). The force and moment resultants are given by Eq. (14.273). The strain energy due to the deformation of the Mindlin plate is given by

$$\pi = \frac{1}{2} \iiint_V \vec{\sigma}^{\mathrm{T}} \vec{\varepsilon}\, dV \tag{14.294}$$

where

$$\vec{\sigma} = \begin{Bmatrix} \sigma_{xx} \\ \sigma_{yy} \\ \sigma_{xy} \\ \sigma_{yz} \\ \sigma_{zx} \end{Bmatrix} \quad \text{and} \quad \vec{\varepsilon} = \begin{Bmatrix} \varepsilon_{xx} \\ \varepsilon_{yy} \\ \varepsilon_{xy} \\ \varepsilon_{yz} \\ \varepsilon_{zx} \end{Bmatrix} \tag{14.295}$$

The stress–strain relations of Eq. (14.272) can be expressed in matrix form as

$$\vec{\sigma} = [B]\vec{\varepsilon} \tag{14.296}$$

where (by using the factor k^2 for σ_{yz} and σ_{zx})

$$[B] = \begin{bmatrix} \frac{E}{1-v^2} & \frac{vE}{1-v^2} & 0 & 0 & 0 \\ \frac{vE}{1-v^2} & \frac{E}{1-v^2} & 0 & 0 & 0 \\ 0 & 0 & G & 0 & 0 \\ 0 & 0 & 0 & k^2 G & 0 \\ 0 & 0 & 0 & 0 & k^2 G \end{bmatrix} \tag{14.297}$$

Using Eq. (14.296), Eq. (14.294) can be written as

$$\pi = \frac{1}{2} \iiint_V \vec{\varepsilon}^{\mathrm{T}} [B]\vec{\varepsilon}\, dV \tag{14.298}$$

Substitution of Eq. (14.297) into Eq. (14.298) leads to

$$\pi = \frac{1}{2} \iiint_V \left(\frac{E}{1-v^2}\varepsilon_{xx}^2 + \frac{vE}{1-v^2}\varepsilon_{yy}^2 + G\varepsilon_{xy}^2 + k^2 G\varepsilon_{yz}^2 + k^2 G\varepsilon_{zx}^2 \right) dV \tag{14.299}$$

Using Eqs. (14.271), Eq. (14.299) can be rewritten as

$$\pi = \frac{1}{2} \int_{z=-h/2}^{h/2} dz \iint_A \left[\frac{E}{1-v}z^2 \left(\frac{\partial \phi_x}{\partial x} \right)^2 + \frac{E}{1-v^2}z^2 \left(\frac{\partial \phi_y}{\partial y} \right)^2 + 2v\frac{E}{1-v^2}\frac{\partial \phi_x}{\partial x}\frac{\partial \phi_y}{\partial y} \right.$$

$$\left. + Gz^2 \left(\frac{\partial \phi_x}{\partial y} + \frac{\partial \phi_y}{\partial x} \right)^2 + k^2 G \left(\phi_y + \frac{\partial w}{\partial y} \right)^2 + k^2 G \left(\phi_x + \frac{\partial w}{\partial x} \right)^2 \right] dA \tag{14.300}$$

where $dA = dx\, dy$ denotes the area of a differential element of the plate. By carrying out integration with respect to z and using Eq. (14.274), Eq. (14.300) can be expressed as

$$\pi = \frac{1}{2} \iint_A \left\{ D \left[\left(\frac{\partial \phi_x}{\partial x} + \frac{\partial \phi_y}{\partial y} \right)^2 - 2(1-v) \left\langle \frac{\partial \phi_x}{\partial x} \frac{\partial \phi_y}{\partial y} - \frac{1}{4} \left(\frac{\partial \phi_x}{\partial y} + \frac{\partial \phi_y}{\partial x} \right)^2 \right\rangle \right] \right.$$
$$\left. + k^2 Gh \left\langle \left(\phi_x + \frac{\partial w}{\partial x} \right)^2 + \left(\phi_y + \frac{\partial w}{\partial y} \right)^2 \right\rangle \right\} dA \qquad (14.301)$$

The kinetic energy T of the plate can be expressed as

$$T = \frac{1}{2} \iiint_V \rho \left[\left(\frac{\partial u}{\partial t} \right)^2 + \left(\frac{\partial v}{\partial t} \right)^2 + \left(\frac{\partial w}{\partial t} \right)^2 \right] dV \qquad (14.302)$$

where ρ is the mass density of the plate and t is the time. Substituting Eq. (14.269) into Eq. (14.302), we obtain

$$T = \frac{\rho h}{2} \iint_A \left\{ \left(\frac{\partial w}{\partial t} \right)^2 + \frac{h^2}{12} \left[\left(\frac{\partial \phi_x}{\partial t} \right)^2 + \left(\frac{\partial \phi_y}{\partial t} \right)^2 \right] \right\} dA \qquad (14.303)$$

The work done by the distributed transverse load $f(x,y,t)$ (load per unit area of the middle surfaces of the plate) can be expressed as

$$W = \iint_A wf\, dA \qquad (14.304)$$

The equations of motion can be derived from Hamilton's principle:

$$\delta \int_{t_1}^{t_2} (T - \pi + W)\, dt = 0 \qquad (14.305)$$

Substituting Eqs. (14.301), (14.303), and (14.304) into Eq. (14.305), we obtain

$$\int_{t_1}^{t_2} \iint_A \left\{ -D \left[\frac{\partial \phi_x}{\partial x} \frac{\partial(\delta\phi_x)}{\partial x} + \frac{\partial \phi_y}{\partial y} \frac{\partial(\delta\phi_y)}{\partial y} + v \frac{\partial \phi_x}{\partial x} \frac{\partial(\delta\phi_y)}{\partial y} + v \frac{\partial \phi_y}{\partial y} \frac{\partial(\delta\phi_x)}{\partial x} \right] \right.$$
$$- \frac{D(1-v)}{2} \left(\frac{\partial \phi_x}{\partial y} + \frac{\partial \phi_y}{\partial x} \right) \left[\frac{\partial(\delta\phi_x)}{\partial y} + \frac{\partial(\delta\phi_y)}{\partial x} \right] - k^2 Gh \left[\left(\phi_x + \frac{\partial w}{\partial x} \right) \left(\delta\phi_x + \frac{\partial(\delta w)}{\partial x} \right) \right.$$
$$\left. + \left(\phi_y + \frac{\partial w}{\partial y} \right) \left(\delta\phi_y + \frac{\partial(\delta w)}{\partial y} \right) \right] + \rho h \frac{\partial w}{\partial t} \frac{\partial(\delta w)}{\partial t} + \frac{\rho h^3}{12} \left[\frac{\partial \phi_x}{\partial t} \frac{\partial(\delta\phi_x)}{\partial t} \right.$$
$$\left. + \frac{\partial \phi_y}{\partial t} \frac{\partial(\delta\phi_y)}{\partial t} \right] + f\delta w \right\} dA\, dt = 0 \qquad (14.306)$$

By performing integration by parts, Eq. (14.306) yields

$$
\int_{t_1}^{t_2} \iint_A \left\{ \left[D \left(\frac{\partial^2 \phi_x}{\partial x^2} + v \frac{\partial^2 \phi_y}{\partial x \, \partial y} \right) + \frac{D(1-v)}{2} \left(\frac{\partial^2 \phi_x}{\partial y^2} + \frac{\partial^2 \phi_y}{\partial x \, \partial y} \right) \right. \right.
$$

$$
\left. - k^2 G h \left(\phi_x + \frac{\partial w}{\partial x} \right) - \frac{\rho h^3}{12} \frac{\partial^2 \phi_x}{\partial t^2} \right] \delta \phi_x
$$

$$
+ \left[D \left(\frac{\partial^2 \phi_y}{\partial y^2} + v \frac{\partial^2 \phi_x}{\partial x \, \partial y} \right) + \frac{D(1-v)}{2} \left(\frac{\partial^2 \phi_y}{\partial x^2} + \frac{\partial^2 \phi_x}{\partial x \, \partial y} \right) \right.
$$

$$
\left. - k^2 G h \left(\phi_y + \frac{\partial w}{\partial y} \right) - \frac{\rho h^3}{12} \frac{\partial^2 \phi_y}{\partial t^2} \right] \delta \phi_y
$$

$$
+ \left[k^2 G h \left(\frac{\partial \phi_x}{\partial x} + \frac{\partial^2 w}{\partial x^2} + \frac{\partial \phi_y}{\partial y} + \frac{\partial^2 w}{\partial y^2} \right) - \rho h \frac{\partial^2 w}{\partial t^2} \right] \delta w + f \delta w \right\} dA \, dt
$$

$$
- \int_{t_1}^{t_2} \oint_C \left\{ \left[D \left(\frac{\partial \phi_x}{\partial x} dy + v \frac{\partial \phi_y}{\partial y} dy \right) - \frac{D(1-v)}{2} \left(\frac{\partial \phi_x}{\partial y} dx + \frac{\partial \phi_y}{\partial x} dx \right) \right] \delta \phi_x \right.
$$

$$
+ \left[-D \left(\frac{\partial \phi_y}{\partial y} dx + v \frac{\partial \phi_x}{\partial x} dx \right) + \frac{D(1-v)}{2} \left(\frac{\partial \phi_x}{\partial y} dy + \frac{\partial \phi_y}{\partial x} dy \right) \right] \delta \phi_y
$$

$$
\left. + k^2 G h \left[\phi_x \, dy + \frac{\partial w}{\partial x} dy - \phi_y \, dx - \frac{\partial w}{\partial y} dx \right] \delta w \right\} dt = 0 \tag{14.307}
$$

where C denotes the boundary of the plate.

Equations of Motion   By equating the coefficients of the various terms involving $\delta \phi_x$, $\delta \phi_y$, and δw under the area integral in Eq. (14.307) to zero, we obtain

$$
D \left(\frac{\partial^2 \phi_x}{\partial x^2} + v \frac{\partial^2 \phi_y}{\partial x \, \partial y} \right) + \frac{D(1-v)}{2} \left(\frac{\partial^2 \phi_x}{\partial y^2} + \frac{\partial^2 \phi_y}{\partial x \, \partial y} \right) - k^2 G h \left(\phi_x + \frac{\partial w}{\partial x} \right) - \frac{\rho h^3}{12} \frac{\partial^2 \phi_x}{\partial t^2} = 0
$$
$$
\tag{14.308}
$$

$$
D \left(\frac{\partial^2 \phi_y}{\partial y^2} + v \frac{\partial^2 \phi_x}{\partial x \, \partial y} \right) + \frac{D(1-v)}{2} \left(\frac{\partial^2 \phi_y}{\partial x^2} + \frac{\partial^2 \phi_x}{\partial x \, \partial y} \right) - k^2 G h \left(\phi_y + \frac{\partial w}{\partial y} \right) - \frac{\rho h^3}{12} \frac{\partial^2 \phi_y}{\partial t^2} = 0
$$
$$
\tag{14.309}
$$

$$
k^2 G h \left(\frac{\partial \phi_x}{\partial x} + \frac{\partial^2 w}{\partial x^2} + \frac{\partial \phi_y}{\partial y} + \frac{\partial^2 w}{\partial y^2} \right) + f - \rho h \frac{\partial^2 w}{\partial t^2} = 0
$$
$$
\tag{14.310}
$$

Equations (14.308)–(14.310) can be rewritten as

$$
\frac{D(1-v)}{2} \nabla^2 \phi_x + \frac{D(1+v)}{2} \frac{\partial}{\partial x} \left(\frac{\partial \phi_x}{\partial x} + \frac{\partial \phi_y}{\partial y} \right) - k^2 G h \left(\phi_x + \frac{\partial w}{\partial x} \right) = \frac{\rho h^3}{12} \frac{\partial^2 \phi_x}{\partial t^2}
$$
$$
\tag{14.311}
$$

$$\frac{D(1-v)}{2}\nabla^2\phi_y + \frac{D(1+v)}{2}\frac{\partial}{\partial y}\left(\frac{\partial\phi_x}{\partial x} + \frac{\partial\phi_y}{\partial y}\right) - k^2 Gh\left(\phi_y + \frac{\partial w}{\partial y}\right) = \frac{\rho h^3}{12}\frac{\partial^2\phi_y}{\partial t^2}$$

(14.312)

$$k^2 Gh\left(\nabla^2 w + \frac{\partial\phi_x}{\partial x} + \frac{\partial\phi_y}{\partial y}\right) + f = \rho h\frac{\partial^2 w}{\partial t^2}$$

(14.313)

Equations (14.311)–(14.313) denote the equations of motion of the Mindlin plate and can be seen to be same as Eqs. (14.281)–(14.283), derived using the dynamic equilibrium approach.

It can be seen that Eqs. (14.311)–(14.313) are coupled in the variables $w(x,y,t)$, $\phi_x(x,y,t)$, and $\phi_y(x,y,t)$. The explicit functions ϕ_x and ϕ_y can be eliminated from Eqs. (14.311)–(14.313), and a single equation of motion in terms of w can be derived as indicated earlier [see Eqs. (14.281)–(14.286)]:

$$\left(\nabla^2 - \frac{\rho}{k^2 G}\frac{\partial^2}{\partial t^2}\right)\left(D\nabla^2 - \frac{\rho h^3}{12}\frac{\partial^2}{\partial t^2}\right)w + \rho h\frac{\partial^2 w}{\partial t^2} = \left(1 - \frac{D}{k^2 Gh}\nabla^2 + \frac{\rho h^2}{12 k^2 G}\frac{\partial^2}{\partial t^2}\right)f$$

(14.314)

Note that in the equations above, the terms containing $k^2 G$ denote the effect of shear deformation, and the terms containing $(\rho h^3/12)(\partial^2/\partial t^2)$ represent the effect of rotary inertia.

General Boundary Conditions   The boundary conditions can be identified by setting the line integral in Eq. (14.307) equal to zero:

$$\oint_C \left[D\left(\frac{\partial\phi_x}{\partial x} + v\frac{\partial\phi_y}{\partial y}\right)\delta\phi_x\, dy - D\left(\frac{\partial\phi_y}{\partial y} + v\frac{\partial\phi_x}{\partial x}\right)\delta\phi_y\, dx\right.$$

$$- \frac{D(1-v)}{2}\left(\frac{\partial\phi_x}{\partial y} + \frac{\partial\phi_y}{\partial x}\right)\delta\phi_x\, dx$$

$$+ \frac{D(1-v)}{2}\left(\frac{\partial\phi_x}{\partial y} + \frac{\partial\phi_y}{\partial x}\right)\delta\phi_y\, dy + k^2 Gh\left(\phi_x + \frac{\partial w}{\partial x}\right)\delta w\, dy$$

$$\left. -k^2 Gh\left(\phi_y + \frac{\partial w}{\partial y}\right)\delta w\, dx\right] = 0$$

(14.315)

Using the expressions for the force and moment resultants given in Eq. (14.273), Eq. (14.315) can be expressed as

$$\oint_C (M_x\delta\phi_x\, dy - M_y\delta\phi_y\, dx - M_{xy}\delta\phi_x\, dx + M_{xy}\delta\phi_y\, dy + Q_x\delta w\, dy - Q_y\delta w\, dx) = 0$$

(14.316)

Boundary Conditions on an Inclined Boundary   Consider an arbitrary boundary of the plate whose normal and tangential directions are denoted n and s, respectively,

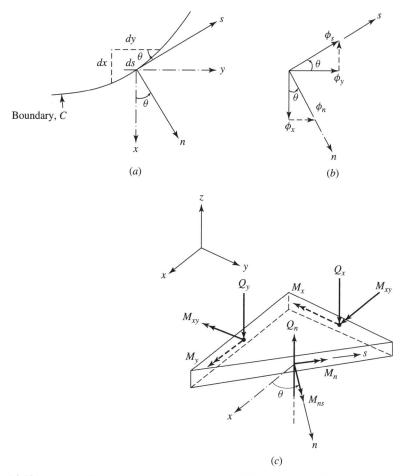

Figure 14.12 (*a*) coordinate systems on a boundary; (**b**) bending rotations at a boundary; (*c*) shear force and moment resultants on an inclined boundary.

as shown in Fig. 14.12(*a*). The (x,y) coordinates and the (n,s) coordinates on the boundary are related as

$$dx = -\sin\theta \, ds \tag{14.317}$$

$$dy = \cos\theta \, ds \tag{14.318}$$

If ϕ_n denotes the rotation of the normal to the midplane in the plane nz (normal plane) and ϕ_s the rotation of the normal to the midplane in the plane sz (tangent plane), the bending rotations ϕ_x and ϕ_y can be related to ϕ_n and ϕ_s as shown in Fig. 14.12(*b*):

$$\phi_x = \phi_n \cos\theta - \phi_s \sin\theta \tag{14.319}$$

$$\phi_y = \phi_n \sin\theta + \phi_s \cos\theta \tag{14.320}$$

The bending moment resultant normal to the boundary, M_n, and the twisting moment resultant at the boundary, M_{ns}, can be expressed in terms of M_x, M_y, and M_{xy} as [see Fig. 14.12(c)]

$$M_n = M_x \cos^2 \theta + M_y \sin^2 \theta + 2M_{xy} \cos \theta \sin \theta \qquad (14.321)$$

$$M_{ns} = (M_y - M_x) \cos \theta \sin \theta + M_{xy}(\cos^2 \theta - \sin^2 \theta) \qquad (14.322)$$

Similarly, the shear force resultant acting on the boundary in the z direction, Q_n, can be expressed in terms of Q_x and Q_y as

$$Q_n = Q_x \cos \theta + Q_y \sin \theta \qquad (14.323)$$

Using Eqs. (14.317) and (14.318), Eq. (14.316) can be expressed as

$$\oint_C (M_x \cos \theta \delta \phi_x + M_y \sin \theta \delta \phi_y + M_{xy} \sin \theta \delta \phi_x + M_{xy} \cos \theta \delta \phi_y$$

$$+ Q_x \cos \theta \delta w + Q_y \sin \theta \delta w) \, ds = 0 \qquad (14.324)$$

Substitution of Eqs. (14.319) and (14.320) into Eq. (14.324) results in

$$\oint_C \{(M_x \cos^2 \theta + M_y \sin^2 \theta + 2M_{xy} \sin \theta \cos \theta)\delta \phi_n$$

$$[(-M_x \cos \theta \sin \theta + M_y \cos \theta \sin \theta) + M_{xy}(\cos^2 \theta - \sin^2 \theta)]\delta \phi_s$$

$$+ (Q_x \cos \theta + Q_y \sin \theta)\delta w\} \, ds = 0 \qquad (14.325)$$

In view of Eqs. (14.321), (14.322), and (14.323), Eq. (14.325) can be written as

$$\oint_C (M_n \delta \phi_n + M_{ns} \delta \phi_s + Q_n \delta w) \, ds = 0 \qquad (14.326)$$

Equation (14.326) will be satisfied when each of the terms under the integral is equal to zero. This implies that the following conditions are to be satisfied along the boundary of the plate:

$$M_n = 0 \qquad (14.327)$$

or

$$\delta \phi_n = 0 \quad (\phi_n \text{ is specified}) \qquad (14.328)$$

$$M_{ns} = 0 \qquad (14.329)$$

or

$$\delta \phi_s = 0 \quad (\phi_s \text{ is specified}) \qquad (14.330)$$

$$Q_n = 0 \qquad (14.331)$$

or

$$\delta w = 0 \quad (w \text{ is specified}) \qquad (14.332)$$

Thus, three boundary conditions are to be specified along a boundary or edge of the plate in Mindlin theory. The boundary conditions corresponding to some common support conditions are as follows:

1. Fixed or clamped boundary:

$$\phi_n = 0, \qquad \phi_s = 0, \qquad w = 0 \tag{14.333}$$

2. Simply supported or hinged boundary:

$$M_n = 0, \qquad \phi_s = 0, \qquad w = 0 \tag{14.334}$$

The support conditions of Eq. (14.334) are sometimes called *hard-type simple support conditions*. In some cases, the following conditions, known as *soft-type simple support conditions*, are used [7]:

$$M_n = 0, \qquad M_{ns} = 0, \qquad w = 0 \tag{14.335}$$

3. Free boundary:

$$M_n = 0, \qquad M_{ns} = 0, \qquad Q_n = 0 \tag{14.336}$$

14.9.3 Free Vibration Solution

For free vibration, $f(x,y,t)$ is set equal to zero and the variables $\phi_x(x,y,t)$, $\phi_y(x,y,t)$, and $w(x,y,t)$ are assumed to be harmonic as [3]

$$\phi_x(x,y,t) = \Phi_x(x,y)e^{i\omega t}$$
$$\phi_y(x,y,t) = \Phi_y(x,y)e^{i\omega t} \tag{14.337}$$
$$w(x,y,t) = W(x,y)e^{i\omega t}$$

so that Eqs. (14.311)–(14.313) become

$$\frac{D}{2}\left[(1-v)\nabla^2\Phi_x + (1+v)\frac{\partial}{\partial x}\tilde{\Phi}\right] - k^2 Gh\left(\Phi_x + \frac{\partial W}{\partial x}\right) + \frac{\rho h^3 \omega^2}{12}\Phi_x = 0 \tag{14.338}$$

$$\frac{D}{2}\left[(1-v)\nabla^2\Phi_y + (1+v)\frac{\partial}{\partial y}\tilde{\Phi}\right] - k^2 Gh\left(\Phi_y + \frac{\partial W}{\partial y}\right) + \frac{\rho h^3 \omega^2}{12}\Phi_y = 0 \tag{14.339}$$

$$k^2 Gh(\nabla^2 W + \tilde{\Phi}) + \rho h \omega^2 W = 0 \tag{14.340}$$

where

$$\tilde{\Phi} = \frac{\partial\Phi_x}{\partial x} + \frac{\partial\Phi_y}{\partial y} \tag{14.341}$$

To find the solution of Eqs. (14.338)–(14.340), $\Phi_x(x,y)$ and $\Phi_y(x,y)$ are expressed in terms of two potentials $\psi(x,y)$ and $H(x,y)$ such that

$$\Phi_x = \frac{\partial\psi}{\partial x} + \frac{\partial H}{\partial y} \tag{14.342}$$

$$\Phi_y = \frac{\partial\psi}{\partial y} - \frac{\partial H}{\partial x} \tag{14.343}$$

where $\psi(x,y)$ and $H(x,y)$ can be shown to correspond to the dilatational and shear components of motion of the plate, respectively. By substituting Eqs. (14.342) and (14.343) into Eqs. (14.338)–(14.340), we obtain

$$\frac{\partial}{\partial x}\left[\nabla^2\psi + \left(Rk_b^4 - \frac{1}{S}\right)\psi - \frac{W}{S}\right] + \frac{1-\nu}{2}\frac{\partial}{\partial y}(\nabla^2 + \delta_3^2)H = 0 \qquad (14.344)$$

$$\frac{\partial}{\partial y}\left[\nabla^2\psi + \left(Rk_b^4 - \frac{1}{S}\right)\psi - \frac{W}{S}\right] - \frac{1-\nu}{2}\frac{\partial}{\partial x}(\nabla^2 + \delta_3^2)H = 0 \qquad (14.345)$$

$$\nabla^2(\psi + W) + Sk_b^4 W = 0 \qquad (14.346)$$

where

$$R = \frac{I}{h} = \frac{h^2}{12} \qquad (14.347)$$

represents the effect of rotary inertia and is equal to the square of the radius of gyration of the cross-sectional area of a unit width of plate,

$$S = \frac{D}{k^2 Gh} \qquad (14.348)$$

denotes the effect of shear deformation,

$$k_b^4 = \frac{\rho h \omega^2}{D} \qquad (14.349)$$

indicates the frequency parameter in classical plate theory, and

$$\delta_3^2 = \frac{2(Rk_b^4 - 1/S)}{1 - \nu} \qquad (14.350)$$

To uncouple Eqs. (14.344)–(14.346), first differentiate Eqs. (14.344) and (14.345) with respect to x and y, respectively, and add the resulting equations to obtain

$$\nabla^2\left[\nabla^2\psi + \left(Rk_b^2 - \frac{1}{S}\right)\psi - \frac{W}{S}\right] = 0 \qquad (14.351)$$

Next differentiate Eqs. (14.344) and (14.345) with respect to y and x, respectively, and subtract the resulting equations to obtain

$$\nabla^2(\nabla^2 + \delta_3^2)H = 0 \qquad (14.352)$$

Equation (14.346) gives

$$\nabla^2\psi = -\nabla^2 W - Sk_b^4 W \qquad (14.353)$$

By substituting Eq. (14.353) into Eq. (14.351), the resulting equation can be expressed as

$$(\nabla^2 + \delta_1^2)(\nabla^2 + \delta_2^2)W = 0 \qquad (14.354)$$

where

$$\delta_1^2, \delta_2^2 = \frac{1}{2}k_b^4\left[R + S \pm \sqrt{(R-S)^2 + \frac{4}{k_b^4}}\right] \qquad (14.355)$$

Thus, the solution of Eq. (14.354) is given by

$$W = W_1 + W_2 \tag{14.356}$$

where W_j satisfies the equation

$$(\nabla^2 + \delta_j^2)W_j = 0, \qquad j = 1, 2 \tag{14.357}$$

It can be verified that Eqs. (14.346) and (14.351) are satisfied by

$$\psi = (\mu - 1)W \tag{14.358}$$

where μ is a constant. Substitution of Eq. (14.358) into Eqs. (14.346) and (14.351) yields two values for μ as

$$\mu_{1,2} = \frac{\delta_{2,1}^2}{Rk_b^4 - 1/S} \tag{14.359}$$

Using Eqs. (14.356) and (14.358), the terms in brackets on the left-hand sides of Eqs. (14.344) and (14.345) can be seen to be zero, so that the equation governing H becomes [from Eqs. (14.344), (14.345), and (14.352)]

$$(\nabla^2 + \delta_3^2)H = 0 \tag{14.360}$$

Thus, the solution of Eqs. (14.338)– (14.340) can be written as

$$\Phi_x = (\mu_1 - 1)\frac{\partial W_1}{\partial x} + (\mu_2 - 1)\frac{\partial W_2}{\partial x} + \frac{\partial H}{\partial y} \tag{14.361}$$

$$\Phi_y = (\mu_1 - 1)\frac{\partial W_1}{\partial y} + (\mu_2 - 1)\frac{\partial W_2}{\partial y} - \frac{\partial H}{\partial x} \tag{14.362}$$

$$W = W_1 + W_2 \tag{14.363}$$

where W_1, W_2, and H are governed by Eqs. (14.357) and (14.360).

14.9.4   Plate Simply Supported on All Four Edges

Let the origin of the coordinate system be located at the center of the plate as shown in Fig. 14.13 so that the boundary conditions can be stated as [20]

$$W = M_x = \Phi_y = 0 \qquad \text{at} \quad x = \pm\frac{a}{2} \tag{14.364}$$

$$W = M_y = \Phi_x = 0 \qquad \text{at} \quad y = \pm\frac{b}{2} \tag{14.365}$$

The solutions of Eqs. (14.357) and (14.360) are assumed as

$$W_1(x,y) = C_1 \sin\alpha_1 x \sin\beta_1 y \tag{14.366}$$

$$W_2(x,y) = C_2 \sin\alpha_2 x \sin\beta_2 y \tag{14.367}$$

$$H(x,y) = C_3 \cos\alpha_3 x \cos\beta_3 y \tag{14.368}$$

with the conditions

$$\alpha_i^2 + \beta_i^2 = \delta_i^2, \qquad i = 1, 2, 3 \tag{14.369}$$

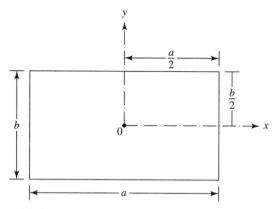

Figure 14.13 Rectangular plate simply supported on all edges.

By substituting Eqs. (14.366)–(14.368) into (14.361)–(14.363), we obtain

$$\Phi_x = C_1(\mu_1 - 1)\alpha_1 \cos \alpha_1 x \sin \beta_1 y + C_2(\mu_2 - 1)\alpha_2 \cos \alpha_2 x \sin \beta_2 y$$
$$- C_3\beta_3 \cos \alpha_3 x \sin \beta_3 y \tag{14.370}$$

$$\Phi_y = C_1(\mu_1 - 1)\beta_1 \sin \alpha_1 x \cos \beta_1 y + C_2(\mu_2 - 1)\beta_2 \sin \alpha_2 x \cos \beta_2 y$$
$$+ C_3\alpha_3 \sin \alpha_3 x \cos \beta_3 y \tag{14.371}$$

$$W = C_1 \sin \alpha_1 x \sin \beta_1 y + C_2 \sin \alpha_2 x \sin \beta_2 y \tag{14.372}$$

Using Eqs. (14.370) and (14.371), the bending moments M_x and M_y [see Eqs. (14.273)] can be expressed as

$$M_x = D[-C_1(\mu_1 - 1)(\alpha_1^2 + v\beta_1^2) \sin \alpha_1 x \sin \beta_1 y$$
$$- C_2(\mu_2 - 1)(\alpha_2^2 + v\beta_2^2) \sin \alpha_2 x \sin \beta_2 y + C_3\alpha_3\beta_3(1 - v) \sin \alpha_3 x \sin \beta_3 y] \tag{14.373}$$

$$M_y = D[-C_1(\mu_1 - 1)(\beta_1^2 + v\alpha_1^2) \sin \alpha_1 x \sin \beta_1 y$$
$$- C_2(\mu_2 - 1)(\beta_2^2 + v\alpha_2^2) \sin \alpha_2 x \sin \beta_2 y - C_3\alpha_3\beta_3(1 - v) \sin \alpha_3 x \sin \beta_3 y] \tag{14.374}$$

In view of Eqs. (14.370)–(14.374), the boundary conditions of Eqs. (14.364) and (14.365) yield

$$\alpha_j = \frac{r_j\pi}{2a}, \qquad j = 1, 2, 3 \tag{14.375}$$

$$\beta_j = \frac{s_j\pi}{2b}, \qquad j = 1, 2, 3 \tag{14.376}$$

where r_j and s_j are even integers. Thus, Eqs. (14.366) and (14.367) denote modes or deflections odd in both x and y. The function H does not represent any mode or deflection but produces rotations with the same type of symmetry obtained from W_1 and W_2. Hence, all the modes or solutions given by Eqs. (14.366)–(14.368) will be odd in both x and y. If we want the modes even in x, we need to interchange $\sin \alpha_j x$ and $\cos \alpha_j x$ in all the three equations (14.366)–(14.368) and r_j are to be taken as odd integers. Similarly, for modes even in y, we need to interchange the terms $\sin \beta_j y$ and $\cos \beta_j y$, with s_j taken as odd integers.

By substituting Eqs. (14.375) and (14.376) into Eqs. (14.369) and using the definitions of δ_j, $j = 1, 2, 3$ [Eqs. (14.355) and (14.350)], we obtain

$$\left(\frac{\omega_j}{\omega}\right)^2 = \frac{1}{2}\left\{1 + \frac{2}{1-\nu}\left[1 + \frac{k^2(1-\nu)}{2}\right]\psi_j^2 + (-1)^j \Omega_j\right\}, \qquad j = 1, 2$$

$$\left(\frac{\omega_3}{\omega}\right)^2 = 1 + \psi_3^2 \tag{14.377}$$

where

$$\psi_j^2 = \frac{h^2}{\pi^2}(\alpha_j^2 + \beta_j^2), \qquad j = 1, 2, 3 \tag{14.378}$$

$$\Omega_j = \left\{\left[1 + \frac{2}{1-\nu}\left(1 + \frac{k^2(1-\nu)}{2}\right)\psi_j^2\right]^2 - \frac{8k^2}{1-\nu}\psi_j^4\right\}^{1/2}, \qquad j = 1, 2, 3 \tag{14.379}$$

For any specific ratios a/b and a/h and the mode number j, the three frequencies ω_j, $j = 1, 2, 3$, satisfy the inequalities $\omega_1 < \omega_3 < \omega_2$, with ω_2 and ω_3 being much larger than ω_1 except for very thick plates. The mode corresponding to ω_1 is associated with W_1 and is most closely related to the one given by the classical plate theory. The mode associated with ω_2 corresponds to W_2 and denotes the thickness shear deformation mode. The mode shape associated with ω_3 contains no transverse deflection except for two components of rotation which are related to one another so as to denote twist about an axis normal to the plate. The types of modes or deflections generated by W_1, W_2, and H are called flexural, thickness–shear, and thickness–twist modes, respectively. The mode shapes corresponding to the three frequency ratios given by Eq. (14.377) are shown in Fig. 14.14.

14.9.5   Circular Plates

Equations of Motion   Considering the dynamic equilibrium of an infinitesimal element shown in Fig. 14.6, the equations of motion of a circular plate, can be derived in terms of the force and moment resultants as

$$\frac{\partial M_r}{\partial r} + \frac{1}{r}\frac{\partial M_{r\theta}}{\partial \theta} + \frac{M_r - M_\theta}{r} - Q_r = \frac{\rho h^3}{12}\frac{\partial^2 \phi_r}{\partial t^2} \tag{14.380}$$

$$\frac{\partial M_r}{\partial r} + \frac{1}{r}\frac{\partial M_\theta}{\partial \theta} + \frac{2}{r}M_{r\theta} - Q_\theta = \frac{\rho h^3}{12}\frac{\partial^2 \phi_\theta}{\partial t^2} \tag{14.381}$$

$$\frac{\partial Q_r}{\partial r} + \frac{1}{r}\frac{\partial Q_\theta}{\partial \theta} + \frac{Q_r}{r} = \rho h\frac{\partial^2 w}{\partial t^2} \tag{14.382}$$

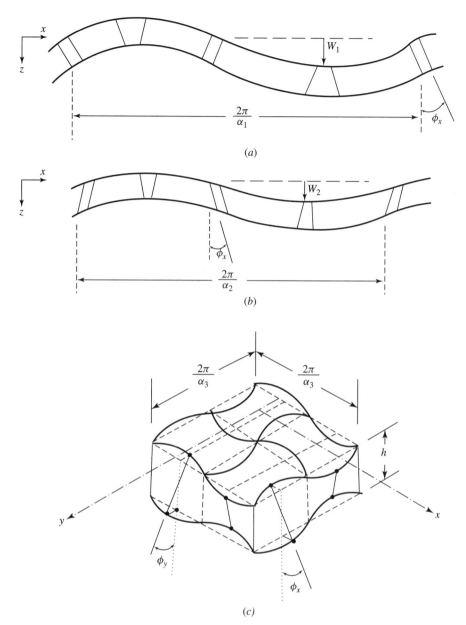

Figure 14.14 Modes of a rectangular plate: (*a*) flexural mode; (*b*) thickness–shear mode; (*c*) thickness–twist mode. (From Ref. 20; reprinted with permission from the publisher.)

where the displacement components have been assumed to be of the form

$$u(r, \theta, t) = z\phi_r(r, \theta, t) \tag{14.383}$$

$$v(r, \theta, t) = z\phi_\theta(r, \theta, t) \tag{14.384}$$

$$w(r, \theta, t) = w(r, \theta, t) \tag{14.385}$$

The moment and force resultants can be expressed in terms of ϕ_r, ϕ_θ and w as

$$M_r = D\left[\frac{\partial \phi_r}{\partial r} + \frac{\nu}{r}\left(\phi_r + \frac{\partial \phi_\theta}{\partial \theta}\right)\right] \tag{14.386}$$

$$M_\theta = D\left[\frac{1}{r}\left(\phi_r + \frac{\partial \phi_\theta}{\partial \theta}\right) + \nu\frac{\partial \phi_r}{\partial r}\right] \tag{14.387}$$

$$M_{r\theta} = \frac{D}{2}(1-\nu)\left[\frac{1}{r}\left(\frac{\partial \phi_r}{\partial \theta} - \phi_\theta\right) + \frac{\partial \phi_\theta}{\partial r}\right] \tag{14.388}$$

$$Q_r = k^2 Gh\left(\phi_r + \frac{\partial w}{\partial r}\right) \tag{14.389}$$

$$Q_\theta = k^2 Gh\left(\phi_\theta + \frac{1}{r}\frac{\partial w}{\partial \theta}\right) \tag{14.390}$$

The equations of motion, Eqs. (14.380)–(14.382), can be written as

$$\frac{D}{2}\left[(1-\nu)\nabla^2\phi_r + (1+\nu)\frac{\partial \Phi}{\partial r}\right] - k^2 Gh\left(\phi_r + \frac{\partial w}{\partial r}\right) = \frac{\rho h^3}{12}\frac{\partial^2 \phi_r}{\partial t^2} \tag{14.391}$$

$$\frac{D}{2}\left[(1-\nu)\nabla^2\phi_\theta + (1+\nu)\frac{\partial \Phi}{\partial \theta}\right] - k^2 Gh\left(\phi_\theta + \frac{1}{r}\frac{\partial w}{\partial \theta}\right) = \frac{\rho h^3}{12}\frac{\partial^2 \phi_\theta}{\partial t^2} \tag{14.392}$$

$$k^2 Gh(\nabla^2 w + \Phi) = \rho h\frac{\partial^2 w}{\partial t^2} \tag{14.393}$$

where

$$\Phi = \frac{\partial \phi_r}{\partial r} + \frac{1}{r}\frac{\partial \phi_\theta}{\partial \theta} \tag{14.394}$$

By assuming harmonic solution as

$$\phi_r(r, \theta, t) = \Phi_r(r, \theta)e^{i\omega t} \tag{14.395}$$

$$\phi_\theta(r, \theta, t) = \Phi_\theta(r, \theta)e^{i\omega t} \tag{14.396}$$

$$w(r, \theta, t) = W(r, \theta)e^{i\omega t} \tag{14.397}$$

Eqs. (14.391)–(14.393) can be expressed as

$$\frac{D}{2}\left[(1-\nu)\nabla^2\Phi_r + (1+\nu)\frac{\partial \tilde{\Phi}}{\partial r}\right] - k^2 Gh\left(\Phi_r + \frac{\partial W}{\partial r}\right) + \frac{\rho h^3\omega^2}{12}\Phi_r = 0 \tag{14.398}$$

$$\frac{D}{2}\left[(1-\nu)\nabla^2\Phi_\theta + (1+\nu)\frac{\partial \tilde{\Phi}}{\partial \theta}\right] - k^2 Gh\left(\Phi_\theta + \frac{1}{r}\frac{\partial W}{\partial \theta}\right) + \frac{\rho h^3\omega^2}{12}\Phi_\theta = 0 \tag{14.399}$$

$$k^2 Gh(\nabla^2 W + \tilde{\Phi}) + \rho h\omega^2 W = 0 \tag{14.400}$$

where

$$\tilde{\Phi} = \frac{\partial \Phi_r}{\partial r} + \frac{1}{r} \frac{\partial \Phi_\theta}{\partial \theta} \tag{14.401}$$

Free Vibration Solution   Using a procedure similar to that of a rectangular plate, the solution of Eqs. (14.398)–(14.400) is expressed as

$$W(r, \theta) = W_1(r, \theta) + W_2(r, \theta) \tag{14.402}$$

$$\Phi_r(r, \theta) = (\mu_1 - 1)\frac{\partial W_1}{\partial r} + (\mu_2 - 1)\frac{\partial W_2}{\partial r} + \frac{1}{r}\frac{\partial H}{\partial \theta} \tag{14.403}$$

$$\Phi_\theta(r, \theta) = (\mu_1 - 1)\frac{1}{r}\frac{\partial W_1}{\partial \theta} + (\mu_2 - 1)\frac{1}{r}\frac{\partial W_2}{\partial \theta} - \frac{\partial H}{\partial r} \tag{14.404}$$

where W_1, W_2 and H are solutions of the equations

$$(\nabla^2 + \delta_j^2)W_j = 0, \qquad j = 1, 2 \tag{14.405}$$

$$(\nabla^2 + \delta_3^2)H = 0 \tag{14.406}$$

with

$$\nabla^2 = \frac{\partial^2}{\partial r^2} + \frac{1}{r}\frac{\partial}{\partial r} + \frac{1}{r^2}\frac{\partial^2}{\partial \theta^2} \tag{14.407}$$

δ_j^2, $j = 1, 2, 3$, given by Eqs. (14.355) and (14.350), and μ_1 and μ_2 by Eq. (14.359). Note that if the effects of rotary inertia and shear deflection are neglected, Eqs. (14.405) and (14.406) reduce to

$$(\nabla^2 + \delta^2)W_1 = 0 \tag{14.408}$$

$$(\nabla^2 - \delta^2)W_2 = 0 \tag{14.409}$$

where

$$\delta^2 = \delta_1^2\big|_{R=S=0} = -\delta_2^2\big|_{R=S=0} = k_b^2 = \left(\frac{\rho h \omega^2}{D}\right)^{1/2} \tag{14.410}$$

The solution of Eqs. (14.405) and (14.406) are expressed in product form as

$$W_j(r, \theta) = R_j(r)\Theta_j(\theta), \qquad j = 1, 2 \tag{14.411}$$

$$H(r, \theta) = R_3(r)\Theta_3(\theta) \tag{14.412}$$

to obtain the following pairs of ordinary differential equations:

$$\frac{d^2 R_j(r)}{dr^2} + \frac{1}{r}\frac{d R_j(r)}{dr} + \left(\delta_j^2 - \frac{m^2}{r^2}\right)R_j(r) = 0, \qquad j = 1, 2, 3 \tag{14.413}$$

$$\frac{d^2 \Theta_j(\theta)}{d\theta^2} + m^2 \Theta_j(\theta) = 0, \qquad j = 1, 2, 3 \tag{14.414}$$

where m^2 is the separation constant. Noting that Eq. (14.413) is Bessel's differential equation, the solutions, $W_j(r, \theta)$ and $H(r, \theta)$, can be expressed as

$$W_j(r, \theta) = \sum_{m=0}^{\infty} W_{jm}(r, \theta) = \sum_{m=0}^{\infty} [A_j^{(m)} J_m(\delta_j r) + B_j^{(m)} Y_m(\delta_j r)] \cos m\theta, \qquad j = 1, 2$$

(14.415)

$$H(r, \theta) = \sum_{m=0}^{\infty} H_m(r, \theta) = \sum_{m=0}^{\infty} [A_3^{(m)} J_m(\delta_3 r) + B_3^{(m)} Y_m(\delta_3 r)] \sin m\theta \qquad (14.416)$$

where $A_j^{(m)}$ and $B_j^{(m)}$, $j = 1, 2, 3$, are constants, J_m and Y_m are Bessel functions of the first and second kind, respectively, of order m, and m corresponds to the number of nodal diameters.

In terms of the solutions given by Eqs. (14.415) and (14.416), the moments M_r and $M_{r\theta}$ and the shear force Q_r (for any particular value of m) can be expressed as

$$
M_r^{(m)}(r, \theta) = D \Bigg[\sum_{i=1}^{2} A_i^{(m)} \left\{ (\sigma_i - 1) \left[J_m''(\delta_i r) + \frac{\nu}{r} J_m'(\delta_i r) - \frac{\nu m^2}{r^2} J_m(\delta_i r) \right] \right\} \cos m\theta
$$
$$
+ \sum_{i=1}^{2} B_i^{(m)} \left\{ (\sigma_i - 1) \left[Y_m''(\delta_i r) + \frac{\nu}{r} Y_m'(\delta_i r) - \frac{\nu m^2}{r^2} Y_m(\delta_i r) \right] \right\} \cos m\theta
$$
$$
+ A_3^{(m)} \left\{ (1 - \nu) \left[\frac{m}{r} J_m'(\delta_3 r) - \frac{m}{r^2} J_m(\delta_3 r) \right] \right\} \cos m\theta
$$
$$
+ B_3^{(m)} \left\{ (1 - \nu) \left[\frac{m}{r} Y_m'(\delta_3 r) - \frac{m}{r^2} Y_m(\delta_3 r) \right] \right\} \cos m\theta \Bigg] \qquad (14.417)
$$

$$
M_{r\theta}^{(m)}(r, \theta) = D(1 - \nu) \Bigg\{ \sum_{i=1}^{2} A_i^{(m)} \left[-\frac{m}{r} J_m'(\delta_i r) + \frac{m}{r^2} J_m(\delta_i r) \right] (\sigma_i - 1)
$$
$$
+ \sum_{i=1}^{2} B_i^{(m)} \left[-\frac{m}{r} Y_m'(\delta_i r) + \frac{m}{r^2} Y_m(\delta_i r) \right] (\sigma_i - 1)
$$
$$
+ A_3^{(m)} \left[-\frac{1}{2} J_m''(\delta_3 r) + \frac{1}{2r} J_m'(\delta_3 r) - \frac{m}{2r^2} J_m(\delta_3 r) \right]
$$
$$
+ B_3^{(m)} \left[-\frac{1}{2} Y_m''(\delta_3 r) + \frac{1}{2r} Y_m'(\delta_3 r) - \frac{m^2}{2r^2} Y_m(\delta_3 r) \right] \Bigg\} \sin m\theta \qquad (14.418)
$$

$$
Q_r^{(m)}(r, \theta) = k^2 G h \Bigg\{ \sum_{i=1}^{2} [A_i^{(m)} \sigma_1 J_m'(\delta_i r) + B_i^{(m)} \sigma_i Y_m'(\delta_i r)]
$$
$$
+ A_3^{(m)} \frac{m}{r} J_m(\delta_3 r) + B_3^{(m)} \frac{m}{r} Y_m(\delta_3 r) \Bigg\} \cos m\theta \qquad (14.419)
$$

The transverse deflection of the plate can be expressed as

$$W^{(m)}(r, \theta) = \left\{ \sum_{i=1}^{2} [A_i^{(m)} J_m(\delta_i r) + B_i^{(m)} Y_m(\delta_i r)] \right\} \cos m\theta \qquad (14.420)$$

The boundary conditions can be stated as follows:

1. Clamped or fixed edge:

$$W = \Phi_r = \Phi_\theta = 0 \qquad (14.421)$$

2. Simply supported edge (hard type):

$$W = \Phi_\theta = M_r = 0 \qquad (14.422)$$

3. Simply supported edge (soft type):

$$W = M_r = M_{r\theta} = 0 \qquad (14.423)$$

4. Free edge:

$$M_r = M_{r\theta} = Q_r = 0 \qquad (14.424)$$

14.9.6   Natural Frequencies of a Clamped Circular Plate

For a solid circular plate, the constants $B_j^{(m)}$, $j = 1, 2, 3$, are set equal to zero in Eqs. (14.415) and (14.416) so as to avoid infinite displacements, slopes, and bending moments at $r = 0$. The natural frequencies can be determined by substituting Eqs. (14.402)–(14.404), (14.415), and (14.416) into the appropriate equation in Eqs. (14.421)–(14.424). If the plate is clamped at the outer boundary, $r = a$, Eqs. (14.421) lead to the frequency equation in the form of a determinantal equation:

$$\begin{vmatrix} C_{11} & C_{12} & C_{13} \\ C_{21} & C_{22} & C_{23} \\ C_{31} & C_{32} & C_{33} \end{vmatrix} = 0 \qquad (14.425)$$

where

$$\begin{aligned}
C_{1j} &= (\mu_j - 1) J_m'(\delta_j a), & j = 1, 2 \\
C_{2j} &= m(\mu_j - 1) J_m(\delta_j a), & j = 1, 2 \\
C_{3j} &= J_m(\delta_j a), & j = 1, 2 \\
C_{13} &= m\, J_m(\delta_3 a) \\
C_{23} &= J_m'(\delta_3 a) \\
C_{33} &= 0
\end{aligned} \qquad (14.426)$$

Some of the natural frequencies given by Eq. (14.425) are shown in Table 14.3.

Table 14.3   Natural Frequencies of a Clamped Circular Plate[a]

Number of nodal diameters, m	Number of nodal circles, n	h/a			Thin plate theory
		0.05	0.10	0.25	
0	1	10.145	9.941	8.807	10.216
	2	38.855	36.479	27.253	39.771
1	1	21.002	20.232	16.521	21.260
	2	58.827	53.890	37.550	60.829
2	0	34.258	32.406	24.670	34.877
	1	80.933	72.368	47.650	84.583

Source: Ref. [7].
[a] Values of $a^2 \sqrt{\rho h \omega^2 / D}$, $\nu = 0.3$, $k^2 = \pi^2/12$.

14.10   PLATE ON AN ELASTIC FOUNDATION

The problem of vibration of a plate on an elastic foundation finds application in several practical situations, such as reinforced concrete pavements of highways and airport runways and foundation slabs of heavy machines and buildings. By assuming a Winkler foundation where the reaction force of the foundation is assumed to be proportional to the deflection, we can express the reaction force, $R(x,y,t)$, as (Fig. 14.15):

$$R = kw(x,y,t) \tag{14.427}$$

where k, a constant known as the *foundation modulus*, can be interpreted as the reaction force of the foundation per unit surface area of the plate per unit deflection of the plate. Since the reaction force acts in a direction opposite to that of the external force $f(x,y,t)$, the equation of motion governing the vibration of a thin plate resting on an elastic foundation can be expressed as

$$\frac{\partial^4 w}{\partial x^4} + 2\frac{\partial^4 w}{\partial x^2 \partial y^2} + \frac{\partial^4 w}{\partial y^4} = \frac{1}{D}\left[f(x,y,t) - kw(x,y,t) - \rho h \frac{\partial^2 w(x,y,t)}{\partial t^2} \right] \tag{14.428}$$

For free vibration, $f(x,y,t) = 0$ and the solution is assumed to be harmonic:

$$w(x,y,t) = W(x,y)\, e^{i\omega t} \tag{14.429}$$

so that Eq. (14.428) becomes

$$\frac{\partial^4 W}{\partial x^4} + 2\frac{\partial^4 W}{\partial x^2 \partial y^2} + \frac{\partial^4 W}{\partial y^4} = \frac{\rho h \omega^2 - k}{D}\, W = 0 \tag{14.430}$$

Defining

$$\tilde{\lambda}^4 = \frac{\rho h \omega^2 - k}{D} \tag{14.431}$$

Eq. (14.430) can be expressed as

$$\nabla^4 W - \tilde{\lambda}^4 W = 0 \tag{14.432}$$

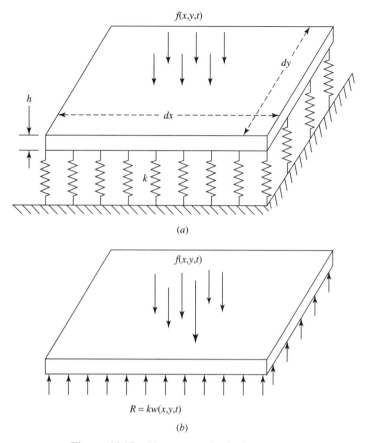

$f(x,y,t)$

dy

h

dx

k

(a)

$f(x,y,t)$

$R = kw(x,y,t)$

(b)

Figure 14.15   Plate on an elastic foundation.

Equation (14.432) can be seen to be similar to Eq. (14.91) and hence the solutions obtained earlier (Section 14.4) are applicable to plates on elastic foundation provided that $\tilde{\lambda}$ is used in place of λ. For example, for a plate on elastic foundation, simply supported on all the edges, the natural frequencies of vibration and the natural modes can be determined using Eqs. (14.105) and (14.107) as

$$\tilde{\lambda}_{mn}^2 = \sqrt{\frac{\rho h \omega_{mn}^2 - k}{D}} = \pi^2 \left[\left(\frac{m}{a} \right)^2 + \left(\frac{n}{b} \right)^2 \right] \tag{14.433}$$

or

$$\omega_{mn} = \left[\frac{D \pi^4}{\rho h} \left[\left(\frac{m}{a} \right)^2 + \left(\frac{n}{b} \right)^2 \right]^2 + \frac{k}{\rho h} \right]^{1/2} \tag{14.434}$$

and

$$w_{mn}(x,y,t) = W_{mn}(x,y) \left(A_{mn} \cos \omega_{mn} t + B_{mn} \sin \omega_{mn} t \right) \tag{14.435}$$

where

$$W_{mn}(x,y) = \sin \frac{m\pi x}{a} \sin \frac{n\pi y}{b} \qquad (14.436)$$

14.11 TRANSVERSE VIBRATION OF PLATES SUBJECTED TO IN-PLANE LOADS

14.11.1 Equation of Motion

To derive the equation of motion governing the transverse vibration of a plate subjected to in-plane loads (see Fig. 14.16) using the equilibrium approach, consider an infinitesimal plate element of sides dx and dy. Let the element be subjected to the time-independent (static) in-plane or membrane loads $N_x(x,y)$, $N_y(x,y)$, and $N_{xy}(x,y) = N_{yx}(x,y)$ per unit length as well as a time-dependent transverse load $f(x,y,t)$ per unit area. The in-plane loads acting on the various sides of the element are shown in Fig. 14.17(a), and the transverse loads and moments acting on the sides of the element are shown in Fig. 14.17(b). The transverse deformation, along with the forces acting on the element, as seen in the xz and yz planes are shown in Fig. 14.17(c) and (d), respectively. The force equilibrium equation in the x direction gives [from Fig. 14.17(c)]

$$\sum F_x = \left(N_x + \frac{\partial N_x}{\partial x}\, dx \right)\, dy\, \cos\theta_1' + \left(N_{yx} + \frac{\partial N_{yx}}{\partial y}\, dy \right)\, dx\, \cos\frac{\theta_1 + \theta_1'}{2}$$
$$- N_x\, dy\, \cos\theta_1 - N_{yx}\, dx\, \cos\frac{\theta_1 + \theta_1'}{2} = 0 \qquad (14.437)$$

where

$$\theta_1' = \theta_1 + d\theta_1 = \theta_1 + \frac{\partial \theta_1}{\partial x}\, dx$$

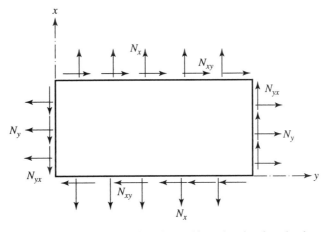

Figure 14.16 Rectangular plate subjected to in-plane loads.

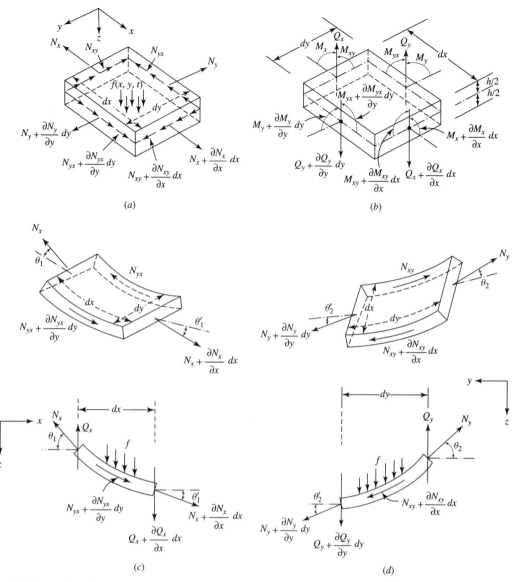

Figure 14.17 (*a*) In-plane loads acting on the sides of an element of a plate; (*b*) moment and shear force resultants acting on the sides on an element of the plate; (*c*) deformation in the *xz* plane; (*d*) deformation in the *yz* plane.

Since the deflections are assumed to be small, θ_1 will be small, so that

$$\cos \theta_1 \approx 1, \qquad \cos \theta_1' \approx 1, \qquad \cos \frac{\theta_1 + \theta_1'}{2} \approx 1 \qquad (14.438)$$

In view of Eq. (14.438) and the fact that $N_{yx} = N_{xy}$, Eq. (14.437) can be simplified as

$$\frac{\partial N_x}{\partial x} + \frac{\partial N_y}{\partial y} = 0 \qquad (14.439)$$

In a similar manner, the force equilibrium equation in the y direction yields [from Fig. 14.17(d)]:

$$
\sum F_y = \left(N_y + \frac{\partial N_y}{\partial y}\, dy \right) dx\, \cos\theta_2' + \left(N_{xy} + \frac{\partial N_{xy}}{\partial x}\, dx \right) dy\, \cos\frac{\theta_2 + \theta_2'}{2}
$$

$$
- N_y\, dx\, \cos\theta_2 - N_{xy}\, dx\, \cos\frac{\theta_2 + \theta_2'}{2} = 0 \qquad (14.440)
$$

where

$$
\theta_2' = \theta_2 + d\theta_2 = \theta_2 + \frac{\partial\theta_2}{\partial y}\, dy
$$

Since the deflections are assumed to be small, θ_2 and θ_2' will also be small, so that

$$
\cos\theta_2 \approx 1, \qquad \cos\theta_2' \approx 1, \qquad \cos\frac{\theta_2 + \theta_2'}{2} \approx 1 \qquad (14.441)
$$

In view of Eq. (14.441), Eq. (14.440) reduces to

$$
\frac{\partial N_y}{\partial y} + \frac{\partial N_{xy}}{\partial x} = 0 \qquad (14.442)
$$

The force equilibrium equation in the z direction can be obtained by considering the projections of all the in-plane and transverse forces as follows [from Fig. 14.17(b), (c), and (d)]:

$$
\sum F_z = \left(N_x + \frac{\partial N_x}{\partial x}\, dx \right) dy\, \sin\theta_1' - N_x\, dy\, \sin\theta_1
$$

$$
+ \left(N_y + \frac{\partial N_y}{\partial y}\, dy \right) dx\, \sin\theta_2' - N_y\, dx\, \sin\theta_2
$$

$$
+ \left(N_{yx} + \frac{\partial N_{yx}}{\partial y}\, dy \right) dx\, \sin\overline{\theta}_1' - N_{yx}\, dx\, \sin\overline{\theta}_1
$$

$$
+ \left(N_{xy} + \frac{\partial N_{xy}}{\partial x}\, dx \right) dy\, \sin\overline{\theta}_2' - N_{xy}\, dy\, \sin\overline{\theta}_2
$$

$$
+ \left(Q_x + \frac{\partial Q_x}{\partial x}\, dx \right) dy - Q_x\, dy
$$

$$
+ \left(Q_y + \frac{\partial Q_y}{\partial y}\, dy \right) dx + f\, dx\, dy = 0 \qquad (14.443)
$$

By using the assumption of small displacements and slopes, we can express

$$\sin\theta_1 \approx \theta_1 \approx \frac{\partial w}{\partial x}, \qquad \sin\theta_1' \approx \theta_1' \approx \theta_1 + \frac{\partial\theta_1}{\partial x}dx = \frac{\partial w}{\partial x} + \frac{\partial^2 w}{\partial x^2}dx$$

$$\sin\theta_2 \approx \theta_2 \approx \frac{\partial w}{\partial y}, \qquad \sin\theta_2' \approx \theta_2' \approx \theta_2 + \frac{\partial\theta_2}{\partial y}dy = \frac{\partial w}{\partial y} + \frac{\partial^2 w}{\partial y^2}dy$$

$$\bar\theta_2 \approx \theta_2 \approx \frac{\partial w}{\partial y}, \qquad \bar\theta_2' \approx \bar\theta_2 + \frac{\partial\bar\theta_2}{\partial x}dx \approx \frac{\partial w}{\partial y} + \frac{\partial^2 w}{\partial x\,\partial y}dx$$

$$\bar\theta_1 \approx \theta_1 \approx \frac{\partial w}{\partial x}, \qquad \bar\theta_1' \approx \bar\theta_1 + \frac{\partial\bar\theta_1}{\partial y}dy \approx \frac{\partial w}{\partial x} + \frac{\partial^2 w}{\partial x\,\partial y}dy$$

$$(14.444)$$

The first two terms in Eq. (14.443) can be rewritten as

$$\left(N_x + \frac{\partial N_x}{\partial x}dx\right)dy\sin\theta_1' - N_x dy\sin\theta_1 \approx \left(N_x + \frac{\partial N_x}{\partial x}dx\right)dy\left(\frac{\partial w}{\partial x} + \frac{\partial^2 w}{\partial x^2}dx\right)$$

$$- N_x\,dy\,\frac{\partial w}{\partial x}$$

$$\approx N_x\frac{\partial^2 w}{\partial x^2}dx\,dy + \frac{\partial N_x}{\partial x}\frac{\partial w}{\partial x}dx\,dy$$

$$(14.445)$$

by neglecting the higher-order term. Similarly, the next pair of terms in Eq. (14.443) can be rewritten as

$$\left(N_y + \frac{\partial N_y}{\partial y}dy\right)dx\sin\theta_2' - N_y dx\sin\theta_2 \approx \left(N_y + \frac{\partial N_y}{\partial y}dy\right)dx\left(\frac{\partial w}{\partial y} + \frac{\partial^2 w}{\partial y^2}dy\right)$$

$$- N_y\,dx\,\frac{\partial w}{\partial y}$$

$$\approx N_y\frac{\partial^2 w}{\partial y^2}dx\,dy + \frac{\partial N_y}{\partial y}\frac{\partial w}{\partial y}dx\,dy$$

$$(14.446)$$

The components of the in-plane shear forces N_{xy} and $(\partial N_{xy}/\partial x)\,dx$, acting on the x edges (i.e ., edges whose normals lie parallel to the x axis) in the z direction can be expressed as

$$\left(N_{xy} + \frac{\partial N_{xy}}{\partial x}\,dx\right)\,dy\,\sin\bar\theta_2' - N_{xy}\,dy\,\sin\bar\theta_2 \qquad (14.447)$$

where the slopes of the deflection surface in the y direction on the x edges, $\bar\theta_2$ and $\bar\theta_2'$, are given by Eq. (14.444):

$$\bar\theta_2 \approx \theta_2 \approx \frac{\partial w}{\partial y}, \qquad \bar\theta_2' \approx \bar\theta_2 + \frac{\partial\bar\theta_2}{\partial x}dx \approx \frac{\partial w}{\partial y} + \frac{\partial^2 w}{\partial x\,\partial y}dx$$

Thus, Eq. (14.447) yields, by neglecting the higher-order term,

$$\frac{\partial N_{xy}}{\partial x} \frac{\partial w}{\partial y} \, dx \, dy + N_{xy} \frac{\partial^2 w}{\partial x \, \partial y} \, dx \, dy \tag{14.448}$$

Similarly, the component of the in-plane shear forces N_{yx} and $(\partial N_{yx}/\partial y) \, dy$, acting on the y edges (i.e ., edges whose normals lie parallel to the y axis) in the z direction can be written as

$$\left(N_{yx} + \frac{\partial N_{yx}}{\partial y} \, dy \right) \, dx \, \sin \overline{\theta}_1' - N_{yx} \, dx \, \sin \overline{\theta}_1 \tag{14.449}$$

where the slopes of the deflection surface in the x direction on the y edges, $\overline{\theta}_1$ and $\overline{\theta}_1'$, are given in Eq. (14.444):

$$\overline{\theta}_1 \approx \theta_1 \approx \frac{\partial w}{\partial x}, \qquad \overline{\theta}_1' \approx \overline{\theta}_1 + \frac{\partial \overline{\theta}_1}{\partial y} \, dy \approx \frac{\partial w}{\partial x} + \frac{\partial^2 w}{\partial x \, \partial y} \, dy$$

Thus, the expression (14.449) can be simplified as

$$N_{yx} \frac{\partial^2 w}{\partial x \, \partial y} \, dx \, dy + \frac{\partial N_{yx}}{\partial y} \frac{\partial w}{\partial x} \, dx \, dy \tag{14.450}$$

Thus, Eq. (14.443) can be expressed as

$$\frac{\partial Q_x}{\partial x} + \frac{\partial Q_y}{\partial y} + f + N_x \frac{\partial^2 w}{\partial x^2} + N_y \frac{\partial^2 w}{\partial y^2} + 2N_{xy} \frac{\partial^2 w}{\partial x \, \partial y}$$

$$+ \left(\frac{\partial N_x}{\partial x} + \frac{\partial N_{yx}}{\partial y} \right) \frac{\partial w}{\partial x} + \left(\frac{\partial N_{xy}}{\partial x} + \frac{\partial N_y}{\partial y} \right) \frac{\partial w}{\partial y} = 0 \tag{14.451}$$

The expressions within the parentheses of Eq.(14.451) are zero in view of Eqs. (14.439) and (14.442) and hence Eq. (14.451) reduces to

$$\frac{\partial Q_x}{\partial x} + \frac{\partial Q_y}{\partial y} + f + N_x \frac{\partial^2 w}{\partial x^2} + N_y \frac{\partial^2 w}{\partial y^2} + 2N_{xy} \frac{\partial^2 w}{\partial x \, \partial y} = 0 \tag{14.452}$$

The in-plane forces do not contribute to any moment along the edges of the element. As such, the moment equilibrium equations about x and y axes lead to Eqs. (3.29) and (3.30). By substituting the shear force resultants in terms of the displacement w [Eqs. (14.5) and (14.6)], Eq. (14.452) yields

$$\frac{\partial^4 w}{\partial x^4} + 2\frac{\partial^4 w}{\partial x^2 \partial y^2} + \frac{\partial^4 w}{\partial y^4} = \frac{1}{D} \left(f + N_x \frac{\partial^2 w}{\partial x^2} + N_y \frac{\partial^2 w}{\partial y^2} + 2N_{xy} \frac{\partial^2 w}{\partial x \, \partial y} \right) \tag{14.453}$$

Finally, by adding the inertia force, the total external force is given by

$$f - \rho h \frac{\partial^2 w}{\partial t^2} \tag{14.454}$$

and the equation of motion for the vibration of a plate subjected to combined in-plane and transverse loads becomes

$$\frac{\partial^4 w}{\partial x^4} + 2\frac{\partial^4 w}{\partial x^2 \partial y^2} + \frac{\partial^4 w}{\partial y^4} = \frac{1}{D} \left(f - \rho h \frac{\partial^2 w}{\partial t^2} + N_x \frac{\partial^2 w}{\partial x^2} + N_y \frac{\partial^2 w}{\partial y^2} + 2N_{xy} \frac{\partial^2 w}{\partial x \, \partial y} \right) \tag{14.455}$$

14.11.2   Free Vibration

For free vibration, f is set equal to zero in Eq. (14.455) and the solution of the resulting equation is assumed to be harmonic with frequency ω:

$$w(x,y,t) = W(x,y)e^{i\omega t} \tag{14.456}$$

Substitution of Eq. (14.456) into Eq. (14.455) with $f = 0$ leads to

$$\frac{\partial^4 W}{\partial x^4} + 2\frac{\partial^4 W}{\partial x^2 \partial y^2} + \frac{\partial^4 W}{\partial y^4} = \frac{1}{D}\left(\rho h\omega^2 W + N_x \frac{\partial^2 W}{\partial x^2} + N_y \frac{\partial^2 W}{\partial y^2} + 2N_{xy}\frac{\partial^2 W}{\partial x\,\partial y}\right) \tag{14.457}$$

Introducing

$$\lambda^4 = \frac{\rho h\omega^2}{D} \tag{14.458}$$

Eq. (14.457) can be rewritten as

$$\nabla^4 W - \lambda^4 W = \frac{1}{D}\left(N_x \frac{\partial^2 W}{\partial x^2} + N_y \frac{\partial^2 W}{\partial y^2} + 2N_{xy}\frac{\partial^2 W}{\partial x\,\partial y}\right) \tag{14.459}$$

14.11.3   Solution for a Simply Supported Plate

We consider the free vibration of a plate simply supported on all edges subjected to the in-plane forces $N_x = N_1$, $N_y = N_2$ and $N_{xy} = 0$, where N_1 and N_2 are constants. For this case, the equation of motion, Eq. (14.459), becomes

$$\nabla^4 W - \lambda^4 W = \frac{1}{D}\left(N_1 \frac{\partial^2 W}{\partial x^2} + N_2 \frac{\partial^2 W}{\partial y^2}\right) \tag{14.460}$$

As in the case of free vibration of a rectangular plate with no in-plane forces, the following solution can be seen to satisfy the boundary conditions of the plate:

$$W(x, y) = \sum_{m,n=1}^{\infty} A_{mn} \sin\frac{m\pi x}{a} \sin\frac{n\pi y}{b} \tag{14.461}$$

where A_{mn} are constants. By substituting Eq. (14.461) into (14.460), we can obtain the frequency equation as

$$\left[\left(\frac{m\pi}{a}\right)^2 + \left(\frac{n\pi}{b}\right)^2\right]^2 + \frac{1}{D}\left[N_1\left(\frac{m\pi}{a}\right)^2 + N_2\left(\frac{n\pi}{b}\right)^2\right] = \lambda^4 = \frac{\rho h\omega_{mn}^2}{D} \tag{14.462}$$

or

$$\omega_{mn}^2 = \frac{D}{\rho h}\left[\left(\frac{m\pi}{a}\right)^2 + \left(\frac{n\pi}{b}\right)^2\right]^2 + \frac{1}{\rho h}\left[N_1\left(\frac{m\pi}{a}\right)^2 + N_2\left(\frac{n\pi}{b}\right)^2\right], \qquad m, n = 1, 2, \ldots \tag{14.463}$$

If N_1 and N_2 are compressive, Eq. (14.463) can be written as

$$\omega_{mn}^2 = \frac{1}{\rho h}\left\{D\left[\left(\frac{m\pi}{a}\right)^2 + \left(\frac{n\pi}{b}\right)^2\right]^2 - N_1\left(\frac{m\pi}{a}\right)^2 - N_2\left(\frac{n\pi}{b}\right)^2\right\}, \qquad m, n = 1, 2, \ldots \tag{14.464}$$

It can be seen that ω_{mn} reduces to zero as the magnitude of N_1 and/or N_2 increases. For example, when $N_2 = 0$, the value of N_1 that makes $\omega_{mn} = 0$, called the *critical* or *buckling load*, can be determined from Eq. (14.464) as

$$(N_1)_{\text{cri}} = -D\left(\frac{a}{m\pi}\right)^2 \left[\left(\frac{m\pi}{a}\right)^2 + \left(\frac{n\pi}{b}\right)^2\right]^2 = -\frac{D\pi^2}{a^2}\left[m + n\left(\frac{n}{m}\right)\left(\frac{a}{b}\right)^2\right]^2 \tag{14.465}$$

where the negative sign represents a compressive load. It is to be noted from Eq. (14.464) that the fundamental natural frequency may not correspond to $m = 1$ and $n = 1$ but depends on the values of N_1, N_2, and a/b. Using Eq. (14.465), the frequency given by Eq. (14.463) can be expressed for $N_2 = 0$ as

$$\omega_{mn}^2 = \left(\frac{m\pi}{a}\right)^2 [N_1 - (N_1)_{\text{cri}}] \tag{14.466}$$

It can be shown that the fundamental frequency in this case always occurs when $n = 1$ and not necessarily when $m = 1$ [17]. By introducing

$$\omega_{\text{ref}}^2 = \frac{4D\pi^4}{\rho ha^4} \tag{14.467}$$

where ω_{ref} denotes the fundamental natural frequency of vibration of a square plate with no in-plane loads, and the square of the frequency ratio, $(\omega_{mn}/\omega_{\text{ref}})^2$, can be written as

$$\left(\frac{\omega_{mn}}{\omega_{\text{ref}}}\right)^2 = \frac{(m\pi/a)^2[N_1 - (N_1)_{\text{cri}}]}{4D\pi^4/\rho ha^4} = \frac{(m\pi/a)^2(N_1)_{\text{cri}}\{[N_1/(N_1)_{\text{cri}}] - 1\}}{4D\pi^4/\rho ha^4} \tag{14.468}$$

Using Eq. (14.465), Eq. (14.468) can be rewritten as

$$\left(\frac{\omega_{mn}}{\omega_{\text{ref}}}\right)^2 = -\frac{m^4}{4}\left[1 + \left(\frac{a}{b}\right)^2\right]^2 \left[\frac{N_1}{(N_1)_{\text{cri}}} - 1\right] \tag{14.469}$$

The variation of the square of the frequency ratio $(\omega_{mn}/\omega_{\text{ref}})^2$ with the load ratio $N_1/(N_1)_{\text{cri}}$ is shown in Fig. 14.18 for different values of m and a/b (with $n = 1$).

14.12   VIBRATION OF PLATES WITH VARIABLE THICKNESS

14.12.1   Rectangular Plates

Let the thickness of the plate vary continuously with no abrupt changes so that it can be represented as $h = h(x, y)$. In this case the expressions given by Eqs. (14.1)–(14.3) for the moment resultants can be used with sufficient accuracy. However, the shear

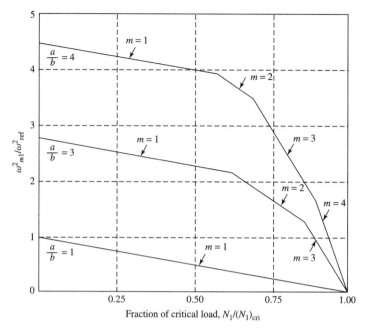

Figure 14.18   Effect of the in-plane force $N_1 = N_x$ on the fundamental frequency of a rectangular plate simply supported on all sides. (From Ref. [17].).

force resultants given by Eqs. (14.5) and (14.6) will be modified as

$$
Q_x = \frac{\partial}{\partial x}\left[-D\left(\frac{\partial^2 w}{\partial x^2} + v\frac{\partial^2 w}{\partial y^2}\right)\right] + \frac{\partial}{\partial y}\left[-(1-v)D\frac{\partial^2 w}{\partial x\,\partial y}\right]
$$

$$
= -D\frac{\partial}{\partial x}\left(\frac{\partial^2 w}{\partial x^2} + v\frac{\partial^2 w}{\partial y^2}\right) - \frac{\partial D}{\partial x}\left(\frac{\partial^2 w}{\partial x^2} + v\frac{\partial^2 w}{\partial y^2}\right)
$$

$$
- (1-v)D\frac{\partial^3 w}{\partial x\,\partial y^2} - (1-v)\frac{\partial D}{\partial y}\frac{\partial^3 w}{\partial x\,\partial y^2} \tag{14.470}
$$

$$
Q_y = \frac{\partial}{\partial x}\left[-D\left(\frac{\partial^2 w}{\partial y^2} + v\frac{\partial^2 w}{\partial x^2}\right)\right] + \frac{\partial}{\partial x}\left[-(1-v)D\frac{\partial^2 w}{\partial x\,\partial y}\right]
$$

$$
= -D\frac{\partial}{\partial y}\left(\frac{\partial^2 w}{\partial y^2} + v\frac{\partial^2 w}{\partial x^2}\right) - \frac{\partial D}{\partial y}\left(\frac{\partial^2 w}{\partial y^2} + v\frac{\partial^2 w}{\partial x^2}\right)
$$

$$
- (1-v)D\frac{\partial^3 w}{\partial x^2\partial y} - (1-v)\frac{\partial D}{\partial x}\frac{\partial^2 w}{\partial x\,\partial y} \tag{14.471}
$$

where

$$
\frac{\partial D}{\partial x} = \frac{\partial}{\partial x}\left[\frac{Eh^3(x,y)}{12(1-v^2)}\right] = \frac{Eh^2(x,y)}{4(1-v^2)}\frac{\partial h(x,y)}{\partial x} \tag{14.472}
$$

$$
\frac{\partial D}{\partial y} = \frac{\partial}{\partial y}\left[\frac{Eh^3(x,y)}{12(1-v^2)}\right] = \frac{Eh^2(x,y)}{4(1-v^2)}\frac{\partial h(x,y)}{\partial y} \tag{14.473}
$$

Finally, by substituting Eqs. (14.470) and (14.471) into Eq. (14.7) yields the equation of motion of a plate with variable thickness as

$$
D\left(\frac{\partial^4 w}{\partial x^4} + 2\frac{\partial^4 w}{\partial x^2 \partial y^2} + \frac{\partial^4 w}{\partial y^4}\right) + 2\frac{\partial D}{\partial x}\frac{\partial}{\partial x}\left(\frac{\partial^2 w}{\partial x^2} + \frac{\partial^2 w}{\partial y^2}\right) + 2\frac{\partial D}{\partial y}\frac{\partial}{\partial y}\left(\frac{\partial^2 w}{\partial x^2} + \frac{\partial^2 w}{\partial y^2}\right)
$$

$$
+ \left(\frac{\partial^2 D}{\partial x^2} + \frac{\partial^2 D}{\partial y^2}\right)\left(\frac{\partial^2 w}{\partial x^2} + \frac{\partial^2 w}{\partial y^2}\right)
$$

$$
- (1-v)\left(\frac{\partial^2 D}{\partial x^2}\frac{\partial^2 w}{\partial y^2} - 2\frac{\partial^2 D}{\partial x \partial y}\frac{\partial^2 w}{\partial x \partial y} + \frac{\partial^2 D}{\partial y^2}\frac{\partial^2 w}{\partial y^2}\right) + \rho h(x,y)\frac{\partial^2 w}{\partial t^2} = f(x,y,t)
$$

$$(14.474)$$

Closed-form solutions of Eq. (14.474) are possible only for very simple forms of variation of thickness of the plate, $h(x,y)$. In general, the solution of Eq. (14.474) can be found using either approximate analytical methods such as the Rayleigh–Ritz and Galerkin methods or numerical methods such as the finite element method.

14.12.2   Circular Plates

As in the case of rectangular plates, the variable thickness of the plate in the case of a circular plate is denoted as $h = h(r, \theta)$ and the bending rigidity as

$$
D = \frac{Eh^3(r, \theta)}{12(1 - v^2)}
$$

$$(14.475)$$

By assuming the expressions of moment resultants given by Eqs. (14.184)–(14.186) to be applicable with sufficient accuracy, the shear force resultants given by Eqs. (14.187a) and (14.188a) will be modified by treating D as a function of r and θ. The resulting expressions when substituted into Eq.(14.170) lead to the equation of motion for the vibration of a circular plate of variable thickness. The resulting equation will be quite lengthy and the closed-form solutions will be almost impossible for a general variation of the thickness of the plate. However, if the plate is assumed to be axisymmetric, the thickness of the plate will vary with r only, so that $h = h(r)$. Although the equation of motion can be derived as in the case of a rectangular plate with variable thickness, the equilibrium approach is used directly in this section. For this, consider the free-body diagram of an element of the circular plate of variable thickness, along with internal force and moment resultants and external force, as shown in Fig. 14.19(a). Note that the shear force resultant Q_θ and the twisting moment resultant $M_{r\theta}$ are not indicated in Fig. 14.19(a), as they will be zero, due to symmetry. The equilibrium of forces in the z direction gives

$$
\left(Q_r + \frac{\partial Q_r}{\partial r}\,dr\right)(r + dr)\,d\theta - Q_r r\,d\theta + fr\,d\theta\,dr = \rho h\frac{\partial^2 w}{\partial t^2}
$$

$$(14.476)$$

Noting that the component of M_θ in the tangential direction is given by [from Fig. 14.19(b)]

$$
M_\theta \sin\frac{d\theta}{2} + M_\theta \sin\frac{d\theta}{2} \approx M_\theta\,d\theta
$$

$$(14.477)$$

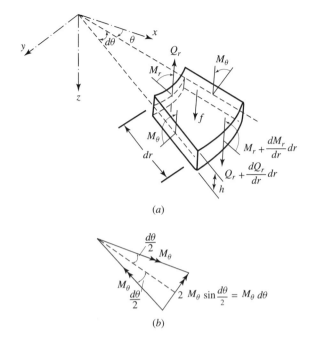

Figure 14.19   Circular plate element.

for small angles $d\theta$, the moment equilibrium about the θ direction leads to

$$\left(M_r + \frac{\partial M_r}{\partial r}\, dr\right)(r + dr)\, d\theta - M_r r\, d\theta - Q_r r\, d\theta\, dr - M_\theta\, dr\, d\theta = 0 \quad (14.478)$$

By dividing Eqs. (14.476) and (14.478) by $dr\, d\theta$ and neglecting small quantities of higher order, we obtain

$$-\frac{1}{r}\frac{\partial}{\partial r}(r Q_r) + \rho h \frac{\partial^2 w(r, t)}{\partial t^2} = f(r, t) \quad (14.479)$$

$$\frac{\partial M_r}{\partial r} + \frac{1}{r}(M_r - M_\theta) - Q_r = 0$$

or

$$Q_r = \frac{\partial M_r}{\partial r} + \frac{1}{r}(M_r - M_\theta) \quad (14.480)$$

For the axisymmetric vibration of plates with variable thickness, the bending moment resultants of Eqs.(14.184) and (14.185) can be expressed as

$$M_r = -D(r)\left(\frac{\partial^2 w}{\partial r^2} + \frac{v}{r}\frac{\partial w}{\partial r}\right) \quad (14.481)$$

$$M_\theta = -D(r)\left(\frac{1}{r}\frac{\partial w}{\partial r} + v\frac{\partial^2 w}{\partial r^2}\right) \quad (14.482)$$

where

$$D(r) = \frac{Eh^3(r)}{12(1 - v^2)} \quad (14.483)$$

Substituting Eqs. (14.481) and (14.482) into Eq. (14.480), we obtain

$$Q_r = -\left[D\frac{\partial^3 w}{\partial r^3} + \left(\frac{\partial D}{\partial r} + \frac{D}{r}\right)\frac{\partial^2 w}{\partial r^2} + \left(\frac{\nu}{r}\frac{\partial D}{\partial r} - \frac{D}{r^2}\right)\frac{\partial w}{\partial r}\right] \tag{14.484}$$

When Eq. (14.484) is substituted for Q_r in Eq. (14.479), we can obtain the equation of motion for the transverse axisymmetric vibration of a circular plate with axisymmetric variation of thickness. An alternative but identical equation of motion can be derived in a simple manner by first specifying Eq. (14.187b) for Q_r for the axisymmetric case as

$$Q_r = -D(r)\frac{\partial}{\partial r}\left(\frac{\partial^2 w}{\partial r^2} + \frac{1}{r}\frac{\partial w}{\partial r}\right) \equiv -D(r)\frac{\partial}{\partial r}\left[\frac{1}{r}\frac{\partial}{\partial r}\left(r\frac{\partial w}{\partial r}\right)\right] \tag{14.485}$$

and then substituting Eq. (14.485) into Eq. (14.479):

$$\frac{1}{r}\frac{\partial}{\partial r}\left\{rD(r)\frac{\partial}{\partial r}\left[\frac{1}{r}\frac{\partial}{\partial r}\left(r\frac{\partial w}{\partial r}\right)\right]\right\} + \rho h\frac{\partial^2 w}{\partial t^2} = f(r,t) \tag{14.486}$$

14.12.3   Free Vibration Solution

According to the procedure used by Conway et al. [12], the external force, $f(r, t)$, is set equal to zero and Eq. (14.486) is rewritten as

$$\frac{\partial}{\partial r}\left\{rD(r)\frac{\partial}{\partial r}\left[\frac{1}{r}\frac{\partial}{\partial r}\left(r\frac{\partial w}{\partial r}\right)\right]\right\} = -\rho hr\frac{\partial^2 w}{\partial t^2} \tag{14.487}$$

For a plate with thickness varying linearly with the radial distance, we have

$$h(r) = h_0 r \tag{14.488}$$

where h_0 is a constant. Equation (14.488) leads to

$$D(r) = D_0 r^3 \tag{14.489}$$

where

$$D_0 = \frac{Eh_0^3}{12(1 - \nu^2)} \tag{14.490}$$

Using Eqs. (14.488)–(14.490), the equation of motion, Eq. (14.487), can be expressed as

$$r^4\frac{\partial^4 w}{\partial r^4} + 8r^3\frac{\partial^3 w}{\partial r^3} + (11 + 3\nu)r^2\frac{\partial^2 w}{\partial r^2} - (2 - 6\nu)\frac{\partial w}{\partial r} = -\frac{12(1 - \nu^2)\rho}{Eh_0^3}r^2\frac{\partial^2 w}{\partial t^2} \tag{14.491}$$

Assuming the value of ν as $\frac{1}{3}$, which is applicable to many materials, Eq. (14.491) can be reduced to

$$\frac{\partial^2}{\partial r^2}\left(r^4\frac{\partial^2 w}{\partial r^2}\right) + \frac{32\rho}{3Eh_0^3}r^2\frac{\partial^2 w}{\partial t^2} = 0 \tag{14.492}$$

Assuming a harmonic solution with frequency ω for free vibration as

$$w(r, t) = W(r)e^{i\omega t} \tag{14.493}$$

we can rewrite Eq. (14.492) as

$$\frac{d^2}{dz^2}\left(z^4 \frac{d^2 W}{dz^2}\right) = z^2 W \tag{14.494}$$

where

$$z = pr \tag{14.495}$$

with

$$p = \left(\frac{32\rho\omega^2}{3Eh_0^3}\right)^{1/2} \tag{14.496}$$

The solution of Eq. (14.494) can be expressed as

$$W(z) = \frac{C_1 J_2(2\sqrt{z})}{z} + \frac{C_2 Y_2(2\sqrt{z})}{z} + \frac{C_3 I_2(2\sqrt{z})}{z} + \frac{C_4 K_2(2\sqrt{z})}{z} \tag{14.497}$$

where J_2 and Y_2 are Bessel functions of the second kind and I_2 and K_2 are modified Bessel functions of the second kind. For specificity, we consider the axisymmetric plate to be clamped at both the inner and outer radii as shown in Fig. 14.20, so that

$$W(z) = \frac{dW}{dz}(z) = 0, \qquad z = z_2 = pR_2, \quad z = z_1 = pR_1 \tag{14.498}$$

Using the relations [13, 18]

$$\frac{d}{dz}[z^{-m/2} J_m(kz^{1/2})] = -\frac{1}{2}kz^{-(m+1)/2} J_{m+1}(kz^{1/2}) \tag{14.499}$$

$$\frac{d}{dz}[z^{-m/2} Y_m(kz^{1/2})] = -\frac{1}{2}kz^{-(m+1)/2} Y_{m+1}(kz^{1/2}) \tag{14.500}$$

$$\frac{d}{dz}[z^{-m/2} I_m(kz^{1/2})] = -\frac{1}{2}kz^{-(m+1)/2} I_{m+1}(kz^{1/2}) \tag{14.501}$$

$$\frac{d}{dz}[z^{-m/2} K_m(kz^{1/2})] = -\frac{1}{2}kz^{-(m+1)/2} K_{m+1}(kz^{1/2}) \tag{14.502}$$

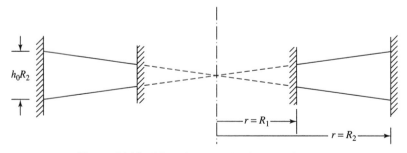

Figure 14.20   Linearly tapered axisymmetric plate.

Table 14.4 Natural Frequencies of Vibration of Axisymmetric Annular Plates Clamped at Both Edges

Mode	Value of the frequency parameter δ for[a]:			
	$R_2/R_1 = 2$	$R_2/R_1 = 3$	$R_2/R_1 = 4$	$R_2/R_1 = 10$
1	16.5	8.04	5.84	3.32
2	45.2	21.9	15.8	8.71
3	88.4	42.6	30.6	16.7
4	146	70.3	50.4	27.3
5	211	104.8	75.0	40.5

Source: Ref. [12].
[a] $\delta = (\omega R_2/h_0)(2\rho/3E)^{1/2}$, $\nu = \frac{1}{3}$.

application of the boundary conditions of Eq. (14.498) in Eq. (14.497) leads to the frequency equation in the form of a determinantal equation:

$$\begin{vmatrix} J_2(\beta) & Y_2(\beta) & I_2(\beta) & K_2(\beta) \\ J_3(\beta) & Y_3(\beta) & -I_3(\beta) & K_3(\beta) \\ J_2(\alpha) & Y_2(\alpha) & I_2(\alpha) & K_2(\alpha) \\ J_3(\alpha) & Y_3(\alpha) & -I_3(\alpha) & K_3(\alpha) \end{vmatrix} = 0 \qquad (14.503)$$

where

$$\beta^2 = 4z_2 = 4R_2 \left(\frac{32\rho\omega^2}{3Eh_0^3} \right)^{1/2} \qquad (14.504)$$

$$\alpha^2 = 4z_1 = 4R_1 \left(\frac{32\rho\omega^2}{3Eh_0^3} \right)^{1/2} \qquad (14.505)$$

The natural frequencies of vibration given by Eq. (14.503) for different values of the thickness ratio R_2/R_1 are given in Table 14.4.

14.13 RECENT CONTRIBUTIONS

Thin Plates A comprehensive review of the various aspects of plate vibration has been given by Liessa [1]. A free vibration analysis of rectangular plates was presented by Gorman [21]. Cote et al. [22] investigated the effects of shear deformation and rotary inertia on the free vibration of a rotating annular plate. Kim and Dickinson studied the flexural vibration of rectangular plates with point supports [23]. Vera et al. [24] dealt with a theoretical analysis of the dynamic behavior of a system made up of a plate with an elastically mounted two-degree-of-freedom system, a study based on an analytical model with Lagrange multipliers.

Plates on Elastic Supports Gorman [25] conducted a comprehensive study of the free vibration of rectangular plates resting on symmetrically distributed uniform elastic edge supports. Xiang et al. [26] investigated the problem of free vibration of a moderately thick rectangular plate with edges elastically restrained against transverse and rotational displacements. The Ritz method, combined with a variational formulation and Mindlin plate theory was used. The applicability of the formulation was illustrated using plates with different combinations of elastically restrained edges and classical boundary conditions.

Plates with Variable Geometry and Properties Gupta and Sharma [27] analyzed the forced motion of a plate of infinite length whose thickness, density, and elastic properties vary in steps along the finite breadth by the eigenfunction method. The numerical results for transverse deflection computed for a clamped–clamped plate subjected to constant or half-sine pulse load were plotted in graphs. Wang presented a power series solution for the vibration of plates with generalized variable thickness [28]. These solutions, represented by the recursive relations of the coefficients of the infinite power series, can be applied to various boundary conditions to obtain the resonant frequency spectra and mode shapes.

Plates on an Elastic Foundation Horvath described different subgrade models of soil–structure interaction analysis [29]. In the Winkler model, the foundation was represented by continuous springs. In the Pasternak model, the shear deformation between the spring elements was also considered by connecting the ends of the springs to the plate with incompressible vertical elements that deform only by the transverse shear. The exact relationship between the ith natural frequency of a simply supported polygonal Mindlin plate resting on a Pasternak foundation and the corresponding natural frequency of a Kirchhoff plate has been derived by Liew et al. [7]. The effect of elastic foundation on the mode shapes in stability and vibration problems of simply supported rectangular plates was considered by Raju and Rao [30]. The results of this work are expected to be useful in the dynamic stability studies of plates resting on elastic foundation.

Numerical solutions of the von Kármán partial differential equation governing the nonlinear dynamic response of circular plates on Winkler, Pasternak, and nonlinear Winkler elastic foundation and subjected to uniformly distributed step loading were presented by Smaill [31]. The effect of the foundation parameters on the central deflection was presented for both clamped and simply supported immovable edge boundary conditions. The nonlinear effects of Pasternak and nonlinear Winkler foundations on the deflection of plates were also determined.

Plates with In-Plane Forces Devarakonda and Bert [32] considered the flexural vibrations of rectangular plates subjected to sinusoidally distributed in-plane compressive loading on two opposite edges. The procedure involves first finding a plane elasticity solution for an in-plane problem satisfying all boundary conditions. Using this solution, flexural vibration analysis was then carried out. The free lateral vibrations of simply supported rectangular plates subject to both direct and shear in-plane forces have been considered by Dickinson [33].

Thick Plates The best shear coefficient and validity of higher spectra in the Mindlin plate theory has been discussed by Stephen [34]. The axisymmetric free vibrations of moderately thick circular plates described by Mindlin theory were analyzed by the differential quadrature method by Liew et al. [35]. The first 15 natural frequencies of vibration were calculated for uniform circular plates with free, simply supported, and clamped edges. A rigid point support at the plate center was also considered.

Martincek [36] used a dynamic resonance method to estimate the elastic properties of materials in a nondestructive manner. The method was based on measured frequencies of natural vibration in test specimens in the shape of circular plates. Exact analytical solutions for the free vibration of sectorial plates with simply supported radial edges have been given by Huang et al. [37]. Wittrick [38] gave analytical, three-dimensional elasticity solutions to some plate problems along with some observations on Mindlin plate theory.

REFERENCES

1. A. W. Leissa, *Vibration of Plates*, NASA SP-160, National Aeronautics and Space Administration, Washington, DC, 1969.

2. K. F. Graff, *Wave Motion in Elastic Solids*, Ohio State University Press, Columbus, OH, 1975.

3. R. D. Mindlin, Influence of rotary inertia and shear on flexural motions of isotropic plates, *Journal of Applied Mechanics*, Vol. 18, pp. 31–38, 1951.

4. R. D. Mindlin and H. Deresiewicz, Thickness-shear and flexural vibrations of a circular disk, *Journal of Applied Physics*, Vol. 25, pp. 1320–1332, 1954.

5. E. H. Mansfield, *The Bending and Stretching of Plates*, Macmillan, New York, 1964.

6. R. Szilard, *Theory and Analysis of Plates: Classical and Numerical Methods*, Prentice-Hall, Englewood Cliffs, NJ, 1974.

7. K. M. Liew, C. M. Wang, Y. Xiang, and S. Kitipornchai, *Vibration of Mindlin Plates: Programming the p-Version Ritz Method,* Elsevier, Amsterdam, 1998.

8. E. Ventsel and T. Krauthammer, *Thin Plates and Shells: Theory, Analysis, and Applications,* Marcel Dekker, New York, 2001.

9. E. B. Magrab, *Vibrations of Elastic Structural Members*, Sijthoff & Noordhoff, Alphen aan den Rijn, The Netherlands, 1979.

10. E. Skudrzyk, *Simple and Complex Vibratory Systems*, Pennsylvania State University Press, University Park, PA, 1968.

11. S. S. Rao and A. S. Prasad, Vibrations of annular plates including the effects of rotatory inertia and transverse shear deformation, *Journal of Sound and Vibration*, Vol. 42, No. 3, pp. 305–324, 1975.

12. H. D. Conway, E. C. H. Becker, and J. F. Dubil, Vibration frequencies of tapered bars and circular plates, *Journal of Applied Mechanics*, Vol. 31, No. 2, pp. 329–331, 1964.

13. K. G. Korenev, *Bessel Functions and Their Applications*, Taylor & Francis, London, 2002.

14. I. N. Sneddon, *Fourier Transforms*, McGraw-Hill, New York, 1951.

15. W. Nowacki, *Dynamics of Elastic Systems*, translated by H. Zorski, Wiley, New York, 1963.

16. R. C. Colwell and H. C. Hardy, The frequencies and nodal systems of circular plates, *Philosophical Magazine*, Ser. 7, Vol. 24, No. 165, pp. 1041–1055, 1937.

17. G. Herrmann, The influence of initial stress on the dynamic behavior of elastic and viscoelastic plates, *Publications of the International Association for Bridge and Structural Engineering*, Vol. 16, pp. 275–294, 1956.

18. A. Jeffrey, *Advanced Engineering Mathematics*, Harcourt/Academic Press, San Diego, CA, 2002.

19. T. Wah, Vibration of circular plates, *Journal of the Acoustical Society of America*, Vol. 34, No. 3, pp. 275–281, 1962.

20. R. D. Mindlin, A. Schacknow, and H. Deresiewicz, Flexural vibrations of rectangular plates, *Journal of Applied Mechanics*, Vol. 23, No. 3, pp. 430–436, 1956.

21. D. J. Gorman, *Free Vibration Analysis of Rectangular Plates*, Elsevier/North-Holland, New York, 1982.

22. A. Cote, N. Atalla, and J. Nicolas, Effects of shear deformation and rotary inertia on the free vibration of a rotating annular plate, *Journal of Vibration and Acoustics*, Vol. 119, No. 4, pp. 641–643, 1997.

23. C. S. Kim and S. M. Dickinson, The flexural vibration of rectangular plates with point supports, *Journal of Sound and Vibration*, Vol. 117, No. 2, pp. 249–261, 1987.

24. S. A. Vera, M. Febbo, C. G. Mendez, and R. Paz, Vibrations of a plate with an attached two degree of freedom system, *Journal of Sound and Vibration*, Vol. 285, No. 1–2, pp. 457–466, 2005.

25. D. J. Gorman, A comprehensive study of the free vibration of rectangular plates resting on symmetrically distributed uniform elastic edge supports, *Journal of Applied Mechanics*, Vol. 56, No. 4 pp. 893–899, 1989.

26. Y. Xiang, K. M. Liew, and S. Kitipornchai, Vibration analysis of rectangular Mindlin plates resting on elastic edge supports, *Journal of Sound and Vibration*, Vol. 204, No. 1, pp. 1–16, 1997.

27. A. P. Gupta and N. Sharma, Forced motion of a stepped semi-infinite plate, *Journal of Sound and Vibration*, Vol. 203, No. 4, pp. 697–705, 1997.

28. J. Wang, Generalized power series solutions of the vibration of classical circular plates with variable thickness, *Journal of Sound and Vibration*, Vol. 202, No. 4, pp. 593–599, 1997.

29. J. S. Horvath, Subgrade models for soil–structure interaction analysis, in *Foundation Engineering: Current Principles and Practices*, American Society of Civil Engineers, Reston, VA, 1989 Vol. 1, pp. 599–612.

30. K. K. Raju and G. V. Rao, Effect of elastic foundation on the mode shapes in stability and vibration problems of simply supported rectangular plates, *Journal of Sound and Vibration*, Vol. 139, No. 1, pp. 170–173, 1990.

31. J. S. Smaill, Dynamic response of circular plates on elastic foundations: linear and non-linear deflection, *Journal of Sound and Vibration*, Vol. 139, No. 3, pp. 487–502, 1990.

32. K. K. V. Devarakonda and C. W. Bert, Flexural vibration of rectangular plates subjected to sinusoidally distributed compressive loading on two opposite sides, *Journal of Sound and Vibration*, Vol. 283, No. 3–5, pp. 749–763, 2005.

33. S. M. Dickinson, Lateral vibration of rectangular plates subject to in-plane forces, *Journal of Sound and Vibration*, Vol. 16, No. 4, pp. 465–472, 1971.

34. N. G. Stephen, Mindlin plate theory: best shear coefficient and higher spectra validity, *Journal of Sound and Vibration*, Vol. 202, No. 4, pp. 539–553, 1997.

35. K. M. Liew, J. B. Han and Z. M. Xiao, Vibration analysis of circular Mindlin plates using the differential quadrature method, *Journal of Sound and Vibration*, Vol. 205, No. 5, pp. 617–630, 1997.

36. G. Martincek, The determination of Poisson's ratio and the dynamic modulus of elasticity from the frequencies of natural vibration in thick circular plates, *Journal of Sound and Vibration*, Vol. 2, No. 2, pp. 116–127, 1965.

37. C. S. Huang, O. G. McGee and A. W. Leissa, Exact analytical solutions for free vibrations of thick sectorial plates with simply supported radial edges, *International Journal of Solids and Structures*, Vol. 31, No. 11, pp. 1609–1631, 1994.

38. W. H. Wittrick, Analytical, three-dimensional elasticity solutions to some plate problems and some observations on Mindlin plate theory, *International Journal of Solids and Structures*, Vol. 23, No. 4, pp. 441–464, 1987.

39. E. Kreyszig, *Advanced Engineering Mathematics*, 8th ed., Wiley, New York, 1999.

40. R. Solecki and R. J. Conant, *Advanced Mechanics of Materials*, Oxford University Press, New York, 2003.

PROBLEMS

14.1 Determine the fundamental frequency of a square plate simply supported on all sides. Assume that the side and thickness of the plate are 500 and 5 mm, respectively, and the material is steel with $E = 207$ GPa, $\nu = 0.291$, and unit weight $= 76.5$ kN/m^3.

14.2 Determine the first two natural frequencies of a steel rectangular plate of sides 750 and 500 mm and thickness 10 mm. Assume the values of E and ν as 207 GPa and 0.291, respectively, with a weight density 76.5 kN/m^3 and the boundary conditions to be simply supported on all sides.

14.3 Derive the frequency equation for the free vibrations of *a* circular plate of radius *a* with a free edge.

14.4 Consider the static deflection of a simply supported rectangular plate subjected to a uniformly distributed load of magnitude f_0. By assuming the solution as

$$w(x,y) = \sum_{m=1}^{\infty} \sum_{n=1}^{\infty} \eta_{mn} W_{mn}(x,y)$$

where $W_{mn}(x,y)$ are the normalized modes (eigenfunctions), determine the values of the constants η_{mn}.

14.5 Derive the frequency equation for a uniform annular plate of inner radius *b* and outer radius *a* when both edges are clamped.

14.6 Consider the equation of motion of a simply supported rectangular plate subjected to a harmonically varying pressure distribution given by Eq. (E14.2.1). Specialize the solution for the following cases: (**a**) static pressure distribution; (**b**) when the forcing frequency is very close to one of the natural frequencies of vibration of the plate, $\Omega \rightarrow \omega_{mn}$; (**c**) static load, P, concentrated at $(x = x_0, y = y_0)$.

14.7 State the boundary conditions of a circular plate of radius *a* supported by linear and torsional springs all around the edge as shown in Fig. 14.21.

14.8 Derive the equation of motion of an infinite plate resting on an elastic foundation that is subjected to a constant force F that moves at a uniform speed c along a straight line passing through the origin using Cartesian coordinates. Assume the mass of the foundation to be negligible.

14.9 Derive the equation of motion of a plate in polar coordinates by considering the free body diagram of an element shown in Fig. 14.6 using the equilibrium approach.

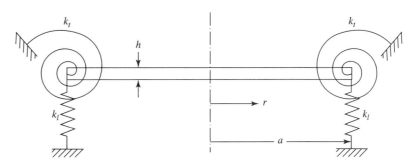

Figure 14.21

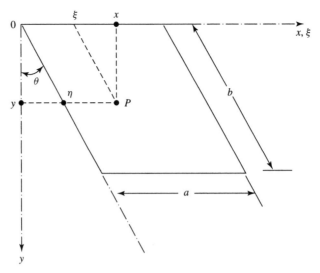

Figure 14.22

14.10 Skew plates find use in the swept wings of air-planes and parallelogram slabs in buildings and bridges. Derive the equation of motion for the vibration of the skew plate shown in Fig. 14.22 using the transformation of coordinates. The oblique coordinates ξ and η are related to x and y as

$$\xi = x - y\tan\theta, \qquad \eta = \frac{y}{\cos\theta}$$

Hint: Transform the Laplacian operator to the oblique coordinate system and derive

$$\nabla^2 = \frac{1}{\cos^2\theta}\left(\frac{\partial^2}{\partial\xi^2} - 2\sin\theta\frac{\partial^2}{\partial\xi\,\partial\eta} + \frac{\partial^2}{\partial\eta^2}\right)$$

14.11 Derive the frequency equation of an annular plate with both the inner and outer edges simply supported. Assume the inner and outer radii of the plate to be b and a, respectively.

14.12 Find the response of a circular plate subjected to a suddenly applied concentrated force F_0 at $r = 0$ at $t = 0$.

14.13 Derive the frequency equation of a rectangular plate with SS–F–SS–F boundary conditions.

14.14 Derive the frequency equation of a circular plate of radius a with a free edge including the effects of rotary inertia and shear deformation.

14.15 Find the first four natural frequencies of vibration of a rectangular plate resting on an elastic foundation with all the edges simply supported. Compare the results with the frequencies of a simply supported plate with no elastic foundation. Size of the plate: $a = 10$, $b = 20$, $h = 0.2$ in; unit weight: 0.283 lb/in^3; $E = 30 \times 10^6$ psi; $\nu = 0.3$, foundation modulus $k = 1000$ lb/in^2-in.

14.16 Find the steady-state response of a simply supported uniform rectangular plate subjected to the force

$$f(x,y,t) = f_0\delta\left(x - \frac{a}{4}, y - \frac{b}{4}\right)\cos\Omega t$$

14.17 Derive Eqs. (14.184)–(14.190) using the coordinate transformation relations.

15

Vibration of Shells

15.1 INTRODUCTION AND SHELL COORDINATES

A thin shell is a three-dimensional body that is bounded by two curved surfaces that are separated by a small distance compared to the radii of curvature of the curved surfaces. The middle surface of the shell is defined by the locus of points that lie midway between the two bounding curved surfaces. The thickness of the shell is denoted by the distance between the bounding surfaces measured along the normal to the middle surface. The thickness of the shell is assumed to be constant. Shells and shell structures find application in several areas of aerospace, architectural, civil, marine, and mechanical engineering. Examples of shells include aircraft fuselages, rockets, missiles, ships, submarines, pipes, water tanks, pressure vessels, boilers, fluid storage tanks, gas cylinders, civil engineering structures, nuclear power plants, concrete arch dams, and roofs of large span buildings.

15.1.1 Theory of Surfaces

The deformation of a thin shell can be described completely by the deformation of its middle (neutral) surface. The undeformed middle surface of a thin shell can be described conveniently by the two independent coordinates α and β shown in Fig. 15.1. In the global Cartesian coordinate system $OXYZ$, the position vector of a typical point in the middle surface, $\vec{r}$, can be expressed in terms of α and β as

$$\vec{r} = \vec{r}(\alpha, \beta) \tag{15.1}$$

which, in Cartesian coordinate system, can be written as

$$X = X(\alpha, \beta)$$

$$Y = Y(\alpha, \beta) \tag{15.2}$$

$$Z = Z(\alpha, \beta)$$

or

$$\vec{r} = X(\alpha, \beta)\vec{i} + Y(\alpha, \beta)\vec{j} + Z(\alpha, \beta)\vec{k} \tag{15.3}$$

where $\vec{i}$, $\vec{j}$, and $\vec{k}$ denote the unit vectors along the X, Y, and Z axes, respectively. The derivatives of the position vector with respect to α and β, denoted as

$$\frac{\partial \vec{r}}{\partial \alpha} = \vec{r}_{,\alpha}$$

$$\frac{\partial \vec{r}}{\partial \beta} = \vec{r}_{,\beta} \tag{15.4}$$

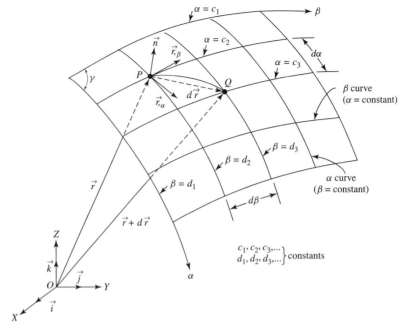

Figure 15.1   Curvilinear coordinates in the middle surface of a shell.

represent tangent vectors at any point of the surface to the α and β coordinate lines as shown in Fig. 15.1.

15.1.2   Distance between Points in the Middle Surface before Deformation

The distance vector between the infinitesimally separated points P and Q lying in the middle surface of the shell (before deformation) can be expressed as

$$d\vec{r} = \frac{\partial \vec{r}}{\partial \alpha}\, d\alpha + \frac{\partial \vec{r}}{\partial \beta}\, d\beta = \vec{r}_{,\alpha}\, d\alpha + \vec{r}_{,\beta}\, d\beta \tag{15.5}$$

The magnitude of the distance vector (ds) is given by

$$(ds)^2 = d\vec{r} \cdot d\vec{r} = \frac{\partial \vec{r}}{\partial \alpha} \frac{\partial \vec{r}}{\partial \alpha}(d\alpha)^2 + 2\,\frac{\partial \vec{r}}{\partial \alpha} \frac{\partial \vec{r}}{\partial \beta}\, d\alpha\, d\beta + \frac{\partial \vec{r}}{\partial \beta} \frac{\partial \vec{r}}{\partial \beta}(d\beta)^2 \tag{15.6}$$

which can be written as

$$(ds)^2 = A^2(d\alpha)^2 + 2AB \cos\gamma\, d\alpha\, d\beta + B^2(d\beta)^2 \tag{15.7}$$

where A and B denote the lengths of the vectors $\vec{r}_{,\alpha}$ and $\vec{r}_{,\beta}$, respectively:

$$A^2 = \vec{r}_{,\alpha} \cdot \vec{r}_{,\alpha} = |\vec{r}_{,\alpha}|^2 \tag{15.8}$$

$$B^2 = \vec{r}_{,\beta} \cdot \vec{r}_{,\beta} = |\vec{r}_{,\beta}|^2 \tag{15.9}$$

and γ indicates the angle between the coordinate curves α and β defined by

$$\frac{\vec{r}_{,\alpha} \cdot \vec{r}_{,\beta}}{A\,B} = \cos\gamma \tag{15.10}$$

Equation (15.7) is called the *fundamental form* or *first quadratic form* of the surface defined by $\vec{r} = \vec{r}(\alpha, \beta)$ and the quantities A^2, $AB \cos \gamma$, and B^2 are called the *coefficients* of the fundamental form. If the coordinates α and β are orthogonal, γ will be $90°$, so that

$$\vec{r}_{,\alpha} \cdot \vec{r}_{,\beta} = 0 \tag{15.11}$$

and Eq. (15.7) reduces to

$$(ds)^2 = A^2(d\alpha)^2 + B^2(d\beta)^2 \tag{15.12}$$

where A and B are called the *Lamé parameters*. Equation (15.12) can be rewritten as

$$(ds)^2 = ds_1^2 + ds_2^2 \tag{15.13}$$

where

$$ds_1 = A \, d\alpha \quad \text{and} \quad ds_2 = B \, d\beta \tag{15.14}$$

denote the lengths of the arc segments corresponding to the increments $d\alpha$ and $d\beta$ in the curvilinear coordinates α and β. It can be seen that Lamé parameters relate a change in the arc length in the middle surface to the changes in the curvilinear coordinates of the shell.

Example 15.1   Determine the Lamé parameters and the fundamental form of the surface for a cylindrical shell.

SOLUTION   The curved surface of a shell is defined by the two lines of principal curvature α and β and the thickness of the shell wall is defined along the z axis. Thus, at each point in the middle surface of the shell, there will be two radii of curvature, one a maximum value and the other a minimum value. In the case of a cylindrical shell, the shell surface is defined by the curvilinear (cylindrical) coordinates $\alpha = x$ and $\beta = \theta$, where α is parallel to the axis of revolution and β is parallel to the circumferential direction as shown in Fig. 15.2. Any point S in the middle surface of the shell is defined by the radius vector

$$\vec{r} = x\vec{i} + R \, \cos \theta \cdot \vec{j} + R \, \sin \theta \cdot \vec{k} \tag{E15.1.1}$$

where R is the radius of the middle surface of the cylinder and $\vec{i}$, $\vec{j}$, and $\vec{k}$ are unit vectors along the Cartesian coordinates x, y, and z, respectively, as shown in Fig. 15.2. The Lamé parameters can be obtained from Eq. (E15.1.1) as

$$\frac{\partial \vec{r}}{\partial \alpha} = \frac{\partial \vec{r}}{\partial x} = \vec{i} \tag{E15.1.2}$$

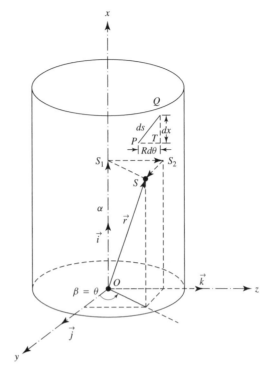

Figure 15.2   Cylindrical shell.

$$A = \left| \frac{\partial \vec{r}}{\partial \alpha} \right| = |\vec{i}| = 1 \tag{E15.1.3}$$

$$\frac{\partial \vec{r}}{\partial \beta} = \frac{\partial \vec{r}}{\partial \theta} = -R \, \sin \theta \vec{j} + R \cos \theta \vec{k} \tag{E15.1.4}$$

$$B = \left| \frac{\partial \vec{r}}{\partial \theta} \right| = [(-R \sin \theta)^2 + (R \cos \theta)^2]^{1/2} = R \tag{E15.1.5}$$

The fundamental form of the surface is given by

$$(ds)^2 = A^2 (d\alpha)^2 + B^2 (d\beta)^2 = (dx)^2 + R^2 (d\theta)^2 \tag{E15.1.6}$$

The interpretation of Eq. (E15.1.6) is shown in Fig. 15.2, where the distance between two infinitesimally separated points P and Q in the middle surface of the shell can be found as the hypotenuse of the right triangle PQT, where the infinitesimal sides $QT = dx$ and $PT = R \, d\theta$ are parallel to the surface coordinates of the shell.

Example 15.2   Determine the Lamé parameters and the fundamental form of the surface for a conical shell.

SOLUTION   Any point S in the middle surface of the shell is defined by the radius vector $\vec{r}$ taken from the top of the cone (origin O in Fig. 15.3) along its generator,

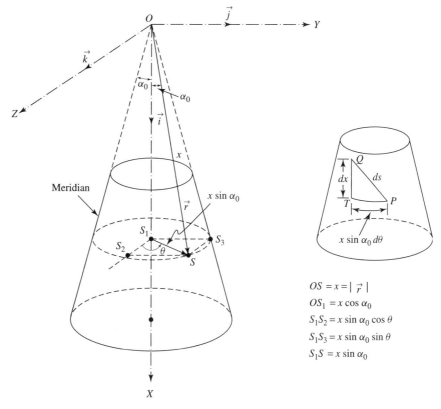

Figure 15.3 Conical shell.

and angle θ formed by a meridional plane passing through the point S with some reference meridional plane (the meridional plane passing through the Z axis is taken as the reference meridional plane in Fig. 15.3). If the length of the vector $\vec{r}$ is x and the semi cone angle is α_0, the radius vector can be expressed in the Cartesian coordinate system XYZ as

$$\vec{r} = x \; \cos\alpha_0 \vec{i} + x \sin\alpha_0 \sin\theta \vec{j} + x \sin\alpha_0 \cos\theta \vec{k} \qquad \text{(E15.2.1)}$$

With $\alpha = x$ and $\beta = \theta$, the Lamé parameters can be obtained from

$$\frac{\partial \vec{r}}{\partial \alpha} = \frac{\partial \vec{r}}{\partial x} = \cos\alpha_0 \vec{i} + \sin\alpha_0 \sin\theta \vec{j} + \sin\alpha_0 \cos\theta \vec{k} \qquad \text{(E15.2.2)}$$

$$A = \left| \frac{\partial \vec{r}}{\partial x} \right| = (\cos\alpha_0^2 + \sin^2\alpha_0 \sin^2\theta + \sin^2\alpha_0 \cos^2\theta)^{1/2} = 1 \qquad \text{(E15.2.3)}$$

$$\frac{\partial \vec{r}}{\partial \beta} = \frac{\partial \vec{r}}{\partial \theta} = x \; \sin\alpha_0 \cos\theta \vec{j} - x \sin\alpha_0 \sin\theta \vec{k} \qquad \text{(E15.2.4)}$$

$$B = \left| \frac{\partial \vec{r}}{\partial \theta} \right| = [(x \sin\alpha_0 \cos\theta)^2 + (-x \sin\alpha_0 \sin\theta)^2]^{1/2} = x \sin\alpha_0 \qquad \text{(E15.2.5)}$$

The fundamental form of the surface is given by

$$(ds)^2 = A^2(d\alpha)^2 + B^2(d\beta)^2 = (dx)^2 + x^2 \sin^2 \alpha_0 (d\theta)^2 \qquad \text{(E15.2.6)}$$

The interpretation of Eq. (E15.2.6) is shown in Fig. 15.3, where the distance between two infinitesimally separated points P and Q in the middle surface of the shell can be found as the hypotenuse of the right triangle PQT with the infinitesimal sides QT and PT given by dx and $x \sin \alpha_0 \, d\theta$, respectively.

Example 15.3   Find the Lamé parameters and the fundamental form of the surface for a spherical shell.

SOLUTION   A typical point S in the middle surface of the shell is defined by the radius vector $\vec{r}$ as shown in Fig. 15.4. Using the spherical coordinates ϕ and θ shown in the figure, the radius vector can be expressed as

$$\vec{r} = R \cos\phi \vec{i} + R \sin\phi \cos\theta \vec{j} + R \sin\phi \sin\theta \vec{k} \qquad \text{(E15.3.1)}$$

where $\vec{i}$, $\vec{j}$, and $\vec{k}$ denote the unit vectors along the Cartesian coordinates X, Y, and Z, respectively, and R is the radius of the spherical shell. Using $\alpha = \phi$ and $\beta = \theta$, the

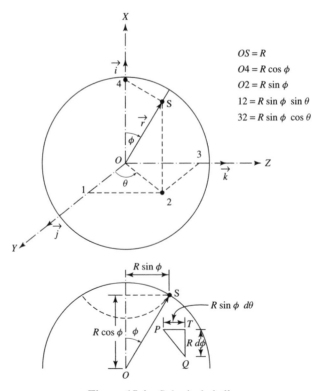

Figure 15.4   Spherical shell.

Lamé parameters can be determined as

$$\frac{\partial \vec{r}}{\partial \alpha} = \frac{\partial \vec{r}}{\partial \phi} = -R \sin \phi \vec{i} + R \cos \phi \cos \theta \vec{j} + R \cos \phi \sin \theta \vec{k} \tag{E15.3.2}$$

$$A = \left| \frac{\partial \vec{r}}{\partial \phi} \right| = [(-R \sin \phi)^2 + (R \cos \phi \cos \theta)^2 + (R \cos \phi \sin \theta)^2]^{1/2} = R \tag{E15.3.3}$$

$$\frac{\partial \vec{r}}{\partial \beta} = \frac{\partial \vec{r}}{\partial \theta} = -R \sin \phi \sin \theta \vec{j} + R \sin \phi \cos \theta \vec{k} \tag{E15.3.4}$$

$$B = \left| \frac{\partial \vec{r}}{\partial \theta} \right| = [(-R \sin \phi \sin \theta)^2 + (R \sin \phi \cos \theta)^2]^{1/2} = R \sin \phi \tag{E15.3.5}$$

The fundamental form of the surface is given by

$$(ds)^2 = A^2 (d\alpha)^2 + B^2 (d\beta)^2 = R^2 (d\phi)^2 + R^2 \sin^2 \phi (d\theta)^2 \tag{E15.3.6}$$

Equation (E15.3.6) implies that the distance between two infinitesimally separated points P and Q in the middle surface of the shell is determined as the hypotenuse of the right triangle PQT with the infinitesimal sides QT and PT given by $R \, d\phi$ and $R \sin \phi \, d\theta$, respectively (see Fig. 15.4).

15.1.3   Distance between Points Anywhere in the Thickness of a Shell before Deformation

Consider two infinitesimally separated points P and Q in the middle surface of the shell and two more points P' and Q' that lie on the normals to the middle surface of the shell, passing through points P and Q, respectively, as shown in Fig. 15.5. Let $\vec{n}$ denote the unit vector normal to the middle surface passing through the point P and z indicate the distance between P and P'. Then the point Q' will be at a distance of $z + dz$ from the middle surface. The position vector of $P'(\vec{R})$ can be denoted as

$$\vec{R}(\alpha, \beta, z) = \vec{r}(\alpha, \beta) + z \, \vec{n}(\alpha, \beta) \tag{15.15}$$

Since the position vector of Q was assumed to be $\vec{r} + d\vec{r}$, the position vector of Q' can be denoted as $\vec{R} + d\vec{R}$, where the differential vector $d\vec{R}$ can be expressed, using Eq. (15.15), as

$$d\vec{R} = d\vec{r} + z \, d\vec{n}(\alpha, \beta) + dz\vec{n} \tag{15.16}$$

with

$$d\vec{n}(\alpha, \beta) = \frac{\partial \vec{n}}{\partial \alpha} \, d\alpha + \frac{\partial \vec{n}}{\partial \beta} \, d\beta \tag{15.17}$$

The magnitude ds' of $d\vec{R}$ is given by

$$\begin{aligned} (ds')^2 = d\vec{R} \cdot d\vec{R} &= (d\vec{r} + z \, d\vec{n} + dz \, \vec{n}) \cdot (d\vec{r} + z \, d\vec{n} + dz \, \vec{n}) \\ &= d\vec{r} \cdot d\vec{r} + z^2 \, d\vec{n} \cdot d\vec{n} + (dz)^2 \vec{n} \cdot \vec{n} \\ &\quad + 2z \, d\vec{r} \cdot d\vec{n} + 2 \, dz \, d\vec{r} \cdot \vec{n} + 2z \, dz \, d\vec{n} \cdot \vec{n} \end{aligned} \tag{15.18}$$

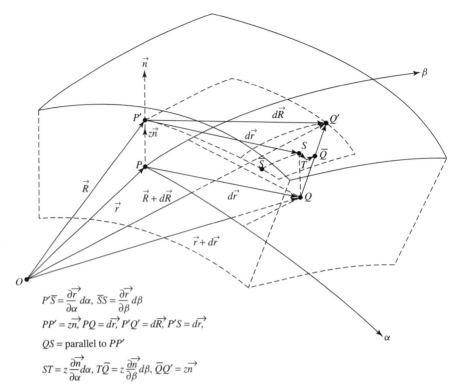

$$P'\overline{S} = \frac{\partial \vec{r}}{\partial \alpha}\,d\alpha,\ \overline{S}S = \frac{\partial \vec{r}}{\partial \beta}\,d\beta$$

$$PP' = z\vec{n},\ PQ = \vec{dr},\ P'Q' = \vec{dR},\ P'S = \vec{dr},$$

$$QS = \text{parallel to } PP'$$

$$ST = z\frac{\partial \vec{n}}{\partial \alpha}d\alpha,\ T\overline{Q} = z\frac{\partial \vec{n}}{\partial \beta}d\beta,\ \overline{Q}Q' = z\vec{n}$$

Figure 15.5   Distance between arbitrary points P' and Q'.

Since $\vec{n}$ is a unit vector, $\vec{n} \cdot \vec{n} = 1$, and because of the orthogonality of the coordinate system, $d\vec{r} \cdot \vec{n} = 0$ and $d\vec{n} \cdot \vec{n} = 0$. Thus, Eq. (15.18) reduces to

$$(ds')^2 = (ds)^2 + z^2\,d\vec{n} \cdot d\vec{n} + (dz)^2 + 2z\,d\vec{r} \cdot d\vec{n} \tag{15.19}$$

The second term on the right-hand side of Eq. (15.19) can be expressed as

$$z^2\,d\vec{n} \cdot d\vec{n} = z^2\left[\frac{\partial \vec{n}}{\partial \alpha} \cdot \frac{\partial \vec{n}}{\partial \alpha}(d\alpha)^2 + \frac{\partial \vec{n}}{\partial \beta} \cdot \frac{\partial \vec{n}}{\partial \beta}(d\beta)^2 + 2\frac{\partial \vec{n}}{\partial \alpha} \cdot \frac{\partial \vec{n}}{\partial \beta}\,d\alpha\,d\beta\right] \tag{15.20}$$

Because of the orthogonality of the coordinates,

$$\frac{\partial \vec{n}}{\partial \alpha} \cdot \frac{\partial \vec{n}}{\partial \beta} = 0 \tag{15.21}$$

in Eq. (15.20). The quantities $z\,(\partial \vec{n}/\partial \alpha)$ and $z\,(\partial \vec{n}/\partial \beta)$ can be expressed in terms of the radii of curvature as

$$\frac{|\partial \vec{r}/\partial \alpha|}{R_\alpha} = \frac{|z(\partial \vec{n}/\partial \alpha)|}{z}, \qquad \frac{|\partial \vec{r}/\partial \beta|}{R_\beta} = \frac{|z(\partial \vec{n}/\partial \beta)|}{z} \tag{15.22}$$

where R_a and R_β denote the radii of curvature of the α and β curves, respectively. When the relations

$$\left|\frac{\partial \vec{r}}{\partial \alpha}\right| = A, \qquad \left|\frac{\partial \vec{r}}{\partial \beta}\right| = B \tag{15.23}$$

are used from Eqs. (15.8) and (15.9), Eqs. (15.22) yield

$$\left| z \, \frac{\partial \vec{n}}{\partial \alpha} \right| = \frac{zA}{R_\alpha}, \qquad \left| z \, \frac{\partial \vec{n}}{\partial \beta} \right| = \frac{zB}{R_\beta} \qquad (15.24)$$

Thus, Eq. (15.20) can be rewritten as

$$z^2 \, d\vec{n} \cdot d\vec{n} = \frac{z^2 A^2}{R_\alpha^2} (d\alpha)^2 + \frac{z^2 B^2}{R_\beta^2} (d\beta)^2 \qquad (15.25)$$

The fourth term on the right-hand side of Eq. (15.19) can be expressed as

$$2z \, d\vec{r} \cdot d\vec{n} = 2z \left(\frac{\partial \vec{r}}{\partial \alpha} d\alpha + \frac{\partial \vec{r}}{\partial \beta} d\beta \right) \cdot \left(\frac{\partial \vec{n}}{\partial \alpha} d\alpha + \frac{\partial \vec{n}}{\partial \beta} d\beta \right)$$

$$= 2z \left[\frac{\partial \vec{r}}{\partial \alpha} \cdot \frac{\partial \vec{n}}{\partial \alpha} (d\alpha)^2 + \frac{\partial \vec{r}}{\partial \beta} \cdot \frac{\partial \vec{n}}{\partial \alpha} d\alpha \, d\beta + \frac{\partial \vec{r}}{\partial \alpha} \cdot \frac{\partial \vec{n}}{\partial \beta} d\alpha \, d\beta + \frac{\partial \vec{r}}{\partial \beta} \cdot \frac{\partial \vec{n}}{\partial \beta} (d\beta)^2 \right] \qquad (15.26)$$

Because of the orthogonality of the coordinates, Eq. (15.26) reduces to

$$2z \, d\vec{r} \cdot d\vec{n} = 2z \, \frac{\partial \vec{r}}{\partial \alpha} \cdot \frac{\partial \vec{n}}{\partial \alpha} (d\alpha)^2 + 2z \, \frac{\partial \vec{r}}{\partial \beta} \cdot \frac{\partial \vec{n}}{\partial \beta} (d\beta)^2 \qquad (15.27)$$

Using Eqs. (15.23) and (15.24), we can express

$$\frac{\partial \vec{r}}{\partial \alpha} \cdot \frac{\partial \vec{n}}{\partial \alpha} (d\alpha)^2 = \left| \frac{\partial \vec{r}}{\partial \alpha} \right| \left| \frac{\partial \vec{n}}{\partial \alpha} \right| (d\alpha)^2 = \frac{A^2}{R_\alpha} (d\alpha)^2$$

$$\frac{\partial \vec{r}}{\partial \beta} \cdot \frac{\partial \vec{n}}{\partial \beta} (d\beta)^2 = \left| \frac{\partial \vec{r}}{\partial \beta} \right| \left| \frac{\partial \vec{n}}{\partial \beta} \right| (d\beta)^2 = \frac{B^2}{R_\beta} (d\beta)^2 \qquad (15.28)$$

Using Eqs. (15.12), (15.25), (15.27) and (15.28) in Eq. (15.19), we obtain

$$(ds')^2 = A^2 \left(1 + \frac{z}{R_\alpha} \right)^2 (d\alpha)^2 + B^2 \left(1 + \frac{z}{R_\beta} \right)^2 (d\beta)^2 + (dz)^2 \qquad (15.29)$$

15.1.4   Distance between Points Anywhere in the Thickness of a Shell after Deformation

Equation (15.29), which gives the distance between two infinitesimally separated points P' and Q' (before deformation), can be rewritten as

$$(ds')^2 = h_{11}(\alpha, \beta, z) \, (d\alpha)^2 + h_{22}(\alpha, \beta, z) \, (d\beta)^2 + h_{33}(\alpha, \beta, z) \, (dz)^2 \qquad (15.30)$$

where

$$h_{11}(\alpha, \beta, z) = A^2 \left(1 + \frac{z}{R_\alpha} \right)^2 \qquad (15.31)$$

$$h_{22}(\alpha, \beta, z) = B^2 \left(1 + \frac{z}{R_\beta} \right)^2 \qquad (15.32)$$

$$h_{33}(\alpha, \beta, z) = 1 \qquad (15.33)$$

Note that Eqs. (15.29) and (15.30) are applicable for a shell in the undeformed condition. When the shell deforms under external loads, a point P' with coordinates (α, β, z)

in the undeformed condition will assume a new position P'' defined by the coordinates $(\alpha + \eta_1, \beta + \eta_2, z + \eta_3)$ after deformation. Similarly, if a point Q' in the neighborhood of P' has coordinates $(\alpha + d\alpha, \beta + d\beta, z + dz)$ in the undeformed condition, it will assume a new position Q'' with coordinates $(\alpha + d\alpha + \eta_1 + d\eta_1, \beta + d\beta + \eta_2 + d\eta_2, z + dz + \eta_3 + d\eta_3)$ after deformation. Let the physical deflections of the point P' along the α, β, and z directions resulting from the deformation of the shell be given by $\bar{u}$, $\bar{v}$ and $\bar{w}$, respectively. Since the coordinates of P' are given by (α, β, z) and of P'' by $(\alpha + \eta_1, \beta + \eta_2, z + \eta_3)$, u_i and η_i are related by Eq. (15.30) as

$$\bar{u} = \sqrt{h_{11}(\alpha, \beta, z)}\,\eta_1, \qquad \bar{v} = \sqrt{h_{22}(\alpha, \beta, z)}\,\eta_2, \qquad \bar{w} = \sqrt{h_{33}(\alpha, \beta, z)}\,\eta_3 \quad (15.34)$$

The distance (ds'') between P'' and Q'', after deformation, can be found as

$$(ds'')^2 = h_{11}(\alpha + \eta_1, \beta + \eta_2, z + \eta_3)(d\alpha + d\eta_1)^2$$
$$+ h_{22}(\alpha + \eta_1, \beta + \eta_2, z + \eta_3)(d\beta + d\eta_2)^2$$
$$+ h_{33}(\alpha + \eta_1, \beta + \eta_2, z + \eta_3)(dz + d\eta_3)^2 \quad (15.35)$$

where, in general, h_{11}, h_{22}, and h_{33} are nonlinear functions of $\alpha + \eta_1$, $\beta + \eta_2$, and $z + \eta_3$. For simplicity, we can linearize them using Taylor's series expansion about the point $P'(\alpha, \beta, z)$ as

$$h_{ii}(\alpha + \eta_1, \beta + \eta_2, z + \eta_3) \approx h_{ii}(\alpha, \beta, z) + \frac{\partial h_{ii}(\alpha, \beta, z)}{\partial \alpha}\eta_1$$
$$+ \frac{\partial h_{ii}}{\partial \beta}(\alpha, \beta, z)\eta_2 + \frac{\partial h_{ii}}{\partial z}(\alpha, \beta, z)\eta_3, \qquad i = 1, 2, 3$$
$$(15.36)$$

Similarly, we approximate the terms $(d\alpha + d\eta_1)^2$, $(d\beta + d\eta_2)^2$, and $(dz + d\eta_3)^2$ in Eq. (15.35) as

$$(d\alpha + d\eta_1)^2 \approx (d\alpha)^2 + 2(d\alpha)(d\eta_1)$$
$$(d\beta + d\eta_2)^2 \approx (d\beta)^2 + 2(d\beta)(d\eta_2)$$
$$(dz + d\eta_3)^2 \approx (dz)^2 + 2(dz)(d\eta_3) \quad (15.37)$$

by neglecting the terms $(d\eta_1)^2$, $(d\eta_2)^2$, and $(d\eta_3)^2$. Since η_i varies with the coordinates α, β, and z, Eq. (15.37) can be expressed as

$$(d\alpha + d\eta_1)^2 \approx (d\alpha)^2 + 2(d\alpha)\left(\frac{\partial \eta_1}{\partial \alpha}\,d\alpha + \frac{\partial \eta_1}{\partial \beta}\,d\beta + \frac{\partial \eta_1}{\partial z}\,dz\right)$$

$$(d\beta + d\eta_2)^2 \approx (d\beta)^2 + 2(d\beta)\left(\frac{\partial \eta_2}{\partial \alpha}\,d\alpha + \frac{\partial \eta_2}{\partial \beta}\,d\beta + \frac{\partial \eta_2}{\partial z}\,dz\right)$$

$$(dz + d\eta_3)^2 \approx (dz)^2 + 2(dz)\left(\frac{\partial \eta_3}{\partial \alpha}\,d\alpha + \frac{\partial \eta_3}{\partial \beta}\,d\beta + \frac{\partial \eta_3}{\partial z}\,dz\right) \quad (15.38)$$

Using Eqs. (15.36) and (15.38), Eq. (15.35) can be expressed as

$$
(ds'')^2 = \left(h_{11} + \frac{\partial h_{11}}{\partial \alpha} \eta_1 + \frac{\partial h_{11}}{\partial \beta} \eta_2 + \frac{\partial h_{11}}{\partial z} \eta_3 \right) \left[(d\alpha)^2 + 2(d\alpha) \left(\frac{\partial \eta_1}{\partial \alpha} d\alpha \right. \right.
$$

$$
\left. \left. + \frac{\partial \eta_1}{\partial \beta} d\beta + \frac{\partial \eta_1}{\partial z} dz \right) \right]
$$

$$
+ \left(h_{22} + \frac{\partial h_{22}}{\partial \alpha} \eta_1 + \frac{\partial h_{22}}{\partial \beta} \eta_2 + \frac{\partial h_{22}}{\partial z} \eta_3 \right) \left[(d\beta)^2 + 2(d\beta) \left(\frac{\partial \eta_2}{\partial \alpha} d\alpha \right. \right.
$$

$$
\left. \left. + \frac{\partial \eta_2}{\partial \beta} d\beta + \frac{\partial \eta_2}{\partial z} dz \right) \right]
$$

$$
+ \left(h_{33} + \frac{\partial h_{33}}{\partial \alpha} \eta_1 + \frac{\partial h_{33}}{\partial \beta} \eta_2 + \frac{\partial h_{33}}{\partial z} \eta_3 \right) \left[(dz)^2 + 2(dz) \left(\frac{\partial \eta_3}{\partial \alpha} d\alpha \right. \right.
$$

$$
\left. \left. + \frac{\partial \eta_3}{\partial \beta} d\beta + \frac{\partial \eta_3}{\partial z} dz \right) \right]
$$

$$
= \left(h_{11} + \frac{\partial h_{11}}{\partial \alpha} \eta_1 + \frac{\partial h_{11}}{\partial \beta} \eta_2 + \frac{\partial h_{11}}{\partial z} \eta_3 \right) (d\alpha)^2
$$

$$
+ 2h_{11}(d\alpha) \left(\frac{\partial \eta_1}{\partial \alpha} d\alpha + \frac{\partial \eta_1}{\partial \beta} d\beta + \frac{\partial \eta_1}{\partial z} dz \right)
$$

$$
+ 2(d\alpha) \left(\frac{\partial h_{11}}{\partial \alpha} \eta_1 + \frac{\partial h_{11}}{\partial \beta} \eta_2 + \frac{\partial h_{11}}{\partial z} \eta_3 \right) \left(\frac{\partial \eta_1}{\partial \alpha} d\alpha + \frac{\partial \eta_1}{\partial \beta} d\beta + \frac{\partial \eta_1}{\partial z} dz \right)
$$

$$
+ \left(h_{22} + \frac{\partial h_{22}}{\partial \alpha} \eta_1 + \frac{\partial h_{22}}{\partial \beta} \eta_2 + \frac{\partial h_{22}}{\partial z} \eta_3 \right) (d\beta)^2
$$

$$
+ 2h_{22}(d\beta) \left(\frac{\partial \eta_2}{\partial \alpha} d\alpha + \frac{\partial \eta_2}{\partial \beta} d\beta + \frac{\partial \eta_2}{\partial z} dz \right)
$$

$$
+ 2(d\beta) \left(\frac{\partial h_{22}}{\partial \alpha} \eta_1 + \frac{\partial h_{22}}{\partial \beta} \eta_2 + \frac{\partial h_{22}}{\partial z} \eta_3 \right) \left(\frac{\partial \eta_2}{\partial \alpha} d\alpha + \frac{\partial \eta_2}{\partial \beta} d\beta + \frac{\partial \eta_2}{\partial z} dz \right)
$$

$$
+ \left(h_{33} + \frac{\partial h_{33}}{\partial \alpha} \eta_1 + \frac{\partial h_{33}}{\partial \beta} \eta_2 + \frac{\partial h_{33}}{\partial z} \eta_3 \right) (dz)^2
$$

$$
+ 2h_{33}(dz) \left(\frac{\partial \eta_3}{\partial \alpha} d\alpha + \frac{\partial \eta_3}{\partial \beta} d\beta + \frac{\partial \eta_3}{\partial z} dz \right)
$$

$$
+ 2(dz) \left(\frac{\partial h_{33}}{\partial \alpha} \eta_1 + \frac{\partial h_{33}}{\partial \beta} \eta_2 + \frac{\partial h_{33}}{\partial z} \eta_3 \right) \left(\frac{\partial \eta_3}{\partial \alpha} d\alpha + \frac{\partial \eta_3}{\partial \beta} d\beta + \frac{\partial \eta_3}{\partial z} dz \right)
$$

$$
\tag{15.39}
$$

The terms involving products of partial derivatives of h_{ii} and η_j on the right-hand side of Eq. (15.39) can be neglected in most practical cases. Hence, Eq. (15.39) can be expressed as

$$
(ds'')^2 \simeq H_{11}(d\alpha)^2 + H_{22}(d\beta)^2 + H_{33}(dz)^2
$$

$$
+ 2H_{12}(d\alpha)(d\beta) + 2H_{23}(d\beta)(dz) + 2H_{13}(d\alpha)(dz) \tag{15.40}
$$

where

$$H_{11} = h_{11} + \frac{\partial h_{11}}{\partial \alpha} \eta_1 + \frac{\partial h_{11}}{\partial \beta} \eta_2 + \frac{\partial h_{11}}{\partial z} \eta_3 + 2 h_{11} \frac{\partial \eta_1}{\partial \alpha} \tag{15.41}$$

$$H_{22} = h_{22} + \frac{\partial h_{22}}{\partial \alpha} \eta_1 + \frac{\partial h_{22}}{\partial \beta} \eta_2 + \frac{\partial h_{22}}{\partial z} \eta_3 + 2 h_{22} \frac{\partial \eta_2}{\partial \beta} \tag{15.42}$$

$$H_{33} = h_{33} + \frac{\partial h_{33}}{\partial \alpha} \eta_1 + \frac{\partial h_{33}}{\partial \beta} \eta_2 + \frac{\partial h_{33}}{\partial z} \eta_3 + 2 h_{33} \frac{\partial \eta_3}{\partial z} \tag{15.43}$$

$$H_{12} = H_{21} = h_{11} \frac{\partial \eta_1}{\partial \beta} + h_{22} \frac{\partial \eta_2}{\partial \alpha} \tag{15.44}$$

$$H_{23} = H_{32} = h_{22} \frac{\partial \eta_2}{\partial z} + h_{33} \frac{\partial \eta_3}{\partial \beta} \tag{15.45}$$

$$H_{13} = H_{31} = h_{11} \frac{\partial \eta_1}{\partial z} + h_{33} \frac{\partial \eta_3}{\partial \alpha} \tag{15.46}$$

Thus, Eq. (15.40) gives the distance between P'' and Q'' after deformation. Points P'' and Q'' indicate the deflected position of points P' and Q' whose coordinates (in the undeformed shell) are given by (α, β, z) and $(\alpha + d\alpha, \beta + d\beta, z + dz)$, respectively.

15.2 STRAIN–DISPLACEMENT RELATIONS

The distance between two arbitrary points after deformation, given by Eq. (15.40), can also be used to define the components of strain in the shell. The normal strain $(\varepsilon_{\alpha\alpha} \equiv \varepsilon_{11})$ along the coordinate direction α is defined as

$$\varepsilon_{11} = \frac{\text{change in length of a fiber originally oriented along the } \alpha \text{ direction}}{\text{original length of the fiber}} \tag{15.47}$$

Since the fiber is originally oriented along the α direction, the coordinates of the end points of the fiber, P' and Q', are given by (α, β, z) and $(\alpha + d\alpha, \beta, z)$, respectively, and Eqs. (15.40) and (15.30) give

$$(ds'')^2 \equiv (ds'')_{11}^2 = H_{11}(d\alpha)^2 \tag{15.48}$$

$$(ds')^2 \equiv (ds')_{11}^2 = h_{11}(d\alpha)^2 \tag{15.49}$$

Using a similar procedure, for fibers oriented originally along the β and z directions, we can write

$$(ds'')_{22}^2 = H_{22}(d\beta)^2 \tag{15.50}$$

$$(ds')_{22}^2 = h_{22}(d\beta)^2 \tag{15.51}$$

$$(ds'')_{33}^2 = H_{33}(dz)^2 \tag{15.52}$$

$$(ds')_{33}^2 = h_{33}(dz)^2 \tag{15.53}$$

Since the normal strain along the coordinate direction $\beta(z)$, denoted by $\varepsilon_{\beta\beta} \equiv \varepsilon_{22}$ (or $\varepsilon_{zz} \equiv \varepsilon_{33}$), is defined as

$$\varepsilon_{22}(\varepsilon_{33}) = \frac{\text{change in length of a fiber originally oriented along the } \beta(z) \text{ direction}}{\text{original length of the fiber}}$$

(15.54)

the normal strains ε_{ii}, $i = 1, 2, 3$, can be expressed as

$$\varepsilon_{ii} = \frac{(ds'')_{ii} - (ds')_{ii}}{(ds')_{ii}} = \frac{(ds'')_{ii}}{(ds')_{ii}} - 1 = \sqrt{\frac{H_{ii}}{h_{ii}}} - 1 = \sqrt{\frac{h_{ii} + H_{ii} - h_{ii}}{h_{ii}}} - 1$$

$$= \sqrt{1 + \frac{H_{ii} - h_{ii}}{h_{ii}}} - 1$$

(15.55)

The quantity $(H_{ii} - h_{ii})/h_{ii}$ is very small compared to 1 and hence the binomial expansion of Eq. (15.55) yields

$$\varepsilon_{ii} = \left(1 + \frac{1}{2}\frac{H_{ii} - h_{ii}}{h_{ii}} + \cdots\right) - 1 \approx \frac{1}{2}\frac{H_{ii} - h_{ii}}{h_{ii}}, \qquad i = 1, 2, 3$$

(15.56)

The shear strain in the $\alpha\beta$ plane (denoted as $\varepsilon_{\alpha\beta} \equiv \varepsilon_{12}$) is defined as

$$\varepsilon_{12} = \text{change in the angle of two mutually perpendicular fibers originally}$$

$$\text{oriented along the } \alpha \text{ and } \beta \text{ directions}$$

(15.57)

which from Fig. 15.6, can be expressed as,

$$\varepsilon_{12} = \frac{\pi}{2} - \theta_{12}$$

(15.58)

where $\theta_{12} \equiv \theta_{\alpha\beta}$. To be specific, consider two fibers, one originally oriented along the α direction [with endpoint coordinates $R = (\alpha, \beta, z)$ and $P = (\alpha + d\alpha, \beta, z)$ and the

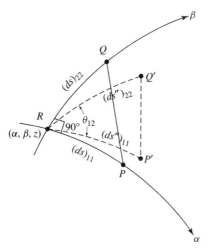

Figure 15.6   Angular change of fibers.

other originally oriented along the β direction (with endpoint coordinates $R = (\alpha, \beta, z)$ and $Q = (\alpha, \beta + d\beta, z)$]. This implies that the distance PQ is given by

$$(ds')^2 \equiv (ds')_{12}^2 = h_{11}(d\alpha)^2 + h_{22}(d\beta)^2 \tag{15.59}$$

By considering points P and Q to be originally located, equivalently, at (α, β, z) and $(\alpha - d\alpha, \beta + d\beta, z)$ instead of at $(\alpha + d\alpha, \beta, z)$ and $(\alpha, \beta + d\beta, z)$, respectively, we can obtain the distance $P'Q'$ as

$$(ds'')^2 \equiv (ds'')_{12}^2 = H_{11}(d\alpha)^2 + H_{22}(d\beta)^2 - 2H_{12}\,d\alpha\,d\beta \tag{15.60}$$

Using a similar procedure in the βz plane (for $\varepsilon_{\beta z} \equiv \varepsilon_{23}$) and $z\alpha$ plane (for $\varepsilon_{z\alpha} \equiv \varepsilon_{31}$), we can obtain

$$(ds')_{23}^2 = h_{22}(d\beta)^2 + h_{33}(dz)^2 \tag{15.61}$$

$$(ds'')_{23}^2 = H_{22}(d\beta)^2 + H_{33}(dz)^2 - 2H_{23}(d\beta)(dz) \tag{15.62}$$

$$(ds')_{31}^2 = h_{33}(dz)^2 + h_{11}(d\alpha)^2 \tag{15.63}$$

$$(ds'')_{31}^2 = H_{33}(dz)^2 + H_{11}(d\alpha)^2 - 2H_{31}(dz)(d\alpha) \tag{15.64}$$

In Fig. 15.6, the sides RQ', $Q'P'$, and RP' are related by the cosine law:

$$(ds'')_{12}^2 = (ds'')_{11}^2 + (ds'')_{22}^2 - 2(ds'')_{11}(ds'')_{22}\cos\theta_{12} \tag{15.65}$$

Equation (15.65) can be expressed in a general form as

$$(ds'')_{ij}^2 = (ds'')_{ii}^2 + (ds'')_{jj}^2 - 2(ds'')_{ii}(ds'')_{jj}\cos\theta_{ij}, \qquad ij = 12, 23, 31 \tag{15.66}$$

Using Eqs. (15.60), (15.48), and (15.50), Eq. (15.65) can be expressed, as

$$H_{11}(d\alpha)^2 + H_{22}(d\beta)^2 - 2H_{12}(d\alpha)(d\beta)$$
$$= H_{11}(d\alpha)^2 + H_{22}(d\beta)^2 - 2(\sqrt{H_{11}}\,d\alpha)(\sqrt{H_{22}}\,d\beta)\cos\theta_{12} \tag{15.67}$$

which can be simplified to obtain

$$\cos\theta_{12} = \frac{H_{12}}{\sqrt{H_{11}H_{22}}} \tag{15.68}$$

Equations (15.58) and (15.68) yield

$$\cos\theta_{12} = \cos\left(\frac{\pi}{2} - \varepsilon_{12}\right) = \sin\varepsilon_{12} = \frac{H_{12}}{\sqrt{H_{11}H_{22}}} \tag{15.69}$$

In most practical cases, shear strains are small, so that the following approximations can be used:

$$\sin\varepsilon_{12} \approx \varepsilon_{12} = \frac{H_{12}}{\sqrt{H_{11}H_{22}}} \approx \frac{H_{12}}{\sqrt{h_{11}h_{22}}} \tag{15.70}$$

A similar procedure can be used for the other shear strains, ε_{23} (in the βz plane) and ε_{31} (in the $z\alpha$ plane). In general, the shear strains can be expressed as

$$\varepsilon_{ij} = \frac{H_{ij}}{\sqrt{h_{ii}h_{jj}}}, \qquad ij = 12, 23, 31 \tag{15.71}$$

Strains in Terms of Displacement Components   The normal strains can be expressed in terms of the displacement components, using Eqs. (15.56), (15.41)–(15.43) and (15.31)–(15.34), as

$$
\varepsilon_{11} = \varepsilon_{\alpha\alpha} = \frac{1}{2A^2 \left(1 + z/R_\alpha\right)^2} \left\{ \frac{\partial}{\partial\alpha} \left[A^2 \left(1 + \frac{z}{R_\alpha}\right)^2 \right] \frac{\bar{u}}{A \left(1 + z/R_\alpha\right)} \right.
$$

$$
+ \frac{\partial}{\partial\beta} \left[A^2 \left(1 + \frac{z}{R_\alpha}\right)^2 \right] \frac{\bar{v}}{B \left(1 + z/R_\beta\right)} + \frac{\partial}{\partial z} \left[A^2 \left(1 + \frac{z}{R_\alpha}\right)^2 \right] \bar{w} \right\}
$$

$$
+ \frac{\partial}{\partial\alpha} \left[\frac{\bar{u}}{A \left(1 + z/R_\alpha\right)} \right]
$$

$$
= \frac{1}{A \left(1 + z/R_\alpha\right)} \left\{ \frac{\partial}{\partial\alpha} \left[A \left(1 + \frac{z}{R_\alpha}\right) \right] \frac{\bar{u}}{A \left(1 + z/R_\alpha\right)} \right.
$$

$$
+ \frac{\partial}{\partial\beta} \left[A \left(1 + \frac{z}{R_\alpha}\right) \right] \frac{\bar{v}}{B \left(1 + z/R_\beta\right)} + \frac{A}{R_\alpha} \bar{w} \right\}
$$

$$
+ \frac{1}{A \left(1 + z/R_\alpha\right)} \frac{\partial\bar{u}}{\partial\alpha} - \frac{\partial}{\partial\alpha} \left[A \left(1 + \frac{z}{R_\alpha}\right) \right] \frac{\bar{u}}{A^2 \left(1 + z/R_\alpha\right)^2} \tag{15.72}
$$

The following partial derivatives, known as *Codazzi conditions* [9], are used to simplify Eq. (15.72):

$$
\frac{\partial}{\partial\beta} \left[A \left(1 + \frac{z}{R_\alpha}\right) \right] = \left(1 + \frac{z}{R_\beta}\right) \frac{\partial A}{\partial\beta} \tag{15.73}
$$

$$
\frac{\partial}{\partial\alpha} \left[B \left(1 + \frac{z}{R_\beta}\right) \right] = \left(1 + \frac{z}{R_\alpha}\right) \frac{\partial B}{\partial\alpha} \tag{15.74}
$$

Thus, the normal strain ε_{11} can be expressed as

$$
\varepsilon_{11} = \varepsilon_{\alpha\alpha} = \frac{1}{A \left(1 + z/R_\alpha\right)} \left(\frac{\partial\bar{u}}{\partial\alpha} + \frac{\bar{v}}{B} \frac{\partial A}{\partial\beta} + \bar{w} \frac{A}{R_\alpha} \right) \tag{15.75}
$$

Using a similar procedure, the normal strains ε_{22} and ε_{33} can be determined as

$$
\varepsilon_{22} = \varepsilon_{\beta\beta} = \frac{1}{B \left(1 + z/R_\beta\right)} \left(\frac{\partial\bar{v}}{\partial\beta} + \frac{\bar{u}}{A} \frac{\partial B}{\partial\alpha} + \bar{w} \frac{B}{R_\beta} \right) \tag{15.76}
$$

$$
\varepsilon_{33} = \varepsilon_{zz} = \frac{\partial\bar{w}}{\partial z} \tag{15.77}
$$

Similarly, using Eqs. (15.71), (15.44)–(15.46), and (15.31)–(15.34), the shear strains can be expressed in terms of displacement components as

$$\varepsilon_{12} = \varepsilon_{\alpha\beta} = \frac{A\left(1 + z/R_\alpha\right)}{B\left(1 + z/R_\beta\right)} \frac{\partial}{\partial\beta}\left[\frac{\bar{u}}{A\left(1 + z/R_\alpha\right)}\right]$$
$$+ \frac{B\left(1 + z/R_\beta\right)}{A\left(1 + z/R_\alpha\right)} \frac{\partial}{\partial\alpha}\left[\frac{\bar{v}}{B\left(1 + z/R_\beta\right)}\right] \tag{15.78}$$

$$\varepsilon_{23} = \varepsilon_{\beta z} = B\left(1 + \frac{z}{R_\beta}\right) \frac{\partial}{\partial z}\left[\frac{\bar{v}}{B\left(1 + z/R_\beta\right)}\right] + \frac{1}{B\left(1 + z/R_\beta\right)} \frac{\partial \bar{w}}{\partial \beta} \tag{15.79}$$

$$\varepsilon_{31} = \varepsilon_{z\alpha} = A\left(1 + \frac{z}{R_\alpha}\right) \frac{\partial}{\partial z}\left[\frac{\bar{v}}{B\left(1 + z/R_\alpha\right)}\right] + \frac{1}{A\left(1 + z/R_\alpha\right)} \frac{\partial \bar{w}}{\partial \alpha} \tag{15.80}$$

15.3 LOVE'S APPROXIMATIONS

In the classical or small displacement theory of thin shells, the following assumptions, originally made by Love [1], are universally accepted to be valid for a *first approximation shell theory*:

1. The thickness of the shell is small compared to its other dimensions, such as the radii of curvature of the middle surface of the shell.
2. The displacements and strains are very small, so that quantities involving second- and higher-order magnitude can be neglected in the strain–displacement relations.
3. The normal stress in the transverse (z) direction is negligibly small compared to the other normal stress components.
4. The normals to the undeformed middle surface of the shell remain straight and normal to the middle surface even after deformation, and undergo no extension or contraction.

The first assumption basically defines a thin shell. For thin shells used in engineering applications, the ratio h/R_{min}, where h is the thickness and R_{min} is the smallest radius of curvature of the middle surface of the shell, is less than $\frac{1}{50}$ and thus the ratio z/R_{min} will be less than $\frac{1}{100}$. Hence, terms of the order h/R_i or z/R_i ($i = \alpha, \beta$) can be neglected compared to unity in the strain–displacement relations

$$\frac{z}{R_\alpha} \ll 1, \qquad \frac{z}{R_\beta} \ll 1 \tag{15.81}$$

The second assumption enables us to make all computations in the undeformed configuration of the shell and ensures that the governing differential equations will be linear. The third assumption assumes that the normal stress $\sigma_{zz} \equiv \sigma_{33} = 0$ in the direction of z. If the outer shell surface is unloaded, σ_{33} will be zero. Even if the outer shell surface is loaded, σ_{33} can be assumed to be negligibly small in most cases. The fourth assumption, also known as *Kirchhoff's hypothesis*, leads to zero transverse shear strains and zero transverse normal strain:

$$\varepsilon_{13} = \varepsilon_{23} = 0 \qquad (15.82)$$

$$\varepsilon_{33} = 0 \qquad (15.83)$$

It is to be noted that assumptions 3 and 4 introduce the following inconsistencies:

(a) The transverse normal stress (σ_{33}) cannot be zero theoretically, especially when the outer surface of the shell is subjected to load.

(b) In thin shell theory, the resultant shear forces, $Q_{13} \equiv Q_{\alpha z}$ (acting on a face normal to the α curve) and $Q_{23} \equiv Q_{\beta z}$ (acting on a face normal to the β curve), are assumed to be present. These resultants can be related to (caused by) the transverse shear stresses, σ_{13} and σ_{23}, in the shell. If transverse shear stresses are present, the transverse shear strains cannot be zero. This violates the assumption in Eq. (15.82).

Despite these inconsistencies, Love's approximations are most commonly used in thin shell theory.

Equations (15.82) imply, from Hooke's law, that the transverse shear stresses are also zero:

$$\sigma_{13} = \sigma_{23} = 0 \qquad (15.84)$$

To satisfy Kirchhoff's hypothesis (fourth assumption), the components of displacement at any point in the thickness of the shell are assumed to be of the following form:

$$\bar{u}(\alpha, \beta, z) = u(\alpha, \beta) + z\theta_1(\alpha, \beta) \qquad (15.85)$$

$$\bar{v}(\alpha, \beta, z) = v(\alpha, \beta) + z\theta_2(\alpha, \beta) \qquad (15.86)$$

$$\bar{w}(\alpha, \beta, z) = w(\alpha, \beta) \qquad (15.87)$$

where u, v, and w denote the components of displacement in the middle surface of the shell along α, β, and z directions and θ_1 and θ_2 indicate the rotations of the normal to the middle surface about the β and α axes, respectively, during deformation:

$$\theta_1 = \frac{\partial \bar{u}(\alpha, \beta, z)}{\partial z} \qquad (15.88)$$

$$\theta_2 = \frac{\partial \bar{v}(\alpha, \beta, z)}{\partial z} \qquad (15.89)$$

Note that Eq. (15.83) will be satisfied by Eq. (15.87) since the transverse displacement is completely defined by the middle surface component, w:

$$\varepsilon_{33} = \frac{\partial \bar{w}}{\partial z} = \frac{\partial w}{\partial z} = 0 \qquad (15.90)$$

Substituting Eqs. (15.85)–(15.87) into Eqs. (15.79) and (15.80), we obtain

$$\varepsilon_{13} = A \left(1 + \frac{z}{R_\alpha}\right) \frac{\partial}{\partial z} \left[\frac{\overline{u}}{A\,(1 + z/R_\alpha)}\right] + \frac{1}{A\,(1 + z/R_\alpha)} \frac{\partial \overline{w}}{\partial \alpha}$$

$$= A \frac{\partial}{\partial z} \left[\frac{u(\alpha,\,\beta) + z\theta_1}{A\,(1 + z/R_\alpha)}\right] + \frac{1}{A} \frac{\partial w(\alpha,\,\beta)}{\partial \alpha}$$

$$= A \left\{\frac{1}{A\,(1 + z/R_\alpha)} \frac{\partial}{\partial z}(u + z\,\theta_1) - \frac{u + z\,\theta_1}{A^2(1 + z/R_\alpha)^2} \frac{\partial}{\partial z}\left[A\left(1 + \frac{z}{R_\alpha}\right)\right]\right\}$$

$$+ \frac{1}{A} \frac{\partial w}{\partial \alpha}$$

$$= \theta_1 - \frac{u}{R_\alpha} + \frac{1}{A} \frac{\partial w}{\partial \alpha} \tag{15.91}$$

$$\varepsilon_{23} = B \left(1 + \frac{z}{R_\beta}\right) \frac{\partial}{\partial z} \left[\frac{\overline{v}}{B\,(1 + z/R_\beta)}\right] + \frac{1}{B\,(1 + z/R_\beta)} \frac{\partial \overline{w}}{\partial \beta}$$

$$= B \frac{\partial}{\partial z} \left[\frac{v(\alpha,\,\beta) + z\,\theta_2}{B\,(1 + z/R_\beta)}\right] + \frac{1}{B} \frac{\partial w(\alpha,\,\beta)}{\partial \beta}$$

$$= B \left\{\frac{1}{B\,(1 + z/R_\beta)} \frac{\partial}{\partial z}(v + z\theta_2) - \frac{v + z\,\theta_2}{B^2\,(1 + z/R_\beta)^2} \frac{\partial}{\partial z}\left[B\left(1 + \frac{z}{R_\beta}\right)\right]\right\}$$

$$+ \frac{1}{B} \frac{\partial w}{\partial \beta}$$

$$= B \left(\frac{\theta_2}{B} - \frac{v + z\,\theta_2}{B^2} \frac{B}{R_\beta}\right) + \frac{1}{B} \frac{\partial w}{\partial \beta}$$

$$= \theta_2 - \frac{v}{R_\beta} + \frac{1}{B} \frac{\partial w}{\partial \beta} \tag{15.92}$$

To satisfy Eqs. (15.82), Eqs. (15.91) and (15.92) are set equal to zero. These give expressions for θ_1 and θ_2:

$$\theta_1 = \frac{u}{R_\alpha} - \frac{1}{A} \frac{\partial w}{\partial \alpha} \tag{15.93}$$

$$\theta_2 = \frac{v}{R_\beta} - \frac{1}{B} \frac{\partial w}{\partial \beta} \tag{15.94}$$

Using Eqs. (15.81), (15.85)–(15.87), (15.93) and (15.94), the strains in the shell, Eqs. (15.75)–(15.80), can be expressed as

$$\varepsilon_{11} = \frac{1}{A} \frac{\partial}{\partial \alpha}(u + z\theta_1) + \frac{v + z\theta_2}{AB} \frac{\partial A}{\partial \beta} + \frac{w}{R_\alpha} \tag{15.95}$$

$$\varepsilon_{22} = \frac{1}{B} \frac{\partial}{\partial \beta}(v + z\,\theta_2) + \frac{u + z\,\theta_1}{AB} \frac{\partial B}{\partial \alpha} + \frac{w}{R_\beta} \tag{15.96}$$

$$\varepsilon_{33} = 0 \tag{15.97}$$

$$\varepsilon_{12} = \frac{A}{B} \frac{\partial}{\partial \beta} \left(\frac{u + z\theta_1}{A} \right) + \frac{B}{A} \frac{\partial}{\partial \alpha} \left(\frac{v + z\theta_2}{B} \right) \qquad (15.98)$$

$$\varepsilon_{23} = 0 \qquad (15.99)$$

$$\varepsilon_{31} = 0 \qquad (15.100)$$

The nonzero components of strain, Eqs. (15.95), (15.96), and (15.98) are commonly expressed by separating the membrane strains (which are independent of z) and bending strains (which are dependent on z) as

$$\varepsilon_{11} = \varepsilon_{11}^0 + zk_{11} \qquad (15.101)$$

$$\varepsilon_{22} = \varepsilon_{22}^0 + zk_{22} \qquad (15.102)$$

$$\varepsilon_{12} = \varepsilon_{12}^0 + zk_{12} \qquad (15.103)$$

where the membrane strains, denoted by the superscript 0, are given by

$$\varepsilon_{11}^0 = \frac{1}{A} \frac{\partial u}{\partial \alpha} + \frac{v}{AB} \frac{\partial A}{\partial \beta} + \frac{w}{R_\alpha} \qquad (15.104)$$

$$\varepsilon_{22}^0 = \frac{1}{B} \frac{\partial v}{\partial \beta} + \frac{u}{AB} \frac{\partial B}{\partial \alpha} + \frac{w}{R_\beta} \qquad (15.105)$$

$$\varepsilon_{12}^0 = \frac{A}{B} \frac{\partial}{\partial \beta} \left(\frac{u}{A} \right) + \frac{B}{A} \frac{\partial}{\partial \alpha} \left(\frac{v}{B} \right) \qquad (15.106)$$

and the parameters k_{11}, k_{22}, and k_{12} are given by

$$k_{11} = \frac{1}{A} \frac{\partial \theta_1}{\partial \alpha} + \frac{\theta_2}{AB} \frac{\partial A}{\partial \beta} \qquad (15.107)$$

$$k_{22} = \frac{1}{B} \frac{\partial \theta_2}{\partial \beta} + \frac{\theta_1}{AB} \frac{\partial B}{\partial \alpha} \qquad (15.108)$$

$$k_{12} = \frac{A}{B} \frac{\partial}{\partial \beta} \left(\frac{\theta_1}{A} \right) + \frac{B}{A} \frac{\partial}{\partial \alpha} \left(\frac{\theta_2}{B} \right) \qquad (15.109)$$

Notes

1. The parameters k_{11} and k_{22} denote the midsurface changes in curvature, and k_{12} denotes the midsurface twist.
2. The strain–displacement relations given by Eqs. (15.101)–(15.109) define the thin shell theories of Love and Timoshenko. Based on the type of approximations used in the strain–displacement relations, other shell theories, those of such as (a) Byrne, Flügge, Goldenveizer, Lurye, and Novozhilov; (b) Reissner, Naghdi, and Berry; (c) Vlasov; (d) Sanders; and (e) Mushtari and Donnell have also been used for the analysis of thin shells [5].

Example 15.4 Simplify the strain–displacement relations given by Eqs. (15.101)–(15.109) for a circular cylindrical shell.

SOLUTION   Noting that $\alpha = x$, $\beta = \theta$, $A = 1$, $B = R$, and the subscripts 1 and 2 refer to x and θ, respectively. $R_\alpha = R_x = \infty$, $R_\beta = R_\theta = R$, Eqs. (15.93) and (15.94) can be expressed for a circular cylindrical shell as

$$\theta_x = -\frac{\partial w}{\partial x} \tag{E15.4.1}$$

$$\theta_\theta = \frac{v}{R} - \frac{1}{R}\frac{\partial w}{\partial \theta} \tag{E15.4.2}$$

The membrane strains are given by Eqs. (15.104)–(15.106):

$$\varepsilon_{xx}^0 = \frac{\partial u}{\partial x} \tag{E15.4.3}$$

$$\varepsilon_{\theta\theta}^0 = \frac{1}{R}\frac{\partial v}{\partial \theta} + \frac{w}{R} \tag{E15.4.4}$$

$$\varepsilon_{x\theta}^0 = \frac{\partial v}{\partial x} + \frac{1}{R}\frac{\partial u}{\partial \theta} \tag{E15.4.5}$$

The parameters k_{11}, k_{22}, and k_{12} can be determined from Eqs. (15.107)–(15.109) as

$$k_{xx} = \frac{\partial \theta_x}{\partial x} = -\frac{\partial^2 w}{\partial x^2} \tag{E15.4.6}$$

$$k_{\theta\theta} = \frac{1}{R}\frac{\partial \theta_\theta}{\partial \theta} = \frac{1}{R^2}\frac{\partial v}{\partial \theta} - \frac{1}{R^2}\frac{\partial^2 w}{\partial \theta^2} \tag{E15.4.7}$$

$$k_{x\theta} = \frac{\partial \theta_\theta}{\partial x} + \frac{1}{R}\frac{\partial \theta_x}{\partial \theta} = \frac{1}{R}\frac{\partial v}{\partial x} - \frac{2}{R}\frac{\partial^2 w}{\partial x\, \partial \theta} \tag{E15.4.8}$$

The total strains in the shell are given by Eqs. (15.101)–(15.103):

$$\varepsilon_{xx} = \varepsilon_{xx}^0 + z k_{xx} = \frac{\partial u}{\partial x} - z\frac{\partial^2 w}{\partial x^2} \tag{E15.4.9}$$

$$\varepsilon_{\theta\theta} = \varepsilon_{\theta\theta}^0 + z k_{\theta\theta} = \frac{1}{R}\frac{\partial v}{\partial \theta} + \frac{w}{R} + \frac{z}{R^2}\frac{\partial v}{\partial \theta} - \frac{z}{R^2}\frac{\partial^2 w}{\partial \theta^2} \tag{E15.4.10}$$

$$\varepsilon_{x\theta} = \varepsilon_{x\theta}^0 + z k_{x\theta} = \frac{\partial v}{\partial x} + \frac{1}{R}\frac{\partial u}{\partial \theta} + \frac{z}{R}\frac{\partial v}{\partial x} - \frac{2z}{R}\frac{\partial^2 w}{\partial x\, \partial \theta} \tag{E15.4.11}$$

In Eqs. (E15.4.1)–(E15.4.11), u, v and w denote the components of displacement along the x, θ, and z directions, respectively, in the midplane of the shell.

Example 15.5   Simplify the strain–displacement relations given by Eqs. (15.101)–(15.109) for a conical shell.

SOLUTION   For a conical shell, $\alpha = x$, $\beta = \theta$, $A = 1$, $B = x\ \sin \alpha_0$, $R_\alpha = R_x = \infty$, and $R_\beta = R_\theta = x\ \tan \alpha_0$. Using x and θ for the subscripts 1 and 2, respectively, θ_x and θ_θ can be obtained from Eqs. (15.93) and (15.94):

$$\theta_x = -\frac{\partial w}{\partial x} \tag{E15.5.1}$$

$$\theta_\theta = \frac{v}{x\ \tan \alpha_0} - \frac{1}{x\ \sin \alpha_0}\frac{\partial w}{\partial \theta} \tag{E15.5.2}$$

The strains in the shell are given by Eqs. (15.101)–(15.103):

$$\varepsilon_{xx} = \varepsilon_{xx}^0 + zk_{xx} \tag{E15.5.3}$$

$$\varepsilon_{\theta\theta} = \varepsilon_{\theta\theta}^0 + zk_{\theta\theta} \tag{E15.5.4}$$

$$\varepsilon_{x\theta} = \varepsilon_{x\theta}^0 + zk_{x\theta} \tag{E15.5.5}$$

where the membrane and bending parts of the strain are given by Eqs. (15.104)–(15.109):

$$\varepsilon_{xx}^0 = \frac{\partial u}{\partial x} \tag{E15.5.6}$$

$$\varepsilon_{\theta\theta}^0 = \frac{1}{x \sin \alpha_0} \frac{\partial v}{\partial \theta} + \frac{u}{x} + \frac{w}{x \tan \alpha_0} \tag{E15.5.7}$$

$$\varepsilon_{x\theta}^0 = \frac{1}{x \sin \alpha_0} \frac{\partial u}{\partial \theta} + \frac{\partial v}{\partial x} - \frac{v}{x} \tag{E15.5.8}$$

$$k_{xx} = \frac{\partial \theta_x}{\partial x} = -\frac{\partial^2 w}{\partial x^2} \tag{E15.5.9}$$

$$
\begin{aligned}
k_{\theta\theta} &= \frac{1}{x \sin \alpha_0} \frac{\partial \theta_\theta}{\partial \theta} + \frac{1}{x} \theta_x \\
&= \frac{\cos \alpha_0}{x^2 \sin^2 \alpha_0} \frac{\partial v}{\partial \theta} - \frac{1}{x^2 \sin^2 \alpha_0} \frac{\partial^2 w}{\partial \theta^2} - \frac{1}{x} \frac{\partial w}{\partial x}
\end{aligned}
\tag{E15.5.10}
$$

$$
\begin{aligned}
k_{x\theta} &= x \frac{\partial}{\partial x} \left(\frac{\theta_\theta}{x} \right) + \frac{1}{x \sin \alpha_0} \frac{\partial \theta_x}{\partial \theta} \\
&= \frac{1}{x \tan \alpha_0} \frac{\partial v}{\partial x} - \frac{2v}{x^2 \tan \alpha_0} - \frac{1}{x \sin \alpha_0} \frac{\partial^2 w}{\partial \theta^2} + \frac{2}{x^2 \sin \alpha_0} \frac{\partial w}{\partial \theta}
\end{aligned}
\tag{E15.5.11}
$$

Note that u, v, and w in Eqs. (E15.5.1)–(E15.5.11) denote the components of displacement along x, θ, and z directions, respectively, in the midplane of the shell.

Example 15.6   Simplify the strain–displacement relations given by Eqs. (15.101)–(15.109) for a spherical shell.

SOLUTION   For a spherical shell, $\alpha = \phi$, $\beta = \theta$, $A = R$, $B = R \sin \phi$, $R_\alpha = R_\phi = R$, and $R_\beta = R_\theta = R$. Using ϕ and θ as subscripts in place of 1 and 2, respectively, Eqs. (15.93) and (15.94) yield

$$\theta_\phi = \frac{1}{R} \left(u - \frac{\partial w}{\partial \phi} \right) \tag{E15.6.1}$$

$$\theta_\theta = \frac{1}{R} \left(v - \frac{1}{\sin \phi} \frac{\partial w}{\partial \theta} \right) \tag{E15.6.2}$$

The membrane strains are given by Eqs. (15.104)–(15.106):

$$\varepsilon^0_{\phi\phi} = \frac{1}{R}\frac{\partial u}{\partial \phi} + \frac{w}{R} \tag{E15.6.3}$$

$$\varepsilon^0_{\theta\theta} = \frac{1}{R \sin \phi}\frac{\partial v}{\partial \theta} + \frac{u \cot \phi}{R} + \frac{w}{R} \tag{E15.6.4}$$

$$\varepsilon^0_{\phi\theta} = \frac{1}{R \sin \phi}\frac{\partial u}{\partial \theta} + \frac{1}{R}\frac{\partial v}{\partial \phi} - \frac{v \cot \phi}{R} \tag{E15.6.5}$$

The total strains are given by Eqs. (15.101)–(15.103):

$$\varepsilon_{\phi\phi} = \varepsilon^0_{\phi\phi} + zk_{\phi\phi} \tag{E15.6.6}$$

$$\varepsilon_{\theta\theta} = \varepsilon^0_{\theta\theta} + zk_{\theta\theta} \tag{E15.6.7}$$

$$\varepsilon_{\phi\theta} = \varepsilon^0_{\phi\theta} + zk_{\phi\theta} \tag{E15.6.8}$$

where $k_{\phi\phi}$, $k_{\theta\theta}$, and $k_{\phi\theta}$ are given by Eqs. (15.107)–(15.109):

$$k_{\phi\phi} = \frac{1}{R^2}\frac{\partial u}{\partial \phi} - \frac{1}{R^2}\frac{\partial^2 w}{\partial \phi^2} \tag{E15.6.9}$$

$$k_{\theta\theta} = \frac{1}{R^2 \sin \phi}\frac{\partial v}{\partial \theta} - \frac{1}{R^2 \sin^2 \phi}\frac{\partial^2 w}{\partial \theta^2} + \frac{\cot \phi}{R^2}u - \frac{\cos \phi}{R^2}\frac{\partial w}{\partial \phi} \tag{E15.6.10}$$

$$k_{\phi\theta} = \frac{1}{R^2 \sin \phi}\frac{\partial u}{\partial \theta} - \frac{1}{R^2 \sin \phi}\frac{\partial^2 w}{\partial \phi \partial \theta} + \frac{1}{R^2}\frac{\partial v}{\partial \phi}$$
$$- \frac{1}{R^2 \sin \phi}\frac{\partial^2 w}{\partial \phi \partial \theta} + \frac{\cos \phi}{R^2 \sin^2 \phi}\frac{\partial w}{\partial \theta} - \frac{\cot \phi}{R^2}v + \frac{\cot \phi}{R^2 \sin \phi}\frac{\partial w}{\partial \theta} \tag{E15.6.11}$$

Note that u, v, and w denote the components of displacement along ϕ, θ, and z directions, respectively, in the mid plane of the shell.

15.4 STRESS–STRAIN RELATIONS

In a three-dimensional isotropic body, such as a thin shell, the stresses are related to the strains by Hooke's law as

$$\varepsilon_{11} = \frac{1}{E}[\sigma_{11} - \nu(\sigma_{22} + \sigma_{33})] \tag{15.110}$$

$$\varepsilon_{22} = \frac{1}{E}[\sigma_{22} - \nu(\sigma_{11} + \sigma_{33})] \tag{15.111}$$

$$\varepsilon_{33} = \frac{1}{E}[\sigma_{33} - \nu(\sigma_{11} + \sigma_{22})] \tag{15.112}$$

$$\varepsilon_{12} = \frac{\sigma_{12}}{G} \tag{15.113}$$

$$\varepsilon_{23} = \frac{\sigma_{23}}{G} \tag{15.114}$$

$$\varepsilon_{13} = \frac{\sigma_{13}}{G} \tag{15.115}$$

where σ_{11}, σ_{22}, and σ_{33} are normal stresses, $\sigma_{12} = \sigma_{21}$, $\sigma_{13} = \sigma_{31}$, and $\sigma_{23} = \sigma_{32}$ are shear stresses. Based on Kirchhoff's hypothesis, we have

$$\varepsilon_{33} = 0 \tag{15.116}$$

$$\varepsilon_{13} = \varepsilon_{23} = 0 \tag{15.117}$$

Equations (15.112) and (15.116) yield

$$\sigma_{33} = \nu(\sigma_{11} + \sigma_{22}) \tag{15.118}$$

But according to Love's third assumption, $\sigma_{33} = 0$, which is in contradiction to Eq. (15.118). As stated earlier, this is an unavoidable contradiction in thin shell theory. Another contradiction is that σ_{13} and σ_{23} cannot be equal to zero because their integrals must be able to balance the transverse shear forces needed for satisfying the equilibrium conditions.

However, the magnitudes of σ_{13} and σ_{23} are usually very small compared to those of σ_{11}, σ_{22} and σ_{12}. Thus, the problem reduces to one of plane stress, described by

$$\varepsilon_{11} = \frac{1}{E}(\sigma_{11} - \nu\sigma_{22}) \tag{15.119}$$

$$\varepsilon_{22} = \frac{1}{E}(\sigma_{22} - \nu\sigma_{11}) \tag{15.120}$$

$$\varepsilon_{12} = \frac{\sigma_{12}}{G} \tag{15.121}$$

Inverting these equations, we are able to express stresses in terms of strains as

$$\sigma_{11} = \frac{E}{1 - \nu^2}(\varepsilon_{11} + \nu\varepsilon_{22}) \tag{15.122}$$

$$\sigma_{22} = \frac{E}{1 - \nu^2}(\varepsilon_{22} + \nu\varepsilon_{11}) \tag{15.123}$$

$$\sigma_{12} = \frac{E}{2(1 + \nu)}\varepsilon_{12} \tag{15.124}$$

Substitution of Eqs. (15.101)–(15.103) into Eqs. (15.122)–(15.124) leads to

$$\sigma_{11} = \frac{E}{1 - \nu^2}[\varepsilon_{11}^0 + \nu\varepsilon_{22}^0 + z(k_{11} + \nu k_{22})] \tag{15.125}$$

$$\sigma_{22} = \frac{E}{1 - \nu^2}[\varepsilon_{22}^0 + \nu\varepsilon_{11}^0 + z(k_{22} + \nu k_{11})] \tag{15.126}$$

$$\sigma_{12} = \frac{E}{2(1 + \nu)} \cdot (\varepsilon_{12}^0 + z k_{12}) \tag{15.127}$$

15.5 FORCE AND MOMENT RESULTANTS

Consider a differential element of the shell isolated from the shell by using four sections normal to its middle surface and tangential to the lines α, $\alpha + d\alpha$, β, and $\beta + d\beta$ as shown in Fig. 15.7. Here the middle surface of the shell is defined by *abco*. The stresses

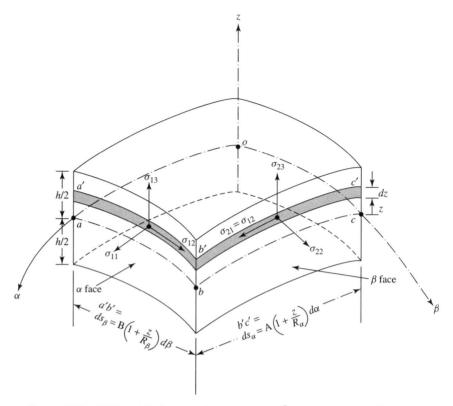

Figure 15.7   Differential element of shell. $ab = ds_\beta^0 = B\,d\beta$, $bc = ds_\alpha^0 = A\,d\alpha$.

acting on the positive faces of the element are also shown in the figure. For the stresses, the first subscript denotes the face on which it is acting (1 for the face normal to the α curve and 2 for the face normal to the β curve), and the second subscript denotes the direction along which it is acting (1 for the α direction and 2 for the β direction).

The force and moment resultants induced due to the various stresses are shown in Figs. 15.8 and 15.9. These resultants can be seen to be similar to those induced for a plate. To find the force resultants, first consider the face of the element shown in Fig. 15.7 that is perpendicular to the α-axis (called the α face). The arc length ab of the intercept of the middle surface with the face is given by

$$ds_\beta^0 = B\,d\beta \tag{15.128}$$

and the arc length $a'b'$ of intercept of a parallel surface located at a distance z from the middle surface is given by

$$ds_\beta = B\left(1 + \frac{z}{R_\beta}\right) d\beta \tag{15.129}$$

The stress σ_{11} multiplied by the elemental area $ds_\beta\,dz$ gives the elemental force dN_{11}. Integration of the elemental force over the thickness of the shell gives the total in-plane

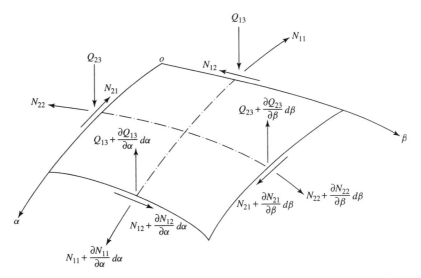

Figure 15.8 Force resultants in a shell (positive directions are indicated).

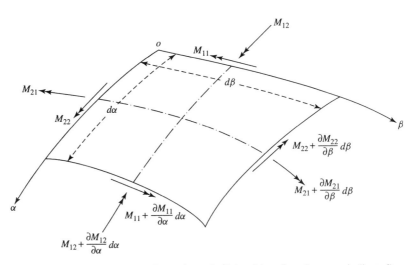

Figure 15.9 Moment resultants in a shell (positive directions are indicated).

normal force acting on the face along the α direction as

$$\text{total force} = \int_{-h/2}^{h/2} \sigma_{11} \, ds_\beta \, dz = \int_{-h/2}^{h/2} \sigma_{11} B \left(1 + \frac{z}{R_\beta}\right) d\beta \, dz \qquad (15.130)$$

The force resultant N_{11} (force per unit length of the middle surface) can be obtained by dividing the total force by the arc length $ds_\beta^0 = B \, d\beta$:

$$N_{11} = \int_{-h/2}^{h/2} \sigma_{11} \left(1 + \frac{z}{R_\beta}\right) dz \qquad (15.131)$$

By neglecting z/R_β as a small quantity compared to 1, Eq. (15.131) gives

$$N_{11} = \int_{-h/2}^{h/2} \sigma_{11}\,dz \tag{15.132}$$

Using Eq. (15.125) in Eq. (15.132), we obtain

$$
\begin{aligned}
N_{11} &= \frac{E}{1 - v^2} \int_{-h/2}^{h/2} [\varepsilon_{11}^0 + v\varepsilon_{22}^0 + z(k_{11} + vk_{22})]\,dz \\
&= \frac{E}{1 - v^2} \left[(\varepsilon_{11}^0 + v\varepsilon_{22}^0)(z)_{-h/2}^{h/2} + (k_{11} + vk_{22})\left(\frac{z^2}{2}\right)_{-h/2}^{h/2} \right] = \frac{Eh}{1 - v^2}(\varepsilon_{11}^0 + v\,\varepsilon_{22}^0)
\end{aligned}
\tag{15.133}
$$

By defining the middle surface or membrane stiffness of the shell (C) as

$$C = \frac{Eh}{1 - v^2} \tag{15.134}$$

Eq. (15.133) can be written as

$$N_{11} = C(\varepsilon_{11}^0 + v\varepsilon_{22}^0) \tag{15.135}$$

The stress σ_{12} acting on the α face multiplied by the elemental area $ds_\beta\,dz$ (or the stress $\sigma_{21} = \sigma_{12}$ acting on the β face multiplied by the elemental area $ds_\alpha\,dz$) gives the elemental force N_{12}. Integration of the elemental force over the thickness of the shell gives the total in-plane shear force, which when divided by the arc length ds_β^0 for the α face and ds_α^0 for the β face, gives the force resultant N_{12} as

$$N_{12} = N_{21} = C\left(\frac{1 - v}{2}\right)\varepsilon_{12}^0 \tag{15.136}$$

The transverse shear force resultant, Q_{13}, acting on the α face due to the shear stress σ_{13}, can be found as

$$Q_{13} = \int_{-h/2}^{h/2} \sigma_{13}\,dz \tag{15.137}$$

Similarly, by integrating σ_{22} on the β face, the force resultant N_{22} can be obtained as

$$N_{22} = C(\varepsilon_{22}^0 + v\varepsilon_{11}^0) \tag{15.138}$$

By integrating the shear stress σ_{23} on the β face, we can find the transverse shear force resultant, Q_{23}, acting on the β face as

$$Q_{23} = \int_{-h/2}^{h/2} \sigma_{23}\,dz \tag{15.139}$$

To find the bending moment resultants, we again consider the α face of the element in Fig. 15.7. The moment of the elemental force ($\sigma_{11}\,ds_\beta\,dz$) about the β line will be $z\sigma_{11}\,ds_\beta\,dz$, and the bending moment resultant M_{11} (moment per unit length of the middle surface) can be found by integrating the elemental moment over the thickness and dividing the result by the arc length $ds_\beta^0 = B\,d\beta$. Thus, we obtain

$$\text{total moment} = \int_{-h/2}^{h/2} \sigma_{11} z B \left(1 + \frac{z}{R_\beta}\right) d\beta\,dz \tag{15.140}$$

$$M_{11} = \int_{-h/2}^{h/2} \sigma_{11} z \left(1 + \frac{z}{R_\beta}\right) dz \tag{15.141}$$

By neglecting the quantity z/R_β in comparison to unity, Eq. (15.141) gives

$$M_{11} = \int_{-h/2}^{h/2} \sigma_{11} z\,dz \tag{15.142}$$

By substituting Eq. (15.125) in Eq. (15.142), we obtain

$$
\begin{aligned}
M_{11} &= \int_{-h/2}^{h/2} \frac{E}{1-\nu^2}[\varepsilon_{11}^0 + \nu\varepsilon_{22}^0 + z(k_{11} + \nu k_{22})]z\,dz \\
&= \frac{E}{1-\nu^2}\left[(\varepsilon_{11}^0 + \nu\varepsilon_{22}^0)\left(\frac{z^2}{2}\right)_{-h/2}^{h/2} + (k_{11} + \nu k_{22})\left(\frac{z^3}{3}\right)_{-h/2}^{h/2}\right] \\
&= \frac{Eh^3}{12(1-\nu^2)}(k_{11} + \nu k_{22}) \\
&= D(k_{11} + \nu k_{22}) \tag{15.143}
\end{aligned}
$$

where D, the bending stiffness of the shell, is defined by

$$D = \frac{Eh^3}{12(1-\nu^2)} \tag{15.144}$$

Using a similar procedure, the moment resultants due to the stresses σ_{22} and $\sigma_{12} = \sigma_{21}$ can be found as

$$M_{22} = D(k_{22} + \nu k_{11}) \tag{15.145}$$

$$M_{12} = M_{21} = D\left(\frac{1-\nu}{2}\right)k_{12} \tag{15.146}$$

Notes

1. Although the shear strains ε_{13} and ε_{23} are assumed to be zero (according to Kirchhoff's hypothesis), we still assume the presence of the transverse shear stresses σ_{13} and σ_{23} and the corresponding shear force resultants, Q_{13} and Q_{23}, as given by Eqs. (15.137) and (15.139).

2. The expressions of the force and moment resultants given in Eqs. (15.132)–(15.146) have been derived by neglecting the quantities z/R_α and z/R_β in comparison to unity. This assumption is made in the thin shell theories of Love, Timoshenko, Reissner, Naghdi, Berry, Sanders, Mushtari, and Donnell.

3. If the quantities z/R_α and z/R_β are not neglected, we will find that $N_{12} \neq N_{21}$ and $M_{12} \neq M_{21}$ (unless $R_\alpha = R_\beta$), although the shear stresses causing them are the same (i.e., $\sigma_{12} = \sigma_{21}$).

Example 15.7   Express the force and moment resultants in terms of displacements for a circular cylindrical shell.

SOLUTION   The force resultants are given in terms of membrane strains by Eqs. (15.135), (15.136), and (15.138). Using the strain–displacement relations given in Example 15.4, the force resultants can be expressed as[1]

$$N_{xx} = C(\varepsilon_{xx}^0 + \nu\varepsilon_{\theta\theta}^0) = C\left(\frac{\partial u}{\partial x} + \frac{\nu}{R}\frac{\partial v}{\partial\theta} + \frac{\nu}{R}w\right) \tag{E15.7.1}$$

$$N_{x\theta} = N_{\theta x} = C\left(\frac{1-\nu}{2}\right)\varepsilon_{x\theta}^0 = C\left(\frac{1-\nu}{2}\right)\left(\frac{\partial v}{\partial x} + \frac{1}{R}\frac{\partial u}{\partial\theta}\right) \tag{E15.7.2}$$

$$N_{\theta\theta} = C(\varepsilon_{\theta\theta}^0 + \nu\varepsilon_{xx}^0) = C\left(\frac{1}{R}\frac{\partial v}{\partial\theta} + \frac{w}{R} + \nu\frac{\partial u}{\partial x}\right) \tag{E15.7.3}$$

where C is given by Eq. (15.134). The moment resultants, given by Eqs. (15.143), (15.145) and (15.146), can be expressed using the strain–displacement relations of Example 15.4 as

$$M_{xx} = D(k_{xx} + \nu k_{\theta\theta}) = D\left(\frac{\partial\theta_x}{\partial x} + \frac{\nu}{R}\frac{\partial\theta_\theta}{\partial\theta}\right) \tag{E15.7.4}$$

$$= D\left(-\frac{\partial^2 w}{\partial x^2} + \frac{\nu}{R^2}\frac{\partial v}{\partial\theta} - \frac{\nu}{R^2}\frac{\partial^2 w}{\partial\theta^2}\right) \tag{E15.7.5}$$

$$M_{\theta\theta} = D(k_{\theta\theta} + \nu k_{xx}) = D\left(\frac{1}{R}\frac{\partial\theta_\theta}{\partial\theta} + \nu\frac{\partial\theta_x}{\partial x}\right) \tag{E15.7.6}$$

$$= D\left(\frac{1}{R^2}\frac{\partial v}{\partial\theta} - \frac{1}{R^2}\frac{\partial^2 w}{\partial\theta^2} - \nu\frac{\partial^2 w}{\partial x^2}\right) \tag{E15.7.7}$$

$$M_{x\theta} = M_{\theta x} = D\left(\frac{1-\nu}{2}\right)k_{x\theta} = D\left(\frac{1-\nu}{2}\right)\left(\frac{\partial\theta_\theta}{\partial x} + \frac{1}{R}\frac{\partial\theta_x}{\partial\theta}\right) \tag{E15.7.8}$$

$$= D\left(\frac{1-\nu}{2}\right)\left(\frac{1}{R}\frac{\partial v}{\partial x} - \frac{2}{R}\frac{\partial^2 w}{\partial x\partial\theta}\right) \tag{E15.7.9}$$

where D is given by Eq. (15.144).

Example 15.8   Express the force and moment resultants in terms of displacements for a circular conical shell.

SOLUTION   Equations (15.135), (15.136), and (15.138) give the force resultants in terms of membrane strains. By using the strain–displacement relations of Example 15.5,

---

[1]For a cylindrical shell, u, v and w denote the components of displacement along the axial (x), circumferential (θ), and radial (z) directions, respectively, as shown in Fig. 15.10.

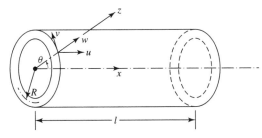

Figure 15.10 Coordinate system of a thin circular cylindrical shell.

the force resultants can be expressed as[2]

$$N_{xx} = C(\varepsilon_{xx}^0 + \nu\varepsilon_{\theta\theta}^0)$$

$$= C\left[\frac{\partial u}{\partial x} + \nu\left(\frac{1}{x\sin\alpha_0}\frac{\partial v}{\partial\theta} + \frac{u}{x} + \frac{w}{x\tan\alpha_0}\right)\right] \qquad (E15.8.1)$$

$$N_{x\theta} = N_{\theta x} = C\left(\frac{1-\nu}{2}\right)\varepsilon_{x\theta}^0$$

$$= C\left(\frac{1-\nu}{2}\right)\left(\frac{1}{x\sin\alpha_0}\frac{\partial u}{\partial\theta} + \frac{\partial v}{\partial x} - \frac{v}{x}\right) \qquad (E15.8.2)$$

$$N_{\theta\theta} = C(\varepsilon_{\theta\theta}^0 + \nu\varepsilon_{xx}^0)$$

$$= C\left(\frac{1}{x\sin\alpha_0}\frac{\partial v}{\partial\theta} + \frac{u}{x} + \frac{w}{x\tan\alpha_0} + \nu\frac{\partial u}{\partial x}\right) \qquad (E15.8.3)$$

where C is given by Eq. (15.134). The moment resultants are given by Eqs. (15.143), (15.145), and (15.146). Using the strain–displacement relations of Example 15.5, the moment resultants can be expressed as

$$M_{xx} = D(k_{xx} + \nu k_{\theta\theta})$$

$$= D\left[-\frac{\partial^2 w}{\partial x^2} + \nu\left(\frac{\cos\alpha_0}{x^2\sin^2\alpha_0}\frac{\partial v}{\partial\theta} - \frac{1}{x^2\sin^2\alpha_0}\frac{\partial^2 w}{\partial\theta^2} - \frac{1}{x}\frac{\partial w}{\partial x}\right)\right] \qquad (E15.8.4)$$

$$M_{\theta\theta} = D(k_{\theta\theta} + \nu k_{xx})$$

$$= D\left(\frac{\cos\alpha_0}{x^2\sin^2\alpha_0}\frac{\partial v}{\partial\theta} - \frac{1}{x^2\sin^2\alpha_0}\frac{\partial^2 w}{\partial\theta^2} - \frac{1}{x}\frac{\partial w}{\partial x} - \nu\frac{\partial^2 w}{\partial x^2}\right) \qquad (E15.8.5)$$

$$M_{x\theta} = M_{\theta x} = D\left(\frac{1-\nu}{2}\right)k_{x\theta}$$

$$= D\left(\frac{1-\nu}{2}\right)\left(\frac{1}{x\tan\alpha_0}\frac{\partial v}{\partial x} - \frac{2\nu}{x^2\tan\alpha_0} - \frac{1}{x\sin\alpha_0}\frac{\partial^2 w}{\partial\theta^2} + \frac{2}{x^2\sin\alpha_0}\frac{\partial w}{\partial\theta}\right) \qquad (E15.8.6)$$

where D is given by Eq. (15.144).

---

[2]For a circular conical shell, u, v, and w denote the components of displacement along the generator (x), circumferential (θ), and radial (z) directions, respectively, as shown in Fig. 15.11.

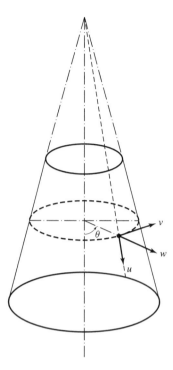

Figure 15.11 Components of displacement in a circular conical shell. u, v, and w are along the x, θ, and z directions.

Example 15.9 Express the force and moment resultants in terms of displacements for a spherical shell.

SOLUTION Using the force resultant–membrane strain relations given by Eqs. (15.135), (15.136) and (15.138), and the strain–displacement relations given in Example 15.6, the force resultants can be expressed as

$$N_{\phi\phi} = C(\varepsilon_{\phi\phi}^0 + \nu\varepsilon_{\theta\theta}^0)$$

$$= C\left[\frac{1}{R}\frac{\partial u}{\partial \phi} + \frac{w}{R} + \nu\left(\frac{1}{R\sin\phi}\frac{\partial v}{\partial \theta} + \frac{u\cot\phi}{R} + \frac{w}{R}\right)\right] \tag{E15.9.1}$$

$$N_{\phi\theta} = N_{\theta\phi} = C\left(\frac{1-\nu}{2}\right)\varepsilon_{\phi\theta}$$

$$= C\left(\frac{1-\nu}{2}\right)\left(\frac{1}{R\sin\phi}\frac{\partial u}{\partial \theta} + \frac{1}{R}\frac{\partial v}{\partial \phi} - \frac{v\cot\phi}{R}\right) \tag{E15.9.2}$$

$$N_{\theta\theta} = C(\varepsilon_{\theta\theta}^0 + \nu\varepsilon_{\phi\phi}^0)$$

$$= C\left[\frac{1}{R\sin\phi}\frac{\partial u}{\partial \theta} + \frac{1}{R}\frac{\partial v}{\partial \phi} - \frac{v\cot\phi}{R} + \nu\left(\frac{u}{R} - \frac{1}{R}\frac{\partial w}{\partial \phi}\right)\right] \tag{E15.9.3}$$

where C is given by Eq. (15.134). Using the moment resultant–bending strain relations given by Eqs. (15.143), (15.145), and (15.146) and the strain–displacement relations given in Example 15.6, the moment resultants can be expressed as[3]

$$M_{\phi\phi} = D(k_{\phi\phi} + \nu k_{\theta\theta})$$

$$= D\left[\frac{1}{R^2} \frac{\partial u}{\partial \phi} - \frac{1}{R^2} \frac{\partial^2 w}{\partial \phi^2} + \nu \left(\frac{1}{R^2 \sin\phi} \frac{\partial v}{\partial \theta} - \frac{1}{R^2 \sin^2\phi} \frac{\partial^2 w}{\partial \theta^2} \right. \right.$$

$$\left. \left. + \frac{\cot\phi}{R^2} u - \frac{\cos\phi}{R^2} \frac{\partial w}{\partial \phi} \right) \right] \tag{E15.9.4}$$

$$M_{\theta\theta} = D(k_{\theta\theta} + \nu k_{\phi\phi})$$

$$= D\left[\frac{1}{R^2 \sin\phi} \frac{\partial v}{\partial \theta} - \frac{1}{R^2 \sin^2\phi} \frac{\partial^2 w}{\partial \theta^2} + \frac{\cot\phi}{R^2} u \right.$$

$$\left. - \frac{\cot\phi}{R^2} \frac{\partial w}{\partial \phi} + \nu \left(\frac{1}{R^2} \frac{\partial u}{\partial \phi} - \frac{1}{R^2} \frac{\partial^2 w}{\partial \phi^2} \right) \right] \tag{E15.9.5}$$

$$M_{\phi\theta} = M_{\theta\phi} = D\left(\frac{1-\nu}{2} \right) k_{\phi\theta}$$

$$= D\left(\frac{1-\nu}{2} \right) \left(\frac{1}{R^2 \sin\phi} \frac{\partial u}{\partial \theta} - \frac{1}{R^2 \sin\phi} \frac{\partial^2 w}{\partial \phi \partial \theta} \right.$$

$$\left. + \frac{1}{R^2} \frac{\partial v}{\partial \phi} - \frac{1}{R^2 \sin\phi} \frac{\partial^2 w}{\partial \phi \partial \theta} + \frac{\cos\phi}{R^2 \sin^2\phi} \frac{\partial w}{\partial \theta} - \frac{\cot\phi}{R^2} v + \frac{\cot\phi}{R^2 \sin\phi} \frac{\partial w}{\partial \theta} \right) \tag{E15.9.6}$$

where D is given by Eq. (15.144).

15.6   STRAIN ENERGY, KINETIC ENERGY, AND WORK DONE BY EXTERNAL FORCES

To derive the equations of motion of the shell using Hamilton's principle, the expressions for strain energy, kinetic energy, and work done by the external forces are required. These expressions are derived in this section.

15.6.1   Strain Energy

The strain energy density or strain energy per unit volume of an elastic body is given by

$$\pi_0 = \tfrac{1}{2}(\sigma_{11}\varepsilon_{11} + \sigma_{22}\varepsilon_{22} + \sigma_{33}\varepsilon_{33} + \sigma_{12}\varepsilon_{12} + \sigma_{23}\varepsilon_{23} + \sigma_{31}\varepsilon_{31}) \tag{15.147}$$

The strain energy (π) of the shell can be found by integrating the strain energy density over the volume of the shell:

$$\pi = \frac{1}{2} \iiint_V (\sigma_{11}\varepsilon_{11} + \sigma_{22}\varepsilon_{22} + \sigma_{33}\varepsilon_{33} + \sigma_{12}\varepsilon_{12} + \sigma_{23}\varepsilon_{23} + \sigma_{31}\varepsilon_{31})\,dV \tag{15.148}$$

---

[3]For a spherical shell, u, v, and w denote the components of displacement along the ϕ, θ, and z directions, respectively, as shown in Fig. 15.12.

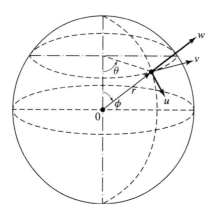

Figure 15.12 Components of displacement in a spherical shell. u, v, and w are along the ϕ, θ, and z directions.

where the volume of an infinitesimal shell element is given by

$$dV = ds_\alpha\, ds_\beta\, dz = \left[A\left(1 + \frac{z}{R_\alpha}\right) d\alpha \right]\left[B\left(1 + \frac{z}{R_\beta}\right) d\beta \right] dz \qquad (15.149)$$

By neglecting the quantities z/R_α and z/R_β in comparison to 1, Eq. (15.149) can be expressed as

$$dV = AB\, d\alpha\, d\beta\, dz \qquad (15.150)$$

Note that the transverse normal stress σ_{33} is assumed to be zero according to Love's third assumption in Eq. (15.148). At the same time, the transverse shear strains ε_{23} and ε_{31} are retained in Eq. (15.148) although they were assumed to be zero earlier [Eqs. (15.99) and (15.100)]. By substituting Eqs. (15.91), (15.92), (15.95), (15.96) and (15.98), respectively, for the strains ε_{13}, ε_{23}, ε_{11}, ε_{22}, and ε_{12}, and Eq. (15.150) for dV into Eq. (15.148), we obtain the strain energy of the shell as

$$
\begin{aligned}
\pi = \frac{1}{2} \int_\alpha \int_\beta \int_z \Bigg\{ & \sigma_{11}\left[\frac{1}{A}\left(\frac{\partial u}{\partial \alpha} + z\frac{\partial \theta_1}{\partial \alpha}\right) + \frac{1}{AB}\frac{\partial A}{\partial \beta}(v + z\theta_2) + \frac{w}{R_\alpha} \right] \\
& + \sigma_{22}\left[\frac{1}{B}\left(\frac{\partial v}{\partial \beta} + z\frac{\partial \theta_2}{\partial \beta}\right) + \frac{1}{AB}\frac{\partial B}{\partial \alpha}(u + z\theta_1) + \frac{w}{R_\beta} \right] \\
& + \sigma_{12}\left[\frac{1}{B}\frac{\partial u}{\partial \beta} - \frac{u}{AB}\frac{\partial A}{\partial \beta} + \frac{z}{B}\frac{\partial \theta_1}{\partial \beta} - \frac{z\theta_1}{AB}\frac{\partial A}{\partial \beta} \right. \\
& \left. + \frac{1}{A}\frac{\partial v}{\partial \alpha} - \frac{v}{AB}\frac{\partial B}{\partial \alpha} + \frac{z}{A}\frac{\partial \theta_2}{\partial \alpha} - \frac{z\theta_2}{AB}\frac{\partial B}{\partial \alpha} \right] + \sigma_{23}\left(\theta_2 - \frac{v}{R_\beta} + \frac{1}{B}\frac{\partial w}{\partial \beta} \right) \\
& + \sigma_{31}\left(\theta_1 - \frac{u}{R_\alpha} + \frac{1}{A}\frac{\partial w}{\partial \alpha} \right) \Bigg\} AB\, d\alpha\, d\beta\, dz \qquad (15.151)
\end{aligned}
$$

Performing the integration with respect to z and using the definitions of the force and moment resultants given in Eqs. (15.135)–(15.139), (15.143), (15.145), and (15.146), the strain energy [Eq. (15.151)] can be expressed as

$$
\begin{aligned}
\pi = \frac{1}{2} \int_\alpha \int_\beta \bigg(&N_{11}B\frac{\partial u}{\partial \alpha} + M_{11}B\frac{\partial \theta_1}{\partial \alpha} + N_{11}\frac{\partial A}{\partial \beta}v + M_{11}\frac{\partial A}{\partial \beta}\theta_2 + N_{11}AB\frac{w}{R_\alpha} \\
&+ N_{22}A\frac{\partial v}{\partial \beta} + M_{22}A\frac{\partial \theta_2}{\partial \beta} + N_{22}\frac{\partial B}{\partial \alpha}u + M_{22}\frac{\partial B}{\partial \alpha}\theta_1 + N_{22}AB\frac{w}{R_\beta} \\
&+ N_{12}A\frac{\partial u}{\partial \beta} - N_{12}\frac{\partial A}{\partial \beta}u + M_{12}A\frac{\partial \theta_1}{\partial \beta} - M_{12}\frac{\partial A}{\partial \beta}\theta_1 \\
&+ N_{12}B\frac{\partial v}{\partial \alpha} - N_{12}\frac{\partial B}{\partial \alpha}v + M_{12}B\frac{\partial \theta_2}{\partial \alpha} - M_{12}\frac{\partial B}{\partial \alpha}\theta_2 \\
&+ Q_{23}AB\theta_2 - Q_{23}AB\frac{v}{R_\beta} + Q_{23}A\frac{\partial w}{\partial \beta} \\
&+ Q_{13}AB\theta_1 - Q_{13}AB\frac{u}{R_\alpha} + Q_{13}B\frac{\partial w}{\partial \alpha} \bigg)\, d\alpha\, d\beta
\end{aligned}
\tag{15.152}
$$

15.6.2   Kinetic Energy

The kinetic energy of an infinitesimal element of the shell can be expressed as

$$
dT = \frac{1}{2}\rho(\dot{\bar u}^2 + \dot{\bar v}^2 + \dot{\bar w}^2)\,dV
\tag{15.153}
$$

where ρ indicates the density of the shell and a dot over a symbol denotes the derivative with respect to time. By substituting Eqs. (15.85)–(15.87) for $\bar u$, $\bar v$, and $\bar w$ into Eq. (15.153), and integrating the resulting expression over the volume of the shell, we obtain the total kinetic energy of the shell as

$$
\begin{aligned}
T = \frac{1}{2}\rho \int_\alpha \int_\beta \int_z [\dot u^2 + \dot v^2 + \dot w^2 + \alpha_3^2(\dot\theta_1^2 + \dot\theta_2^2) \\
+ 2\alpha_3(\dot u\dot\theta_1 + \dot v\dot\theta_2)]AB\left(1 + \frac{z}{R_\alpha}\right)\left(1 + \frac{z}{R_\beta}\right)d\alpha\, d\beta\, dz
\end{aligned}
\tag{15.154}
$$

By neglecting the terms z/R_α and z/R_β in comparison to 1 in Eq. (13.154) and performing the integration with respect to z between the limits $-h/2$ to $h/2$ yields

$$
T = \frac{1}{2}\rho \int_\alpha \int_\beta \left[h(\dot u^2 + \dot v^2 + \dot w^2) + \frac{h^3}{12}(\dot\theta_1^2 + \dot\theta_2^2) \right] AB\, d\alpha\, d\beta
\tag{15.155}
$$

15.6.3   Work Done by External Forces

The work done by external forces (W) indicates the work done by the components of the distributed loads f_α, f_β, and f_z along the α, β, and z directions, respectively, (shown in Fig. 15.13), and the force and moment resultants acting on the boundaries of the shell defined by constant values of α and β (shown in Fig. 15.14). All the external

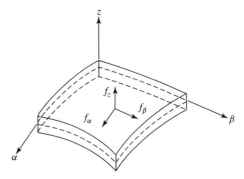

Figure 15.13 Distributed loads f_α, f_β, and f_z acting along the α, β, and z directions in the middle surface of a shell.

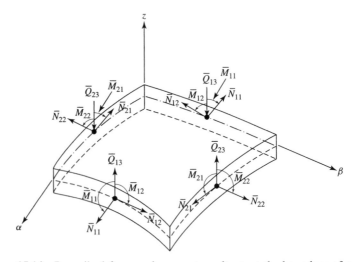

Figure 15.14 Prescribed force and moment resultants at the boundary of a shell.

loads are assumed to act in the middle surface of the shell. The work done (W_d) by the components of the distributed load is given by

$$W_d = \int_\alpha \int_\beta (f_\alpha u + f_\beta v + f_z w)\, ds_\alpha^0\, ds_\beta^0$$

$$= \int_\alpha \int_\beta (f_\alpha u + f_\beta v + f_z w) AB\, d\alpha\, d\beta \tag{15.156}$$

The work done by the boundary force and moment resultants (W_b) can be expressed as

$$W_b = \int_\alpha (\overline{N}_{22}v + \overline{N}_{21}u + \overline{Q}_{23}w + \overline{M}_{22}\theta_2 + \overline{M}_{21}\theta_1) A\, d\alpha$$

$$+ \int_\beta (\overline{N}_{11}u + \overline{N}_{12}v + \overline{Q}_{13}w + \overline{M}_{11}\theta_1 + \overline{M}_{12}\theta_2) B\, d\beta \tag{15.157}$$

where a bar over a symbol denotes a prescribed or specified quantity. Thus, the total work done (W) by the external forces is given by

$$W = W_d + W_b \tag{15.158}$$

15.7   EQUATIONS OF MOTION FROM HAMILTON'S PRINCIPLE

The generalized Hamilton's principle can be stated as

$$\delta \int_{t_1}^{t_2} L \, dt = \delta \int_{t_1}^{t_2} (T - \pi + W) \, dt = 0 \tag{15.159}$$

where L denotes the Lagrangian, and the kinetic energy T, strain energy π, and work done by external forces, W, are given by Eqs. (15.155), (15.152), and (15.158), respectively. Equation (15.159) is rewritten as

$$\int_{t_1}^{t_2} (\delta T - \delta \pi + \delta W) \, dt = 0 \tag{15.160}$$

For convenience, integrals of the variations of T, π, and W are evaluated individually as described next.

15.7.1   Variation of Kinetic Energy

From Eq. (15.155), $\int_{t_1}^{t_2} \delta T \, dt$ can be written as

$$
\int_{t_1}^{t_2} \delta T \, dt = \rho h \int_{t_1}^{t_2} \int_{\alpha} \int_{\beta} \Big[\dot{u} \, \delta \dot{u} + \dot{v} \, \delta \dot{v} + \dot{w} \, \delta \dot{w}
$$
$$
+ \frac{h^2}{12} (\dot{\theta}_1 \, \delta \dot{\theta}_1 + \dot{\theta}_2 \, \delta \dot{\theta}_2) \Big] AB \, d\alpha \, d\beta \, dt \tag{15.161}
$$

The integration of individual terms on the right-hand side of Eq. (15.161) can be carried by parts. For example, the first term can be evaluated as

$$
\int_{t_1}^{t_2} \int_{\alpha} \int_{\beta} \dot{u} \delta \dot{u} \, d\alpha \, d\beta \, dt = \int_{\alpha} \int_{\beta} d\alpha \, d\beta \int_{t_1}^{t_2} \frac{\partial u}{\partial t} \frac{\partial (\delta u)}{\partial t} \, dt
$$
$$
= \int_{\alpha} \int_{\beta} d\alpha \, d\beta \left(- \int_{t_1}^{t_2} \frac{\partial^2 u}{\partial t^2} \delta u \, dt + \frac{\partial u}{\partial t} \delta u \Big|_{t_1}^{t_2} \right) \tag{15.162}
$$

Since the displacement component, u, is specified at the initial and final times (t_1 and t_2), its variation is zero at t_1 and t_2, and hence Eq. (15.162) becomes

$$
\int_{t_1}^{t_2} \int_{\alpha} \int_{\beta} \dot{u} \delta \dot{u} \, d\alpha \, d\beta \, dt = - \int_{t_1}^{t_2} \int_{\alpha} \int_{\beta} \ddot{u} \, \delta u \, d\alpha \, d\beta \, dt \tag{15.163}
$$

Thus, Eq. (15.161) can be expressed as

$$
\int_{t_1}^{t_2} \delta T \, dt = -\rho h \int_{t_1}^{t_2} \int_{\alpha} \int_{\beta} \Big[\ddot{u} \, \delta u + \ddot{v} \, \delta v + \ddot{w} \, \delta w
$$
$$
+ \frac{h^2}{12} (\ddot{\theta}_1 \, \delta \theta_1 + \ddot{\theta}_2 \, \delta \theta_2) \Big] AB \, d\alpha \, d\beta \, dt \tag{15.164}
$$

Note that the terms involving $\ddot{\theta}_1$ and $\ddot{\theta}_2$ denote the effect of rotatory inertia. By neglecting the effect of rotatory inertia (as in the case of Euler–Bernoulli or thin beam theory), Eq. (15.164) can be expressed as

$$\int_{t_1}^{t_2} \delta T \, dt = -\rho h \int_{t_1}^{t_2} \int_\alpha \int_\beta (\ddot{u}\,\delta u + \ddot{v}\,\delta v + \ddot{w}\,\delta w) AB \, d\alpha \, d\beta \, dt \qquad (15.165)$$

15.7.2   Variation of Strain Energy

From Eq. (15.152), $\int_{t_1}^{t_2} \delta\pi \, dt$ can be written as

$$
\begin{aligned}
\int_{t_1}^{t_2} \delta\pi \, dt = \int_{t_1}^{t_2} \int_\alpha \int_\beta &\left[N_{11} B \frac{\partial(\delta u)}{\partial\alpha} + M_{11} B \frac{\partial(\delta\theta_1)}{\partial\alpha} \right. \\
&+ N_{11} \frac{\partial A}{\partial\beta} \delta v + M_{11} \frac{\partial A}{\partial\beta} \delta\theta_2 + N_{11} \frac{AB}{R_\alpha} \delta w \\
&+ N_{22} A \frac{\partial(\delta v)}{\partial\beta} + M_{22} A \frac{\partial(\delta\theta_2)}{\partial\beta} + N_{22} \frac{\partial B}{\partial\alpha} \delta u \\
&+ M_{22} \frac{\partial B}{\partial\alpha} \delta\theta_1 + N_{22} \frac{AB}{R_\beta} \delta w \\
&+ N_{12} A \frac{\partial(\delta u)}{\partial\beta} - N_{12} \frac{\partial A}{\partial\beta} \delta u + M_{12} A \frac{\partial(\delta\theta_1)}{\partial\beta} \\
&- M_{12} \frac{\partial A}{\partial\beta} \delta\theta_1 + N_{12} B \frac{\partial(\delta v)}{\partial\alpha} - N_{12} \frac{\partial B}{\partial\alpha} \delta v \\
&+ M_{12} B \frac{\partial(\delta\theta_2)}{\partial\alpha} - M_{12} \frac{\partial B}{\partial\alpha} \delta\theta_2 + Q_{23} AB \delta\theta_2 \\
&- Q_{23} \frac{AB}{R_\beta} \delta v + Q_{23} A \frac{\partial(\delta w)}{\partial\beta} + Q_{13} AB \delta\theta_1 \\
&\left. - Q_{13} \frac{AB}{R_\alpha} \delta u + Q_{13} B \frac{\partial(\delta w)}{\partial\alpha} \right] d\alpha \, d\beta \, dt \qquad (15.166)
\end{aligned}
$$

The terms involving partial derivatives of variations, δu, δv, δw, $\delta\theta_1$, and $\delta\theta_2$, in Eq. (15.166) can be evaluated using integration by parts. For example, the first term can be evaluated as

$$\int_{t_1}^{t_2} \int_\alpha \int_\beta N_{11} B \frac{\partial(\delta u)}{\partial\alpha} \, d\alpha \, d\beta \, dt$$

$$= \int_{t_1}^{t_2} \left[-\int_\alpha \int_\beta \frac{\partial}{\partial\alpha}(N_{11}B)\,\delta u \, d\alpha \, d\beta + \int_\beta N_{11} B \delta u \, d\beta \right] dt \qquad (15.167)$$

where the second term on the right-hand side of Eq. (15.167) denotes the contribution to the boundary condition. Thus, Eq. (15.166) can be expressed as

$$
\int_{t_1}^{t_2} \delta\pi \, dt = \int_{t_1}^{t_2} \int_\alpha \int_\beta \left[-\frac{\partial}{\partial\alpha}(N_{11}B)\,\delta u - \frac{\partial}{\partial\alpha}(M_{11}B)\delta\theta_1 \right.
$$

$$
+ N_{11} \frac{\partial A}{\partial\beta} \delta v + M_{11} \frac{\partial A}{\partial\beta} \delta\theta_2 + N_{11} \frac{AB}{R_\alpha} \delta w
$$

$$-\frac{\partial}{\partial\beta}(N_{22}A)\,\delta v - \frac{\partial}{\partial\beta}(M_{22}A)\,\delta\theta_2 + N_{22}\frac{\partial B}{\partial\alpha}\,\delta u$$

$$+ M_{22}\frac{\partial B}{\partial\alpha}\,\delta\theta_1 + N_{22}\frac{AB}{R_\beta}\,\delta w - \frac{\partial}{\partial\beta}(N_{12}A)\,\delta u$$

$$- N_{12}\frac{\partial A}{\partial\beta}\,\delta u - \frac{\partial}{\partial\beta}(M_{12}A)\,\delta\theta_1 - M_{12}\frac{\partial A}{\partial\beta}\,\delta\theta_1$$

$$- \frac{\partial}{\partial\alpha}(N_{12}B)\,\delta v - N_{12}\frac{\partial B}{\partial\alpha}\,\delta v - \frac{\partial}{\partial\alpha}(M_{12}B)\,\delta\theta_2$$

$$- M_{12}\frac{\partial B}{\partial\alpha}\,\delta\theta_2 + Q_{23}AB\,\delta\theta_2 - Q_{23}\frac{AB}{R_\beta}\,\delta v - \frac{\partial}{\partial\beta}(Q_{23}A)\,\delta w$$

$$+ Q_{13}AB\,\delta\theta_1 - Q_{13}\frac{AB}{R_\alpha}\,\delta u - \frac{\partial}{\partial\alpha}(Q_{13}B)\,\delta w\Bigg]\,d\alpha\,d\beta\,dt$$

$$+ \int_{t_1}^{t_2}\int_\alpha (N_{22}A\,\delta v + M_{22}A\,\delta\theta_2 + N_{12}A\,\delta u + M_{12}A\,\delta\theta_1 + Q_{23}A\,\delta w)\,d\alpha\,dt$$

$$+ \int_{t_1}^{t_2}\int_\beta (N_{11}B\,\delta u + M_{11}B\,\delta\theta_1 + N_{12}B\,\delta v + M_{12}B\,\delta\theta_2 + Q_{13}B\,\delta w)\,d\beta\,dt$$

$$(15.168)$$

15.7.3   Variation of Work Done by External Forces

From Eq. (15.158), $\int_{t_1}^{t_2}\delta W\,dt$ can be expressed as

$$\int_{t_1}^{t_2}\delta W\,dt = \int_{t_1}^{t_2}(\delta W_d + \delta W_b)\,dt = \int_{t_1}^{t_2}\Bigg[\int_\alpha\int_\beta (f_\alpha\,\delta u + f_\beta\,\delta v + f_z\,\delta w)\,AB\,d\alpha\,d\beta$$

$$+ \int_\alpha (\overline{N}_{22}\,\delta v + \overline{N}_{21}\,\delta u + \overline{Q}_{23}\,\delta w + \overline{M}_{22}\,\delta\theta_2 + \overline{M}_{21}\,\delta\theta_1)A\,d\alpha$$

$$+ \int_\beta (\overline{N}_{11}\,\delta u + \overline{N}_{12}\,\delta v + \overline{Q}_{13}\,\delta w + \overline{M}_{11}\,\delta\theta_1 + \overline{M}_{12}\,\delta\theta_2)B\,d\beta\Bigg]\,dt$$

$$(15.169)$$

15.7.4   Equations of Motion

Finally, by substituting Eqs. (15.165), (15.168) and (15.169) into Eq. (15.160), Hamilton's principle can be expressed as follows:

$$\int_{t_1}^{t_2}\int_\alpha\int_\beta\Bigg\{\Bigg[\frac{\partial}{\partial\alpha}(N_{11}B) + \frac{\partial}{\partial\beta}(N_{21}A) + N_{12}\frac{\partial A}{\partial\beta}$$

$$- N_{22}\frac{\partial B}{\partial\alpha} + Q_{13}\frac{AB}{R_\alpha} + (f_\alpha - \rho h\ddot{u})AB\Bigg]\,\delta u$$

$$+ \Bigg[\frac{\partial(N_{12}B)}{\partial\alpha} + \frac{\partial(N_{22}A)}{\partial\beta} + N_{21}\frac{\partial B}{\partial\alpha} - N_{11}\frac{\partial A}{\partial\beta} + Q_{23}\frac{AB}{R_\beta} + (f_\beta - \rho h\ddot{v})AB\Bigg]\,\delta v$$

$$+ \left[\frac{\partial(Q_{13}B)}{\partial\alpha} + \frac{\partial(Q_{23}A)}{\partial\beta} - \frac{N_{11}}{R_\alpha}AB - \frac{N_{22}AB}{R_\beta} + (f_z - \rho h \ddot{w})AB \right] \delta w$$

$$+ \left[\frac{\partial}{\partial\alpha}(M_{11}B) + \frac{\partial}{\partial\beta}(M_{21}A) - M_{22}\frac{\partial B}{\partial\alpha} + M_{12}\frac{\partial A}{\partial\beta} - Q_{13}AB \right] \delta\theta_1$$

$$+ \left[\frac{\partial}{\partial\alpha}(M_{12}B) + \frac{\partial}{\partial\beta}(M_{22}A) - M_{11}\frac{\partial A}{\partial\beta} + M_{21}\frac{\partial B}{\partial\alpha} - Q_{23}AB \right] \delta\theta_2 \Bigg\} \, d\alpha \, d\beta \, dt$$

$$+ \int_{t_1}^{t_2} \int_\alpha [(\overline{N}_{21} - N_{21})\,\delta u + (\overline{N}_{22} - N_{22})\,\delta v + (\overline{Q}_{23} - Q_{23})\,\delta w$$

$$+ (\overline{M}_{22} - M_{22})\,\delta\theta_2 + (\overline{M}_{21} - M_{21})\,\delta\theta_1]A \, d\alpha \, dt$$

$$+ \int_{t_1}^{t_2} \int_\beta [(\overline{N}_{11} - N_{11})\delta u + (\overline{N}_{12} - N_{12})\,\delta v + (\overline{Q}_{13} - Q_{13})\,\delta w$$

$$+ (\overline{M}_{11} - M_{11})\,\delta\theta_1 + (\overline{M}_{12} - M_{12})\,\delta\theta_2]B \, d\beta \, dt = 0 \tag{15.170}$$

To satisfy Eq. (15.170), the terms involving the triple and double integrals must be set equal to zero individually. By setting the term involving the triple integral equal to zero, we can obtain the equations of motion of the shell. When the terms involving the double integrals are set equal to zero individually, we can derive the boundary conditions of the shell. First we set the triple integral term equal to zero.

$$\int_{t_1}^{t_2} \int_\alpha \int_\beta \left\{ \left[\frac{\partial}{\partial\alpha}(N_{11}B) + \frac{\partial}{\partial\beta}(N_{21}A) + N_{12}\frac{\partial A}{\partial\beta} \right. \right.$$

$$\left. -N_{22}\frac{\partial B}{\partial\alpha} + Q_{13}\frac{AB}{R_\alpha} + (f_\alpha - \rho h \ddot{u})AB \right] \delta u$$

$$+ \left[\frac{\partial(N_{12}B)}{\partial\alpha} + \frac{\partial(N_{22}A)}{\partial\beta} + N_{21}\frac{\partial B}{\partial\alpha} - N_{11}\frac{\partial A}{\partial\beta} + Q_{23}\frac{AB}{R_\beta} + (f_\beta - \rho h \ddot{v})AB \right] \delta v$$

$$+ \left[\frac{\partial(Q_{13}B)}{\partial\alpha} + \frac{\partial(Q_{23}A)}{\partial\beta} - \frac{N_{11}}{R_\alpha}AB - \frac{N_{22}AB}{R_\beta} + (f_z - \rho h \ddot{w})AB \right] \delta w$$

$$+ \left[\frac{\partial}{\partial\alpha}(M_{11}B) + \frac{\partial}{\partial\beta}(M_{21}A) - M_{22}\frac{\partial B}{\partial\alpha} + M_{12}\frac{\partial A}{\partial\beta} - Q_{13}AB \right] \delta\theta_1$$

$$+ \left[\frac{\partial}{\partial\alpha}(M_{12}B) + \frac{\partial}{\partial\beta}(M_{22}A) - M_{11}\frac{\partial A}{\partial\beta} + M_{21}\frac{\partial B}{\partial\alpha} \right.$$

$$\left. -Q_{23}AB \right] \delta\theta_2 \Bigg\} \, d\alpha \, d\beta \, dt = 0 \tag{15.171}$$

In Eq. (15.171), the variations of displacements, δu, δv, δw, $\delta\theta_1$, and $\delta\theta_2$, are arbitrary and hence their coefficients must be equal to zero individually. This yields the following equations, also known as *Love's equations*:

$$-\frac{\partial(N_{11}B)}{\partial\alpha} - \frac{\partial(N_{21}A)}{\partial\beta} - N_{12}\frac{\partial A}{\partial\beta} + N_{22}\frac{\partial B}{\partial\alpha} - AB\frac{Q_{13}}{R_\alpha} + AB\rho h \ddot{u} = AB f_\alpha$$

$$\tag{15.172}$$

$$-\frac{\partial(N_{12}B)}{\partial\alpha} - \frac{\partial(N_{22}A)}{\partial\beta} - N_{21}\frac{\partial B}{\partial\alpha} + N_{11}\frac{\partial A}{\partial\beta} - AB\frac{Q_{23}}{R_\beta} + AB\rho h\ddot{v} = ABf_\beta \quad (15.173)$$

$$-\frac{\partial(Q_{13}B)}{\partial\alpha} - \frac{\partial(Q_{23}A)}{\partial\beta} + AB\left(\frac{N_{11}}{R_\alpha} + \frac{N_{22}}{R_\beta}\right) + AB\ddot{w} = ABf_z \quad (15.174)$$

$$\frac{\partial(M_{11}B)}{\partial\alpha} + \frac{\partial(M_{21}A)}{\partial\beta} + M_{12}\frac{\partial A}{\partial\beta} - M_{22}\frac{\partial B}{\partial\alpha} - Q_{13}AB = 0 \quad (15.175)$$

$$\frac{\partial(M_{12}B)}{\partial\alpha} + \frac{\partial(M_{22}A)}{\partial\beta} + M_{12}\frac{\partial B}{\partial\alpha} - M_{11}\frac{\partial A}{\partial\beta} - Q_{23}AB = 0 \quad (15.176)$$

Here Eqs. (15.172)–(15.174) denote the equations of motion of the shell for motions in the α, β, and z directions, respectively, and Eqs. (15.175) and (15.176) indicate how the transverse shear force resultants Q_{13} and Q_{23} are related to the various moment resultants.

15.7.5   Boundary Conditions

Next, each of the terms involving the double integral is set equal to zero in Eq. (15.170) to obtain

$$\int_{t_1}^{t_2}\int_\alpha [(\overline{N}_{21} - N_{21})\,\delta u + (\overline{N}_{22} - N_{22})\,\delta v + (\overline{Q}_{23} - Q_{23})\,\delta w$$

$$+ (\overline{M}_{21} - M_{21})\,\delta\theta_1 + (\overline{M}_{22} - M_{22})\,\delta\theta_2]A\,d\alpha\,dt = 0 \quad (15.177)$$

$$\int_{t_1}^{t_2}\int_\beta [(\overline{N}_{11} - N_{11})\,\delta u + (\overline{N}_{12} - N_{12})\,\delta v + (\overline{Q}_{13} - Q_{13})\,\delta w$$

$$+ (\overline{M}_{11} - M_{11})\,\delta\theta_1 + (\overline{M}_{12} - M_{12})\,\delta\theta_2]B\,d\beta\,dt = 0 \quad (15.178)$$

It appears that Eqs. (15.177) and (15.178) will be satisfied only if either the variation of the displacement component ($\delta u, \delta v, \delta w, \delta\theta_1,$ or $\delta\theta_2$) or its coefficient is zero in each of these equations:

$$\left.\begin{array}{l} \overline{N}_{21} - N_{21} = 0 \quad \text{or} \quad \delta u = 0 \\ \overline{N}_{22} - N_{22} = 0 \quad \text{or} \quad \delta v = 0 \\ \overline{Q}_{23} - Q_{23} = 0 \quad \text{or} \quad \delta w = 0 \\ \overline{M}_{21} - M_{21} = 0 \quad \text{or} \quad \delta\theta_1 = 0 \\ \overline{M}_{22} - M_{22} = 0 \quad \text{or} \quad \delta\theta_2 = 0 \end{array}\right\} \text{at } \beta = \overline{\beta} = \text{constant} \quad (15.179)$$

$$\left.\begin{array}{l} \overline{N}_{11} - N_{11} = 0 \quad \text{or} \quad \delta u = 0 \\ \overline{N}_{12} - N_{12} = 0 \quad \text{or} \quad \delta v = 0 \\ \overline{Q}_{13} - Q_{13} = 0 \quad \text{or} \quad \delta w = 0 \\ \overline{M}_{11} - M_{11} = 0 \quad \text{or} \quad \delta\theta_1 = 0 \\ \overline{M}_{12} - M_{12} = 0 \quad \text{or} \quad \delta\theta_2 = 0 \end{array}\right\} \text{at } \alpha = \overline{\alpha} = \text{constant} \quad (15.180)$$

Equations (15.179) and (15.180) indicate that there are 10 boundary conditions for the problem. However, from the force resultant/moment resultant–stress relations, stress–strain relations, and strain-displacement relations, we can find that the three equations of motion, Eqs. (15.172)–(15.174), together involve partial derivatives of order 8 involving spatial variables. This indicates that we can have only four boundary conditions on any edge. Before developing the actual boundary conditions, we assume that the boundaries coincide with the coordinate curves, that is, the α and β coordinates.

Let a boundary of the shell be defined by the line $\beta = \overline{\beta} = $ constant. The five force resultants (including the moment resultants) acting on this boundary are given by N_{21}, N_{22}, Q_{23}, M_{21}, and M_{22}. The deformation on this boundary is characterized by the five displacements (including the slopes), u, v, w, θ_1, and θ_2, which correspond to N_{21}, N_{22}, Q_{23}, M_{21}, and M_{22}, respectively. Equations (15.179) indicate that either one of the five force resultants or its corresponding displacement must be prescribed on the boundary, $\beta = \overline{\beta} = $ constant. However, this is not true because the five quantities indicated (either force resultants or the displacements) are not independent. For example, the slope θ_2 is related to the displacements w and v in order to preserve the normal to the middle surface after deformation to satisfy Kirchhoff's hypothesis (Love's fourth assumption). Thus, the number of independent displacements (and hence the corresponding generalized forces) will only be four. Hence, only four boundary conditions need to be prescribed on each edge of the shell. To identify the four boundary conditions for the edge, $\beta = \overline{\beta} = $ constant, we rewrite Eq. (15.177) by expressing θ_1 in terms of u and w using Eq. (15.93) as

$$
\int_{t_1}^{t_2} \int_{\alpha} \{ (\overline{N}_{21} - N_{21}) \, \delta u + (\overline{N}_{22} - N_{22}) \, \delta v + (\overline{Q}_{23} - Q_{23}) \, \delta w
$$
$$
+ (\overline{M}_{21} - M_{21}) \left[\frac{\delta u}{R_\alpha} - \frac{1}{A} \frac{\partial}{\partial \alpha} (\delta w) \right] + (\overline{M}_{22} - M_{22}) \, \delta \theta_2 \} \, A \, d\alpha \, dt = 0 \quad (15.181)
$$

The term involving $\partial(\delta w)/\partial \alpha$ can be integrated by parts as

$$
\int_{t_1}^{t_2} \int_{\alpha} (\overline{M}_{21} - M_{21}) \frac{\partial(\delta w)}{\partial \alpha} \, d\alpha \, dt
$$
$$
= \int_{t_1}^{t_2} \left[(\overline{M}_{21} - M_{21}) \, \delta w |_\alpha - \int_\alpha \frac{\partial}{\partial \alpha} (\overline{M}_{21} - M_{21}) \, \delta w \, d\alpha \right] dt \quad (15.182)
$$

where the first term on the right-hand side is equal to zero since $M_{21} = \overline{M}_{21}$ along the edge on which α varies. Using the resulting equation (15.182) in Eq. (15.181) and collecting the coefficients of the variations $\delta u, \delta v, \delta w$, and $\delta \theta_2$, we obtain

$$
\int_{t_1}^{t_2} \int_{\alpha} \left\{ \left[\left(\overline{N}_{21} + \frac{\overline{M}_{21}}{R_\alpha} \right) - \left(N_{21} + \frac{M_{21}}{R_\alpha} \right) \right] \delta u + (\overline{N}_{22} - N_{22}) \, \delta v \right.
$$
$$
+ \left[\left(\overline{Q}_{23} + \frac{1}{A} \frac{\partial \overline{M}_{21}}{\partial \alpha} \right) - \left(Q_{23} + \frac{1}{A} \frac{\partial M_{21}}{\partial \alpha} \right) \right] \delta w
$$
$$
\left. + (\overline{M}_{22} - M_{22}) \, \delta \theta_2 \right\} A \, d\alpha \, dt = 0 \quad (15.183)
$$

A similar procedure can be used with Eq. (15.178) by expressing θ_2 in terms of v and w using Eq. (15.94) to obtain

$$\int_{t_1}^{t_2} \int_\beta \left\{ (\overline{N}_{11} - N_{11}) \, \delta u + \left[\left(\overline{N}_{12} + \frac{\overline{M}_{12}}{R_\beta} \right) - \left(N_{12} + \frac{M_{12}}{R_\beta} \right) \right] \delta v \right.$$

$$+ \left[\left(\overline{Q}_{13} + \frac{1}{B} \frac{\partial \overline{M}_{12}}{\partial \beta} \right) - \left(Q_{13} + \frac{1}{B} \frac{\partial M_{12}}{\partial \beta} \right) \right] \delta w$$

$$\left. + (\overline{M}_{11} - M_{11}) \, \delta\theta_1 \right\} B \, d\beta \, dt = 0 \tag{15.184}$$

Defining the effective inplane shear force resultants F_{12} and F_{21} as

$$F_{12} = N_{12} + \frac{M_{12}}{R_\beta} \tag{15.185}$$

$$F_{21} = N_{21} + \frac{M_{21}}{R_\alpha} \tag{15.186}$$

and the effective transverse shear force resultants V_{13} and V_{23} as

$$V_{13} = Q_{13} + \frac{1}{B} \frac{\partial M_{12}}{\partial \beta} \tag{15.187}$$

$$V_{23} = Q_{23} + \frac{1}{A} \frac{\partial M_{21}}{\partial \alpha} \tag{15.188}$$

Eqs. (15.183) and (15.184) can be expressed as

$$\int_{t_1}^{t_2} \int_\alpha [(\overline{F}_{21} - F_{21}) \, \delta u + (\overline{N}_{22} - N_{22}) \, \delta v$$

$$+ (\overline{V}_{23} - V_{23}) \, \delta w + (\overline{M}_{22} - M_{22}) \, \delta\theta_2] A \, d\alpha \, dt = 0 \tag{15.189}$$

$$\int_{t_1}^{t_2} \int_\beta [(\overline{N}_{11} - N_{11}) \, \delta u + (\overline{F}_{12} - F_{12}) \, \delta v + (\overline{V}_{13} - V_{13}) \, \delta w$$

$$+ (\overline{M}_{11} - M_{11}) \, \delta\theta_1] B \, d\beta \, dt = 0 \tag{15.190}$$

Equations (15.189) and (15.190) will be satisfied only when each of the displacement variations or its coefficient will be zero. Noting that the variation in a displacement will be zero only when the displacement is prescribed, the boundary conditions can be stated as follows:

$$\left. \begin{array}{lll} F_{21} = \overline{F}_{21} & \text{or} & u = \overline{u} \\ N_{22} = \overline{N}_{22} & \text{or} & v = \overline{v} \\ V_{23} = \overline{V}_{23} & \text{or} & w = \overline{w} \\ M_{22} = \overline{M}_{22} & \text{or} & \theta_2 = \overline{\theta}_2 \end{array} \right\} \quad \text{at} \quad \beta = \overline{\beta} = \text{constant} \tag{15.191}$$

$$\left. \begin{array}{lll} N_{11} = \overline{N}_{11} & \text{or} & u = \overline{u} \\ F_{12} = \overline{F}_{12} & \text{or} & v = \overline{v} \\ V_{13} = \overline{V}_{13} & \text{or} & w = \overline{w} \\ M_{11} = \overline{M}_{11} & \text{or} & \theta_1 = \overline{\theta}_1 \end{array} \right\} \quad \text{at} \quad \alpha = \overline{\alpha} = \text{constant} \tag{15.192}$$

where $\bar{u}, \bar{v}, \bar{w}, \bar{\theta}_1$, and $\bar{\theta}_2$ denote the prescribed values of u, v, w, θ_1, and θ_2, respectively. Equations (15.191) and (15.192) represent the four independent boundary conditions to be satisfied in solving the equations of motion of shells. The boundary conditions for some of the commonly encountered edges can be stated as follows:

1. For the edge defined by $\beta = \bar{\beta} = $ constant:

(a) Clamped or fixed edge:

$$u = 0 \qquad v = 0, \qquad w = 0, \qquad \theta_2 = 0 \qquad (15.193)$$

(b) Hinged or simply supported edge with the support free to move in the normal direction:

$$u = 0, \qquad v = 0, \qquad M_{22} = 0, \qquad V_{23} = 0 \qquad (15.194)$$

(c) Hinged or simply supported edge with no motion permitted in the normal direction:

$$u = 0, \qquad v = 0, \qquad w = 0, \qquad M_{22} = 0 \qquad (15.195)$$

(d) Free edge:

$$N_{22} = 0, \qquad F_{21} = 0, \qquad V_{23} = 0, \qquad M_{22} = 0 \qquad (15.196)$$

2. For the edge defined by $\alpha = \bar{\alpha} = $ constant:

(a) Clamped or fixed edge:

$$u = 0, \qquad v = 0, \qquad w = 0, \qquad \theta_1 = 0 \qquad (15.197)$$

(b) Hinged or simply supported edge with the support free to move in the normal direction:

$$u = 0, \qquad v = 0, \qquad M_{11} = 0, \qquad V_{13} = 0 \qquad (15.198)$$

(c) Hinged or simply supported edge with no motion permitted in the normal direction:

$$u = 0, \qquad v = 0, \qquad w = 0, \qquad M_{11} = 0 \qquad (15.199)$$

(d) Free edge:

$$N_{11} = 0, \qquad F_{12} = 0, \qquad V_{13} = 0, \qquad M_{11} = 0 \qquad (15.200)$$

15.8 CIRCULAR CYLINDRICAL SHELLS

For a cylindrical shell, x, θ, and z are used as the independent coordinates, as shown in Fig. 15.10. The components of displacement parallel to the x, θ and z directions are denoted as u, v and w, respectively. The radius of the shell is assumed to be R. The parameters of the shell are given by (see Example 15.1):

$$\alpha = x, \qquad A = 1, \qquad \beta = \theta, \qquad B = R \qquad (15.201)$$

The radius of curvatures of the x and θ lines are given by

$$R_\alpha = R_x = \infty, \qquad R_\beta = R_\theta = R \qquad (15.202)$$

15.8.1   Equations of Motion

The equations of motion, Eqs. (15.172)–(15.174) reduce to

$$\frac{\partial N_{x,x}}{\partial x} + \frac{1}{R}\frac{\partial N_{\theta x}}{\partial \theta} + f_x = \rho h \ddot{u} \tag{15.203}$$

$$\frac{\partial N_{x\theta}}{\partial x} + \frac{1}{R}\frac{\partial N_{\theta\theta}}{\partial \theta} + \frac{Q_{\theta z}}{R} + f_\theta = \rho h \ddot{v} \tag{15.204}$$

$$\frac{\partial Q_{xz}}{\partial x} + \frac{1}{R}\frac{\partial Q_{\theta z}}{\partial \theta} - \frac{N_{\theta\theta}}{R} + f_z = \rho h \ddot{w} \tag{15.205}$$

The relations between the transverse shear force resultants and the moment resultants, Eqs. (15.175) and (15.176), reduce to

$$Q_{xz} = \frac{\partial M_{xx}}{\partial x} + \frac{1}{R}\frac{\partial M_{\theta x}}{\partial \theta} \tag{15.206}$$

$$Q_{\theta z} = \frac{\partial M_{x\theta}}{\partial x} + \frac{1}{R}\frac{\partial M_{\theta\theta}}{\partial \theta} \tag{15.207}$$

Using Eqs. (E15.7.5)–(E15.7.7), Q_{xz} and $Q_{\theta z}$ [Eqs. (15.206) and (15.207)] can be expressed as

$$Q_{xz} = D\left(\frac{\partial^2 \theta_x}{\partial x^2} + \frac{1-\nu}{2R^2}\frac{\partial^2 \theta_x}{\partial \theta^2} + \frac{1+\nu}{2R}\frac{\partial^2 \theta_\theta}{\partial x\,\partial\theta}\right) \tag{15.208}$$

$$Q_{\theta z} = D\left(\frac{1-\nu}{2}\frac{\partial^2 \theta_\theta}{\partial x^2} + \frac{1}{R^2}\frac{\partial^2 \theta_\theta}{\partial \theta^2} + \frac{1+\nu}{2R}\frac{\partial^2 \theta_x}{\partial x\,\partial\theta}\right) \tag{15.209}$$

Substituting Eqs. (E15.4.1) and (E15.4.2) into Eqs. (15.208) and (15.209) leads to

$$Q_{xz} = D\left(-\frac{\partial^3 w}{\partial x^3} + \frac{1+\nu}{2R^2}\frac{\partial^2 v}{\partial x\,\partial\theta} - \frac{1}{R^2}\frac{\partial^3 w}{\partial x\,\partial\theta^2}\right) \tag{15.210}$$

$$Q_{\theta z} = D\left(\frac{1-\nu}{2R}\frac{\partial^2 v}{\partial x^2} + \frac{1}{R^3}\frac{\partial^2 v}{\partial \theta^2} - \frac{1}{R^3}\frac{\partial^3 w}{\partial \theta^3} - \frac{1}{R}\frac{\partial^3 w}{\partial x^2\,\partial\theta}\right) \tag{15.211}$$

Finally, using Eqs. (E15.7.1)–(E15.7.3) and (E15.7.5)–(E15.7.7) and Eqs. (15.210) and (15.211), the equations of motion, Eqs. (15.203)–(15.205), can be expressed in terms of the displacement components u, v, and w as

$$C\left(\frac{\partial^2 u}{\partial x^2} + \frac{1-\nu}{2R^2}\frac{\partial^2 u}{\partial \theta^2} + \frac{\nu}{R}\frac{\partial w}{\partial x} + \frac{1+\nu}{2R}\frac{\partial^2 v}{\partial x\,\partial\theta}\right) + f_x = \rho h \ddot{u} \tag{15.212}$$

$$C\left(\frac{1-\nu}{2}\frac{\partial^2 v}{\partial x^2} + \frac{1}{R^2}\frac{\partial^2 v}{\partial \theta^2} + \frac{1}{R^2}\frac{\partial w}{\partial \theta} + \frac{1+\nu}{2R}\frac{\partial^2 u}{\partial x\,\partial\theta}\right)$$
$$+ D\left(\frac{1-\nu}{2R^2}\frac{\partial^2 v}{\partial x^2} + \frac{1}{R^4}\frac{\partial^2 v}{\partial \theta^2} - \frac{1}{R^4}\frac{\partial^3 w}{\partial \theta^3} - \frac{1}{R^2}\frac{\partial^3 w}{\partial x^2\,\partial\theta}\right) + f_\theta = \rho h \ddot{v} \tag{15.213}$$

$$D\left(-\frac{\partial^4 w}{\partial x^4} + \frac{1}{R^2}\frac{\partial^3 v}{\partial x^2\,\partial\theta} - \frac{2}{R^2}\frac{\partial^4 w}{\partial x^2\,\partial\theta^2} - \frac{1}{R^4}\frac{\partial^4 w}{\partial \theta^4} + \frac{1}{R^4}\frac{\partial^3 v}{\partial \theta^3}\right)$$
$$- C\left(\frac{1}{R^2}\frac{\partial v}{\partial \theta} + \frac{w}{R^2} + \frac{\nu}{R}\frac{\partial u}{\partial x}\right) + f_z = \rho h \ddot{w} \tag{15.214}$$

15.8.2   Donnell–Mushtari–Vlasov Theory

The equations of motion of a cylindrical shell, Eqs. (15.212)–(15.214), can be simplified using the Donnell–Mushtari–Vlasov (DMV) theory. The following assumptions are made in the DMV theory in the context of vibration of cylindrical shells:

1. The contribution of in-plane displacements u and v to the bending strain parameters k_{11}, k_{22}, and k_{12} [in Eqs. (E15.4.6)–(E15.4.8)] is negligible.
2. The effect of the shear term $(1/R)Q_{\theta z}$ in the equation of motion corresponding to v [Eq. (15.204)] is negligible. This is equivalent to neglecting the term involving D in Eq. (15.213).

The equations of motion corresponding to the DMV theory can be expressed as follows:

$$\frac{\partial^2 u}{\partial x^2} + \frac{1-\nu}{2R^2}\frac{\partial^2 u}{\partial \theta^2} + \frac{\nu}{R}\frac{\partial w}{\partial x} + \frac{1+\nu}{2R}\frac{\partial^2 v}{\partial x \partial \theta} = \frac{(1-\nu^2)\rho}{E}\frac{\partial^2 u}{\partial t^2} \tag{15.215}$$

$$\frac{1-\nu}{2}\frac{\partial^2 v}{\partial x^2} + \frac{1}{R^2}\frac{\partial^2 v}{\partial \theta^2} + \frac{1}{R^2}\frac{\partial w}{\partial \theta} + \frac{1+\nu}{2R}\frac{\partial^2 u}{\partial x \partial \theta} = \frac{(1-\nu^2)\rho}{E}\frac{\partial^2 v}{\partial t^2} \tag{15.216}$$

$$-\left(\frac{\nu}{R}\frac{\partial u}{\partial x} + \frac{1}{R^2}\frac{\partial v}{\partial \theta} + \frac{w}{R^2}\right)$$

$$-\frac{h^2}{12}\left(\frac{\partial^4 w}{\partial x^4} + \frac{2}{R^2}\frac{\partial^4 w}{\partial x^2 \partial \theta^2} + \frac{1}{R^4}\frac{\partial^4 w}{\partial \theta^4}\right) = \frac{(1-\nu^2)\rho}{E}\frac{\partial^2 w}{\partial t^2} \tag{15.217}$$

15.8.3   Natural Frequencies of Vibration According to DMV Theory

Let the circular cylindrical shell, of length l, be simply supported on its edges as shown in Fig. 15.15(a). The boundary conditions of the shell, simply supported at $x = 0$ and $x = l$, can be stated as follows:

$$v(0, \theta, t) = 0 \tag{15.218}$$

$$w(0, \theta, t) = 0 \tag{15.219}$$

$$N_{xx}(0, \theta, t) = C\left(\frac{\partial u}{\partial x} + \frac{\nu}{R}\frac{\partial v}{\partial \theta} + \frac{\nu}{R}w\right)(0, \theta, t) = 0 \tag{15.220}$$

$$M_{xx}(0, \theta, t) = D\left(-\frac{\partial^2 w}{\partial x^2} + \frac{\nu}{R^2}\frac{\partial v}{\partial \theta} - \frac{\nu}{R^2}\frac{\partial^2 w}{\partial \theta^2}\right)(0, \theta, t) = 0 \tag{15.221}$$

$$v(l, \theta, t) = 0 \tag{15.222}$$

$$w(l, \theta, t) = 0 \tag{15.223}$$

$$N_{xx}(l, \theta, t) = C\left(\frac{\partial u}{\partial x} + \frac{\nu}{R}\frac{\partial v}{\partial \theta} + \frac{\nu}{R}w\right)(l, \theta, t) = 0 \tag{15.224}$$

$$M_{xx}(l, \theta, t) = D\left(-\frac{\partial^2 w}{\partial x^2} + \frac{\nu}{R^2}\frac{\partial v}{\partial \theta} - \frac{\nu}{R^2}\frac{\partial^2 w}{\partial \theta^2}\right)(l, \theta, t) = 0 \tag{15.225}$$

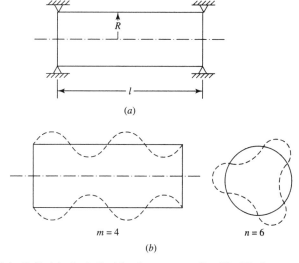

Figure 15.15 (*a*) Cylindrical shell (simply supported); (*b*) Displacement pattern during vibration.

The solution of the equations of motion corresponding to DMV theory, Eqs. (15.215)–(15.217), is assumed to be in the following form:

$$u(x, \theta) = \sum_m \sum_n A_{mn} \cos \frac{m \pi x}{l} \cos n\theta \cos \omega t \tag{15.226}$$

$$v(x, \theta) = \sum_m \sum_n B_{mn} \sin \frac{m \pi x}{l} \sin n\theta \cos \omega t \tag{15.227}$$

$$w(x, \theta) = \sum_m \sum_n C_{mn} \sin \frac{m \pi x}{l} \cos n\theta \cos \omega t \tag{15.228}$$

where m is the number of half-waves of displacement in the length of shell, n is the number of half-waves of displacement in the circumference of the shell [see Fig. 15.15(*b*)], and A_{mn}, B_{mn}, and C_{mn} are constants. Note that the assumed solution [Eqs. (15.226)–(15.228)] satisfies the boundary conditions [Eqs. (15.218)–(15.225)] as well as the periodicity condition in the circumferential (θ) direction. Substituting Eqs. (15.226)–(15.228) into the equations of motion, Eqs. (15.215)–(15.217), yields the following equations:

$$(-\lambda^2 - a_1 n^2 + \Omega)A_{mn} + (a_2 \lambda n)B_{mn} + (\nu \lambda)C_{mn} = 0 \tag{15.229}$$

$$(a_2 \lambda n)A_{mn} + (-a_1 \lambda^2 - n^2 + \Omega)B_{mn} + (-n)C_{mn} = 0 \tag{15.230}$$

$$(\nu \lambda)A_{mn} + (-n)B_{mn} + (-1 - \lambda^4 \mu - 2\lambda^2 n^2 \mu - n^4 \mu + \Omega)C_{mn} = 0 \tag{15.231}$$

where

$$\lambda = \frac{m \pi R}{l} \tag{15.232}$$

$$\mu = \frac{h^2}{12R^2} \tag{15.233}$$

$$\Omega = \frac{(1 - v^2)R^2\rho}{E}\omega^2 \tag{15.234}$$

$$a_1 = \frac{1 - v}{2} \tag{15.235}$$

$$a_2 = \frac{1 + v}{2} \tag{15.236}$$

For a nontrivial solution of the constants A_{mn}, B_{mn}, and C_{mn}, the determinant of their coefficient matrix in Eqs. (15.229)–(15.231) must be zero. This leads to

$$\begin{vmatrix} -\lambda^2 - a_1 n^2 + \Omega & a_2\lambda n & v\lambda \\ a_2\lambda n & -a_1\lambda^2 - n^2 + \Omega & -n \\ v\lambda & -n & -1 - \lambda^4\mu - 2\lambda^2 n^2\mu - n^4\mu + \Omega) \end{vmatrix} = 0 \tag{15.237}$$

The expansion of Eq. (15.237) leads to the frequency equation

$$\Omega^3 + b_1\Omega^2 + b_2\Omega + b_3 = 0 \tag{15.238}$$

where

$$b_1 = -\lambda^2(1 + a_1 + 2n^2\mu) - n^2(1 + a_1) - \lambda^4\mu - n^4\mu - 1 \tag{15.239}$$

$$\begin{aligned} b_2 = {} & \lambda^6\mu(1 + a_1) + \lambda^4(a_1 + 3n^2\mu + 3a_1 n^2\mu) \\ & + \lambda^2(1 + n^2 + a_1^2 n^2 - a_2^2 n^2 - v^2 + a_1 + 3a_1 n^4\mu + 3n^4\mu) \\ & + n^6\mu(1 + a_1) + n^4 a_1 + n^2 a_1 \end{aligned} \tag{15.240}$$

$$\begin{aligned} b_3 = {} & -\lambda^8 a_1\mu - \lambda^6 n^2\mu(1 + 2a_1 - a_2^2 + a_1^2) \\ & - \lambda^4(a_1 + 2a_1 n^4\mu + 2a_1^2 n^4\mu + 2n^4\mu - 2a_2^2 n^4\mu - a_2 v^2) \\ & - \lambda^2[n^6\mu(1 + a_1^2 + 2a_1 - a_2^2) + n^2(-a_2^2 + a_1^2 + 2a_2 v - v^2)] - n^8 a_1\mu \end{aligned} \tag{15.241}$$

It can be seen that Eq. (15.238) is a cubic equation in Ω and that the coefficients b_1, b_2, and b_3 depend on the material properties (E, v and ρ), geometry (R, l, and h), and the vibration mode (m and n). It can be shown that all the roots of Eq. (15.238) are always real and the positive square roots of Ω can be used to find the natural frequencies of the shell. Thus, there will be three values of the natural frequency (ω) for any specific combination of values of m and n. For any specific natural frequency of vibration (ω), the ratio between the amplitudes, (e.g., B_{mn}/A_{mn} and C_{mn}/A_{mn}) can be computed from any two of the equations among Eqs. (15.229)–(15.231). These ratios provide the ratios between the amplitudes of the longitudinal, tangential, and normal displacements of the shell for any specific natural frequency ω.

15.8.4 Natural Frequencies of Transverse Vibration According to DMV Theory

It has been observed that, of the three natural frequencies given by Eq. (15.238), the frequency corresponding to the transverse mode of vibration will have the smallest

Table 15.1 Natural Frequencies of Transverse Vibration of a Cylindrical Shell ($h = 0.1$in.)

Mode shape		Natural frequency (rad/s)	
m	n	From Eq. (15.238)	From Eq. (15.243)
1	1	5, 187.4575	6, 567.4463
1	2	2, 298.6262	2, 494.8096
1	3	1, 295.3844	1, 360.8662
2	1	11, 446.168	13, 839.495
2	2	6, 632.3569	7, 386.4487
2	3	3, 984.2207	4, 267.3979
3	1	15, 160.400	16, 958.043
3	2	10, 466.346	11, 527.878
3	3	7, 072.3271	7, 612.5889

value. Hence the square and cubic terms of Ω will be comparatively small and can be neglected in Eq. (15.238). This gives the approximate value of Ω as

$$\Omega \approx -\frac{b_3}{b_2} \tag{15.242}$$

or

$$\omega^2 = \frac{E\Omega}{(1 - \nu^2)R^2\rho} \approx -\frac{Eb_3}{(1 - \nu^2)R^2\rho b_2} \tag{15.243}$$

Example 15.10 Find the natural frequencies of vibration of a circular cylindrical shell simply supported at $x = 0$ and $x = l$, using DMV theory for the following data: $E = 30 \times 10^6$ psi, $\nu = 0.3$, $\rho = 7.324 \times 10^{-4}$ lb-sec^2/in^4, $R = 10$ in, $l = 40$ in, and $h = 0.1$ in

SOLUTION   The natural frequencies computed using DMV theory [Eqs. (15.238) and (15.243)] are given in Table 15.1. Only the smallest natural frequency given by Eq. (15.238), corresponding to the transverse mode, is given. It can be seen that the natural frequencies given by Eq. (15.243) are slightly larger than the corresponding values given by the DMV theory.

15.8.5 Natural Frequencies of Vibration According to Love's Theory

Consider a cylindrical shell of radius R, length l, and simply supported at both ends, $x = 0$ and $x = l$. The boundary conditions are given by Eqs. (13.218)–(13.225). In the absence of external forces ($f_x = f_\theta = f_z = 0$), the equations of motion, Eqs. (13.212)–(13.214), reduce to

$$C\left(\frac{\partial^2 u}{\partial x^2} + \frac{1 - \nu}{2R^2}\frac{\partial^2 u}{\partial \theta^2} + \frac{\nu}{R}\frac{\partial w}{\partial x} + \frac{1 + \nu}{2R}\frac{\partial^2 v}{\partial x \partial \theta}\right) = \rho h \ddot{u} \tag{15.244}$$

$$C\left(\frac{1 - \nu}{2}\frac{\partial^2 v}{\partial x^2} + \frac{1}{R^2}\frac{\partial^2 v}{\partial \theta^2} + \frac{1}{R^2}\frac{\partial w}{\partial \theta} + \frac{1 + \nu}{2R}\frac{\partial^2 u}{\partial x \partial \theta}\right)$$
$$+ \frac{D}{R^2}\left(\frac{1 - \nu}{2}\frac{\partial^2 v}{\partial x^2} + \frac{1}{R^2}\frac{\partial^2 v}{\partial \theta^2} - \frac{1}{R^2}\frac{\partial^3 w}{\partial \theta^3} - \frac{\partial^3 w}{\partial x^2 \partial \theta}\right) = \rho h \ddot{v} \tag{15.245}$$

$$C\left(-\frac{1}{R^2}\frac{\partial v}{\partial \theta} - \frac{w}{R^2} - \frac{v}{R}\frac{\partial u}{\partial x}\right)$$

$$+ \frac{D}{R^2}\left(-R^2\frac{\partial^4 w}{\partial x^4} + \frac{\partial^3 v}{\partial x^2 \partial \theta} - 2\frac{\partial^4 w}{\partial x^2 \partial \theta^2} - \frac{1}{R^2}\frac{\partial^4 w}{\partial \theta^4} + \frac{1}{R^2}\frac{\partial^3 v}{\partial \theta^3}\right) = \rho h \ddot{w} \quad (15.246)$$

The solution is assumed to be harmonic during free vibration as

$$u(x, \theta, t) = U(x, \theta)e^{i\omega t} \qquad (15.247)$$

$$v(x, \theta, t) = V(x, \theta)e^{i\omega t} \qquad (15.248)$$

$$w(x, \theta, t) = W(x, \theta)e^{i\omega t} \qquad (15.249)$$

where ω is the frequency of vibration. In view of Eqs. (15.247)–(15.249), the boundary conditions, Eqs. (15.218)–(15.225), can be stated as

$$V(0, \theta) = 0 \qquad (15.250)$$

$$W(0, \theta) = 0 \qquad (15.251)$$

$$\left(-\frac{\partial^2 W}{\partial x^2} + \frac{v}{R^2}\frac{\partial V}{\partial \theta} - \frac{v}{R^2}\frac{\partial^2 W}{\partial \theta^2}\right)(0, \theta) = 0 \qquad (15.252)$$

$$\left(\frac{\partial U}{\partial x} + \frac{v}{R}\frac{\partial V}{\partial \theta} + \frac{v}{R}W\right)(0, \theta) = 0 \qquad (15.253)$$

$$V(l, \theta) = 0 \qquad (15.254)$$

$$W(l, \theta) = 0 \qquad (15.255)$$

$$\left(-\frac{\partial^2 W}{\partial x^2} + \frac{v}{R^2}\frac{\partial V}{\partial \theta} - \frac{v}{R^2}\frac{\partial^2 W}{\partial \theta^2}\right)(l, \theta) = 0 \qquad (15.256)$$

$$\left(\frac{\partial U}{\partial x} + \frac{v}{R}\frac{\partial V}{\partial \theta} + \frac{v}{R}W\right)(l, \theta) = 0 \qquad (15.257)$$

Substitution of Eqs. (15.247)–(15.249) into Eqs. (15.244)–(15.246) leads to

$$C\left(\frac{\partial^2 U}{\partial x^2} + \frac{1-v}{2R^2}\frac{\partial^2 U}{\partial \theta^2} + \frac{v}{R}\frac{\partial W}{\partial x} + \frac{1+v}{2R}\frac{\partial^2 V}{\partial x \partial \theta}\right) + \rho h\omega^2 U = 0 \qquad (15.258)$$

$$C\left(\frac{1-v}{2}\frac{\partial^2 V}{\partial x^2} + \frac{1}{R^2}\frac{\partial^2 V}{\partial \theta^2} + \frac{1}{R^2}\frac{\partial W}{\partial \theta} + \frac{1+v}{2R}\frac{\partial^2 U}{\partial x \partial \theta}\right)$$

$$+ \frac{D}{R^2}\left(\frac{1-v}{2}\frac{\partial^2 V}{\partial x^2} + \frac{1}{R^2}\frac{\partial^2 V}{\partial \theta^2} - \frac{1}{R^2}\frac{\partial^3 W}{\partial \theta^3} - \frac{\partial^3 W}{\partial x^2 \partial \theta}\right) + \rho h\omega^2 V = 0 \quad (15.259)$$

$$C\left(-\frac{1}{R^2}\frac{\partial V}{\partial \theta} - \frac{W}{R^2} - \frac{v}{R}\frac{\partial U}{\partial x}\right) + \frac{D}{R^2}\left(-R^2\frac{\partial^4 W}{\partial x^4} + \frac{\partial^3 V}{\partial x^2 \partial \theta} - 2\frac{\partial^4 W}{\partial x^2 \partial \theta^2}\right.$$

$$\left. - \frac{1}{R^2}\frac{\partial^4 W}{\partial \theta^4} + \frac{1}{R^2}\frac{\partial^3 V}{\partial \theta^3}\right) + \rho h\omega^2 W = 0 \qquad (15.260)$$

The following solutions are assumed to satisfy the boundary conditions of Eqs. (15.250)–(15.257):

$$U(x, \theta) = C_1 \cos \frac{m\pi x}{l} \cos n(\theta - \phi_0) \tag{15.261}$$

$$V(x, \theta) = C_2 \sin \frac{m\pi x}{l} \sin n(\theta - \phi_0) \tag{15.262}$$

$$W(x, \theta) = C_3 \sin \frac{m\pi x}{l} \cos n (\theta - \phi_0) \tag{15.263}$$

where C_1, C_2, and C_3 are constants and ϕ_0 is the phase angle. Using Eqs. (15.261)–(15.263), Eqs. (15.258)–(15.260) can be expressed as

$$C_1 \left[-C \left(\frac{m\pi}{l} \right)^2 - C \frac{1 - \nu}{2R^2} (n^2) + \rho h \omega^2 \right]$$

$$+ C_2 \left(C \frac{1 + \nu}{2R} \frac{m\pi}{l} n \right) + C_3 \left(C \frac{\nu}{R} \frac{m\pi}{l} \right) = 0 \tag{15.264}$$

$$C_1 \left(C \frac{1 + \nu}{2} \frac{m\pi}{l} \frac{n}{R} \right)$$

$$+ C_2 \left[-C \frac{1 - \nu}{2} \left(\frac{m\pi}{l} \right)^2 - C \frac{n^2}{R^2} - \frac{D}{R^2} \frac{1 - \nu}{2} \left(\frac{m\pi}{l} \right)^2 - \frac{D}{R^2} \frac{n^2}{R^2} + \rho h \omega^2 \right]$$

$$+ C_3 \left[-C \frac{n}{R^2} - \frac{D}{R^2} \frac{n^3}{R^2} - \frac{D}{R^2} \left(\frac{m\pi}{l} \right)^2 n \right] = 0 \tag{15.265}$$

$$C_1 \left(C \frac{\nu}{R} \frac{m\pi}{l} \right) + C_2 \left[-C \frac{n}{R^2} - \frac{D}{R^2} \left(\frac{m\pi}{l} \right)^2 n - \frac{D}{R^2} \frac{n^3}{R^2} \right]$$

$$+ C_3 \left[-\frac{C}{R^2} - D \left(\frac{m\pi}{l} \right)^4 - \frac{D}{R^2} (2) \left(\frac{m\pi}{l} \right)^2 n^2 \right.$$

$$\left. - \frac{D}{R^2} \frac{n^4}{R^2} + \rho h \omega^2 \right] = 0 \tag{15.266}$$

Equations (15.264)–(15.266) can be written in matrix form as

$$\begin{bmatrix} \rho h \omega^2 - d_{11} & d_{12} & d_{13} \\ d_{21} & \rho h \omega^2 - d_{22} & d_{23} \\ d_{31} & d_{32} & \rho h \omega^2 - d_{33} \end{bmatrix} \begin{Bmatrix} C_1 \\ C_2 \\ C_3 \end{Bmatrix} = \begin{Bmatrix} 0 \\ 0 \\ 0 \end{Bmatrix} \tag{15.267}$$

where

$$d_{11} = C \left(\frac{m\pi}{l} \right)^2 + C \frac{1 - \nu}{2} \left(\frac{n}{R} \right)^2 \tag{15.268}$$

$$d_{12} = d_{21} = C \frac{1 + \nu}{2} \frac{m\pi}{l} \frac{n}{R} \tag{15.269}$$

$$d_{13} = d_{31} = C \frac{\nu}{R} \frac{m\pi}{l} \tag{15.270}$$

$$d_{22} = C \frac{1 - \nu}{2} \left(\frac{m\pi}{l} \right)^2 + C \left(\frac{n}{R} \right)^2 + \frac{D}{R^2} \frac{1 - \nu}{2} \left(\frac{m\pi}{l} \right)^2 + \frac{D}{R^2} \left(\frac{n}{R} \right)^2 \tag{15.271}$$

$$d_{23} = d_{32} = -\frac{Cn}{R^2} - \frac{Dn}{R^2}\left(\frac{m\pi}{l}\right)^2 - \frac{Dn}{R^2}\left(\frac{n}{R}\right)^2 \tag{15.272}$$

$$d_{33} = \frac{C}{R^2} + D\left(\frac{m\pi}{l}\right)^4 + 2D\left(\frac{m\pi}{l}\right)^2\left(\frac{n}{R}\right)^2 + D\left(\frac{n}{R}\right)^4 \tag{15.273}$$

For a nontrivial solution of C_1, C_2, and C_3, the determinant of their coefficient matrix in Eq. (15.267) must be equal to zero. This leads to the frequency equation:

$$\begin{vmatrix} \rho h\omega^2 - d_{11} & d_{12} & d_{13} \\ d_{12} & \rho h\omega^2 - d_{22} & d_{23} \\ d_{13} & d_{23} & \rho h\omega^2 - d_{33} \end{vmatrix} = 0 \tag{15.274}$$

or

$$\omega^6 + p_1\omega^4 + p_2\omega^2 + p_3 = 0 \tag{15.275}$$

where

$$p_1 = \frac{1}{\rho h}(d_{11} + d_{22} + d_{33}) \tag{15.276}$$

$$p_2 = \frac{1}{\rho^2 h^2}(d_{11}d_{22} + d_{22}d_{33} + d_{11}d_{33} - d_{12}^2 - d_{23}^2 - d_{13}^2) \tag{15.277}$$

$$p_3 = \frac{1}{\rho^3 h^3}(d_{11}d_{23}^2 + d_{22}d_{13}^2 + d_{33}d_{12}^2 + 2d_{12}d_{23}d_{13} - d_{11}d_{22}d_{33}) \tag{15.278}$$

It can be seen that the frequency equation, Eq. (15.275) is a cubic equation in ω^2 (as in the case of DMV theory).

Example 15.11   Find the natural frequencies of transverse vibration of a circular cylindrical shell simply supported at $x = 0$ and $x = l$ using Love's theory for the following data: $E = 30 \times 10^6$ psi, $v = 0.3$, $\rho = 7.324 \times 10^{-4}$ lb-sec^2/in^4, $R = 10$ in., $l = 40$ in., and $h = 0.1$ in.

SOLUTION   The natural frequencies corresponding to transverse vibration of the shell are given by the smallest roots of Eq. (15.275). The values of ω obtained from Eq. (15.275) for different combinations of m and n are given in Table 15.2. These results can be compared with the values given by the DMV theory in Example 15.10.

Table 15.2   Natural Frequencies of Transverse Vibration of a Cylindrical Shell ($h = 0.1$ in.)

Mode shape		Natural frequency	Mode shape		Natural frequency
m	n	from Eq. (15.275) (rad/s)	m	n	from Eq. (15.275) (rad/s)
1	1	4,958.2515	2	3	4,105.0059
1	2	2,375.8223	3	1	12,989.213
1	3	1,321.9526	3	2	10,086.031
2	1	9,878.3359	3	3	7,181.8760
2	2	6,595.9062			

15.9 EQUATIONS OF MOTION OF CONICAL AND SPHERICAL SHELLS

In this section, the general equations of motion of a shell are specialized for circular conical and spherical shells. The equations of motion are obtained by substituting the proper values of α, β, A, B, R_α, and R_β into Love's equations of motion, Eqs. (15.172)–(15.176).

15.9.1 Circular Conical Shells

Using $\alpha = x$, $\beta = \theta$, $A = 1$, $B = x \sin \alpha_0$, $R_\alpha = R_x = \infty$, and $R_\beta = R_\theta = x \tan \alpha_0$ (see Example 15.5) and N_{xx}, $N_{x\theta}$, $N_{\theta x}$, $N_{\theta\theta}$, M_{xx}, $M_{x\theta}$, $M_{\theta x}$, $M_{\theta\theta}$, Q_{xz}, and $Q_{\theta z}$ for N_{11}, N_{12}, N_{21}, N_{22}, M_{11}, M_{12}, M_{21}, M_{22}, Q_{13}, and Q_{23}, respectively, in Eqs. (15.172)–(15.176), we obtain the equations governing the vibration of a conical shell as follows:

$$\frac{\partial N_{xx}}{\partial x} + \frac{1}{x \sin \alpha_0} \frac{\partial N_{\theta x}}{\partial \theta} + \frac{1}{x}(N_{xx} - N_{\theta\theta}) + f_x = \rho h \frac{\partial^2 u}{\partial t^2} \tag{15.279}$$

$$\frac{\partial N_{x\theta}}{\partial x} + \frac{2}{x} N_{\theta x} + \frac{1}{x \sin \alpha_0} \frac{\partial N_{\theta\theta}}{\partial \theta} + \frac{1}{x \tan \alpha_0} Q_{\theta z} + f_\theta = \rho h \frac{\partial^2 v}{\partial t^2} \tag{15.280}$$

$$\frac{\partial Q_{xz}}{\partial x} + \frac{1}{x} Q_{xz} + \frac{1}{x \sin \alpha_0} \frac{\partial Q_{\theta z}}{\partial \theta} - \frac{1}{x \tan \alpha_0} N_{\theta\theta} + f_z = \rho h \frac{\partial^2 w}{\partial t^2} \tag{15.281}$$

$$\frac{\partial M_{xx}}{\partial x} + \frac{M_{xx}}{x} + \frac{1}{x \sin \alpha_0} \frac{\partial M_{\theta x}}{\partial \theta} - \frac{M_{\theta\theta}}{x} - Q_{xz} = 0 \tag{15.282}$$

$$\frac{\partial M_{x\theta}}{\partial x} + \frac{2}{x} M_{\theta x} + \frac{1}{x \sin \alpha_0} \frac{\partial M_{\theta\theta}}{\partial \theta} - Q_{\theta z} = 0 \tag{15.283}$$

where u, v, and w denote the components of displacement along the x, θ, and z directions, respectively. By using the expressions of N_{xx}, $N_{x\theta} = N_{\theta x}$, $N_{\theta\theta}$, M_{xx}, $M_{x\theta} = M_{\theta x}$, and $M_{\theta\theta}$ in terms of the displacements u, v, and w given by Eqs. (E15.8.1)–(E15.8.8) the equations of motion, Eqs. (15.279)–(15.281) can be expressed in terms of u, v, and w (see Problem 15.1).

15.9.2 Spherical Shells

Using $\alpha = \phi$, $\beta = \theta$, $A = R$, $B = R \sin \phi$, $R_\alpha = R_\phi = R$, and $R_\beta = R_\theta = R$ (see Example 15.6) and $N_{\phi\phi}$, $N_{\phi\theta}$, $N_{\theta\phi}$, $N_{\theta\theta}$, $M_{\phi\phi}$, $M_{\phi\theta}$, $M_{\theta\phi}$, $M_{\theta\theta}$, $Q_{\phi z}$, and $Q_{\theta z}$ for N_{11}, N_{12}, N_{21}, N_{22}, M_{11}, M_{12}, M_{21}, M_{22}, Q_{13}, and Q_{23}, respectively, in Eqs. (15.172)–(15.176), we obtain the equations governing the vibration of a spherical shell as follows:

$$\frac{\partial}{\partial \phi}(N_{\phi\phi} \sin \phi) + \frac{\partial N_{\theta\phi}}{\partial \theta} - N_{\theta\theta} \cos \phi + Q_{\phi z} \sin \phi$$

$$+ R f_\phi \sin \phi = R \sin \phi \, \rho h \frac{\partial^2 u}{\partial t^2} \tag{15.284}$$

$$\frac{\partial}{\partial \phi}(N_{\phi\theta} \sin \phi) + \frac{\partial N_{\theta\theta}}{\partial \theta} + N_{\theta\phi} \cos \phi + Q_{\theta z} \sin \phi$$

$$+ R f_\theta \sin \phi = R \sin \phi \, \rho h \frac{\partial^2 v}{\partial t^2} \tag{15.285}$$

$$\frac{\partial}{\partial \phi} (Q_{\phi z} \ \sin \phi) + \frac{\partial Q_{\theta z}}{\partial \theta} - (N_{\phi\phi} + N_{\theta\theta}) \ \sin \phi$$

$$+ Rf_z \ \sin \phi = R \ \sin \phi \rho h \ \frac{\partial^2 w}{\partial t^2} \tag{15.286}$$

$$\frac{\partial}{\partial \phi} (M_{\phi\phi} \ \sin \phi) + \frac{\partial M_{\theta\phi}}{\partial \theta} - M_{\theta\theta} \cos \phi - Q_{\phi z} \ R \ \sin \phi = 0 \tag{15.287}$$

$$\frac{\partial}{\partial \phi} (M_{\theta\phi} \ \sin \phi) + \frac{\partial M_{\theta\theta}}{\partial \theta} + M_{\theta\phi} \cos \phi - Q_{\theta z} \ R \ \sin \phi = 0 \tag{15.288}$$

where u, v, and w denote the components of displacement along the ϕ, θ, and z directions, respectively. By using the expressions of $N_{\phi\phi}$, $N_{\phi\theta} = N_{\theta\phi}$, $N_{\theta\theta}$, $M_{\phi\phi}$, $M_{\phi\theta} = M_{\theta\phi}$, and $M_{\theta\theta}$ in terms of the displacements u, v, and w given by Eqs. (E15.9.1)–(E15.9.8) the equations of motion, Eqs. (15.284)–(15.286), can be expressed in terms of u, v, and w (see Problem 15.2).

15.10   EFFECT OF ROTARY INERTIA AND SHEAR DEFORMATION

In the vibration of thick shells, the effects of shear deformation and rotary inertia play an important role. In this section, the equations of motion of a shell are derived by including the effects of shear deformation and rotary inertia using an approach similar to that of Timoshenko beams and Mindlin plates [5,14,15]. The effects of shear deformation and rotary inertia become increasingly important as the thickness of the shell (or the value of h/R_α or h/R_β) increases. These effects can be significant even for thin shells in higher modes. Thus, the effects of shear deformation and rotary inertia can be significant when dealing with short wavelengths, especially those that have the same order as the thickness of the shell or less. As in the case of beams and plates, the effect of shear deformation is incorporated through generalization of the strain–displacement relations, and the effect of rotary inertia is incorporated through the basic equations of motion (in the dynamic equilibrium approach).

15.10.1   Displacement Components

The displacement components of an arbitrary point in the shell are assumed to be given by

$$\overline{u}(\alpha, \beta, z) = u(\alpha, \beta) + z\psi_1(\alpha, \beta) \tag{15.289}$$

$$\overline{v}(\alpha, \beta, z) = v(\alpha, \beta) + z\psi_2(\alpha, \beta) \tag{15.290}$$

$$\overline{w}(\alpha, \beta, z) = w(\alpha, \beta) \tag{15.291}$$

where ψ_1 and ψ_2 denote the total angular rotations, including the angular rotations due to shear, of the normal to the middle surface about the β and α axes, respectively. Note that ψ_1 and ψ_2 are different from θ_1 and θ_2 used in Eqs. (15.85) and (15.86). θ_1 and θ_2 were used to denote the rotations, with no shear deformation, of the normal to the middle surface, about the β and α axes, during deformation. Equations (15.289)–(15.291) assume that straight lines normal to the middle surface remain straight lines after deformation, even if they no longer are normal. This assumption is consistent with the Timoshenko beam and Mindlin plate theories.

15.10.2   Strain–Displacement Relations

When shear deformation is considered, the fourth assumption in Love's theory (Kirch-hoff's hypothesis), which states that normals remain normal, will not be considered. This implies that Eq. (15.82) is to be dropped and the shear strains $\varepsilon_{\alpha z}$ and $\varepsilon_{\beta z}$ will no longer be zero. Thus, ψ_1 and ψ_2 will no longer be related to u, v, and w [as in Eqs. (15.93) and (15.94)]; they need to be treated as independent variables in addition to u, v, and w. At the same time, Love's third assumption, which states that the transverse normal stress σ_{zz} is negligible, is included. Thus, the normal stress σ_{zz} is ignored and the transverse shear strains ε_{13} and ε_{23} are retained in the analysis. The strain–displacement relations in curvilinear coordinates are given by Eqs. (15.75)–(15.80).

By substituting Eqs. (15.289)–(15.291) into Eqs. (15.75)–(15.80), and imposing the assumption of Eq. (15.81), the strain–displacement relations can be expressed as

$$\varepsilon_{11} = \varepsilon_{11}^0 + z k_{11} \tag{15.292}$$

$$\varepsilon_{22} = \varepsilon_{22}^0 + z k_{22} \tag{15.293}$$

$$\varepsilon_{12} = \varepsilon_{12}^0 + z k_{12} \tag{15.294}$$

$$\varepsilon_{23} = \psi_2 - \frac{v}{R_\beta} + \frac{1}{B}\frac{\partial w}{\partial \beta} \tag{15.295}$$

$$\varepsilon_{13} = \psi_1 - \frac{u}{R_\alpha} + \frac{1}{A}\frac{\partial w}{\partial \alpha} \tag{15.296}$$

$$\varepsilon_{33} = 0 \tag{15.297}$$

where ε_{11}^0, ε_{22}^0, and ε_{12}^0 denote membrane strains (which are independent of z) and k_{11}, k_{22}, and k_{12} indicate curvatures given by

$$\varepsilon_{11}^0 = \frac{1}{A}\frac{\partial u}{\partial \alpha} + \frac{v}{AB}\frac{\partial A}{\partial \beta} + \frac{w}{R_\alpha} \tag{15.298}$$

$$\varepsilon_{22}^0 = \frac{1}{B}\frac{\partial v}{\partial \beta} + \frac{u}{AB}\frac{\partial B}{\partial \alpha} + \frac{w}{R_\beta} \tag{15.299}$$

$$\varepsilon_{12}^0 = \frac{B}{A}\frac{\partial}{\partial \alpha}\left(\frac{v}{B}\right) + \frac{A}{B}\frac{\partial}{\partial \beta}\left(\frac{u}{A}\right) \tag{15.300}$$

$$k_{11} = \frac{1}{A}\frac{\partial \psi_1}{\partial \alpha} + \frac{\psi_2}{AB}\frac{\partial A}{\partial \beta} \tag{15.301}$$

$$k_{22} = \frac{1}{B}\frac{\partial \psi_2}{\partial \beta} + \frac{\psi_1}{AB}\frac{\partial B}{\partial \alpha} \tag{15.302}$$

$$k_{12} = \frac{B}{A}\frac{\partial}{\partial \alpha}\left(\frac{\psi_2}{B}\right) + \frac{A}{B}\frac{\partial}{\partial \beta}\left(\frac{\psi_1}{A}\right) \tag{15.303}$$

Notes

1. The shear strains ε_{23} and ε_{13} denote the shear strains in the middle surface of the shell.

2. Equations (15.295) and (15.296) are set equal to zero in Love's shell theory, and ψ_1 and ψ_2 (θ_1 and θ_2 in the notation of Love's shell theory) are expressed in terms of u, v, and w.

15.10.3   Stress–Strain Relations

The stress–strain relations are given by Eqs. (15.122)–(15.124) and (15.113)–(15.115). The transverse shear stresses given by Eqs. (15.114) and (15.115) are valid only at the middle surface of the shell. These stresses must diminish to zero at the free surface of the shell (at $z = \pm h/2$). The average values of the transverse shear stresses, denoted $\overline{\sigma}_{13}$ and $\overline{\sigma}_{23}$, are defined as

$$\overline{\sigma}_{13} = k\sigma_{13} = kG\varepsilon_{13} \tag{15.304}$$

$$\overline{\sigma}_{23} = k\sigma_{23} = kG\varepsilon_{23} \tag{15.305}$$

where k is a constant less than unity, called the *shear coefficient*.

15.10.4   Force and Moment Resultants

Force and moment resultants due to σ_{11}, σ_{22}, and σ_{12} are given by Eqs. (15.135), (15.136), (15.138), (15.143), (15.145), and (15.146):

$$N_{11} = C(\varepsilon_{11}^0 + v\varepsilon_{22}^0) \tag{15.306}$$

$$N_{22} = C(\varepsilon_{22}^0 + v\varepsilon_{11}^0) \tag{15.307}$$

$$N_{12} = N_{21} = \frac{1-v}{2} C\varepsilon_{12}^0 \tag{15.308}$$

$$M_{11} = D(k_{11} + vk_{22}) \tag{15.309}$$

$$M_{22} = D(k_{22} + vk_{11}) \tag{15.310}$$

$$M_{12} = M_{21} = \frac{1-v}{2} Dk_{12} \tag{15.311}$$

where ε_{11}^0, ε_{22}^0, ε_{12}^0, k_{11}, k_{22}, and k_{12}, are given by Eqs. (15.298)–(15.303). The transverse shear force resultants Q_{13} and Q_{23} due to σ_{13} and σ_{23}, respectively, are defined by [see Eqs. (15.137) and (15.139)]

$$Q_{13} = \int_{-h/2}^{h/2} \sigma_{13} \, dz \tag{15.312}$$

$$Q_{23} = \int_{-h/2}^{h/2} \sigma_{23} \, dz \tag{15.313}$$

In view of Eqs. (15.304) and (15.305), Eqs. (15.312) and (15.313) can be written as

$$Q_{13} = \overline{\sigma}_{13}h = kGh\varepsilon_{13} \tag{15.314}$$

$$Q_{23} = \overline{\sigma}_{23}h = kGh\varepsilon_{23} \tag{15.315}$$

Noting that the shear strains ε_{13} and ε_{23} can be expressed in terms of u, v, w, ψ_1 and ψ_2, using Eqs. (15.295) and (15.296), Eqs. (15.314) and (15.315) can be expressed as

$$Q_{13} = kGh \left(\psi_1 - \frac{u}{R_\alpha} + \frac{1}{A} \frac{\partial w}{\partial \alpha} \right) \tag{15.316}$$

$$Q_{23} = kGh \left(\psi_2 - \frac{v}{R_\beta} + \frac{1}{B} \frac{\partial w}{\partial \beta} \right) \tag{15.317}$$

15.10.5   Equations of Motion

The generalized Hamilton's principle is used to derive the equations of motion:

$$\delta \int_{t_1}^{t_2} L \, dt = \delta \int_{t_1}^{t_2} (T - \pi + W) \, dt = 0 \tag{15.318}$$

The variation of the kinetic energy term in Eq. (15.318), including the effect of rotary inertia, is given by Eq. (15.164) by replacing θ_i by $\psi_i (i = 1, 2)$:

$$\int_{t_1}^{t_2} \delta T \, dt = -\rho h \int_{t_1}^{t_2} \int_\alpha \int_\beta [\ddot{u} \, \delta u + \ddot{v} \, \delta v + \ddot{w} \, \delta w$$

$$+ \frac{h^2}{12} (\ddot{\psi}_1 \, \delta \psi_1 + \ddot{\psi}_2 \, \delta \psi_2) \Big] AB \, d\alpha \, d\beta \, dt \tag{15.319}$$

The strain energy term in Eq. (15.318) can be obtained from Eq. (15.168), by replacing θ_i by $\psi_i (i = 1, 2)$ and expressing Q_{13} and Q_{23} in terms of ε_{13} and ε_{23} using Eqs. (15.314) and (15.315), as

$$\int_{t_1}^{t_2} \delta \pi \, dt = \int_{t_1}^{t_2} \int_\alpha \int_\beta \left[-\frac{\partial}{\partial \alpha} (N_{11} B) \, \delta u - \frac{\partial}{\partial \alpha} (M_{11} B) \, \delta \psi_1 \right.$$

$$+ N_{11} \frac{\partial A}{\partial \beta} \delta v + M_{11} \frac{\partial A}{\partial \beta} \delta \psi_2 + N_{11} \frac{AB}{R_\alpha} \delta w$$

$$- \frac{\partial}{\partial \beta} (N_{22} A) \, \delta v - \frac{\partial}{\partial \beta} (M_{22} A) \, \delta \psi_2 + N_{22} \frac{\partial B}{\partial \alpha} \delta u$$

$$+ M_{22} \frac{\partial B}{\partial \alpha} \delta \psi_1 + N_{22} \frac{AB}{R_\beta} \delta w - \frac{\partial}{\partial \beta} (N_{12} A) \, \delta u$$

$$- N_{12} \frac{\partial A}{\partial \beta} \delta u - \frac{\partial}{\partial \beta} (M_{12} A) \, \delta \psi_1 - M_{12} \frac{\partial A}{\partial \beta} \delta \psi_1$$

$$- \frac{\partial}{\partial \alpha} (N_{12} B) \, \delta v - N_{12} \frac{\partial B}{\partial \alpha} \delta v - \frac{\partial}{\partial \alpha} (M_{12} B) \, \delta \psi_2$$

$$- M_{12} \frac{\partial B}{\partial \alpha} \delta \psi_2 + kGh\varepsilon_{23} AB \, \delta \psi_2 - kGh\varepsilon_{23} \frac{AB}{R_\beta} \delta v$$

$$- \frac{\partial}{\partial \beta} (kGh\varepsilon_{23} A) \, \delta w + kGh\varepsilon_{13} AB \, \delta \psi_1$$

$$\left. - kGh\varepsilon_{13} \frac{AB}{R_\alpha} \delta u - \frac{\partial}{\partial \alpha} (kGh\varepsilon_{13} B) \, \delta w \right] d\alpha \, d\beta \, dt$$

$$+ \int_{t_1}^{t_2} \int_{\alpha} (N_{22} A \, \delta v + M_{22} A \delta \psi_2 + N_{21} A \, \delta u + M_{21} A \, \delta \psi_1$$

$$+ kGh\varepsilon_{23} A \, \delta w) \, d\alpha \, dt$$

$$+ \int_{t_1}^{t_2} \int_{\beta} (N_{11} B \, \delta u + M_{11} B \, \delta \psi_1 + N_{12} B \, \delta v + M_{12} B \delta \psi_2$$

$$+ kGh\varepsilon_{13} B \, \delta w) \, d\beta \, dt \tag{15.320}$$

The term related to the work done by the external forces in Eq. (15.318) can be obtained from Eq. (15.169) by replacing θ_i by $\psi_i \, (i = 1, 2)$. By substituting Eqs. (15.319), (15.320) and the modified Eq. (15.169) into Eq. (15.318), the equations of motion corresponding to the variables u, v, w, ψ_1, and ψ_2 can be identified as [14]

$$- \frac{\partial}{\partial \alpha} (N_{11} B) - \frac{\partial}{\partial \beta} (N_{21} A) - N_{12} \frac{\partial A}{\partial \beta} + N_{22} \frac{\partial B}{\partial \alpha}$$

$$- \frac{AB}{R_\alpha} \varepsilon_{13} kGh + AB\rho h \ddot{u} = AB f_\alpha \tag{15.321}$$

$$- \frac{\partial}{\partial \alpha} (N_{12} B) - \frac{\partial}{\partial \beta} (N_{22} A) - N_{21} \frac{\partial B}{\partial \alpha} + N_{11} \frac{\partial A}{\partial \beta}$$

$$- \frac{AB}{R_\beta} \varepsilon_{23} kGh + AB\rho h \ddot{v} = AB f_\beta \tag{15.322}$$

$$- kGh \frac{\partial}{\partial \alpha} (\varepsilon_{13} B) - kGh \frac{\partial}{\partial \beta} (\varepsilon_{23} A)$$

$$+ AB \left(\frac{N_{11}}{R_\alpha} + \frac{N_{22}}{R_\beta} \right) + AB\rho h \ddot{w} = AB f_z \tag{15.323}$$

$$\frac{\partial}{\partial \alpha} (M_{11} B) + \frac{\partial}{\partial \beta} (M_{21} A) + M_{12} \frac{\partial A}{\partial \beta} - M_{22} \frac{\partial B}{\partial \alpha}$$

$$- kGh\varepsilon_{13} AB - AB \frac{\rho h^3}{12} \ddot{\psi}_1 = 0 \tag{15.324}$$

$$\frac{\partial}{\partial \alpha} (M_{12} B) + \frac{\partial}{\partial \beta} (M_{22} A) + M_{21} \frac{\partial B}{\partial \alpha} - M_{11} \frac{\partial A}{\partial \beta}$$

$$- kGh\varepsilon_{23} AB - AB \frac{\rho h^3}{12} \ddot{\psi}_2 = 0 \tag{15.325}$$

15.10.6 Boundary Conditions

The boundary conditions of the shell on the edges $\alpha = \bar{\alpha} = $ constant and $\beta = \bar{\beta} = $ constant can be identified as follows:

$$\left. \begin{array}{lll} N_{22} = \overline{N}_{22} & \text{or} \quad v = \bar{v} \\ M_{22} = \overline{M}_{22} & \text{or} \quad \psi_2 = \overline{\psi}_2 \\ N_{21} = \overline{N}_{21} & \text{or} \quad u = \bar{u} \\ Q_{23} = kGh\varepsilon_{23} = \overline{Q}_{23} & \text{or} \quad w = \bar{w} \\ M_{21} = \overline{M}_{21} & \text{or} \quad \psi_1 = \overline{\psi}_1 \end{array} \right\} \quad \text{at} \quad \beta = \bar{\beta} = \text{constant} \tag{15.326}$$

$$\left.\begin{array}{ll} N_{11} = \overline{N}_{11} & \text{or} \quad u = \overline{u} \\ N_{12} = \overline{N}_{12} & \text{or} \quad v = \overline{v} \\ Q_{13} = kGh\varepsilon_{13} = \overline{Q}_{13} & \text{or} \quad w = \overline{w} \\ M_{11} = \overline{M}_{11} & \text{or} \quad \psi_1 = \overline{\psi}_1 \\ M_{12} = \overline{M}_{12} & \text{or} \quad \psi_2 = \overline{\psi}_2 \end{array}\right\} \quad \text{at} \quad \alpha = \overline{\alpha} = \text{constant} \qquad (15.327)$$

15.10.7   Vibration of Cylindrical Shells

For a cylindrical shell, x, θ, and z are used as the independent coordinates. From the fundamental form of the surface, we find that $\alpha = x$, $\beta = \theta$, $A = 1$, $B = R$, $R_\alpha = R_x = \infty$, and $R_\beta = R_\theta = R$(see Example 15.4). In this case, the equations of motion, Eqs. (15.321)–(15.325), become (using the notation ψ_x for ψ_1 and ψ_θ for ψ_2)

$$R\frac{\partial N_{xx}}{\partial x} + \frac{\partial N_{\theta x}}{\partial \theta} = R\rho h \ddot{u} - Rf_x \qquad (15.328)$$

$$R\frac{\partial N_{x\theta}}{\partial x} + \frac{\partial N_{\theta\theta}}{\partial \theta} + Q_{\theta z} = R\rho h \ddot{v} - Rf_\theta \qquad (15.329)$$

$$R\frac{\partial Q_{xz}}{\partial x} + \frac{\partial Q_{\theta z}}{\partial \theta} - N_{\theta\theta} = R\rho h \ddot{w} - Rf_z \qquad (15.330)$$

$$R\frac{\partial M_{xx}}{\partial x} + \frac{\partial M_{\theta x}}{\partial \theta} - RQ_{xz} = R\frac{\rho h^3}{12}\ddot{\psi}_x \qquad (15.331)$$

$$R\frac{\partial M_{x\theta}}{\partial x} + \frac{\partial M_{\theta\theta}}{\partial \theta} - RQ_{\theta z} = R\frac{\rho h^3}{12}\ddot{\psi}_\theta \qquad (15.332)$$

The force and moment resultants can be expressed in terms of the displacement components, from Eqs. (15.306)–(15.311), (15.316) and (15.317), as:

$$N_x = C\left[\frac{\partial u}{\partial x} + v\left(\frac{1}{R}\frac{\partial v}{\partial \theta} + \frac{w}{R}\right)\right] \qquad (15.333)$$

$$N_\theta = C\left(\frac{1}{R}\frac{\partial v}{\partial \theta} + \frac{w}{R} + v\frac{\partial u}{\partial x}\right) \qquad (15.334)$$

$$N_{x\theta} = N_{\theta x} = \frac{1-v}{2}C\left(\frac{1}{R}\frac{\partial u}{\partial \theta} + \frac{\partial v}{\partial x}\right) \qquad (15.335)$$

$$M_x = D\left(\frac{\partial \psi_x}{\partial x} + \frac{v}{R}\frac{\partial \psi_\theta}{\partial \theta}\right) \qquad (15.336)$$

$$M_\theta = D\left(\frac{1}{R}\frac{\partial \psi_\theta}{\partial \theta} + v\frac{\partial \psi_x}{\partial x}\right) \qquad (15.337)$$

$$M_{x\theta} = M_{\theta x} = \frac{1-v}{2}D\left(\frac{1}{R}\frac{\partial \psi_x}{\partial \theta} + \frac{\partial \psi_\theta}{\partial x}\right) \qquad (15.338)$$

$$Q_x = kGh\left(\frac{\partial w}{\partial x} + \psi_x\right) \qquad (15.339)$$

$$Q_\theta = kGh\left[\frac{1}{R}\frac{\partial w}{\partial \theta} - \left(\frac{v}{R} - \psi_\theta\right)\right] \qquad (15.340)$$

where C and D are given by Eqs. (15.134) and (15.144). By substituting Eqs. (15.333)–(15.340) into Eqs. (15.328)–(15.332), we obtain the equations of motion in terms of the displacement components as

$$\frac{\partial^2 u}{\partial x^2} + \frac{1-v}{2}\frac{1}{R^2}\frac{\partial^2 u}{\partial \theta^2} + \frac{1+v}{2}\frac{1}{R}\frac{\partial^2 v}{\partial x \partial \theta} + \frac{v}{R}\frac{\partial w}{\partial x} = \frac{\rho(1-v^2)}{E}\frac{\partial^2 u}{\partial t^2} \tag{15.341}$$

$$\frac{1}{R^2}\frac{\partial^2 v}{\partial \theta^2} + \frac{1-v}{2}\frac{\partial^2 v}{\partial x^2} + \frac{1+v}{2}\frac{1}{R}\frac{\partial^2 u}{\partial x \partial \theta} + \frac{1}{R^2}\frac{\partial w}{\partial \theta} + \frac{\bar{k}}{R}\left(\psi_\theta - \frac{v}{R} + \frac{1}{R}\frac{\partial w}{\partial \theta}\right)$$

$$= \frac{\rho(1-v^2)}{E}\frac{\partial^2 v}{\partial t^2} \tag{15.342}$$

$$\bar{k}\left(\nabla^2 w + \frac{\partial \psi_x}{\partial x} + \frac{1}{R}\frac{\partial \psi_\theta}{\partial \theta} - \frac{1}{R^2}\frac{\partial v}{\partial \theta}\right) - \frac{1}{R}\left(\frac{1}{R}\frac{\partial v}{\partial \theta} + \frac{w}{R} + v\frac{\partial u}{\partial x}\right)$$

$$= \frac{\rho(1-v^2)}{E}\frac{\partial^2 w}{\partial t^2} \tag{15.343}$$

$$\left(\frac{\partial^2 \psi_x}{\partial x^2} + \frac{1-v}{2R^2}\frac{\partial^2 \psi_x}{\partial \theta^2} + \frac{1+v}{2R}\frac{\partial^2 \psi_\theta}{\partial x \partial \theta}\right) - \frac{12\bar{k}}{h^2}\left(\frac{\partial w}{\partial x} + \psi_x\right)$$

$$= \frac{\rho(1-v^2)}{E}\frac{\partial^2 \psi_x}{\partial t^2} \tag{15.344}$$

$$\left(\frac{1}{R^2}\frac{\partial^2 \psi_\theta}{\partial \theta^2} + \frac{1-v}{2}\frac{\partial^2 \psi_\theta}{\partial x^2} + \frac{1+v}{2R}\frac{\partial^2 \psi_x}{\partial x \partial \theta}\right) - \frac{12\bar{k}}{h^2}\left(\frac{1}{R}\frac{\partial w}{\partial \theta} + \psi_\theta - \frac{v}{R}\right)$$

$$= \frac{\rho(1-v^2)}{E}\frac{\partial^2 \psi_\theta}{\partial t^2} \tag{15.345}$$

where

$$\bar{k} = \frac{1-v}{2}k \tag{15.346}$$

$$\nabla^2 w = \frac{\partial^2 w}{\partial x^2} + \frac{1}{R^2}\frac{\partial^2 w}{\partial \theta^2} \tag{15.347}$$

15.10.8   Natural Frequencies of Vibration of Cylindrical Shells

Consider a cylindrical shell of radius R, length l, and simply supported at both ends, $x = 0$ and $x = l$. The boundary conditions of the shell can be expressed as

$$v(0, \theta, t) = 0 \tag{15.348}$$

$$w(0, \theta, t) = 0 \tag{15.349}$$

$$M_{xx}(0, \theta, t) = D\left(\frac{\partial \psi_x}{\partial x} + \frac{v}{R}\frac{\partial \psi_\theta}{\partial \theta}\right)(0, \theta, t) = 0 \tag{15.350}$$

$$N_{xx}(0, \theta, t) = C\left[\frac{\partial u}{\partial x} + v\left(\frac{1}{R}\frac{\partial v}{\partial \theta} + \frac{w}{R}\right)\right](0, \theta, t) = 0 \tag{15.351}$$

$$\psi_\theta(0, \theta, t) = 0 \tag{15.352}$$

$$v(l, \theta, t) = 0 \tag{15.353}$$

$$w(l, \theta, t) = 0 \tag{15.354}$$

$$M_{xx}(l, \theta, t) = D \left(\frac{\partial \psi_x}{\partial x} + \frac{v}{R} \frac{\partial \psi_\theta}{\partial \theta} \right) (l, \theta, t) = 0 \tag{15.355}$$

$$N_{xx}(l, \theta, t) = C \left[\frac{\partial u}{\partial x} + v \left(\frac{1}{R} \frac{\partial v}{\partial \theta} + \frac{w}{R} \right) \right] (l, \theta, t) = 0 \tag{15.356}$$

$$\psi_\theta(l, \theta, t) = 0 \tag{15.357}$$

For free vibration, the solution is assumed to be harmonic as

$$u(x, \theta, t) = U(x, \theta) e^{i\omega t} \tag{15.358}$$

$$v(x, \theta, t) = V(x, \theta) e^{i\omega t} \tag{15.359}$$

$$w(x, \theta, t) = W(x, \theta) e^{i\omega t} \tag{15.360}$$

$$\psi_x(x, \theta, t) = \Psi_x(x, \theta) e^{i\omega t} \tag{15.361}$$

$$\psi_\theta(x, \theta, t) = \Psi_\theta(x, \theta) e^{i\omega t} \tag{15.362}$$

where ω is the frequency of vibration. In view of Eqs. (15.358)–(15.362), the boundary conditions of Eqs. (15.348)–(15.357) can be restated as

$$V(0, \theta) = 0 \tag{15.363}$$

$$W(0, \theta) = 0 \tag{15.364}$$

$$\left(\frac{\partial \Psi_x}{\partial x} + \frac{v}{R} \frac{\partial \Psi_\theta}{\partial \theta} \right) (0, \theta) = 0 \tag{15.365}$$

$$\left[\frac{\partial U}{\partial x} + v \left(\frac{1}{R} \frac{\partial V}{\partial \theta} + \frac{W}{R} \right) \right] (0, \theta) = 0 \tag{15.366}$$

$$\Psi_\theta(0, \theta) = 0 \tag{15.367}$$

$$V(l, \theta) = 0 \tag{15.368}$$

$$W(l, \theta) = 0 \tag{15.369}$$

$$\left(\frac{\partial \Psi_x}{\partial x} + \frac{v}{R} \frac{\partial \Psi_\theta}{\partial \theta} \right) (l, \theta) = 0 \tag{15.370}$$

$$\left[\frac{\partial U}{\partial x} + v \left(\frac{1}{R} \frac{\partial V}{\partial \theta} + \frac{W}{R} \right) \right] (l, \theta) = 0 \tag{15.371}$$

$$\Psi_\theta(l, \theta) = 0 \tag{15.372}$$

The following solutions are assumed to satisfy the boundary conditions of Eqs. (15.363)–(15.372):

$$U(x, \theta) = C_1 \cos \frac{m\pi x}{l} \cos n(\theta - \phi) \tag{15.373}$$

$$V(x, \theta) = C_2 \sin \frac{m\pi x}{l} \sin n(\theta - \phi) \tag{15.374}$$

$$W(x, \theta) = C_3 \sin \frac{m\pi x}{l} \cos n(\theta - \phi) \tag{15.375}$$

$$\Psi_x(x, \theta) = C_4 \cos \frac{m\pi x}{l} \cos n(\theta - \phi) \tag{15.376}$$

$$\Psi_\theta(x, \theta) = C_5 \sin \frac{m\pi x}{l} \sin n(\theta - \phi) \tag{15.377}$$

Substituting Eqs. (15.358)–(15.362) and (15.373)–(15.377) into the equations of motion, Eqs. (15.341)–(15.345), we obtain

$$
\begin{bmatrix}
d_{11} - \Omega^2 & d_{12} & d_{13} & 0 & 0 \\
d_{21} & d_{22} - \Omega^2 & d_{23} & 0 & d_{25} \\
d_{31} & d_{32} & d_{33} - \Omega^2 & d_{34} & d_{35} \\
0 & 0 & d_{43} & d_{44} - \Omega^2 & d_{45} \\
0 & d_{52} & d_{53} & d_{54} & d_{55} - \Omega^2
\end{bmatrix}
\begin{Bmatrix}
C_1 \\ C_2 \\ C_3 \\ C_4 \\ C_5
\end{Bmatrix}
=
\begin{Bmatrix}
0 \\ 0 \\ 0 \\ 0 \\ 0
\end{Bmatrix}
\tag{15.378}
$$

where

$$d_{11} = \left(\frac{m\pi}{l}\right)^2 + \frac{1-\nu}{2}\left(\frac{n}{R}\right)^2 \tag{15.379}$$

$$d_{12} = d_{21} = -\frac{1+\nu}{2}\frac{n}{R}\frac{m\pi}{l} \tag{15.380}$$

$$d_{13} = d_{31} = -\frac{\nu}{R}\frac{m\pi}{l} \tag{15.381}$$

$$d_{22} = \frac{1-\nu}{2}\left(\frac{m\pi}{l}\right)^2 + \left(\frac{n}{R}\right)^2 + \frac{\bar{k}}{R^2} \tag{15.382}$$

$$d_{23} = d_{32} = \frac{1}{R}\frac{n}{R} + \frac{n}{R}\frac{\bar{k}}{R} \tag{15.383}$$

$$d_{25} = d_{52} = -\frac{\bar{k}}{R} \tag{15.384}$$

$$d_{33} = \bar{k}\left\{\left(\frac{m\pi}{l}\right)^2 + \left(\frac{n}{R}\right)^2\right\} + \frac{1}{R^2} \tag{15.385}$$

$$d_{34} = d_{43} = \bar{k}\frac{m\pi}{l} \tag{15.386}$$

$$d_{35} = d_{53} = -\bar{k}\frac{n}{R} \tag{15.387}$$

$$d_{44} = \frac{12\bar{k}}{h^2} + \frac{1-\nu}{2}\left(\frac{n}{R}\right)^2 + \left(\frac{m\pi}{l}\right)^2 \tag{15.388}$$

$$d_{45} = d_{54} = -\frac{1-\nu}{2}\frac{n}{R}\frac{m\pi}{l} \tag{15.389}$$

$$d_{55} = \frac{12\bar{k}}{h^2} + \frac{1-\nu}{2}\left(\frac{m\pi}{l}\right)^2 + \left(\frac{n}{R}\right)^2 \tag{15.390}$$

where

$$\Omega^2 = \frac{\rho(1 - \nu^2)}{E} \omega^2 \tag{15.391}$$

and $\bar{k}$ is given by Eq. (15.346).

For a nontrivial solution of $C_1, C_2, \ldots, C_5$, the determinant of the coefficient matrix in Eq. (15.378) must be zero. This yields the frequency equation as a fifth-degree polynomial equation in Ω^2. The roots of this polynomial equation give the natural frequencies of the cylindrical shell ω^2. For every combination of m and n, there will be five distinct natural frequencies. The mode shapes of the shell can be determined by first substituting each natural frequency into the matrix equations (15.378), and then solving for any four constants among $C_1, C_2, \ldots, C_5$ in terms of the remaining constant. For example, by selecting C_5 as the independent constant, we can find the values of C_1/C_5, C_2/C_5, C_3/C_5, and C_4/C_5.

15.10.9   Axisymmetric Modes

In the particular case of axisymmetric modes ($n = 0$), the five equations of motion will be uncoupled into two sets: one consisting of three equations involving u, w, and θ_x and the other consisting of two equations involving v and θ_θ. Thus, the first set of equations leads to a cubic frequency equation and describes flexural or radial modes and the second set leads to a quadratic frequency equation and describes circumferential modes. For $n = 0$, Eqs. (15.379)–(15.390) reduce to

$$d_{11} = \left(\frac{m\pi}{l}\right)^2 \tag{15.392}$$

$$d_{12} = d_{21} = 0 \tag{15.393}$$

$$d_{13} = d_{31} = -\frac{\nu}{R}\frac{m\pi}{l} \tag{15.394}$$

$$d_{22} = \frac{1 - \nu}{2}\left(\frac{m\pi}{l}\right)^2 + \frac{\bar{k}}{R^2} \tag{15.395}$$

$$d_{23} = d_{32} = 0 \tag{15.396}$$

$$d_{25} = d_{52} = -\frac{\bar{k}}{R} \tag{15.397}$$

$$d_{33} = \bar{k}\left(\frac{m\pi}{l}\right)^2 + \frac{1}{R^2} \tag{15.398}$$

$$d_{34} = d_{43} = \bar{k}\frac{m\pi}{l} \tag{15.399}$$

$$d_{35} = d_{53} = 0 \tag{15.400}$$

$$d_{44} = \frac{12\bar{k}}{h^2} + \left(\frac{m\pi}{l}\right)^2 \tag{15.401}$$

$$d_{45} = d_{54} = 0 \tag{15.402}$$

$$d_{55} = \frac{12\bar{k}}{h^2} + \frac{1 - \nu}{2}\left(\frac{m\pi}{l}\right)^2 \tag{15.403}$$

In this case, Eq. (15.378) can be written as

$$\begin{bmatrix} d_{11} - \Omega^2 & 0 & d_{13} & 0 & 0 \\ 0 & d_{22} - \Omega^2 & 0 & 0 & d_{25} \\ d_{31} & 0 & d_{33} - \Omega^2 & d_{34} & 0 \\ 0 & 0 & d_{43} & d_{44} - \Omega^2 & 0 \\ 0 & d_{52} & 0 & 0 & d_{55} - \Omega^2 \end{bmatrix} \begin{Bmatrix} C_1 \\ C_2 \\ C_3 \\ C_4 \\ C_5 \end{Bmatrix} = \begin{Bmatrix} 0 \\ 0 \\ 0 \\ 0 \\ 0 \end{Bmatrix} \quad (15.404)$$

This equation can be rewritten as a system of two uncoupled matrix equations as

$$\begin{bmatrix} d_{11} - \Omega^2 & d_{13} & 0 \\ d_{31} & d_{33} - \Omega^2 & d_{34} \\ 0 & d_{43} & d_{44} - \Omega^2 \end{bmatrix} \begin{Bmatrix} C_1 \\ C_3 \\ C_4 \end{Bmatrix} = \begin{Bmatrix} 0 \\ 0 \\ 0 \end{Bmatrix} \quad (15.405)$$

$$\begin{bmatrix} d_{22} - \Omega^2 & d_{25} \\ d_{52} & d_{55} - \Omega^2 \end{bmatrix} \begin{Bmatrix} C_2 \\ C_5 \end{Bmatrix} = \begin{Bmatrix} 0 \\ 0 \end{Bmatrix} \quad (15.406)$$

Equations (15.405) and (15.406) lead to the following cubic and quadratic frequency equations, respectively:

$$\Omega^6 - (d_{11} + d_{33} + d_{44})\Omega^4 + (d_{11}d_{33} + d_{11}d_{44} + d_{33}d_{44} - d_{13}^2 - d_{34}^2)\Omega^2$$
$$+ (d_{11}d_{33}d_{44} - d_{11}d_{34}^2 - d_{44}d_{13}^2) = 0 \quad (15.407)$$

$$\Omega^4 - (d_{22} + d_{55})\Omega^2 + (d_{22}d_{55} - d_{25}^2) = 0 \quad (15.408)$$

Notes

1. The equations of motion of a circular cylindrical shell, considering the effects of rotary inertia and shear deformation, were derived by several investigators [15–17, 23]. Naghdi and Cooper derived a set of equations that are more general than those presented in Sections 15.10.8 and 15.10.9 (see Problem 15.14).

2. The equations of motion for the axisymmetric vibration of cylindrical shells and resulting equations for free vibration, Eq. (15.406), were derived as a special case of the more general equations by Naghdi and Cooper [15] and from general shell theory equations by Soedel [14].

Example 15.12 Find the natural frequencies of axisymmetric vibration of a circular cylindrical shell simply supported at $x = 0$ and $x = l$, considering the effects of rotary inertia and shear deformation for the following data: $E = 30 \times 10^6$ psi, $G = 12 \times 10^6$ psi, $\nu = 0.3$, $\rho = 7.32 \times 10^{-4}$ lb-sec^2/in.^{4}, $R = 10$ in., $l = 40$ in., $h = 0.1$ in., and $k = \frac{5}{6}$.

SOLUTION The natural frequencies of axisymmetric vibration of the cylindrical shell can be determined from the roots of Eqs. (15.407) and (15.408). The smallest values of the natural frequencies found for different modes ($m = 1, 2, \ldots, 10$) are given in Table 15.3. If the smallest natural frequencies of the cylindrical shell are to be determined by neglecting the effect of shear deformation, we need to set the value of k equal to 0 in Eqs. (15.407) and (15.408)

Table 15.3 Smallest Values of Natural Frequencies
Given by Eqs. (15.407) and (15.408)

	Smallest value of natural frequency, ω_1 (rad/sec)
m	With shear deformation and rotary inertia
1	24,057.074
2	29,503.338
3	36,832.332
4	45,135.996
5	53,966.332
6	63,102.639
7	72,429.156
8	81,880.961
9	91,419.211
10	101,019.41

15.11 RECENT CONTRIBUTIONS

A complete solution for the dynamic response due to time-dependent mechanical and/or thermal loading of spherical and cylindrical shells with arbitrary time-dependent boundary conditions was presented by Pilkey [19]. The solution was obtained in the form of a series expansion of the products of modes of free vibration and a generalized coordinate. The generalized Fourier transform was used to determine the generalized coordinate which contains all physically admissible boundary conditions. The free flexural vibrations of cylindrical shells stiffened by equidistant ring frames was investigated by Wah [20] using finite difference calculus. The theory accounts for both in-plane flexural and torsional vibration of the ring stiffeners.

Tables of natural frequencies and graphs of representative mode shapes of harmonic elastic waves propagating in an infinitely long isotropic hollow cylinder have been presented by Armenakas et al. [21]. The free vibration problem of a homogeneous isotropic thick cylindrical shell or panel subjected to a certain type of simply supported edge conditions was investigated by Soldatos and Hadjigeorgiou [22]. The governing equations of three-dimensional linear elasticity were employed and solved using an iterative approach, which in practice leads to the prediction of the exact frequencies of vibration. In the case of a flat or a complete shell, excellent agreement was found between the results given by this approach and those given by other exact analysis methods.

The role of median surface curvature in large-amplitude flexural vibrations of thin shells was studied by Prathap and Pandalai [23]. It was shown that whereas the nonlinear behavior of flat plates and straight bars is generally of a hardening type, the behavior of thin structural elements involving finite curvature of the undeformed median surface in one or both principal axis directions may be of the hardening or softening type, depending on the structural parameters as well as on whether the shell is open or closed.

The free vibration of circular cylindrical shells with axially varying thickness was considered by Sivadas and Ganesan [24]. The free vibration of noncircular cylindrical shells with circumferentially varying thickness was discussed by Suzuki and

Leissa [25]. Soedel summarized the vibration analysis of shells in Ref. [26]. The free vibration of thin cylindrical shells, consisting of two sections of different thicknesses but with a common mean radius, was investigated by Warburton and Al-Najafi [27]. They determined the natural frequencies and mode shapes using two types of ring finite elements as well as by solving the shell equations using the boundary conditions and the continuity condition at the intersection of the two thicknesses. Sharma and Johns [28] determined the circumferential mode vibration characteristics of clamped-free circular cylindrical shells experimentally.

A two-dimensional higher-order shell theory was applied to the free vibration problem of simply supported cylindrical shell subjected to axial stresses by Matsunaga [29]. Using a power series expansion of displacement components, the dynamical equations were derived, including the effects of rotary inertia and shear deformation, from Hamilton's principle. Thin-walled regular polygonal prismatic shells were used in several practical applications, such as honeycomb cores of sandwich plates, guide supports of welding frameworks, and high piers of highway bridges. The free vibration of regular polygonal prismatic shells has been presented by Liang et al. [30] using a novel plate model and beam model on the basis of geometric symmetry. Analytical solutions were obtained by combining the vibration theories of Euler beams and thin-walled plates.

REFERENCES

1. A. E. H. Love, *A Treatise on the Mathematical Theory of Elasticity*, 4th ed., Dover, New York, 1944.

2. W. Flügge, *Stresses in Shells*, Springer-Verlag, Berlin, 1962.

3. A. L. Goldenveizer, *Theory of Thin Shells*, translated by G. Herrmann, Pergamon Press, New York, 1961.

4. H. Kraus, *Thin Elastic Shells*, Wiley, New York, 1967.

5. A. W. Leissa, *Vibration of Shells*, NASA SP-288, National Aeronautics and Space Administration, Washington, D C, 1973.

6. W. Soedel, *Vibration of Shells and Plates*, 2nd ed., Marcel Dekker, New York, 1993.

7. A. C. Ugural, *Stresses in Plates and Shells*, McGraw-Hill, New York, 1981.

8. S. P. Timoshenko and S. Woinowsky-Krieger, *Theory of Plates and Shells*, McGraw-Hill, New York, 1959.

9. E. Ventsel and T. Krauthammer, *Thin Plates and Shells: Theory, Analysis, and Applications*, Marcel Dekker, New York, 2001.

10. V. V. Novozhilov, *Thin Shell Theory*, translated by P. G. Lowe, Noordhoff, Groningen, The Netherlands, 1964.

11. L. H. Donnell, *Beams, Plates, and Shells*, McGraw-Hill, New York, 1976.

12. V. Z. Vlasov, *General Theory of Shells and Its Applications in Engineering*, NASA-TT-F99, National Aeronautics and Space Administration, Washington, DC, 1964.

13. N. A. Kilchevskiy, *Fundamentals of the Analytical Mechanics of Shells*, NASA-TT-F292, National Aeronautics and Space Administration, Washington, DC, 1965.

14. W. Soedel, On the vibration of shells with Timoshenko–Mindlin type shear deflections and rotatory inertia, *Journal of Sound and Vibration*, Vol. 83, No. 1, pp. 67–79, 1982.

15. P. M. Naghdi and R. M. Cooper, Propagation of elastic waves in cylindrical shells, including the effects of transverse shear and rotatory inertia, *Journal of the Acoustical Society of America*, Vol. 28, No. 1, pp. 56–63, 1956.

16. T. C. Lin and G. W. Morgan, A study of axisymmetric vibrations of cylindrical shells as affected by rotary inertia and transverse shear, *Journal of Applied Mechanics*, Vol. 23, No. 2, pp. 255–261, 1956.

17. I. Mirsky and G. Herrmann, Nonaxially symmetric motions of cylindrical shells, *Journal of the Acoustical Society of America*, Vol. 29, No. 10, pp. 1116–1123, 1957.

18. I. Mirsky, Vibrations of orthotropic, thick, cylindrical shells, *Journal of the Acoustical Society of America*, Vol. 36, No. 1, pp. 41–51, 1964.

19. W. D. Pilkey, Mechanically and/or thermally generated dynamic response of thick spherical and cylindrical shells with variable material properties, *Journal of Sound and Vibration*, Vol. 6, No. 1, pp. 105–109, 1967.

20. T. Wah, Flexural vibrations of ring-stiffened cylindrical shells, *Journal of Sound and Vibration*, Vol. 3, No.3 pp. 242–251, 1966.

21. A. E. Armenakas, D. C. Gazis, and G. Herrmann, *Free Vibrations of Circular Cylindrical Shells*, Pergamon Press, Oxford, 1969.

22. K. P. Soldatos and V. P. Hadjigeorgiou, Three-dimensional solution of the free vibration problem of homogeneous isotropic cylindrical shells and panels, *Journal of Sound and Vibration*, Vol. 137, No. 3, pp. 369–384, 1990.

23. G. Prathap and K. A. V. Pandalai, The role of median surface curvature in large amplitude flexural vibrations of thin shells, *Journal of Sound and Vibration*, Vol. 60, No. 1, pp. 119–131, 1978.

24. K. R. Sivadas and N. Ganesan, Free vibration of circular cylindrical shells with axially varying thickness, *Journal of Sound and Vibration*, Vol. 147, No.1, pp. 73–85, 1991.

25. K. Suzuki and A. W. Leissa, Free vibrations of non-circular cylindrical shells having circumferentially varying thickness, *Journal of Applied Mechanics,* Vol. 52, No. 1, pp. 149–154, 1985.

26. W. Soedel, Shells, in *Encyclopedia of Vibration*, S. Braun, D. Ewins, and S. S. Rao, Eds., Academic Press, San Diego, CA, 2002, Vol. 3, pp. 1155–1167.

27. G. B. Warburton and A. M. J. Al-Najafi, Free vibration of thin cylindrical shells with a discontinuity in the thickness, *Journal of Sound and Vibration*, Vol. 9, No. 3, pp. 373–382, 1969.

28. C. B. Sharma and D. J. Johns, Natural frequencies of clamped-free circular cylindrical shells, *Journal of Sound and Vibration*, Vol. 21, No. 3, pp. 317–318, 1972.

29. H. Matsunaga, Free vibration of thick circular cylindrical shells subjected to axial stresses, *Journal of Sound and Vibration*, Vol. 211, No. 1, pp. 1–17, 1998.

30. S. Liang, H. L. Chen, and T. X. Liang, An analytical investigation of free vibration for a thin-walled regular polygonal prismatic shell with simply supported odd/even number of sides, *Journal of Sound and Vibration*, Vol. 284, No. 1–2, pp. 520–530, 2005.

PROBLEMS

15.1 Derive the equations of motion of a conical shell in terms of the components of displacement u, v, and w from Eqs. (15.279)–(15.281).

15.2 Derive the equations of motion of a spherical shell in terms of the components of displacement u, v, and w from Eqs. (15.284)–(15.286).

15.3 Specialize the equations of motion of a cylindrical shell given in Section 15.8.1 to the case of a rectangular plate. [*Hint:* Use $R \, d\theta = y$ and $1/R = 0$.]

15.4 Specialize the equations of motion of a cylindrical shell given in Section 15.8.1 to the case of a circular ring.

15.5 Derive the frequency equation (15.238) using the solution given by Eqs. (15.226)–(15.228) in the equations of motion, Eqs. (15.215)– (15.217).

15.6 Using Donnell–Mushtari–Vlasov theory, find the natural frequencies of vibration corresponding to m, $n = 1, 2, 3$ for an aluminum circular cylindrical shell with simple supports at $x = 0$ and $x = l$. $E = 71$ GPa, $\nu = 0.334$, unit weight $= 26.6$ kN/m^3, $R = 0.2$ m, $l = 2$ m, and $h = 2$ mm.

15.7 Derive the frequency equation (15.275) using the solution given by Eqs. (15.261)–(15.263) in the equations of motion, Eqs. (15.258)–(15.260).

15.8 Using Love's theory, find the natural frequencies of vibration of the cylindrical shell described in Problem 15.6.

15.9 Solve Problem 15.6 considering the material of the shell as steel with $E = 207$ GPa, $\nu = 0.292$, unit weight $= 76.5$ kN/m^3, $R = 0.2$ m, $l = 2$ m, and $h = 2$ mm.

15.10 Using Love's theory, find the natural frequencies of vibration corresponding to m, $n = 1, 2, 3$ for a steel circular cylindrical shell with simple supports at $x = 0$ and $x = l$. $E = 207$ GPa, $\nu = 0.292$, unit weight $= 76.5$ kN/m^3, $R = 0.2$ m, $l = 2$ m, and $h = 2$ mm.

15.11 (a) Derive the equation of motion for the axisymmetric vibrations of a thin cylindrical shell from the general equations.

(b) Assuming the transverse deflection $w(x, t)$ to be

$$w(x, t) = \sum_{m=1}^{\infty} C_n \sin \frac{n\pi x}{l} \sin \omega t$$

where x is the axial direction, l is the length, and ω is the frequency of axisymmetric vibration of a cylindrical shell simply supported at $x = 0$ and $x = l$, find an expression for the natural frequency ω.

15.12 Consider a cylindrical shell simply supported at $x = 0$ and $x = l$, with radius 6 in., length 15 in., wall thickness 0.025 in., Poisson's ratio 0.3, Young's modulus 30×10^6 psi, and unit weight 0.283 lb/in^3. Find the number of half-waves in the circumference corresponding to the minimum natural frequency of vibration according to DMV theory.

15.13 Show that the solution given by Eqs. (15.373)–(15.377) satisfies the boundary conditions of Eqs. (15.363) –(15.372).

15.14 According to Naghdi and Cooper [15], the force and moment resultants in a cylindrical shell, by considering the effects of rotary inertia and shear deformation, are given by

$$N_x = C \left[\frac{\partial u}{\partial x} + \nu \left(\frac{1}{R} \frac{\partial v}{\partial \theta} + \frac{w}{R} \right) + \frac{h^2}{12R} \frac{\partial \psi_x}{\partial x} \right]$$

$$N_\theta = C \left[\frac{1}{R} \frac{\partial v}{\partial \theta} + \frac{w}{R} + \nu \frac{\partial u}{\partial x} - \frac{h^2}{12R^2} \frac{\partial \psi_\theta}{\partial \theta} \right]$$

$$N_{x\theta} = \frac{1 - \nu}{2} C \left(\frac{1}{R} \frac{\partial u}{\partial \theta} + \frac{\partial v}{\partial x} + \frac{h^2}{12R} \frac{\partial \psi_x}{\partial \theta} \right)$$

$$N_{\theta x} = \frac{1-\nu}{2} C \left[\frac{1}{R} \frac{\partial u}{\partial \theta} + \frac{\partial v}{\partial x} + \frac{h^2}{12R} \left(\frac{1}{R^2} \frac{\partial u}{\partial \theta} - \frac{1}{R} \frac{\partial \psi_x}{\partial \theta} \right) \right]$$

$$M_x = D \left(\frac{\partial \psi_x}{\partial x} + \frac{\nu}{R} \frac{\partial \psi_\theta}{\partial \theta} + \frac{1}{R} \frac{\partial u}{\partial x} \right)$$

$$M_\theta = D \left[\frac{1}{R} \frac{\partial \psi_\theta}{\partial \theta} + \nu \frac{\partial \psi_x}{\partial x} - \frac{1}{R^2} \left(\frac{\partial v}{\partial \theta} + w \right) \right]$$

$$M_{x\theta} = \frac{1 - \nu}{2} D \left(\frac{1}{R} \frac{\partial \psi_x}{\partial \theta} + \frac{\partial \psi_\theta}{\partial x} + \frac{1}{R} \frac{\partial v}{\partial x} \right)$$

$$M_{\theta x} = \frac{1 - \nu}{2} D \left(\frac{1}{R} \frac{\partial \psi_x}{\partial \theta} + \frac{\partial \psi_\theta}{\partial x} - \frac{1}{R^2} \frac{\partial u}{\partial \theta} \right)$$

$$Q_x = kGh \left(\frac{\partial w}{\partial x} + \psi_x \right)$$

$$Q_\theta = kGh \left[\frac{1}{R} \frac{\partial w}{\partial \theta} - \left(\frac{v}{R} - \psi_\theta \right) \right]$$

Express the equations of motion, Eqs. (15.328)–(15.332), in terms of the displacement components u, v, w, ψ_x, and ψ_θ.

15.15 A cylindrical shell is loaded by a concentrated harmonic force $f(x, \theta, t) = f_0 \sin \Omega t$ in the radial direction at the point $x = x_0$ and $\theta = \theta_0$. Determine the amplitude of the resulting forced vibration.

16

Elastic Wave Propagation

16.1 INTRODUCTION

Any localized disturbance in a medium will be transmitted to other parts of the medium through the phenomenon of wave propagation. The spreading of ripples in a water pond, the transmission of sound in air, and the propagation of seismic tremors in Earth are examples of waves in different media. We consider wave propagation only in solid bodies in this chapter. Although the propagation of a disturbance in a solid takes place at a microscopic level, through the interaction of atoms of the solid, we consider only the physics of wave propagation by treating properties such as density and elastic constants of the solid body to be continuous functions that represent the averages of microscopic quantities.

In solid bodies, compression and shear waves can occur. In compression waves, the compressive and tensile stresses are transmitted through the motion of particles in the direction of the wave motion. In shear waves, shear stress is transmitted through the motion of particles in a direction transverse to the direction of wave propagation. Three types of waves can occur in a solid body: elastic waves, viscoelastic waves, and plastic waves. In elastic waves, the stresses in the material obey Hooke's law. In viscoelastic waves, viscous as well as elastic stresses act, and in plastic waves, the stresses exceed the yield stress of the material. We consider only elastic waves in this chapter.

Elastic waves in deformable bodies play an important role in many practical applications. For example, oil and gas deposits are detected and Earth's geological structure is studied with the help of waves transmitted through the soil. The waves generated by Earth's tremors are used to detect and study earthquakes. Properties of materials are determined by measuring the behavior of waves transmitted through them. Some recent medical diagnosis and therapy procedures are based on a study of elastic waves transmitted through the human body.

16.2 ONE-DIMENSIONAL WAVE EQUATION

The one-dimensional wave equation is given by

$$c^2 \frac{\partial^2 \phi}{\partial x^2} = \frac{\partial \phi}{\partial t^2} \tag{16.1}$$

where $\phi = \phi(x, t)$ is the dependent variable and x and t are the independent variables. Equation (16.1) represents the equation of motion for the free lateral vibration of strings,

Table 16.1 Physical Significance of ϕ and c in Eq. (16.1)

Type of problem	Significance of ϕ	Significance of c [a]
Lateral vibration of strings	Lateral or transverse displacement of a string	$c = \sqrt{P/\rho}$ $P =$ tension $\rho =$ mass per unit length
Longitudinal vibration of bars	Longitudinal or axial displacement of the cross section of a bar	$c = \sqrt{E/\rho}$ $E =$ Young's modulus $\rho =$ mass density
Torsional vibration of rods	Angular rotation of the cross section of a rod	$c = \sqrt{G/\rho}$ $G =$ shear modulus $\rho =$ mass density

[a] c has dimensions of linear velocity.

longitudinal vibration of bars, and torsional vibration of rods. The study of propagation of waves in a taut string is useful in the manufacture of thread and in understanding the characteristics of many musical instruments and the dynamics of electrical transmission lines. The longitudinal waves in a bar have application in seismic studies. Waves transmitted through the Earth are used to detect and study earthquakes. The physical significances of ϕ and c in different problems are given in Table 16.1.

16.3 TRAVELING-WAVE SOLUTION

16.3.1 D'Alembert's Solution

D'Alembert derived the solution of Eq. (16.1) in 1747 in a form that provides considerable insight into the phenomenon of wave propagation. According to his approach, the general solution of Eq. (16.1) is obtained by introducing two new independent variables, ξ and η as

$$\xi = x - ct \tag{16.2}$$

$$\eta = x + ct \tag{16.3}$$

By expressing the dependent variable w^1 in terms of ξ and η (instead of x and t), we can obtain the following relationships using Eqs. (16.2) and (16.3):

$$\frac{\partial \xi}{\partial \xi} = 1, \qquad \frac{\partial \xi}{\partial t} = -c, \qquad \frac{\partial \eta}{\partial x} = 1, \qquad \frac{\partial \eta}{\partial t} = c \tag{16.4}$$

$$\frac{\partial w}{\partial x} = \frac{\partial w}{\partial \xi}\frac{\partial \xi}{\partial x} + \frac{\partial w}{\partial \eta}\frac{\partial \eta}{\partial x} = \frac{\partial w}{\partial \xi} + \frac{\partial w}{\partial \eta} \tag{16.5}$$

$$\frac{\partial^2 w}{\partial x^2} = \frac{\partial^2 w}{\partial \xi^2}\frac{\partial \xi}{\partial x} + \frac{\partial^2 w}{\partial \xi \partial \eta}\frac{\partial \eta}{\partial x} + \frac{\partial^2 w}{\partial \xi \partial \eta}\frac{\partial \xi}{\partial x} + \frac{\partial^2 w}{\partial \eta^2}\frac{\partial \eta}{\partial x} = \frac{\partial^2 w}{\partial \xi^2} + 2\frac{\partial^2 w}{\partial \xi \partial \eta} + \frac{\partial^2 w}{\partial \eta^2} \tag{16.6}$$

---

[1] The dependent variable, ϕ, in Eq. (16.1) is denoted as w in this section.

$$\frac{\partial w}{\partial t} = \frac{\partial w}{\partial \xi}\frac{\partial \xi}{\partial t} + \frac{\partial w}{\partial \eta}\frac{\partial \eta}{\partial t} = -c\frac{\partial w}{\partial \xi} + c\frac{\partial w}{\partial \eta} \tag{16.7}$$

$$\frac{\partial^2 w}{\partial t^2} = -c\left(\frac{\partial^2 w}{\partial \xi^2}\frac{\partial \xi}{\partial t} + \frac{\partial^2 w}{\partial \xi \partial \eta}\frac{\partial \eta}{\partial t}\right) + c\left(\frac{\partial^2 w}{\partial \xi \partial \eta}\frac{\partial \xi}{\partial t} + \frac{\partial^2 w}{\partial \eta^2}\frac{\partial \eta}{\partial t}\right)$$

$$= c^2\frac{\partial^2 w}{\partial \xi^2} - 2c^2\frac{\partial^2 w}{\partial \xi \partial \eta} + c^2\frac{\partial^2 w}{\partial \eta^2} \tag{16.8}$$

Substituting Eqs. (16.6) and (16.8) into Eq. (16.1), we obtain the one-dimensional wave equation in the form

$$\frac{\partial^2 w}{\partial \xi \partial \eta} = 0 \tag{16.9}$$

This equation can be integrated twice to obtain its general solution. Integration of Eq. (16.9) with respect to η gives

$$\frac{\partial w}{\partial \xi} = h(\xi) \tag{16.10}$$

where $h(\xi)$ is an arbitrary function of ξ. Integration of Eq. (16.10) with respect to ξ yields

$$w = \int h(\xi)\,d\xi + g(\eta) \tag{16.11}$$

where $g(\eta)$ is an arbitrary function of η. By defining

$$f(\xi) = \int h(\xi)\,d\xi \tag{16.12}$$

the solution of the one-dimensional wave equation can be expressed as

$$w(\xi, \eta) = f(\xi) + g(\eta)$$

$$= f(x - ct) + g(x + ct) \tag{16.13}$$

The solution given by Eq. (16.13) is D'Alembert's solution [see Eq. (8.35)]. Note that f and g in Eq. (16.13) are arbitrary functions of integration which can be determined from the known initial conditions of the problem. To interpret the solution given by Eq. (16.13), assume that the term $g(x + ct)$ is zero and the term $f(x - ct)$ is nonzero. Assume that at $t = 0$, the function $f(x - ct) = f(x)$ denotes a triangular profile as shown in Fig. 16.1 with the peak of $f(x)$, equal to $f(0)$, occurring at $x = 0$. At a later time, the peak $f(0)$ occurs when the value of the argument of the function $f(x - ct)$ is zero. Thus, the peak occurs when $x_2 - ct_2 = 0$ or $t_2 = x_2/c$, as shown in Fig. 16.1. Using a similar argument, every point in the triangular profile can be shown to propagate in the positive x direction with a constant velocity c. It follows that the function $f(x - ct)$ represents a wave that propagates undistorted with velocity c in the positive direction of the x axis. In a similar manner, the function $g(x + ct)$ can be shown to represent a wave that propagates undistorted with velocity c in the negative direction of the x axis. Note that the shape of the disturbances, $f(x - ct)$ and $g(x + ct)$, which is decided by the initial conditions specified, remains the same during wave propagation.

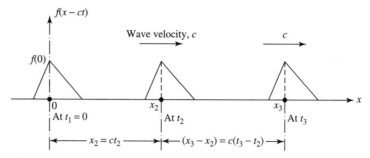

Figure 16.1   Propagation of the wave $f(x - ct)$ with no distortion.

16.3.2   Two-Dimensional Problems

For a two-dimensional problem (membrane), the equation of motion is given by [see Eq. (13.2)]

$$c^2 \left(\frac{\partial^2 w}{\partial x^2} + \frac{\partial^2 w}{\partial y^2} \right) = \frac{\partial^2 w}{\partial t^2} \tag{16.14}$$

where $w(x, y, t)$ denotes the transverse displacement of the membrane. The solution of Eq. (16.14) is given by [3]

$$w(x, y, t) = f(lx + my - ct) + g(lx + my + ct) \tag{16.15}$$

where f and g denote plane waves propagating in a direction whose direction cosines are given by l and m. The transverse displacement w can be shown to be constant along the line $lx + my = $ constant at each instant of time t.

16.3.3   Harmonic Waves

By redefining the variables ξ and η as

$$\xi = -\frac{1}{c}(x - ct) = t - \frac{x}{c}, \qquad \eta = \frac{1}{c}(x + ct) = t + \frac{x}{c} \tag{16.16}$$

the solution of the one-dimensional wave equation can be expressed as

$$w(x, t) = f \left(t - \frac{x}{c} \right) + g \left(t + \frac{x}{c} \right) \tag{16.17}$$

where f and g denote, respectively, the forward- and backward-propagating waves. A wave whose profile (or shape or displacement configuration) is sinusoidal is called a *harmonic wave*. In general, harmonic waves moving in the positive and negative x directions can be represented, respectively, as

$$w(x, t) = \begin{cases} A \sin \omega \left(t - \dfrac{x}{c} \right) \quad \text{or} \quad A \cos \omega \left(t - \dfrac{x}{c} \right) & (16.18) \\[2mm] A \sin \omega \left(t + \dfrac{x}{c} \right) \quad \text{or} \quad A \cos \omega \left(t + \dfrac{x}{c} \right) & (16.19) \end{cases}$$

where A denotes the amplitude of the wave. By defining the wavelength λ as

$$\lambda = \frac{2\pi c}{\omega} \tag{16.20}$$

we note that

$$\sin \omega \left(t \pm \frac{x+\lambda}{c} \right) = \sin \left[\omega \left(t \pm \frac{x}{c} \right) + 2\pi \right] = \sin \omega \left(t \pm \frac{x}{c} \right) \tag{16.21}$$

$$\cos \omega \left(t \pm \frac{x+\lambda}{c} \right) = \cos \left[\omega \left(t \pm \frac{x}{c} \right) + 2\pi \right] = \cos \omega \left(t \pm \frac{x}{c} \right) \tag{16.22}$$

Thus, the wave profile given by Eq. (16.18) or (16.19) repeats itself at regular intervals of the wavelength ($x = \lambda$). The reciprocal of the wavelength, known as the wave number (n),

$$n = \frac{1}{\lambda} \tag{16.23}$$

denotes the number of cycles of the wave per unit length. The period of the wave (τ) denotes the time required for a complete cycle to pass through a fixed point x so that

$$\tau = \frac{\lambda}{c} = \frac{2\pi}{\omega} \tag{16.24}$$

The frequency of the wave ($\hat{f}$) is defined as the *reciprocal* of the period:

$$\hat{f} = \frac{1}{\tau} = \frac{c}{\lambda} \tag{16.25}$$

It can be seen that the frequency ($\hat{f}$) and the wavelength (λ) are related as

$$\hat{f}\lambda = c \tag{16.26}$$

16.4   WAVE MOTION IN STRINGS

16.4.1   Free Vibration and Harmonic Waves

The boundaries of a string of finite length invariably introduce complications in wave propagation due to the phenomenon of reflections. Hence, we first consider a long, infinite or semi-infinite string, where the problem of boundary reflections need not be considered. Using the separation-of-variables approach, the solution of the wave equation (free vibration solution of the string) can be obtained as [see Eq. (8.87)]

$$w(x, t) = \left(A \cos \frac{\omega}{c} x + B \sin \frac{\omega}{c} x \right) (C \cos \omega t + D \sin \omega t)$$

$$= A_1 \cos \frac{\omega}{c} x \cos \omega t + A_2 \cos \frac{\omega}{c} x \sin \omega t + A_3 \sin \frac{\omega}{c} x \cos \omega t$$

$$+ A_4 \sin \frac{\omega}{c} x \sin \omega t \tag{16.27}$$

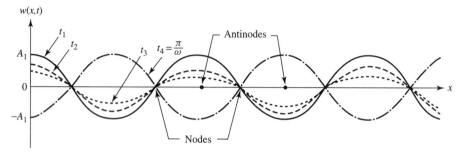

Figure 16.2   Vibration pattern of a string given by Eq. (16.28) (standing wave).

where $A_1 = AC$, $A_2 = AD$, $A_3 = BC$, and $A_4 = BD$ are the new constants. Consider a typical term in Eq. (16.27):

$$w_1(x, t) = A_1 \cos \frac{\omega}{c} x \cos \omega t \qquad (16.28)$$

The deflections of the string at successive instants of time, given by Eq. (16.28), are shown in Fig. 16.2. It can be seen that certain points (called *nodes*) on the string undergo zero vibration amplitude, whereas other points (called *antinodes*) will attain maximum amplitude. The nodes and antinodes occur at regular spacings along the string and remain fixed in that position for all time as indicated in Fig. 16.2. This type of vibration is called a *stationary* or *standing wave*. The solution given by Eq. (16.27) can also be expressed, using trigonometric identities, as

$$w(x, t) = B_1 \sin \left(\frac{\omega x}{c} + \omega t \right) + B_2 \sin \left(\frac{\omega x}{c} - \omega t \right)$$
$$+ B_3 \cos \left(\frac{\omega x}{c} + \omega t \right) + B_4 \cos \left(\frac{\omega x}{c} - \omega t \right) \qquad (16.29)$$

where B_1, B_2, B_3, and B_4 are constants. The term

$$w_1(x, t) = B_4 \cos \left(\frac{\omega x}{c} - \omega t \right) = B_4 \cos \omega \left(\frac{x}{c} - t \right) \qquad (16.30)$$

can be seen to denote a wave propagating in the positive x direction. We can see that as time (t) progresses, larger values of x are needed to maintain a constant (e.g., zero) value of the argument $\omega[(x/c) - t]$. In this case, the deflection pattern of the string at successive instants of time, given by Eq. (16.30), appears as shown in Fig. 16.3. It can be observed that the constant $(\omega/2\pi c)$ is the *wave number*, the argument $\omega[(x/c) - t]$ is the phase ϕ, $2\pi c/\omega$ is the *wavelength*, and c is the *phase velocity*, which indicates the propagation velocity of the constant phase ϕ. Note that Eq. (16.30), with a constant value of ϕ, indicates a wave of infinite length that has no wavefront or beginning.

A propagation velocity (in the positive x direction) can be associated with motion only when the phase is considered. Consider another term in Eq. (16.29) to represent $w_1(x, t)$ [instead of Eq. (16.30)]:

$$w_1(x, t) = B_3 \cos \omega \left(\frac{x}{c} + t \right) \qquad (16.31)$$

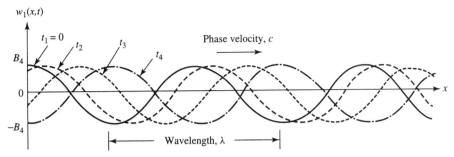

Figure 16.3   Vibration pattern of a string given by Eq. (16.30) (propagating wave).

We can see that Eq. (16.31) represents a wave propagating in the negative direction. The solutions given by the other terms of Eq. (16.29) are similar in nature. Thus, the solution given by Eq. (16.29) denotes a propagating wave solution, whereas the solution given by Eq. (16.27) represents a standing-wave solution. In fact, the standing-wave solution of Eq. (16.27) can be obtained from constructive and destructive interference of waves propagating to the right and left. The validity of this aspect can be seen by considering the sum of two waves of equal amplitude propagating in different directions as

$$w(x, t) = A \cos \omega \left(\frac{x}{c} + t \right) + A \cos \omega \left(\frac{x}{c} - t \right)$$

$$= 2A \cos \frac{\omega x}{c} \cos \omega t \tag{16.32}$$

where the term on the right-hand side represents a standing wave.

16.4.2   Solution in Terms of Initial Conditions

If the initial displacement and velocity of the string are specified as $U(x)$ and $V(x)$, respectively, we have

$$w(x, 0) = U(x)$$
$$\frac{\partial w}{\partial t}(x, 0) = V(x) \tag{16.33}$$

Using the general solution of Eq. (16.13) at $t = 0$, we obtain

$$f(x) + g(x) = U(x) \tag{16.34}$$

$$\frac{\partial w}{\partial t}(x, 0) = \frac{\partial f}{\partial (x - ct)} \frac{\partial (x - ct)}{\partial t} + \frac{\partial g}{\partial (x + ct)} \frac{\partial (x + ct)}{\partial t} = V(x)$$

or

$$-cf'(x) + cg'(x) = V(x) \tag{16.35}$$

where a prime denotes differentiation with respect to the argument. Integration of Eq.(16.35) yields

$$f(x) - g(x) = -\frac{1}{c} \int_{x_0}^{x} V(y) \, dy \tag{16.36}$$

where x_0 denotes an arbitrary lower limit introduced to eliminate the constant of integration. The solution of Eqs. (16.34) and (16.36) yields

$$f(x) = \frac{1}{2}U(x) - \frac{1}{2c}\int_{x_0}^{x} V(y)\,dy \tag{16.37}$$

$$g(x) = \frac{1}{2}U(x) + \frac{1}{2c}\int_{x_0}^{x} V(y)\,dy \tag{16.38}$$

To express the general solution for $t \neq 0$, we replace x by $x - ct$ in Eq. (16.37) and $x + ct$ in Eq. (16.38) and add the results to obtain

$$w(x, t) = f(x - ct) + g(x + ct)$$

$$= \frac{1}{2}U(x - ct) - \frac{1}{2c}\int_{x_0}^{x-ct} V(y)\,dy + \frac{1}{2}U(x + ct) + \frac{1}{2c}\int_{x_0}^{x+ct} V(y)\,dy \tag{16.39}$$

Noting that

$$-\int_{x_0}^{x-ct} V(y)\,dy + \int_{x_0}^{x+ct} V(y)\,dy = \int_{x-ct}^{x+ct} V(y)\,dy \tag{16.40}$$

and defining

$$\int_{x-ct}^{x+ct} V(y)\,dy = R(x + ct) - R(x - ct) \tag{16.41}$$

the displacement solution can be expressed as

$$w(x, t) = \frac{1}{2}[U(x - ct) + U(x + ct)] + \frac{1}{2c}[R(x - ct) + R(x + ct)] \tag{16.42}$$

It can be seen that the motion, given by Eq. (16.42), consists of identical disturbances propagating to the left and the right with separate contributions from the initial displacement and initial velocity.

16.4.3   Graphical Interpretation of the Solution

To interpret the solution given by Eq. (16.42) graphically, consider a simple case with zero initial velocity [with $V(x) = 0$ so that $R(x - ct) = R(x + ct) = 0$]:

$$w(x, t) = \tfrac{1}{2}[U(x - ct) + U(x + ct)] \tag{16.43}$$

This solution denotes the sum of waves propagating to the right and left which have the same shape as the initial displacement $U(x)$ but have one-half its magnitude. Let the initial displacement, at $t = 0$, be assumed as

$$w(x, t) = \tfrac{1}{2}[U(x) + U(x)] = U(x) \equiv p(x) \tag{16.44}$$

with a peak value of $2p_0$ as shown in Fig. 16.4(a). The displacement distribution at any particular time can be obtained by superposing the waves propagating to the right

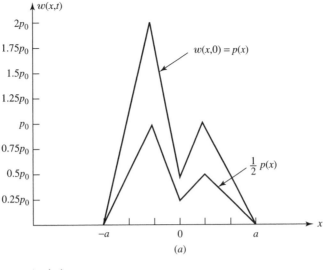

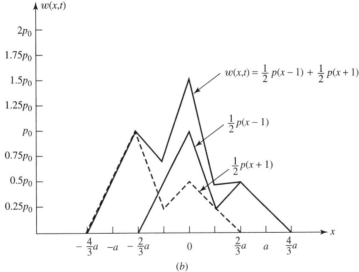

Figure 16.4   Propagation of displacement disturbance with time: (a) at $ct = 0$; (b) at $ct = a/3$; (c) at $ct = \frac{2}{3}a$; (d) at $ct = a$; (e) at $ct = \frac{4}{3}a$.

and left as shown in Fig. 16.4(b)(d). Notice that the separation of the two waves at any value of ct would be twice the distance traveled by one wave in the time ct. As such, for all values of $ct > a$, the two waves will not overlap and travel with the same shape as the initial disturbance but with one–half its magnitude (see Fig. 16.4e).

Next, consider the case with initial velocity, $V(x)$, and zero initial displacement $[U(x) = 0]$, so that the solution becomes

$$w(x, t) = \frac{1}{2c}[-R(x - ct) + R(x + ct)] = \frac{1}{2c}\int_{x-ct}^{x+ct} V(y)\,dy \qquad (16.45)$$

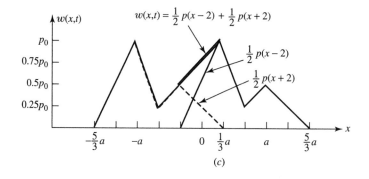

(c)

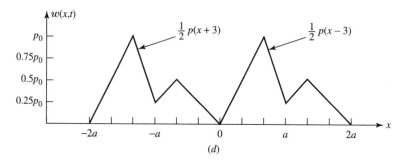

(d)

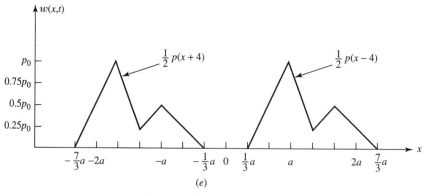

(e)

Figure 16.4   (*continued*)

The graphical interpretation of Eq. (16.45) is based on the integral of the initial velocity distribution. At any specific position x and time t, the solution given by Eq. (16.45) depends on the initial velocity distribution from $x - ct$ to $x + ct$. For example, if the initial velocity is assumed to be a constant as

$$\frac{\partial w}{\partial t}(x, 0) = V_0 \equiv q_0 \tag{16.46}$$

the displacement given by Eq. (16.45) at different instants of time will appear as shown in Fig. 16.5. It can be seen that when $t < a/c$, the maximum displacement is given by $V_0 t$ in the interval $|x| \leq a - ct$. In the ranges $a - ct \leq x \leq a + ct$ and $-(a -$

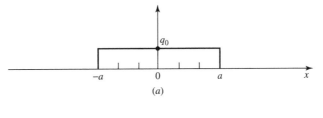

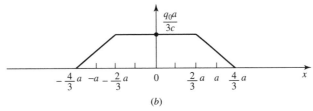

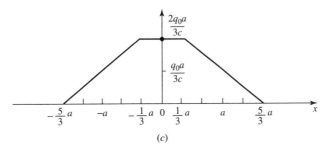

Figure 16.5 Displacement given by Eq. (16.45) at (*a*) $t = 0$, (*b*) $ct = \frac{a}{3}$, (*c*) $ct = \frac{2}{3}a$, (*d*) $ct = a$ and (*e*) $ct = \frac{4}{3}a$.

$ct) \leq x \leq -(a + ct)$, the displacement decreases linearly to zero. When $t > a/c$, the maximum displacement is given by $V_0 a/c$ in the interval $-(ct - a) \leq x \leq (ct - a)$. As $ct \to \infty$, the string gets displaced at a uniform distance $V_0 a/c$ from its original position.

16.5   REFLECTION OF WAVES IN ONE-DIMENSIONAL PROBLEMS

The investigation of propagation of waves in one-dimensional problems such as strings and bars with infinite domain does not require consideration of interaction of waves at boundaries. We consider the reflection of waves at fixed and free boundaries in this section. For simplicity, an intuitive approach is presented instead of a mathematical approach.

16.5.1   Reflection at a Fixed or Rigid Boundary

Consider a semi-infinite bar fixed at $x = 0$. The boundary condition at the fixed end is given by

$$u(0, t) = 0 \qquad (16.47)$$

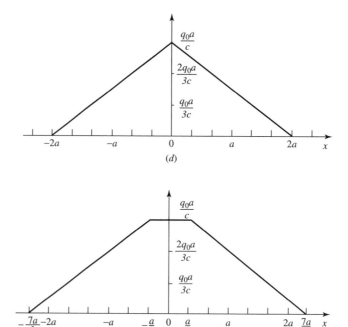

Figure 16.5 (*continued*)

where u is the axial displacement. An image displacement pulse system is introduced to satisfy the boundary condition of Eq. (16.47). Consider a wave approaching the rigid boundary at $x = 0$ from the right as shown in Fig. 16.6(a). Imagine the rigid boundary at $x = 0$ as removed and extend the bar to infinity. Now assume an "image" wave to the original propagating wave as shown in Fig. 16.6(b). The image wave is placed symmetrically with respect to $x = 0$, is opposite in sense to the original propagating wave, and propagates to the right. As the original and image waves approach $x = 0$, they interact as shown in Fig. 16.6(c). As they pass, their displacements will mutually cancel at $x = 0$, yielding $u(0, t) = 0$ always. Thus, the rigid boundary condition of the semi-infinite bar is always satisfied by the image wave system in the infinite bar. After some time, the interaction stage will be completed and the image wave propagates into $x > 0$, while the original "real" wave propagates into $x < 0$ as shown in Fig. 16.6(d). Thus, the reflected wave propagates along the positive x axis with the sign of the wave reversed.

16.5.2   Reflection at a Free Boundary

Consider a semi-infinite bar free at $x = 0$. Since the longitudinal stress at a free end is zero, the boundary condition is given by

$$\sigma_{xx}(0, t) = E \frac{\partial u}{\partial x}(0, t) = 0$$

or

$$\frac{\partial u}{\partial x}(0, t) = 0 \tag{16.48}$$

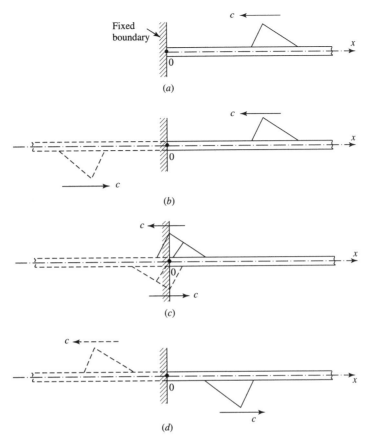

Figure 16.6 Sequence of events during reflection of a wave at a rigid boundary. (From Refs. [1] and [4].)

Consider a wave approaching a free boundary from the right as shown in Fig. 16.7(a). Imagine the free boundary at $x = 0$ as removed and extend the bar to negative infinity. Now assume a "mirror image" wave to the original propagating wave as shown in Fig. 16.7(b). The image wave is symmetrically placed with respect to $x = 0$ and propagates to the right. As the original and image waves approach $x = 0$, they interact as shown in Fig. 16.7(c). As they pass, their slopes will mutually cancel at $x = 0$ yielding always. $\partial u(0, t)/\partial x = 0$. Thus, the free boundary condition of the semi-infinite bar is always satisfied by the image wave system in the infinite bar. As time passes, the interaction stage will be completed and the image wave propagates into $x > 0$ while the original real wave propagates into $x < 0$ as shown in Fig. 16.7(d). Note that the sign of the original wave remains unchanged after completion of the reflection process.

16.6 REFLECTION AND TRANSMISSION OF WAVES AT THE INTERFACE OF TWO ELASTIC MATERIALS

Consider two elastic half-spaces made of two different materials bonded together at their bounding surfaces as shown in Fig. 16.8(a), where the plane $x = 0$ denotes the

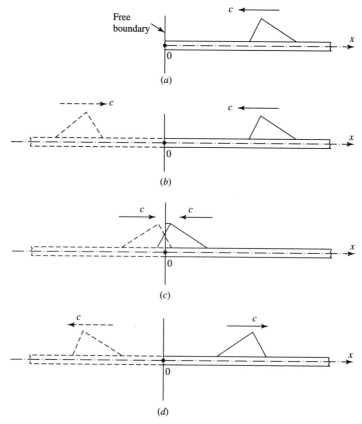

Figure 16.7   Sequence of events during reflection of a wave at a free boundary. (From Refs. [1] and [4].)

interface between the two materials. The two materials are designated 1 and 2. Let a specified incident wave propagate in the positive x direction in the half-space 1. When the wave reaches the interface between the two materials, it gives rise to a reflected wave propagating in material 1 in the negative x direction and a transmitted wave propagating in material 2 in the positive x direction as shown in Fig. 16.8(b). Denoting the incident wave as

$$p(\xi_1) \equiv p\left(t - \frac{x}{c_1}\right) \tag{16.49}$$

and the reflected wave as

$$r(\eta_1) \equiv r\left(t + \frac{x}{c_1}\right) \tag{16.50}$$

the displacement in material 1 can be expressed as

$$u_1 = p(\xi_1) + r(\eta_1) = p\left(t - \frac{x}{c_1}\right) + r\left(t + \frac{x}{c_1}\right) \tag{16.51}$$

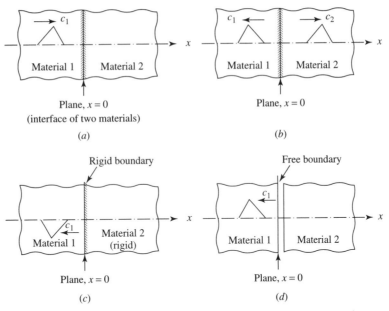

Figure 16.8 Reflection and transmission of waves at the interface of materials: (*a*) Incident wave; (*b*) reflected and transmitted waves; (*c*) reflection at a rigid boundary; (*d*) reflection at a free boundary.

By considering only the transmitted wave (s), the displacement in material 2 can be written as

$$u_2 = s(\xi_2) \equiv s\left(t - \frac{x}{c_2}\right) \tag{16.52}$$

Since the incident wave is specified, the function r is known. However, the functions s and p corresponding to the reflected and transmitted waves are to be determined using the following conditions:

1. The displacements in the two half-spaces must be equal at the interface:

$$u_1(0, t) = u_2(0, t) \tag{16.53}$$

2. The normal stress in the two materials must be equal at the interface:

$$\sigma_{xx_1}(0, t) = \sigma_{xx_2}(0, t) \tag{16.54}$$

Using Eq. (A.16), Eq. (16.54) can be rewritten as

$$(\lambda + 2\mu)_1 \frac{\partial u_1}{\partial x}(0, t) = (\lambda + 2\mu)_2 \frac{\partial u_2}{\partial x}(0, t)$$

or, equivalently,

$$\rho_1 c_1^2 \frac{\partial u_1}{\partial x}(0, t) = \rho_2 c_2^2 \frac{\partial u_2}{\partial x}(0, t) \tag{16.55}$$

where ρ_1 and ρ_2 are the densities and c_1 and c_2 are the velocities of dilatational (compressional) wave propagation in materials 1 and 2, respectively. Equations (16.51) and (16.52) yield

$$\frac{\partial u_1}{\partial x} = \frac{\partial p(\xi_1)}{\partial \xi_1}\frac{\partial \xi_1}{\partial x} + \frac{\partial r(\eta_1)}{\partial \eta_1}\frac{\partial \eta_1}{\partial x} = -\frac{1}{c_1}\frac{dp(\xi_1)}{d\xi_1} + \frac{1}{c_1}\frac{dr(\eta_1)}{d\eta_1} \tag{16.56}$$

$$\frac{\partial u_2}{\partial x} = \frac{\partial s(\xi_2)}{\partial \xi_2}\frac{\partial \xi_2}{\partial x} = -\frac{1}{c_2}\frac{ds(\xi_2)}{d\xi_2} \tag{16.57}$$

The boundary conditions, Eqs. (16.53) and (16.54), can be expressed, using Eqs. (16.51), (16.52), (16.56), and (16.57), as

$$p(t) + r(t) = s(t) \tag{16.58}$$

$$-\frac{dp(t)}{dt} + \frac{dr(t)}{dt} = -a\frac{ds(t)}{dt} \tag{16.59}$$

where

$$a = \frac{\rho_2 c_2}{\rho_1 c_1} \tag{16.60}$$

depends on the properties of the two materials. Equation (16.59) can be integrated to obtain

$$-p(t) + r(t) = -as(t) + A \tag{16.61}$$

where A is an integration constant. By neglecting A, Eqs. (16.58) and (16.61) can be solved to obtain $r(t)$ and $s(t)$ in terms of the known function $p(t)$ as

$$r\left(t + \frac{x}{c_1}\right) = \frac{1 - a}{1 + a}p\left(t + \frac{x}{c_1}\right) \tag{16.62}$$

$$s\left(t - \frac{x}{c_2}\right) = \frac{2}{1 + a}p\left(t - \frac{x}{c_2}\right) \tag{16.63}$$

Note that if the two materials are the same, $a = 1$ and there will be no reflected wave, but there will be a transmitted wave that is identical to the incident wave.

16.6.1   Reflection at a Rigid Boundary

Consider an elastic material 1 (first half-space) bonded to a rigid material 2 (second half-space) as shown in Fig. 16.8(c). Let a specified incident wave propagate in the positive x direction in the half-space 1. The wave reflected from the rigid boundary (interface) can be found from Eq. (16.62) using the relation $\rho_2 c_2 \to \infty$ for material 2 so that $a \to \infty$:

$$r\left(t + \frac{x}{c_1}\right) = -p\left(t + \frac{x}{c_1}\right) \tag{16.64}$$

Equation (16.63) shows that there will be no transmitted wave in material 2, and Eq. (16.64) indicates that the reflected wave is identical in form to the incident wave but propagates in the negative x direction. The reflection process at a rigid boundary is shown graphically in Fig. 16.8(c).

16.6.2 Reflection at a Free Boundary

Consider an elastic half-space 1 (material 1) with a free boundary at $x = 0$ as shown in Fig. 16.8(d). Let a specified incident wave propagate in the positive x direction in material 1. The wave reflected from the free boundary can be found from Eq. (16.62) using the relation $\rho_2 c_2 \to 0$ for material 2 so that $a \to 0$:

$$r\left(t + \frac{x}{c_1}\right) = p\left(t + \frac{x}{c_1}\right) \tag{16.65}$$

Equation (16.65) indicates that the reflected wave is identical in form to the incident wave and propagates in the positive direction. The reflection process at a free boundary is shown graphically in Fig. 16.8(d). Note that the processes of reflection at fixed and free boundaries are similar to those shown for a bar in Figs. 16.6 and 16.7.

16.7 COMPRESSIONAL AND SHEAR WAVES

16.7.1 Compressional or P Waves

Consider a half-space with the x axis pointing into the material and the yz plane forming the boundary of the half-space as shown in Fig. 16.9(a). The half-space can be disturbed and compressional waves can be generated by either a displacement or a normal stress boundary condition, as indicated in Fig. 16.9(b) and (c). In Fig. 16.9(b), the half-space is assumed to be initially undisturbed with $u(x, t) = 0$ for $t \leq 0$, and the boundary is then given a uniform displacement in the x direction so that

$$u(0, t) = r(t) \tag{16.66}$$

where $r(t)$ is a known function of time that is zero for $t \leq 0$. The motion resulting from this displacement boundary condition (with no motion of the material in the y and z directions) can be described by the components of displacement:

$$u = u(x, t)$$
$$v = 0 \tag{16.67}$$
$$w = 0$$

For this one-dimensional motion, the equation of motion, Eq. (A.31), reduces to

$$\frac{\partial^2 u}{\partial t^2}(x, t) = \alpha^2 \frac{\partial^2 u(x, t)}{\partial x^2} \tag{16.68}$$

where

$$\alpha = \left(\frac{\lambda + 2\mu}{\rho_0}\right)^{1/2} \tag{16.69}$$

ρ_0 is the density of the material in the reference state, and λ and μ are Lamé constants, related to Young's modulus E and Poisson's ratio ν of an isotropic linear elastic material, as

$$\lambda = \frac{\nu E}{(1 + \nu)(1 - 2\nu)} \tag{16.70}$$

$$\mu = \frac{E}{2(1 + \nu)} \equiv G \tag{16.71}$$

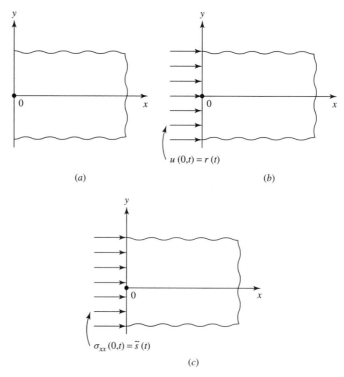

(a)   (b)

(c)

Figure 16.9   (a) Undisturbed half-space; (b) displacement boundary condition; (c) normal stress boundary condition.

and G is the shear modulus. Equation (16.68) is known as a *one-dimensional wave equation* and α, in Eq. (16.69), denotes the wave velocity. Thus, the one-dimensional motion of the material, given by Eq. (16.67), is governed by Eq. (16.68). The solution or waves given by Eq. (16.68) are called *compressional* or *P waves* and α is called the *compressional* or *P wave speed* [1, 4]. The displacements of a set of material points due to a compressional wave at a specific time t are shown in Fig. 16.10. Note that the material points move only in the x direction and their motions depend on x and t only. In Fig. 16.9(c), the boundary of the initially undisturbed half-space is given a uniform normal stress in the x direction so that

$$\sigma_{xx}(0, t) = \tilde{s}(t) \tag{16.72}$$

where $\tilde{s}(t)$ is a known function of time that is zero for $t \leq 0$. Equation (16.72), called the *normal stress boundary condition*, can be expressed in terms of the displacement u, using the stress–strain and strain–displacement relations, as

$$\sigma_{xx}(0, t) = (\lambda + 2\mu)\varepsilon_{xx}(0, t) = (\lambda + 2\mu)\frac{\partial u(0, t)}{\partial x} = \tilde{s}(t) \tag{16.73}$$

or

$$\frac{\partial u}{\partial x}(0, t) = s(t) \tag{16.74}$$

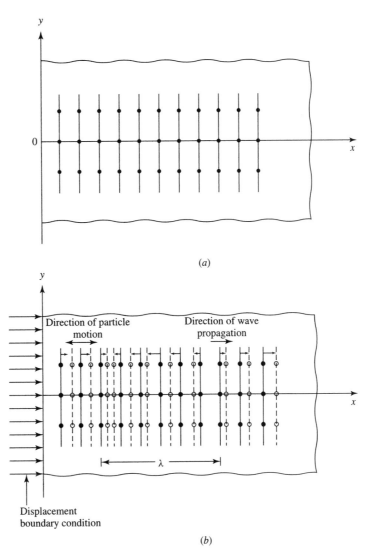

Figure 16.10 (*a*) Set of material points in the undisturbed half-space; (*b*) displacements of the set of material points due to a compressional wave.

where

$$s(t) = \frac{1}{\lambda + 2\mu}\tilde{s}(t)$$ (16.75)

In this case also, compressional waves are generated.

16.7.2 Shear or S Waves

The half-space shown in Fig. 16.11(*a*) can be disturbed and shear waves can be generated either by a displacement or a shear stress boundary condition, as indicated in

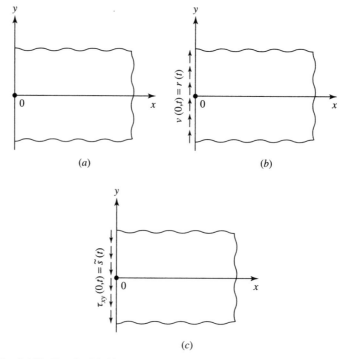

Figure 16.11 (a) Undisturbed half-space; (b) displacement boundary condition; (c) shear stress boundary condition.

Fig. 16.11(b) and (c). In Fig. 16.11(b), the initially undisturbed half-space is given a uniform displacement in the y direction, so that

$$v(0, t) = r(t) \tag{16.76}$$

where $r(t)$ is a known function of time that is zero for $t \leq 0$. The motion resulting from this displacement boundary condition (with no motions along the x and z directions) can be described by

$$u = 0$$
$$v = v(x, t) \tag{16.77}$$
$$w = 0$$

For this one-dimensional motion, the equation of motion, Eq. (A.33), reduces to

$$\frac{\partial v(x, t)}{\partial t^2} = \beta^2 \frac{\partial^2 v(x, t)}{\partial x^2} \tag{16.78}$$

where

$$\beta = \left(\frac{\mu}{\rho_0} \right)^{1/2} \tag{16.79}$$

Equation (16.78) is also called a one-dimensional wave equation, and β, given by Eq. (16.79), denotes the wave speed. The solution or waves given by Eq. (16.78) are called *shear* or *S waves* and β is called the *shear* or *S-wave speed* [1, 4]. The displacements of a set of material points due to a shear wave at a specific time t are shown in Fig. 16.12. Note that the material points move only in the y direction and their motions depend on x and t only. In Fig. 16.11(c), the boundary of the initially undisturbed half-space is given a uniform shear stress in the y direction, so that

$$\tau_{xy}(0, t) = \tilde{s}(t) \tag{16.80}$$

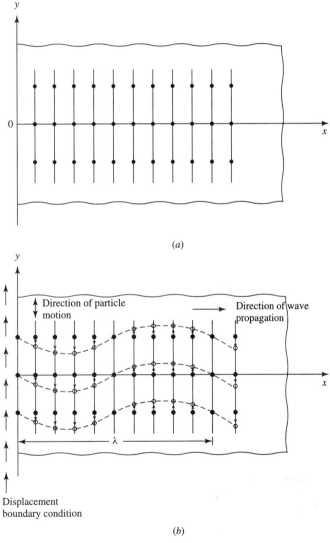

(a)

(b)

Figure 16.12   (*a*) Set of material points in the undisturbed half-space; (*b*) displacements of the set of material points due to a shear wave.

where $\tilde{s}(t)$ is a known function of time that is zero for $t \geq 0$. Equation (16.80), called the *shear stress boundary condition*, can be expressed in terms of the displacement v, using the stress–strain and strain–displacement relations, as

$$\tau_{xy}(0, t) = \mu \varepsilon_{xy}(0, t) = \mu \left[\frac{\partial u}{\partial y}(0, t) + \frac{\partial v}{\partial x}(0, t) \right]$$

$$= \mu \frac{\partial v}{\partial x}(0, t) = \tilde{s}(t) \qquad (16.81)$$

or

$$\frac{\partial v}{\partial x}(0, t) = s(t) \qquad (16.82)$$

where

$$s(t) = \frac{1}{\mu} \tilde{s}(t) \qquad (16.83)$$

In this case also, shear waves are generated.

16.8 FLEXURAL WAVES IN BEAMS

The equation of motion for the transverse motion of a thin uniform beam, according to Euler–Bernoulli theory, is given by

$$\frac{\partial^4 w(x, t)}{\partial x^4} + \frac{1}{c^2} \frac{\partial^2 w(x, t)}{\partial t^2} = 0 \qquad (16.84)$$

where

$$c = \sqrt{\frac{EI}{\rho A}} \qquad (16.85)$$

It can be observed that Eq. (16.84) differs from the one-dimensional wave equation, Eq. (16.1), studied earlier in terms of the following:

1. Equation (16.84) contains a fourth derivative with respect to x instead of the second derivative.

2. The constant c does not have the dimensions of velocity; its dimensions are in^2/sec and not the in./sec required for velocity.

Thus, the general solution of the wave equation,

$$w(x, t) = f(x - ct) + g(x + ct) \qquad (16.86)$$

will not be a solution of Eq. (16.84). As such, we will not be able to state that the motion given by Eq. (16.84) consists of waves traveling at constant velocity and without alteration of shape. Consider the solution of Eq. (16.84) for an infinitely long beam in the form of a harmonic wave traveling with velocity v in the positive x direction:

$$w(x, t) = A \cos \frac{2\pi}{\lambda}(x - vt) \equiv A \cos(kx - \omega t) \qquad (16.87)$$

where A is a constant, λ is the wavelength, v is the phase velocity, k is the wave number, and ω is the circular frequency of the wave, with the following interrelationships:

$$\omega = 2\pi f = kv \tag{16.88}$$

$$k = \frac{2\pi}{\lambda} \tag{16.89}$$

Substitution of Eq. (16.87) into Eq. (16.84) yields the velocity, also called the *wave velocity* or *phase velocity*, as

$$v = \frac{2\pi}{\lambda}c = \frac{2\pi}{\lambda}\sqrt{\frac{EI}{\rho A}} \tag{16.90}$$

Thus, unlike in the case of transverse vibration of a string, the velocity of propagation of a harmonic flexural wave is not a constant but varies inversely as the wavelength. The material or medium in which the wave velocity v depends on the wavelength is called a *dispersive medium*. Physically, it implies that a nonharmonic flexural pulse (of arbitrary shape) can be considered as the superposition of a number of harmonic waves of different wavelengths. Since each of the component harmonic waves has different phase velocity, a flexural pulse of arbitrary shape cannot propagate along the beam without dispersion, which results in a change in the shape of the pulse.

A pulse composed of several or a group of harmonic waves is called a *wave packet*, and the velocity with which the group of waves travel is called the *group velocity* [4, 5]. The group velocity, denoted by v_g, is the velocity with which the energy is propagated, and its physical interpretation can be seen by considering a wave packet composed of two simple harmonic waves of equal amplitude but slightly different frequencies $\omega + \Delta\omega$ and $\omega - \Delta\omega$. The waves can be described as

$$w_1(x, t) = A\cos(k_1 x - \omega_1 t) \tag{16.91}$$

$$w_2(x, t) = A\cos(k_2 x - \omega_2 t) \tag{16.92}$$

where A denotes the amplitude and

$$k_1 = k + \Delta k \tag{16.93}$$

$$k_2 = k - \Delta k \tag{16.94}$$

$$\omega_1 = \omega + \Delta\omega \tag{16.95}$$

$$\omega_2 = \omega - \Delta\omega \tag{16.96}$$

$$\Delta\omega = \tfrac{1}{2}(\omega_1 - \omega_2) \tag{16.97}$$

$$\Delta k = \tfrac{1}{2}(k_1 - k_2) \tag{16.98}$$

The wave packet can be represented by adding the two waves:

$$w_p = w_1(x, t) + w_2(x, t) = A\left[\cos(k_1 x - \omega_1 t) + \cos(k_2 x - \omega_2 t)\right] \tag{16.99}$$

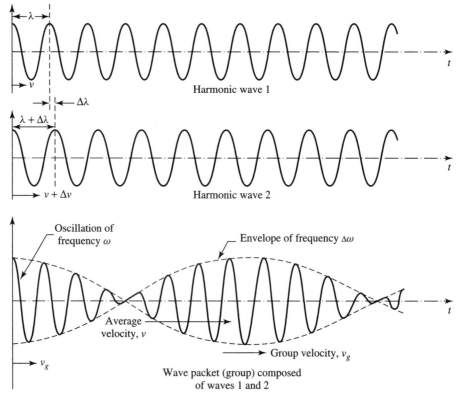

Figure 16.13 Wave packet and group velocity.

Equation (16.99) can be rewritten as

$$w_p(x, t) = 2A \cos(\Delta kx - \Delta\omega t) \cos(kx - \omega t) \tag{16.100}$$

The wave packet, given by Eq. (16.100), is shown graphically in Fig. 16.13. It can be seen that it contains a high-frequency cosine term (at frequency ω) and a low-frequency cosine term (at frequency $\Delta\omega$). The high-frequency oscillatory motion is called the *carrier wave* and moves at a velocity v, known as the *phase velocity*, given by

$$v = \frac{\omega}{k} \tag{16.101}$$

while the low-frequency oscillatory motion propagates at a velocity v_g, known as the *group velocity*, given by

$$v_g = \frac{\Delta\omega}{\Delta k} \tag{16.102}$$

The wave motion given by Eq. (16.100) is called an *amplitude-modulated carrier* and is shown in Fig. 16.13. It can be seen that the low-frequency term acts as a modulator on the carrier (denoted by the high-frequency term). Accordingly, the factor $\cos(kx - \omega t)$ represents the carrier wave and the factor $\cos(\Delta kx - \Delta\omega t)$ indicates the envelope moving at the group velocity. The behavior is similar to that of beats observed in coupled oscillators (see Section 1.9.2).

The group velocity can be represented in an alternative form using the basic relation $\omega = kv$ as

$$v_g = \lim_{\Delta k \to 0} \frac{\Delta \omega}{\Delta k} = \frac{d\omega}{dk} = \frac{d(kv)}{dk} = v + k\frac{dv}{dk} \tag{16.103}$$

where v is considered a function of k. Noting that $k = 2\pi/\lambda$, Eq. (16.103) can also be expressed in terms of the wavelength λ as

$$v_g = v - \lambda\frac{dv}{d\lambda} \tag{16.104}$$

In the case of propagation of flexural waves in beams, the group velocity is given by

$$v_g = v - \lambda\frac{dv}{d\lambda} = v - \lambda\left(-\frac{2\pi c}{\lambda^2}\right) = v + \frac{2\pi c}{\lambda} = 2v = \frac{4\pi}{\lambda}\sqrt{\frac{EI}{\rho A}} \tag{16.105}$$

Equation (16.105) shows that for flexural waves in beams, the group velocity is twice the wave velocity.

16.9 WAVE PROPAGATION IN AN INFINITE ELASTIC MEDIUM

In this section the elastic wave propagation in solid bodies is considered. Since the body is infinite, boundary interactions of waves need not be considered. The equations governing waves in infinite media are derived from the basic equations of elasticity. Two basic types of waves, dilatational and distortional waves, can propagate in an infinite medium. These two types of waves can exist independent of one another.

16.9.1 Dilatational Waves

To see the simplest types of waves generated in the solid body, differentiate the equations of motion (A.31), (A.33), and (A.34) with respect to x, y, and z, respectively, and add the resulting equations to obtain

$$(\lambda + 2\mu)\nabla^2\Delta = \rho\frac{\partial^2\Delta}{\partial t^2} \tag{16.106}$$

or

$$\frac{\partial^2\Delta}{\partial t^2} = c_1^2\nabla^2\Delta \tag{16.107}$$

where c_1, called the *velocity of wave propagation*, is given by

$$c_1 = \left(\frac{\lambda + 2\mu}{\rho}\right)^{1/2} \tag{16.108}$$

Equation (16.106) is called a wave equation that governs propagation of the dilatation Δ through the medium with a velocity c.

16.9.2 Distortional Waves

The dilatation (Δ) can be eliminated from the equations of motion (A.31), (A.33), and (A.34) to obtain the equation governing distortional waves. For this, differentiate Eqs. (A.31) and (A.34) with respect to z and y, respectively, and subtract the resulting equations one from the other to obtain

$$\mu\nabla^2\left(-\frac{\partial v}{\partial z} + \frac{\partial w}{\partial y}\right) = \rho\frac{\partial^2}{\partial t^2}\left(-\frac{\partial v}{\partial z} + \frac{\partial w}{\partial y}\right) \tag{16.109}$$

Noting that the quantity in parentheses denotes twice the rotation ω_x, Eq. (16.109) can be rewritten as

$$\mu\nabla^2\omega_x = \rho\frac{\partial^2\omega_x}{\partial t^2} \tag{16.110}$$

By using a similar procedure, the other two equations of motion can be written in terms of rotations ω_y and ω_z as

$$\mu\nabla^2\omega_y = \rho\frac{\partial^2\omega_y}{\partial t^2} \tag{16.111}$$

$$\mu\nabla^2\omega_z = \rho\frac{\partial^2\omega_z}{\partial t^2} \tag{16.112}$$

Each of Eqs. (16.110)–(16.112) denotes a wave equation that governs the propagation of rotational or equivoluminal or distortional or shear waves through the elastic medium with a velocity of wave propagation given by

$$c_2 = \left(\frac{\mu}{\rho}\right)^{1/2} \tag{16.113}$$

It can be seen from Eqs. (16.108) and (16.113) that the velocity of rotational waves is smaller than the velocity of dilatational waves. The dilatational and rotational waves denote the two basic types of wave motion possible in an infinite elastic medium.

16.9.3 Independence of Dilatational and Distortional Waves

In general, a wave consists of dilatation and rotation simultaneously [2, 4]. The dilatation propagates with velocity c_1 and the rotation propagates with velocity c_2. By setting dilatation Δ equal to zero in Eqs. (A.31), (A.33), and (A.34) one obtains the equations of motion:

$$\mu\nabla^2 u = \rho\frac{\partial^2 u}{\partial t^2} \tag{16.114}$$

$$\mu\nabla^2 v = \rho\frac{\partial^2 v}{\partial t^2} \tag{16.115}$$

$$\mu\nabla^2 w = \rho\frac{\partial^2 w}{\partial t^2} \tag{16.116}$$

Equations (16.114)–(16.116) describe distortional waves that propagate with a velocity c_2 given by Eq. (16.113). In fact, the motion implied by Eqs. (16.114)–(16.116) is that of rotation. Thus, the wave equations (A.31), (A.33), and (A.34), in the absence of dilatation, reduce to those obtained for rotation, Eqs. (16.110)–(16.112). Next, we eliminate rotations ω_x, ω_y and ω_z from the equations of motion. For this, we set $\omega_x = \omega_y = \omega_z = 0$:

$$\frac{\partial w}{\partial y} - \frac{\partial v}{\partial z} = \frac{\partial u}{\partial z} - \frac{\partial w}{\partial x} = \frac{\partial v}{\partial x} - \frac{\partial u}{\partial y} = 0 \tag{16.117}$$

and define a potential function ϕ such that

$$u = \frac{\partial \phi}{\partial x}, \qquad v = \frac{\partial \phi}{\partial y}, \qquad w = \frac{\partial \phi}{\partial z} \tag{16.118}$$

Using Eq. (16.118), the rotation ω_x, for example, can be seen to be zero:

$$\omega_x = \frac{1}{2}\left(\frac{\partial w}{\partial y} - \frac{\partial v}{\partial z}\right) = \frac{1}{2}\left(\frac{\partial^2 \phi}{\partial y \partial z} - \frac{\partial^2 \phi}{\partial z \partial y}\right) = 0 \tag{16.119}$$

Using Eq. (16.118), the dilatation can be written as

$$\Delta = \frac{\partial u}{\partial x} + \frac{\partial v}{\partial y} + \frac{\partial w}{\partial z} = \nabla^2 \phi \tag{16.120}$$

Differentiation of Eq. (16.120) with respect to x gives

$$\frac{\partial \Delta}{\partial x} = \frac{\partial}{\partial x}(\nabla^2 \phi) = \nabla^2\left(\frac{\partial \phi}{\partial x}\right) = \nabla^2 u \tag{16.121}$$

and using a similar procedure leads to

$$\frac{\partial \Delta}{\partial y} = \nabla^2 v \tag{16.122}$$

$$\frac{\partial \Delta}{\partial z} = \nabla^2 w \tag{16.123}$$

Substitution of Eqs. (16.121), (16.122), and (16.123) into the equations of motion, (A.31), (A.33), and (A.34) respectively, yields

$$(\lambda + 2\mu)\nabla^2 u = \rho\frac{\partial^2 u}{\partial t^2} \tag{16.124}$$

$$(\lambda + 2\mu)\nabla^2 v = \rho\frac{\partial^2 v}{\partial t^2} \tag{16.125}$$

$$(\lambda + 2\mu)\nabla^2 w = \rho\frac{\partial^2 w}{\partial t^2} \tag{16.126}$$

Equations (16.124)–(16.126) describe dilatational waves that propagate with a velocity c_1 given by Eq. (16.108).

Notes

1. If the displacement components u, v, and w depend only on one coordinate, such as x, instead of x, y, and z, the resulting waves are called *plane waves* that propagate in the direction of x. Assuming that

$$u = u(x, t), \qquad v = v(x, t) \qquad w = w(x, t) \tag{16.127}$$

the dilatation is given by

$$\Delta = \frac{\partial u}{\partial x} \tag{16.128}$$

and the equations of motion (A.31), (A.33) and (A.34) reduce to

$$(\lambda + 2\mu)\frac{\partial^2 u}{\partial x^2} = \rho \frac{\partial^2 u}{\partial t^2} \tag{16.129}$$

$$\mu \frac{\partial^2 v}{\partial x^2} = \rho \frac{\partial^2 v}{\partial t^2} \tag{16.130}$$

$$\mu \frac{\partial^2 w}{\partial x^2} = \rho \frac{\partial^2 w}{\partial t^2} \tag{16.131}$$

Equations (16.129) describes a longitudinal wave in which the direction of motion is parallel to the direction of propagation of the wave. This wave is a *dilatational wave* and propagates with a velocity c_1. Each of Eqs. (16.130) and (16.131) describes a transverse wave in which the direction of motion is perpendicular to the direction of propagation of the wave. These are *distortional waves* and propagate with a velocity c_2.

2. *Dilatational waves* are also called *irrotational, longitudinal,* or *primary* (P) waves, and *distortional waves* are also known as *equivoluminal, rotational, shear* or *secondary* (S) *waves*.

3. All the wave equations derived in this section have the general form

$$c^2 \nabla^2 \eta = \rho \frac{\partial^2 \eta}{\partial t^2} \tag{16.132}$$

Consider a specific case in which the deformation takes place along the x coordinate so that Eq. (16.132) can be written as

$$c^2 \frac{\partial^2 \eta}{\partial x^2} = \frac{\partial^2 \eta}{\partial t^2} \tag{16.133}$$

The general solution (or D'Alembert's solution) of Eq. (16.133) can be expressed as

$$\eta(x, t) = f(x - ct) + g(x + ct) \tag{16.134}$$

where f denotes the wave traveling in the $+x$ direction and g indicates the wave traveling in the $-x$ direction. If the disturbance occurs only at one point in the elastic medium, the deformation depends only on the radial distance (r) from the point. The radial distance r can be expressed in terms of the Cartesian coordinates x, y, and z as

$$r = \sqrt{x^2 + y^2 + z^2} \tag{16.135}$$

and hence the derivatives of η with respect to x, y and z can be expressed as

$$\frac{\partial^2 \eta}{\partial x^2} = \frac{x^2}{r^2} \frac{\partial^2 \eta}{\partial r^2} + \frac{r^2 - x^2}{r^3} \frac{\partial \eta}{\partial r} \tag{16.136}$$

$$\frac{\partial^2 \eta}{\partial y^2} = \frac{y^2}{r^2} \frac{\partial^2 \eta}{\partial r^2} + \frac{r^2 - y^2}{r^3} \frac{\partial \eta}{\partial r} \tag{16.137}$$

$$\frac{\partial^2 \eta}{\partial z^2} = \frac{z^2}{r^2} \frac{\partial^2 \eta}{\partial r^2} + \frac{r^2 - z^2}{r^3} \frac{\partial \eta}{\partial r} \tag{16.138}$$

Addition of Eqs. (16.136)–(16.138) yields

$$\nabla^2 \eta = \frac{\partial^2 \eta}{\partial r^2} + \frac{2}{r} \frac{\partial \eta}{\partial r} \tag{16.139}$$

and hence Eq. (16.132) can be written as

$$c^2 \left(\frac{\partial^2 \eta}{\partial r^2} + \frac{2}{r} \frac{\partial \eta}{\partial r} \right) = \nabla^2 \eta \tag{16.140}$$

or

$$c^2 \frac{\partial^2 (r\eta)}{\partial r^2} = \frac{\partial^2 (r\eta)}{\partial t^2} \tag{16.141}$$

Equation (16.141) is known as the *spherical wave equation*, whose general solution can be expressed as

$$r\eta = f(r - ct) + g(r + ct) \tag{16.142}$$

In Eq. (16.142), f denotes a diverging spherical wave and g indicates a converging spherical wave from the point of disturbance. The amplitude of the wave (η) is, in general, inversely proportional to the radial distance r from the point of disturbance.

16.10   RAYLEIGH OR SURFACE WAVES

As seen earlier, two types of waves, dilatational and distortional waves, can exist in an isotropic infinite elastic medium. When there is a boundary, as in the case of an elastic half-space, a third type of waves, whose effects are confined close to the bounding surface, may exist. These waves were first investigated by Rayleigh [7] and hence are called *Rayleigh* or *surface waves*. The effect of surface waves decreases rapidly along the depth of the material, and their velocity of propagation is smaller than those of P and S waves. The discovery of surface waves was closely related to seismological studies, where it is observed that earthquake tremors usually consist of two minor disturbances, corresponding to the arrival of P and S waves followed closely by a third tremor that causes significant damage. The third wave (surface wave) is found to be associated with significant energy that is dissipated less rapidly than the P and S waves and is essentially confined to the ground surface. To study these surface waves, consider a semi-infinite elastic medium bounded by the xy plane, with the z axis pointing into the material, as shown in Fig. 16.14. Assume a wave that is propagating on the bounding surface along the x direction with its crests parallel to the y axis. Hence, all the components of

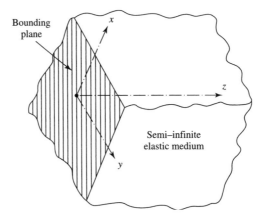

Figure 16.14 Semi-infinite medium.

displacement u, v, and w are independent of y. For this case, the equations of motion (A.31), (A.33) and (A.34) can be written as

$$(\lambda + \mu)\frac{\partial \Delta}{\partial x} + \mu \nabla^2 u = \rho \frac{\partial^2 u}{\partial t^2} \tag{16.143}$$

$$\mu \nabla^2 v = \rho \frac{\partial^2 v}{\partial t^2} \tag{16.144}$$

$$(\lambda + \mu)\frac{\partial \Delta}{\partial z} + \mu \nabla^2 w = \rho \frac{\partial^2 w}{\partial t^2} \tag{16.145}$$

where

$$\Delta = \frac{\partial u}{\partial x} + \frac{\partial w}{\partial z} \tag{16.146}$$

$$\nabla^2 = \frac{\partial^2}{\partial x^2} + \frac{\partial^2}{\partial z^2} \tag{16.147}$$

Note that Eq. (16.144) describes distortional waves that are not confined just to the bounding surface. Also, Eqs. (16.143) and (16.145) do not contain v. Hence, we assume that $v = 0$ and consider only Eqs. (16.143) and (16.145) in the analysis. Define two potential functions ϕ and ψ such that

$$u = \frac{\partial \phi}{\partial x} + \frac{\partial \psi}{\partial z} \tag{16.148}$$

$$w = \frac{\partial \phi}{\partial z} - \frac{\partial \psi}{\partial x} \tag{16.149}$$

Since v and y are not considered, dilatation Δ and rotation ω_y are given by

$$\Delta = \frac{\partial u}{\partial x} + \frac{\partial w}{\partial z} = \nabla^2 \phi \tag{16.150}$$

$$\omega_y = \frac{1}{2}\left(\frac{\partial u}{\partial z} - \frac{\partial w}{\partial x}\right) = \frac{1}{2}\nabla^2 \psi \tag{16.151}$$

$$\omega_x = \frac{1}{2}\left(\frac{\partial w}{\partial y} - \frac{\partial v}{\partial z}\right) = 0 \tag{16.152}$$

$$\omega_z = \frac{1}{2}\left(\frac{\partial v}{\partial x} - \frac{\partial u}{\partial y}\right) = 0 \tag{16.153}$$

It can be seen that the definitions of ϕ and ψ given by Eqs. (16.148) and (16.149) separated the effects of dilatation and rotation since ϕ is associated with dilatation only and ψ with rotation only in Eqs. (16.150)–(16.153). Substitution of Eqs. (16.148), (16.149), and (16.150) into Eqs. (16.153) and (16.145) permits the equations of motion to be written in terms of ϕ and ψ as

$$(\lambda + 2\mu)\frac{\partial}{\partial x}(\nabla^2\phi) + \mu\frac{\partial}{\partial z}(\nabla^2\psi) = \rho\frac{\partial}{\partial x}\left(\frac{\partial^2\phi}{\partial t^2}\right) + \rho\frac{\partial}{\partial z}\left(\frac{\partial^2\psi}{\partial t^2}\right) \tag{16.154}$$

$$(\lambda + 2\mu)\frac{\partial}{\partial z}(\nabla^2\phi) - \mu\frac{\partial}{\partial x}(\nabla^2\psi) = \rho\frac{\partial}{\partial z}\left(\frac{\partial^2\phi}{\partial t^2}\right) - \rho\frac{\partial}{\partial x}\left(\frac{\partial^2\psi}{\partial t^2}\right) \tag{16.155}$$

Equations (16.154) and (16.155) will be satisfied if ϕ and ψ are the solutions of the equations (can be verified by direct substitution):

$$(\lambda + 2\mu)\nabla^2\phi = \rho\frac{\partial^2\phi}{\partial t^2} \quad \text{or} \quad c_1^2\nabla^2\phi = \frac{\partial^2\phi}{\partial t^2} \tag{16.156}$$

$$\mu\nabla^2\psi = \rho\frac{\partial^2\psi}{\partial t^2} \quad \text{or} \quad c_2^2\nabla^2\psi = \frac{\partial^2\psi}{\partial t^2} \tag{16.157}$$

where c_1 and c_2 denote the velocities of dilatational and distortional waves given by Eqs. (16.108) and (16.113), respectively. Next, the wave described by Eqs. (16.156) and (16.157) is assumed to be harmonic or sinusoidal propagating in the positive x direction with velocity c at a frequency ω. The velocity can be expressed as

$$c = \frac{\omega}{n} \equiv \omega\lambda \tag{16.158}$$

where n is the wave number and λ is the wavelength ($n = 1/\lambda$). The solutions of Eqs. (16.156) and (16.157) are assumed to be of the following complex harmonic form:

$$\phi = Z_1(z)e^{i(\omega t - nx)} \tag{16.159}$$

$$\psi = Z_2(z)e^{i(\omega t - nx)} \tag{16.160}$$

where the functions $Z_1(z)$ and $Z_2(z)$ indicate the dependence of the amplitudes of ϕ and ψ on the coordinate z. When Eq. (16.159) is used in Eq. (16.156), we obtain

$$c_1^2\left(\frac{\partial^2}{\partial x^2} + \frac{\partial^2}{\partial z^2}\right)Z_1(z)e^{i(\omega t - nx)} = \frac{\partial^2}{\partial t^2}[Z_1(z)e^{i(\omega t - nx)}]$$

or

$$\frac{d^2Z_1}{dz^2} - \left(n^2 - \frac{\omega^2}{c_1^2}\right)Z_1 = 0 \tag{16.161}$$

The solution of this second-order ordinary differential equation can be expressed as

$$Z_1(z) = \tilde{c}_1 e^{-\alpha z} + \tilde{c}_2 e^{\alpha z} \tag{16.162}$$

where

$$\alpha = \left(n^2 - \frac{\omega^2}{c_1^2} \right)^{1/2} \tag{16.163}$$

For real values of the amplitude of ϕ, α must be positive. For positive values of α, the second term of Eqs. (16.162), and hence the amplitude of the wave, increases with increasing values of z. This is not physically possible from the point of view of conservation of energy. Hence, $\tilde{c}_2$ must be zero in Eq. (16.162) for a surface wave. By substituting Eq. (16.160) into Eq. (16.157) and proceeding as in the case of ϕ, we obtain the solution of $Z_2(z)$ as

$$Z_2(z) = c_3 e^{-\beta z} + c_4 e^{\beta z} \tag{16.164}$$

where $c_4 = 0$ and

$$\beta = \left(n^2 - \frac{\omega^2}{c_2^2} \right)^{1/2} \tag{16.165}$$

Using Eqs. (16.162) and (16.164) in Eqs. (16.159) and (16.160), the solutions of ϕ and ψ of Eqs. (16.156) and (16.157) representing a harmonic wave propagating in the positive z direction with frequency ω and wave number n and decaying with increasing values of z can be expressed as

$$\phi = \tilde{c}_1 e^{-\alpha z + i(\omega t - nx)} \tag{16.166}$$

$$\psi = c_3 e^{-\beta z + i(\omega t - nx)} \tag{16.167}$$

where α and β are given by Eqs. (16.163) and (16.165), respectively, and $\tilde{c}_1$ and c_3 are constants to be determined from the boundary conditions. The free surface conditions applicable to the boundary surface ($z = 0$) of the elastic half-space are

$$\sigma_{zz} = \sigma_{xz} = \sigma_{yz} = 0 \tag{16.168}$$

Using the stress–strain relations of Eq. (A.16) and the strain–displacement relations of Eq. (A.2), the normal stress σ_{zz} can be expressed as

$$\sigma_{zz} = \lambda \Delta + 2\mu \frac{\partial u}{\partial x} \tag{16.169}$$

where Δ is given by Eq. (16.150). Using Eqs. (16.148) and (16.149) for u and w, Eq. (16.169) can be written as

$$\sigma_{zz} = \lambda \left(\frac{\partial^2 \phi}{\partial x^2} + \frac{\partial^2 \psi}{\partial x \partial z} + \frac{\partial^2 \phi}{\partial z^2} - \frac{\partial^2 \psi}{\partial x \partial z} \right) + 2\mu \left(\frac{\partial^2 \phi}{\partial z^2} - \frac{\partial^2 \psi}{\partial x \partial z} \right)$$

or

$$\sigma_{zz} = (\lambda + 2\mu) \frac{\partial^2 \phi}{\partial z^2} + \lambda \frac{\partial^2 \phi}{\partial x^2} - 2\mu \frac{\partial^2 \psi}{\partial x \partial z} \tag{16.170}$$

Substitution of Eqs. (16.166) and (16.167) into Eq. (16.170) gives

$$\sigma_{zz} = [(\lambda + 2\mu)\alpha^2 - \lambda n^2]\tilde{c}_1 e^{-\alpha z + i(\omega t - nx)} - (2\mu i\beta n)c_3 e^{-\beta z + i(\omega t - nx)} \qquad (16.171)$$

The condition $\sigma_{zz}(z = 0) = 0$ can be written as

$$[(\lambda + 2\mu)\alpha^2 - \lambda n^2]\tilde{c}_1 - (2\mu i\beta n)c_3 = 0 \qquad (16.172)$$

The shear stress σ_{xz} can be expressed using Eqs. (A.16) and (A.8) as

$$\sigma_{xz} = \mu\varepsilon_{xz} = \mu\left(\frac{\partial u}{\partial z} + \frac{\partial w}{\partial x}\right) \qquad (16.173)$$

Using Eqs. (16.148) and (16.149) for u and w, Eq. (16.173) can be written as

$$\sigma_{xz} = \mu\left(2\frac{\partial^2\phi}{\partial x\partial z} + \frac{\partial^2\psi}{\partial z^2} - \frac{\partial^2\psi}{\partial x^2}\right) \qquad (16.174)$$

Substitution of Eqs. (16.166) and (16.167) into Eq. (16.174) yields

$$\sigma_{xz} = \mu[2in\alpha\tilde{c}_1 e^{-\alpha z + i(\omega t - nx)} + (\beta^2 + n^2)c_3 e^{-\beta z + i(\omega t - nx)}] \qquad (16.175)$$

The condition $\sigma_{xz}(z = 0) = 0$ can be expressed as

$$(2in\alpha)\tilde{c}_1 + (\beta^2 + n^2)c_3 = 0 \qquad (16.176)$$

Finally the shear stress σ_{yz} can be expressed, using Eq. (A.16), as

$$\sigma_{yz} = \mu\varepsilon_{yz} = \mu\left(\frac{\partial w}{\partial y} + \frac{\partial v}{\partial z}\right) \qquad (16.177)$$

Nothing that u and w are independent of y and v is assumed to be zero, Eq. (16.177) gives $\sigma_{yz} = 0$ throughout the elastic half-space. Hence, the boundary condition $\sigma_{yz}(z = 0) = 0$ is satisfied automatically. Equations (16.172) and (16.176) represent two simultaneous homogeneous algebraic equations with $\tilde{c}_1$ and c_3 as unknowns. For a nontrivial solution of $\tilde{c}_1$ and c_3, the determinant of their coefficient matrix is set equal to zero to obtain

$$[(\lambda + 2\mu)\alpha^2 - \lambda n^2](\beta^2 + n^2) + 2\mu i\beta n \cdot 2in\alpha = 0$$

or

$$[(\lambda + 2\mu)\alpha^2 - \lambda n^2](\beta^2 + n^2) = 4\alpha\beta\mu n^2 \qquad (16.178)$$

Squaring both sides of Eq. (16.178) and using Eqs. (16.163) and (16.165) for α and β, respectively, we obtain

$$\left[-(\lambda + 2\mu)\frac{\omega^2}{c_1^2} + 2\mu n^2\right]^2\left(2n^2 - \frac{\omega^2}{c_2^2}\right)^2 = 16\mu^2 n^4\left(n^2 - \frac{\omega^2}{c_2^2}\right)\left(n^2 - \frac{\omega^2}{c_1^2}\right) \qquad (16.179)$$

Dividing both sides of Eq. (16.179) by $\mu^2 n^8$ leads to

$$\left(2 - \frac{\lambda + 2\mu}{\mu} \frac{\omega^2}{c_1^2 n^2}\right)^2 \left(2 - \frac{\omega^2}{c_2^2 n^2}\right)^2 = 16\left(1 - \frac{\omega^2}{c_1^2 n^2}\right)\left(1 - \frac{\omega^2}{c_2^2 n^2}\right) \tag{16.180}$$

Note the following relationship from Eqs. (16.108) and (16.113) and Eqs. (A.22) and (A.23):

$$\frac{c_2^2}{c_1^2} = \frac{\mu}{\lambda + 2\mu} = \frac{E}{2(1+v)}\left\{\frac{1}{[vE/(1+v)(1-2v)] + 2E/2(1+v)}\right\} = \frac{1-2v}{2-2v} \tag{16.181}$$

Introducing the notation

$$\frac{c_2^2}{c_1^2} = \frac{1-2v}{2-2v} = \gamma^2 \tag{16.182}$$

$$p^2 = \frac{\omega^2}{c_2^2 n^2} \tag{16.183}$$

we can express

$$\frac{\omega^2}{c_1^2 n^2} = \gamma^2 \frac{\omega^2}{c_2^2 n^2} = \gamma^2 p^2 \tag{16.184}$$

Substituting Eqs. (16.181)–(16.184) into Eq. (16.180) gives

$$(2 - p^2)^2 (2 - p^2)^2 = 16(1 - \gamma^2 p^2)(1 - p^2)$$

or

$$p^2[p^6 - 8p^4 + (24 - 16\gamma^2)p^2 + (16\gamma^2 - 16)] = 0 \tag{16.185}$$

Since $p^2 \neq 0$, Eq. (16.185) leads to

$$p^6 - 8p^4 + 8(3 - 2\gamma^2)p^2 + 16(\gamma^2 - 1) = 0 \tag{16.186}$$

To investigate the roots of Eq. (16.186), we first note that it is a cubic equation in p^2, and second, that the roots depend on Poisson's ratio [see Eq. (16.182)]. Since p is defined as the ratio of the velocity of the surface wave to that of the distortion wave in Eq. (16.183), the velocity of propagation of the surface wave will be a constant for any given material. The roots of Eq. (16.186) can be real, imaginary, or complex, depending on the range of Poisson's ratio [1, 4]:

1. For v less than about 0.263; all three roots (p^2) are real.
2. For v about 0.263; one root is real and two roots are complex conjugates.
3. Complex roots will not be valid in the present case since this will result in the attenuations of ϕ and ψ with respect to time, as if damping were present, which is not the case in the present problem [1, 4].

Example 16.1   Investigate the variation of the velocity of propagation of surface waves for different values of the Poisson's ratio (ν).

SOLUTION   First we consider the values of $\nu = 0.25$. It can be verified that this value of ν corresponds to a surface wave with $\lambda = \mu$ and is often used for rock. For $\nu = 0.25$, Eq. (16.182) gives

$$\gamma^2 = \frac{1 - 2\nu}{2 - 2\nu} = \frac{1}{3} \tag{E16.1.1}$$

and Eq. (16.186) becomes

$$p^6 - 8p^4 - \frac{56}{3}p^2 - \frac{32}{3} = 0 \tag{E16.1.2}$$

The roots of Eq. (E16.1.2) are given by

$$p^2 = 4, \quad 2 + \frac{2}{\sqrt{3}}, \quad 2 - \frac{2}{\sqrt{3}} \tag{E16.1.3}$$

and the corresponding values of α and β, given by Eqs. (16.163) and (16.165), can be found as follows:

$$\frac{\alpha^2}{n^2} = 1 - \frac{\omega^2}{c_1^2 n^2} = 1 - \frac{\omega^2}{c_2^2 n^2}\frac{c_2^2}{c_1^2} = 1 - \gamma^2 p^2 = 1 - \frac{p^2}{3} \tag{E16.1.4}$$

$$\frac{\beta^2}{n^2} = 1 - \frac{\omega^2}{c_2^2 n^2} = 1 - p^2 \tag{E16.1.5}$$

For $p^2 = 4$:

$$\frac{\alpha^2}{n^2} = 1 - \frac{4}{3} = -\frac{1}{3}, \quad \frac{\alpha}{n} = \text{imaginary}$$

$$\frac{\beta^2}{n^2} = 1 - 4 = -3, \quad \frac{\beta}{n} = \text{imaginary}$$

For $p^2 = 2 + 2/\sqrt{3}$:

$$\frac{\alpha^2}{n^2} = 1 - \frac{2}{3} - \frac{2}{3\sqrt{3}} \approx -0.051567, \quad \frac{\alpha}{n} = \text{imaginary}$$

$$\frac{\beta^2}{n^2} = 1 - 2 - \frac{2}{\sqrt{3}} \approx -2.15470, \quad \frac{\beta}{n} = \text{imaginary}$$

For $p^2 = 2 - 2/\sqrt{3}$:

$$\frac{\alpha^2}{n^2} = 1 - \frac{2}{3} + \frac{2}{3\sqrt{3}} \approx 0.71823, \quad \frac{\alpha}{n} \approx 0.84748$$

$$\frac{\beta^2}{n^2} = 1 - 2 + \frac{2}{\sqrt{3}} \approx 0.15470, \quad \frac{\beta}{n} \approx 0.39332$$

It can be seen that the larger two values of p^2 in Eq. (E16.1.3) give imaginary values of α/n and β/n which violate the requirement of $\alpha/n > 0$ and $\beta/n > 0$. Hence, the only valid value of p^2 yields

$$\frac{\omega}{c_2 n} = p = \left(2 - \frac{2}{\sqrt{3}}\right)^{1/2} = 0.91940$$

or

$$\frac{\omega}{n} = 0.91940 c_2 \tag{E16.1.6}$$

The variations of $\omega/c_2 n$ (and $c_2/c_1 = 1/\gamma$) for different values of Poisson's ratio (ν) are shown graphically in Fig. 16.15. Since the velocity of propagation, $c = \omega/n$, is independent of the frequency for any given material, the surface wave propagates with no dispersion at a velocity slightly smaller than the velocity of propagation of distortion or S waves.

Example 16.2 Derive expressions for the variations of u and w with the depth of material (z).

SOLUTION The displacement components u and w can be expressed, using Eqs. (16.148), (16.149), (16.166), and (16.167) as

$$u = -\tilde{c}_1 in e^{-\alpha z + i(\omega t - nx)} - c_3 \beta e^{-\beta z + i(\omega t - nx)} \tag{E16.2.1}$$

$$w = -\tilde{c}_1 \alpha e^{-\alpha z + i(\omega t - nx)} - c_3 in e^{-\beta z + i(\omega t - nx)} \tag{E16.2.2}$$

Equation (16.176) gives

$$c_3 = -\frac{2in\alpha}{\beta^2 + n^2} \tilde{c}_1 \tag{E16.2.3}$$

$$u = \tilde{c}_1 \left(-in e^{-\alpha z} + \frac{2in\alpha\beta}{\beta^2 + n^2} e^{-\beta z}\right) e^{i(\omega t - nx)} \tag{E16.2.4}$$

$$w = \tilde{c}_1 \alpha \left(-e^{-\alpha z} + \frac{2n^2}{\beta^2 + n^2} e^{-\beta z}\right) e^{i(\omega t - nx)} \tag{E16.2.5}$$

The physical displacements u and w are given by the real parts of Eqs. (E16.2.4) and (E16.2.5) as

$$u = \tilde{c}_1 n \left(-e^{-\alpha z} - \frac{2\alpha\beta}{\beta^2 + n^2} e^{-\beta z}\right) \sin(\omega t - nx) \tag{E16.2.6}$$

$$w = \tilde{c}_1 \alpha \left(-e^{-\alpha z} + \frac{2n^2}{\beta^2 + n^2} e^{-\beta z}\right) \cos(\omega t - nx) \tag{E16.2.7}$$

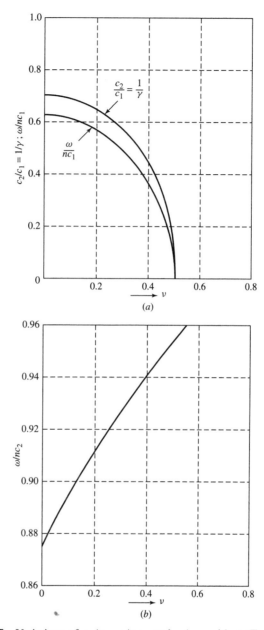

Figure 16.15   Variations of c_2/c_1, ω/nc_1, and ω/nc_2 with ν. (From Ref. [8].)

16.11   RECENT CONTRIBUTIONS

Comprehensive Studies   A brief history of the study of wave propagation in solids and a presentation of wave motion in elastic strings, rods, beams, membranes, plates, shells, semi-infinite media, and infinite media have been given by Graff [4]. Kolsky summarized the basic developments in the field of stress waves in Ref. [9].

Longitudinal Waves   A theory of propagation of longitudinal stress waves in a cylindrical rod with several step changes in the cross-sectional area was developed by Beddone [10]. The analysis obtained a transient solution of the one-dimensional wave equation by means of Laplace transform methods based on the concepts of traveling waves and reflection and transmission coefficients.

Wave Propagation in Periodic Structures   The problem of free coupled longitudinal and flexural waves of a periodically supported beam was studied by Lee and Yeen [11]. It was shown that the characteristic or dispersion equation can be factorized into product form, which simplifies the analysis and classification of the dynamic nature of the system. Sen Gupta [12] studied the propagation of flexural waves in doubly periodic structures consisting of the repetition of a basic unit that is a periodic structure in itself. The analysis is simplified by introducing a direct and a cross-chain receptance for multispan structures and by utilizing the concept of the equivalent internal restraint.

Wave Propagation Under Moving Loads   Ju used the three-dimensional finite element method to simulate the soil vibrations due to high-speed trains moving across bridges in Ref. [13]. He first analyzed a bridge system passed by trains. Then the pier forces and moments calculated were applied to a pile cap to simulate wave propagation in the soil.

Waves Through Plate or Beam Junctions   In a study of elastic wave transmission through plate–beam junctions by Langley and Heron [14], a generic plate–beam junction was considered to be composed of an arbitrary number of plates either coupled through a beam or coupled directly along a line. The effects of shear deformation, rotary inertia, and warping were included in the analysis of the beam, and due allowance was made for offsets between the plate attachment lines and the shear axis of the beam.

Vibration Analysis Using a Wave Equation   Langley showed that the vibrations of beams and plates may be analyzed in the frequency domain by using a wave equation instead of the conventional differential equations of motion provided that certain assumptions are made regarding the response of the system in the vicinity of a structural discontinuity [15].

Measurement of Wave Intensity   Halkyard and Mace [16] presented a Fourier series approach to the measurement of flexural wave intensity in plates. The approach is based on the fact that in regions sufficiently remote from excitation and discontinuities, the flexural motion of a plate can be expressed as the sum of plane propagating waves.

Lamb Waves   With the aim of clarifying the manner in which the dispersion curves for real- and imaginary-valued Lamb modes in a free, isotropic, elastic plate vary with the Poisson ratio, Freedman presented a set of Lamb mode spectra at fixed values of the Poisson ratio covering its full range [17].

REFERENCES

1. A. Bedford and D. S. Drumheller, *Introduction to Elastic Wave Propagation*, Wiley, Chichester, West Jeessey, England, 1994.
2. S. K.Clark, *Dynamics of Continuous Elements*, Prentice-Hall, Englewood Cliffs, NJ, 1972.
3. E.Volterra and E. C. Zachmanoglou, *Dynamics of Vibrations*, Charles E. Merrill, Columbus, OH, 1965.

4. K. F. Graff, *Wave Motion in Elastic Solids*, Ohio State University Press, Columbus, OH 1975.

5. L. Meirovitch, *Analytical Methods in Vibrations*, Macmillan, New York, 1967.

6. H. J. Pain, *The Physics of Vibrations and Waves*, 4th ed, Wiley, Chichester, West Jeessey, England, 1993.

7. J. W. S. Rayleigh, On waves propagated along the plane surface of an elastic solid, *Proceedings of the London Mathematical Society*, Vol. 17, pp. 4–11, 1887.

8. A. Knopoff, On Rayleigh wave velocities, *Bulletin of the Seismological Society of America*, Vol. 42, pp. 307–308, 1952.

9. H. Kolsky, *Stress Waves in Solids*, Dover, New York, 1963.

10. Beddone, Propagation of elastic stress waves in a necked rod, *Journal of Sound and Vibration*, Vol. 2, No. 2, pp. 150–164, 1965.

11. S. Y. Lee and W. F. Yeen, Free coupled longitudinal and flexural waves of a periodically supported beam, *Journal of Sound and Vibration*, Vol. 142, No. 2, pp. 203–211, 1990.

12. G. Sen Gupta, Propagation of flexural waves in doubly-periodic structures, *Journal of Sound and Vibration*, Vol. 20, No. 1, pp. 39–49, 1972.

13. S. H. Ju, Finite element analysis of wave propagations due to high-speed train across bridges, *International Journal for Numerical Methods in Engineering*, Vol. 54, No. 9, pp. 1391–1408, 2002.

14. R. S. Langley and K. H. Heron, Elastic wave transmission through plate/beam junctions, *Journal of Sound and Vibration*, Vol. 143, No. 2, pp. 241–253, 1990.

15. R. S. Langley, Analysis of beam and plate vibrations by using the wave equation, *Journal of Sound and Vibration*, Vol. 150, No. 1, pp. 47–65, 1991.

16. C. R. Halkyard and B. R. Mace, A Fourier series approach to the measurement of flexural wave intensity in plates, *Journal of Sound and Vibration*, Vol. 203, No. 1, pp. 101–126, 1997.

17. A. Freedman, The variation, with the Poisson ratio, of Lamb modes in a free plate, I: General spectra, *Journal of Sound and Vibration*, Vol. 137, No. 2, pp. 209–230, 1990.

PROBLEMS

16.1 Determine the velocity of propagation of longitudinal, torsional, and bending waves in a steel bar with $E = 207$ GPa, $G = 79.3$ GPa, and a unit weight of 76.5 kN/m^3.

16.2 Consider a bar with density 7500 kg/m^3 and Lamé constants $\lambda = 1.15 \times 10^{11}$ Pa and $\mu = 0.75 \times 10^{11}$ Pa. Determine the compressional and shear wave velocities.

16.3 Consider the following initial conditions for a long, transversely vibrating string:

$$w(x, 0) = \begin{cases} 2, & -1 < x < 1 \\ 0, & |x| > 1 \end{cases}$$

$$\dot{w}(x, 0) = 0$$

Show graphically the propagation of the wave along the string at different instants of time.

16.4 (a) Derive the equation of motion of a taut string resting on a foundation of elastic modulus k.

(b) Derive the condition to be satisfied between γ and ω for the propagation of a wave of the form $w(x, t) = Ae^{\gamma x - \omega t}$ in the string.

16.5 Derive the equation of motion for the longitudinal vibration of an inhomogeneous bar for which Young's modulus E and density ρ are given by $E = E_0(1 + \varepsilon_1 x^3)$ and $\rho = \rho_0(1 + \varepsilon_2 x)$.

16.6 The initial conditions of an unbounded elastic material are given by

$$u(x, 0) = \begin{cases} 0, & x < 0 \\ A \sin \pi x, & 0 \le x \le 1 \\ 0, & x > 1 \end{cases}$$

$$\dot{u}(x, 0) = 0$$

where A is a constant. Plot the displacement of the material at $t = 0, t = 1/(2c), t = 1/c, t = 2/c$, and $t = 4/c$.

16.7 **(a)** Show that the ratio of velocities of P and S waves depends only on Poisson's ratio of the material.

(b) Determine the ratio of velocities of P and S waves for the following materials:

Aluminum, $\nu = 0.334$	Copper, $\nu = 0.326$
Brass, $\nu = 0.324$	Glass, $\nu = 0.245$
Carbon steel, $\nu = 0.292$	Stainless steel, $\nu = 0.305$
Cast iron, $\nu = 0.211$	

16.8 Consider a thin steel beam of circular cross section with Young's modulus 207 GPa, radius 1 cm, and unit weight 76.5 kN/m^3. Determine the flexural wave velocity and the group velocity in the beam.

16.9 Consider a thick beam for which the equation of motion, according to Timoshenko theory, is given by

$$\frac{EI}{\rho A}\frac{\partial^4 w}{\partial x^4} - \frac{I}{A}\left(1 + \frac{E}{kG}\right)\frac{\partial^4 w}{\partial x^2 \partial t^2} + \frac{\partial^2 w}{\partial t^2} + \frac{\rho I}{kAG}\frac{\partial^4 w}{\partial t^4} = 0$$
(P16.1)

Assuming the propagation of harmonic waves in an infinite beam with solution

$$w(x, t) = Ae^{i(\gamma x - \omega t)}$$
(P16.2)

where A is a constant and γ is the wave number, the substitution of Eq. (P16.2) into Eq. (P16.1) leads to

$$\frac{EI}{\rho A}\gamma^4 - \frac{I}{A}\left(1 + \frac{E}{kG}\right)\gamma^4 c^2 - \gamma^2 c^2 + \frac{\rho I}{kAG}\gamma^4 c^4 = 0$$
(P16.3)

with

$$\omega = \gamma c$$
(P16.4)

By considering the cases $\gamma \to \infty$ and $\gamma \to 0$ in Eq. (P16.3) separately, derive expressions for determining the wave velocity c.

16.10 The potential and kinetic energy densities of a string are given by

$$v(x, t) = \frac{P}{2}\left(\frac{\partial w}{\partial x}\right)^2, \qquad k(x, t) = \frac{\rho}{2}\left(\frac{\partial w}{\partial t}\right)^2$$

The wave propagating in the positive x direction can be expressed as

$$w(x, t) = f(x - ct)$$

where $c = \sqrt{P/\rho}$ is the phase velocity. Show that the potential and kinetic energy densities are equal for a propagating wave.

17

Approximate Analytical Methods

17.1 INTRODUCTION

The exact solutions of problems associated with the free and forced vibration of continuous systems have been considered in earlier chapters. Exact solutions are usually represented by an infinite series expressed in terms of the normal or principal modes of vibration. In many practical applications, the solution of the vibration problem is dominated by the first few low-frequency modes, and the effect of high-frequency modes is negligible. In such cases the solution may be expressed in terms of a finite number of normal modes or in terms of assumed polynomials that describe the deformation shape of the continuous system. Exact solutions are possible only in relatively few simple cases of continuous systems. The exact solutions are particularly difficult to find for two- and three-dimensional problems. Exact solutions are often desirable because they provide valuable insight into the behavior of the system through ready access to the natural frequencies and mode shapes.

Most of the continuous systems considered in earlier chapters have uniform mass and stiffness distributions and simple boundary conditions. However, some vibration problems may pose insurmountable difficulties either because the governing differential equation is difficult to solve or the boundary conditions may be extremely difficult or impossible to satisfy. In such cases we may be satisfied with an approximate solution of the vibration problem. Several methods are available for finding the approximate solutions of vibration problems. The approximate methods can be classified into two categories. The first category is based on the expansion of the solution in the form of a finite series consisting of known functions multiplied by unknown coefficients. Depending on the particular method used, the known functions can be comparison functions, admissible functions, or functions that satisfy the differential equation but not the boundary conditions. If a series is assumed to consist of n functions, the corresponding eigenvalue problem will yield n eigenvalues and the corresponding eigenfunctions.

The second category of methods is based on a simple lumping of system properties. For example, the mass of a system can be assumed to be concentrated at certain points, known as *stations*, and the segments between consecutive stations, called *fields*, are assumed to be massless with uniform stiffness distribution. This model, with n stations, can be used to derive an algebraic eigenvalue problem of size n whose solution yields n eigenvalues and the corresponding eigenvectors.

The first class of methods is more analytical in nature and the second is more intuitive in nature. All the approximate methods basically convert a problem described by partial differential equations into a problem described by a set of ordinary differential equations. This essentially converts a differential eigenvalue problem into an algebraic eigenvalue problem. There are two classes of methods that are based on series expansions: Rayleigh–Ritz methods and weighted residual methods. In this chapter we consider the Rayleigh and Rayleigh–Ritz method and the closely related assumed modes method as well as several weighted residual methods, including the Galerkin, collocation, and least squares methods.

17.2   RAYLEIGH'S QUOTIENT

The expression for Rayleigh's quotient and the stationary property of Rayleigh's quotient can be discussed conveniently by considering a specific system. Consider the torsional vibration of a shaft. In the absence of external torque, the expressions for the potential and kinetic energies of the shaft are given by

$$\pi_p(t) = \frac{1}{2} \int_0^l GI_p(x) \left[\frac{\partial \theta(x,t)}{\partial x} \right]^2 dx \tag{17.1}$$

$$T(t) = \frac{1}{2} \int_0^l \rho I_p(x) \left[\frac{\partial \theta(x,t)}{\partial t} \right]^2 dx \tag{17.2}$$

The angular displacement is given by

$$\theta(x,t) = X(x)f(t) \tag{17.3}$$

where $X(x)$ is a trial function used to denote the maximum angular displacement at point x and $f(t)$ indicates the harmonic time dependence,

$$f(t) = e^{i\omega t} \tag{17.4}$$

where ω is the frequency of vibration. Substituting Eqs. (17.3) and (17.4) into Eqs. (17.1) and (17.2), the potential and kinetic energies can be expressed as

$$\pi_p(t) = \frac{e^{i\omega t}}{2} \int_0^l GI_p(x) \left[\frac{dX(x)}{dx} \right]^2 dx \tag{17.5}$$

$$T(t) = \frac{e^{i\omega t}}{2} (-\omega^2) \int_0^l \rho I_p(x)[X(x)]^2 dx \tag{17.6}$$

Equating the maximum values of potential and kinetic energies leads to $\pi_{p\max} = T_{\max}$ or

$$\int_0^l GI_p(x) \left[\frac{dX(x)}{dx} \right]^2 dx = \omega^2 \int_0^l \rho I_p(x)[X(x)]^2 dx \tag{17.7}$$

Rayleigh's quotient, R, is defined as

$$R(X(x)) = \lambda = \omega^2 = \frac{\int_0^l GI_p(x)[dX(x)/dx]^2 \, dx}{\int_0^l \rho I_p(x)[X(x)]^2 \, dx} \tag{17.8}$$

It can be seen that the value of Rayleigh's quotient depends on the trial function $X(x)$ used. To investigate the variation of R with different trial functions $X(x)$, the trial function $X(x)$ is expressed as a combination of the normal modes of the shaft, $\Theta_i(x)$, using the expansion theorem as

$$X(x) = \sum_{i=1}^{\infty} c_i \Theta_i(x) \qquad (17.9)$$

where c_i are unknown coefficients. Substituting Eq. (17.9) into Eq. (17.8) gives

$$R(c_1, c_2, \cdots) = \lambda = \omega^2 = \frac{\int_0^l GI_p(x) \sum_{i=1}^{\infty} c_i [d\Theta_i(x)/dx] \sum_{j=1}^{\infty} c_j [d\Theta_j(x)/dx] \, dx}{\int_0^l \rho I_p(x) \sum_{i=1}^{\infty} c_i \Theta_i(x) \sum_{j=1}^{\infty} c_j \Theta_j(x) \, dx}$$

$$= \frac{\sum_{i=1}^{\infty} \sum_{j=1}^{\infty} c_i c_j \int_0^l GI_p(x)[d\Theta_i(x)/dx][d\Theta_j(x)/dx] \, dx}{\sum_{i=1}^{\infty} \sum_{j=1}^{\infty} c_i c_j \int_0^l \Theta_i(x)\Theta_j(x) \, dx} \qquad (17.10)$$

The orthogonality conditions of the normal modes of the shaft are given by [similar to Eqs. (E10.3.8) and (E10.3.9)]

$$\int_0^l \rho I_p(x)\Theta_i(x)\Theta_j(x) \, dx = \delta_{ij}, \qquad i, j = 1, 2, \ldots \qquad (17.11)$$

$$\int_0^l GI_p(x)\frac{d\Theta_i(x)}{dx}\frac{d\Theta_j(x)}{dx} \, dx = \lambda_i \delta_{ij}, \qquad i, j = 1, 2, \ldots \qquad (17.12)$$

where $\lambda_i = \omega_i^2$ is the ith eigenvalue of the shaft. Using Eqs. (17.11) and (17.12), Eq. (17.10) can be expressed as

$$R(c_1, c_2, \ldots) = \frac{\sum_{i=1}^{\infty} \sum_{j=1}^{\infty} c_i c_j \lambda_i \delta_{ij}}{\sum_{i=1}^{\infty} \sum_{j=1}^{\infty} c_i c_j \delta_{ij}} = \frac{\sum_{i=1}^{\infty} c_i^2 \lambda_i}{\sum_{i=1}^{\infty} c_i^2} = \frac{c_k^2 \lambda_k + \sum_{i=1, i \neq k}^{\infty} c_i^2 \lambda_i}{c_k^2 + \sum_{i=1, i \neq k}^{\infty} c_i^2}$$

$$(17.13)$$

If the trial function $X(x)$ closely resembles any of the eigenfunctions $\Theta_k(x)$, it implies that all the coefficients c_i other than c_k are small compared to c_k, so that we can write

$$c_i = \varepsilon_i c_k, \qquad i = 1, 2, \ldots, k-1, k+1, \ldots \qquad (17.14)$$

where ε_i are small numbers. By substituting Eq. (17.14) into Eq. (17.13) and dividing the numerator and denominator by c_k^2 and neglecting terms in ε_i of order greater than 2, we obtain

$$R(c_1, c_2, \ldots) = \frac{\lambda_k + \sum_{i=1, i \neq k}^{\infty} \varepsilon_i^2 \lambda_i}{1 + \sum_{i=1, i \neq k}^{\infty} \varepsilon_i^2} = \left(\lambda_k + \sum_{\substack{i=1 \\ i \neq k}}^{\infty} \varepsilon_i^2 \lambda_i \right) \left(1 + \sum_{\substack{i=1 \\ i \neq k}}^{\infty} \varepsilon_i^2 \right)^{-1}$$

$$\approx \left(\lambda_k + \sum_{\substack{i=1 \\ i \neq k}}^{\infty} \varepsilon_i^2 \lambda_i \right) \left(1 - \sum_{\substack{i=1 \\ i \neq k}}^{\infty} \varepsilon_i^2 \right) \approx \lambda_k + \sum_{\substack{i=1 \\ i \neq k}}^{\infty} (\lambda_i - \lambda_k)\varepsilon_i^2 \qquad (17.15)$$

Note that if the trial function $X(x)$ differs from the kth normal mode $\Theta_k(x)$ by a small quantity of order 1 in ε [i.e., $X(x) = \Theta_k(x) + O(\varepsilon)$], Rayleigh's quotient differs from the kth eigenvalue by a small quantity of order 2 in ε [i.e., $R(c_1, c_2, \ldots) = \lambda_k + O(\varepsilon^2)$]. This implies that Rayleigh's quotient has a stationary value at an eigenfunction $\Theta_k(x)$, and the stationary value is the corresponding eigenvalue λ_k. The stationary value is actually a minimum value at the fundamental or first eigenvector, $\Theta_1(x)$. To see this, let $k = 1$ in Eq. (17.15) and write

$$R(X(x)) = \lambda_1 + \sum_{i=2}^{\infty}(\lambda_i - \lambda_1)\varepsilon_i^2 \qquad (17.16)$$

Since the eigenvalues satisfy the relation

$$\lambda_1 \le \lambda_2 \le \lambda_3 \le \cdots \qquad (17.17)$$

Eq. (17.16) leads to

$$R(X(x)) \ge \lambda_1 \qquad (17.18)$$

which shows that Rayleigh's quotient is never smaller than the first eigenvalue. Equation (17.18) can also be interpreted as follows:

$$\lambda_1 = \omega_1^2 = \min R(X(x)) = R(\Theta_1(x)) \qquad (17.19)$$

which implies that the lowest eigenvalue is the minimum value that Rayleigh's quotient can assume and that the minimum value of R occurs at the fundamental eigenfunction, $X(x) = \Theta_1(x)$. Equation (17.19) denotes *Rayleigh's principle*, which can be stated as follows: The frequency of vibration of a conservative system vibrating about an equilibrium position has a stationary value in the neighborhood of a natural mode. This stationary value, in fact, is a minimum value in the neighborhood of the fundamental natural mode.

17.3   RAYLEIGH'S METHOD

In most structural and mechanical systems, the fundamental or lowest natural frequency is the most important. For a quick estimate of the dynamic response of the system, especially during the preliminary design studies, the fundamental natural frequency is used. In such cases, Rayleigh's method can be used most conveniently to find the approximate fundamental natural frequency of vibration of a system without having to solve the governing differential equation of motion. The method is based on Rayleigh's principle. It can be used for a discrete or continuous conservative system.

In Rayleigh's method, we choose a trial function, $X(x)$, that resembles closely the first natural mode, $\Theta_1(x)$, substitute it into the Rayleigh's quotient of the system, similar to Eq. (17.10), carry out the integrations involved, and find the value of $R = \lambda = \omega^2$. Because of the fact that Rayleigh's quotient has a minimum at the fundamental or lowest eigenfunction, the fundamental natural frequency of the system can be taken as $\omega = \sqrt{R}$. Because of the stationarity of Rayleigh's quotient, the method gives remarkably good estimates of the fundamental frequency even if the trial function does not resemble the true fundamental eigenfunction too closely. In fact, the natural frequency thus computed will be one order of magnitude closer to the true fundamental natural frequency ω_1 than the trial function $X(x)$ is to the true fundamental natural mode $\Theta_1(x)$. Of course, the closer the trial function resembles the first eigenfunction, the better the estimate of the fundamental frequency.

Usually, selecting a suitable trial function for use in Rayleigh's quotient is not difficult. For example, the static deflection curve under self-weight can be used for bar, beam, plate, or shell structures. Even if the system has nonuniform mass and stiffness distributions, the static deflection curve of the system found by assuming a uniform mass and stiffness distributions can be used in Rayleigh's quotient. Similarly, even if the system has complex boundary conditions, the static deflection curve, found with simple boundary conditions, can be used in Rayleigh's quotient. For example, for simplicity, a free end condition can be used instead of a spring-supported end condition of a beam. The strain and kinetic energies required in defining Rayleigh's quotient are given in Table 17.1 for some uniform structural elements.

Example 17.1   Determine the fundamental frequency of transverse vibration of a uniform beam fixed at both ends (Fig. 17.1) using Rayleigh's method. Use the following trial functions for approximating the fundamental mode shape:

(a)

$$X(x) = C\left(1 - \cos\frac{2\pi x}{l}\right) \tag{E17.1.1}$$

where C is a constant. This function is selected to satisfy the boundary conditions of the beam: $X(0) = 0$, $dX(0)/dx = 0$, $X(l) = 0$, and $dX(l)/dx = 0$.

(b)

$$X(x) = C(x^2)(l - x)^2 \tag{E17.1.2}$$

with $C = w_0/24EI$. This function is the static deflection curve of a fixed–fixed beam under a self-weight of w_0 per unit length.

SOLUTION   The expressions for the strain and kinetic energies of a uniform beam are given by

$$\pi = \frac{1}{2}EI\int_0^l \left[\frac{\partial^2 w(x, t)}{\partial x^2}\right]^2 dx \tag{E17.1.3}$$

$$T = \frac{1}{2}\rho A\int_0^l \left[\frac{\partial w(x, t)}{\partial t}\right]^2 dx \tag{E17.1.4}$$

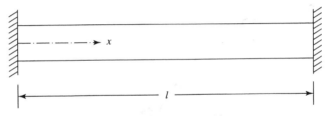

Figure 17.1   Fixed–fixed beam.

Table 17.1 Strain and Kinetic Energies of Some Structural Elements

Element or member	Strain energy, π	Kinetic energy, T	
		General	Maximum value in harmonic motion with frequency ω_n
1. String undergoing transverse motion	$\dfrac{P}{2}\displaystyle\int_0^l \left(\dfrac{\partial w}{\partial x}\right)^2 dx$ P = tension w = transverse deflection l = length	$\dfrac{\rho}{2}\displaystyle\int_0^l \left(\dfrac{\partial w}{\partial t}\right)^2 dx$ ρ = mass per unit length	$\dfrac{\rho\omega_n^2}{2}\displaystyle\int_0^l W^2\, dx$ W = amplitude of transverse deflection
2. Bar in tension or compression	$\dfrac{AE}{2}\displaystyle\int_0^l \left(\dfrac{\partial u}{\partial x}\right)^2 dx$ A = cross-sectional area E = Young's modulus u = axial displacement l = length	$\dfrac{\rho A}{2}\displaystyle\int_0^l \left(\dfrac{\partial u}{\partial t}\right)^2 dx$ ρ = density	$\dfrac{\rho A\omega_n^2}{2}\displaystyle\int_0^l U^2\, dx$ U = amplitude of axial displacement
3. Rod in torsion	$\dfrac{GJ}{2}\displaystyle\int_0^l \left(\dfrac{\partial \theta}{\partial x}\right)^2 dx$ GJ = torsional stiffness θ = angular deflection l = length	$\dfrac{\rho J}{2}\displaystyle\int_0^l \left(\dfrac{\partial \theta}{\partial t}\right)^2 dx$ ρ = density J = polar moment of inertia of cross section	$\dfrac{\rho J\omega_n^2}{2}\displaystyle\int_0^l \Theta^2\, dx$ Θ = amplitude of angular deflection
4. Beam in bending	$\dfrac{EI}{2}\displaystyle\int_0^l \left(\dfrac{\partial^2 w}{\partial x^2}\right)^2 dx$ E = Young's modulus I = moment of inertia of cross section w = transverse deflection l = length	$\dfrac{\rho A}{2}\displaystyle\int_0^l \left(\dfrac{\partial w}{\partial t}\right)^2 dx$ ρ = density A = area of cross section	$\dfrac{\rho A\omega_n^2}{2}\displaystyle\int_0^l W^2\, dx$ W = amplitude of transverse deflection

Table 17.1.1 (*continued*)

| | | Kinetic energy, T | |
Element or member	Strain energy, π	General	Maximum value in harmonic motion with frequency ω_n
5. Plate in bending (rectangular plate)	$\dfrac{D}{2} \iint\limits_A \left\{ \left(\dfrac{\partial^2 w}{\partial x^2} + \dfrac{\partial^2 w}{\partial y^2} \right)^2 - 2(1-\nu) \right.$ $\left. \times \left[\dfrac{\partial^2 w}{\partial x^2}\dfrac{\partial^2 w}{\partial y^2} - \left(\dfrac{\partial^2 w}{\partial x \partial y} \right)^2 \right] \right\} dx\,dy$ $D = \dfrac{Eh^3}{12(1-\nu^2)} = \text{bending stiffness}$ $E = \text{Young's modulus}$ $h = \text{plate thickness}$ $\nu = \text{Poisson's ratio}$	$\dfrac{\rho h}{2} \iint\limits_A \left(\dfrac{\partial w}{\partial t} \right)^2 dx\,dy$ $\rho = \text{density}$ $w = \text{transverse deflection}$ $A = \text{area of plate}$	$\dfrac{\rho h \omega_n^2}{2} \iint\limits_A W^2\,dx\,dy$ $W = \text{amplitude of transverse deflection}$
6. (a) Plate in bending (circular plate)	$\dfrac{D}{2} \int_0^{2\pi} \int_0^R \left\{ \left(\dfrac{\partial^2 w}{\partial r^2} + \dfrac{1}{r}\dfrac{\partial w}{\partial r} + \dfrac{1}{r^2}\dfrac{\partial^2 w}{\partial \theta^2} \right)^2 \right.$ $-2(1-\nu)\dfrac{\partial^2 w}{\partial r^2}\left(\dfrac{1}{r}\dfrac{\partial w}{\partial r} + \dfrac{1}{r^2}\dfrac{\partial^2 w}{\partial \theta^2} \right)$ $\left. +2(1-\nu)\left[\dfrac{\partial}{\partial r}\left(\dfrac{1}{r}\dfrac{\partial w}{\partial \theta} \right) \right]^2 \right\} r\,d\theta\,dr$ $R = \text{radius of plate}$	$\pi \rho h \int_0^R \left(\dfrac{\partial w}{\partial t} \right)^2 r\,dr$ $\rho = \text{density}$ $h = \text{thickness of plate}$	$\pi \rho h \omega_n^2 \int_0^R W^2 r\,dr$ $W = \text{amplitude of transverse deflection, } w$
(b) Circular plate with axisymmetric deflection	$\pi D \int_0^R \left[\left(\dfrac{\partial^2 w}{\partial r^2} + \dfrac{1}{r}\dfrac{\partial w}{\partial r} \right)^2 \right.$ $\left. -2(1-\nu)\dfrac{\partial^2 w}{\partial r^2}\dfrac{1}{r}\dfrac{\partial w}{\partial r} \right] r\,dr$	(Same as above)	(Same as above)

653

Table 17.1.1 (*continued*)

Element or member	Strain energy, π	Kinetic energy, T General	Maximum value in harmonic motion with frequency ω_n
7. Rectangular membrane	$\dfrac{P}{2} \displaystyle\iint_A \left[\left(\dfrac{\partial w}{\partial x}\right)^2 + \left(\dfrac{\partial w}{\partial y}\right)^2 \right] dx\,dy$ P = tension w = transverse deflection A = area of membrane	$\dfrac{\rho}{2} \displaystyle\iint_A \left(\dfrac{\partial w}{\partial t}\right)^2 dx\,dy$ ρ = mass per unit area	$\dfrac{\rho \omega_n^2}{2} \displaystyle\iint_A W^2\, dx\,dy$ W = amplitude of transverse deflection
8. Circular membrane (with axisymmetric motion)	$\dfrac{P}{2} \displaystyle\int_0^R \left(\dfrac{\partial W}{\partial r}\right)^2 \cdot 2\pi r\,dr$ P = tension R = radius	$\dfrac{\rho}{2} \displaystyle\int_0^R \left(\dfrac{\partial W}{\partial t}\right)^2 \cdot 2\pi r\,dr$ ρ = mass per unit area	$\dfrac{\rho \omega_n^2}{2} \displaystyle\int_0^R W^2 \cdot 2\pi r\,dr$ W = amplitude of transverse deflection

where the transverse deflection function, $w(x, t)$, can be assumed to be harmonic:

$$w(x, t) = X(x) \cos \omega t \qquad (E17.1.5)$$

where ω is the frequency of vibration. Rayleigh's quotient for a beam bending is defined by

$$\text{maximum strain energy} = \text{maximum kinetic energy}$$

Substitution of Eq. (E17.1.5) into Eqs. (E17.1.3) and (E17.1.4) leads to

$$\pi_{\max} = \frac{1}{2} E I \int_0^l \left[\frac{d^2 X(x)}{dx^2} \right]^2 dx \qquad (E17.1.6)$$

$$T_{\max} = \frac{1}{2} \rho A \omega^2 \int_0^l [X(x)]^2 \, dx \qquad (E17.1.7)$$

Equating $\pi_{\max}$ and $T_{\max}$, Rayleigh's quotient can be derived as

$$R(X(x)) = \omega^2 = \frac{\frac{1}{2} E I \int_0^l \left(d^2 X / dx^2 \right)^2 dx}{\frac{1}{2} \rho A \int_0^l [X(x)]^2 \, dx} \qquad (E17.1.8)$$

(a) In this case,

$$X(x) = C \left(1 - \cos \frac{2\pi x}{l} \right) \qquad (E17.1.9)$$

$$\frac{d^2 X(x)}{dx^2} = C \left(\frac{2\pi}{l} \right)^2 \cos \frac{2\pi x}{l} \qquad (E17.1.10)$$

$$\int_0^l \left(\frac{d^2 X}{dx^2} \right)^2 dx = C^2 \left(\frac{2\pi}{l} \right)^4 \int_0^l \cos^2 \frac{2\pi x}{l} \, dx = C^2 \left(\frac{2\pi}{l} \right)^4 \frac{l}{2} = \frac{8 C^2 \pi^4}{l^3} \qquad (E17.1.11)$$

$$\int_0^l [X(x)]^2 = C^2 \int_0^l \left(1 - \cos \frac{2\pi x}{l} \right)^2 dx = \frac{3 C^2 l}{2} \qquad (E17.1.12)$$

Thus, Eq. (E17.1.8) gives

$$R = \omega^2 = \frac{\frac{1}{2} E I (8 C^2 \pi^4 / l^3)}{\frac{1}{2} \rho A (3 C^2 l / 2)} = \frac{16 \pi^4}{3} \frac{E I}{\rho A l^4}$$

or

$$\omega = 22.792879 \sqrt{\frac{E I}{\rho A l^4}} \qquad (E17.1.13)$$

(b) In this case,

$$X(x) = C x^2 (l - x)^2 \qquad (E17.1.14)$$

$$\frac{d^2 X(x)}{dx^2} = 2 C (6 x^2 - 6 l x + l^2) \qquad (E17.1.15)$$

$$\int_0^l \left(\frac{d^2X}{dx^2}\right)^2 dx = 4C^2 \int_0^l (6x^2 - 6lx + l^2)^2 \, dx = \tfrac{4}{5}C^2l^5 \qquad \text{(E17.1.16)}$$

$$\int_0^l (X(x))^2 \, dx = C^2 \int_0^l (x^4 - 2lx^3 + x^2l^2)^2 \, dx = \tfrac{1}{630}C^2l^9 \qquad \text{(E17.1.17)}$$

Thus, Eq. (E17.1.8) gives

$$R = \omega^2 = \frac{\tfrac{1}{2}EI\left(\tfrac{4}{5}C^2l^5\right)}{\tfrac{1}{2}\rho A\left(\tfrac{1}{630}C^2l^9\right)} = 504\frac{EI}{\rho Al^4}$$

or

$$\omega = 22.449944\sqrt{\frac{EI}{\rho Al^4}} \qquad \text{(E17.1.18)}$$

The exact fundamental natural frequency of the beam is given by (see Fig. 11.3)

$$\omega_1^2 = (\beta_1 l)^2 \sqrt{\frac{EI}{\rho Al^4}} = (4.730041)^2\sqrt{\frac{EI}{\rho Al^4}} = 22.373288\sqrt{\frac{EI}{\rho Al^4}}$$
$$\text{(E17.1.19)}$$

and the exact fundamental natural mode is given by [see Eq. (11.70)]

$$W_1(x) = C\left[\sinh \beta_1 x - \sin \beta_1 x + \frac{\sinh \beta_1 l - \sin \beta_1 l}{\cos \beta_1 l - \cosh \beta_1 l}(\cosh \beta_1 x - \cos \beta_1 x)\right]$$
$$\text{(E17.1.20)}$$

where C is a constant. It can be seen that the fundamental natural frequencies given by Rayleigh's method are very close to the exact value and larger than the exact value by only 1.875410% in the first case [with Eq. (E17.1.9)] and 0.342623% in the second case [with (Eq. (E17.1.14))].

Example 17.2   Determine the fundamental frequency of transverse vibration of the uniform beam shown in Fig. 17.2. The beam is fixed at $x = 0$ and carries a concentrated mass m and rests on a linear spring of stiffness k at $x = l$. Use Rayleigh's method with

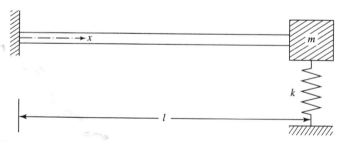

Figure 17.2   Beam fixed at one end and carrying a spring-supported mass at the other.

the trial function $X(x) = Cx^2(3l - x)$, where $C = F/6EI$, which corresponds to the deflection shape of a cantilever beam fixed at $x = 0$ and subjected to a concentrated force F and $x = l$.

SOLUTION   Considering the strain energy due to the deflection of the spring and the kinetic energy due to the motion of the end mass, the strain and kinetic energies of a uniform beam can be expressed as

$$\pi = \frac{1}{2}EI \int_0^l \left[\frac{\partial^2 w(x, t)}{\partial x^2} \right]^2 dx + \frac{1}{2}k[w(l, t)]^2 \tag{E17.2.1}$$

$$T = \frac{1}{2}\rho A \int_0^l \left[\frac{\partial w(x, t)}{\partial t} \right]^2 dx + \frac{1}{2}m \left[\frac{\partial w(l, t)}{\partial t} \right]^2 \tag{E17.2.2}$$

where the transverse deflection function, $w(x, t)$, is assumed to be harmonic:

$$w(x, t) = X(x) \cos \omega t \tag{E17.2.3}$$

where ω is the frequency of vibration. Substitution of Eq. (E17.2.3) into Eqs. (E17.2.1) and (E17.2.2) yields

$$\pi_{\max} = \frac{1}{2}EI \int_0^l \left[\frac{d^2 X(x, t)}{dx^2} \right]^2 dx + \frac{1}{2}kX^2(l) \tag{E17.2.4}$$

$$T_{\max} = \frac{1}{2}\rho A \omega^2 \int_0^l [X(x)]^2 dx + \frac{1}{2}m\omega^2 X^2(l) \tag{E17.2.5}$$

By equating $\pi_{\max}$ and $T_{\max}$, Rayleigh's quotient can be derived as

$$R(X(x)) = \omega^2 = \frac{\frac{1}{2}EI \int_0^l \left(d^2 X/dx^2 \right)^2 dx + \frac{1}{2}kX^2(l)}{\frac{1}{2}\rho A \int_0^l [X(x)]^2 dx + \frac{1}{2}m X^2(l)} \tag{E17.2.6}$$

Using

$$X(x) = Cx^2(3l - x) \tag{E17.2.7}$$

$$\frac{d^2 X(x)}{dx^2} = 6C(l - x) \tag{E17.2.8}$$

we can find

$$\int_0^l \left(\frac{d^2 X}{dx^2} \right)^2 dx = \int_0^l [6C(l - x)]^2 dx = 12C^2 l^3 \tag{E17.2.9}$$

$$\frac{1}{2}kX^2(l) = \frac{1}{2}kC^2(2l^3)^2 = 2kC^2 l^6 \tag{E17.2.10}$$

$$\int_0^l X^2(x) dx = \int_0^l C^2 x^4(3l - x)^2 dx = \frac{33C^2 l^7}{35} \tag{E17.2.11}$$

$$\frac{1}{2}mX^2(l) = \frac{1}{2}mC^2(2l^3)^2 = 2mC^2 l^6 \tag{E17.2.12}$$

Thus, Eq. (E17.2.6) gives

$$R = \omega^2 = \frac{\frac{1}{2}EI(12C^2l^3) + 2kC^2l^6}{\frac{1}{2}\rho A\left(33C^2l^7/35\right) + 2mC^2l^6} = \frac{6EI + 2kl^3}{\frac{33}{70}\rho Al^4 + 2\,ml^3} \qquad \text{(E17.2.13)}$$

Thus, the fundamental frequency of vibration is given by

$$\omega = \left(\frac{420EI + 140kl^3}{33\rho Al^4 + 140\,ml^3}\right)^{1/2} \qquad \text{(E17.2.14)}$$

Note: If k and m are set equal to zero, the beam becomes a cantilever beam and the fundamental frequency given by Rayleigh's method, Eq. (E17.2.14), reduces to

$$\omega = 3.567530\sqrt{\frac{EI}{\rho Al^4}} \qquad \text{(E17.2.15)}$$

This value can be compared with the exact value

$$\omega_1 = (1.875104)^2\sqrt{\frac{EI}{\rho Al^4}} = 3.516150\sqrt{\frac{EI}{\rho Al^4}}$$

Thus, the frequency given by Rayleigh's method is larger than the exact value by 1.461257%.

Example 17.3   Find the fundamental natural frequency of longitudinal vibration of the tapered bar fixed at $x = 0$ and connected to a linear spring of stiffness k at $x = l$ shown in Fig. 17.3 using Rayleigh's method. Assume the variation of the cross-sectional area of the bar to be $A(x) = A_0\,(1 - x/2l)$ and use the trail function $X(x) = C\sin(\pi x/2l)$ for the mode shape.

SOLUTION   The expressions for the strain and kinetic energies of a uniform bar, including the strain energy due to the deformation of the spring at $x = l$, can be

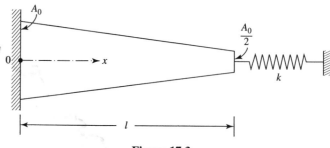

Figure 17.3

expressed as

$$\pi = \frac{1}{2}E \int_0^l A(x) \left[\frac{\partial u(x,t)}{\partial x}\right]^2 dx + \frac{1}{2}ku^2(l,t) \qquad (E17.3.1)$$

$$T = \frac{1}{2}\rho \int_0^l A(x) \left[\frac{\partial u(x,t)}{\partial t}\right]^2 dx \qquad (E17.3.2)$$

where the longitudinal deflection function, $u(x,t)$, is assumed to be harmonic:

$$u(x,t) = X(x)\cos\omega t \qquad (E17.3.3)$$

where ω is the frequency of vibration. Substitution of Eq. (E17.3.3) into Eqs. (E17.3.1) and (E17.3.2) leads to

$$\pi_{max} = \frac{1}{2}E \int_0^l A_0 \left(1 - \frac{x}{2l}\right)\left[\frac{dX(x)}{dx}\right]^2 dx + \frac{1}{2}kX^2(l) \qquad (E17.3.4)$$

$$T_{max}^* = \frac{1}{2}\rho \int_0^l A_0 \left(1 - \frac{x}{2l}\right) X^2(x)\, dx \qquad (E17.3.5)$$

Using

$$X(x) = C\sin\frac{\pi x}{2l} \qquad (E17.3.6)$$

$$\frac{dX}{dx}(x) = \frac{C\pi}{2l}\cos\frac{\pi x}{2l} \qquad (E17.3.7)$$

we can obtain

$$\int_0^l A_0 \left(1 - \frac{x}{2l}\right)\left(\frac{C\pi}{2l}\right)^2 \cos^2\frac{\pi x}{2l}\, dx + \frac{1}{2}k(C)^2 = \frac{A_0 C^2 \pi^2}{8l}\left(\frac{3}{4} + \frac{1}{\pi^2}\right) + \frac{k}{2}C^2 \qquad (E17.3.8)$$

$$\int_0^l A_0 \left(1 - \frac{x}{2l}\right) C^2 \sin^2\frac{\pi x}{2l}\, dx = A_0 C^2 \frac{l}{2}\left(\frac{3}{4} - \frac{1}{\pi^2}\right) \qquad (E17.3.9)$$

Rayleigh's quotient is given by

$$R = \omega^2 = \frac{\pi_{max}}{T_{max}^*} = \frac{\left(EA_0 C^2 \pi^2/16l\right)\left(\frac{3}{4} + 1/\pi^2\right) + kC^2/2}{(\rho A_0 C^2 l/4)\left(\frac{3}{4} - 1/\pi^2\right)}$$

$$= \frac{1}{\rho A_0 l^2}(3.238212EA_0 + 3.063189kl) \qquad (E17.3.10)$$

Thus, the natural frequency of vibration is given by

$$\omega = \left[\frac{1}{\rho A_0 l^2}(3.238212EA_0 + 3.063189kl)\right]^{1/2} \qquad (E17.3.11)$$

Example 17.4 Estimate the fundamental frequency of axisymmetric transverse vibration of a circular plate of radius a fixed along the edge $r = a$ using Rayleigh's method with the trial function

$$W(r) = c\left(1 - \frac{r^2}{a^2}\right)^2 \tag{E17.4.1}$$

where c is a constant.

SOLUTION   The trial solution of Eq. (E17.4.1) satisfies the boundary conditions of the plate at the edge $r = a$. The strain and kinetic energies of a circular plate of thickness h in axisymmetric vibration are given by (from Table 17.1)

$$V = \pi D \int_0^a \left[\left(\frac{\partial^2 w}{\partial r^2} + \frac{1}{r}\frac{\partial w}{\partial r}\right)^2 - 2(1 - v)\frac{\partial^2 w}{\partial r^2}\frac{1}{r}\frac{\partial w}{\partial r}\right] r \, dr \tag{E17.4.2}$$

$$T = \pi \rho h \int_0^a \left(\frac{\partial w}{\partial t}\right)^2 r \, dr \tag{E17.4.3}$$

If the plate is fixed at $r = a$, the integral of the second term on the right-hand side of Eq. (E17.4.2) will be zero and hence the equation reduces to

$$V = \pi D \int_0^a \left(\frac{\partial^2 w}{\partial r^2} + \frac{1}{r}\frac{\partial w}{\partial r}\right)^2 r \, dr \tag{E17.4.4}$$

Assuming harmonic variation of $w(r, t)$ as

$$w(r, t) = W(r) \cos \omega t \tag{E17.4.5}$$

where ω is the frequency of vibration, we can obtain

$$V_{\max} = \pi D \int_0^a \left(\frac{d^2 W}{dr^2} + \frac{1}{r}\frac{dW}{dr}\right)^2 r \, dr \tag{E17.4.6}$$

$$T^*_{\max} = \pi \rho h \int_0^a [W(r)]^2 r \, dr \tag{E17.4.7}$$

Rayleigh's quotient, $R(W(r))$, is given by

$$R(W(r)) = \omega^2 = \frac{V_{\max}}{T^*_{\max}} \tag{E17.4.8}$$

From Eq. (E17.4.1), we have

$$\frac{dW}{dr} = c\left(-\frac{4r}{a^2} + \frac{4r^3}{a^4}\right) \tag{E17.4.9}$$

$$\frac{d^2 W}{dr^2} = c\left(-\frac{4}{a^2} + \frac{12r^2}{a^4}\right) \tag{E17.4.10}$$

Equations (E17.4.6) and (E17.4.7) can be evaluated as

$$V_{max} = \pi D \int_0^a \left[c \left(-\frac{4}{a^2} + \frac{12r^2}{a^4} \right) + \frac{c}{r} \left(-\frac{4r}{a^2} + \frac{4r^3}{a^4} \right) \right]^2 r \, dr$$

$$= \frac{32\pi D c^2}{3a^2} \tag{E17.4.11}$$

$$T^*_{max} = \pi \rho h \int_0^a \left[c \left(1 - \frac{r^2}{a^2} \right)^2 \right]^2 r \, dr = \frac{\pi \rho h a^2 c^2}{10} \tag{E17.4.12}$$

Thus, the estimate of the fundamental natural frequency of the plate can be found from Eqs. (E17.4.8), (E17.4.11), and (E17.4.12) as

$$\omega^2 \simeq \frac{V_{max}}{T^*_{max}} = \frac{32\pi D c^2 / 3a^2}{\pi \rho h a^2 c^2 / 10} = 106.6667 \frac{D}{\rho h a^4}$$

or

$$\omega \simeq 10.3279 \sqrt{\frac{D}{\rho h a^4}} \tag{E17.4.13}$$

17.4 RAYLEIGH–RITZ METHOD

As stated earlier, exact solution of eigenvalue problems of many continuous systems is difficult, sometimes impossible, either because of nonuniform stiffness and mass distributions or because of complex boundary conditions. At the same time, information about the natural frequencies of the system may be required for the dynamic analysis and design of the system. For most systems, only the first few natural frequencies and associated natural modes greatly influence the dynamic response, and the contribution of higher natural frequencies and the corresponding mode shapes is negligible. If only the fundamental natural frequency of the system is required, Rayleigh's method can be used conveniently. If a small number of lowest natural frequencies of the system is required, the Rayleigh–Ritz method can be used. The Rayleigh–Ritz method can be considered as an extension of Rayleigh's method. The method is based on the fact that Rayleigh's quotient gives an upper bound for the first eigenvalue, $\lambda_1 = \omega_1^2$:

$$R(X(x)) \geq \lambda_1 \tag{17.20}$$

where the equality sign holds if and only if the trial function $X(x)$ coincides with the first eigenfunction of the system. In the Rayleigh–Ritz method, the shape of deformation of the continuous system, $X(x)$, is approximated using a trial family of admissible functions that satisfy the geometric boundary conditions of the problem:

$$X(x) = \sum_{i=1}^{n} c_i \phi_i(x) \tag{17.21}$$

where $c_1, c_2, \ldots, c_n$ are unknown (constant) coefficients, also called *Ritz coefficients*, and $\phi_1(x), \phi_2(x), \ldots, \phi_n(x)$ are the known trial family of admissible functions. The functions $\phi_i(x)$ can be a set of assumed mode shapes, polynomials, or even eigenfunctions. When Eq. (17.21) is substituted into the expression for Rayleigh's quotient, R, Rayleigh's quotient becomes a function of the unknown coefficients $c_1, c_2, \ldots, c_n$:

$$R = R(c_1, c_2, \ldots, c_n) \tag{17.22}$$

The coefficients $c_1, c_2, \ldots, c_n$ are selected to minimize Rayleigh's quotient using the necessary conditions:

$$\frac{\partial R}{\partial c_i} = 0, \qquad i = 1, 2, \ldots, n \tag{17.23}$$

Equation (17.23) denotes a system of n algebraic homogeneous linear equations in the unknowns $c_1, c_2, \ldots, c_n$. For the coefficients $c_1, c_2, \ldots, c_n$ to have a nontrivial solution, the determinant of the coefficient matrix is set equal to zero. This yields the frequency equation in the form of a polynomial in ω^2 of order n. The roots of the frequency equation provide the approximate natural frequencies of the system $\omega_1, \omega_2, \ldots, \omega_n$. Using the approximate natural frequency ω_i in Eq. (17.23), the corresponding approximate mode shape $c_1^{(i)}, c_2^{(i)}, \ldots, c_n^{(i)}$ can be determined (for $i = 1, 2, \ldots, n$). It can be seen that a continuous system which has an infinite number of degrees of freedom is represented by a discrete model, through Eq. (17.21), having only n degrees of freedom. The accuracy of the method depends on the value of n and the choice of the trial functions, $\phi_i(x)$, used in the approximation, Eq. (17.21). By using a larger value of n, the approximation can be made more accurate. Similarly, by using the trial functions $\phi_i(x)$, which are closer to the true eigenfunctions of the continuous system, the approximation can be improved.

The fundamental natural frequency given by the Rayleigh–Ritz method will be higher than the true natural frequency. The reason is that the approximation of a continuous system with infinitely many degrees of freedom by an n-degree-of-freedom system amounts to imposing the constraints

$$c_{n+1} = c_{n+2} = \cdots = 0 \tag{17.24}$$

on the system [in Eq. (17.21)]. The addition of constraints is equivalent to increasing the stiffness of the system, and hence the estimated frequency will be higher than the true fundamental frequency, When a larger number of trial functions are used, the number of constraints will be less and hence the fundamental natural frequency given by the Rayleigh–Ritz method, although higher than the true value, will be closer to the true value.

If Rayleigh's quotient is expressed as

$$R = \omega^2 = \frac{\pi_{max}}{T_{max}^*} = \frac{N}{D} \tag{17.25}$$

where $N = \pi_{max}$ and $D = T_{max}^*$ denote, respectively, the maximum strain energy and reference kinetic energy of the system. The reference kinetic energy (T_{max}^*) is related to the maximum kinetic energy (T_{max}) as

$$T_{max} = \omega^2 T_{max}^* \tag{17.26}$$

The maximum strain energy and the reference kinetic energy can be expressed as

$$\pi_{\max} = N = \frac{1}{2} \sum_{i=1}^{n} \sum_{j=1}^{n} c_i c_j k_{ij} = \frac{1}{2} \vec{c}^{\mathrm{T}} [k] \vec{c} \tag{17.27}$$

$$T^*_{\max} = D = \frac{1}{2} \sum_{i=1}^{n} \sum_{j=1}^{n} c_i c_j m_{ij} = \frac{1}{2} \vec{c}^{\mathrm{T}} [m] \vec{c} \tag{17.28}$$

where $[k] = [k_{ij}]$ is the stiffness matrix, $[m] = [m_{ij}]$ is the mass matrix,

$$\vec{c} = \begin{Bmatrix} c_1 \\ c_2 \\ \vdots \\ c_n \end{Bmatrix} \tag{17.29}$$

and the stiffness and mass coefficients k_{ij} and m_{ij} can be evaluated in terms of the stiffness and mass distributions of the system. For example, in the case of longitudinal vibration of bars, k_{ij} and m_{ij} are given by

$$k_{ij} = \int_0^l EA \frac{d\phi_i}{dx} \frac{d\phi_j}{dx} \, dx \tag{17.30}$$

$$m_{ij} = \int_0^l \rho A \phi_i \phi_j \, dx \tag{17.31}$$

and in the case of transverse vibration of beams, k_{ij} and m_{ij} are given by

$$k_{ij} = \int_0^l EI \frac{d^2\phi_i}{dx^2} \frac{d^2\phi_j}{dx^2} \, dx \tag{17.32}$$

$$m_{ij} = \int_0^l \rho A \phi_i \phi_j \, dx \tag{17.33}$$

Using Eqs. (17.27) and (17.28), Eq, (17.25) can be expressed as

$$R(c_1, c_2, \ldots, c_n) = \frac{N(c_1, c_2, \ldots, c_n)}{D(c_1, c_2, \ldots, c_n)} \tag{17.34}$$

since the numerator N and denominator D depend on the coefficients $c_1, c_2, \ldots, c_n$. The necessary conditions for the minimum of Rayleigh's quotient are

$$\frac{\partial R}{\partial c_i} = \frac{D(\partial N / \partial c_i) - N(\partial D / \partial c_i)}{D^2} = 0, \qquad i = 1, 2, \ldots, n$$

or

$$\frac{1}{D} \left(\frac{\partial N}{\partial c_i} - \frac{N}{D} \frac{\partial D}{\partial c_i} \right) = \frac{1}{D} \left(\frac{\partial N}{\partial c_i} - \lambda^{(n)} \frac{\partial D}{\partial c_i} \right) = 0, \qquad i = 1, 2, \ldots, n \tag{17.35}$$

where $\lambda^{(n)}$ is the eigenvalue and the superscript n indicates that the eigenvalue problem corresponds to a series of n terms in Eq. (17.21). Equation (17.35) can be written in matrix form as

$$\frac{\partial R}{\partial \vec{c}} = \frac{1}{D}\left(\frac{\partial N}{\partial \vec{c}} - \lambda^{(n)}\frac{\partial D}{\partial \vec{c}}\right) = \vec{0} \tag{17.36}$$

where $\partial N/\partial \vec{c}$ and $\partial D/\partial \vec{c}$ can be expressed, using Eqs. (17.27) and (17.28), as

$$\frac{\partial N}{\partial \vec{c}} = \vec{c}^{\mathrm{T}}[k] \tag{17.37}$$

$$\frac{\partial D}{\partial \vec{c}} = \vec{c}^{\mathrm{T}}[m] \tag{17.38}$$

and hence Eq. (17.36) leads to

$$\frac{\partial N}{\partial \vec{c}} - \lambda^{(n)}\frac{\partial D}{\partial \vec{c}} = 0$$

or

$$[[k] - \lambda^{(n)}[m]] \, \vec{c} = \vec{0} \tag{17.39}$$

Equation (17.39) denotes an algebraic eigenvalue problem of order n. For a nontrivial solution of the vector $\vec{c}$, the determinant of the coefficient matrix must be zero:

$$|[k] - \lambda^{(n)}[m]| = 0 \tag{17.40}$$

Equation (17.40) denotes the frequency equation, which upon expansion results in a polynomial in $\lambda^{(n)}$ of order n. The roots of the polynomial, $\lambda_i^{(n)}$, $i = 1, 2, \ldots, n$, give the natural frequencies as

$$\omega_i = \sqrt{\lambda_i^{(n)}}, \qquad i = 1, 2, \ldots, n \tag{17.41}$$

For each natural frequency ω_i, the corresponding vector of Ritz coefficients $\vec{c}^{(i)}$ can be determined to within an arbitrary constant by solving the linear simultaneous homogeneous equations:

$$[[k] - \lambda_i^{(n)}[k]] \, \vec{c}^{(i)} = \vec{0} \tag{17.42}$$

which can be written in scalar form as

$$\begin{bmatrix} k_{11} - \lambda_i^{(n)}m_{11} & k_{12} - \lambda_i^{(n)}m_{12} & \cdots & k_{1n} - \lambda_i^{(n)}m_{1n} \\ k_{21} - \lambda_i^{(n)}m_{21} & k_{22} - \lambda_i^{(n)}m_{22} & \cdots & k_{2n} - \lambda_i^{(n)}m_{2n} \\ \vdots & & & \\ k_{n1} - \lambda_i^{(n)}m_{n1} & k_{n2} - \lambda_i^{(n)}m_{n2} & \cdots & k_{nn} - \lambda_i^{(n)}m_{nn} \end{bmatrix} \begin{Bmatrix} c_1^{(i)} \\ c_2^{(i)} \\ \vdots \\ c_n^{(i)} \end{Bmatrix} = \begin{Bmatrix} 0 \\ 0 \\ \vdots \\ 0 \end{Bmatrix},$$

$$i = 1, 2, \ldots, n \tag{17.43}$$

Once $\vec{c}^{(i)} = \begin{Bmatrix} c_1^{(i)} \\ c_2^{(i)} \\ \vdots \\ c_n^{(i)} \end{Bmatrix}$ is determined, the eigenvector or normal mode associated with the

frequency i can be determined using Eq. (17.21) as

$$X^{(i)}(x) = \sum_{j=1}^{n} c_j^{(i)} \phi_j(x) = c_1^{(i)} \phi_1(x) + c_2^{(i)} \phi_2(x) + \cdots + c_n^{(i)} \phi_n(x), \qquad i = 1, 2, \ldots, n$$

(17.44)

Note   Once the natural frequencies and the corresponding mode shapes are determined using Eqs. (17.41) and (17.44), the dynamic or vibration response of the continuous system (with an infinite number of degrees of freedom) can be represented by an equivalent n-degree-of-freedom system. The vibration response of the continuous system can be expressed, using the separation-of-variables technique, as

$$u(x, t) = \sum_{i=1}^{n} X^{(i)}(x)\eta_i(t)$$

(17.45)

where $\eta_i(t), i = 1, 2, \ldots, n$, are the time-dependent coefficients or generalized coordinates.

Example 17.5   Determine the first three natural frequencies of longitudinal vibration of the tapered bar fixed at $x = 0$ and free at $x = l$ shown in Fig. 17.4 using the Rayleigh–Ritz method. Assume the variation of the cross-sectional area of the bar as $A(x) = A_0(1 - x/2l)$. Use the following functions as trial functions:

$$\phi_1(x) = \frac{x}{l}, \qquad \phi_2(x) = \frac{x^2}{l^2}, \qquad \phi_3(x) = \frac{x^3}{l^3}$$

(E17.5.1)

SOLUTION   For the fixed–free bar, the geometric boundary condition is given by

$$u(0) = 0$$

(E17.5.2)

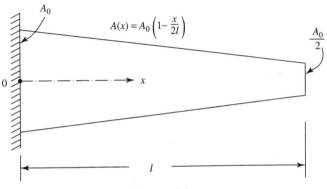

Figure 17.4

and the natural boundary condition by

$$\frac{\partial u}{\partial x}(l) = 0 \tag{E17.5.3}$$

Note that the trial functions in Eq. (E17.5.1) are admissible functions since they satisfy only the geometric boundary condition, Eq. (E17.5.2), but not the natural boundary condition, Eq. (E17.5.3). Assuming the longitudinal deflection function, $u(x, t)$, to be harmonic as

$$u(x, t) = X(x) \cos \omega t \tag{E17.5.4}$$

where ω is the frequency of vibration, the Rayleigh quotient of the bar can be expressed as

$$R = \frac{\pi_{max}}{T^*_{max}} \equiv \frac{\frac{1}{2} \int_0^l EA(x)[dX(x)/dx]^2 \, dx}{\frac{1}{2} \int_0^l \rho A(x)[X(x)]^2 \, dx} \tag{E17.5.5}$$

Using

$$X(x) = \sum_{i=1}^{3} c_i \phi_i(x) = c_1 \frac{x}{l} + c_2 \frac{x^2}{l^2} + c_3 \frac{x^3}{l^3} \tag{E17.5.6}$$

$$\frac{dX(x)}{dx} = \frac{c_1}{l} + \frac{2c_2 x}{l^2} + \frac{3c_3 x^2}{l^3} \tag{E17.5.7}$$

π_{max} and T^*_{max} can be evaluated as follows:

$$\pi_{max} = \frac{E}{2} \int_0^l A_0 \left(1 - \frac{x}{2l}\right) \left(\frac{c_1}{l} + \frac{2c_2 x}{l^2} + \frac{3c_3 x^2}{l^3}\right)^2 dx$$

$$= \frac{EA_0}{2l} \left(\frac{3}{4}c_1^2 + \frac{5}{6}c_2^2 + \frac{21}{20}c_3^2 + \frac{4}{3}c_1 c_2 + \frac{5}{4}c_1 c_3 + \frac{9}{5}c_2 c_3\right) \tag{E17.5.8}$$

$$T^*_{max} = \frac{\rho}{2} \int_0^l A_0 \left(1 - \frac{x}{2l}\right) \left(\frac{c_1}{l}x + \frac{c_2}{l^2}x^2 + \frac{c_3}{l^3}x^3\right)^2 dx$$

$$= \frac{\rho A_0 l}{2} \left(\frac{5}{24}c_1^2 + \frac{7}{60}c_2^2 + \frac{9}{112}c_3^2 + \frac{3}{10}c_1 c_2 + \frac{4}{21}c_2 c_3 + \frac{7}{30}c_1 c_3\right) \tag{E17.5.9}$$

Rayleigh's quotient is given by

$$R = \omega^2 = \frac{\pi_{max}(c_1, c_2, c_3)}{T^*_{max}(c_1, c_2, c_3)} \tag{E17.5.10}$$

The necessary conditions for the minimization of R are given by

$$\frac{\partial R}{\partial c_i} = \frac{T^*_{max}(\partial \pi_{max}/\partial c_i) - \pi_{max}(\partial T^*_{max}/\partial c_i)}{T^{*2}_{max}} = 0$$

or

$$\frac{\partial \pi_{\max}}{\partial c_i} - \frac{\pi_{\max}}{T^*_{\max}} \frac{\partial T^*_{\max}}{\partial c_i} = \frac{\partial \pi_{\max}}{\partial c_i} - \omega^2 \frac{\partial T^*_{\max}}{\partial c_i} = 0, \qquad i = 1, 2, 3 \qquad \text{(E17.5.11)}$$

Using

$$\frac{\partial \pi_{\max}}{\partial c_1} = \frac{EA_0}{2l} \left(\frac{3}{2} c_1 + \frac{4}{3} c_2 + \frac{5}{4} c_3 \right) \qquad \text{(E17.5.12)}$$

$$\frac{\partial \pi_{\max}}{\partial c_2} = \frac{EA_0}{2l} \left(\frac{5}{3} c_2 + \frac{4}{3} c_1 + \frac{9}{5} c_3 \right) \qquad \text{(E17.5.13)}$$

$$\frac{\partial \pi_{\max}}{\partial c_3} = \frac{EA_0}{2l} \left(\frac{21}{10} c_3 + \frac{5}{4} c_1 + \frac{9}{5} c_2 \right) \qquad \text{(E17.5.14)}$$

$$\frac{\partial T^*_{\max}}{\partial c_1} = \frac{\rho A_0 l}{2} \left(\frac{5}{12} c_1 + \frac{3}{10} c_2 + \frac{7}{30} c_3 \right) \qquad \text{(E17.5.15)}$$

$$\frac{\partial T^*_{\max}}{\partial c_2} = \frac{\rho A_0 l}{2} \left(\frac{7}{30} c_2 + \frac{3}{10} c_1 + \frac{4}{21} c_3 \right) \qquad \text{(E17.5.16)}$$

$$\frac{\partial T^*_{\max}}{\partial c_3} = \frac{\rho A_0 l}{2} \left(\frac{9}{56} c_3 + \frac{4}{21} c_2 + \frac{7}{30} c_1 \right) \qquad \text{(E17.5.17)}$$

Eq. (E17.5.11) can be expressed as

$$\frac{EA_0}{2l} \begin{bmatrix} \frac{3}{2} & \frac{4}{3} & \frac{5}{4} \\ \frac{4}{3} & \frac{5}{3} & \frac{9}{5} \\ \frac{5}{4} & \frac{9}{5} & \frac{21}{10} \end{bmatrix} \begin{Bmatrix} c_1 \\ c_2 \\ c_3 \end{Bmatrix} = \omega^2 \frac{\rho A_0 l}{2} \begin{bmatrix} \frac{5}{12} & \frac{3}{10} & \frac{7}{30} \\ \frac{3}{10} & \frac{7}{30} & \frac{4}{21} \\ \frac{7}{30} & \frac{4}{21} & \frac{9}{56} \end{bmatrix} \begin{Bmatrix} c_1 \\ c_2 \\ c_3 \end{Bmatrix} \qquad \text{(E17.5.18)}$$

or

$$[k]\, \vec{c} = \lambda\, [m]\vec{c} \qquad \text{(E17.5.19)}$$

where

$$[k] = \begin{bmatrix} \frac{3}{2} & \frac{4}{3} & \frac{5}{4} \\ \frac{4}{3} & \frac{5}{3} & \frac{9}{5} \\ \frac{5}{4} & \frac{9}{5} & \frac{21}{10} \end{bmatrix} \qquad \text{(E17.5.20)}$$

$$[m] = \begin{bmatrix} \frac{5}{12} & \frac{3}{10} & \frac{7}{30} \\ \frac{3}{10} & \frac{7}{30} & \frac{4}{21} \\ \frac{7}{30} & \frac{4}{21} & \frac{9}{56} \end{bmatrix} \qquad \text{(E17.5.21)}$$

$$\lambda = \frac{\rho l^2 \omega^2}{E} \qquad \text{(E17.5.22)}$$

The solution of Eq. (E17.5.19) is given by

$$\vec{\lambda} = \begin{Bmatrix} 3.2186 \\ 24.8137 \\ 100.8460 \end{Bmatrix} \qquad \text{(E17.5.23)}$$

The mode shapes can be found as

$$\vec{c}^{(1)} = \begin{Bmatrix} -0.8710 \\ -0.2198 \\ 0.4394 \end{Bmatrix}, \qquad \vec{c}^{(2)} = \begin{Bmatrix} 0.3732 \\ -0.8370 \\ 0.4001 \end{Bmatrix}, \qquad \vec{c}^{(3)} = \begin{Bmatrix} -0.2034 \\ 0.7603 \\ -0.6170 \end{Bmatrix} \qquad \text{(E17.5.24)}$$

Example 17.6 Solve Example 17.5 using the following as trial functions:

$$\phi_1(x) = \sin \frac{\pi x}{2l}, \qquad \phi_2(x) = \sin \frac{3\pi x}{2l}, \qquad \phi_3(x) = \sin \frac{5\pi x}{2l} \qquad \text{(E17.6.1)}$$

SOLUTION   It can be seen that the trial functions in Eq. (E17.6.1) are comparison functions since they satisfy both the geometric boundary condition, Eq. (E17.5.2), and the natural boundary condition, Eq. (E17.5.3). As in Example 17.5, the longitudinal deflection function, $u(x, t)$, is assumed to be harmonic as

$$u(x, t) = X(x) \cos \omega t \qquad \text{(E17.6.2)}$$

where ω denotes the natural frequency of vibration. Rayleigh's quotient of the bar can be expressed as

$$R = \frac{\pi_{max}}{T^*_{max}} \equiv \frac{\frac{1}{2} \int_0^l EA(x)[dX(x)/dx]^2\, dx}{\frac{1}{2} \int_0^l \rho A(x)[X(x)]^2\, dx} \qquad \text{(E17.6.3)}$$

Using

$$X(x) = \sum_{i=1}^3 c_i \phi_i(x) = c_1 \sin \frac{\pi x}{2l} + c_2 \sin \frac{3\pi x}{2l} + c_3 \sin \frac{5\pi x}{2l} \qquad \text{(E17.6.4)}$$

$$\frac{dX(x)}{dx} = \frac{c_1 \pi}{2l} \cos \frac{\pi x}{2l} + \frac{c_2 3\pi}{2l} \cos \frac{3\pi x}{2l} + \frac{c_3 5\pi}{2l} \cos \frac{5\pi x}{2l} \qquad \text{(E17.6.5)}$$

π_{max} and T^*_{max} can be evaluated as follows:

$$\pi_{max} = \frac{E}{2} \int_0^l A_0 \left(1 - \frac{x}{2l}\right) \left(\frac{c_1 \pi}{2l} \cos \frac{\pi x}{2l} + \frac{c_2 3\pi}{2l} \cos \frac{3\pi x}{2l} + \frac{c_3 5\pi}{2l} \cos \frac{5\pi x}{2l}\right)^2 dx$$

$$= \frac{EA_0}{2} \left[c_1^2 \left(\frac{3\pi^2}{32l} + \frac{1}{8l}\right) + c_2^2 \left(\frac{27\pi^2}{32l} + \frac{1}{8l}\right) + c_3^2 \left(\frac{75\pi^2}{32l} + \frac{1}{8l}\right) \right.$$

$$\left. + c_1 c_2 \left(\frac{3}{4l}\right) + c_2 c_3 \left(\frac{15}{4l}\right) + c_1 c_3 \left(\frac{5}{36l}\right) \right] \qquad \text{(E17.6.6)}$$

$$T^*_{max} = \frac{\rho}{2} \int A_0 \left(1 - \frac{x}{2l}\right) \left(c_1 \sin \frac{\pi x}{2l} + c_2 \sin \frac{3\pi x}{2l} + c_3 \sin \frac{5\pi x}{2l}\right)^2 dx$$

$$= \frac{\rho A_0 l}{2} \left[c_1^2 \left(\frac{3}{8} - \frac{1}{2\pi^2}\right) + c_2^2 \left(\frac{3}{8} - \frac{1}{18\pi^2}\right) \right.$$

$$\left. + c_3^2 \left(\frac{3}{8} - \frac{1}{50\pi^2}\right) + c_1 c_2 \left(\frac{1}{\pi^2}\right) - c_1 c_3 \left(\frac{1}{9\pi^2}\right) + c_2 c_3 \left(\frac{1}{\pi^2}\right) \right] \qquad \text{(E17.6.7)}$$

The Rayleigh's quotient is given by

$$R = \omega^2 = \frac{\pi_{max}(c_1, c_2, c_3)}{T^*_{max}(c_1, c_2, c_3)} \tag{E17.6.8}$$

The necessary conditions for the minimization of R are given by

$$\frac{\partial R}{\partial c_i} = \frac{T^*_{max}(\partial \pi_{max}/\partial c_i) - \pi_{max}(\partial T^*_{max}/\partial c_i)}{T^{*^2}_{max}} = 0 \tag{E17.6.9}$$

or

$$\frac{\partial \pi_{max}}{\partial c_i} - \omega^2 \frac{\partial T^*_{max}}{\partial c_i} = 0, \qquad i = 1, 2, 3 \tag{E17.6.10}$$

Using

$$\frac{\partial \pi_{max}}{\partial c_1} = \frac{E A_0}{2} \left[c_1 \left(\frac{3\pi^2}{16l} + \frac{1}{4l} \right) + c_2 \left(\frac{3}{4l} \right) + c_3 \left(\frac{5}{36l} \right) \right] \tag{E17.6.11}$$

$$\frac{\partial \pi_{max}}{\partial c_2} = \frac{E A_0}{2} \left[c_2 \left(\frac{27\pi^2}{16l} + \frac{1}{4l} \right) + c_1 \left(\frac{3}{4l} \right) + c_3 \left(\frac{15}{4l} \right) \right] \tag{E17.6.12}$$

$$\frac{\partial \pi_{max}}{\partial c_3} = \frac{E A_0}{2} \left[c_3 \left(\frac{75\pi^2}{16l} + \frac{1}{4l} \right) + c_2 \left(\frac{15}{4l} \right) + c_1 \left(\frac{5}{36l} \right) \right] \tag{E17.6.13}$$

$$\frac{\partial T^*_{max}}{\partial c_1} = \frac{\rho A_0 l}{2} \left[c_1 \left(\frac{3}{4} - \frac{1}{\pi^2} \right) + c_2 \left(\frac{1}{\pi^2} \right) - c_3 \left(\frac{1}{9\pi^2} \right) \right] \tag{E17.6.14}$$

$$\frac{\partial T^*_{max}}{\partial c_2} = \frac{\rho A_0 l}{2} \left[c_2 \left(\frac{3}{4} - \frac{1}{9\pi^2} \right) + c_1 \left(\frac{1}{\pi^2} \right) + c_3 \left(\frac{1}{\pi^2} \right) \right] \tag{E17.6.15}$$

$$\frac{\partial T^*_{max}}{\partial c_3} = \frac{\rho A_0 l}{2} \left[c_3 \left(\frac{3}{4} - \frac{1}{25\pi^2} \right) - c_1 \left(\frac{1}{9\pi^2} \right) + c_2 \left(\frac{1}{\pi^2} \right) \right] \tag{E17.6.16}$$

Eqs. (E17.6.9) can be expressed as

$$\frac{E A_0}{2l}
\begin{bmatrix}
\dfrac{3\pi^2}{16} + \dfrac{1}{4} & \dfrac{3}{4} & \dfrac{5}{36} \\[2ex]
\dfrac{3}{4} & \dfrac{27\pi^2}{16} + \dfrac{1}{4} & \dfrac{15}{4} \\[2ex]
\dfrac{5}{36} & \dfrac{15}{4} & \dfrac{75\pi^2}{16} + \dfrac{1}{4}
\end{bmatrix}
\begin{Bmatrix} c_1 \\ c_2 \\ c_3 \end{Bmatrix}$$

$$= \frac{\omega^2 \rho A_0 l}{2}
\begin{bmatrix}
\dfrac{3}{4} - \dfrac{1}{\pi^2} & \dfrac{1}{\pi^2} & -\dfrac{1}{9\pi^2} \\[2ex]
\dfrac{1}{\pi^2} & \dfrac{3}{4} - \dfrac{1}{9\pi^2} & \dfrac{1}{\pi^2} \\[2ex]
-\dfrac{1}{9\pi^2} & \dfrac{1}{\pi^2} & \dfrac{3}{4} - \dfrac{1}{25\pi^2}
\end{bmatrix}
\begin{Bmatrix} c_1 \\ c_2 \\ c_3 \end{Bmatrix} \tag{E17.6.17}$$

or

$$[k]\vec{c} = \lambda[m]\vec{c} \qquad \text{(E17.6.18)}$$

where

$$[k] = \begin{bmatrix} \dfrac{3\pi^2}{16} + \dfrac{1}{4} & \dfrac{3}{4} & \dfrac{5}{36} \\[3mm] \dfrac{3}{4} & \dfrac{27\pi^2}{16} + \dfrac{1}{4} & \dfrac{15}{4} \\[3mm] \dfrac{5}{36} & \dfrac{15}{4} & \dfrac{75\pi^2}{16} + \dfrac{1}{4} \end{bmatrix} \qquad \text{(E17.6.19)}$$

$$[m] = \begin{bmatrix} \dfrac{3}{4} - \dfrac{1}{\pi^2} & \dfrac{1}{\pi^2} & -\dfrac{1}{9\pi^2} \\[3mm] \dfrac{1}{\pi^2} & \dfrac{3}{4} - \dfrac{1}{9\pi^2} & \dfrac{1}{\pi^2} \\[3mm] -\dfrac{1}{9\pi^2} & \dfrac{1}{\pi^2} & \dfrac{3}{4} - \dfrac{1}{25\pi^2} \end{bmatrix} \qquad \text{(E17.6.20)}$$

$$\lambda = \dfrac{\omega^2 \rho l^2}{E} \qquad \text{(E17.6.21)}$$

The solution of Eq. (E17.6.17) is given by

$$\vec{\lambda} = \begin{Bmatrix} 3.2189 \\ 23.0627 \\ 62.7291 \end{Bmatrix} \qquad \text{(E17.6.22)}$$

The mode shapes can be computed as

$$\vec{c}^{(1)} = \begin{Bmatrix} -0.9996 \\ 0.0288 \\ 0.0017 \end{Bmatrix}, \qquad \vec{c}^{(2)} = \begin{Bmatrix} 0.1237 \\ -0.9912 \\ 0.0461 \end{Bmatrix}, \qquad \vec{c}^{(3)} = \begin{Bmatrix} -0.0356 \\ 0.0948 \\ -0.9949 \end{Bmatrix} \qquad \text{(E17.6.23)}$$

17.5 ASSUMED MODES METHOD

The assumed modes method is closely related to the Rayleigh–Ritz method. In fact, the discrete model obtained with the assumed modes method is identical to the one obtained with the Rayleigh–Ritz method. The main difference between the two methods is that the Rayleigh–Ritz method is commonly used to solve the eigenvalue problem, whereas the assumed modes method is generally used to solve the forced vibration problem. In the assumed modes method, the solution of the vibration problem of the continuous system is assumed in the form of a series composed of a linear combination of admissible functions ϕ_i, which are functions of the spatial coordinates, multiplied by time-dependent generalized coordinates, $\eta_i(t)$. Thus, for a one-dimensional continuous system, the displacement solution is assumed to be

$$w(x, t) = \sum_{i=1}^{n} \phi_i(x)\eta_i(t) \tag{17.46}$$

where $\phi_i(x)$ are known trial functions that satisfy the geometric boundary conditions (admissible functions) and $\eta_i(t)$ are unknown functions of time, also called generalized coordinates. For a forced vibration problem, the expressions of strain energy (π), kinetic energy (T), and virtual work of nonconservative forces, δW_{nc}, are expressed in terms of the assumed modes solution, Eq. (17.46), and then Lagrange's equations are used to derive the equations of motion of the equivalent n-degree-of-freedom discrete system of the continuous system. For specificity, consider a tapered bar under longitudinal vibration subjected to the distributed load, $f(x, t)$ per unit length. The strain energy, kinetic energy, and virtual work of nonconservative forces of the system are given by

$$\pi(t) = \frac{1}{2} \int_0^l E A(x) \left[\frac{\partial u(x, t)}{\partial x} \right]^2 dx \tag{17.47}$$

$$T(t) = \frac{1}{2} \int_0^l \rho A(x) \left[\frac{\partial u}{\partial t}(x, t) \right]^2 dx \tag{17.48}$$

$$\delta W_{nc} = \int_0^l f(x, t) \delta u(x, t) \, dx \tag{17.49}$$

By substituting Eq. (17.46) into Eqs. (17.47)–(17.49), we obtain

$$\pi(t) = \frac{1}{2} \int_0^l E A(x) \sum_{i=1}^{n} \frac{d\phi_i(x)}{dx} \eta_i(t) \sum_{j=1}^{n} \frac{d\phi_j(x)}{dx} \eta_j(t) \, dx$$

$$= \frac{1}{2} \sum_{i=1}^{n} \sum_{j=1}^{n} \eta_i(t)\eta_j(t) \left[\int_0^l E A(x) \frac{d\phi_i(x)}{dx} \frac{d\phi_j(x)}{dx} \, dx \right] = \frac{1}{2} \sum_{i=1}^{n} \sum_{j=1}^{n} k_{ij} \eta_i(t)\eta_j(t) \tag{17.50}$$

where k_{ij} denote the symmetric stiffness coefficient, given by

$$k_{ij} = k_{ji} = \int_0^l E A(x) \frac{d\phi_i(x)}{dx} \frac{d\phi_j(x)}{dx} \, dx, \qquad i, j = 1, 2, \ldots, n \tag{17.51}$$

$$T(t) = \int_0^l \rho A(x) \sum_{i=1}^{n} \phi_i(x)\dot{\eta}_i(t) \sum_{j=1}^{n} \phi_j(x)\dot{\eta}_j(t) \, dx$$

$$= \frac{1}{2} \sum_{i=1}^{n} \sum_{j=1}^{n} \dot{\eta}_i(t)\dot{\eta}_j(t) \left[\int_0^l \rho A(x)\phi_i(x)\phi_j(x) \, dx \right] = \frac{1}{2} \sum_{i=1}^{n} \sum_{j=1}^{n} m_{ij} \dot{\eta}_i(t)\dot{\eta}_j(t) \tag{17.52}$$

where $\dot{\eta}_i(t) = d\eta_i(t)/dt$ and m_{ij} indicate the symmetric mass coefficients, given by

$$m_{ij} = m_{ji} = \int_0^l \rho A(x)\phi_i(x)\phi_j(x)\,dx, \qquad i, j = 1, 2, \ldots, n \tag{17.53}$$

$$\delta W_{nc}(t) = \int_0^l f(x, t) \sum_{i=1}^n \phi_i(x)\delta\eta_i(t)\,dx = \sum_{i=1}^n Q_{i_{nc}}(t)\delta\eta_i(t) \tag{17.54}$$

where $Q_{i_{nc}}(t)$ denotes the generalized nonconservative force corresponding to the generalized coordinate $\eta_i(t)$, given by

$$Q_{i_{nc}}(t) = \int_0^l f(x, t)\phi_i(x)\,dx, \qquad i = 1, 2, \ldots, n \tag{17.55}$$

The Lagrange equations can be expressed as

$$\frac{d}{dt}\left(\frac{\partial T}{\partial \dot{\eta}_i}\right) - \frac{\partial T}{\partial \eta_i} + \frac{\partial \pi}{\partial \eta_i} = Q_{i_{nc}}, \qquad i = 1, 2, \ldots, n \tag{17.56}$$

Substituting Eqs. (17.50), (17.52), and (17.55) into Eq. (17.56), and noting that $\partial T/\partial \eta_i = 0$, $i = 1, 2, \ldots, n$, we can derive the equations of motion of the discretized system as

$$\sum_{j=1}^n m_{ij}\ddot{\eta}_j(t) + \sum_{j=1}^n k_{ij}\eta_j(t) = Q_{i_{nc}}(t), \qquad i = 1, 2, \ldots, n \tag{17.57}$$

Equations (17.57) can be expressed in matrix form as

$$[m]\ddot{\vec{\eta}}(t) + [k]\vec{\eta}(t) = \vec{Q}(t) \tag{17.58}$$

where

$$\vec{\eta}(t) = \begin{Bmatrix} \eta_1(t) \\ \eta_2(t) \\ \vdots \\ \eta_n(t) \end{Bmatrix}, \qquad \ddot{\vec{\eta}}(t) = \begin{Bmatrix} \dfrac{d^2\eta_1(t)}{dt^2} \\ \dfrac{d^2\eta_2(t)}{dt^2} \\ \vdots \\ \dfrac{d^2\eta_n(t)}{dt^2} \end{Bmatrix}, \qquad \vec{Q}(t) = \begin{Bmatrix} Q_{1_{nc}}(t) \\ Q_{2_{nc}}(t) \\ \vdots \\ Q_{n_{nc}}(t) \end{Bmatrix}$$

Notes

1. If $\vec{Q}(t)$ is set equal to $\vec{0}$, Eq. (17.58) denotes the equivalent n-degree-of-freedom free vibration equation of the continuous system:

$$[m]\ddot{\vec{\eta}}(t) + [k]\vec{\eta}(t) = \vec{0} \tag{17.59}$$

If all $\eta_i(t)$ are assumed to be harmonic in Eq. (17.59) as

$$\eta_i(t) = X_i \cos \omega t \tag{17.60}$$

where X_i denotes the amplitude of $\eta_i(t)$ and ω indicates the frequency of vibration, the resulting equations define the eigenvalue problem of the discretized system:

$$\omega^2[m]\vec{X} = [k]\vec{X} \qquad (17.61)$$

2. If the same trial functions $\phi_i(t)$ used in Eq. (17.46) are used, the eigenvalue problem, Eq. (17.39), given by the Rayleigh–Ritz methods will be identical to the one given by the assumed modes methods, Eq. (17.61).

17.6   WEIGHTED RESIDUAL METHODS

The Rayleigh and Rayleigh–Ritz methods of solving the eigenvalue problem are based on the stationarity of Rayleigh's quotient. These methods can be classified as variational methods because Rayleigh's quotient is related to the variational methods. There is another class of methods, known as *weighted residual methods*, for solving vibration problems. The Galerkin, collocation, subdomain collocation, and least squares methods fall into the category of weighted residual methods. The weighted residual methods work directly with the governing differential equation and boundary conditions of a problem.

Let the eigenvalue problem of the continuous system be stated by the differential equation

$$AW = \lambda BW \qquad \text{in} D \qquad (17.62)$$

with the boundary conditions

$$E_j W = 0, \qquad j = 1, 2, \ldots, p \quad \text{on} S \qquad (17.63)$$

where A, B, and E_j are linear differential operators, W is the eigenfunction or normal mode (or displacement pattern), λ is the eigenvalue, p is the number of boundary conditions, D is the domain, and S is the boundary of the system. In all the weighted residual methods, a trial solution, $\overline{\phi}$, is assumed for the problem. In general, the trial solution does not satisfy the governing equation, Eq. (17.62), and hence a measure of error is defined: for example, for a one-dimensional problem involving x as

$$R(\overline{\phi}, x) = A\overline{\phi} - \lambda B\overline{\phi} \qquad (17.64)$$

where $R(\overline{\phi}, x)$ is called the *residual*. It can be observed that if the trial function $\overline{\phi}(x)$ happens to be an eigenfunction $W_i(x)$ and λ the eigenvalue λ_i, the residual will be zero.

17.7   GALERKIN'S METHOD

The Galerkin method is the most widely used weighted residual method. In this method, solution of the eigenvalue problem is assumed in the form of a series of n comparison functions which satisfy all the boundary conditions of the problem:

$$\overline{\phi}^{(n)}(x) = \sum_{i=1}^{n} c_i \phi_i(x) \qquad (17.65)$$

where the c_i are coefficients to be determined and the $\phi_i(x)$ are known comparison functions. When Eq. (17.65) is substituted into the differential equation (17.62), the resulting error or residual is defined as

$$R = A\overline{\phi}^{(n)} - \lambda^{(n)} B\overline{\phi}^{(n)} \tag{17.66}$$

where $\lambda^{(n)}$ is the estimate of the eigenvalue obtained with an n-term trial solution, Eq. (17.65). Note that the residual will be zero from the boundary conditions, Eqs. (17.63), since the trial solution is composed of comparison functions which satisfy all the boundary conditions. In the Galerkin method, the selection of the coefficients of the trial solution is based on the criterion of making the residual small.

Specifically, we multiply the residual by the comparison functions $\phi_1(x), \phi_2(x), \dots, \phi_n(x)$, in sequence, integrate the product over the domain of the system, and equate the result to zero:

$$\int_0^l R(\overline{\phi}^{(n)})\phi_i(x)\,dx = 0, \qquad i = 1, 2, \dots, n \tag{17.67}$$

It can be seen that in Eq. (17.67), the integral of the weighted residual is set equal to zero, with the functions $\phi_i(x)$ serving as weighting functions. Upon integration, Eqs. (17.67) denote a set of linear homogeneous algebraic equations in the unknown coefficients $c_1, c_2, \dots, c_n$, and the eigenvalue, $\lambda^{(n)}$. These equations are known as *Galerkin equations*. They represent an algebraic eigenvalue problem of order n. The solution of the algebraic eigenvalue problem yields n eigenvalues $\lambda_1, \lambda_2, \dots, \lambda_n$ and the corresponding eigenvectors $\vec{c}^{(1)}, \vec{c}^{(2)}, \dots, \vec{c}^{(n)}$ (each within a multiplicative constant), where

$$\vec{c}^{(i)} = \begin{Bmatrix} c_1^{(i)} \\ c_2^{(i)} \\ \vdots \\ c_n^{(i)} \end{Bmatrix}, \qquad i = 1, 2, \dots, n \tag{17.68}$$

The Rayleigh and Rayleigh–Ritz methods are applicable to only conservative systems. However, the Galerkin method is more general and is applicable to both conservative and nonconservative systems.

Example 17.7 Find the natural frequencies of vibration of a fixed–fixed beam of length L, bending stiffness EI, and mass per unit length m (Fig. 17.1) using the Galerkin method with the following trial (comparison) functions:

$$\phi_1(x) = \cos\frac{2\pi x}{L} - 1 \tag{E17.7.1}$$

$$\phi_2(x) = \cos\frac{4\pi x}{L} - 1 \tag{E17.7.2}$$

SOLUTION   The equation governing the free vibration of a beam is given by

$$\frac{d^4W}{dx^4} - \beta^4 W = 0 \tag{E17.7.3}$$

where

$$\beta^4 = \frac{m\omega^2}{EI} \tag{E17.7.4}$$

with ω denoting the natural frequency of vibration of the beam. Using the trial functions of Eqs. (E17.7.1) and (E17.7.2), the approximate solution of the beam vibration problem is assumed as

$$\overline{W}(x) = c_1 \phi_1(x) + c_2 \phi_2(x) = c_1 \left(\cos \frac{2\pi x}{L} - 1 \right) + c_2 \left(\cos \frac{4\pi x}{L} - 1 \right) \quad \text{(E17.7.5)}$$

Substitution of Eq. (E17.7.5) into Eq. (E17.7.3) gives the residual as

$$R(c_1, c_2) = c_1 \left[\left(\frac{2\pi}{L} \right)^4 - \beta^4 \right] \cos \frac{2\pi x}{L} + c_1 \beta^4 + c_2 \left[\left(\frac{4\pi}{L} \right)^4 - \beta^4 \right] \cos \frac{4\pi x}{L} + c_2 \beta^4$$

$$\text{(E17.7.6)}$$

The Galerkin method gives

$$\int_{x=0}^{L} R\phi_i \, dx = 0, \quad i = 1, 2 \quad \text{(E17.7.7)}$$

which can be expressed, using Eqs. (E17.7.6), (E17.7.1), and (E17.7.2), as

$$\int_{x=0}^{L} \left(\cos \frac{2\pi x}{L} - 1 \right) \left\{ c_1 \left[\left(\frac{2\pi}{L} \right)^4 - \beta^4 \right] \cos \frac{2\pi x}{L} + c_1 \beta^4 \right.$$

$$\left. + c_2 \left[\left(\frac{4\pi}{L} \right)^4 - \beta^4 \right] \cos \frac{4\pi x}{L} + c_2 \beta^4 \right\} dx = 0 \quad \text{(E17.7.8)}$$

$$\int_{x=0}^{L} \left(\cos \frac{4\pi x}{L} - 1 \right) \left\{ c_1 \left[\left(\frac{2\pi}{L} \right)^4 - \beta^4 \right] \cos \frac{2\pi x}{L} + c_1 \beta^4 \right.$$

$$\left. + c_2 \left[\left(\frac{4\pi}{L} \right)^4 - \beta^4 \right] \cos \frac{4\pi x}{L} + c_2 \beta^4 \right\} dx = 0 \quad \text{(E17.7.9)}$$

or

$$c_1 \left\{ \frac{1}{2} \left[\left(\frac{2\pi}{L} \right)^4 - \beta^4 \right] - \beta^4 \right\} - c_2 \beta^4 = 0 \quad \text{(E17.7.10)}$$

$$-c_1 \beta^4 + c_2 \left\{ \frac{1}{2} \left[\left(\frac{4\pi}{L} \right)^4 - \beta^4 \right] - \beta^4 \right\} = 0 \quad \text{(E17.7.11)}$$

For a nontrivial solution of Eqs. (E17.7.10) and (E17.7.11), the determinant of the coefficient matrix of c_1 and c_2 must be zero. This gives

$$\begin{vmatrix} \frac{1}{2} \left[\left(\frac{2\pi}{L} \right)^4 - \beta^4 \right] - \beta^4 & -\beta^4 \\ -\beta^4 & \frac{1}{2} \left[\left(\frac{4\pi}{L} \right)^4 - \beta^4 \right] - \beta^4 \end{vmatrix} = 0 \quad \text{(E17.7.12)}$$

Simplification of Eq. (E17.7.12) yields the frequency equation:

$$(\beta L)^8 - 15,900(\beta L)^4 + 7,771,000 = 0 \tag{E17.7.13}$$

The solution of Eq. (E17.7.13) is

$$\beta L = 4.741 \quad \text{or} \quad 11.140 \tag{E17.7.14}$$

Thus, the first two natural frequencies of the beam are given by

$$\omega_1 = \frac{22.48}{L^2}\sqrt{\frac{EI}{m}}$$

$$\omega_2 = \frac{124.1}{L^2}\sqrt{\frac{EI}{m}} \tag{E17.7.15}$$

The eigenvectors corresponding to ω_1 and ω_2 can be obtained by solving Eq. (E17.7.10) or (E17.7.11) with the appropriate value of β. The results are as follows. For ω_1:

$$\begin{Bmatrix} c_1 \\ c_2 \end{Bmatrix}^{(1)} = \begin{Bmatrix} 23.0 \\ 1.0 \end{Bmatrix} \tag{E17.7.16}$$

For ω_2:

$$\begin{Bmatrix} c_1 \\ c_2 \end{Bmatrix}^{(2)} = \begin{Bmatrix} -0.69 \\ 1.00 \end{Bmatrix} \tag{E17.7.17}$$

Example 17.8 Derive the equations of motion for the free vibration of a viscously damped tapered beam using the Galerkin method. The governing equation is given by

$$\rho A(x)\frac{\partial^2 w(x,t)}{\partial t^2} + d(x)\frac{\partial w(x,t)}{\partial t} + \frac{\partial^2}{\partial x^2}\left[EI(x)\frac{\partial^2 w(x,t)}{\partial x^2}\right] = 0, \qquad 0 < x < l \tag{E17.8.1}$$

with two boundary conditions at $x = 0$ as well as at $x = l$. In Eq. (E17.8.1), the term $d(x)[\partial w(x,t)/\partial t]$ denotes the viscous damping force per unit length of the beam.

SOLUTION   The transverse deflection function $w(x,t)$ is assumed to be of the form

$$w(x,t) = W(x)e^{\lambda t} \tag{E17.8.2}$$

where λ is complex and $W(x)$ is the deflection shape (or mode shape). When Eq. (E17.8.2) is substituted into Eq. (E17.8.1), we obtain

$$\rho A(x)\lambda^2 W(x) + d(x)\lambda W(x) + \frac{d^2}{dx^2}\left[EI(x)\frac{d^2 W(x)}{dx^2}\right] = 0, \qquad 0 < x < l \tag{E17.8.3}$$

with two boundary conditions at each end. We assume the solution, $W(x)$, in the form of a series of n comparison functions $\phi_1(x), \phi_2(x), \ldots, \phi_n(x)$, to be

$$W(x) = \sum_{i=1}^{n} c_i \phi_i(x) \tag{E17.8.4}$$

where each function $\phi_i(x)$ satisfies all the boundary conditions of the beam. Since the assumed function $W(x)$ does not satisfy the differential equation (E17.8.3), the residual is defined as

$$R = \lambda^2 \rho A(x) \sum_{i=1}^{n} c_i \phi_i(x) + \lambda d(x) \sum_{i=1}^{n} c_i \phi_i(x) + \sum_{i=1}^{n} c_i \frac{d^2}{dx^2} \left[EI(x) \frac{d^2 \phi_i(x)}{dx^2} \right] \tag{E17.8.5}$$

Multiplying Eq. (E17.8.5) by $\phi_j(x)$, integrating the product $R\phi_j(x)$ from 0 to l, and setting the result equal to zero, we obtain

$$\lambda^2 \sum_{i=1}^{n} c_i \int_0^l \rho A(x)\phi_i(x)\phi_j(x)\,dx + \lambda \sum_{i=1}^{n} c_i \int_0^l d(x)\phi_i(x)\phi_j(x)\,dx$$

$$+ \sum_{i=1}^{n} c_i \int_0^l \frac{d^2}{dx^2}\left[EI(x)\frac{d^2\phi_i(x)}{dx^2} \right]\phi_j(x)\,dx = 0, \qquad j = 1, 2, \ldots, n \tag{E17.8.6}$$

Defining the symmetric stiffness, damping, and mass coefficients k_{ij}, d_{ij}, and m_{ij}, respectively, as

$$k_{ij} = \int_0^l \frac{d^2}{dx^2}\left[EI(x)\frac{d^2\phi_i(x)}{dx^2} \right]\phi_j(x)\,dx \tag{E17.8.7}$$

$$d_{ij} = \int_0^l d(x)\phi_i(x)\phi_j(x)\,dx \tag{E17.8.8}$$

$$m_{ij} = \int_0^l \rho A(x)\phi_i(x)\phi_j(x)\,dx \tag{E17.8.9}$$

Eq. (E17.8.6) can be written as

$$\lambda^2 \sum_{i=1}^{n} m_{ij}c_i + \lambda \sum_{i=1}^{n} d_{ij}c_i + \sum_{i=1}^{n} k_{ij}c_i = 0, \qquad j = 1, 2, \ldots, n \tag{E17.8.10}$$

or in matrix form as

$$\lambda^2 [m]\vec{c} + \lambda[d]\vec{c} + [k]\vec{c} = \vec{0} \tag{E17.8.11}$$

Solution of Eq. (E17.8.11)   Three approaches can be used for solving Eq. (E11.8.11).

Approach 1: Direct solution: For a nonzero solution of $\vec{c}$ in Eq. (E17.8.11), we must have

$$| \lambda^2[m] + \lambda[d] + [k] | = 0 \tag{E17.8.12}$$

Equation (E17.8.12) leads to a polynomial equation in λ of order $2n$, whose solution yields the roots $\lambda_i, i = 1, 2, \ldots, 2n$. For each λ_i, Eq. (E11.8.11) can be solved to find the corresponding vector $\vec{c}^{(i)}$. This procedure is, in general, tedious and inconvenient to handle; hence the following procedures are commonly used to solve Eq. (E17.8.11).

Approach 2: Proportional damping: In this approach, the damping matrix is assumed to be given by a linear combination of the mass and stiffness matrices:

$$[d] = \alpha[m] + \beta[k] \tag{E17.8.13}$$

where α and β are known constants. This type of damping is known as *proportional damping* because $[c]$ is proportional to $[m]$ and $[k]$. As in the case of a multidegree-of-freedom system, the modal matrix of the corresponding undamped discretized system, $[X]$, is defined as

$$[X] = \begin{bmatrix} \vec{c}_1 \vec{c}_2 \cdots \vec{c}_n \end{bmatrix} \tag{E17.8.14}$$

where $\vec{c}^{(i)}, i = 1, 2, \ldots, n$, denote the modal vectors that satisfy the undamped eigenvalue problem:

$$[k]\vec{c} = \lambda[m]\vec{c} \tag{E17.8.15}$$

where $\lambda = \omega^2$ and ω is the natural frequency of the discretized undamped system.

Assuming that the modal vectors are normalized with respect to the mass matrix, we have

$$[X]^{\mathrm{T}} [m] [X] = [I] \tag{E17.8.16}$$

$$[X]^{\mathrm{T}}[k][X] = [\Lambda] = \begin{bmatrix} \lambda_1 & & & 0 \\ & \lambda_2 & & \\ & & \cdots & \\ 0 & & & \lambda_n \end{bmatrix} \tag{E17.8.17}$$

where $\lambda_i = \omega_i^2, i = 1, 2, \ldots, n$. Substituting Eq. (E17.8.13) into Eq. (E17.8.11) and using the transformation

$$\vec{c} = [X]\vec{p} \tag{E17.8.18}$$

we obtain

$$\lambda^2[m][X]\vec{p} + \lambda(\alpha[m] + \beta[k])[X]\vec{p} + [k][X]\vec{p} = \vec{0} \tag{E17.8.19}$$

Premultiplication of Eq. (E17.8.19) by $[X]^{\mathrm{T}}$ and use of Eqs. (E17.8.16) and (E17.8.17) results in

$$\lambda^2\vec{p} + \lambda(\alpha[I] + \beta[\Lambda])\vec{p} + [\Lambda]\vec{p} = \vec{0} \tag{E17.8.20}$$

Defining

$$
\alpha[I] + \beta[\Lambda] = [\gamma] \equiv
\begin{bmatrix}
2\zeta_1\omega_1 & & & 0 \\
& 2\zeta_2\omega_2 & & \\
& & \ddots & \\
0 & & & 2\zeta_n\omega_n
\end{bmatrix}
\tag{E17.8.21}
$$

where ζ_i is called the *damping ratio* in mode i, Eq. (E17.8.20) can be written as

$$
(\lambda^2[I] + \lambda[\gamma] + [\Lambda])\vec{p} = \vec{0}
\tag{E17.8.22}
$$

or, in scalar form,

$$
\lambda^2 + 2\zeta_i\omega_i\lambda + \omega_i^2 = 0, \qquad i = 1, 2, \ldots, n
\tag{E17.8.23}
$$

Each equation in (E17.8.23) denotes a quadratic equation in the eigenvalue λ which can be solved to find the ith eigenvalue, λ_i, of the proportionally damped continuous system. Substituting this eigenvalue λ_i along with Eq. (E17.8.13), into Eq. (E17.8.11) and solving the resulting linear equations gives the eigenvectors $\vec{c}^{(i)}$ of the proportionally damped system. In general, the eigenvalues and the eigenvectors of the proportionally damped system occur in complex-conjugate pairs.

Approach 3: General viscous damping: In the case of general viscous damping, damping will not be proportional and hence the undamped modal matrix $[X]$ will not diagonalize the damping matrix $[d]$. In such a case, we transform the eigenvalue problem, Eq. (E17.8.11), as indicated by the following steps:

(a) Define the identity

$$
\lambda\vec{c} = \lambda\vec{c}
\tag{E17.8.24}
$$

(b) Rewrite Eq. (E17.8.11) as

$$
\lambda^2\vec{c} = -\lambda[m]^{-1}[d]\vec{c} - [m]^{-1}[k]\vec{c}
\tag{E17.8.25}
$$

(c) Define a vector $\vec{y}$ of dimension $2n$ as:

$$
\vec{y} = \begin{Bmatrix} \vec{c} \\ \lambda\vec{c} \end{Bmatrix}
\tag{E17.8.26}
$$

(d) Combine Eqs. (E17.8.24) and (E17.8.25) as

$$
\lambda \begin{Bmatrix} \vec{c} \\ \lambda\vec{c} \end{Bmatrix} = \begin{bmatrix} [0] & [I] \\ -[m]^{-1}[k] & -[m]^{-1}[d] \end{bmatrix} \begin{Bmatrix} \vec{c} \\ \lambda\vec{c} \end{Bmatrix}
\tag{E17.8.27}
$$

or

$$
\lambda\vec{y} = [B]\vec{y}
\tag{E17.8.28}
$$

where $[B]$ denotes a $2n \times 2n$ matrix:

$$[B] = \begin{bmatrix} [0] & [I] \\ -[m]^{-1}[k] & -[m]^{-1}[d] \end{bmatrix} \qquad \text{(E17.8.29)}$$

The transformed algebraic eigenvalue problem of order $2n$ defined by Eq. (E17.8.28), is solved to find the eigenvalues λ_i and the corresponding mode shapes $\vec{y}^{(i)}$, $i = 1, 2, \ldots, 2n$.

17.8   COLLOCATION METHOD

In the Galerkin method, the integral of the weighted residual over the domain of the problem is set equal to zero where the weighting function is the same as one of the comparison functions used in the series solution. The collocation method is similar to the Galerkin method except that the weighting functions are spatial Dirac delta functions. Thus, for a one-dimensional eigenvalue problem, an approximate solution is assumed in the form of a linear sum of trial functions $\phi_i(x)$ as

$$\overline{\phi}^{(n)}(x) = \sum_{i=1}^{n} c_i \phi_i(x) \qquad (17.69)$$

where the c_i are unknown coefficients and the $\phi_i(x)$ are the trial functions. Depending on the nature of the trial functions used, the collocation method may be classified in one of the following three types:

1. *Boundary method:* used when the functions $\phi_i(x)$ satisfy the governing differential equation over the domain but not all the boundary conditions of the problem.
2. *Interior method:* used when the functions $\phi_i(x)$ satisfy all the boundary conditions but not the governing differential equation of the problem.
3. *Mixed method:* used when the functions $\phi_i(x)$ do not satisfy either the governing differential equation or the boundary conditions of the problem.

When the integral of the weighted residual is set equal to zero, the collocation method yields

$$\int_0^l \delta(x - x_i) R(\overline{\phi}^{(n)}(x)) \, dx = 0, \qquad i = 1, 2, \ldots, n \qquad (17.70)$$

where δ is the Dirac delta function and x_i, $i = 1, 2, \ldots, n$, are the known collocation points where the residual is specified to be equal to zero. Due to the sampling property of the Dirac delta function, Eqs. (17.70) require no integration and hence can be expressed as

$$R(\overline{\phi}^{(n)}(x_i)) = 0, \qquad i = 1, 2, \ldots, n \qquad (17.71)$$

This amounts to setting the residue at $x_1, x_2, \ldots, x_n$ equal to zero. Equations (17.71) denote a system of n homogeneous algebraic equations in the unknown coefficients

$c_1, c_2, \ldots, c_n$ and the parameter λ. In fact, they represent an algebraic eigenvalue problem of order n. It can be seen that the selection of the collocation points $x_1, x_2, \ldots, x_n$ is important in obtaining a well-conditioned system of equations and a convergent solution. The locations of the collocation points should be selected as evenly as possible in the domain and/or boundary of the system to avoid ill-conditioning of the resulting equations.

To see the nature of the eigenvalue problem, consider the problem of longitudinal vibration of a tapered bar. The governing differential equation is given by [see Eq. (9.14) with harmonic variation of $u(x, t)$]:

$$\frac{d}{dx}\left[E A(x)\frac{dU(x)}{dx}\right] - \lambda \rho A(x)U(x) = 0 \qquad (17.72)$$

By assuming the trial functions $\phi_i(x)$ in Eq. (17.69) as comparison functions, substituting the assumed solution into Eq. (17.72), and setting the residual equal to zero at x_i, we obtain

$$\sum_{j=1}^{n} c_j \left\{ \frac{d}{dx}\left[E A(x)\frac{d\phi_j(x)}{dx}\right] - \lambda \rho A(x)\phi_j(x) \right\}\bigg|_{x=x_i} = 0, \qquad i = 1, 2, \ldots, n \qquad (17.73)$$

or

$$\sum_{j=1}^{n} k_{ij}c_j = \lambda \sum_{j=1}^{n} m_{ij}c_j, \qquad i = 1, 2, \ldots, n \qquad (17.74)$$

where λ is the eigenvalue of the problem and k_{ij} and m_{ij} denote the stiffness and mass coefficients, respectively, defined by

$$k_{ij} = \frac{d}{dx}\left[E A(x_i)\frac{d\phi_j(x_i)}{dx}\right] \qquad (17.75)$$

$$m_{ij} = \rho A(x_i)\phi_j(x_i) \qquad (17.76)$$

Equations (17.74) denote an algebraic eigenvalue problem which can be expressed in matrix form as

$$[k]\vec{c} = \lambda[m]\vec{c} \qquad (17.77)$$

where

$$\vec{c} = \begin{Bmatrix} c_1 \\ c_2 \\ \vdots \\ c_n \end{Bmatrix}$$

is an n-dimensional vector of the coefficients, and $[k] = [k_{ij}]$ and $[m] = [m_{ij}]$ are the stiffness and mass matrices of order $n \times n$.

It can be seen that the main advantage of the collocation method is its simplicity. Evaluation of the stiffness and mass coefficients involves no integrations. The main disadvantage of the method is that the stiffness and mass matrices, defined by Eqs. (17.75) and (17.76), are not symmetric although the system is conservative. Hence, the solution of the nonsymmetric eigenvalue problem, Eq. (17.77), is not simple. In general, we need to find both the right and left eigenvectors of the system in order to find the system response.

Example 17.9 Find the natural frequencies of transverse vibration of a uniform fixed–fixed beam shown in Fig. 17.1 using the collocation method with the approximate solution

$$W(x) \equiv X(x) = c_1\phi_1(x) + c_2\phi_2(x) \tag{E17.9.1}$$

where

$$\phi_1(x) = 1 - \cos\frac{2\pi x}{l} \tag{E17.9.2}$$

$$\phi_2(x) = 1 - \cos\frac{4\pi x}{l} \tag{E17.9.3}$$

SOLUTION   It can be seen that the trial functions satisfy all the boundary (geometric) conditions of the beam:

$$\phi_i(x = 0) = \phi_i(x = l) = 0, \qquad i = 1, 2 \tag{E17.9.4}$$

$$\frac{d\phi_i}{dx}(x = 0) = \frac{d\phi_i}{dx}(x = l) = 0, \qquad i = 1, 2 \tag{E17.9.5}$$

Since the assumed solution has two unknown coefficients, we need to use two collocation points. The collocation points are chosen as $x_1 = l/4$ and $x_2 = l/2$. The eigenvalue problem is defined by the differential equation

$$EI\frac{d^4 W(x)}{dx^4} - \rho A\omega^2 W(x) = 0 \tag{E17.9.6}$$

When Eq. (E17.9.1) is substituted in Eq. (E17.9.6), the residual is given by

$$R(X(x)) = \frac{d^4 X(x)}{dx^4} - \lambda X(x) \tag{E17.9.7}$$

where

$$\lambda = \frac{\rho A\omega^2}{EI} \tag{E17.9.8}$$

Using Eqs. (E17.9.1)–(E17.9.3), the residual can be expressed as

$$\begin{aligned}
R &= \frac{d^4}{dx^4}\left[c_1\left(1 - \cos\frac{2\pi x}{l}\right) + c_2\left(1 - \cos\frac{4\pi x}{l}\right)\right] \\
&\quad - \lambda\left[c_1\left(1 - \cos\frac{2\pi x}{l}\right) + c_2\left(1 - \cos\frac{4\pi x}{l}\right)\right] \\
&= -c_1\left(\frac{2\pi}{l}\right)^4\cos\frac{2\pi x}{l} - c_2\left(\frac{4\pi}{l}\right)^4\cos\frac{4\pi x}{l} \\
&\quad - \lambda\left[c_1\left(1 - \cos\frac{2\pi x}{l}\right) + c_2\left(1 - \cos\frac{4\pi x}{l}\right)\right]
\end{aligned} \tag{E17.9.9}$$

By setting the residual equal to zero at $x_1 = l/4$ and $x_2 = l/2$, Eq. (E17.9.9) gives

$$c_1\left[-\left(\frac{2\pi}{l}\right)^4 \cos\frac{\pi}{2} - \lambda\left(1 - \cos\frac{\pi}{2}\right)\right] + c_2\left[-\left(\frac{4\pi}{l}\right)^4 \cos\pi - \lambda(1 - \cos\pi)\right] = 0$$
(E17.9.10)

$$c_1\left[-\left(\frac{2\pi}{l}\right)^4 \cos\pi - \lambda(1 - \cos\pi)\right] + c_2\left[-\left(\frac{4\pi}{l}\right)^4 \cos2\pi - \lambda(1 - \cos2\pi)\right] = 0$$
(E17.9.11)

Equations (E17.9.10) and (E17.9.11) can be simplified as

$$c_1(-\lambda) + c_2\left[\left(\frac{4\pi}{l}\right)^4 - 2\lambda\right] = 0 \qquad (E17.9.12)$$

$$c_1\left[\left(\frac{2\pi}{l}\right)^4 - 2\lambda\right] + c_2\left[-\left(\frac{4\pi}{l}\right)^4\right] = 0 \qquad (E17.9.13)$$

By setting the determinant of the coefficient matrix in Eqs. (E17.9.12) and (E17.9.13) equal to zero, we obtain the frequency equation as

$$\begin{vmatrix} -\lambda & \left(\frac{4\pi}{l}\right)^4 - 2\lambda \\ \left(\frac{2\pi}{l}\right)^4 - 2\lambda & -\left(\frac{4\pi}{l}\right)^4 \end{vmatrix} = 0 \qquad (E17.9.14)$$

which can be simplified as

$$\lambda^2 - \lambda\left(\frac{\pi}{l}\right)^4 (200) + 1024\left(\frac{\pi^2}{l^2}\right)^4 = 0 \qquad (E17.9.15)$$

The roots of Eq. (E17.9.15) are given by

$$\lambda_{1,2} = 5.258246\left(\frac{\pi}{l}\right)^4, \qquad 194.741754\left(\frac{\pi}{l}\right)^4 \qquad (E17.9.16)$$

Using Eqs (E17.9.8) and (E17.9.16), the natural frequencies can be obtained as

$$\omega_1 = 22.6320\sqrt{\frac{EI}{\rho Al^4}}, \qquad \omega_2 = 137.730878\sqrt{\frac{EI}{\rho Al^4}} \qquad (E17.9.17)$$

The exact values of the first two natural frequencies of a fixed–fixed beam are given by

$$\omega_1 = 22.3729\sqrt{\frac{EI}{\rho Al^4}} \quad \text{and} \quad \omega_2 = 61.6696\sqrt{\frac{EI}{\rho Al^4}} \qquad (E17.9.18)$$

It can be seen that the first and the second natural frequencies given by the collocation method are larger by 1.1581% and 123.3367%, respectively.

Note: If we use the points $x_1 = l/4$ and $x_2 = 3l/4$ or $x_1 = l/3$ and $x_2 = 2l/3$ as collocation points, it would not be possible to compute the natural frequencies. Because of symmetry, use of the points $x_1 = l/4$ and $x_2 = l/2$ in the half-beam would be sufficient.

The mode shapes can be determined, using (E17.9.12), as

$$c_2 = \left\{ \frac{\lambda}{\left(\frac{4\pi}{l}\right)^4 - 2\lambda} \right\} c_1 \qquad \text{(E17.9.19)}$$

Thus, the first mode shape, corresponding to $\lambda_1 = 5.258246(\pi^4/l^4)$, is given by

$$c_2^{(1)} = 0.0214199 c_1^{(1)}$$

or

$$\begin{Bmatrix} c_1 \\ c_2 \end{Bmatrix}^{(1)} = \begin{Bmatrix} 1 \\ 0.0214199 \end{Bmatrix} c_1^{(1)} \qquad \text{(E17.9.20)}$$

The second mode shape, corresponding to $\lambda = 194.741754(\pi^4/l^4)$, is given by

$$c_2^{(2)} = -1.458920 c_1^{(2)}$$

or

$$\begin{Bmatrix} c_1 \\ c_2 \end{Bmatrix}^{(2)} = \begin{Bmatrix} 1 \\ -1.458920 \end{Bmatrix} c_1^{(2)} \qquad \text{(E17.9.21)}$$

17.9   SUBDOMAIN METHOD

In this method, the domain of the problem, D, is subdivided into n smaller subdomains $D_i (i = 1, 2, \ldots, n)$, so that

$$D = \sum_{i=1}^{n} D_i \qquad (17.78)$$

and the integral of the residual over each subdomain is set equal to zero:

$$\int_{D_i} R(\overline{\phi}(x)) \, dx = 0, \qquad i = 1, 2, \ldots, n \qquad (17.79)$$

where $\overline{\phi}(x)$ denotes the assumed solution in the form of a linear sum of trial functions (such as comparison functions), Eq. (17.69). Equations (17.79) indicate that the average value of the residual in each subdomain is zero. Obviously, in this method, negative errors can cancel positive errors to give least net error, although the sum of the absolute value of the errors is very large. The subdomain method can be interpreted as a weighted residual method where the weighting functions, $\psi_i(x)$, are defined as

$$\psi_i(x) = \begin{cases} 1, & \text{if } x \text{ is in } D_i \\ 0, & \text{otherwise} \end{cases} \qquad (17.80)$$

The stiffness and mass matrices given by the subdomain method are also nonsymmetric, and hence the resulting algebraic eigenvalue problem is difficult to solve.

Example 17.10 Find the natural frequencies of transverse vibration of a uniform fixed–fixed beam using the subdomain method with the approximate solution

$$X(x) = c_1\phi_1(x) + c_2\phi_2(x) \tag{E17.10.1}$$

where

$$\phi_1(x) = 1 - \cos\frac{2\pi x}{l} \tag{E17.10.2}$$

$$\phi_2(x) = 1 - \cos\frac{4\pi x}{l} \tag{E17.10.3}$$

SOLUTION   As seen in Example 17.9, the trial functions satisfy all the (geometric) boundary conditions of the beam. Since the assumed solution has two unknown coefficients, we need to use two subdomains. Because of the symmetry, we choose the subdomains in the first half of the beam as $D_1 = (0, l/4)$ and $D_2 = (l/4, l/2)$. For the approximate solution given by Eq. (E17.10.1), the residual is given by [Eq. (E17.9.9)]:

$$
R = -c_1 \left(\frac{2\pi}{l}\right)^4 \cos\frac{2\pi x}{l} - c_2 \left(\frac{4\pi}{l}\right)^4 \cos\frac{4\pi x}{l}
$$
$$
- \lambda \left[c_1 \left(1 - \cos\frac{2\pi x}{l}\right) + c_2 \left(1 - \cos\frac{4\pi x}{l}\right) \right] \tag{E17.10.4}
$$

By setting the integral of the residual over the two subdomains equal to zero, we obtain

$$
\int_0^{l/4} R\,dx = \int_0^{l/4} \left[-c_1 \left(\frac{2\pi}{l}\right)^4 \cos\frac{2\pi x}{l} - c_2 \left(\frac{4\pi}{l}\right)^4 \cos\frac{4\pi x}{l} \right.
$$
$$
\left. - \lambda c_1 + \lambda c_1 \cos\frac{2\pi x}{l} - \lambda c_2 + \lambda c_2 \cos\frac{4\pi x}{l} \right] dx = 0
$$

or

$$
c_1 \left[\left(\frac{2\pi}{l}\right)^3 - \lambda\frac{l}{4} - \lambda\frac{l}{2\pi} \right] + c_2 \left(-\lambda\frac{l}{4}\right) = 0 \tag{E17.10.5}
$$

and

$$
\int_{l/4}^{l/2} R\,dx = \int_{l/4}^{l/2} \left[-c_1 \left(\frac{2\pi}{l}\right)^4 \cos\frac{2\pi x}{l} - c_2 \left(\frac{4\pi}{l}\right)^4 \cos\frac{4\pi x}{l} \right.
$$
$$
\left. - \lambda c_1 + \lambda c_1 \cos\frac{2\pi x}{l} - \lambda c_2 + \lambda c_2 \cos\frac{4\pi x}{l} \right] dx = 0
$$

or

$$
c_1 \left[-\left(\frac{2\pi}{l}\right)^3 - \lambda\frac{l}{4} + \lambda\frac{l}{2\pi} \right] + c_2 \left(-\lambda\frac{l}{4}\right) = 0 \tag{E17.10.6}
$$

For a nontrivial solution of c_1 and c_2 in Eqs. (E17.10.5) and (E17.10.6), the determinant of the coefficient matrix must be zero:

$$\begin{vmatrix} \left(\frac{2\pi}{l}\right)^3 - \lambda\frac{l}{4} - \lambda\frac{l}{2\pi} & -\lambda\frac{l}{4} \\ -\left(\frac{2\pi}{l}\right)^3 - \lambda\frac{l}{4} + \lambda\frac{l}{2\pi} & -\lambda\frac{l}{4} \end{vmatrix} = 0 \qquad (E17.10.7)$$

Equation (E17.10.7) can be simplified as

$$\lambda^2\left(\frac{l^2}{4\pi}\right) - \lambda\left[\left(\frac{2\pi}{l}\right)^3\frac{l}{2}\right] = 0 \qquad (E17.10.8)$$

The roots of Eq. (E17.10.8) are

$$\lambda_1 = 0, \qquad \lambda_2 = \frac{16\pi^4}{l^4} \qquad (E17.10.9)$$

The mode shapes corresponding to λ_1 and λ_2 can be determined using Eq. (E17.10.5) as

$$c_2 = \frac{(2\pi/l)^3 - \lambda\,(l/4 + l/2\pi)}{\lambda\,(l/4)}c_1 \qquad (E17.10.10)$$

Thus, for $\lambda_1 = 0$, the mode shape is given by

$$\vec{c}^{(1)} = \left\{\begin{array}{c} c_1 \\ c_2 \end{array}\right\}^{(1)} = \left\{\begin{array}{c} 0 \\ 1 \end{array}\right\}c_2^{(1)} \qquad (E17.10.11)$$

For $\lambda_2 = 16\pi^4/l^4$, the mode shape is given by

$$\vec{c}^{(2)} = \left\{\begin{array}{c} c_1 \\ c_2 \end{array}\right\}^{(2)} = \left\{\begin{array}{c} 1.0 \\ 183.4405/l^3 \end{array}\right\}c_1^{(2)} \qquad (E17.10.12)$$

17.10 LEAST SQUARES METHOD

The least squares method can be considered as a variational method as well as a weighted residual method. Because the method is also applicable to problems for which a classical variational principle does not exist, it is considered more as a weighted residual method. Basically, the least squares method minimizes the integral of the square of the residual over the domain:

$$\int_D R^2\,dD = \text{minimum} \qquad (17.81)$$

where R is the residual of the governing differential equation and D is the domain of the problem. Assuming the approximate solution in the form of Eq. (17.69), Eq. (17.81) can be expressed as

$$\int_0^l R^2(\overline{\phi}(x))\,dx = \int_0^l R^2(c_1, c_2, \ldots, c_n)\,dx = \text{minimum} \qquad (17.82)$$

The minimization is carried with respect to the unknown coefficients $c_1, c_2, \ldots, c_n$. The necessary conditions for the minimum of the integral in Eq. (17.82) are given by

$$\frac{\partial}{\partial c_i}\left(\int_0^l R^2\,dx\right) = 2\int_0^l R\frac{\partial R}{\partial c_i}\,dx = 0$$

or

$$\int_0^l R\frac{\partial R}{\partial c_i}\,dx = 0, \qquad i = 1, 2, \ldots, n \tag{17.83}$$

Equation (17.83) indicates that the least squares method is a weighted residual method where the weighting functions, $\psi_i(x)$, are given by

$$\psi_i(x) = \frac{\partial R}{\partial c_i}, \qquad i = 1, 2, \ldots, n \tag{17.84}$$

To see the nature of the algebraic eigenvalue problem given by the method of least squares, consider the problem of the longitudinal vibration of a tapered bar:

$$\frac{d}{dx}\left[EA(x)\frac{dU(x)}{dx}\right] - \lambda\rho A(x)U(x) = 0 \tag{17.85}$$

When the approximate solution given by Eq. (17.69) is used, the residual of Eq. (17.85) can be expressed as

$$R = \sum_{j=1}^n c_j\left\{\frac{d}{dx}\left[EA(x)\frac{d\phi_j(x)}{dx}\right] - \lambda\rho A(x)\phi_j(x)\right\} \tag{17.86}$$

and hence

$$\frac{\partial R}{\partial c_i} = \frac{d}{dx}\left[EA(x)\frac{d\phi_i(x)}{dx}\right] - \lambda\rho A(x)\phi_i(x) \tag{17.87}$$

Using Eqs. (17.86) and (17.87), Eq. (17.83) can be expressed as

$$\int_0^l \sum_{j=1}^n c_j\left\{\frac{d}{dx}\left[EA(x)\frac{d\phi_j(x)}{dx}\right] - \lambda\rho A(x)\phi_j(x)\right\}$$
$$\cdot\left\{\frac{d}{dx}\left[EA(x)\frac{d\phi_i(x)}{dx}\right] - \lambda\rho A(x)\phi_i(x)\right\}dx = 0$$

or

$$\sum_{j=1}^n c_j\left\{\int_0^l \frac{d}{dx}\left[EA(x)\frac{d\phi_j(x)}{dx}\right]\frac{d}{dx}\left[EA(x)\frac{d\phi_i(x)}{dx}\right]dx\right.$$
$$- \lambda\int_0^l \frac{d}{dx}\left[EA(x)\frac{d\phi_j(x)}{dx}\right]\rho A(x)\phi_i(x)\,dx$$
$$- \lambda\int_0^l \rho A(x)\phi_j(x)\frac{d}{dx}\left[EA(x)\frac{d\phi_i(x)}{dx}\right]dx$$
$$\left.+ \lambda^2\int_0^l \rho^2 A^2(x)\phi_i(x)\phi_j(x)\,dx\right\} = 0, \qquad i = 1, 2, \cdots, n \tag{17.88}$$

Defining the $n \times n$ matrices

$$k_{ij} = \int_0^l \frac{d}{dx}\left[EA(x)\frac{d\phi_i(x)}{dx}\right]\frac{d}{dx}\left[EA(x)\frac{d\phi_j(x)}{dx}\right]dx \qquad (17.89)$$

$$m_{ij} = \int_0^l \rho^2 A^2(x)\phi_i(x)\phi_j(x)\,dx \qquad (17.90)$$

$$h_{ij} = \int_0^l \rho A(x)\phi_i(x)\frac{d}{dx}\left[EA(x)\frac{d\phi_j}{dx}\right]dx \qquad (17.91)$$

Eq. (17.88) can be expressed as

$$\sum_{j=1}^n k_{ij}c_j - \lambda \sum_{j=1}^n h_{ij}c_j - \lambda \sum_{j=1}^n h_{ji}c_j + \lambda^2 \sum_{j=1}^n m_{ij}c_j = 0, \qquad i = 1, 2, \ldots, n \qquad (17.92)$$

or, equivalently, in matrix form as

$$[[k] - \lambda([h] + [h]^T) + \lambda^2[m]]\vec{c} = \vec{0} \qquad (17.93)$$

Equation (17.93) denotes a quadratic eigenvalue problem because it involves both λ and λ^2. This equation can be seen to be similar to the one corresponding to the eigenvalue problem of a damped system, Eq. (E17.8.11).

To reduce Eq. (17.93) to the form of a standard eigenvalue problem, define the following vectors and matrices:

$$\vec{b}_{2n \times 1} = \left\{ \begin{array}{c} \lambda\vec{c} \\ \vec{c} \end{array} \right\} \qquad (17.94)$$

$$[A]_{2n \times 2n} = \left[\begin{array}{cc} [h] + [h]^T & -[k] \\ -[k] & [0] \end{array} \right] \qquad (17.95)$$

$$[B]_{2n \times 2n} = \left[\begin{array}{cc} [m] & [0] \\ [0] & -[k] \end{array} \right] \qquad (17.96)$$

Using Eqs. (17.94), (17.95), and (17.96), Eq. (17.93) can be rewritten as

$$[A]\vec{b} = \lambda[B]\vec{b} \qquad (17.97)$$

which can be seen to be a standard matrix eigenvalue problem of order $2n$. Thus, the least squares method requires the solution of an eigenvalue problem of twice the order of that required by other methods, such as the Rayleigh–Ritz and Galerkin methods.

Example 17.11 Find the first two natural frequencies and mode shapes of a fixed–fixed uniform beam using the least squares method. The free vibration equation of a uniform beam is given by

$$\frac{d^4W}{dx^4} - \lambda W = 0, \qquad 0 \le x \le l \qquad (E17.11.1)$$

where

$$\lambda = \beta^4 = \frac{m\omega^2}{EI} \tag{E17.11.2}$$

Assume an approximate solution for $W(x)$ as

$$\tilde{W}(x) = c_1 \left(1 - \cos \frac{2\pi x}{l} \right) + c_2 \left(1 - \cos \frac{4\pi x}{l} \right) \tag{E17.11.3}$$

SOLUTION   The least squares method requires that

$$\int_0^l R^2(\tilde{W}(x)) \, dx \rightarrow \text{minimum} \tag{E17.11.4}$$

where $R(\tilde{W}(x))$ is the residual. The conditions for the minimum in Eq. (E17.11.4) are given by

$$\int_0^l R(\tilde{W}(x)) \frac{\partial R}{\partial c_i} (\tilde{W}(x)) \, dx = 0, \qquad i = 1, 2 \tag{E17.11.5}$$

With the assumed solution, the residual becomes

$$R(\tilde{W}(x)) = \frac{d^4 \tilde{W}}{dx^4} - \lambda \tilde{W} = -c_1 \left(\frac{2\pi}{l} \right)^4 \cos \frac{2\pi x}{l} - c_2 \left(\frac{4\pi}{l} \right)^4 \cos \frac{4\pi x}{l}$$

$$- \beta^4 \left[c_1 \left(1 - \cos \frac{2\pi x}{l} \right) + c_2 \left(1 - \cos \frac{4\pi x}{l} \right) \right] \tag{E17.11.6}$$

The necessary conditions, given by Eq. (E17.11.5), can be expressed as follows.
For $i = 1$:

$$\int_0^l R(\tilde{W}(x)) \frac{\partial R(\tilde{W}(x))}{\partial c_1} \, dx = \int_0^l \left[c_1 \left(\frac{2\pi}{l} \right)^4 \cos \frac{2\pi x}{l} + c_2 \left(\frac{4\pi}{l} \right)^4 \cos \frac{4\pi x}{l} \right.$$

$$+ c_1 \beta^4 \left(1 - \cos \frac{2\pi x}{l} \right) + c_2 \beta^4 \left(1 - \cos \frac{4\pi x}{l} \right) \right]$$

$$\cdot \left[\left(\frac{2\pi}{l} \right)^4 \cos \frac{2\pi x}{l} + \beta^4 \left(1 - \cos \frac{2\pi x}{l} \right) \right] dx = 0$$

$$\tag{E17.11.7}$$

Upon integration and simplification, Eq. (E17.11.7), yields

$$c_1 \left(\frac{128\pi^8}{l^7} - \frac{16\pi^4 \beta^4}{l^3} + \frac{3}{2} l\beta^8 \right) + c_2(l\beta^8) = 0 \tag{E17.11.8}$$

For $i = 2$:

$$\int_0^l R(\tilde{W}(x)) \frac{\partial R(\tilde{W}(x))}{\partial c_2} dx = \int_0^l \left[c_1 \left(\frac{2\pi}{l}\right)^4 \cos\frac{2\pi x}{l} + c_2 \left(\frac{4\pi}{l}\right)^4 \cos\frac{4\pi x}{l} \right.$$

$$+ c_1\beta^4 \left(1 - \cos\frac{2\pi x}{l}\right) + c_2\beta^4 \left(1 - \cos\frac{4\pi x}{l}\right) \Bigg]$$

$$\cdot \left[\left(\frac{4\pi}{l}\right)^4 \cos\frac{4\pi x}{l} + \beta^4 \left(1 - \cos\frac{4\pi x}{l}\right) \right] dx = 0$$

$$(E17.11.9)$$

Upon integration and simplification, Eq. (E17.11.9), yields

$$c_1(l\beta^8) + c_2 \left(\frac{2^{15}\pi^8}{l^7} - \frac{4^4\pi^4\beta^4}{l^3} + \frac{3}{2}l\beta^8\right) = 0 \qquad (E17.11.10)$$

For a nontrivial solution of c_1 and c_2 in Eqs. (E17.11.8) and (E17.11.10), the determinant of their coefficient matrix must be equal to zero. This gives

$$\begin{bmatrix} \dfrac{128\pi^8}{l^7} - \dfrac{16\pi^4\beta^4}{l^3} + \dfrac{3}{2}l\beta^8 & l\beta^8 \\ l\beta^8 & \dfrac{2^{15}\pi^8}{l^7} - \dfrac{4^4\pi^4\beta^4}{l^3} + \dfrac{3}{2}l\beta^8 \end{bmatrix} \begin{Bmatrix} c_1 \\ c_2 \end{Bmatrix} = \begin{Bmatrix} 0 \\ 0 \end{Bmatrix}$$

$$(E17.11.11)$$

which upon simplification gives the frequency equation

$$5l^{16}\beta^{16} - 1632\pi^4 l^{12}\beta^{12} + 213,760\pi^8 l^8\beta^8 - 2,228,224\pi^{12}l^4\beta^4 + 2^{24}\pi^{16} = 0$$

$$(E17.11.12)$$

Defining

$$\tilde{\lambda} = \frac{l^4\beta^4}{2^4\pi^4} \qquad (E17.11.13)$$

Eq. (E17.11.12) can be rewritten as

$$5\tilde{\lambda}^4 - 102\tilde{\lambda}^3 + 835\tilde{\lambda}^2 - 544\tilde{\lambda} + 256 = 0 \qquad (E17.11.14)$$

The roots of Eq. (E17.11.14) are given by (using MATLAB)

$$\tilde{\lambda}_{1,2} = 9.8671 \pm 7.4945i, \qquad \tilde{\lambda}_{3,4} = 0.3329 \pm 0.4719i \qquad (E17.11.15)$$

The natural frequencies can be computed using Eqs. (E17.11.2) and (E17.11.13) as

$$\tilde{\lambda}_i = \frac{l^4\beta_i^4}{2^4\pi^4} = \frac{ml^4 \omega_i^2}{EI \cdot 2^4\pi^4}$$

or

$$\omega_i = 39.478602\sqrt{\tilde{\lambda}_i} \left(\frac{EI}{ml^4}\right)^{1/2}, \qquad i = 1, 2, 3, 4 \qquad (E17.11.16)$$

Noting that

$$\sqrt{\tilde{\lambda}_{1,2}} = 3.3360 \pm 1.1233i \tag{E17.11.17}$$

$$\sqrt{\tilde{\lambda}_{3,4}} = 0.6747 \pm 0.3497i \tag{E17.11.18}$$

Eq. (E17.11.16) gives natural frequencies as

$$\omega_1 = (131.700617 + 44.346314i) \left(\frac{EI}{ml^4} \right)^{1/2} \tag{E17.11.19}$$

$$\omega_2 = (131.700617 - 44.346314i) \left(\frac{EI}{ml^4} \right)^{1/2} \tag{E17.11.20}$$

$$\omega_3 = (26.636213 + 13.805667i) \left(\frac{EI}{ml^4} \right)^{1/2} \tag{E17.11.21}$$

$$\omega_4 = (26.636213 - 13.805667i) \left(\frac{EI}{ml^4} \right)^{1/2} \tag{E17.11.22}$$

To find the mode shapes, Eq. (E17.11.11), is rewritten in terms of $\tilde{\lambda}$ as

$$\begin{bmatrix} \frac{1}{2} - \tilde{\lambda} + \frac{3}{2}\tilde{\lambda}^2 & \tilde{\lambda}^2 \\ \tilde{\lambda}^2 & 128 - 16\tilde{\lambda} + \frac{3}{2}\tilde{\lambda}^2 \end{bmatrix} \begin{Bmatrix} c_1 \\ c_2 \end{Bmatrix} = \begin{Bmatrix} 0 \\ 0 \end{Bmatrix} \tag{E17.11.23}$$

The first equation of (E17.11.23) can be written in scalar form as

$$c_1 \left(\frac{1}{2} - \tilde{\lambda} + \frac{3}{2}\tilde{\lambda}^2 \right) + c_2(\tilde{\lambda}^2) = 0 \tag{E17.11.24}$$

or

$$c_2 = \left(-\frac{3}{2} + \frac{1}{\tilde{\lambda}} - \frac{1}{2\tilde{\lambda}^2} \right) c_1 \tag{E17.11.25}$$

By substituting the value of $\tilde{\lambda}_i$ given by Eq. (E17.11.15) into Eq. (E17.11.25), the corresponding eigenvector $\vec{c}^{(i)}$ can be expressed as

$$\vec{c}^{(i)} = \begin{Bmatrix} c_1^{(i)} \\ c_2^{(i)} \end{Bmatrix} = \begin{Bmatrix} 1.0 + 0.0i \\ -\frac{3}{2} + \frac{1}{\tilde{\lambda}} - \frac{1}{2\tilde{\lambda}^2} \end{Bmatrix} c_1^{(i)}, \quad i = 1, 2, 3, 4 \tag{E17.11.26}$$

Thus, the eigenvectors are as follows: For $\tilde{\lambda}_1 = 9.8671 + 7.4945i$,

$$\vec{c}^{(1)} = \begin{Bmatrix} c_1^{(1)} \\ c_2^{(1)} \end{Bmatrix} = \begin{Bmatrix} 1.0 + 0.0i \\ -1.4366 - 0.0457i \end{Bmatrix} c_1^{(1)} \tag{E17.11.27}$$

For $\tilde{\lambda}_2 = 9.8671 - 7.4945i$,

$$\vec{c}^{(2)} = \begin{Bmatrix} c_1^{(2)} \\ c_2^{(2)} \end{Bmatrix} = \begin{Bmatrix} 1.0 + 0.0i \\ -1.4366 + 0.0457i \end{Bmatrix} c_1^{(2)} \tag{E17.11.28}$$

For $\tilde{\lambda}_3 = 0.3329 + 0.4719i$,

$$\vec{c}^{(3)} = \left\{ \begin{array}{c} c_1^{(3)} \\ c_2^{(3)} \end{array} \right\} = \left\{ \begin{array}{c} 1.0 + 0.0i \\ 0.0010 - 0.0026i \end{array} \right\} c_1^{(3)} \tag{E17.11.29}$$

For $\tilde{\lambda}_4 = 0.3329 - 0.4719i$,

$$\vec{c}^{(4)} = \left\{ \begin{array}{c} c_1^{(4)} \\ c_2^{(4)} \end{array} \right\} = \left\{ \begin{array}{c} 1.0 + 0.0i \\ 0.0010 + 0.0026i \end{array} \right\} c_1^{(4)} \tag{E17.11.30}$$

Thus, the natural frequencies of the beam are given by Eqs. (E17.11.19)–(E17.11.22) and the corresponding mode shapes by Eqs. (E17.11.27)–(E17.11.30)

Notes
1. The eigenvalues (and the natural frequencies) and the corresponding eigenvectors in the least squares method will be complex conjugates. Although it is difficult to interpret the complex eigenvalues and eigenvectors, usually the imaginary parts are small and can be neglected.
2. The least squares method leads to a quadratic eigenvalue problem. That is, the size of the eigenvalue problem will be twice that of the problem in the Rayleigh–Ritz method.
3. The matrices involved in the least squares method are more difficult to compute.
4. In view of the foregoing features, the least squares method is not as popular as the other methods, such as the Rayleigh–Ritz and Galerkin methods for solving eigenvalue problems. However, the least squares method works well for equilibrium problems, as indicated in the following example.

Example 17.12 Find the deflection of a fixed–fixed uniform beam subject to a uniformly distributed load f_0 per unit length using the least squares method. The governing differential equation for the deflection of a beam is given by

$$\frac{d^4 w}{dx^4} - f_0 = 0, \qquad 0 \le x \le l \tag{E17.12.1}$$

where

$$f_0 = \frac{p_0}{EI} \tag{E17.12.2}$$

Assume an approximate solution for $w(x)$, using comparison functions, as

$$\tilde{w}(x) = c_1 x^2 (x - l)^2 + c_2 x^3 (x - l)^2 \tag{E17.12.3}$$

SOLUTION The least squares method requires the minimization of the function

$$I = \int_0^l R^2(\tilde{w}(x)) \, dx \tag{E17.12.4}$$

The necessary conditions for the minimum of I are given by

$$\int_0^l R(\tilde{w}(x)) \frac{\partial R(\tilde{w}(x))}{\partial c_i} \, dx = 0, \qquad i = 1, 2 \tag{E17.12.5}$$

With the assumed solution of Eq. (E17.12.3), the residual and its partial derivatives are given by

$$R(\tilde{w}(x)) = 24c_1 + (120x - 48l)c_2 - f_0 \qquad (E17.12.6)$$

$$\frac{\partial R}{\partial c_1} = 24 \qquad (E17.12.7)$$

$$\frac{\partial R}{\partial c_2} = 120x - 48l \qquad (E17.12.8)$$

Thus, Eq. (E17.12.5) can be expressed as

$$\int_0^l [24c_1 + (120x - 48l)c_2 - f_0](24)\, dx = 0 \qquad (E17.12.9)$$

$$\int_0^l [24c_1 + (120x - 48l)c_2 - f_0](120x - 48l)\, dx = 0 \qquad (E17.12.10)$$

Equations (E17.12.9) and (E17.12.10) can be evaluated to obtain

$$24c_1 + 12lc_2 = f_0 \qquad (E17.12.11)$$
$$24c_1 - 80lc_2 = f_0 \qquad (E17.12.12)$$

The solution of Eqs. (E17.12.11) and (E17.12.12) gives

$$c_1 = \frac{f_0}{24}, \qquad c_2 = 0 \qquad (E17.12.13)$$

Thus, the deflection of the beam, in view of Eqs. (E17.12.3) and (E17.12.13), becomes

$$\tilde{w}(x) = c_1 x^2(x - l)^2 = \frac{f_0}{24} x^2(x - l)^2 \qquad (E17.12.14)$$

It can be seen that this solution coincides with the exact solution.

17.11 RECENT CONTRIBUTIONS

Dunkerley's Method   The basic idea behind Dunkerley's method of finding the smallest natural frequency of a multidegree-of-freedom elastic system was extended by Levy [6]. to determine all the frequencies of the system simultaneously. The method is found to converge fast when the frequencies are not close to each other. The method is demonstrated with the help of several lumped-parameter systems. Badrakhan presented the application of Rayleigh's method to an unconstrained system [7].

Frequency in Terms of Static Deflection   Radhakrishnan et al. [8] developed a method to estimate the fundamental frequency of a plate through the finite element solution of its static deflection under a uniformly distributed load without the associated eigenvalue problem. The results computed in the case of a clamped rectangular plate with a central circular hole were found to be in reasonable agreement with experimental results. The method is useful for determining the fundamental frequency of elastic plates of arbitrary geometry and boundary conditions. Bert [9] proposed the simple relation

$$\omega = C \left(\frac{g}{\delta} \right)^{1/2} \tag{17.98}$$

where C is a dimensionless constant and g is the acceleration due to gravity for estimating the natural frequency (ω) in terms of the static deflection (δ) of a linear system. Nagaraj [10] showed that Eq. (17.98) also holds for a rotating Timoshenko beam if C is selected properly. A variational formulation of the Rayleigh–Ritz method was presented by Bhat [11]. The stationarity of the natural frequencies was investigated with respect to coefficients in the linear combination of the assumed deflection shape as well as natural modes.

Beams   The frequencies of beams carrying multiple masses using Rayleigh's method were considered by Low [12]. The solution methods for frequencies of three mass-loaded beams are presented with both the transcendental characteristic equation and Rayleigh estimation. Gladwell [13] presented a method of finding the natural frequencies and principal modes of undamped free vibration of a plane frame consisting of a rectangular grid of uniform beams. A general method of finding a set of assumed modes for use in the Rayleigh–Ritz method was given. The resulting equations, expressed in matrix form, were solved for the case of a simple frame, to illustrate the method.

Membranes   The dynamic stability of a flat sag cable subjected to an axial periodic load was investigated by Takahashi using the Galerkin method [14]. The results include unstable regions for various sag-to-span ratios and ratios of wave speeds. The transient response of hanging curtains clamped at three edges was considered by Yamada and his associates [15]. A hanging curtain was replaced by an equivalent membrane for deriving the equation of motion. The free vibration of the membrane was analyzed theoretically, and its transient response when subjected to a rectangularly varying point force was studied using Galerkin's method.

Plates   The use of two-dimensional orthogonal plate functions as admissible deflection functions in the study of flexural vibration of skew plates by the Rayleigh–Ritz method was presented by Liew and Lam [16]. Free vibration analysis of triangular and trapezoidal plates was considered by the superposition technique [17, 18]. The superposition method was extended by Gorman for the free vibration solution of rectangular plates resting on uniform elastic edge supports [19]. The elastic edge supports were assumed to be uniform elastic rotational and translational supports of any stiffness magnitudes in terms of eight stiffness coefficients. The vibrations of circular plates with thickness varying in a discontinuous fashion were studied by Avalos et al. [20]. The free vibration of a solid circular plate free at its edge and attached to a Winkler foundation was considered by Salari et al. [21]. The free vibrations of a solid circular plate of linearly varying thickness attached to a Winkler foundation were considered by Laura et al. [22] using linear analysis and the Rayleigh–Schmidt method.

Avalos et al. [20] presented general approximate solution for vibrating circular plates with stepped thickness over a concentrated circular region. Approximate values of the fundamental frequencies of vibration of circular plates with discontinuous variations of thickness in a nonconcentric fashion were determined by Laura et al. [23]. The Ritz method and Rayleigh's optimization procedure were used in finding the solution of plates whose edges are elastically restrained against rotation.

REFERENCES

1. S. H. Crandall, *Engineering Analysis: A Survey of Numerical Procedures*, McGraw-Hill, New York, 1956.

2. S. S. Rao, *Applied Numerical Methods for Engineers and Scientists*, Prentice Hall, Upper Saddle River, NJ, 2002.

3. S. S. Rao, *The Finite Element Method in Engineering*, 3rd ed., Butterworth-Heinemann, Boston, 1999.

4. S. S. Rao, *Mechanical Vibrations*, 4th ed., Prentice Hall, Upper Saddle River, NJ, 2004.

5. J. N. Reddy, *Energy and Variational Methods in Applied Mechanics*, Wiley, New York, 1984.

6. C. Levy, An iterative technique based on the Dunkerley method for determining the natural frequencies of vibrating systems, *Journal of Sound and Vibration*, Vol. 150, No. 1, pp. 111–118, 1991.

7. F. Badrakhan, On the application of Rayleigh's method to an unconstrained system, *Journal of Sound and Vibration*, Vol. 162, No.1, pp. 190–194, 1993.

8. G. Radhakrishnan, M. K. Sundaresan, and B. Nageswara Rao, Fundamental frequency of thin elastic plates, *Journal of Sound and Vibration*, Vol. 209, No. 2, pp. 373–376, 1998.

9. C. W. Bert, Application of a version of the Rayleigh technique to problems of bars, beams, columns, membranes and plates, *Journal of Sound and Vibration*, Vol. 119, No. 2, pp. 317–326, 1987.

10. V. T. Nagaraj, Relationship between fundamental natural frequency and maximum static deflection for rotating Timoshenko beams, *Journal of Sound and Vibration*, Vol. 201, No. 3, pp. 404–406, 1997.

11. R. B. Bhat, Nature of stationarity of the natural frequencies at the natural modes in the Rayleigh–Ritz method, *Journal of Sound and Vibration*, Vol. 203, No. 2, pp. 251–263, 1997.

12. K. H. Low, Frequencies of beams carrying multiple masses: Rayleigh estimation versus eigenanalysis solutions, *Journal of Sound and Vibration*, Vol. 268, No. 4, pp. 843–853, 2003.

13. G. M. L. Gladwell, The vibration of frames, *Journal of Sound and Vibration*, Vol. 1, No. 4, pp. 402–425, 1964.

14. K. Takahashi, Dynamic stability of cables subjected to an axial periodic load, *Journal of Sound and Vibration*, Vol. 144, No. 2, pp. 323–330, 1991.

15. G. Yamada, Y. Kobayashi, and H. Hamaya, Transient response of a hanging curtain, *Journal of Sound and Vibration*, Vol. 130, No. 2, pp. 223–235, 1989.

16. K. M. Liew and K. Y. Lam, Application of two-dimensional orthogonal plate function to flexural vibration of skew plates, *Journal of Sound and Vibration*, Vol. 139, No. 2, pp. 241–252, 1990.

17. H. T. Saliba, Transverse free vibration of fully clamped symmetrical trapezoidal plates, *Journal of Sound and Vibration*, Vol. 126, No. 2, pp. 237–247, 1988.

18. H. T. Saliba, Transverse free vibration of simply supported right triangular thin plates: a highly accurate simplified solution, *Journal of Sound and Vibration*, Vol. 139, No. 2, pp. 289–297, 1990.

19. D. J. Gorman, A general solution for the free vibration of rectangular plates resting on uniform elastic edge supports, *Journal of Sound and Vibration*, Vol. 139, No. 2, pp. 325–335, 1990.

20. D. Avalos, P. A. A. Laura, and H. A. Larrondo, Vibrating circular plates with stepped thickness over a concentric circular region: a general approximate solution, *Journal of the Acoustical Society of America*, Vol. 84, No. 4, pp. 1181–1185, 1988.

21. M. Salari, C. W. Bert, and A. G. Striz, Free vibrations of a solid circular plate free at its edge and attached to a Winkler foundation, *Journal of Sound and Vibration*, Vol. 118, No. 1, pp. 188–191, 1987.

22. P. A. A. Laura, R. H. Gutierrez, R. Carnicer, and H. C. Sanzi, Free vibrations of a solid circular plate of linearly varying thickness and attached to a Winkler foundation, *Journal of Sound and Vibration*, Vol. 144, No. 1, pp. 149–161, 1991.

23. P. A. A. Laura, R. H. Gutierrez, A. Bergmann, R. Carnicer, and H. C. Sanzi, Vibrations of circular plates with discontinuous variation of the thickness in a non-concentric fashion, *Journal of Sound and Vibration*, Vol. 144, No. 1, pp. 1–8, 1991.

PROBLEMS

17.1 The eigenvalue problem for finding the natural frequencies of vibration of a taut string, shown in Fig. 17.5, is given by

$$\frac{d^2 W}{dy^2} + \lambda W = 0, \qquad 0 < y < 1$$

with the boundary conditions

$$W(y) = 0 \qquad \text{at} \quad y = 0 \text{ and } y = 1$$

where

$$\lambda = \frac{\rho L^2 \omega^2}{P}$$

ρ is the mass per unit length, L is the length, P is the tension, ω is the natural frequency, y is the nondimensional length $= x/L$, and W is the transverse deflection shape of the string. Find the natural frequency of vibration of the string using the Galerkin method with the following trial solution:

$$W(y) = c_1 y(1 - y)$$

17.2 Solve Problem 17.1 and find the natural frequencies of the string using the trial solution

$$W(y) = c_1 y(1 - y) + c_2 y(1 - y)^2$$

17.3 Solve Problem 17.1 and find the natural frequencies of the string using the trial solution

$$W(y) = c_1 y(1 - y) + c_2 y(1 - y)^2 + c_3 y(1 - y)^3$$

17.4 Find the natural frequency of vibration of the string described in Problem 17.1 using the subdomain collocation method with the trial solution

$$W(y) = c_1 y(1 - y)$$

Assume that the subdomain is defined by $y = 0$ to $\frac{1}{8}$.

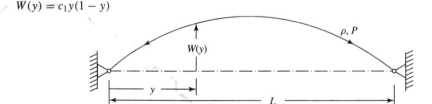

Figure 17.5

17.5 Find the natural frequencies of vibration of the string described in Problem 17.1 using the subdomain collocation method with the trail solution

$$W(y) = c_1 y(1-y) + c_2 y(1-y)^2$$

Assume that the subdomains are defined by $y_1 = \left(0, \frac{1}{8}\right)$ and $y_2 = \left(\frac{3}{8}, \frac{5}{8}\right)$.

17.6 Find the natural frequencies of vibration of the string described in Problem 17.1 using the subdomain collocation method with the trail solution

$$W(y) = c_1 y(1-y) + c_2 y(1-y)^2 + c_3 y(1-y)^3$$

Assume that the subdomains are defined as $y_1 = \left(0, \frac{1}{8}\right)$, $y_2 = \left(\frac{3}{8}, \frac{5}{8}\right)$, and $y_3 = \left(\frac{7}{8}, 1\right)$.

17.7 Consider the eigenvalue problem of the taut string described in Problem 17.1. Find the natural frequency of vibration of the string using the least squares method with the trial solution

$$W(y) = c_1 y(1-y)$$

17.8 Solve Problem 17.7 and find the natural frequencies of the string with the trial solution

$$W(y) = c_1 y(1-y) + c_2 y(1-y)^2$$

17.9 Solve Problem 17.7 and find the natural frequencies of the string with the trial solution

$$W(y) = c_1 y(1-y) + c_2 y(1-y)^2 + c_3 y(1-y)^3$$

17.10 Find the natural frequency of vibration of the string described in Problem 17.1 using the collocation method with the trail solution

$$W(y) = c_1 y(1-y)$$

Assume the collocation point to be $y_1 = \frac{1}{8}$.

17.11 Find the natural frequencies of vibration of the string described in Problem 17.1 using the collocation method with the trail solution

$$W(y) = c_1 y(1-y) + c_2 y(1-y)^2$$

Assume the collocation points to be $y_1 = \frac{1}{8}$ and $y_2 = \frac{1}{2}$.

17.12 Find the natural frequencies of vibration of the string described in Problem 17.1 using the collocation method with the trail solution

$$W(y) = c_1 y(1-y) + c_2 y(1-y)^2 + c_3 y(1-y)^3$$

Assume the collocation points to be $y_1 = \frac{1}{8}$, $y_2 = \frac{1}{2}$, and $y_3 = \frac{5}{8}$.

17.13 Rayleigh's quotient corresponding to the transverse vibration of a string, shown in Fig. 17.5, is given by

$$R(W(x)) = \lambda = \frac{\int_0^1 (dW/dy)^2\, dy}{\int_0^1 W^2\, dy}$$

where

$$\lambda = \frac{\rho l^2\, \omega^2}{P}$$

ρ is the mass per unit length, l is the length, P is the tension, ω is the natural frequency, y is the nondimensional length $= x/l$, and W is the transverse deflection shape of the string. Find the natural frequencies of vibration of the string using the Rayleigh–Ritz method with the following trial solution:

$$W(y) = c_1 y(1-y) + c_2 y(1-y)^2$$

17.14 Solve Problem 17.13 with the trial solution

$$W(y) = c_1 y(1-y) + c_2 y(1-y)^2 + c_3 y(1-y)^3$$

17.15 Find the natural frequency of the transverse vibration of the string described in Problem 17.13 using Rayleigh's method with the trial solution

$$W(y) = c_1 y(1-y)$$

17.16 The natural frequencies of vibration of a tapered bar in axial vibration are governed by the equation

$$\frac{d}{dx}\left[EA(x)\frac{dU(x)}{dx}\right] + m(x)U(x)\omega^2 = 0$$

where E is Young's modulus, $A(x)$ is the cross-sectional, area, $U(x)$ is the axial displacement shape, $m(x)$ is the mass per unit length, and ω is the natural frequency. Find the natural frequencies of axial vibration of the bar shown in Fig. 17.6 using the Galerkin method with the trial solution

$$U(x) = c_1 \sin\frac{\pi x}{2l} + c_2 \sin\frac{3\pi x}{2l}$$

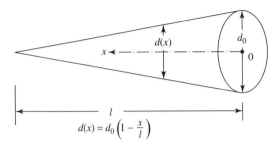

$$d(x) = d_0\left(1 - \frac{x}{l}\right)$$

Figure 17.6

17.17 Rayleigh's quotient corresponding to the axial vibration of a nonuniform bar is given by

$$R(U(x)) = \omega^2 = \frac{\int_0^l EA(x)[dU(x)/dx]^2\,dx}{\int_0^l m(x)[U(x)]^2\,dx}$$

Find the natural frequency of vibration of the tapered bar shown in Fig. 17.6 using Rayleigh's method with the trial solution

$$U(x) = c_1 \sin\frac{\pi x}{2l}$$

17.18 Rayleigh's quotient corresponding to the axial vibration of a nonuniform bar is given by

$$R(U(x)) = \omega^2 = \frac{\int_0^l EA(x)[dU(x)/dx]^2\,dx}{\int_0^l m(x)[U(x)]^2\,dx}$$

Find the natural frequencies of vibration of the tapered bar shown in Fig. 17.6 using the Rayleigh–Ritz method with the trial solution

$$U(x) = c_1 \sin\frac{\pi x}{2l} + c_2 \sin\frac{3\pi x}{2l}$$

17.19 Find the natural frequencies of axial vibration of the tapered bar described in Problem 17.16 and shown in Figure 17.6 using the subdomain collocation method with the trial solution

$$U(x) = c_1 \sin\frac{\pi x}{2l} + c_2 \sin\frac{3\pi x}{2l}$$

Assume the subdomains for collocation as $y_1 = (0, l/4)$ and $y_2 = (l/4, l/2)$.

17.20 Find the natural frequencies of axial vibration of the tapered bar described in Problem 17.16 and shown in Figure 17.6 using the collocation method with the trial solution

$$U(x) = c_1 \sin\frac{\pi x}{2l} + c_2 \sin\frac{3\pi x}{2l}$$

Assume the collocation points to be $y_1 = l/4$ and $y_2 = l/2$.

17.21 The natural frequencies of transverse vibration of a tapered beam are governed by the equation

$$\frac{d^2}{dx^2}\left[EI(x)\frac{d^2\,W(x)}{dx^2}\right] - \rho A(x)W(x)\omega^2 = 0$$

where $W(x)$ is the deflection shape, E is the Young's modulus, $I(x)$ is the area moment of inertia of the cross section, ρ is the density, $A(x)$ is the cross-sectional area, and ω is the natural frequency. Find the natural frequencies of vibration of the tapered beam shown in Fig. 17.7 using the Galerkin method with the trial solution

$$W(x) = c_1\left(1 - \frac{x}{l}\right)^4 + c_2\frac{x}{l}\left(1 - \frac{x}{l}\right)^4$$

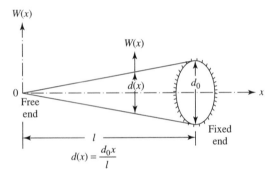

Figure 17.7

17.22 Find the natural frequencies of transverse vibration of the tapered beam described in Problem 17.21 and shown in Figure 17.7 using the collocation method with the trial solution

$$W(x) = c_1\left(1 - \frac{x}{l}\right)^4 + c_2\frac{x}{l}\left(1 - \frac{x}{l}\right)^4$$

Assume the collocation points to be $x_1 = l/4$ and $x_2 = l/2$.

17.23 Find the natural frequencies of transverse vibration of the tapered beam described in Problem 17.21 and shown in Figure 17.7 using the subdomain collocation method with the trial solution

$$W(x) = c_1\left(1 - \frac{x}{l}\right)^4 + c_2\frac{x}{l}\left(1 - \frac{x}{l}\right)^4$$

Assume the subdomains to be $(0, l/4)$ and $(l/4, l/2)$.

17.24 Rayleigh's quotient corresponding to the transverse vibration of a nonuniform beam is given by

$$R(W(x)) = \omega^2 = \frac{\int_0^l EI(x)[\frac{d^2 W(x)}{dx^2}]^2\, dx}{\int_0^l \rho A(x)[W(x)]^2\, dx}$$

Find the natural frequency of the tapered beam shown in Fig. 17.7 using Rayleigh's method with the trial solution

$$W(x) = c_1 \left(1 - \frac{x}{l}\right)^4$$

17.25 Find the natural frequencies of transverse vibration of the tapered beam shown in Fig. 17.7 using the Rayleigh–Ritz method with the trial solution

$$W(x) = c_1 \left(1 - \frac{x}{l}\right)^4 + c_2 \frac{x}{l} \left(1 - \frac{x}{l}\right)^4$$

17.26 Consider a fixed–free beam in the form of a wedge with width b, maximum depth d, and length l, as shown in Fig. 17.8. Estimate the fundamental natural frequency of vibration of the beam using Rayleigh's method with the following function for transverse deflection:

$$W(x) = c \left(1 - \cos \frac{\pi x}{2l}\right)$$

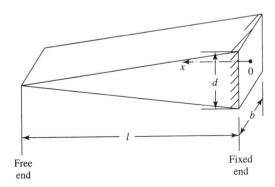

Figure 17.8

17.27 Estimate the fundamental natural frequency of vibration of a uniform fixed–fixed beam using Rayleigh's method. Assume the deflection function to be the same as the static deflection shape under self-weight:

$$W(x) = c(x^4 - 2lx^3 + l^2x^2)$$

where $c = \rho A/24EI$.

17.28 Find the natural frequencies of axisymmetric transverse vibration of a circular plate of thickness h and radius R clamped at the edge $r = R$ using the Rayleigh–Ritz method with the trial solution

$$W(r) = c_1 \left(1 - \frac{r^2}{R^2}\right)^2 + c_2 \left(1 - \frac{r^2}{R^2}\right)^3$$

where c_1 and c_2 are constants.

17.29 Estimate the fundamental natural frequency of transverse vibration of a rectangular plate of thickness h and dimensions a and b clamped on all four edges using Rayleigh's method with the trial solution

$$W(x, y) = c \sin \frac{\pi x}{a} \sin \frac{\pi y}{b}$$

where c is a constant.

17.30 Estimate the fundamental natural frequency of transverse vibration of a rectangular membrane of dimensions a and b under uniform tension P clamped at the edges using Rayleigh's method with the trial solution

$$W(x, y) = c \sin \frac{\pi x}{a} \sin \frac{\pi y}{b}$$

where c is a constant.

A

Basic Equations of Elasticity

A.1 STRESS

The state of stress at any point in a loaded body is defined completely in terms of the nine components of stress: $\sigma_{xx}, \sigma_{yy}, \sigma_{zz}, \sigma_{xy}, \sigma_{yx}, \sigma_{yz}, \sigma_{zy}, \sigma_{zx}$, and σ_{xz}, where the first three are the normal components and the latter six are the components of shear stress. The equations of internal equilibrium in terms of the nine components of stress can be derived by considering the equilibrium of moments and forces acting on the elemental volume shown in Fig. A.1. The equilibrium of moments about the x, y, and z axes, assuming that there are no body moments, leads to the relations

$$\sigma_{yx} = \sigma_{xy}, \qquad \sigma_{zy} = \sigma_{yz}, \qquad \sigma_{xz} = \sigma_{zx} \qquad (A.1)$$

Equations (A.1) show that the state of stress at any point can be defined completely by the six components $\sigma_{xx}, \sigma_{yy}, \sigma_{zz}, \sigma_{xy}, \sigma_{yz}$, and σ_{zx}.

A.2 STRAIN–DISPLACEMENT RELATIONS

The deformed shape of an elastic body under any given system of loads can be described completely by the three components of displacement u, v, and w parallel to the directions x, y, and z, respectively. In general, each of these components u, v, and w is a function of the coordinates x, y, and z. The strains and rotations induced in the body can be expressed in terms of the displacements u, v, and w. We shall assume the deformations to be small in this work. To derive the expressions for the normal strain components ε_{xx} and ε_{yy} and the shear strain component ε_{xy}, consider a small rectangular element $OACB$ whose sides (of lengths dx and dy) lie parallel to the coordinate axes before deformation. When the body undergoes deformation under the action of external load and temperature distribution, the element $OACB$ also deforms to the shape $O'A'C'B'$, as shown in Fig. A.2. We can observe that the element $OACB$ has two basic types of deformation, one of change in length and the other of angular distortion.

Since the normal strain is defined as change in length divided by original length, the strain components ε_{xx} and ε_{yy} can be found as

$$\varepsilon_{xx} = \frac{\text{change in length of the fiber } OA \text{ which lies in the } x \text{ direction before deformation}}{\text{original length of the fiber}}$$

$$= \frac{\{dx + [u + (\partial u/\partial x)dx] - u\} - dx}{dx} = \frac{\partial u}{\partial x} \qquad (A.2)$$

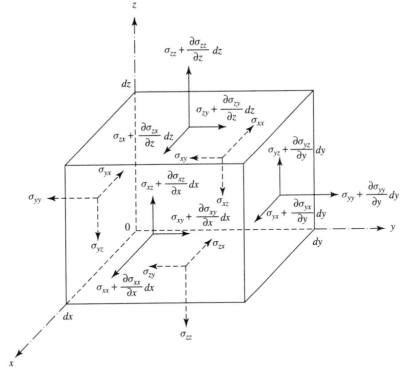

Figure A.1 Stresses on an element of size $dx\,dy\,dz$.

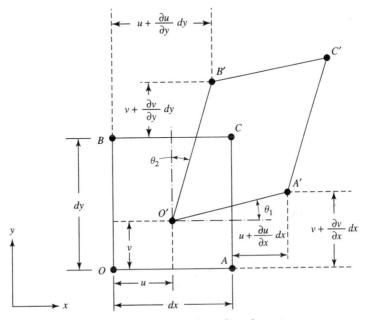

Figure A.2 Deformation of an element.

$$\varepsilon_{yy} = \frac{\text{change in length of the fiber } OB \text{ which lies in the } y \text{ direction before deformation}}{\text{original length of the fiber } OB}$$

$$= \frac{\{dy + [v + (\partial v/\partial y)dy] - v\} - dy}{dy} = \frac{\partial v}{\partial y} \tag{A.3}$$

The shear strain is defined as the decrease in the right angle between fibers OA and OB, which were at right angles to each other before deformation. Thus, the expression for the shear strain ε_{xy} can be obtained as

$$\varepsilon_{xy} = \theta_1 + \theta_2 \approx \frac{[v + (\partial v/\partial x)dx] - v}{dx + [u + (\partial u/\partial x)dx] - u} + \frac{[u + (\partial u/\partial y)dy] - u}{dy + [v + (\partial v/\partial y)dy] - v} \tag{A.4}$$

If the displacements are assumed to be small, ε_{xy} can be expressed as

$$\varepsilon_{xy} = \frac{\partial u}{\partial y} + \frac{\partial v}{\partial x} \tag{A.5}$$

The expressions for the remaining normal strain component ε_{zz} and shear strain components ε_{yz} and ε_{zx} can be derived in a similar manner as

$$\varepsilon_{zz} = \frac{\partial w}{\partial z} \tag{A.6}$$

$$\varepsilon_{yz} = \frac{\partial w}{\partial y} + \frac{\partial v}{\partial z} \tag{A.7}$$

$$\varepsilon_{zx} = \frac{\partial u}{\partial z} + \frac{\partial w}{\partial x} \tag{A.8}$$

A.3   ROTATIONS

Consider the rotation of a rectangular element of sides dx and dy as a rigid body by a small angle, as shown in Fig. A.3. Noting that $A'D$ and $C'E$ denote the displacements of A and C along the y and $-x$ axes, the rotation angle α can be expressed as

$$\alpha = \frac{\partial v}{\partial x} = -\frac{\partial u}{\partial y} \tag{A.9}$$

Of course, the strain in the element will be zero during rigid-body movement. If both rigid-body displacements and deformation or strain occur, the quantity

$$\omega_z = \frac{1}{2}\left(\frac{\partial v}{\partial x} - \frac{\partial u}{\partial y}\right) \tag{A.10}$$

can be seen to represent the average of angular displacement of dx and the angular displacement of dy, and is called *rotation about the z axis*. Thus, the rotations of an elemental body about the x, y, and z axes can be expressed as

$$\omega_x = \frac{1}{2}\left(\frac{\partial w}{\partial y} - \frac{\partial v}{\partial z}\right) \tag{A.11}$$

$$\omega_y = \frac{1}{2}\left(\frac{\partial u}{\partial z} - \frac{\partial w}{\partial x}\right) \tag{A.12}$$

$$\omega_z = \frac{1}{2}\left(\frac{\partial v}{\partial x} - \frac{\partial u}{\partial y}\right) \tag{A.13}$$

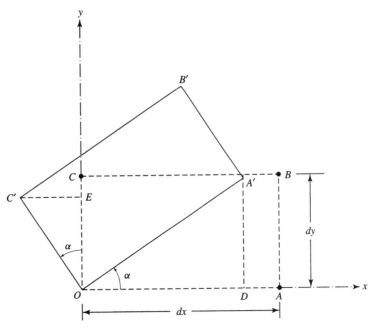

Figure A.3 Rotation of an element.

A.4 STRESS–STRAIN RELATIONS

The *stress–strain relations*, also known as the *constitutive relations*, of an anisotropic elastic material are given by the generalized Hooke's law, based on the experimental observation that strains are linearly related to the applied load within the elastic limit. The six components of stress at any point are related to the six components of strain linearly as

$$
\left\{
\begin{array}{c}
\sigma_{xx} \\
\sigma_{yy} \\
\sigma_{zz} \\
\sigma_{yz} \\
\sigma_{zx} \\
\sigma_{xy}
\end{array}
\right\}
=
\left[
\begin{array}{ccccc}
C_{11} & C_{12} & C_{13} & \cdots & C_{16} \\
C_{21} & C_{22} & C_{23} & \cdots & C_{26} \\
C_{31} & C_{32} & C_{33} & \cdots & C_{36} \\
\cdot & \cdot & \cdot & \cdots & \cdot \\
\cdot & \cdot & \cdot & \cdots & \cdot \\
C_{61} & C_{62} & C_{63} & \cdots & C_{66}
\end{array}
\right]
\left\{
\begin{array}{c}
\varepsilon_{xx} \\
\varepsilon_{yy} \\
\varepsilon_{zz} \\
\varepsilon_{yz} \\
\varepsilon_{zx} \\
\varepsilon_{xy}
\end{array}
\right\}
\tag{A.14}
$$

where the C_{ij} denote one form of elastic constants of the particular material. Equation (A.14) has 36 elastic constants. However, for real materials, the condition for the elastic energy to be a single-valued function of the strain requires the constants C_{ij} to be symmetric; that is, $C_{ij} = C_{ji}$. Thus, there are only 21 different elastic constants in Eq. (A.14) for an anisotropic material.

For an isotropic material, the elastic constants are invariant, that is, independent of the orientation of the x, y, and z axes. This reduces to two the number of independent elastic constants in Eq. (A.14). The two independent elastic constants, called *Lamé's elastic constants*, are commonly denoted as λ and μ. The Lamè constants are related to C_{ij} as follows:

$$
\begin{aligned}
C_{11} &= C_{22} = C_{33} = \lambda + 2\mu \\
C_{12} &= C_{21} = C_{31} = C_{13} = C_{32} = C_{23} = \lambda \\
C_{44} &= C_{55} = C_{66} = \mu \\
\text{all other } C_{ij} &= 0
\end{aligned}
\tag{A.15}
$$

Equation (A.14) can be rewritten for an elastic isotropic material as

$$
\begin{aligned}
\sigma_{xx} &= \lambda\Delta + 2\mu\varepsilon_{xx} \\
\sigma_{yy} &= \lambda\Delta + 2\mu\varepsilon_{yy} \\
\sigma_{zz} &= \lambda\Delta + 2\mu\varepsilon_{zz} \\
\sigma_{yz} &= \mu\varepsilon_{yz} \\
\sigma_{zx} &= \mu\varepsilon_{zx} \\
\sigma_{xy} &= \mu\varepsilon_{xy}
\end{aligned}
\tag{A.16}
$$

where

$$
\Delta = \varepsilon_{xx} + \varepsilon_{yy} + \varepsilon_{zz}
\tag{A.17}
$$

denotes the dilatation of the body and denotes the change in the volume per unit volume of the material. Lamé's constants λ and μ are related to Young's modulus E, shear modulus G, bulk modulus K, and Poisson's ratio ν as follows:

$$
E = \frac{\mu(3\lambda + 2\mu)}{\lambda + \mu}
\tag{A.18}
$$

$$
G = \mu
\tag{A.19}
$$

$$
K = \lambda + \tfrac{2}{3}\mu
\tag{A.20}
$$

$$
\nu = \frac{\lambda}{2(\lambda + \mu)}
\tag{A.21}
$$

or

$$
\lambda = \frac{\nu E}{(1 + \nu)(1 - 2\nu)}
\tag{A.22}
$$

$$
\mu = \frac{E}{2(1 + \nu)} = G
\tag{A.23}
$$

A.5   EQUATIONS OF MOTION IN TERMS OF STRESSES

Due to the applied loads (which may be dynamic), stresses will develop inside an elastic body. If we consider an element of material inside the body, it must be in dynamic equilibrium due to the internal stresses developed. This leads to the equations of motion of a typical element of the body. The sum of all forces acting on the element shown in Fig. A.1 in the x direction is given by

$$
\begin{aligned}
\sum F_x &= \left(\sigma_{xx} + \frac{\partial\sigma_{xx}}{\partial x}dx\right)dy\,dz - \sigma_{xx}\,dy\,dz + \left(\sigma_{xy} + \frac{\partial\sigma_{xy}}{\partial y}dy\right)dx\,dz - \sigma_{xy}\,dy\,dz \\
&\quad + \left(\sigma_{zx} + \frac{\partial\sigma_{zx}}{\partial z}dz\right)dx\,dy - \sigma_{zx}\,dx\,dy \\
&= \frac{\partial\sigma_{xx}}{\partial x}dx\,dy\,dz + \frac{\partial\sigma_{xy}}{\partial y}dx\,dy\,dz + \frac{\partial\sigma_{zx}}{\partial z}dx\,dy\,dz
\end{aligned}
\tag{A.24}
$$

According to Newton's second law of motion, the net force acting in the x direction must be equal to mass times acceleration in the x direction:

$$\sum F_x = \rho \, dx \, dy \, dz \, \frac{\partial^2 u}{\partial t^2} \tag{A.25}$$

where ρ is the density, u is the displacement, and $\partial^2 u / \partial t^2$ is the acceleration parallel to the x axis. Equations (A.24) and (A.25) lead to the equation of motion in the x direction. A similar procedure can be used for the y and z directions. The final equations of motion can be expressed as

$$\frac{\partial \sigma_{xx}}{\partial x} + \frac{\partial \sigma_{xy}}{\partial y} + \frac{\partial \sigma_{zx}}{\partial z} = \rho \, \frac{\partial^2 u}{\partial t^2} \tag{A.26}$$

$$\frac{\partial \sigma_{xy}}{\partial x} + \frac{\partial \sigma_{yy}}{\partial y} + \frac{\partial \sigma_{yz}}{\partial z} = \rho \, \frac{\partial^2 v}{\partial t^2} \tag{A.27}$$

$$\frac{\partial \sigma_{zx}}{\partial x} + \frac{\partial \sigma_{yz}}{\partial y} + \frac{\partial \sigma_{zz}}{\partial z} = \rho \, \frac{\partial^2 w}{\partial t^2} \tag{A.28}$$

where u, v, and w denote the components of displacement parallel to the x, y, and z axes, respectively. Note that the equations of motion are independent of the stress–strain relations or the type of material.

A.6 EQUATIONS OF MOTION IN TERMS OF DISPLACEMENTS

Using Eqs. (A.16), the equation of motion, Eq. (A.26), can be expressed as

$$\frac{\partial}{\partial x} (\lambda \, \Delta + 2\mu\varepsilon_{xx}) + \frac{\partial}{\partial y} (\mu \, \varepsilon_{xy}) + \frac{\partial}{\partial z} (\mu\varepsilon_{xz}) = \rho \, \frac{\partial^2 u}{\partial t^2} \tag{A.29}$$

Using the strain–displacement relations given by Eqs. (A.2), (A.4), and (A.8), Eq. (A.29) can be written as

$$\frac{\partial}{\partial x} \left(\lambda \, \Delta + 2\mu \, \frac{\partial u}{\partial x} \right) + \frac{\partial}{\partial y} \left[\mu \left(\frac{\partial v}{\partial x} + \frac{\partial u}{\partial y} \right) \right] + \frac{\partial}{\partial z} \left[\mu \left(\frac{\partial w}{\partial x} + \frac{\partial u}{\partial z} \right) \right] = \rho \, \frac{\partial^2 u}{\partial t^2} \tag{A.30}$$

which can be rewritten as

$$(\lambda + \mu) \frac{\partial \Delta}{\partial x} + \mu\nabla^2 u = \rho \, \frac{\partial^2 u}{\partial t^2} \tag{A.31}$$

where Δ is the dilatation and ∇^2 is the Laplacian operator:

$$\nabla^2 = \frac{\partial^2}{\partial x^2} + \frac{\partial^2}{\partial y^2} + \frac{\partial^2}{\partial z^2} \tag{A.32}$$

Using a similar procedure, the other two equations of motion, Eqs. (A.27) and (A.28), can be expressed as

$$(\lambda + \mu)\frac{\partial \Delta}{\partial y} + \mu \nabla^2 v = \rho \frac{\partial^2 v}{\partial t^2} \tag{A.33}$$

$$(\lambda + \mu)\frac{\partial \Delta}{\partial z} + \mu \nabla^2 w = \rho \frac{\partial^2 w}{\partial t^2} \tag{A.34}$$

The equations of motion, Eqs. (A.31), (A.33), and (A.34), govern the propagation of waves as well as the vibratory motion in elastic bodies.

B

Laplace and Fourier Transforms

Table B.1 Laplace Transforms

Serial number	$f(t)$	$F(s) = L[f(t)]$ $= \int_0^\infty e^{-st} f(t)dt$
1	1	$\dfrac{1}{s}$
2	t	$\dfrac{1}{s^2}$
3	$t^n,\ n = 1, 2, \ldots$	$\dfrac{n!}{s^{n+1}}$
4	$t^a,\ a > -1$	$\dfrac{\Gamma(a+1)}{s^{a+1}},\ s > a$
5	e^{at}	$\dfrac{1}{s-a},\ s > a$
6	$t^n e^{at},\ n = 1, 2, \ldots$	$\dfrac{n!}{(s-a)^{n+1}},\ s > a$
7	$\sin at$	$\dfrac{a}{s^2 + a^2}$
8	$\cos at$	$\dfrac{s}{s^2 + a^2}$
9	$t\sin at$	$\dfrac{2as}{(s^2 + a^2)^2}$
10	$t\cos at$	$\dfrac{s^2 - a^2}{(s^2 + a^2)^2}$
11	$e^{at}\sin bt$	$\dfrac{b}{(s-a)^2 + b^2}$
12	$e^{at}\cos bt$	$\dfrac{s-a}{(s-a)^2 + b^2}$
13	$\sinh at$	$\dfrac{a}{s^2 - a^2}$
14	$\cosh at$	$\dfrac{s}{s^2 - a^2}$

Table B.1 (*continued*)

Serial number	$f(t)$	$F(s) = L[f(t)]$ $= \int_0^\infty e^{-st} f(t)dt$
15	$t \sinh at$	$\dfrac{2as}{(s^2 - a^2)^2}$
16	$t \cosh at$	$\dfrac{s^2 + a^2}{(s^2 - a^2)^2}$
17	$H(t - a)$	$\dfrac{e^{-as}}{s}, s \geq a$
18	$\delta(t - a)$	e^{-as}
19	$af_1(t) + bf_2(t)$	$aF_1(s) + bF_2(s)$
20	$f(at)$	$\dfrac{1}{a} F\left(\dfrac{s}{a}\right)$
21	$e^{at} f(t)$	$F(s - a)$
22	$f'(t)$	$sF(s) - f(0)$
23	$f^{(n)}(t)$	$s^n F(s) - s^{n-1} f(0) - s^{n-2} f'(0)$ $- \cdots - f^{(n-1)}(0)$
24	$\displaystyle\int_0^t f(u)\,du$	$\dfrac{F(s)}{s}$
25	$\displaystyle\int_0^t f(u)g(t - u)\,du$	$F(s)G(s)$
26		$\dfrac{\pi\tau}{\tau^2 s^2 + \pi^2} \coth \dfrac{\tau s}{2}$
27		$\dfrac{1}{\tau s^2} \tanh \dfrac{\tau s}{2}$

Table B.1   (*continued*)

Serial number	$f(t)$	$F(s) = L[f(t)]$ $= \int_0^\infty e^{-st} f(t)dt$
28		$\dfrac{1}{s} \tanh \dfrac{\tau s}{2}$
29		$\dfrac{1}{\tau s^2} - \dfrac{e^{-\tau s}}{s(1 - e^{-\tau s})}$
30		$\dfrac{e^{-as}}{s}$
31		$\dfrac{e^{-as}}{s}(1 - e^{-\tau s})$
32	$f(t) = \begin{cases} \sin \dfrac{\pi t}{a}, & 0 \le t \le a \\ 0, & t > a \end{cases}$	$\dfrac{\pi a(1 + e^{-as})}{a^2 s^2 + \pi^2}$

Table B.2 Fourier Transform Pairs

Serial number	$f(x)$	$F(\omega) = \dfrac{1}{\sqrt{2\pi}}\displaystyle\int_{-\infty}^{\infty} f(x)e^{-i\omega x}\,dx$
1	$\begin{cases} 1, & \lvert x\rvert < a \\ 0, & \lvert x\rvert > a \end{cases}\quad (a>0)$	$\sqrt{\dfrac{2}{\pi}}\,\dfrac{\sin a\omega}{\omega}$
2	$\begin{cases} 1, & a < x < b \\ 0, & \text{otherwise} \end{cases}$	$\dfrac{1}{\sqrt{2\pi}}\,\dfrac{e^{-ia\omega} - e^{-ib\omega}}{i\omega}$
3	$\begin{cases} e^{-ax}, & x > 0 \\ 0, & x < 0 \end{cases}\quad (a>0)$	$\dfrac{1}{\sqrt{2\pi}}\,\dfrac{1}{a+i\omega}$
4	$\begin{cases} e^{ax}, & b < x < c \\ 0, & \text{otherwise} \end{cases}\quad (a>0)$	$\dfrac{1}{\sqrt{2\pi}}\,\dfrac{e^{(a+i\omega)c} - e^{(a+i\omega)b}}{a-i\omega}$
5	$e^{-a\lvert x\rvert},\ a>0$	$\sqrt{\dfrac{2}{\pi}}\,\dfrac{a}{a^2+\omega^2}$
6	$xe^{-a\lvert x\rvert},\ a>0$	$-\sqrt{\dfrac{2}{\pi}}\,\dfrac{2ia\omega}{(a^2+\omega^2)^2}$
7	$\begin{cases} e^{iax}, & \lvert x\rvert < b \\ 0, & \lvert x\rvert > b \end{cases}$	$\sqrt{\dfrac{2}{\pi}}\,\dfrac{\sin b(\omega-a)}{\omega-a}$
8	$e^{-a^2 x^2},\ a>0$	$\dfrac{1}{\sqrt{2}\,a}\,e^{-(\omega^2/4a^2)}$
9	$J_0(ax),\ a>0$	$\sqrt{\dfrac{2}{\pi}}\,\dfrac{H(a-\lvert\omega\rvert)}{(a^2-\omega^2)^{1/2}}$
10	$\delta(x-a),\ a\ \text{real}$	$\dfrac{1}{\sqrt{2\pi}}\,e^{-ia\omega}$
11	$\dfrac{\sin ax}{x}$	$\sqrt{\dfrac{\pi}{2}}\,H(a-\lvert\omega\rvert)$
12	e^{iax}	$\sqrt{2\pi}\,\delta(\omega-a)$
13	$H(x)$	$\sqrt{\dfrac{\pi}{2}}\left[\dfrac{1}{i\pi\omega} + \delta(\omega)\right]$
14	$H(x-a)$	$\sqrt{\dfrac{\pi}{2}}\left[\dfrac{e^{-i\omega a}}{i\pi\omega} + \delta(\omega)\right]$
15	$H(x) - H(-x)$	$\sqrt{\dfrac{2}{\pi}}\,\dfrac{-i}{\omega}$
16	$\dfrac{1}{x}$	$-i\sqrt{\dfrac{\pi}{2}}\,\operatorname{sgn}\omega$
17	$\dfrac{1}{x^n}$	$-i\sqrt{\dfrac{\pi}{2}}\left[\dfrac{(-i\omega)^{n-1}}{(n-1)!}\,\operatorname{sgn}\omega\right]$
18	$\dfrac{1}{x-a}$	$-i\sqrt{\dfrac{\pi}{2}}\,e^{-i\omega a}\,\operatorname{sgn}\omega$
19	$\dfrac{1}{(x-a)^n}$	$-i\sqrt{\dfrac{\pi}{2}}\,e^{-ia\omega}\,\dfrac{(-i\omega)^{n-1}}{(n-1)!}\,\operatorname{sgn}\omega$

Table B.3   Fourier Cosine Transforms

Serial number	f(x)	$F(\omega) = \sqrt{\dfrac{2}{\pi}} \displaystyle\int_0^\infty \cos(\omega x)\, f(x)dx$
1	$e^{-ax},\ a > 0$	$\sqrt{\dfrac{2}{\pi}}\dfrac{a}{a^2 + \omega^2}$
2	xe^{-ax}	$\sqrt{\dfrac{2}{\pi}}\dfrac{a^2 - \omega^2}{(a^2 + \omega^2)^2}$
3	$e^{-a^2 x^2}$	$\dfrac{1}{\sqrt{2}\,a} e^{-\omega^2/4a^2}$
4	$H(a - x)$	$\sqrt{\dfrac{2}{\pi}}\dfrac{\sin a\omega}{\omega}$
5	$x^{a-1},\ 0 < a < 1$	$\sqrt{\dfrac{2}{\pi}}\,\Gamma(a)\dfrac{1}{\omega^a}\cos\dfrac{a\pi}{2}$
6	$\cos ax^2$	$\dfrac{1}{2\sqrt{a}}\left(\cos\dfrac{\omega^2}{4a} + \sin\dfrac{\omega^2}{4a}\right)$
7	$\sin ax^2$	$\dfrac{1}{2\sqrt{a}}\left(\cos\dfrac{\omega^2}{4a} - \sin\dfrac{\omega^2}{4a}\right)$

Table B.4   Fourier Sine Transforms

Serial number	f(x)	$F(\omega) = \sqrt{\dfrac{2}{\pi}} \displaystyle\int_0^\infty \sin(\omega x)\, f(x)dx$
1	$e^{-ax},\ a > 0$	$\sqrt{\dfrac{2}{\pi}}\dfrac{\omega}{a^2 + \omega^2}$
2	$xe^{-ax},\ a > 0$	$\sqrt{\dfrac{2}{\pi}}\dfrac{2a\omega}{(a^2 + \omega^2)^2}$
3	$x^{a-1},\ 0 < a < 1$	$\sqrt{\dfrac{2}{\pi}}\dfrac{\Gamma(a)}{\omega^a}\sin\dfrac{\pi a}{2}$
4	$\dfrac{1}{\sqrt{x}}$	$\dfrac{1}{\sqrt{\omega}},\ \omega > 0$
5	$x^{-1}e^{-ax},\ a > 0$	$\sqrt{\dfrac{2}{\pi}}\tan^{-1}\dfrac{\omega}{a},\ \omega > 0$
6	$xe^{-a^2 x^2}$	$2^{-3/2}\dfrac{\omega}{a^3} e^{-\omega^2/4a^2}$
7	$\dfrac{x}{a^2 + x^2}$	$\sqrt{\dfrac{\pi}{2}}\,e^{-a\omega}$

Index